Geometric Algebra Computing

Eduardo Bayro-Corrochano • Gerik Scheuermann
Editors

Geometric Algebra Computing

in Engineering and Computer Science

Springer

Editors
Prof. Eduardo Bayro-Corrochano
Dept. Electrical Eng. &
Computer Science
CINVESTAV
Unidad Guadalajara
Av. Científica 1145
45015 Colonia el Bajío,
Zapopan, JAL
Mexico
edb@gdl.cinvestav.mx
http://www.gdl.cinvestav.mx/edb

Prof. Dr. Gerik Scheuermann
Inst. Informatik
Universität Leipzig
04009 Leipzig
Germany
scheuermann@informatik.uni-leipzig.de

ISBN 978-1-4471-5768-7 ISBN 978-1-84996-108-0 (eBook)
DOI 10.1007/978-1-84996-108-0
Springer London Dordrecht Heidelberg New York

British Library Cataloguing in Publication Data
A catalogue record for this book is available from the British Library

Printed on acid-free paper

Springer is part of Springer Science+Business Media (www.springer.com)

Preface

This book presents new results on applications of geometric algebra. The time when researchers and engineers were starting to realize the potential of quaternions for applications in electrical, mechanic, and control engineering passed a long time ago. Since the publication of *Space-Time Algebra* by David Hestenes (1966) and *Clifford Algebra to Geometric Calculus: A Unified Language for Mathematics and Physics* by David Hestenes and Garret Sobczyk (1984), consistent progress in the applications of geometric algebra has taken place. Particularly due to the great developments in computer technology and the Internet, researchers have proposed new ideas and algorithms to tackle a variety of problems in the areas of computer science and engineering using the powerful language of geometric algebra. In this process, pioneer groups started the conference series entitled "Applications of Geometric Algebra in Computer Science and Engineering" (AGACSE) in order to promote the research activity in the domain of the application of geometric algebra. The first conference, AGACSE'1999, organized by Eduardo Bayro-Corrochano and Garret Sobczyk, took place in Ixtapa-Zihuatanejo, Mexico, in July 1999. The contributions were published in *Geometric Algebra with Applications in Science and Engineering*, Birkhäuser, 2001. The second conference, ACACSE'2001, was held in the Engineering Department of the Cambridge University on 9–13 July 2001 and was organized by Leo Dorst, Chris Doran, and Joan Lasenby. The best conference contributions appeared as a book entitled *Applications of Geometric Algebra in Computer Science and Engineering*, Birkhäuser, 2002. The third conference, AGACSE'2008, took place in August 2008 in Grimma, Leipzig, Germany. The conference chairs, Eduardo Bayro-Corrochano and Gerik Sheuermann, edited this book using selected contributions that were peer-reviewed by at least two reviewers.

In the history of science, theories would have not been developed at all without essential mathematical concepts. In various periods of the history of mathematics and physics, there is clear evidence of stagnation, and it is only thanks to new mathematical developments that astonishing progress has taken place. Furthermore, researchers unavoidably cause fragmented knowledge in their various attempts to combine different mathematical systems. We realize that each mathematical system brings about some parts of geometry; however, together, they constitute a system that is highly redundant due to an unnecessary multiplicity of representations for geometric concepts. In contrast, in the geometric algebra language, most of the standard matter taught in engineering and computer science can be advantageously reformulated without redundancies and in a highly condensed fashion.

This book presents a selection of articles about the theory and applications of the advanced mathematical language *geometric algebra* which greatly helps to express the ideas and concepts and to develop algorithms in the broad domains of computer science and engineering. The contributions are organized in seven parts.

The *first part* presents screw theory in geometric algebra, the parameterization of 3D conformal transformations in conformal geometric algebra, and an overview of applications of geometric algebra. The *second part* includes thorough studies on Cliffor–Fourier transforms: the two-dimensional Clifford windowed Fourier transform; the cylindrical Fourier transform; applications of the 3D geometric algebra Fourier transform in graphics engineering; the 4D Clifford–Fourier transform for color image processing; and the use of the Hilbert transforms in Clifford analysis for signal processing. In the *third part*, self-organizing geometric neural networks are utilized for 2D contour and 3D surface reconstruction in medical image processing. The clustering and classification are handled using geometric neural networks and associative memories designed in the conformal geometric algebra. This part concludes with a retrospective of the quaternion wavelet transform, including an application for stereo vision. The *fourth part* for computer vision starts with a new cone-pixel camera using a convex hull and twists in conformal geometric algebra. The next work introduces a model-based approach for global self-localization using active stereo vision and Gaussian spheres. In the *fifth part*, the geometric characterization of M-conformal mappings is discussed, and a study of fluid flow problems is carried out in depth using quaternionic analysis. The *sixth part* shows the impressive space group visualizer for all 230 3D groups using the software packet for geometric algebra computations CLUCalc. The second author studies geometric algebra formalism as an alternative to distributed representation models; here convolutions are replaced by geometric products, and, as a result, a natural language for visualization of higher concepts is proposed. Another author studies computational complexity reductions using Clifford algebras and shows that graph problems of complexity class NP are polynomial in the number of Clifford operations required. The *seventh part* includes new developments in efficient geometric algebra computing: The first author presents an efficient blade factorization algorithm to produce faster implementations of the *Join*; with the software packet GALOOP, the second author symbolically reduces involved formulas of conformal geometric algebra, generating suitable code for computing using hardware accelerators. Another chapter shows applications of Grobner bases in robotics, formulated in the language of Clifford algebras, in engineering to the theory of curves, including Fermat and Bezier cubics, and in the interpolation of functions used in finite element theory.

We are very thankful to all book contributors, who are working persistently to advance the applications of geometric algebra. We do hope that the reader will find this collection of contributions in a broad scope of the areas of engineering and computer science very stimulating and encouraging. We hope that, as a result, we will see our community growing and benefitting from new and promising scientific contributions. Finally, we thank also for the support to this book project given by CINVESTAV Unidad Guadalajara and CONACYT Project 2007-1 82084.

CINVESTAV, Guadalajara, México
Universität Leipzig,
Institut für Informatik, Germany

Eduardo Bayro-Corrochano
Gerik Sheuermann

Contents

Part IV Computer Vision

Part V Conformal Mapping and Fluid Analysis

Contributors

Rafał Abłamowicz Department of Mathematics, Tennessee Technological University, Box 5054, Cookeville, TN 38505, USA, rablamowicz@tntech.edu

Sriwulan Adji School of Mathematical Sciences, Universiti Sains Malaysia, 11800 Penang, Malaysia, wulan@cs.usm.my

Tamim Asfour Humanoids and Intelligence Systems Lab, Karlsruhe Institute of Technology, Adenauerring 2, 76131 Karlsruhe, Germany, asfour@ira.uka.de

Mawardi Bahri School of Mathematical Sciences, Universiti Sains Malaysia, 11800 Penang, Malaysia, mawardibahri@gmail.com

Ricardo Barron Center of Computing Research, Juan de Dios Batiz s/n Col. Nueva Industrial Vallejo Gustavo A. Madero, C.P. 07738, Mexico DF, Mexico, rbarron@cic.ipn.mx

Thomas Batard Lab. MIA, La Rochelle University, La Rochelle, France, tbatar01@univ-lr.fr

Eduardo Bayro-Corrochano Dept. Electrical Eng. & Computer Science, CINVESTAV, Unidad Guadalajara, Av. Científica 1145, 45015 Colonia el Bajío, Zapopan, JAL, Mexico, edb@gdl.cinvestav.mx

Ryad Benosman Institut des Systèmes Intelligents et de Robotique (ISIR), 4 place Jussieu, Paris, France, benosman@isir.fr

Michel Berthier Lab. MIA, La Rochelle University, La Rochelle, France, michel.berthier@univ-lr.fr

Fred Brackx Clifford Research Group, Department of Mathematical Analysis, Ghent University, Galglaan 2, 9000 Gent, Belgium, fb@cage.ugent.be; Clifford Research Group, Faculty of Engineering, Ghent University, Galglaan 2, 9000 Gent, Belgium, Freddy.Brackx@UGent.be

Benjamin Cruz Center of Computing Research, Juan de Dios Batiz s/n Col. Nueva Industrial Vallejo Gustavo A. Madero, C.P. 07738, Mexico DF, Mexico, benjamincruz@sagitario.cic.ipn.mx

Thibaud Debaecker Institut des Systèmes Intelligents et de Robotique (ISIR), 4 place Jussieu, Paris, France, thibaud.debaecker@isir.jussieu.fr

Ruediger Dillmann ITEC, Universitaet Karlsruhe (TH), Karlsruhe, Germany, dillmann@ira.uka.de; Karlsruhe Institute of Technology, Adenauerring 2, 76131, Karlsruhe, Germany

Leo Dorst Intelligent Systems Laboratory, University of Amsterdam, Science Park 107, 1098 XG, Amsterdam, The Netherlands, L.Dorst@uva.nl

Daniel Fontijne University of Amsterdam, Kruislaan 403, 1098 SJ, Amsterdam, The Netherlands, fontijne@science.uva.nl

Takeshi Furuhashi Nagoya University, Furou 1-1, Chikusa, Nagoya, Japan, furuhashi@cse.nagoya-u.ac.jp

David Gonzalez-Aguirre Humanoids and Intelligence Systems Lab, Karlsruhe Institute of Technology, Adenauerring 2, 76131 Karlsruhe, Germany, gonzalez@ira.uka.de

K. Gürlebeck Institut für Mathematik/Physik, Bauhaus-Universität Weimar, Coudraystr. 13B, 99421 Weimar, Germany, klaus.guerlebeck@uni-weimar.de, guerlebe@fossi.uni-weimar.de

David Hestenes Arizona State University, Tempe, AZ 85287, USA, hestenes@asu.edu

Dietmar Hildenbrand Interactive Graphics Systems Group, University of Technology Darmstadt, Darmstadt, Germany, dhilden@gris.informatik.tu-darmstadt.de

Eckhard M.S. Hitzer Department of Applied Physics, University of Fukui, Bunkyo 3-9-1, 910-8507 Fukui, Japan, hitzer@mech.u-fukui.ac.jp

Daisuke Ichikawa Department of Applied Physics, University of Fukui, Bunkyo 3-9-1, 910-8507 Fukui, Japan, ichikawa@hello.apphy.fukui-u.ac.jp

Sio H. Ieng Institut des Systèmes Intelligents et de Robotique (ISIR), 4 place Jussieu, Paris, France, ieng@isir.fr

Bram Knock Clifford Research Group, Faculty of Engineering, Ghent University, Galglaan 2, 9000 Gent, Belgium, bdk@cage.UGent.be

Andreas Koch Embedded Systems and Applications Group, University of Technology Darmstadt, Darmstadt, Germany, koch@esa.informatik.tu-darmstadt.de

Hongbo Li Mathematics Mechanization Key Laboratory, Academy of Mathematics and Systems Science, Chinese Academy of Sciences, Beijing 100080, China, hli@mmrc.iss.ac.cn

Wei Lu Institute of Image Communication and Information Processing, Department of Electronic Engineering, Shanghai Jiao Tong University, Shanghai 200240, China, learza@sjtu.edu.cn

J. Morais Institut für Mathematik/Physik, Bauhaus-Universität Weimar, Coudraystr. 13B, 99421 Weimar, Germany, jmorais@mat.ua.pt

Agnieszka Patyk Faculty of Applied Physics and Mathematics, Gdańsk University of Technology, 80-233 Gdańsk, Poland, patyk@mif.pg.gda.pl; Centrum Leo Apostel (CLEA), Vrije Universiteit Brussel, 1050 Brussels, Belgium

Christian Perwass Institute for Informatics, University of Kiel, 24118 Kiel, Germany, christian@perwass.de

Minh Tuan Pham Nagoya University, Furou 1-1, Chikusa, Nagoya, Japan, minhtuan@cmplx.cse.nagoya-u.ac.jp

Joachim Pitt Interactive Graphics Systems Group, University of Technology Darmstadt, Darmstadt, Germany,

Wieland Reich Institut fuer Informatik, Universität Leipzig, 04009 Leipzig, Germany, reich@informatik.uni-leipzig.de

Jorge Rivera-Rovelo Universidad Anahuac Mayab, Carr. Merida-Progreso Km. 15.5, Merida, Mexico, jorge.rivera@unimayab.edu.mx

Alyn Rockwood Geometric Modeling and Scientific Visualization Research Center, King Abdullah University of Science and Technology, Thuwal, Saudi Arabia, alynrock@yahoo.com

Christophe Saint-Jean Lab. MIA, La Rochelle University, La Rochelle, France, christophe.saintjean@univ-lr.fr

Hennie Schepper Clifford Research Group, Faculty of Engineering, Ghent University, Galglaan 2, 9000 Gent, Belgium, Hennie.DeSchepper@UGent.be

Nele Schepper Clifford Research Group, Department of Mathematical Analysis, Ghent University, Galglaan 2, 9000 Gent, Belgium, nds@cage.ugent.be

Gerik Scheuermann Institut fuer Informatik, Universität Leipzig, 04009 Leipzig, Germany, scheuermann@informatik.uni-leipzig.de

René Schott IECN and LORIA Université Henri Poincaré-Nancy I, BP 239, 54506 Vandoeuvre-lès-Nancy, France, schott@loria.fr

Frank Sommen Clifford Research Group, Department of Mathematical Analysis, Ghent University, Galglaan 2, 9000 Gent, Belgium, fs@cage.ugent.be

Li Song Institute of Image Communication and Information Processing, Department of Electronic Engineering, Shanghai Jiao Tong University, Shanghai 200240, China, song_li@sjtu.edu.cn

Humberto Sossa Center of Computing Research, Juan de Dios Batiz s/n Col. Nueva Industrial Vallejo Gustavo A. Madero, C.P. 07738, Mexico DF, Mexico, hsossa@cic.ipn.mx

W. Sprößig Fakultaet fuer Mathematik und Informatik, TU Bergakademie Freiberg, Prueferstraße 9, 09596 Freiberg, Germany, sproessig@math.tu-freiberg.de

G. Stacey Staples Department of Mathematics and Statistics, Southern Illinois University Edwardsville, Edwardsville, IL 62026-1653, USA, sstaple@siue.edu

Kanta Tachibana Kogakuin University, Nishi-Shinjuku 1-24-2, Tokyo, 163-8677, Japan, kanta@cc.kogakuin.ac.jp

Leonardo Traversoni Ciencias Básicas e Ingenieria, Univ. Autonoma Met. (Iztapalapa), 09340 Mexico, Mexico, ltd@xanum.uam.mx

Yi Xu Institute of Image Communication and Information Processing, Department of Electronic Engineering, Shanghai Jiao Tong University, Shanghai 200240, China, xuyi@sjtu.edu.cn

Xiaokang Yang Institute of Image Communication and Information Processing, Department of Electronic Engineering, Shanghai Jiao Tong University, Shanghai 200240, China, xkyang@sjtu.edu.cn

Tomohiro Yoshikawa Nagoya University, Furou 1-1, Chikusa, Nagoya, Japan, yoshikawa@cse.nagoya-u.ac.jp

Part I
Geometric Algebra

New Tools for Computational Geometry and Rejuvenation of Screw Theory

David Hestenes

Abstract *Conformal Geometric Algebraic* (CGA) provides ideal mathematical tools for construction, analysis, and integration of classical *Euclidean, Inversive & Projective Geometries*, with practical applications to computer science, engineering, and physics. This paper is a comprehensive introduction to a CGA tool kit. Synthetic statements in classical geometry translate directly to coordinate-free algebraic forms. Invariant and covariant methods are coordinated by *conformal splits*, which are readily related to the literature using methods of matrix algebra, biquaternions, and screw theory. Designs for a complete system of powerful tools for the mechanics of linked rigid bodies are presented.

1 Introduction

Euclidean geometry supplies essential conceptual underpinnings for physics and engineering. It was recognized and reported only recently that *conformal geometric algebra* (CGA) provides an ideal algebraic arena for all aspects of Euclidean geometry, from theoretical formulation and analysis to practical design and computation [1]. I am pleased to say that response to that announcement has been rapid and enthusiastic, with extensive applications ranging from computer science and robotics to crystallography reported in these proceedings and elsewhere.

My purpose here is to set forth the central ideas and results of this "conceptual revolution" as a convenient summary for practitioners and an outline for beginners. For historical context and perspective on the relevant scientific literature, I comment on where the ideas have come from and on opportunities for further development. I take this opportunity to make minor changes and corrections to my previous accounts [1, 2], as well as to clarify and emphasize important points that have been generally overlooked. In particular, I recommend special attention to the different

D. Hestenes (✉)
Arizona State University, Tempe, AZ 85287, USA
e-mail: hestenes@asu.edu

E. Bayro-Corrochano, G. Scheuermann (eds.), *Geometric Algebra Computing*,
DOI 10.1007/978-1-84996-108-0_1, © Springer-Verlag London Limited 2010

advantages and roles of invariant and covariant approaches to Euclidean geometry and to the prospects for developing and applying Screw Theory.

The adaptation of CGA to serve the purposes of Euclidean geometry is *a fundamental problem in the design of mathematics*. Its objective is a mathematical system that facilitates geometric modeling and analysis, optimizes computational efficiency, and incorporates all aspects of rigid body mechanics. Mathematical invention is most effective when its purpose is clear.

2 Universal Geometric Algebra

The geometric concept of vector as a directed number has a long historical development culminating in the invention of *geometric algebra* [3]. To define it we begin with the standard notion of a *real vector space* $\mathbb{R}^{r,s}\{a, b, c, \ldots\}$ with *dimension* $r + s = n$. On reflection one can see that the concepts of vector addition and scalar multiplication introduced in this way are insufficient to characterize relative directions among vectors. That deficiency is rectified by introducing an associative *geometric product* defined by the simple rule

$$a^2 = \pm|a|^2, \tag{1}$$

where the real number $|a| = 0$ is the *magnitude* of the vector a, and the sign is its *signature*. The vector is said to be *null* if $a^2 = |a|^2 = 0$. The vector space $\mathbb{R}^{r,s}$ is presumed to have nondegenerate signature $\{r, s\}$, where r/s are maximal subspaces of vectors with *positive/negative* signature. Of course, the signature is not defined without the geometric product, which specifies a multiplicative relation between vectors and scalars. Finally, from the vector space $\mathbb{R}^{r,s}$, the geometric product generates the *real geometric algebra* (GA) $\mathbb{G}^{r,s} = \mathbb{G}(\mathbb{R}^{r,s})$ with elements $\{A, G, M, \ldots\}$ called *multivectors*.

Though this completes the definition of GA as an algebraic system, it is only the beginning for *development of GA as a mathematical language*. Development proceeds by creating a system of definitions and theorems that facilitate algebraic encoding and analysis of geometric concepts. A systematic research program to do precisely that was inaugurated in [4] and greatly extended in [5]. I am pleased to say that many others have joined me in this enterprise, including [6–8] to name only three. These references suffice to show that GA has a broader and deeper range of applications than any other mathematical system, including matrix algebra.

Let me briefly review the basic definitions and theorems needed for most applications of GA: For a pair of vectors, a symmetric *inner product* $a \cdot b$ and antisymmetric *outer product* $a \wedge b$ can be defined implicitly by

$$ab = a \cdot b + a \wedge b \quad \text{and} \quad ba = b \cdot a + b \wedge a = a \cdot b - a \wedge b. \tag{2}$$

It is easy to prove that $a \cdot b$ is scalar valued, while the quantity $a \wedge b$, called a *bivector* or 2-vector, is a new algebraic entity that can be interpreted geometrically as an oriented area.

The antisymmetric outer product can be generalized iteratively to define *k-vectors* by

$$a \wedge A_k = \frac{1}{2}\big(aA_k + (-1)^k A_k a\big), \tag{3}$$

which generates a $(k+1)$-*vector* from k-*vector* A_k. It follows that the outer product of k-vectors is the completely antisymmetric part of their geometric product:

$$a_1 \wedge a_2 \wedge \cdots \wedge a_k = \langle a_1 a_2 \cdots a_k \rangle_k, \tag{4}$$

where the angle bracket means k-*vector part*, and k is its *grade*. This product vanishes if and only if the vectors are linearly dependent. Consequently, the maximal grade for nonzero k-*vectors* is $k = n$. It follows that every multivector A can be expanded into its k-vector parts and the entire algebra can be decomposed into k-vector subspaces:

$$\mathbb{G}^{r,s} = \sum_{k=0}^{n} \mathbb{G}_k^{r,s} = \left\{ A = \sum_{k=0}^{n} \langle A \rangle_k \right\}. \tag{5}$$

This is called a *grading* of the algebra. Note that the grading is generated from primitive elements, the vectors or 1-vectors in this case, with the scalars regarded as elements with grade 0. As seen below, alternative gradings are appropriate for geometric subalgebras.

The inner product can also be generalized, leading to the very useful formula

$$a \cdot (a_1 \wedge a_2 \wedge \cdots \wedge a_k) = \sum_{j=1}^{k} (-1)^{j+1} a \cdot a_j (a_1 \wedge \cdots \wedge \breve{a}_j \wedge \cdots \wedge a_k), \tag{6}$$

where $\breve{a}_j$ indicates a missing factor in the outer product. This formula shows that the inner product is a grade-lowering operator, while (3) shows that the outer product is grade-raising.

Reversing the order of multiplication is called *reversion*, as expressed by

$$(a_1 a_2 \cdots a_k)^{\sim} \equiv a_k \cdots a_2 a_1, \quad \text{and} \quad (a_1 \wedge a_2 \wedge \cdots \wedge a_k)^{\sim} = a_k \wedge \cdots \wedge a_2 \wedge a_1, \tag{7}$$

and the *reverse* of an arbitrary multivector is defined by

$$\widetilde{A} = \sum_{k=0}^{n} \big\langle \widetilde{A} \big\rangle_k \equiv \sum_{k=0}^{n} (-1)^{k(k-1)/2} \langle A \rangle_k. \tag{8}$$

Similarly *(space) inversion* (regarded as the result of reversing the sign of all vectors in geometric products) is defined by

$$A^{\#} \equiv \sum_{k=0}^{n} (-1)^k \langle A \rangle_k = \langle A \rangle_+ - \langle A \rangle_-, \tag{9}$$

where $\langle A \rangle_\pm$ are the parts of $A = \langle A \rangle_+ + \langle A \rangle_-$ with *even/odd parity*, respectively.

The unit *n-vector* (or *pseudoscalar*) is so important that it is given a special symbol and defined (up to a sign) by the properties

$$\mathrm{I} = \langle \mathrm{I} \rangle_n, \qquad \widetilde{\mathrm{I}}\mathrm{I} = (-1)^s, \qquad a \wedge \mathrm{I} = 0 \quad \text{for every vector } a. \tag{10}$$

Every multivector A has a *dual* defined by $A^* = A\mathrm{I}^{-1}$. This leads to the basic theorem relating inner and outer products by duality:

$$a \cdot A^* = a \cdot \left(A\mathrm{I}^{-1}\right) = (a \wedge A)\mathrm{I}^{-1} = (a \wedge A)^*. \tag{11}$$

The algebraic system described to this point may be referred to as *universal GA*, because of its broad applicability. That distinguishes it from specialized versions of GA, such as the *spacetime algebra* [4, 6] tailored to the geometry of spacetime.

Names and nomenclature are of great importance in science and mathematics, as they can suppress or reveal deep conceptual distinctions. In this regard, it is important to mark crucial conceptual differences between *Geometric Algebra* (GA) and *Clifford Algebra* (CA) that are often overlooked in the literature. Though they have a common root in the work of W.K. Clifford, CA has been cultivated by mathematicians up to present times as one among many formal algebraic systems with little attention to its geometric meaning. In contrast, the systematic development of GA as a universal geometric language is a more recent development that is still underway. Since Clifford himself proposed the name *Geometric Algebra*, I believe he would be embarrassed to have it named after him; for he was well aware of its universal geometric import, and he attributed to Hermann Grassmann [9] the chief role in its creation. Indeed, Clifford's development of the algebra hardly progressed beyond the rudiments, though it is clear that he had deep insights into the geometric and algebraic issues [10]. No doubt the history of Geometric Algebra would have been quite different if not for Clifford's tragic early death.

Here are some important differences in viewpoint between CA and GA. CA is typically defined with complex numbers instead of the reals as scalars, whereas GA contends that this obscures geometric meaning without providing greater generality. CA is often defined as an ideal in tensor algebra, whereas GA defines tensors as multilinear functions of vector variables [5]. CA is often characterized as the "algebra of a quadratic form," with $a \cdot b$ interpreted as a metric tensor. In contrast, GA contends that the inner product should be regarded as a *contraction* (or grade-lowering operation) as originally conceived by Grassmann. Any metric tensor can then be defined as a scalar-valued bilinear function $g(a,b) = a \cdot \underline{g}(b)$, where $\underline{g}$ is a linear operator, with $a \cdot b$ as the most important special case. Many further differences between GA and CA are obvious in applications.

3 Group Theory with Geometric Algebra

A multivector G that can be expressed as a geometric product $G = n_k \ldots n_2 n_1$ of nonnull vectors is called a *versor*. Obviously it has a multiplicative inverse $G^{-1} =$

$n_k^{-1}\dots n_2^{-1}n_1^{-1}$ and even or odd parity given by $G^{\#} = (-1)^k G$. The multiplicative group of all versors in $\mathbb{G}^{r,s}$ is the *Pin group*

$$\mathrm{Pin}(r, s) = \{G \mid GG^{-1} = 1\}, \tag{12}$$

within which the subgroup of all versors with even parity is the *Spin group*

$$\mathrm{Spin}(r, s) = \{G \mid G^{\#} = G\}. \tag{13}$$

The *orthogonal group* on $\mathbb{R}^{r,s}$ is the group of all *isometries*, that is, linear transformations G that preserve the magnitude of each vector a. In GA it has the elegant representation

$$\mathrm{O}(r, s) = \{\underline{G} \mid \underline{G}(a) = G^{\#}aG^{-1}\}. \tag{14}$$

The versors in $\mathrm{Pin}(r, s)$ are thus generators of the orthogonal group. The versors of $\mathrm{Spin}(r, s)$ generate the *special orthogonal group* $\mathrm{SO}(r, s)$, a subgroup of $\mathrm{O}(r, s)$ sometimes called the rotation group.

This GA approach to isometries has considerable advantages over matrix representations: First, it is completely coordinate-free. Second, it reduces group composition of isometries $\underline{G}_2\underline{G}_1 = \underline{G}_3$ to simple multiplication of versors $G_2G_1 = G_3$. At the same time, it establishes direct connection between the orthogonal group and its *covering* by the pin group. Third, it facilitates reduction of isometries to their irreducible elements, namely, reflection in a hyperplane determined by its vector normal n_i:

$$\underline{G}(a) = n_i^{\#}an_i^{-1} = -n_i a n_i^{-1}. \tag{15}$$

It is a simple matter then to prove the important

Cartan–Dieudonné Theorem *Every isometry of* $\mathbb{R}^{r,s}$ *can be reduced to at most* $n = r + s$ *reflections in hyperplanes.*

As usual, credit for the theorem could probably be more fairly attributed to others, most notably to Lipschitz (1880), who pioneered the approach to isometries presented here. The best practice may be to give it a descriptive name such as "*Isometry Reduction Theorem*."

Evidently the GA approach can be profitably extended to the whole of group representation theory [5]. The *classical groups* have been treated in [11]. We will be most interested below in the *conformal group* $\mathrm{C}(r, s)$, which has a representation in GA specified by the isomorphism

$$\mathrm{C}(r, s) \cong \mathrm{O}(r + 1, s + 1). \tag{16}$$

This representation is so useful that we shall refer to $\mathbb{G}^{r+1,s+1}$ as *Conformal Geometric Algebra* (CGA).

This completes our summary of universal GA and its relation to group theory. In the following we concentrate on practical applications to Euclidean geometry, knowing full well that our results are readily generalized to spaces of arbitrary dimension and signature.

4 Euclidean Geometry with Conformal GA

The conformal model of Euclidean 3-space $\mathbb{E}^3$ is embedded in the CGA $\mathbb{G}^{4,1} = \mathbb{G}(\mathbb{R}^{4,1})$ as follows: First, we identify Euclidean points with vectors in the null cone

$$\mathbb{N}^{4,1} \equiv \{x \in \mathbb{R}^{4,1} \mid x^2 = 0\}. \tag{17}$$

Next, we reduce the remaining degrees of freedom from four to three by choosing a point at infinity $e \equiv x_\infty$ and normalizing all points to the hyperplane $\{x \mid e \cdot x = -1,\ e^2 = 0\}$. Thus, we have

$$\mathbb{E}^3 \cong \mathbb{N}^{4,1} \equiv \{x \in \mathbb{R}^{4,1} \mid x^2 = 0,\ x \cdot e = -1\}. \tag{18}$$

Finally, we confirm this as a model of Euclidean space by verifying that

$$|x_2 - x_1|^2 = (x_2 - x_1)^2 = -2x_2 \cdot x_1 \tag{19}$$

correctly determines the *Euclidean distance* $|x_2 - x_1|$ between any two points. The argument is completed below.

The amazing fact about this embedding of $\mathbb{E}^3$ in CGA is that it automatically imbues all elements of $\mathbb{G}^{4,1}$ with rich geometric meaning and thereby facilitates formulation, analysis, and computation in all aspects of Euclidean geometry. It has two major advantages:

First, it unites the conceptual advantages of classical synthetic geometry with the analytic power of algebra in providing direct algebraic representations of basic geometric objects and their properties.

Second, it enlists the apparatus of the conformal versor groups for multiplicative, coordinate- free representation of Euclidean symmetries and transformations. Specifically, the invariance group of the Euclidean metric (19) is the *Euclidean group* $E(3) = \{\underline{G}\}$, defined as a subgroup of the conformal group $\mathrm{C}(3,0) \cong \mathrm{O}(4,1)$ by the constraint

$$\underline{G}(e) = G^{\#} e G^{-1} = e. \tag{20}$$

This group includes reflections. Its restriction to rigid displacements by requiring $G^{\#} = G$ is the *Special Euclidean group* SE(3). The rest of this paper is an elaboration of these two points with specific recommendations for notation, representation, and method. The subject is young and fluid, so the setting of standards for practice is still open. Of course, I cannot cover everything. For further details and explanation, I refer the serious student to [7], which provides the most thorough exposition of CGA to date, with due emphasis on geometric visualization. Comparison with the present account shows where I think that exposition can be improved.

In CGA the basic *Geometric Objects* $\{O = C, L, S, P\}$ of 3D Euclidean geometry can be defined as follows:

- A *Circle* C is determined by three points:

$$C = x_1 \wedge x_2 \wedge x_3. \tag{21}$$

- A *Line* L is a circle through the point at infinity:

$$L = x_1 \wedge x_2 \wedge e. \tag{22}$$

- A *Sphere* S is determined by four points:

$$S = x_1 \wedge x_2 \wedge x_3 \wedge x_4. \tag{23}$$

- A *Plane* P is a sphere through the point at infinity:

$$P = x_1 \wedge x_2 \wedge x_3 \wedge e. \tag{24}$$

- A *Point* x lies on object O if and only if:

$$x \wedge O = 0. \tag{25}$$

Note the distinction between a geometric object O (defined algebraically) and the set of points $\mathbb{O}$ it determines, as expressed by

$$\mathbb{L}\text{ine} = \{x \mid x \wedge L = 0\}, \qquad \mathbb{P}\text{lane} = \{x \mid x \wedge P = 0\}. \tag{26}$$

In this respect we follow Euclid in introducing points and lines as distinct objects with properties specified by a system of axioms. The idea of defining a line as a set of points emerged in the 19th century with "containment" replacing the geometric concept of "incidence" as the basic relation between points and lines. From our perspective, the limitations of that idea are clear, so we are prepared to use the concepts of set theory but not to confuse them with concepts of geometry.

Of course our concept of geometric object goes beyond Euclid's, most notably in assigning to each an *orientation* (algebraic sign) and a *weight* or *magnitude* (e.g. *length*, *area*, *volume*). Thus, interchanging the product of points in (21)–(24) reverses the sign, hence orientation, of the objects.

For many purposes, the *dual representation* for a geometric object $O^* = O\mathrm{I}^{-1}$ is most convenient. From the duality of inner an outer products (11) it follows that the intersection with a point (25) is then expressed by

$$x \cdot O^* = 0. \tag{27}$$

For a plane, the dual $P^* = P\mathrm{I}^{-1} = n$ is a vector *normal* to the plane (note the use of lower-case letters for vectors). The equation $x \cdot n = 0$ has the familiar form of an equation for a plane through the origin of a vector space, but in this case it applies to any plane in $\mathbb{E}^3$. For the normal, n specifies a location as well as an orientation for the plane. Moreover, the separation of $\mathbb{E}^3$ into disjoint subsets can be neatly expressed by the inequality $x \cdot n > 0$ for points *in front of* the plane, and $x \cdot n < 0$ for points *behind* the plane.

The *intersection* of two planes $P_1 = n_1\mathrm{I}$ and $P_2 = n_2\mathrm{I}$ is a line specified by

$$P_1^* \cdot P_2 = n_1 \cdot P_2 = n_1 \cdot (n_2\mathrm{I}) = (n_1 \wedge n_2)\mathrm{I}. \tag{28}$$

Obviously, this vanishes if the planes are parallel. Moreover, as will be evident later, with the normalization $n_1^2 = n_2^2 = 1$, the magnitude $|n_1 \wedge n_2|$ is the *sine* of the dihedral angle between the intersecting planes.

Similar expressions for the mutual intersections of lines, planes, circles and spheres are discussed in [7].

5 Invariant Euclidean Geometry

There are two different ways to formulate the equations of spacetime physics: (1) *covariant formulations* expressed with respect to one inertial frame and related to other frames by Lorentz transformations, and (2) *invariant formulations* independent of any reference frame choice. Experts prefer to work with invariants, because they are invariably simpler than covariants. However, beginners are usually introduced to a covariant approach, mainly because of educational tradition.

In precise analogy, there are *covariant formulations* of Euclidean geometry that depend on designating an arbitrary point as origin, and *invariant formulations* that do not. The conformal model supports both approaches, so we should examine their respective advantages and how they are related.

The characterization of geometric objects in the preceding section is already an invariant formulation. Let us consider it more closely for extension to an invariant treatment of any topic in Euclidean geometry. We have seen that CGA supplies sufficient algebraic structure to define basic geometric objects. Now note that the structure of CGA suggests a somewhat different approach to geometric primitives than the classical one.

The primitive algebraic objects are vectors. In CGA there are four types of vector with distinct geometric meanings:

$$\begin{aligned} &\mathbb{P}\text{oints:} \quad \{x \mid x^2 = 0,\ x \cdot e = -1\}, \\ &\mathbb{P}\text{lanes:} \quad \{n \mid n^2 > 0,\ n \cdot e = 0\}, \\ &\mathbb{S}\text{pheres:} \quad \{s \mid s^2 = \pm\rho^2,\ s \cdot e = -1\}. \end{aligned} \tag{29}$$

This suggests that the dual form for a plane $n = P^*$ should be regarded as more fundamental than the 4-vector form in (24). It is also algebraically much more convenient, especially for generating translations, as shown below.

According to (29), there are two types of sphere with radius ρ, corresponding to the two signs in the square of the sphere vector s. The length of the sphere vector is scaled so that its square gives the radius directly. The two sphere types are designated as *real* and *imaginary* for positive and negative square respectively. A *real sphere* $s = S^*$ is the dual of the 4-vector sphere S in (23). It is of interest to note that the *center* c of a real sphere can be obtained by a suitably scaled reflection from the point at infinity:

$$c = -\frac{1}{2} ses = -\frac{1}{2}(2e \cdot s - es)s = s + \frac{1}{2}\rho^2 e. \tag{30}$$

Fig. 1

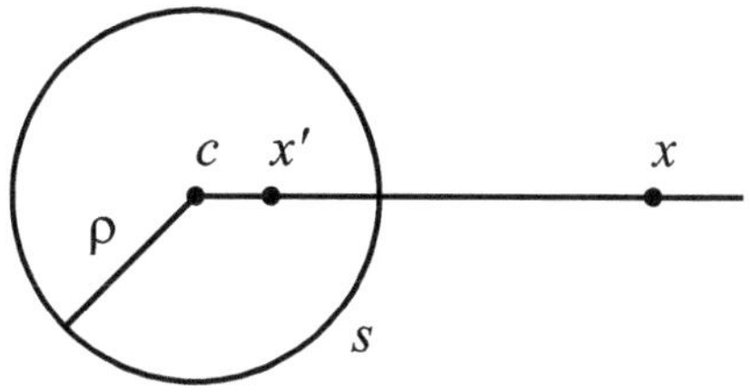

An easy check verifies that c does indeed have the properties of a point. Moreover, this gives us a natural measure for distance from a point to a sphere:

$$2s \cdot x = 2\left(c - \frac{1}{2}\rho^2 e\right) \cdot x = \rho^2 - |x - c|^2. \tag{31}$$

Thus, the point x is *inside*, *on*, or *outside* the sphere when $s \cdot x$ is *positive*, *zero*, or *negative* respectively. The order is reversed by changing the sign (orientation) of s.

Note that the reflection (30) has been defined with respect to the radius of the sphere instead of unity. This kind of reflection is called *inversion in a sphere*. Applied to an arbitrary point, it gives a new point,

$$x' = -\frac{1}{2} sxs, \tag{32}$$

and a little algebra reveals the distance inversion

$$\left|x' - c\right|^2 = \frac{\rho^4}{|x - c|^2}, \tag{33}$$

as illustrated in Fig. 1. Though sphere inversion does not preserve Euclidean distance, it is a powerful means for geometric design and analysis.

The geometric significance of *imaginary spheres* is more subtle, and the reader is referred to [7] for a discussion. The main point of interest here is that all four vector types in (29) are needed for computational Euclidean geometry. Subject to scaling, these constitute all the vectors in the CGA $\mathbb{G}^{4,1} = \mathbb{G}(\mathbb{R}^{4,1})$. We can conclude then that CGA supplies precisely the algebraic structure needed for Euclidean geometry without anything superfluous.

Geometry can be regarded as a system of relations among points. Accordingly, the most basic is the relation between two points described by the vector $n_{21} = x_2 - x_1$. This vector is so important that it deserves a name. I like the venerable old term *chord*, especially when it is relating points in a geometric figure or physical object. Of course, it serves as a *displacement vector* in other contexts. However, it has another geometric property unique to CGA; it is the *perpendicular bisector* of the interval between the two points. The fact that it is the normal for a plane is confirmed immediately by $n_{21} \cdot e = 0$. The fact that it is the bisecting plane is confirmed from the equation $n_{21} \cdot x = 0$, which as illustrated in Fig. 2, implies (using (19)) that

$$|x - x_1|^2 = -2x \cdot x_1 = |x - x_2|^2.$$

Fig. 2

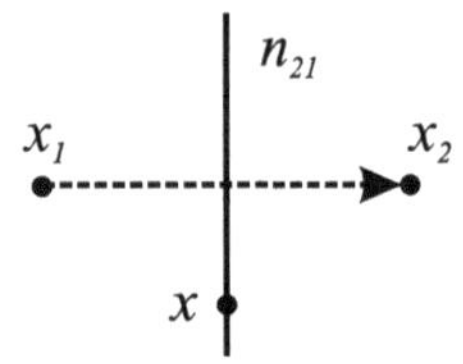

Fig. 3

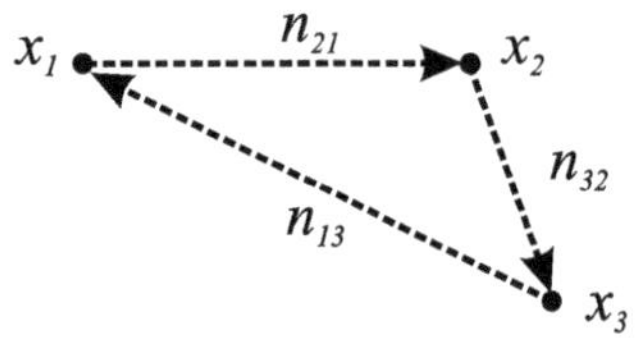

The chords of a *triangle* are especially significant (Fig. 3), for they determine the basic properties of the Euclidean metric:

$$|x_i - x_j|^2 = n_{ij}^2 \geq 0.$$

The chords are related by the triangle equation

$$n_{21} + n_{32} + n_{13} = 0.$$

This gives us immediately the familiar *law of cosines*,

$$n_{21}^2 + n_{32}^2 + 2n_{32} \cdot n_{21} = n_{13}^2,$$

which determines the basic triangle inequalities for Euclidean distances.

Versor products among the chords generate all the reflection and rotation symmetries of a triangle and, consequently, values for all the vertex angles. For example, the versor $n_{32}n_{21}$ is "complementary" to a rotation "about the vertex x_2," and vanishing of its scalar part reduces the law of cosines to the Pythagorean theorem. Note that the chord n_{ij} generates a reflection that takes point x_i to x_j or vise versa. Hence, the versor product of successive chords generates a walk of reflections along any sequence of points, which may return to the initial point if the path is closed, as in a walk around a triangle.

There is much more to be derived from the invariant approach to Euclidean geometry. For example, according to (21) and (24), the outer product of three vertices in a triangle determines its *circumcircle* and the plane in which it lies. Of course, all this applies to 2D as well as 3D geometry. It would interesting to work out what insights and simplifications it brings to the great theorems of classical geometry, such as the *nine circle theorem*. Indeed, the results may even have practical value in applications to mechanical engineering, as we see in later sections.

Finally, to complete our discussion of two point geometry, we note that the sum

$$s_{21} = x_2 + x_1 = c_{21} + \frac{1}{2}\rho_{21}^2 e \tag{34}$$

is a sphere with center c_{21} and poles at the two points. However, in contrast to the real sphere (30), it is an imaginary sphere. Its role in Euclidean geometry remains to be worked out.

6 Projective Geometry

Projective geometry is useful in many applications—in computer vision, for example—but its methodology stands apart from the rest of mathematics. To make the conceptual assets of projective geometry readily available, we need to incorporate them into the algebraic design of CGA. As Dieudonné has famously declared, projective geometry is nothing but linear algebra. Accordingly, let us consider a generic, nonsingular linear transformation that leaves the point at infinity invariant (up to a scale factor σ at least):

$$\underline{f} : x \mapsto x' = \underline{f}(x) \quad \text{with } \underline{f}(e) = \sigma e. \tag{35}$$

The trouble with this is that it need not preserve the null property of points, so we have

$$\underline{f} : x^2 = 0 \mapsto \left[\underline{f}(x)^2\right] = \underline{f}(x) \cdot \underline{f}(x) = x \cdot \overline{f}\underline{f}(x) \neq 0.$$

Note: the underbar notation $\underline{f}$ denotes a linear operator, while the overbar $\overline{f}$ denotes its adjoint. To solve this problem, Anthony Lasenby [12] has proposed that we extend the notion of points to include planes regarded as boundary points at ∞. Thus, we extend our model of Euclidean space to include two kinds of points:

$$\begin{aligned} &\textit{interior points}: && \left\{x \mid x^2 = 0,\ x \cdot e = -1\right\}, \\ &\textit{boundary points}: && \left\{n \mid n^2 = 1,\ n \cdot e = 0\right\}, \end{aligned}$$

where the boundary points, like the interior points, are normalized to make them unique. The set of boundary points can thus be regarded as a $\mathbb{P}$lane (of directions) at ∞. Indeed, each boundary point can be regarded as the intersection of parallel lines at ∞, as parallel lines have a common direction. This is an old idea dating back to Kepler.

Now, projective geometry suppresses metrical notions of scale while maintaining the geometric concept of *incidence*, as expressed by (28). As that equation requires use of the pseudoscalar and duality, we must extend our notion of projective transformations to accommodate them. Happily, GA provides a natural way to do precisely that.

A great advantage of GA is that it enables natural extension of a linear transformation on vectors to the entire algebra. This extension is called an *outermorphism* [5–7, 13] because it preserves the outer product (hence grade), as expressed by

$$\underline{f}(x \wedge M) = \underline{f}(x) \wedge \underline{f}(M). \tag{36}$$

It follows that the pseudoscalar is an eigenblade of the outermorphism, with the determinant as its eigenvalue:

$$\underline{f}(\mathrm{I}) = (\det \underline{f})\mathrm{I}. \tag{37}$$

This prepares us for the fundamental theorem [13]:

$$(\det \underline{f})\underline{f}\big(A^* \cdot B\big) = \underline{f}(A)^* \cdot \underline{f}(B), \tag{38}$$

which can be made to look more elegant by absorbing $(\det \underline{f})$ in A^* defined as the dual with respect to the transformed pseudoscalar (37). This theorem could fairly be called the *Incidence Theorem*, because it expresses the fact that outermorphisms preserve the incidence property (28).

This fundamental theorem of linear algebra has been almost totally overlooked in the literature, presumably because it is not so naturally expressed in standard formalisms.

These developments invite us to employ the pseudoscalar to define "complex objects" such as

$$N = x + \mathrm{I}n, \tag{39}$$

where x is an interior point, and n is a boundary point. As explained in a later section, this object can interpreted as a *point in a plane* if $n \cdot x = 0$ or, equivalently, $N^2 = -1$. All this suggests that we should define *projective transformations* by extending outermorphisms to include duality transformations.

I believe that we now have all the necessary ingredients to incorporate projective geometry smoothly into CGA. There remains the large task of reformulating the classical results of projective geometry. Since modern works have already formulated many of these results in terms of linear algebra [14], the task should be fairly straightforward. I recommend it as a good topic for a doctoral thesis.

7 Covariant Euclidean Geometry with Conformal Splits

The most widely used model of Euclidean geometry by far is the vector space model based on the isomorphism of Euclidean space to a real vector space with Euclidean inner product:

$$\mathbb{E}^3 \cong \mathbb{R}^3\{\mathbf{x}\}. \tag{40}$$

The most effective means of exploiting this model is through its geometric algebra:

$$\mathbb{G}^3 = \mathbb{G}\big(\mathbb{R}^3\big) = \{\alpha + \mathbf{a} + i\mathbf{b} + i\beta\}, \tag{41}$$

where i is the unit right-handed pseudoscalar, and the geometric product of vectors articulates perfectly with the standard dot and cross products:

$$\mathbf{ab} = \mathbf{a} \cdot \mathbf{b} + \mathbf{a} \wedge \mathbf{b} = \mathbf{a} \cdot \mathbf{b} + i\mathbf{a} \times \mathbf{b}. \tag{42}$$

Efficient methods for applying $\mathbb{G}^3$ to any aspect of mechanics are well developed with many innovative features [15]. In particular, details of the quaternion theory of rotations are thoroughly worked out and smoothly articulated with standard vector methods and matrix representations.

These results even articulate smoothly with the arcane literature on applications of complex quaternions to geometry and mechanics. For it is evident in (41) that complex quaternions are isomorphic to multivectors in $\mathbb{G}^3$, though practitioners have not realized that their unit imaginary can be interpreted geometrically as a pseudoscalar.

Despite all these advantages, the algebra $\mathbb{G}^3$ suffers from the drawback of all vector space models, namely, that the vector space (40) singles out the origin as a preferred point. In other words, it introduces an asymmetry that is not inherent in the concept of Euclidean space. Happily, that can be remedied by embedding the vector space model in the conformal model, or better, by factoring it out of the conformal model. We consider two ways to do that.

The first way is a *conformal split* of CGA into a commuting product of subalgebras:

$$\mathbb{G}^{4,1} = \mathbb{G}^3 \otimes \mathbb{G}^{1,1}. \tag{43}$$

The split is defined geometrically by choosing one point e_0 as origin and noting that every other point x lies on the bundle of lines through that point. This defines a mapping of points into trivectors:

$$\mathbf{x} = x \wedge e_0 \wedge e, \tag{44}$$

which we identify with the vectors in (40). Thus, with a *regrading* of trivectors as vectors, we generate $\mathbb{G}^3$ as a subalgebra of $\mathbb{G}^{4,1}$.

The other subalgebra $\mathbb{G}^{1,1} = \mathbb{G}(\mathbb{R}^{1,1})$ in (43) is generated from the null vectors $\{e_0, e\}$. Its pseudoscalar is a bivector of sufficient importance to merit a special symbol:

$$E = e_0 \wedge e \quad \text{with} \quad E^2 = (e \cdot e_0)^2 = 1. \tag{45}$$

We examine this algebra more fully later on. For now, it suffices to note that its content, though not its structure, depends on the arbitrary choice of the origin point e_0. It is *covariant* in the sense that it changes with a change of origin. I have dubbed it *conformal split*, because it is deeply analogous to the spacetime split [4, 6], which is so useful in spacetime physics. The spacetime split is generated by selecting a timelike vector rather than a null vector as here. Otherwise, the structure and utility of the splits are quite comparable.

The nature of the conformal split may be clarified by examining a basis for $\mathbb{R}^{4,1}$:

$$\{e, e_0, e_1, e_2, e_3\} \quad \text{with } e_j \cdot e_k = \delta_{jk} \text{ for } j,k = 1,2,3 \text{ and } e \cdot e_k = 0 = e_0 \cdot e_k. \tag{46}$$

This generates a basis for $\mathbb{R}^3$:

$$\left\{\sigma_k = e_k \wedge e_0 \wedge e = e_k(e_0 \wedge e) = e_k E = E e_k\right\} \tag{47}$$

and a pseudoscalar

$$i = \boldsymbol{\sigma}_1\boldsymbol{\sigma}_2\boldsymbol{\sigma}_3 = (e_1E)(e_2E)(e_3E) = e_1e_2e_3E = \mathrm{I}. \tag{48}$$

Thus, the pseudoscalar for $\mathbb{G}^3$ is identical to the pseudoscalar for $\mathbb{G}^{4,1}$. It is an invariant of the conformal split!

An alternative to the conformal split is the *additive split*:

$$\mathbb{G}^{4,1} = \mathbb{G}(\mathbb{R}^3_+ \oplus \mathbb{R}^{1,1}) \equiv \mathbb{G}^3_+ \oplus \mathbb{G}^{1,1}, \tag{49}$$

defined by choosing $\{e_1, e_2, e_3\}$ from (46) as a basis for $\mathbb{R}^3_+$. Unlike the basis (47), the basis in this case is not algebraically associated with lines through a point, and the pseudoscalar $\mathrm{I}_3 = e_1e_2e_3 = \mathrm{I}E$ is not an invariant. Furthermore, the $\boldsymbol{\sigma}_k$ commute with e, while the e_k do not. Consequently, the additive split is not as convenient as the conformal split. Even so, it has its place, most notably in modeling a rigid body, as we shall see.

To demonstrate the felicity of the conformal split for relating invariant forms for geometric objects to standard vector space forms, results for the most basic geometric objects (point, line, plane) are given here.

The mapping (44) of point x to vector $\mathbf{x} = x \wedge E$ can be inverted. The slickest way to do that is to use the geometric product thus:

$$xE = x \wedge E + x \cdot E \quad \text{with } x \cdot E = x \cdot (e_0 \wedge e) = (x \cdot e_0)e + e_0. \tag{50}$$

Multiplying the first equation by its reverse, we get

$$0 = (x \wedge E)^2 - (x \cdot E)^2; \quad \text{whence } \mathbf{x}^2 = (x \cdot E)^2 = -2x \cdot e_0. \tag{51}$$

Inserting this back into (50), we solve to get

$$x = \left(\mathbf{x} - \frac{1}{2}\mathbf{x}^2 e + e_0\right)E = E\left(\mathbf{x} + \frac{1}{2}\mathbf{x}^2 e - e_0\right) = \mathbf{x}E + \frac{1}{2}\mathbf{x}^2 e + e_0. \tag{52}$$

This can be regarded as the conformal split of point x with respect to point e_0. Illustrations of the two representations for points are superimposed in Fig. 4, which may be misleading because the points are related by projection.

The conformal split of a line L through points x and a gives

$$L = x \wedge a \wedge e = \mathbf{x} \wedge \mathbf{a}e + (\mathbf{a} - \mathbf{x}) = (\mathbf{d}e + 1)\mathbf{n}. \tag{53}$$

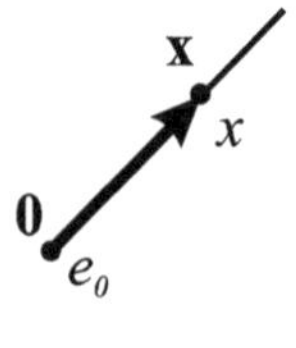

Fig. 4

Fig. 5

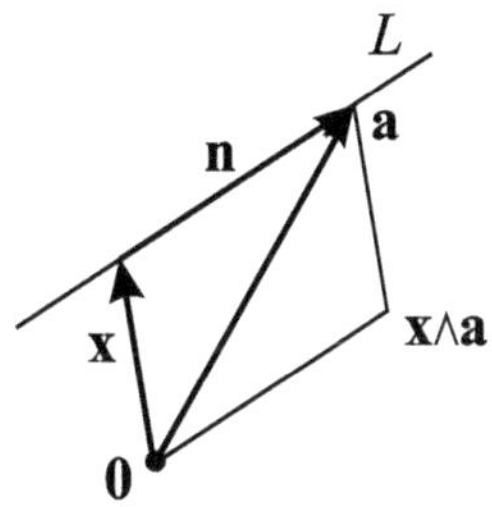

Fig. 6

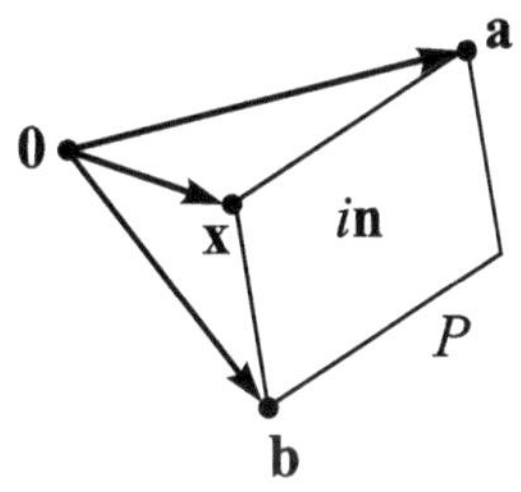

Note that this represents the line in terms of its Plücker coordinates, which consists of a vector, bivector pair for the line tangent, and moment with respect to the origin (Fig. 5):

$$\textit{tangent}: \quad \mathbf{n} = \mathbf{a} - \mathbf{x}, \tag{54}$$

$$\textit{moment}: \quad \mathbf{x} \wedge \mathbf{a} = \mathbf{x} \wedge (\mathbf{a} - \mathbf{x}) = \mathbf{dn}. \tag{55}$$

The directed distance from origin to line is given by the *directance*:

$$\mathbf{d} = (\mathbf{x} \wedge \mathbf{a})\mathbf{n}^{-1} = (\mathbf{x} \wedge \mathbf{n})\mathbf{n}^{-1} = \mathbf{x} - \big(\mathbf{x} \cdot \mathbf{n}^{-1}\big)\mathbf{n}. \tag{56}$$

The conformal split of a plane P through points x, a, b gives

$$P = x \wedge a \wedge b \wedge e = \mathbf{x} \wedge \mathbf{a} \wedge \mathbf{b}\, e + (\mathbf{a} - \mathbf{x}) \wedge (\mathbf{b} - \mathbf{x}) E. \tag{57}$$

Its *Plücker coordinates* consists of the bivector–trivector pair (Fig. 6):

$$\textit{tangent}: \quad (\mathbf{a} - \mathbf{x}) \wedge (\mathbf{b} - \mathbf{x}) = \mathbf{x} \wedge \mathbf{a} + \mathbf{a} \wedge \mathbf{b} + \mathbf{b} \wedge \mathbf{x} = i\mathbf{n}, \tag{58}$$

$$\textit{moment}: \quad \mathbf{x} \wedge \mathbf{a} \wedge \mathbf{b} = \mathbf{x} \wedge \big[(\mathbf{a} - \mathbf{x}) \wedge (\mathbf{b} - \mathbf{x})\big] = \mathbf{x} \wedge (i\mathbf{n}) = i(\mathbf{x} \cdot \mathbf{n}). \tag{59}$$

The dual form for the plane is:

$$P = i(\mathbf{xn}e + \mathbf{n}E) = in. \tag{60}$$

More explicitly, split of the plane normal n (Fig. 2) gives us

$$n = x_2 - x_1 = (\mathbf{x}_2 - \mathbf{x}_1)E + \frac{1}{2}(\mathbf{x}_2^2 - \mathbf{x}_1^2)e$$
$$= (\mathbf{x}_2 - \mathbf{x}_1)E + \frac{1}{2}(\mathbf{x}_2 + \mathbf{x}_1) \cdot (\mathbf{x}_2 - \mathbf{x}_1)e = \mathbf{n}E + \mathbf{c} \cdot \mathbf{n}e. \tag{61}$$

The invariant forms for geometric objects are obviously much simpler than the split forms. Therefore it is preferable to work with invariant forms directly. However, the split forms are essential for relating results to the literature, so we will be using them for that purpose below.

8 Rigid Displacements

From (20) it follows that every *rigid displacement* $\underline{D}$ is a linear transformation of the form

$$\underline{D} : x \mapsto x' = \underline{D}(x) = Dx\widetilde{D}, \tag{62}$$

where its generator D is a versor of even parity that commutes with the point at infinity and is normalized to unity; that is,

$$D^{\#} = D, \qquad De = eD, \qquad DD^{-1} = D\widetilde{D} = 1. \tag{63}$$

With respect to any chosen point e_0, the displacement can decomposed into a rotation $\underline{R}$ followed by a translation $\underline{T}$ or vice versa. This defines a *conformal split* of the displacement as follows:

$$\underline{D} = \underline{T}\,\underline{R} = \underline{R}'\underline{T}, \tag{64}$$

where the rotations satisfy

$$\underline{R}(e_0) = Re_0\widetilde{R} = e_0, \qquad \underline{R}(e_0') = R'e_0'\widetilde{R}' = e_0', \tag{65}$$

and the translation

$$\underline{T}(e_0) = Te_0\widetilde{T} = e_0' = \underline{D}(e_0) \tag{66}$$

is determined solely by the endpoints e_0 and e_0'. For given D and any choice of e_0, the translation can be computed, and the rotation is determined by $R = \widetilde{T}D$. As all displacements can be generated from reflections in planes, let us consider the various possibilities along with their conformal splits.

Reflection in a plane with unit normal n and $e \cdot n = 0$ is specified by

$$x' = \underline{n}(x) = -nxn = x - 2x \cdot nn. \tag{67}$$

If the plane passes through a point c, we have $c \cdot n = 0$ and the conformal split

$$n = \mathbf{n}E + \mathbf{c}\mathbf{n}e, \quad \text{whence } x \cdot n = \mathbf{n}(\mathbf{x} - \mathbf{c}). \tag{68}$$

Fig. 7

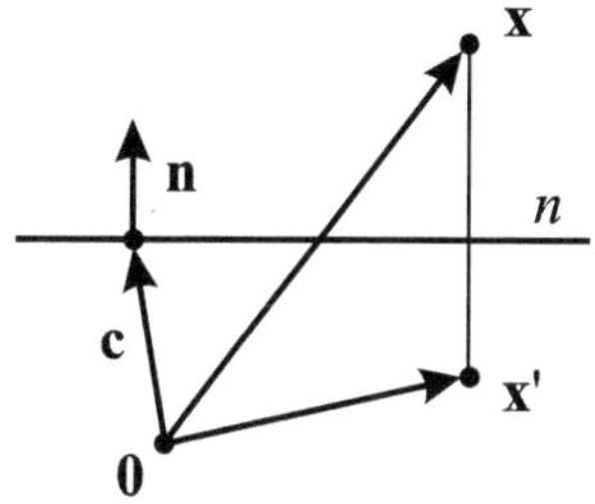

Fig. 8

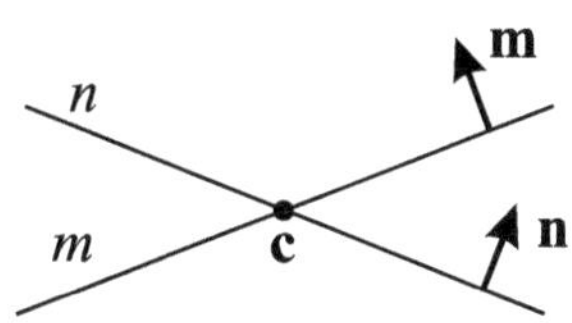

Fig. 9

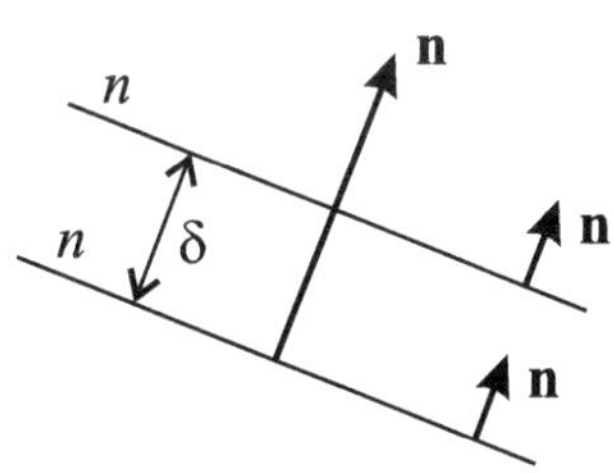

Whence,

$$\mathbf{x}' = \mathbf{x} - 2(\mathbf{x} - \mathbf{c}) \cdot \mathbf{n}\mathbf{n}, \tag{69}$$

as shown in Fig. 7.

Rotation by planes n and m intersecting through a point c (Fig. 8) is generated by

$$\begin{aligned} R_c = mn &= (\mathbf{m}E + \mathbf{m} \cdot \mathbf{c}e)(\mathbf{n}E + \mathbf{n} \cdot \mathbf{c}e) \\ &= \mathbf{mn} + e(\mathbf{m} \wedge \mathbf{n}) \cdot \mathbf{c} = R + e(R \times \mathbf{c}) = T_c^{-1} R T_c, \end{aligned} \tag{70}$$

where T_c generates the translation from origin e_0 to c, $R = \mathbf{mn}$, and we have used the *commutator product*, defined by

$$A \times B = \frac{1}{2}(AB - BA). \tag{71}$$

Note that R_c in (70) can be identified with R' in (65) if $c = e_0'$ and $T_c^{-1} = T$ in (66).

Translation through parallel planes n and m is generated by

$$T_a = mn = (\mathbf{n}E + 0)(\mathbf{n}E + \delta e) = 1 + \frac{1}{2}\mathbf{a}e, \tag{72}$$

where $\mathbf{a} = 2\mathbf{n}\delta$, and, without loss of generality, one plane is presumed to pass through the origin (Fig. 9).

Now we can use (72) and (61) to evaluate $T = T_a$ in (66), with the result

$$a = e_0' - e_0 = \mathbf{a} + \frac{1}{2}\mathbf{a}^2 e. \tag{73}$$

9 Framing a Rigid Body

The *position* and *attitude* of a rigid body in space is uniquely determined by specifying the positions of four points, say $\{x, x_1, x_2, x_3\}$, embedded in the body. Identifying position with the *base point* x, the attitude can be represented by the *body frame* $\{e_k = x_k - x\}$, as illustrated in Fig. 10.

And it is most convenient to orthonormalize the body frame, so $e_j \cdot e_k = \delta_{jk}$. The body frame represents attitude by a set of three vectors. A promising alternative representation in terms of a single geometric object has been proposed by Selig [16] following ideas of Engels. He defines a *Flag* geometrically as *a point on a line in a plane*. CGA gives it the elegant algebraic form,

$$F = x + L + P = x + \mathrm{I}Q, \tag{74}$$

where line L and plane P are defined by

$$L = x \wedge x_1 \wedge e = x \wedge (x_1 - x) \wedge e = x \wedge e_1 \wedge e = \mathrm{I}n_2 n_3,$$
$$P = x \wedge x_1 \wedge x_2 \wedge e = x \wedge e_1 \wedge e_2 \wedge e = e_2 \wedge L = \mathrm{I}n_3,$$

and their combined dual forms are given by

$$Q = n_3 + n_2 n_3 = (1 + n_2)n_3, \tag{75}$$

where $n_j \cdot n_k = \delta_{jk}$. As the n_k are the normals for intersecting planes, they are represented in Fig. 11 by arrows extending symmetrically to each side of the base point.

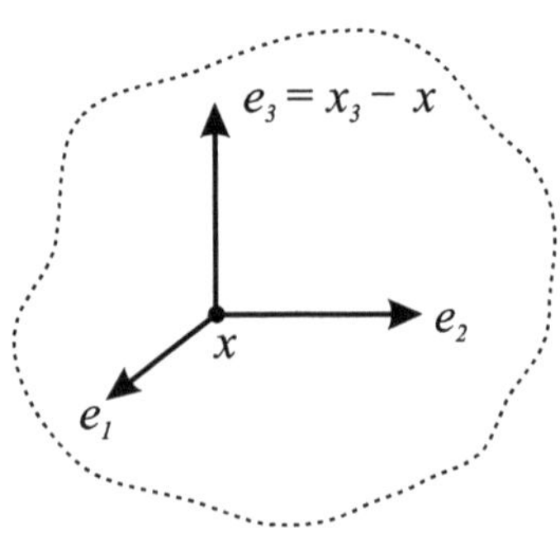

Fig. 10

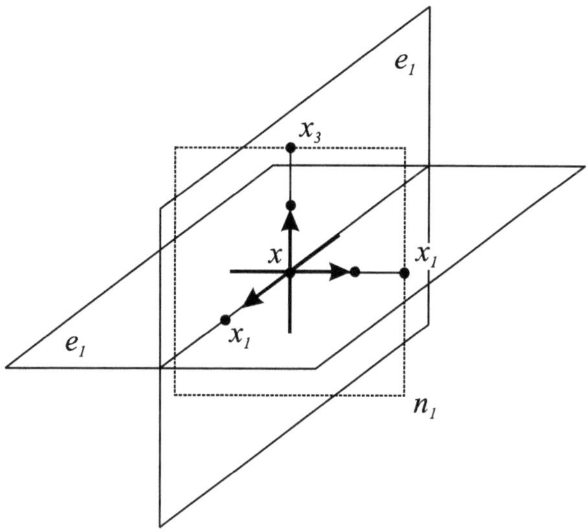

Fig. 11

Of course, the fact that the base point lies on the intersection of line and plane is expressed by

$$x \wedge (L + P) = x \wedge (\mathrm{I}Q) = 0, \quad \text{or dually by } x \cdot Q = 0. \tag{76}$$

Lasenby [12] arrived at Q in a different way, and, noting that $Q^2 = 0$, he identified it with the mysterious absolute conic of projective geometry.

As a related connection to projective geometry, note that the "complex vector" $N = x + \mathrm{I}n$ introduced in (39) is a flag without the line component. There are many other possibilities to explore, such as introducing vectors representing spheres instead of planes.

It seems simplest to work with dual forms for line and plane. This suggests that we consider a *dual flag* defined by

$$F^* = x + Q = x + n_3 + n_2 n_3, \quad \text{with } x \cdot F^* = 0. \tag{77}$$

This looks unsymmetrical, as the plane n_1 is not explicitly represented, though it is determined indirectly by the intersection of the base point with the other planes. Here is a more symmetrical representation for the rigid body:

$$Q' \equiv n_3 + n_2 n_3 + n_1 n_2 n_3, \quad \text{with } x \cdot Q' = 0. \tag{78}$$

Note that this representation is a graded sum of nested subspaces with pseudoscalar $\mathrm{I}_3 = n_1 n_2 n_3$ intrinsic to the body—a practical instance of the additive split (49).

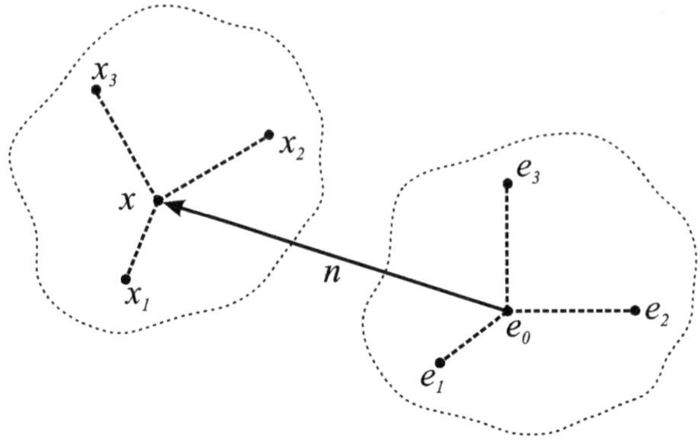

Fig. 12

10 Rigid Body Kinematics

Motion of a rigid body is a one-parameter family of displacements, which we describe by a time-dependent versor function $D = D(t)$. As illustrated in Fig. 12, this determines the evolution of body points from some reference positions e_k to instantaneous positions

$$x_k(t) = De_kD^{-1} = De_k\widetilde{D}. \tag{79}$$

From the versor character of D it can be proved that its derivative must satisfy

$$\dot{D} = \frac{1}{2}VD \quad \text{and} \quad \dot{D}^{-1} = -\frac{1}{2}D^{-1}V, \tag{80}$$

where the velocity $V = V(t)$ is a bivector, as expressed algebraically by $\widetilde{V} = -V = \langle V \rangle_2$. Using (80) to differentiate (79), we get equations of motion for the body points:

$$\dot{x}_k = C \cdot x_k. \tag{81}$$

However, there is no need to integrate this system of three equations, as the body motion is completely determined by integrating the *displacement equation* (80). Indeed, the equation of motion for D is independent of any designation of specific body points, although selection of a base point is necessary to separate rotational and translational components of the motion. To decompose the (generalized) velocity V into rotational and translational parts, we introduce a conformal split defined by

$$D = RT, \qquad De_0D^{-1} = Te_0T^{-1} = e_0 + n, \qquad T = 1 + \frac{1}{2}ne. \tag{82}$$

Derivatives of rotation and translation versors have the form

$$\dot{R} = -\frac{1}{2}\omega R, \qquad \dot{T} = \frac{1}{2}\dot{n}e = \frac{1}{2}\dot{x}e = \frac{1}{2}\dot{x}eT = \frac{1}{2}\dot{\mathbf{x}}e, \tag{83}$$

where $\boldsymbol{\omega}$ is the rotational velocity of the body, and $\dot{\mathbf{x}} = \dot{x} \wedge E$. Hence

$$\dot{D} = \dot{R}T + R\dot{T} = \frac{1}{2}\left(-i\boldsymbol{\omega} + Re\dot{\mathbf{x}}R^{-1}\right)RT = \frac{1}{2}VD,$$

so the velocity has the split form

$$V = -i\boldsymbol{\omega} + e\mathbf{v}, \quad \text{with } \mathbf{v} = R\dot{\mathbf{x}}R^{-1}. \tag{84}$$

One can write $\mathbf{v} = \dot{\mathbf{x}}$ by adopting the instantaneous initial condition $R(t) = 1$ (as done implicitly in most references on kinematics), although that complicates further differentiation of $\mathbf{v}$ if needed. (See the section on rotating systems in [15] for further discussion of this point.) Also, the negative sign for rotational velocity in (83) and (84) is dictated by the convention that the rotation is right handed around the oriented $\boldsymbol{\omega}$ axis [15]. Now consider the effect of shifting the initial base point from the origin e_0 to

$$e_0' = T_0 e_0 \widetilde{T_0} = e_0 + r_0, \quad \text{where } T_0 = 1 + \frac{1}{2}\mathbf{r}_0 e. \tag{85}$$

This determines a shift in base point trajectory to

$$x' = De_0'\widetilde{D} = D'e_0\widetilde{D'}, \tag{86}$$

where

$$D' = DT_0 = T_r D, \quad \text{with } T_r = RT_0\widetilde{R} = 1 + \frac{1}{2}\mathbf{r}e. \tag{87}$$

Differentiating, we have

$$\dot{D}' = \dot{T}_r D + T_r \dot{D} = \frac{1}{2}e\dot{\mathbf{r}}T_r D + \frac{1}{2}VDT_0 = \frac{1}{2}(e\dot{\mathbf{r}} + V)D' = \frac{1}{2}V'D'.$$

Thus, we have proved that a shift in base point induces a shift in velocity:

$$V = -i\boldsymbol{\omega} + e\mathbf{v} \mapsto V' = V + e\dot{\mathbf{r}} = -i\boldsymbol{\omega} + e(\mathbf{v} + \boldsymbol{\omega} \times r). \tag{88}$$

This result is the kinematic version of *Chasles' Theorem* [15]. Note that the rotational part is independent of the base point shift.

The base point need not be located within the rigid body, so at a given time the vector $\mathbf{r}$ can be specified freely. In particular, one can specify

$$\boldsymbol{\omega} \cdot \mathbf{r} = 0 \quad \text{since } \mathbf{r} = \boldsymbol{\omega}^{-1} \times \mathbf{v} = i\left(\mathbf{v} \wedge \boldsymbol{\omega}^{-1}\right) \tag{89}$$

to put V' in the form of a *screw*:

$$V' = -i\boldsymbol{\omega} + eh\boldsymbol{\omega} = \boldsymbol{\omega}(eh - i) \quad \text{with } \textit{pitch } h = \mathbf{v}\boldsymbol{\omega}^{-1} = \mathbf{v} \cdot \boldsymbol{\omega}/\boldsymbol{\omega}^2. \tag{90}$$

For positive pitch, this velocity generates an infinitesimal translation along the axis of a right-handed rotation.

From (79) and (82) it follows that chords $n_k = e_k - e_0 = T(e_k - e_0)T^{-1}$ are invariant under translations. Hence, the evolution of chords and products of chords is simply a rotation, as described by

$$n'_k = Dn_k\widetilde{D} = Rn_k\widetilde{R} \quad \text{and} \quad n'_j n'_k = Dn_j n_k\widetilde{D} = Rn_j n_k\widetilde{R}. \tag{91}$$

Likewise, evolution of the dual flag Q^* in (79) is described by

$$Q^* \to Q^*(t) = DQ^*\widetilde{D} = RQ^*\widetilde{R}. \tag{92}$$

Note that the form of these rotations is independent of base point, though the value of R is not, as described explicitly by (70).

11 Rigid Body Dynamics

In [15] GA is employed for a completely coordinate-free derivation and analysis of the equations for a rigid body. The results are summarized here for embedding in a more compact and deeper formulation with CGA. Then [15] can be consulted for help with detailed applications.

For a rigid body with total *mass* m and *inertial tensor* $\underline{I}$, the *momentum* p and *rotational* (or *angular*) *momentum* $\mathbf{l}$ are defined by

$$\mathbf{p} = m\mathbf{v} \quad \text{and} \quad \mathbf{l} = \underline{I}(\boldsymbol{\omega}), \tag{93}$$

where $\mathbf{v}$ is the velocity of a freely chosen base point (not necessarily the center of mass), and the inertia tensor depends on the choice of base point, as determined by the *parallel axis theorem* (see [15], where the structure of inertia tensors is discussed in detail).

Translational and rotational motions are then determined (respectively) by *Newton's Force Law* and *Euler's Torque Law*:

$$\dot{\mathbf{p}} = \mathbf{f} = \sum_k \mathbf{f}_k \quad \text{and} \quad \dot{\mathbf{l}} = \underline{I}(\dot{\boldsymbol{\omega}}) + \boldsymbol{\omega} \times \underline{I}(\boldsymbol{\omega}) = \boldsymbol{\Gamma} = \sum_k \boldsymbol{\Gamma}_k, \tag{94}$$

where the *net force* $\mathbf{f}$ is the sum of forces applied to specified body points, and the *net torque* $\boldsymbol{\Gamma}$ is the sum of applied torques.

Now, to combine $\mathbf{p}$ and $\mathbf{l}$ into a generalized *comomentum* P that is linearly related to the velocity V in (88), we introduce the generalized *mass operator* $\underline{M}$ defined by

$$P = \underline{M}V = m\mathbf{v}e_0 - i\underline{I}\boldsymbol{\omega} = \mathbf{p}e_0 - i\mathbf{l}. \tag{95}$$

The appearance of e_0 instead of e in this expression requires some explanation. For the moment, it suffices to note that it yields the standard expression for total *kinetic energy*:

$$K = \frac{1}{2}V \cdot \widetilde{P} = -\frac{1}{2}V \cdot \underline{M}V = \frac{1}{2}(\boldsymbol{\omega} \cdot \mathbf{l} + \mathbf{v} \cdot \mathbf{p}). \tag{96}$$

Next, using (95), we combine the two conservation laws (94) into a single equation of motion for a rigid body:

$$\dot{P} = W \quad \text{where } W = \mathbf{f}e_0 - i\boldsymbol{\Gamma} \tag{97}$$

is called a *wrench* or *coforce*. From this we easily derive the standard expression for *power* driving change in kinetic energy:

$$\dot{K} = V \cdot \widetilde{W} = \boldsymbol{\omega} \cdot \boldsymbol{\Gamma} + \mathbf{vf}. \tag{98}$$

Finally, we consider the effect of shifting the base point as specified by (85) and (86). That shift induces shifts in the comomentum and applied wrench:

Parallel axis theorem

$$P \mapsto P' = P + i\mathbf{r} \times \mathbf{p} = \mathbf{p}e_0 - i(\mathbf{l} - \mathbf{r} \times \mathbf{p}), \tag{99}$$

$$W \mapsto W' = W + i\mathbf{r} \times \mathbf{f} = e_0\mathbf{f} - i(\boldsymbol{\Gamma} - \mathbf{r} \times \mathbf{f}). \tag{100}$$

The comomentum shift expresses the *parallel axis theorem*, while the corresponding shift in torque is sometimes called *Poinsot's theorem*. As a check for consistency with Chasles' theorem (88), we verify shift invariance of the kinetic energy:

$$\begin{aligned} 2K &= V' \cdot \widetilde{P'} = -(V + e\boldsymbol{\omega} \times r) \cdot (P + i\mathbf{r} \times \mathbf{p}) \\ &= \boldsymbol{\omega} \cdot (\mathbf{l} - \mathbf{r} \times \mathbf{p}) + (\mathbf{v} + \boldsymbol{\omega} \times \mathbf{r}) \cdot \mathbf{p} = \boldsymbol{\omega} \cdot \mathbf{l} + \mathbf{v} \cdot \mathbf{p} = V \cdot \widetilde{P}. \end{aligned}$$

This completes our transcription of rigid body dynamics into a single invariant equation of motion (97). As indicated by the appearance of e_0 in (95) and (97) and verified by the shift equations (99) and (100), separation of the motion into rotational and translational components is a conformal split that depends on the choice of base point. In applications there is often an optimal choice of base point, such as the point of contact of interacting rigid bodies, as in the rigid body linkages discussed below.

This is a good place to reflect on what makes CGA mechanics so compact and efficient. In writing my mechanics book [15], I noted that momentum is a vector quantity, while angular momentum is a bivector quantity, so I combined them by defining a "complex velocity"

$$V = \mathbf{v} + i\boldsymbol{\omega}, \tag{101}$$

with corresponding definitions for complex momentum and force. That led to a composite equation of motion just as compact as (97). However, it was more a curiosity than an advantage, because you had to take it apart to use it. The trouble was, as I fully understood only with the development of CGA, that the complex velocity (101) does not conform to the structure of the Euclidean group. That defect is remedied by the simple expedient of introducing the null element e to change the definition of velocity from (101) to (84).

The resulting velocity (84) is a bivector in CGA. To make that explicit, we note that the first term is a bivector because it is the dual of a trivector: $i\boldsymbol{\omega} = \mathrm{I}\boldsymbol{\omega} \wedge E = \mathrm{I} \cdot (\boldsymbol{\omega} \wedge e_0 \wedge e)$, while the second term is the contraction of a trivector by a vector:

$e\mathbf{v} = (v \wedge E)e = (v \wedge e_0 \wedge e) \cdot e$. Thus, (84) is a generic form for bivectors generating Euclidean displacements. Of course, that fact was implicit already in the definition of V in the displacement equation (80). It has been reiterated here to confirm consistency with the conformal split. As we see next, the notion of V as bivector generator of the Euclidean group is the foundation of Screw Theory. That is what makes the equation of motion (97) so significant.

12 Screw Theory

Screw theory was developed in the latter part of the nineteenth century [17] from applications of geometry and mechanics to the design of mechanisms and machines. When formulated within the standard matrix algebra of today, the concepts of screw theory seem awkward or even a bit *screwy*! Consequently, applications of screw theory, deep and useful though they be, have remained outside the mainstream of mechanical engineering.

Here we cast screw theory in terms of CGA to secure its rightful place in the Kingdom of Euclidean geometry and facilitate access to its rich literature. To the extent that the conformal model becomes a standard for applications of Euclidean geometry, this will surely promote a rejuvenation of screw theory.

The foundations for screw theory in rigid body mechanics have been laid in the preceding sections. Here we concentrate on explicating the screw concept in relation to displacements. For constant V, the displacement equation (80) integrates immediately to the solution

$$D(t) = e^{\frac{1}{2}Vt} = e^{\frac{1}{2}\underline{T}_r V't} = T_r e^{\frac{1}{2}V't} T_r^{-1}, \tag{102}$$

where (87) to (90) have been used in the form

$$\begin{aligned} V &= \underline{T}_r V' = T_r V' T_r^{-1} = -i T_r \boldsymbol{\omega} T_r^{-1} + eh\boldsymbol{\omega} \\ &= -i\boldsymbol{\omega} + e(h\boldsymbol{\omega} + \mathbf{r} \times \boldsymbol{\omega}) = -i\boldsymbol{\omega} + e\mathbf{v} \end{aligned} \tag{103}$$

to exhibit the conformal split and shift of base point. The translation versor, which, of course, generates a fixed displacement, also has an exponential form:

$$T_r = 1 + \frac{1}{2}\mathbf{r}e = e^{\frac{1}{2}\mathbf{r}e}, \qquad T_r^{-1} = e^{-\frac{1}{2}\mathbf{r}e} = T_{-r}. \tag{104}$$

As expressed by (103), the rotation rate $\boldsymbol{\omega}$ is invariant under a base point shift. As $\boldsymbol{\omega} = \omega \wedge E$ is a trivector representing a line through the origin, the motion generated by $D(t)$ in (102) is a steady screw motion with constant pitch along that line. The translations in (102) can be understood as translating the line through a given base point to one through the origin, unfolding the screw displacement, and then translating it back to the original base point.

To consolidate our concepts it is helpful to introduce nomenclature that conforms to the screw theory literature as closely as possible. In general, any Euclidean displacement versor can be given the exponential form:

$$D = e^{\frac{1}{2}S}, \quad \text{where } S = i\mathbf{m} + e\mathbf{n}. \tag{105}$$

The versor D is called a *twistor*, while its generator S is called a *twist* or a *screw*. The term "screw" is often restricted to the case where $\mathbf{n}$ and $\mathbf{m}$ are collinear and $\mathbf{m}^2 = 1$. The line determined by $\mathbf{m}$ is called the *screw axis* or *axode*.

As implicitly shown in (62) and (63), the multiplicative group of twistors is a double covering of the Special Euclidean group:

$$\mathrm{SE}(3) = \{\text{rigid displacements } \underline{D}\} \cong_2 \{\text{twistors } D\}. \tag{106}$$

The set of all twists constitutes an algebra of bivectors:

$$\mathrm{se}(3) \equiv \textit{Lie algebra of } \mathrm{SE}(3) = \{S_k = i\mathbf{m}_k + e\mathbf{n}_k\}. \tag{107}$$

This algebra is closed under the commutator product:

$$S_1 \times S_2 = \frac{1}{2}(S_1S_2 - S_2S_1) = i(\mathbf{m}_2 \times \mathbf{m}_1) + e(\mathbf{n}_2 \times \mathbf{m}_1 - \mathbf{n}_1 \times \mathbf{m}_2). \tag{108}$$

Let us summarize important general properties of this algebra.

The representation of a Lie group by action on its Lie algebra is called the adjoint representation [16]. In this case, we have

$$S_k' = \underline{U}(S_k) = US_kU^{-1} = Ad_U S_k, \quad \text{where } \{\underline{U}\} = \mathrm{SE}(3). \tag{109}$$

This transformation preserves the geometric product $S_1S_2 = S_1 \cdot S_2 + S_1 \times S_2 + S_1 \wedge S_2$; that is,

$$S_1'S_2' = \underline{U}(S_1S_2) = U(S_1 \cdot S_2 + S_1 \times S_2 + S_1 \wedge S_2)U^{-1}. \tag{110}$$

Separating parts of different grade, we see first that the commutator product is *covariant*, as expressed by

$$S_1' \times S_2' = \underline{U}(S_1 \times S_2) = U(S_1 \times S_2)U^{-1}. \tag{111}$$

Second, the scalar part is an obvious invariant:

$$S_1' \cdot S_2' = S_1 \cdot S_2 = -\mathbf{m}_1 \cdot \mathbf{m}_2, \tag{112}$$

known as the *Killing form* for the group. Finally, the remaining term is a pseudoscalar invariant

$$S_1' \wedge S_2' = \underline{U}(S_1 \wedge S_2) = S_1 \wedge S_2 = ie(\mathbf{m}_1 \cdot \mathbf{n}_2 + \mathbf{m}_2 \cdot \mathbf{n}_1), \tag{113}$$

because it is proportional to the *Euclidean pseudoscalar* $\mathrm{I}_e = ie$, which is invariant because i and e are invariant. This concept and result is unique to CGA, so it merits further discussion.

The pseudoscalar $\mathrm{I}_e = ie$ squares to $I_e^2 = -e^2 = 0$, so it cannot be used for an invertible duality mapping. However, we can define a conjugate pseudoscalar $\mathrm{I}_e^* \equiv ie_0$ so that $\mathrm{I}_e^* \cdot \mathrm{I}_e = \langle \mathrm{I}_e^* \mathrm{I}_e \rangle = -e \cdot e_0 = 1$. Then we can express the invariant (113) as a scalar:

$$(S_1 \wedge S_2) \cdot (ie_0) = \mathbf{m}_1 \cdot \mathbf{n}_2 + \mathbf{m}_2 \cdot \mathbf{n}_1 = S_1 \cdot S_2^*, \tag{114}$$

where a *dual screw* = *coscrew* has been defined by

$$S_k^* \equiv S_k \cdot (ie_0) = \langle S_k ie_0 \rangle_2 = -i\mathbf{n}_k - e_0 \mathbf{m}_k. \tag{115}$$

This equivalent to the *reciprocal screw* first introduced by Ball [17]. Comparison with (97) shows that *wrenches are coscrews*! Of course, the dual defined here should not be confused with the dual introduced in (11). The asterisk notation has been used for both to emphasize their conceptual commonality.

Finally, we note that for a single screw, the *pitch* h appears as a ratio of the invariants (114) and (112):

$$h = -\frac{1}{2}\frac{S \cdot S^*}{S \cdot S} = \mathbf{n} \cdot \mathbf{m}^{-1}. \tag{116}$$

Screw theory is basically about the generators of displacements. The simplicity of its formulation within CGA belies the richness and complexity of its applications in mechanical engineering, for which the serious student must consult the literature.

As guides to the screw theory literature, I recommend two books. The first book [16] concerns use of modern mathematical concepts and notation, which can be compared to the approach taken here. The second book [18] is by two long-time practitioners of screw theory. Though it is designed as a textbook for the ill-prepared engineering student of today, it provides a mature perspective on current status with an authoritative entrée to the literature.

Screw theory literature going back to the nineteenth century contains many gems that can be recovered by reformulation within CGA. Here is another worthy topic for doctoral research. It requires more than historical study, for many of the gems need polishing—there are unsolved problems to be addressed in the new light of CGA.

To facilitate translations from the literature, relations between CGA and matrix algebra are established next.

13 Conformal Split and Matrix Representation

Besides its invariance and incredible compactness, one great advantage of the CGA formulation of rigid body mechanics in the preceding sections is the ease of relating it to alternative formulations by a conformal split. In this section we consider the

Fig. 13

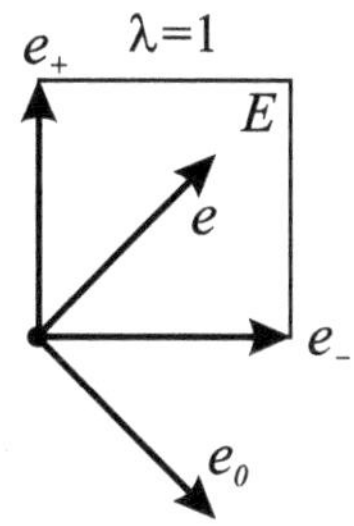

conformal split in more detail, especially to clarify and facilitate connections to the vast literature on mechanical systems.

As defined in (43), one factor in the conformal split is the geometric algebra $\mathbb{G}^{1,1} = \mathbb{G}(\mathbb{R}^{1,1})$. The vector space $\mathbb{R}^{1,1}$ is sometimes referred to as 2D Minkowski space to emphasize its similarity to 4D Minkowski space $\mathbb{R}^{3,1}$, which is a standard model for spacetime in relativistic physics [4, 6]. It can be generated from a *null basis* $\{e, e_0 \mid e^2 = e_0^2 = 0, e \cdot e_0 = -1\}$ or from an equivalent *orthonormal basis* $\{e_\pm = \frac{1}{\sqrt{2}}(\lambda e \mp \lambda^{-1} e_0), \lambda \neq 0, e_\pm^2 = \pm 1\}$.

Though the orthonormal basis is more familiar to most readers, as we have seen already, the null basis is more significant geometrically. It generates a basis $\{1, e, e_0, E\}$ (depicted in Fig. 13) for the entire algebra $\mathbb{G}^{1,1}$, with the basic properties

$$E^2 = 1, \qquad e_0 e = E - 1, \qquad Ee = -eE = e, \qquad e_0 E = -E e_0 = e_0. \quad (117)$$

These properties have been used many times in previous sections.

The algebra of *dual numbers* $\mathbb{D} = \{\alpha + e\beta\}$ is an important subalgebra of $\mathbb{G}^{1,1}$. However, it was first proposed by Clifford as an extension of the real numbers analogous to complex numbers, with the null unit e replacing the imaginary unit i. He introduced it as an extension of scalars in quaternions to form what he called biquaternions [10, 16]. We can regard it as an extension of the algebra $\mathbb{G}^3$ to $\mathbb{G}^3 \otimes \mathbb{D}$. Clifford clearly recognized the geometric significance of this extension for incorporating the additivity and commutativity properties of translations. In terms of translation versors, these properties are expressed by

$$T_{\mathbf{a}} T_{\mathbf{b}} = \left(1 + \frac{1}{2}\mathbf{a}e\right)\left(1 + \frac{1}{2}\mathbf{b}e\right) = 1 + \frac{1}{2}(\mathbf{a} + \mathbf{b})e = T_{\mathbf{a}+\mathbf{b}} = T_{\mathbf{b}} T_{\mathbf{a}}. \quad (118)$$

Clifford's biquaternions have been used to represent translations and screws by many authors since. However, we have seen in the preceding section that the dual numbers must be extended to the entire algebra $\mathbb{G}^{1,1}$ to accommodate coscrews and screw invariants. That has been done in the literature primarily by employing matrices in the following way. The algebra $\mathbb{G}^{1,1}$ is isomorphic to the algebra $\mathbb{M}_2(\mathbb{R})$ of real 2×2 matrices. That is readily established by exhibiting the isomorphism of basis elements:

$$e_+ \simeq \begin{bmatrix} 0 & 1 \\ 1 & 0 \end{bmatrix}, \qquad e_- \simeq \begin{bmatrix} 0 & 1 \\ -1 & 0 \end{bmatrix},$$
$$E \simeq \begin{bmatrix} 1 & 0 \\ 0 & -1 \end{bmatrix}, \qquad 1 \simeq \begin{bmatrix} 1 & 0 \\ 0 & 1 \end{bmatrix}. \tag{119}$$

Accordingly, every multivector M in $\mathbb{G}^{1,1}$ has a matrix representation $\hat{M}$ explicitly given by

$$M = \frac{1}{2}A(1+E)+B(e_+ + e_-)+C(e_+ - e_-)+D(1-E) \simeq \hat{M} = \begin{bmatrix} A & B \\ C & D \end{bmatrix}, \tag{120}$$

where the matrix elements are real numbers. This representation is readily generalized by allowing the matrix elements to have values in other algebras, $\mathbb{G}^3$ in particular. Thus, we arrive at the isomorphism:

$$\mathbb{G}^{4,1} = \mathbb{G}^3 \otimes \mathbb{G}^{1,1} \simeq \mathbb{M}_2(\mathbb{G}^3). \tag{121}$$

Properties of this isomorphism are surprisingly rich and have been thoroughly studied in [13]. That enabled a critique of the matrix representation for the conformal group, which contributed to developing the invariant formulation in CGA introduced in [1].

The matrix algebra $\mathbb{M}_2(\mathbb{G}^3)$ has been much used in screw theory with the elements of $\mathbb{G}^3$ interpreted as complex quaternions. More often, it has been used with the elements of $\mathbb{G}^3$ represented as 3×3 matrices or column vectors. The alternatives are best explained by a representative example.

With an obvious change of notation, we can write (103) for the change in *screw coordinates* induced by a shift $\mathbf{r} = \mathbf{x}_Q - \mathbf{x}_P$ from base point P to point Q in the form

$$V_Q = e\mathbf{v}_Q - i\boldsymbol{\omega} = \underline{T}_r V_P = e(\mathbf{v}_P - \mathbf{r} \times \boldsymbol{\omega}) - i\boldsymbol{\omega}. \tag{122}$$

This has a matrix representation $\hat{V}_Q = \hat{\underline{T}}_r \hat{V}_P$ with the explicit form

$$\begin{bmatrix} \mathbf{v}_Q \\ \boldsymbol{\omega} \end{bmatrix} = \begin{bmatrix} 1 & -\mathbf{r}\times \\ 0 & 1 \end{bmatrix} \begin{bmatrix} \mathbf{v}_P \\ \boldsymbol{\omega} \end{bmatrix} = \begin{bmatrix} \mathbf{v}_P - \mathbf{r} \times \boldsymbol{\omega} \\ \boldsymbol{\omega} \end{bmatrix}. \tag{123}$$

Similarly, we can write (100) for the induced change of *coscrew coordinates* in the form

$$W_Q = -e_0\mathbf{f} + i\boldsymbol{\Gamma}_Q = \underline{T}_r^* W_P = -e_0\mathbf{f} + i(\boldsymbol{\Gamma}_P + \mathbf{r} \times \mathbf{f}). \tag{124}$$

Its matrix representation $\hat{W}_Q = \hat{\underline{T}}_r^* \hat{W}_P$ has the explicit form

$$\begin{bmatrix} \mathbf{f} \\ \boldsymbol{\Gamma}_Q \end{bmatrix} = \begin{bmatrix} 1 & 0 \\ -\mathbf{r}\times & 1 \end{bmatrix} \begin{bmatrix} \mathbf{f} \\ \boldsymbol{\Gamma}_P \end{bmatrix} = \begin{bmatrix} \mathbf{f} \\ \Gamma_P + \mathbf{r} \times \mathbf{f} \end{bmatrix}. \tag{125}$$

These four equations suffice to show how any equation in the literature on screw theory or robotics can be translated into CGA and vice versa. For example, in [18] the screws in (123) and coscrews in (125) are represented as 6-component column vectors. This example reveals a significant drawback of the matrix representations: the matrices do not encode the distinction between screws and coscrews, which, in contrast, is explicitly expressed by the distinction between e and e_0 in (122) and (124). Expressed in more general terms: the matrix representations suppress the geometric meaning of matrix elements, which is explicitly encoded in the algebra $\mathbb{G}^{1,1}$. Furthermore, the matrix representation (121) implicitly forces one to adopt a conformal split, which means, as we have seen, that one is forced into a covariant rather than invariant approach to geometry. Nevertheless the isomorphism $\mathbb{G}^{1,1} \simeq \mathbb{M}_2(\mathbb{R})$ is essential for relating CGA to the robotics literature.

14 Linked Rigid Bodies & Robotics

The potential for application of CGA to robotics is best illustrated by a simple example. Figure 14 depicts a kinematic chain with three segments in a *reference pose*:

$$x_0 = e_0 + a + b + c. \tag{126}$$

Rotations at its three joints are specified by *twistors* $\{R_1, R_2, R_3\}$. Rotation at the first joint gives

$$x_1 = e_0 + a + b + R_1 c R_1^{-1} = e_0 + a + b + c_1. \tag{127}$$

Subsequent or concurrent rotation at the second joint gives

$$x_2 = e_0 + a + R_2\big(b + R_1 c R_1^{-1}\big)R_2^{-1} = e_0 + a + b_2 + c_{21}. \tag{128}$$

Finally, the net result of rotations at all the joints is *general pose*:

$$x = e_0 + R_3\big[a + R_2\big(b + R_1 c R_1^{-1}\big)R_2^{-1}\big]R_3^{-1} = e_0 + a_3 + b_{32} + c_{321}. \tag{129}$$

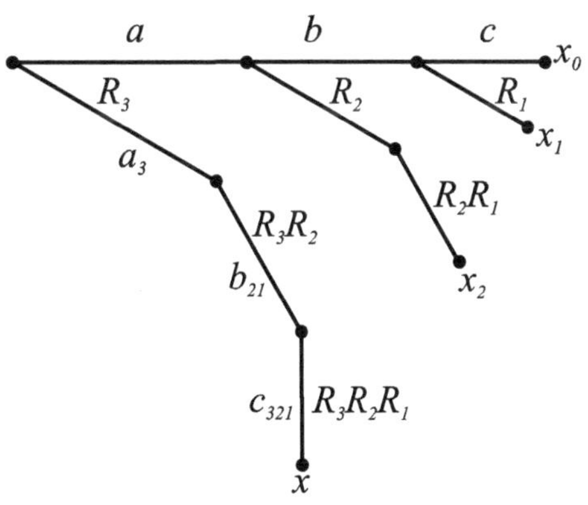

Fig. 14

A conformal split with the fixed point e_0 gives the position vector for the endpoint:

$$\mathbf{x} = x \wedge E = R_3\big[\mathbf{a} + R_2\big(\mathbf{b} + R_1\mathbf{c}R_1^{-1}\big)R_2^{-1}\big]R_3^{-1} = \mathbf{a}_3 + \mathbf{b}_{32} + \mathbf{c}_{321}. \tag{130}$$

Of course, the rotations need not be confined to a plane (as presumed in Fig. 14 for simplicity of illustration). Restrictions on the range of motion at each joint are encoded in the twistors. For example, a rotation R_1 with one degree of freedom can be given the angular form

$$r_1 = \exp\left(-\frac{1}{2}\mathbf{n}_1\alpha_1\right), \quad \text{where } 0 \le \alpha_1 \le \pi, \tag{131}$$

and unit vector $\mathbf{n}_1$ is the direction of the joint axis for a right-handed rotation. Kinematics of the chain can be described a follows. Irrespective of how the joints are characterized, the twistors satisfy equations of the form

$$\dot{R}_k = -\frac{1}{2}i\omega_k R_k. \tag{132}$$

Hence, for $R_{32} = R_3R_2$, we have

$$\dot{R}_{32} = -\frac{1}{2}i\omega_{32}R_{32} \quad \text{with } \omega_{32} = \omega_3 + R_3\omega_2R_3^{-1}. \tag{133}$$

Finally, with $\omega_{321} = \omega_3 + R_3\omega_2R_3^{-1} + R_3R_2\omega_1R_2^{-1}R_3^{-1}$, for the derivative of the end point position vector (130), we get

$$\dot{\mathbf{x}} = \omega_3 \times \mathbf{a}_3 + \omega_{32} \times \mathbf{b}_{32} + \omega_{321} \times \mathbf{c}_{321}. \tag{134}$$

This is only the beginning for application of CGA to robotics, but we have all the theoretical machinery we need for any task. To incorporate dynamics we introduce the inertia properties of each body with (95) and the applied wrenches with (97). Moreover, CGA offers a promising approach to modeling complex interactions between bodies, such as viscoelastic coupling at joints [2].

The next phase in the development of CGA *robotics* is detailed applications to specific problems. This development is already underway by attendees at this conference and others in the GA community. However, I see a need for more systematic mining of the robotics literature to incorporate established problems, results, and methods in CGA and promote broader interaction within the engineering community. I thought about offering suggestions for literature to consult. But the robotics literature is so vast and variable in complexity and quality that I fear my suggestions could be as misleading as helpful. Consequently, I add only one recent reference [19] to those I have already mentioned.

To sum up, CGA provides a powerful mathematical framework for robotics R&D with the twin goals of (1) simplicity and clarity in mathematical formulation, (2) efficiency and speed in computation.

Acknowledgement I thank Arvid Halma and Leo Dorst for assistance in preparing the document.

References

1. Hestenes, D.: Old Wine in New Bottles: A new algebraic framework for computational geometry. In: Bayro-Corrochano, E., Sobczyk, G. (eds.) Advances in Geometric Algebra with Applications in Science and Engineering, pp. 1–14. Birkhäuser, Basel (2001)
2. Hestenes, D., Fasse, E.: Homogeneous rigid body mechanics with elastic coupling. In: Dorst, L., Doran, C., Lasenby, J. (eds.) Applications of Geometric Algebra in Computer Science and Engineering, pp. 197–212. Birkhäuser, Basel (2002)
3. Hestenes, D.: A unified language for mathematics and physics. In: Chisholm, J.S.R., Common, A.K. (eds.) Clifford Algebras and their Applications in Mathematica Physics, pp. 1–23. Reidel, Dordrecht (1986)
4. Hestenes, D.: Space-Time Algebra. Gordon & Breach, New York (1966)
5. Hestenes, D., Sobczyk, G.: CLIFFORD ALGEBRA TO GEOMETRIC CALCULUS, a unified language for mathematics and physics. Kluwer, Dordrecht (1984). Paperback (1985). Fourth printing 1999
6. Doran, C., Lasenby, A.: Geometric Algebra for Physicists. Cambridge University Press, Cambridge (2002)
7. Dorst, L., Fontijne, D., Mann, S.: Geometric Algebra for Computer Science. Morgan Kaufmann Publ., Elsevier, San Mateo, Amsterdam (2007/2009)
8. Li, H.: Invariant Algebras and Geometric Reasoning. World Scientific, Singapore (2008)
9. Hestenes, D.: Grassmann's vision. In: Schubring, G. (ed.) Hermann Günther Grassmann (1809–1877)—Visionary Scientist and Neohumanist Scholar, pp. 191–201. Kluwer, Dordrecht (1996)
10. Clifford, W.K.: Mathematical Papers. Macmillan, London (1882). Ed. by R. Tucker. Reprinted by Chelsea, New York (1968)
11. Doran, C., Hestenes, D., Sommen, F., Van Acker, N.: Lie groups as spin groups. J. Math. Phys. **34**, 3642–3669 (1993)
12. Lasenby, A.: Recent applications of conformal geometric algebra. In: Li, H., et al. (ed.) Computer Algebra and Geometric Algebra with Applications. LNCS, vol. 3519, pp. 298–328. Springer, Berlin (2005)
13. Hestenes, D.: The design of linear algebra and geometry. Acta Appl. Math. **23**, 65–93 (1991)
14. Onishchik, A., Sulanke, R.: Projective and Cayley–Klein Geometries. Springer, Berlin (2006)
15. Hestenes, D.: New Foundations for Classical Mechanics. Kluwer, Dordrecht (1986). Paperback (1987). Second edition (1999)
16. Selig, J.: Geometrical Methods in Robotics. Springer, Berlin (1996)
17. Ball, R.S.: A Treatise on the Theory of Screws. Cambridge University Press, Cambridge (1900). Reprinted in paperback (1998)
18. Davidson, J., Hunt, K.: Robots and Screw Theory: Applications of Kinematics and Statics to Robotics. Oxford University Press, London (2004)
19. Featherstone, R., Orin, D.: Dynamics. In: Siciliano, B., Khatib, O. (eds.) Springer Handbook of Robotics, pp. 35–65. Springer, Berlin (2008)

Tutorial: Structure-Preserving Representation of Euclidean Motions Through Conformal Geometric Algebra

Leo Dorst

Abstract A new and useful set of homogeneous coordinates has been discovered for the treatment of Euclidean geometry. They render Euclidean motions not merely linear (as the classical homogeneous coordinates do), but even turn them into orthogonal transformations, through a clever choice of metric in two (not one) additional dimensions.

To take full advantage of this new possibility, a good representation of orthogonal transformations is required. We find that multiple reflections, while classically giving unwieldy expressions involving the dot product, become practical by introducing the more fundamental geometric product (which has the dot product merely as its symmetric part). We obtain a sandwiching operation between products of vectors as our representation of motions, which is not only easily concatenated, but also incorporates the computational advantages of complex numbers and quaternions in a real manner. The antisymmetric part of the geometric product produces a spanning operation that permits the construction of lines, planes, spheres and tangents from vectors. Since the sandwiching operation distributes over this construction, 'objects' are fully integrated with 'motions', in a structure preserving manner.

Additional techniques such as duality (permitting a universal intersection operation), and the rewriting of operators logarithmically (to obtain quantities that can be interpolated linearly) complete the techniques of what is ultimately a very convenient geometric algebra. It easily incorporates general conformal transformations, and can be implemented to run almost as efficiently as classical homogeneous coordinates. The resulting high-level programming language naturally integrates quantitative computation with the automatic administration of geometric data structures.

Rather than the usual introductions which plow through Clifford algebra before they reach this very useful 'conformal model' (if they do at all), this tutorial does the reverse. It structures the new concepts in a manner that shows how each ad-

L. Dorst (✉)
Intelligent Systems Laboratory, University of Amsterdam, Science Park 107, 1098 XG, Amsterdam, The Netherlands
e-mail: L.Dorst@uva.nl

E. Bayro-Corrochano, G. Scheuermann (eds.), *Geometric Algebra Computing*,
DOI 10.1007/978-1-84996-108-0_2,

ditional sophistication is related to what went before, and how it extends its expressive and computational power. Throughout, the concepts and techniques will be illustrated by interactive visualization software (GAviewer, freely available at www.geometricalgebra.net).

1 Introduction

"Doing geometry" in computer science or engineering requires at least the following ingredients in a practical computational framework:

- *descriptive primitives*: such as points, lines, planes, circles, spheres, tangents
- *basic constructions*: connections, intersections, parametric specification
- *motions*: translation, rotation, reflection, projection
- *properties*: size, location, orientation
- *practical numerics*: approximation, estimation, interpolation, linearization

These ingredients should interweave seamlessly. Notably, the framework should be structure preserving, in the sense that *constructions and properties of primitives should be covariant under motions.* For instance, when moving a circle determined by three points, it should not be necessary to decompose the circle back into the points, move those, and then reconstruct; rather, the circle should be a basic element of computation with an associated motion operator (which should moreover preferably be identical to that for points). Also, all ingredients should be specifiable in a sufficiently high-level programming language, which avoids coordinates as specification level though it may revert to them when executing the operations. The usual linear algebra tools have neither of these desirable properties, not even when using homogeneous coordinates. Yet a practical computational framework exists that can do all of the above. It is called "conformal geometric algebra" (CGA), and this chapter briefly exposes its essential structure. We will explain all elements of Fig. 1, and more.

2 Conformal Geometric Algebra

2.1 Trick 1: Representing Euclidean Points in Minkowski Space

Let us focus on a 3D space in which we want to perform Euclidean motions. We can consider it as a 3D vector space and use a position vector $\mathbf{x}$ to denote a point X relative to an (arbitrary) origin. This is naive practice, and not very convenient, since Euclidean motions are then not even linear transformations. A commonly used improvement is the *homogeneous model*, in which the space is augmented with an extra dimension e_o, and the point X at $\mathbf{x}$ represented as $e_o + \mathbf{x}$. Now Euclidean motions are linear transformations but still not structure preserving. More is required.

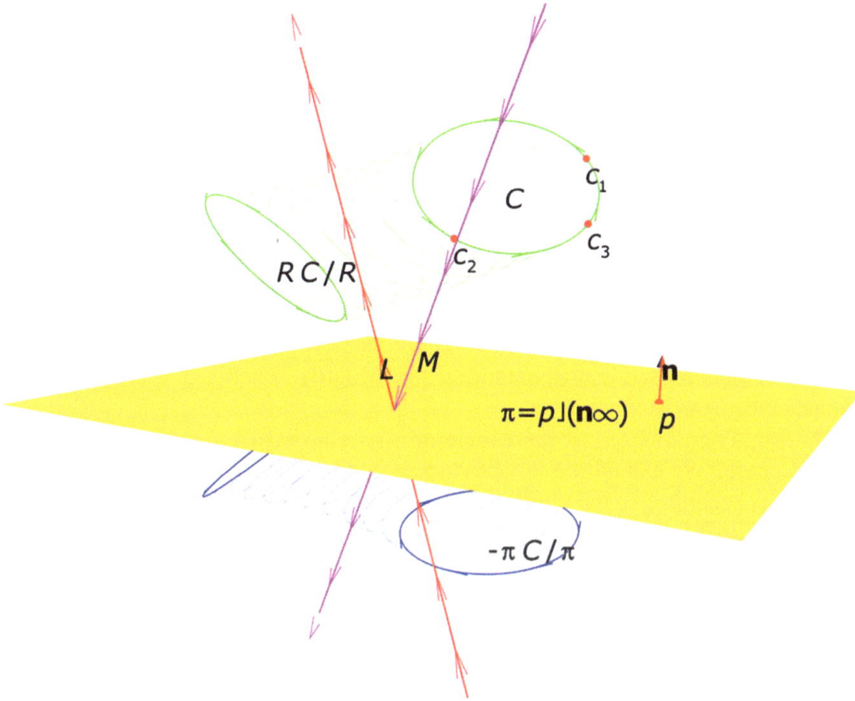

Fig. 1 An example of the ease of CGA (from [2]). A circle C is generated from three points c_1, c_2, c_3 as $C = c_1 \wedge c_2 \wedge c_3$. A line is given as a 3-blade L. The circle C is to be rotated around the line L, producing RCR^{-1}, with R specified as $R = \exp(L^*\phi/2)$. The rotation is interpolated in k steps using $R^{1/k}$. Then the whole scene is reflected in the plane π given by a normal vector $\mathbf{n}$ and a point p on it as $\pi = p \cdot (\mathbf{n} \wedge e_\infty)$; any element X is reflected as $X \mapsto (-1)^{\text{grade}(X)}\pi X \pi^{-1}$. In appropriate software such as [3], these coordinate-free formulas are the literal specification of a computer program producing the scene

In CGA, we introduce *two extra dimensions* for representational purposes, thus constructing a five-dimensional space. We introduce two basis vectors for these extra dimensions, e_o and e_∞, and the specific metric given below. As we will see, the null vectors in this extended space (i.e., the vectors x satisfying $x \cdot x = 0$ in the chosen metric) represent weighted points in the Euclidean space (though one usually employs unit weight points satisfying $x \cdot e_\infty = 0$). Such vectors representing points have algebraic properties to construct other elements in a coordinate-free, invariant manner, as explained in the paper by Hestenes [5] elsewhere in this volume.

In the present introductory paper, we mostly prefer to use an explicit expression for such a vector x representing a point X, relating it to the "classical" Euclidean position vector $\mathbf{x}$ of the point relative to the chosen origin through

$$x = e_o + \mathbf{x} + \frac{1}{2}\|\mathbf{x}\|^2 e_\infty. \tag{1}$$

Table 1

·	e_o	$\mathbf{x}$	e_∞
e_o	0	0	−1
$\mathbf{x}$	0	$\mathbf{x} \cdot \mathbf{x}$	0
e_∞	−1	0	0

If we ignore the component for e_∞, we recognize in $e_o + \mathbf{x}$ just the homogeneous model. In that model, the extra dimension e_o represents *the point at the origin*; and the same interpretation holds in CGA (set $\mathbf{x} = \mathbf{0}$). We see that the term with e_∞ dominates as $\mathbf{x}$ gets large. In fact, e_∞ can be interpreted consistently as *the point at infinity* which is used in mathematics to "compactify" Euclidean space to remove special cases from its algebra.

The Big Trick of CGA is the choice of a specific metric for the 5D representational space. We extend the dot product $\mathbf{x} \cdot \mathbf{x}$ for Euclidean vectors to the new dimensions according to the multiplication table (Table 1), where the bold elements are purely Euclidean and borrow the 3D Euclidean dot product. This table shows that the usual Euclidean metric holds for the bold vectors, but a strange metric applies to the two additional dimensions e_o and e_∞, which are moreover "orthogonal" to the Euclidean part of the representational space since they have dot product zero with Euclidean vectors. (In fact, the full 5D space now has a Minkowski metric, as can be seen by considering the alternative basis vectors $\sigma_+ = e_o - e_\infty/2$ and $\sigma_- = e_o + e_\infty/2$ that have squared norms of $+1$ and -1, respectively. For more on this basis, see [5].)

This metric is introduced to give a sensible real world meaning to the dot product of two point representatives x and y:

$$\begin{aligned} x \cdot y &= \left(e_o + \mathbf{x} + \frac{1}{2}\|\mathbf{x}\|^2 e_\infty\right) \cdot \left(e_o + \mathbf{y} + \frac{1}{2}\|\mathbf{y}\|^2 e_\infty\right) \\ &= \left(0 + 0 - \frac{1}{2}\|\mathbf{y}\|^2\right) + (0 + \mathbf{x} \cdot \mathbf{y} + 0) + \left(-\frac{1}{2}\|\mathbf{x}\|^2 + 0 + 0\right) \\ &= -\frac{1}{2}(\mathbf{x} - \mathbf{y}) \cdot (\mathbf{x} - \mathbf{y}) = -\frac{1}{2}\|\mathbf{x} - \mathbf{y}\|^2. \end{aligned} \tag{2}$$

The dot product in conformal space therefore encodes the (squared) Euclidean distance of the original points! Since points have distance zero to themselves, they are represented by null vectors; and since Euclidean transformations should preserve the inter-point distance, they should preserve the dot product.

Euclidean transformations are represented as orthogonal transformations

in CGA.[1] This is more specific than their representation as a certain strange class of linear transformations in the usual homogeneous model, and it permits us to design

[1]We have simplified slightly; the general representation of a point at $\mathbf{x}$ in CGA is a scalar multiple of x in (1); the scalar factor is the scalar $-e_\infty \cdot x$ (as you may verify), and this can be consistently

a more effective computational framework tailored to this property. Matrices are actually not that great for representing orthogonal transformations, but fortunately there is something better, as we will see in the next section.

First, let us determine what the vectors in the 5D representation space signify geometrically. Suppose that we want to represent a sphere with center C and radius ρ in Euclidean space. A point X on such a sphere would satisfy $\|\mathbf{x}-\mathbf{c}\|^2=\rho^2$ (using Euclidean vectors). Using (2), this can be written in terms of the dot product of the representative vectors x and c as $x\cdot c=-\frac{1}{2}\rho^2$; and using $-e_\infty\cdot x=1$, we can even group into $x\cdot(c-\frac{1}{2}\rho^2 e_\infty)=0$. The vector $\sigma=\alpha(c-\frac{1}{2}\rho^2 e_\infty)$ is the most general vector we can make in the conformal space (it has five parameters), and written in this form we recognize it as representing a sphere with center c, radius ρ, and "weight" α through the equation $x\cdot\sigma=0$. You may verify that $\|\sigma\|^2=\alpha^2\rho^2$ (even "imaginary spheres" with $\rho^2<0$ are included) and that a point is just a sphere with radius zero, represented by a null vector (for which $\|x\|^2=0$). A plane is the degenerate case of a sphere, and it is represented by a vector of the form $\pi=\alpha(\mathbf{n}+\delta e_\infty)$ (which has no e_o-component and therefore satisfies $\pi\cdot e_\infty=0$). Here $\mathbf{n}$ is the unit normal vector of the plane, δ is its oriented distance from the origin, and α a weight. So:

the vectors in conformal space represent weighted spheres and planes.

In this tutorial, we will mostly use unit weights, focusing on the merely geometrical aspects of the representation. In our *notation*, we will use bold for the elements of the conformal model that are in its n-D Euclidean subspace, and nonbold for elements residing in the full $(n+2)$-D representational space or its geometric algebra. Since there is a clear correspondence between elements of Euclidean geometry and their conformal representation, we will drop the distinction between X and x, and talk about a point x at location $\mathbf{x}$.

2.2 *Trick 2: Orthogonal Transformations as Multiple Reflections in a Sandwiching Representation*

In mathematics, the Cartan–Dieudonné theorem states that all orthogonal transformations can be represented as multiple reflections. In linear algebra, this fact is not used much, since reflections are represented awkwardly and therefore unsuitable as atoms of representation. If we want to reflect a Euclidean vector $\mathbf{x}$ in a plane through the origin with normal vector $\mathbf{a}$, this is the linear transformation

$$\mathbf{x}\mapsto\mathbf{x}-2(\mathbf{x}\cdot\mathbf{a})\mathbf{a}/(\mathbf{a}\cdot\mathbf{a}). \tag{3}$$

---

interpreted as the *weight* of the point. The squared distance between weighted points is computed by normalizing first as $(x/(-e_\infty\cdot x))\cdot(y/(-e_\infty\cdot y))$. Euclidean transformations should then not affect this formula; this implies that they are the specific orthogonal transformations that preserve the special vector e_∞.

It does not look elementary at all, and within linear algebra the dot products cannot be simplified.

We now introduce a clever trick: we consider the dot product (which is symmetric) as merely the symmetrical part of a more fundamental product between vectors. That product (invented by Clifford in 1872) is called the *geometric product* and denoted by a space. So we rewrite:

$$\mathbf{a}\cdot\mathbf{x} = \frac{1}{2}(\mathbf{a}\mathbf{x}+\mathbf{x}\mathbf{a}). \tag{4}$$

This more fundamental product is defined to be bilinear and associative but not necessarily commutative. We see that $\|\mathbf{x}\|^2 = \mathbf{x}\cdot\mathbf{x} = \mathbf{x}\mathbf{x} = \mathbf{x}^2$, so that the square of a vector under the geometric product is a scalar. We extend the geometric product to scalars (and later to other elements). Scalars commute under the geometric product, so $\alpha x = x\alpha$ for vector x and scalar α. A vector $\mathbf{x}$ has a unique inverse $\mathbf{x}^{-1}$ under the geometric product, defined through $\mathbf{x}\mathbf{x}^{-1} = 1 = \mathbf{x}^{-1}\mathbf{x}$ and therefore found explicitly as

$$\textit{inverse of a vector:}\quad \mathbf{x}^{-1} = \mathbf{x}/\left(\mathbf{x}^2\right).$$

Now we see how this simplifies the reflection representation:

$$\begin{aligned}\textit{reflection in origin hyperplane with normal } \mathbf{a}\text{:}\quad \mathbf{x} &\mapsto \mathbf{x} - 2(\mathbf{x}\cdot\mathbf{a})\mathbf{a}/(\mathbf{a}\cdot\mathbf{a})\\ &= \mathbf{x} - (\mathbf{x}\mathbf{a}+\mathbf{a}\mathbf{x})\mathbf{a}^{-1}\\ &= -\mathbf{a}\mathbf{x}\mathbf{a}^{-1}. \end{aligned} \tag{5}$$

The reflection of $\mathbf{x}$ in the origin hyperplane with normal vector $\mathbf{a}$ is therefore simply a "sandwiching" of $\mathbf{x}$ by $\mathbf{a}$ and $\mathbf{a}^{-1}$ (with a minus sign). In this form, the fundamental nature of reflections for the representation of transformations is more obvious.

You may rightly object that we have not really reflected a point x, but only its Euclidean part $\mathbf{x}$. Let us try to extend the formula to the point x, using the explicit representation (1). Postulating distributivity of the geometric product, we get $-\mathbf{a}x\mathbf{a}^{-1} = -\mathbf{a}(e_o + \mathbf{x} + \frac{1}{2}\|\mathbf{x}\|^2 e_\infty)\mathbf{a}^{-1} = -\mathbf{a}e_o\mathbf{a}^{-1} - \mathbf{a}\mathbf{x}\mathbf{a}^{-1} - \frac{1}{2}\|\mathbf{x}\|^2\mathbf{a}e_\infty\mathbf{a}^{-1}$. Evaluating this requires computing what $-\mathbf{a}e_o\mathbf{a}^{-1}$ and $-\mathbf{a}e_\infty\mathbf{a}^{-1}$ are. We realize from definition (4) and the dot product table that $-\mathbf{a}e_o\mathbf{a}^{-1} = -(\mathbf{a}e_o)\mathbf{a}^{-1} = -(2\mathbf{a}\cdot e_o - e_o\mathbf{a})\mathbf{a}^{-1} = 0 + e_o\mathbf{a}\mathbf{a}^{-1} = e_o$. Of course, you would expect this geometrically: the point at the origin does not change after the reflection. Similarly for e_∞, as you may verify. Further realize that $\|-\mathbf{a}\mathbf{x}\mathbf{a}^{-1}\|^2 = (-\mathbf{a}\mathbf{x}\mathbf{a}^{-1})(-\mathbf{a}\mathbf{x}\mathbf{a}^{-1}) = \mathbf{a}\mathbf{x}\mathbf{x}\mathbf{a}^{-1} = \mathbf{x}^2(\mathbf{a}\mathbf{a}^{-1}) = \|\mathbf{x}^2\|$—obviously, since reflection is an orthogonal transformation. Combining all this, we find $-\mathbf{a}x\mathbf{a}^{-1} = e_o - \mathbf{a}\mathbf{x}\mathbf{a}^{-1} + \frac{1}{2}\|-\mathbf{a}\mathbf{x}\mathbf{a}^{-1}\|^2 e_\infty$, which is precisely the representation of a point at the reflected location. Therefore a point x is reflected by transfer of the Euclidean formula (3), as $x \mapsto -\mathbf{a}x\mathbf{a}^{-1}$. This structural principle may be illustrated as the commutative diagram

$$
\begin{array}{ccccc}
\boxed{\text{position } \mathbf{x}} & \xrightarrow{\text{normal vector}} & \mathbf{a} & \xrightarrow{\text{as reflector}} & \boxed{\text{position } -\mathbf{a}\mathbf{x}\mathbf{a}^{-1}} \\
\text{embed in CGA}\big\downarrow & & \text{embed in CGA}\big\downarrow & & \text{embed in CGA}\big\downarrow \\
\boxed{\text{point } x} & \xrightarrow{\text{origin plane}} & \mathbf{a} & \xrightarrow{\text{as reflector}} & \boxed{\text{point } -\mathbf{a}x\mathbf{a}^{-1}}
\end{array}
$$

If we perform a second reflection in another origin hyperplane, with normal vector **b**, this should be the mapping

$$x \mapsto = -\mathbf{b}\big(-\mathbf{a}x\mathbf{a}^{-1}\big)\mathbf{b}^{-1} = (\mathbf{ba})x(\mathbf{ba})^{-1},$$

using the associativity of the geometric product in the rewriting. Geometrically, a double reflection is a rotation (see Fig. 2), so the operator (**ba**) represents a rotation operator (in an axis through the origin, determined as the intersection of the planes **a** and **b**). In this manner, we can generate all orthogonal transformations as sandwiching products by elements that are themselves the geometric product of vectors. These elements are called *versors*. A delightful property of versors is that they do not only apply to vectors, but also directly to other geometric elements like lines and circles. Let us first make those geometric elements part of our algebra.

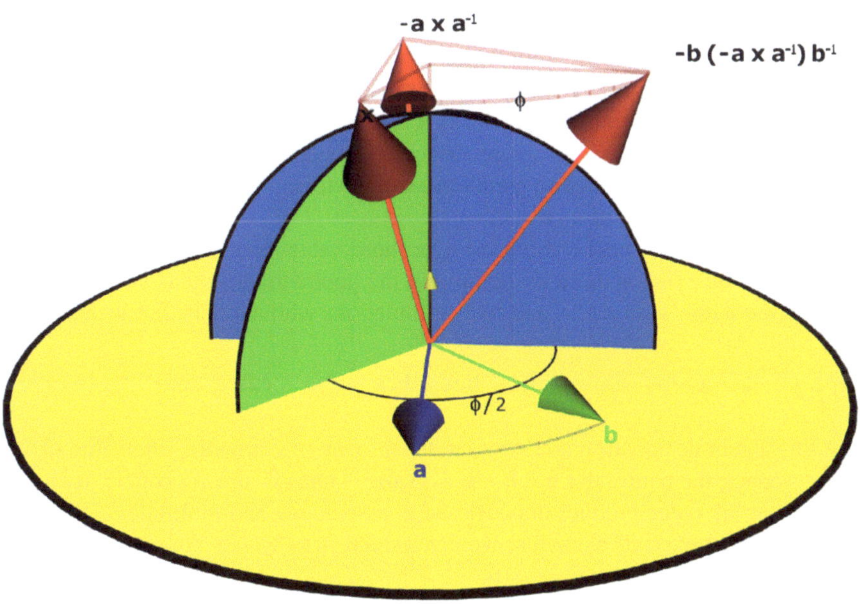

Fig. 2 A reflection in two successive planes is equivalent to a rotation over double their separating angle, around the line of their intersection (in 3D)

2.3 Trick 3: Constructing Elements by Anti-Symmetry

When we introduced the geometric product for vectors, we used only its symmetric part (that was the dot product). But of course there is an anti-symmetric part as well. Let us denote that by $\wedge$ and call it the *outer product*. For vectors, it is defined as

$$\mathbf{x} \wedge \mathbf{a} = \frac{1}{2}(\mathbf{xa} - \mathbf{ax}).$$

It is clear that $\mathbf{x} \wedge \mathbf{a} = -\mathbf{a} \wedge \mathbf{x}$, so that $\mathbf{x} \wedge \mathbf{x} = 0$.

To interpret this new element $\mathbf{x} \wedge \mathbf{a}$ geometrically, let us use some classical linear algebra and take $\mathbf{x}$ and $\mathbf{a}$ as direction vectors. If we take an orthonormal basis $\{\mathbf{e}_1, \mathbf{e}_2\}$ in the plane spanned by $\mathbf{x}$ and $\mathbf{a}$, and choose it such that $\mathbf{x} = \|\mathbf{x}\|\mathbf{e}_1$, then $\mathbf{a}$ can be written as $\mathbf{a} = \|\mathbf{a}\|(\cos(\phi)\mathbf{e}_1 + \sin(\phi)\mathbf{e}_2)$ with ϕ the angle from $\mathbf{x}$ to $\mathbf{a}$. We evaluate:

$$\begin{aligned}\mathbf{x} \wedge \mathbf{a} &= \big(\|\mathbf{x}\|\mathbf{e}_1\big) \wedge \big(\|\mathbf{a}\|\big(\cos(\phi)\mathbf{e}_1 + \sin(\phi)\mathbf{e}_2\big)\big)\\ &= \|\mathbf{x}\|\,\|\mathbf{a}\|\big(\cos(\phi)\mathbf{e}_1 \wedge \mathbf{e}_1 + \sin(\phi)\mathbf{e}_1 \wedge \mathbf{e}_2\big)\\ &= \|\mathbf{x}\|\,\|\mathbf{a}\|\sin(\phi)\mathbf{e}_1 \wedge \mathbf{e}_2,\end{aligned}$$

for being the sum of two bilinear products, the outer product is itself bilinear. We recognize in $\|\mathbf{x}\|\,\|\mathbf{a}\|\sin(\phi)$ the signed area of the oriented parallelogram spanned by $\mathbf{x}$ and $\mathbf{a}$ (in that order) and can therefore interpret $\mathbf{e}_1 \wedge \mathbf{e}_2$ as the algebraic specification of the unit area element in the $(\mathbf{e}_1, \mathbf{e}_2)$-plane. We call this a *unit 2-blade*.We then interpret the 2-blade $\mathbf{x} \wedge \mathbf{a}$ of direction vectors as the full specification of the geometric area element spanned by $\mathbf{x}$ and $\mathbf{a}$ (in that order) in terms of its magnitude, orientation, and geometrical attitude (i.e., spatial stance). Only the shape is not determined, for you can easily verify that, for instance, $\mathbf{x} \wedge (\mathbf{a} + \lambda\mathbf{x}) = \mathbf{x} \wedge \mathbf{a}$ so that $\mathbf{x}$ and $\mathbf{a} + \lambda\mathbf{x}$ span the same element as $\mathbf{x}$ and $\mathbf{a}$. For parallel direction vectors $\mathbf{x}$ and $\mathbf{a}$, the outer product $\mathbf{x} \wedge \mathbf{a}$ is zero, so the commutativity relationship $\mathbf{xa} = \mathbf{ax}$ is the algebraic way of expressing parallelness of vectors. Orthogonality of vectors is expressed as $\mathbf{xa} = -\mathbf{ax}$, or $\mathbf{x} \cdot \mathbf{a} = 0$.

The outer product can be extended over more vector terms, always as the anti-symmetric sum. This is done by permuting the geometric products and endowing even permutations with a plus and odd permutations with a minus. For instance:

$$\mathbf{a} \wedge \mathbf{b} \wedge \mathbf{c} = \frac{1}{3!}(\mathbf{abc} - \mathbf{bac} + \mathbf{bca} - \mathbf{cba} + \mathbf{cab} - \mathbf{acb})$$

(but this algebraic equation is a very inefficient way of computing the value of the outer product; the equivalent $\mathbf{a} \wedge \mathbf{b} \wedge \mathbf{c} = \frac{1}{2}(\mathbf{abc} - \mathbf{cba})$ is already better). It can be shown that the outer product thus defined is associative and multilinear. To make it fully defined over all elements, we can extend it to scalars simply by defining $\alpha \wedge \mathbf{a} = \alpha\mathbf{a}$ for scalar α and vector $\mathbf{a}$.

The outer product of k vector factors is called a k*-blade*, and the number of vector factors k is called its *grade*. Geometrically, a k-blade is a quantitative representation of a weighted, oriented k-dimensional subspace of the space its vectors reside in, and its signed magnitude is an oriented hypervolume. For instance, if you would compute the outer product of three direction vectors in 3D space, you would find that the coordinates of the vectors combine to a familiar signed scalar multiple of the unit volume: $\mathbf{a} \wedge \mathbf{b} \wedge \mathbf{c} = \det([\![\mathbf{a}\,\mathbf{b}\,\mathbf{c}]\!])\mathbf{e}_1 \wedge \mathbf{e}_2 \wedge \mathbf{e}_3$. This volume is zero when the vectors are co-planar, and therefore $\mathbf{x} \wedge (\mathbf{a} \wedge \mathbf{b}) = 0$ can be solved for $\mathbf{x}$ as $\mathbf{x} = \lambda\mathbf{a} + \mu\mathbf{b}$. Again, the 2-blade $\mathbf{a} \wedge \mathbf{b}$ is seen to be a single computational element representing the plane spanned by the direction vectors $\mathbf{a}$ and $\mathbf{b}$.

In the conformal model, the outer product of vectors representing points a and b takes on a different geometric interpretation, even though its algebra is the same. In CGA, the blade *$a \wedge b$ represents an oriented point-pair*, in the sense that the set of points x satisfying $x \wedge a \wedge b = 0$ is either $x = a$ or $x = b$. (Comparing to the derivation just given, we do get $x = \lambda a + \mu b$, as before, but to be a point in CGA, x has to satisfy $x \cdot x = 0$ by (2), as do a and b. Some algebra then leads to $\lambda\mu(a \cdot b) = 0$, and this implies $\lambda = 0$ and/or $\mu = 0$.) Similarly, *$a \wedge b \wedge c$ represents the oriented circle through the points a, b, and c*, and the outer product of four points $a \wedge b \wedge c \wedge d$ represents an oriented *sphere*. We call these elements *rounds*. If the points are in degenerate positions, or if one of them is the point at infinity e_∞, an oriented *flat* results (in 3D, these are: a line $a \wedge b \wedge e_\infty$, a plane $a \wedge b \wedge c \wedge e_\infty$, or a "flat point" $a \wedge e_\infty$). Showing these facts without too much computation requires the technique of dual representation, introduced next.

2.4 Trick 4: Dual Specification of Elements Permits Intersection

A subspace can be characterized by the outer product, but it is often convenient to take a "dual" approach, not specifying the vectors in it but the vectors orthogonal to it. We have already seen this for spheres: the orthogonality demand $x \cdot (c - \frac{1}{2}\rho^2 e_\infty) = 0$ solves for x lying on a sphere with center c and radius ρ. Duality is a fundamental concept of geometric algebra and requires no more than complementation relative to the volume of the vector space, through division.

An n-dimensional vector space cannot have nonzero blades of a grade exceeding n. A nonzero blade of the maximum grade n is called a *pseudoscalar* for the space. It is common to normalize this to a unit pseudoscalar and to denote it by $\mathbf{I}_n$ or I_n. The choice of the sign of the unit pseudoscalar amounts to choosing a reference orientation for the space. In a 3D Euclidean space of direction vectors with an orthonormal basis, $\mathbf{I}_3 = \mathbf{e}_1 \wedge \mathbf{e}_2 \wedge \mathbf{e}_3 (= \mathbf{e}_1\mathbf{e}_2\mathbf{e}_3)$ picks the standard "right-handed" orientation. In the conformal model space, a suitable pseudoscalar is $I_{4,1} = e_o \wedge \mathbf{I}_3 \wedge e_\infty$. The inverse of the unit pseudoscalar in 3D Euclidean space is $\mathbf{I}_3^{-1} = -\mathbf{I}_3$ (verify that $\mathbf{I}_3\mathbf{I}_3^{-1} = 1$!). In the conformal space, $I_{4,1}^{-1} = e_o \wedge \mathbf{I}_3^{-1} \wedge e_\infty = -I_{4,1}$.

One can find the blade representing the orthogonal complement of any subspace through right-dividing its blade A by the pseudoscalar, as AI_n^{-1}. This is called the *dual* of A and denoted A^*:

$$\textit{dualization:} \quad A^* = AI_n^{-1}. \tag{6}$$

For instance, the dual of the 2-blade $\mathbf{e}_1 \wedge \mathbf{e}_2$ in 3D-space is $(\mathbf{e}_1 \wedge \mathbf{e}_2)(\mathbf{e}_1 \wedge \mathbf{e}_2 \wedge \mathbf{e}_3)^{-1} = -(\mathbf{e}_1\mathbf{e}_2)(\mathbf{e}_1\mathbf{e}_2\mathbf{e}_3) = \mathbf{e}_3$. This is indeed the normal vector of the $(\mathbf{e}_1, \mathbf{e}_2)$-plane using the right-hand rule. The familiar 3D cross product of vectors can be made in CGA as $\mathbf{x} \times \mathbf{a} = (\mathbf{x} \wedge \mathbf{a})\mathbf{I}_3^{-1}$, though its use should be avoided.

Duality permits us to intersect subspaces. Let us denote the *intersection* (or meet) of blades A and B as $A \cap B$; then we can define it in terms of outer product and dual as

$$\textit{dual specification of meet:} \quad (A \cap B)^* = B^* \wedge A^*, \tag{7}$$

where the duality is to be taken relative to the smallest-grade blade containing both A and B (this is known as their join, and the intersection as their meet). If one simply takes duality relative to the full space, a meet can become zero in degenerate situations. (More about these operations and their efficient implementation in [4].)

An extension of the inner product beyond vector arguments can be developed as a product in its own right, with its own set of algebraic rules. When done properly, it is consistent with the rest of the framework in the sense that

$$\textit{extension of inner product:} \quad A \cdot B \equiv \left(A \wedge B^*\right)^{-*}, \tag{8}$$

with duality relative to a blade containing the join (one usually takes the pseudoscalar I_n).[2] This inner product has properties like

$$x \cdot (a \wedge b) = (x \cdot a)b - (x \cdot b)a. \tag{9}$$

The inner product is especially convenient to define orthogonal projection of subspaces as

$$\textit{orthogonal projection of } X \textit{ onto } B: \quad X \mapsto \left(X \cdot B^{-1}\right) \cdot B.$$

For flats, this corresponds to the usual orthogonal projection but it is more general: for instance, projecting a line onto a sphere produces a great circle.

Knowing duality also permits us to interpret elements like $\mathbf{a} \wedge \mathbf{b}$. In CGA, $\mathbf{a}$ and $\mathbf{b}$ are the dual representations of planes through the origin, for the points on these planes satisfy $x \cdot \mathbf{a} = 0$ and $x \cdot \mathbf{b} = 0$. Therefore by (7), the 2-blade $\mathbf{a} \wedge \mathbf{b}$ should be the dual representation of their intersection line. Points x on that line should then satisfy $x \cdot (\mathbf{a} \wedge \mathbf{b}) = 0$, and expanding according to (9) shows that this indeed holds. You may verify that the point at infinity e_∞ is on the line $(\mathbf{a} \wedge \mathbf{b})^{-*}$.

We now have enough to show that in CGA, $S = a \wedge b \wedge c \wedge d$ represents the sphere through the four points a, b, c, d. The geometry is illustrated in Fig. 3. By antisymmetry of $\wedge$, we can subtract any factor from the others without changing the value of S. We use a to produce $S = a \wedge (b - a) \wedge (c - a) \wedge (d - a)$. To find out what $(b - a)$ represents, solve $x \cdot (b - a) = 0$. This evaluates to $x \cdot a = x \cdot b$, and because of (2), this means that x has the same distance to a and b. So $(b - a)$

[2] This inner product is called the *left contraction* and denoted "⌋" in [2]. It differs in details from the inner product used in [1].

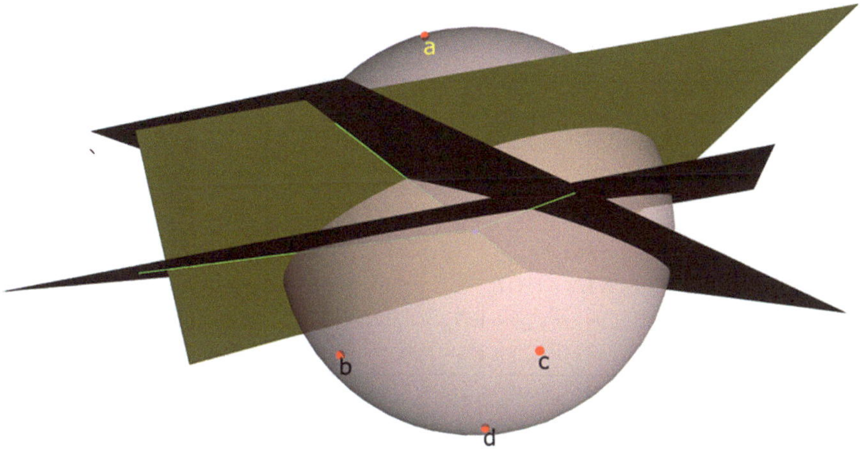

Fig. 3 The proof that $a \wedge b \wedge c \wedge d$ represents a sphere involves the intersection of the midplanes $b - a$, $c - a$, and $d - a$

is the dual representation of the midplane between a and b. Therefore $(b - a) \wedge (c - a) \wedge (d - a)$ is the dual representation of the intersection of three midplanes. These planes intersect in two points: the center of the sphere m and the point at infinity e_∞, so $(b - a) \wedge (c - a) \wedge (d - a)$ is proportional to $(m \wedge e_\infty)^*$. Then we find $S \propto a \wedge (m \wedge e_\infty)^* = (a \cdot (m \wedge e_\infty))^* = (m - \frac{1}{2}\rho^2 e_\infty)^*$ with $a \cdot m = -\frac{1}{2}\rho^2$. So indeed S is the dual of a dual sphere representation and therefore a sphere. This also gives a very compact way to compute center and radius of a sphere given by four points: they are simply the appropriate components of $(a \wedge b \wedge c \wedge d)^*$.

3 Bonus: The Elements of Euclidean Geometry as Blades

Closure of the operations of outer product and duality produces a suite of blades representing recognizable elements of Euclidean geometry. We have seen many examples of this already, and the full list is given in Table 2 from [2] (where n is the dimension of the Euclidean space, $\mathbf{E}$ a purely Euclidean element of appropriate grade, $\mathbf{E}^\star$ denotes the Euclidean dual $\mathbf{E}\mathbf{I}_n^{-1}$, and $T_\mathbf{p}$ denotes the translation versor over $\mathbf{p}$, see (11)). Care has been taken to orient the blades and their duals consistently.

The square of a normalized round gives its radius squared, and this may be negative. Such "imaginary rounds" occur naturally, for instance, when intersecting two spheres that are further apart than the sum of their radii. Because only the squared radius occurs in the conformal model, these elements are tractable in a real algebra. Tangents are in fact rounds of zero radius, indicative of their infinitesimal size. A tangent 2-blade occurs, for instance, as the grade 3 element that is the meet of two touching spheres. In this context, a weighted point may be viewed as a localized tangent scalar.

Table 2

Element	*Standard form X*	*Defining properties*
Direction	$\mathbf{E} \wedge e_\infty$	$e_\infty \wedge X = 0$; $e_\infty \cdot X = 0$
Dual direction	$-\mathbf{E}^\star \wedge e_\infty$	$e_\infty \wedge X = 0$; $e_\infty \cdot X = 0$
Flat	$T_\mathbf{p}(e_o \wedge \mathbf{E} \wedge e_\infty)T_\mathbf{p}^{-1}$	$e_\infty \wedge X = 0$; $e_\infty \cdot X \neq 0$
Dual flat	$T_\mathbf{p}(\mathbf{E}^\star(-1)^{n-\text{grade}(\mathbf{E})})T_\mathbf{p}^{-1}$	$e_\infty \wedge X \neq 0$; $e_\infty \cdot X = 0$
Tangent	$T_\mathbf{p}(e_o \wedge \mathbf{E})T_\mathbf{p}^{-1}$	$e_\infty \wedge X \neq 0$; $e_\infty \cdot X \neq 0$; $X^2 = 0$
Dual tangent	$T_\mathbf{p}(e_o \wedge \mathbf{E}^\star(-1)^n)T_\mathbf{p}^{-1}$	$e_\infty \wedge X \neq 0$; $e_\infty \cdot X \neq 0$; $X^2 = 0$
Round	$T_\mathbf{p}((e_o + \frac{1}{2}\rho^2 e_\infty) \wedge \mathbf{E})T_\mathbf{p}^{-1}$	$e_\infty \wedge X \neq 0$; $e_\infty \cdot X \neq 0$; $X^2 \neq 0$
Dual round	$T_\mathbf{p}((e_o - \frac{1}{2}\rho^2 e_\infty) \wedge \mathbf{E}^\star(-1)^n)T_\mathbf{p}^{-1}$	$e_\infty \wedge X \neq 0$; $e_\infty \cdot X \neq 0$; $X^2 \neq 0$

It is especially notable that the various uses and meanings of "vector with direction $\mathbf{u}$" from applied linear algebra get their own "algebraic data structures":

- *a point at location* $\mathbf{u}$ is represented by the CGA vector $e_o + \mathbf{u} + \frac{1}{2}\mathbf{u}^2 e_\infty$
- *a free vector* is represented by the translation invariant 2-blade $\mathbf{u} \wedge e_\infty$
- *a normal vector* is the vector $p \cdot (\mathbf{u} \wedge e_\infty)$ and can shift on a localized plane
- *a force vector* is represented by the 3-blade $p \wedge \mathbf{u} \wedge e_\infty$ and can shift along a line
- *a tangent vector* $\mathbf{u}$ at p is the localized 2-blade $p \cdot (p \wedge \mathbf{u} \wedge e_\infty)$

All these automatically move appropriately under Euclidean versors, without a programmer needing to specify that they should (by giving them their own "classes" and "methods," as is required in common practice in classical software, even when based on homogeneous coordinates).

4 Bonus: Euclidean Motions Through Sandwiching

We have seen how all orthogonal transformations can be made as multiple reflections and that a single reflection is represented by an invertible vector $\mathbf{a}$ as the transformation $x \mapsto -\mathbf{a}x\mathbf{a}^{-1}$. Now that we know what the vectors in the conformal model represent, we can easily generate the versors for common motions. Euclidean motions are generated by multiple reflections in planes, and we have seen that those are dually represented by vectors of the form $\pi = \mathbf{n} + \delta e_\infty$ that satisfy $e_\infty \cdot \pi = 0$.

- *Rotation in a plane through the origin*: If we take two unit dual planes at the origin $\mathbf{n}_1$ and $\mathbf{n}_2$ with a relative angle of $\phi/2$ from $\mathbf{n}_1$ to $\mathbf{n}_2$, the double reflection first in $\mathbf{n}_1$ and then in $\mathbf{n}_2$ is represented as

$$R_{\mathbf{I}\phi} = \mathbf{n}_2\mathbf{n}_1 = \mathbf{n}_2 \cdot \mathbf{n}_1 + \mathbf{n}_2 \wedge \mathbf{n}_1 = \cos(\phi/2) - \mathbf{I}\sin(\phi/2). \tag{10}$$

When used in a sandwiching operation, this is a rotation over the angle ϕ around the dual line given by the unit 2-blade $\mathbf{I}$ (proportional to $\mathbf{n}_1 \wedge \mathbf{n}_2$). Such a 2-blade has the property $\mathbf{I}^2 = -1$. To show this, introduce an orthonormal basis $\{\mathbf{e}_1, \mathbf{e}_2\}$, write $\mathbf{I} = \mathbf{e}_1 \wedge \mathbf{e}_2 = \mathbf{e}_1\mathbf{e}_2$, and compute using the associativity property of the geometric product: $(\mathbf{e}_1 \wedge \mathbf{e}_2)(\mathbf{e}_1 \wedge \mathbf{e}_2) = (\mathbf{e}_1\mathbf{e}_2)(\mathbf{e}_1\mathbf{e}_2) = -\mathbf{e}_2\mathbf{e}_1\mathbf{e}_1\mathbf{e}_2 = -\mathbf{e}_2\mathbf{e}_2 = -1$. In this real geometric algebra, we therefore naturally get elements that square to -1. In 3D, there is a basis for 2-blades consisting of the elements $\mathbf{I} = \mathbf{e}_1 \wedge \mathbf{e}_2$, $\mathbf{J} = \mathbf{e}_2 \wedge \mathbf{e}_3$, and $\mathbf{K} = \mathbf{e}_3 \wedge \mathbf{e}_1$, each squaring to -1 and having multiplicative relationships like $\mathbf{IJ} = -\mathbf{JI} = -\mathbf{K}$. These are of course isomorphic to the elementary quaternions which have proven so useful for 3D rotation computations. In geometric algebra, they are introduced in a real manner as products of vectors, fully integrated with the real elements they operate on. We will soon see that they can rotate any element, and derive the versor for a rotation around a general line in Sect. 6.

- *Translation*: A translation over a vector $\mathbf{t}$ is generated by reflection in two dual planes separated by a vector $\mathbf{t}/2$, resulting in the element: $(\mathbf{t} + \frac{1}{2}\mathbf{t} \cdot \mathbf{t}e_\infty)\mathbf{t} = \mathbf{t}^2(1 - \mathbf{t}e_\infty/2)$. Since a scalar multiple generates the same motion in the sandwiching product with the inverse, we prefer to define

$$\textit{versor for translation over } \mathbf{t}\text{:} \quad T_\mathbf{t} \equiv 1 - \mathbf{t}e_\infty/2. \tag{11}$$

  You can check that the point representation (1) is indeed related to the point at the origin e_o by translation over $\mathbf{x}$, since $x = T_\mathbf{x} e_o T_\mathbf{x}^{-1}$.
- *General rigid body motion*: A general rigid body motion can be constructed in the usual manner as a rotation followed by a translation. In CGA, an alternative is to make it directly as the reflection in two lines, which produces a screw motion (see [2]).
- *Uniform scaling*: Although not strictly a rigid body motion, the Euclidean similarity transformation of uniform scaling can be made by subsequent reflection in two dual spheres at the origin such as $e_o - \frac{1}{2}\rho_1^2 e_\infty$ and $e_o - \frac{1}{2}\rho_2^2 e_\infty$. After some simplification, the scaling versor for a uniform scaling by e^γ is found to be

$$S_\gamma \equiv \cosh(\gamma/2) + \sinh(\gamma/2) e_o \wedge e_\infty.$$

More versors can be generated by reflection in spheres, notably for the conformal operation of a *transversion*—details may be found elsewhere [2].

5 Bonus: Structure Preservation and the Transfer Principle

All constructions of elements were based on the linear combinations of geometric products, since the other products are ultimately expressible in that manner. Therefore, when we act on them with a versor V in the sandwiching product, all constructions transform covariantly. For the outer product, this means that equations hold like the following:

$$V(a \wedge b)V^{-1} = \left(VaV^{-1}\right) \wedge \left(VbV^{-1}\right).$$

The same structure-preserving property holds for *all* operations we introduced, be they spanning, inner product, or duality (relative to a transformed pseudoscalar). In words: "*the transformation of a construction equals the construction of the transformed elements*." This fact is very convenient, for it implies that we can simply construct something at the origin and then move it into place to find the general form (hence our preference for origin-based specification in the table above). And composite elements move by the same versor as points do: the translation versor $T_{\mathbf{t}}$ universally translates points, lines, planes, spheres, or tangent elements. As we mentioned, there is no longer any need for data structures distinguishing between "position vectors" which feel translations and "direction vectors" which do not; all is automatically administrated in the algebraic behavior of the corresponding elements. This is an enormous advantage relative to the classical homogeneous model for the development of structural code, either by hand or using a code generator [3].

This principle is also extremely useful in derivations. Let us, for instance, use it to prove the general formula for the reflection of a line Λ in a dual plane π as $\Lambda \mapsto -\pi\Lambda\pi^{-1}$, simply from the 1-D direction reflection formula (5). A line Λ_0 with direction $\mathbf{u}$ through the origin is given as $\Lambda_0 = e_o \wedge \mathbf{u} \wedge e_\infty$, and a dual plane π_0 through the origin with normal vector $\mathbf{n}$ as $\pi_0 = \mathbf{n}$. The reflection of the direction $\mathbf{u}$ is affected by (5) as $\mathbf{u} \mapsto \mathbf{u}' \equiv -\mathbf{nun}^{-1} = -\pi_0\mathbf{u}\pi_0{}^{-1}$. The reflected line is then $\Lambda_0' = e_o \wedge \mathbf{u}' \wedge e_\infty$. Now we note that due to the algebraic commutation (i.e., the geometric orthogonality) of the bold Euclidean and the nonbold extra dimensions e_o and e_∞, we have $-\pi_0 e_o \pi_0{}^{-1} = -\mathbf{n}e_o\mathbf{n}^{-1} = e_o$ and $-\pi_0 e_\infty \pi_0{}^{-1} = -\mathbf{n}e_\infty\mathbf{n}^{-1} = e_\infty$. Therefore we can "pull out" the reflection operator to act on the whole line Λ_0 by (5):

$$\begin{aligned}\Lambda_0' &= \left(-\pi_0 e_o \pi_0{}^{-1}\right) \wedge \left(-\pi_0 \mathbf{u} \pi_0{}^{-1}\right) \wedge \left(-\pi_0 e_\infty \pi_0{}^{-1}\right)\\ &= -\pi_0(e_o \wedge \mathbf{u} \wedge e_\infty)\pi_0{}^{-1} = -\pi_0\Lambda_0\pi_0{}^{-1}.\end{aligned}$$

This is still only true at the origin, but we can move this construction by a motion versor V to an arbitrary location. All elements change to their general form $\pi = V\pi_0 V^{-1}$, $\Lambda = V\Lambda_0 V^{-1}$, and the reflection transformation preserves its structure since $V\Lambda_0' V^{-1} = (V\pi_0 V^{-1})(V\Lambda_0 V^{-1})(V\pi_0 V^{-1}) = \pi\Lambda\pi^{-1}$. Therefore the general reflection formula of a line in a plane is simply

$$\textit{reflection of a line } \Lambda \textit{ in the dual plane } \pi: \quad \Lambda \mapsto -\pi\Lambda\pi^{-1}.$$

This includes all aspects of location, direction, and orientation. Note that this computation reflects a general line in a general plane without computing its intersection point—try doing that using linear algebra! (If you need the intersection point of line and plane, it is $\pi \cdot \Lambda$, by straightforward application of the universal meet operation (7) and duality (6), (8).)

$$
\begin{array}{ccccc}
\boxed{\text{direction } \mathbf{u}} & \xrightarrow{\text{normal vector}} & \mathbf{a} & \xrightarrow{\text{as reflector}} & \boxed{\text{direction } \mathbf{u} - 2(\mathbf{u}\cdot\mathbf{a})\mathbf{a}/\|\mathbf{a}\|^2)} \\
\text{embed in GA}\downarrow & & \text{embed in GA}\downarrow & & \text{embed in GA}\downarrow \\
\boxed{\text{direction } \mathbf{u}} & \xrightarrow{\text{normal vector}} & \mathbf{a} & \xrightarrow{\text{as reflector}} & \boxed{\text{direction } \mathbf{u}' = -\mathbf{a}\mathbf{u}\mathbf{a}^{-1}} \\
\text{embed in CGA}\downarrow & & \text{embed in CGA}\downarrow & & \text{embed in CGA}\downarrow \\
\boxed{\text{origin line } \Lambda_o = e_o \wedge \mathbf{u} \wedge e_\infty} & \xrightarrow{\text{dual } e_o\text{-plane}} & \pi_o & \xrightarrow{\text{as reflector}} & \boxed{\text{origin line } \Lambda_o' = -\pi_o \Lambda_o \pi_o^{-1}} \\
\text{Euclidean versor}\downarrow & & \text{Euclidean versor}\downarrow & & \text{Euclidean versor}\downarrow \\
\boxed{\text{general line } \Lambda = V\Lambda_o V^{-1}} & \xrightarrow{\text{dual plane}} & \pi & \xrightarrow{\text{as reflector}} & \boxed{\text{general line } \Lambda' = -\pi\Lambda\pi^{-1}} \\
\text{conformal versor}\downarrow & & \text{conformal versor}\downarrow & & \text{conformal versor}\downarrow \\
\boxed{\text{general circle } K = V\Lambda V^{-1}} & \xrightarrow{\text{dual sphere}} & \sigma & \xrightarrow{\text{as invertor}} & \boxed{\text{general circle } K' = -\sigma K\sigma^{-1}}
\end{array}
$$

We can even apply an arbitrary conformal versor and change the reflecting dual plane π into a dual sphere σ, and the line L into a circle K; the result is a spherical inversion operation. (As a further extension, another application of the structure preservation property shows that the reflection in σ of a general element X is $X \mapsto (-1)^{\text{grade}(X)}\sigma X\sigma^{-1}$.)

The conformal model renders all transitions trivial in this transfer, all the way from a reflection of a Euclidean direction vector at the origin to the inversion of a general circle in a general sphere. Such is the power of a structure-preserving framework!

6 Trick 5: Exponential Representation of Versors

Even-graded versors, made by an even number of reflections, represent motions that can be performed continuously and in small amounts. In Euclidean and Minkowski spaces, *all even-graded versors can be written as the exponentials of bivectors*. The bivector specification of an even versor is often more directly related to the geometry of the situation than the "product of vectors" method.

As an example of the exponential rewriting, take the rotation $R_{\mathbf{I}\phi}$ over the angle ϕ, parallel to the $\mathbf{I}$-plane as treated in (10),

$$R_{\mathbf{I}\phi} = \cos(\phi/2) - \sin(\phi/2)\mathbf{I} = e^{-\mathbf{I}\phi/2}.$$

It is the property $\mathbf{I}^2 = -1$ that makes the exponential rewriting permitted:

$$\begin{aligned} e^{-\mathbf{I}\phi/2} &= 1 + \frac{1}{1!}(-\mathbf{I}\phi/2)^1 + \frac{1}{2!}(-\mathbf{I}\phi/2)^2 + \cdots \\ &= \left(1 - \frac{1}{2!}(\phi/2)^2 + \cdots\right) + \left(\frac{1}{1!}(\phi/2)^1 - \frac{1}{3!}(\phi/2)^3 + \cdots\right)\mathbf{I} \\ &= \cos(\phi/2) - \sin(\phi/2)\mathbf{I}. \end{aligned}$$

The translation versor of (11) can also be written in this exponential form; but since it involves the bivector $\mathbf{t} \wedge e_\infty$, the expansion truncates after two terms (fundamentally due to $e_\infty^2 = 0$):

$$T_\mathbf{t} = 1 - \mathbf{t} \wedge e_\infty/2 = e^{-\mathbf{t}\wedge e_\infty/2}.$$

A rotation around a general 3D unit line Λ over ϕ is now generated by the versor:

$$\textit{rotation around } \Lambda \textit{ over } \phi\textit{:} \quad R_{\Lambda,\phi} = e^{\Lambda^*\phi/2}$$

Proof This follows from the simply derived structural property

$$V \exp(B) V^{-1} = \exp\bigl(V B V^{-1}\bigr)$$

and the transfer property applied as follows. First recognize that the rotation axis of the origin rotation $R_{\mathbf{I}\phi}$ is the line $\Lambda_0 = \mathbf{I}^* = -\mathbf{I}^{-*}$, so the origin rotation is $\exp(\Lambda_0{}^*\phi/2)$. Then transfer this by a translation T to the actual location of the desired axis Λ, which changes $\Lambda_0{}^*$ to $T(\Lambda_0{}^*)T^{-1} = (T\Lambda_0 T^{-1})/(T I_{4,1} T^{-1}) = \Lambda^*$ since the pseudoscalar $I_{4,1}$ involved in the dualization is translation invariant. Done. □

General rigid body motions can of course also be made, for instance, by the usual method of combining an origin rotation with a translation. You find that the result can be written as the exponential of a general conformal bivector on the basis $\{\mathbf{e}_1 \wedge \mathbf{e}_2, \mathbf{e}_2 \wedge \mathbf{e}_3, \mathbf{e}_3 \wedge \mathbf{e}_1, \mathbf{e}_1 \wedge e_\infty, \mathbf{e}_2 \wedge e_\infty, \mathbf{e}_3 \wedge e_\infty\}$, giving the six degrees of freedom required. Since this space of bivectors is linear, it can be used for motion interpolation. To interpolate between two poses characterized by the versors M_0 and M_1, find their bivectors $B_0 = \log(M_0)$ and $B_1 = \log(M_1)$. Now apply a standard vector interpolation method to smoothly change B_0 into B_1 through intermediate bivectors B_i; then use the versors $\exp(B_i)$ to generate the interpolated poses. To execute this procedure, one needs to find the bivector corresponding to a given versor; such "versor logarithms" may be found in [2].

Linearization of versor motions for extrapolation or estimation is also possible and requires geometric calculus. When performed (see [1]), the first order change in an element X that is moved by a changing versor $V(\tau)$ from a standard element X_0 as $X(\tau) = V(\tau)X_0V(\tau)^{-1}$ is

$$X(\tau + d\tau) = X(\tau) + \bigl(\Omega(\tau)X(\tau) - X(\tau)\Omega(\tau)\bigr)$$

$$\text{with } \Omega(\tau) = \left(\frac{d}{d\tau}V(\tau)\right)V(\tau)^{-1}.$$

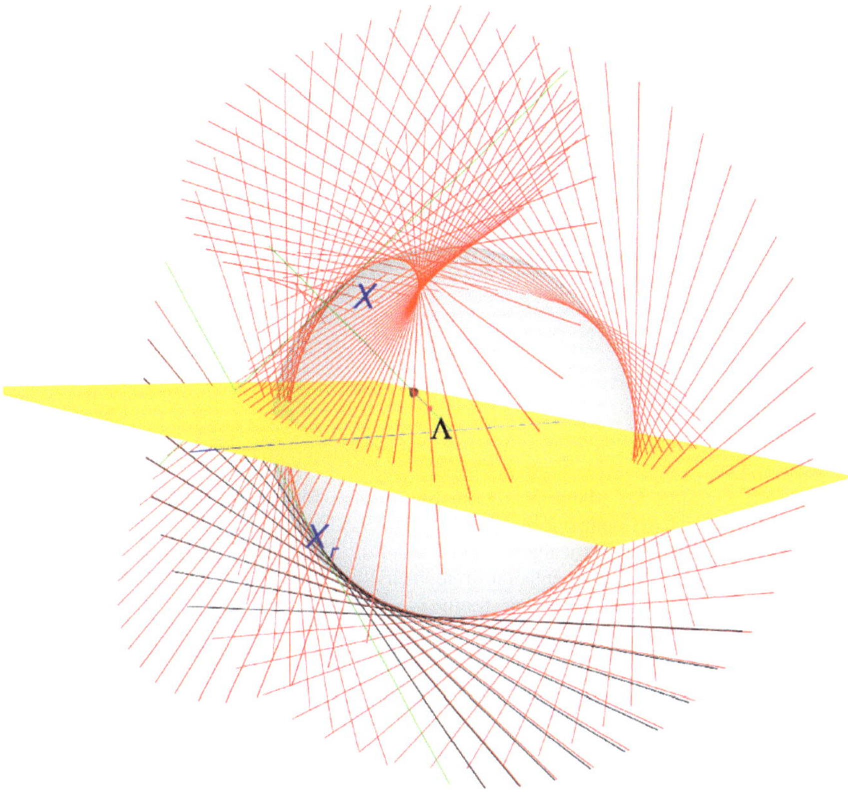

Fig. 4 The mirror Π rotates ϕ round a line Λ, and a line X is reflected in it. Using a local first-order linearization of the reflection versor, one can derive the perturbation of the reflected line to second order (*in black*) to be the rotation with versor $\exp(-\phi((\Lambda \cdot \Pi)/\Pi)^*)$, i.e., around the projection of Λ onto Π with angle $2\phi\cos(\Pi, \Lambda)$. For details, see [2]

If V is normalized, Ω is a bivector, and its commutator product with $X(\tau)$ preserves the grade. This linearization of geometrical perturbations is very useful in applications, see Fig. 4. The full geometric calculus is truly powerful, and one can differentiate relative to an arbitrary element of the algebra (such as a blade or a versor). We cannot treat that here, and the reader is referred to introductions like [2] and [1].

7 Trick 6: Sparse Implementation at Compiler Level

Implementation of CGA may seem to be expensive. After all, to treat a 3D space, we embed into a 5D representational space and use the geometric algebra of that, which involves a 2^5-D basis of constructible elements of all grades. Yet the use we make of this space is restricted, and the elements are therefore somehow sparse.

Ultimately, the main purpose of the algebraic organization is to keep track automatically of the administration of the meaning of the coordinates of points, lines, planes, spheres, etc., simultaneous with performing the quantitative computations. That is in a sense a Boolean selection task of the algebra, which one would intuitively expect not to be too expensive. Indeed it has proved possible to limit the overhead of the use of CGA to about 10% relative to the best available coordinate code programmed classically. For the computer science techniques that achieve this, consult [2] and [3]. A warning: before you start using CGA in commercial applications, be aware that it is covered by a *US patent* [6].

References

1. Doran, C., Lasenby, A.: Geometric Algebra for Physicists. Cambridge University Press, Cambridge (2000)
2. Dorst, L., Fontijne, D., Mann, S.: Geometric Algebra for Computer Science, An Object-Oriented Approach to Geometry. Morgan Kaufman, San Mateo (2007). Second revised printing 2009. See also www.geometricalgebra.net
3. Fontijne, D.: Efficient implementation of geometric algebra. Ph.D. Thesis, U. of Amsterdam (2007). See also www.science.uva.nl/~fontijne
4. Fontijne, D., Dorst, L.: Efficient algorithms for factorization and join of blades. This volume
5. Hestenes, D.: New tools for computational geometry and the rejuvenation of screw theory. This volume
6. Hestenes, D., Rockwood, A., Li, H.: System for encoding and manipulating models of objects. US Patent 6,853,964, granted 8 February 2005

Engineering Graphics in Geometric Algebra

Alyn Rockwood and Dietmar Hildenbrand

Abstract We illustrate the suitability of geometric algebra for representing structures and developing algorithms in computer graphics, especially for engineering applications. A number of example applications are reviewed. Geometric algebra unites many underpinning mathematical concepts in computer graphics such as vector algebra and vector fields, quaternions, kinematics and projective geometry, and it easily deals with geometric objects, operations, and transformations. Not only are these properties important for computational engineering, but also for the computational point-of-view they provide. We also include the potential of geometric algebra for optimizations and highly efficient implementations.

1 Introduction

Computer graphics relies heavily on geometric models and methods. Geometric algebra is a mathematical framework to easily describe geometric concepts and operations. It allows us to develop algorithms fast and in an intuitive way. Geometric algebra is based on the work of Hermann Grassmann (see the conference [45] celebrating his 200th birthday in 2009) and William Clifford [14, 15]. Pioneering work has been done by David Hestenes, who first applied geometric algebra to problems in mechanics and physics [24, 26].

A. Rockwood (✉)
Geometric Modeling and Scientific Visualization Research Center, King Abdullah University of Science and Technology, Thuwal, Saudi Arabia
e-mail: alynrock@yahoo.com

A. Rockwood
e-mail: alyn.rockwood@kaust.edu.sa

E. Bayro-Corrochano, G. Scheuermann (eds.), *Geometric Algebra Computing*,
DOI 10.1007/978-1-84996-108-0_3,

2 Benefits of Geometric Algebra for Computational Engineering

We first highlight some properties of geometric algebra that make it advantageous for graphics engineering applications.

2.1 Unification of Mathematical Systems

In the wide range of engineering applications many different mathematical systems are currently used. One notable advantage of geometric algebra is that it subsumes mathematical systems like vector algebra, complex analysis, quaternions, Plucker coordinates, and tensor analysis. Applications described in Sect. 3 will illustrate this advantage.

2.2 Uniform Handling of Different Geometric Primitives

Conformal geometric algebra, the geometric algebra of conformal space we focus on, is able to treat different geometric objects such as points, vectors, lines, circles, spheres, and planes as the same entities algebraically. Consider the spheres of Fig. 1, for instance. These spheres are simply represented by

$$S = P - \frac{1}{2}r^2 e_\infty \tag{1}$$

based on their center point P, their radius r, and the basis vector e_∞ which represents the point at infinity. The circle of intersection of the spheres is then easily computed using the outer product to operate on the spheres as simply as if they were vectors:

$$Z = S_1 \wedge S_2. \tag{2}$$

This way of computing with geometric algebra clearly benefits applications like kinematics, pose estimation, and other computer graphics applications as seen in Sect. 3.

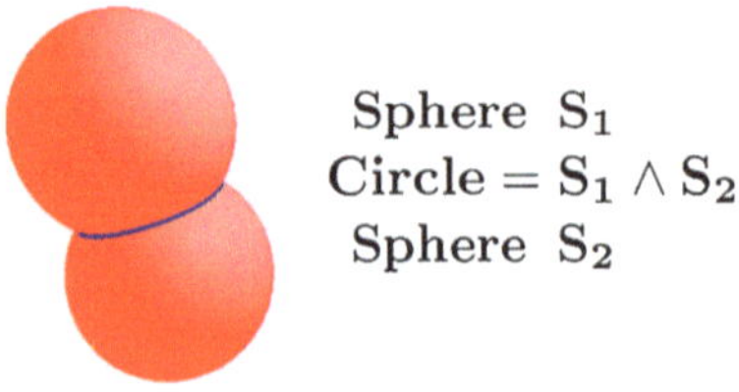

Fig. 1 Spheres and circles are basic entities of geometric algebra. Operations like the intersection of two spheres are easily expressed

2.3 Simplified Rigid Body Motion

Rigid body motions in geometric algebra can be described with one compact linear expression, the so-called *screw*

$$\mathrm{S} = \mathrm{i}\mathbf{m} + e_\infty \mathbf{n} \tag{3}$$

with the Euclidean pseudoscalar $\mathrm{i} = e_1 \wedge e_2 \wedge e_3$ includes both rotational and linear parts described with the 3D vectors **m** and **n** (see [25]). The combinations of rotational and linear velocities, forces and torques are also described with the help of one linear expression.

One result of this property is the improvement of Finite Element methods [9].

2.4 Curl, Vorticity and Rotation

The vector algebra concepts of curl, vorticity, and rotation as expressed in geometric algebra are defined in any dimension, whereas the cross-product in classical vector algebra is restricted to three dimensions. Thus geometric calculus enables vector algebra applications to be considered in any dimensions [13].

2.5 More Efficient Implementations

Geometric algebra as a mathematical language often suggests a clearer structure and greater elegance in understanding methods and formulae. This regularly results in more efficiency and lower runtime performance for derived algorithms. In Sect. 5 we present a dramatically improved optimization approach for kinematics. We will see there that geometric algebra inherently has a large potential for creating optimizations leading to more highly efficient implementations.

3 Some Applications

Computer graphics and thc related areas of robotics and computer vision are active areas of research in geometric algebra. In this section we survey some of these applications in more detail.

For about a decade, researchers at the University of Cambridge, UK, have applied geometric algebra to a number of graphics related projects. They started with ideas in computer vision. Lasenby et al. [31, 32] and Perwass et al. [37, 41, 42] present some applications dealing with structure and motion estimation as well as with the trifocal tensor. Rigid-body pose and position interpolation, mesh deformation, and catadioptric cameras articles using geometric algebra are presented by Cameron et al. [12] and Wareham et al. [54, 55]. Geomerics [53] is a start-up company in Cambridge specializing in simulation software for physics and lighting, which presented its new technology allowing real-time radiosity in videogames utilizing commodity graphics processing hardware. The technology is based on geometric algebra wavelet technology.

Dorst et al. [16–18, 33, 34] at the University of Amsterdam, the Netherlands, are applying their fundamental research on geometric algebra mainly to 3D computer vision. Zaharia et al. [56] investigated modeling and visualization of 3D polygonal mesh surfaces using geometric algebra. Currently D. Fontijne is primarily focusing on the efficient implementation of geometric algebra. He investigated the performance and elegance of five models of 3D Euclidean geometry in a ray tracing, an archetypical computer graphics application [22]. It summarized the investigation by noting that 5D conformal space was the most elegant, but required appropriate hardware to become the most efficient as current hardware supported the 4D affine model. Along this line, research into hardware for geometric algebra continues. The Amsterdam group developed a code generator for geometric algebras [23]. Also, there is a book with applications of geometric algebra edited by Dorst et al. [19]. A new book [20] was published recently, which dedicates its major portion to the issue of geometric algebra calculation.

The first time geometric algebra was introduced to a wider Computer Graphics audience was through a couple of courses at the SIGGRAPH conferences 2000 and 2001 (see [35]).

Bayro-Corrochano et al. from Guadalajara, Mexico, are primarily dealing with the application of geometric algebra in the field of computer vision, robot vision and kinematics. They are using geometric algebra, for instance, for tasks like visual guided grasping, camera self-localization, and reconstruction of shape and motion [3]. Their methods for geometric neural computing are used for tasks like pattern recognition [1, 8]. Registration, the task of finding correspondences between two point sets, is solved based on geometric algebra methods in [47]. Some of their kinematics algorithms can be found in [7] for the 4D motor algebra and in the conformal geometric algebra papers [5, 6] dealing with inverse kinematics, fixation, and grasping as well as with kinematics and differential kinematics of binocular robot heads. Books from Bayro-Corrochano et al. with geometric algebra applications are, for instance, [2] and [4].

At the University of Kiel, Germany, Sommer et al. [51] are applying geometric algebra to robot vision, e.g., Rosenhahn et al. [48, 49] concerning pose estimation and Sommer et al. [52] regarding the twist representation of free-form objects. Perwass et al. are applying conformal geometric algebra to uncertain geometry with circles, spheres, and conics [40] to geometry and kinematics with uncertain data [44] or concerning the inversion camera model [43]. There is a book with applications of geometric algebra edited by Sommer [50] and a new book about the application of geometric algebra in engineering applications by Christian Perwass [38]. Sven Buchholz, together with Kanta Tachibana from the university of Nagoya and Eckhard Hitzer from the university of Fukui, Japan, do some interesting research dealing for instance with neural networks based on geometric algebra [10, 11].

In addition to these examples, there are many other applications like geometric algebra Fourier transforms for the visualization and analysis of vector fields [21] or classification and clustering of spatial patterns with geometric algebra [46] showing a wide area of possibilities of advantageously using this mathematical system in engineering applications.

4 The Geometric Primitives in More Detail

Here, we look into some details of the basic geometric primitives of conformal geometric algebra as introduced in Sect. 2.2 and listed in Table 1. We especially look into the representations of spheres and planes and will see that planes are specific spheres with infinite radius. Increasing the radius of a sphere to infinity, the resulting plane is described by

$$\pi = \mathbf{n} + de_\infty \tag{4}$$

with $\mathbf{n}$ being the 3D unit normal vector of the plane, and d the distance of the plane from the origin. This limit process can be used in order to *fit the best suitable object into a set of points*, whether it is a plane or a sphere. A locally estimated sphere can be used in order to describe local curvature of point clouds [27], while an estimation of a plane describes vanishing curvature.

4.1 Planes as a Limit of Spheres

Spheres and planes, both, are vectors in conformal geometric algebra. In this section, we will see how a sphere

$$S = \mathbf{s} + \frac{1}{2}(\mathbf{s}^2 - r^2)e_\infty + e_0 \tag{5}$$

Table 1 List of the basic geometric primitives provided by the 5D conformal geometric algebra. The bold characters represent 3D entities ($\mathbf{x}$ is a 3D point, $\mathbf{n}$ is a 3D normal vector, and $\mathbf{x}^2$ is the scalar product of the 3D vector $\mathbf{x}$). The two additional basis vectors e_0 and e_∞ represent the origin and infinity. Based on the outer product, circles and lines can be described as intersections of two spheres, respectively two planes. The parameter r represents the radius of the sphere and the parameter d the distance of the plane to the origin

Entity	Representation
Point	$P = \mathbf{x} + \frac{1}{2}\mathbf{x}^2 e_\infty + e_0$
Sphere	$s = P - \frac{1}{2}r^2 e_\infty$
Plane	$\pi = \mathbf{n} + de_\infty$
Circle	$z = s_1 \wedge s_2$
Line	$l = \pi_1 \wedge \pi_2$

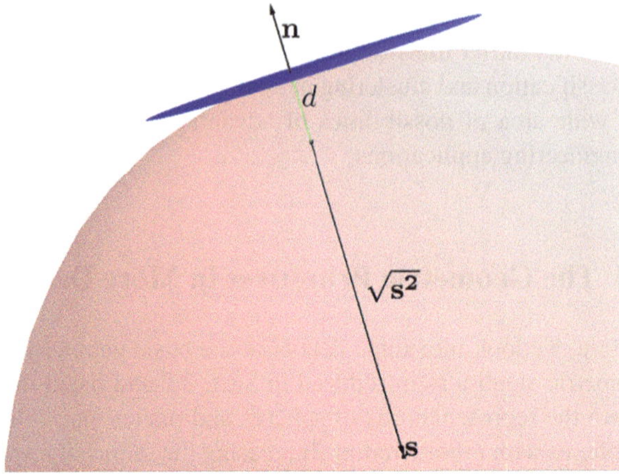

Fig. 2 A sphere with a center point $\mathbf{s}$ (in the opposite direction of a normal vector $\mathbf{n}$) going to infinity (and always adapting its radius), at the end, results in a plane with normal vector $\mathbf{n}$ and distance d to the origin

with Euclidean center point $\mathbf{s}$ and radius r degenerates to a plane as the result of a limit process.

According to the construction of Fig. 2, the minimum distance from the origin to the sphere, having its center in the opposite direction of a normal vector $\mathbf{n}$, is

$$d = r - \sqrt{\mathbf{s}^2}, \tag{6}$$

and the radius is the sum of the length of the 3D vector $\mathbf{s}$ and d,

$$r = \sqrt{\mathbf{s}^2} + d, \tag{7}$$

or

$$r^2 = \mathbf{s}^2 + 2d\sqrt{\mathbf{s}^2} + d^2. \tag{8}$$

Now, the sphere can be written as

$$S = \mathbf{s} + \frac{1}{2}\left(\mathbf{s}^2 - \mathbf{s}^2 - 2d\sqrt{\mathbf{s}^2} - d^2\right)e_\infty + e_0 \tag{9}$$

or equivalently

$$S = \mathbf{s} + \frac{1}{2}\left(-2d\sqrt{\mathbf{s}^2} - d^2\right)e_\infty + e_0. \tag{10}$$

Now, we introduce S' as a scaled version of the algebraic expression of sphere S representing geometrically the same sphere as

$$S' = -\frac{S}{\sqrt{\mathbf{s}^2}} = -\frac{\mathbf{s}}{\sqrt{\mathbf{s}^2}} + \frac{1}{2}\left(2d + \frac{d^2}{\sqrt{\mathbf{s}^2}}\right)e_\infty - \frac{e_0}{\sqrt{\mathbf{s}^2}}. \tag{11}$$

Since the ratio of the 3D vector **s** and its length $\sqrt{\mathbf{s}^2}$ correspond to the negative normal vector **n** (see the construction in Fig. 2),

$$\lim_{\mathbf{s}^2\to\infty} -\frac{S}{\sqrt{\mathbf{s}^2}} = \mathbf{n} + \lim_{\mathbf{s}^2\to\infty} \frac{1}{2}\left(2d + \frac{d^2}{\sqrt{\mathbf{s}^2}}\right)e_\infty - \lim_{\mathbf{s}^2\to\infty} \frac{e_0}{\sqrt{\mathbf{s}^2}}. \tag{12}$$

This is equivalent to

$$\lim_{\mathbf{s}^2\to\infty} -\frac{S}{\sqrt{\mathbf{s}^2}} = \mathbf{n} + de_\infty, \tag{13}$$

which is a representation of a plane with normal vector **n** and distance d to the origin.

4.2 Distances Based on the Inner Product

Points, planes, and spheres are represented as vectors (as listed in Table 1). We will see that the inner products of these vectors describe distances between these geometric objects. The inner product between a vector P and a vector S is defined by

$$P \cdot S = (\mathbf{p} + p_4 e_\infty + p_5 e_o) \cdot (\mathbf{s} + s_4 e_\infty + s_5 e_o). \tag{14}$$

This corresponds to

$$\begin{aligned} P \cdot S = \mathbf{p}\cdot\mathbf{s} &+ s_4 \underbrace{\mathbf{p}\cdot e_\infty}_{0} + s_5 \underbrace{\mathbf{p}\cdot e_o}_{0} \\ &+ p_4 \underbrace{e_\infty \cdot \mathbf{s}}_{0} + p_4 s_4 \underbrace{e_\infty^2}_{0} + p_4 s_5 \underbrace{e_\infty \cdot e_o}_{-1} \\ &+ p_5 \underbrace{e_o \cdot \mathbf{s}}_{0} + p_5 s_4 \underbrace{e_o \cdot e_\infty}_{-1} + p_5 s_5 \underbrace{e_o^2}_{0} \end{aligned}$$

and, based on the rules of conformal geometric algebra, to

$$P \cdot S = \mathbf{p}\cdot\mathbf{s} - p_5 s_4 - p_4 s_5 \tag{15}$$

or

$$P \cdot S = p_1 s_1 + p_2 s_2 + p_3 s_3 - p_5 s_4 - p_4 s_5. \tag{16}$$

4.2.1 Distances Between Points

In the case of P and S being points, we get

$$p_4 = \frac{1}{2}\mathbf{p}^2, \qquad p_5 = 1,$$

$$s_4 = \frac{1}{2}\mathbf{s}^2, \qquad s_5 = 1.$$

The inner product of these points is according to (15)

$$\begin{aligned}
P \cdot S &= \mathbf{p} \cdot \mathbf{s} - \frac{1}{2}\mathbf{s}^2 - \frac{1}{2}\mathbf{p}^2 \\
&= p_1 s_1 + p_2 s_2 + p_3 s_3 - \frac{1}{2}\left(s_1^2 + s_2^2 + s_3^2\right) - \frac{1}{2}\left(p_1^2 + p_2^2 + p_3^2\right) \\
&= -\frac{1}{2}\left(s_1^2 + s_2^2 + s_3^2 + p_1^2 + p_2^2 + p_3^2 - 2p_1 s_1 - 2p_2 s_2 - 2p_3 s_3\right) \\
&= -\frac{1}{2}\left((s_1 - p_1)^2 + (s_2 - p_2)^2 + (s_3 - p_3)^2\right) \\
&= -\frac{1}{2}(\mathbf{s} - \mathbf{p})^2.
\end{aligned}$$

We recognize that the square of the Euclidean distance of the inhomogenous points corresponds to the inner product of the homogenous points multiplied by -2:

$$(\mathbf{s} - \mathbf{p})^2 = -2(P \cdot S). \tag{17}$$

4.2.2 Distance Between Points and Planes

For a vector P representing a point, we get

$$p_4 = \frac{1}{2}\mathbf{p}^2, \qquad p_5 = 1.$$

For a vector S representing a plane with normal vector $\mathbf{n}$ and distance d, we get

$$\mathbf{s} = \mathbf{n}, \qquad s_4 = d, \qquad s_5 = 0.$$

The inner product of point and plane is according to (15)

$$P \cdot S = \mathbf{p} \cdot \mathbf{n} - d, \tag{18}$$

representing the Euclidean distance of a point and a plane.

4.2.3 Distance Between Point and Sphere

We will see now that the inner product of a point and a sphere can be used as a measure of distance between a point and a sphere even if it does not correspond to the minimal Euclidean distance between them.

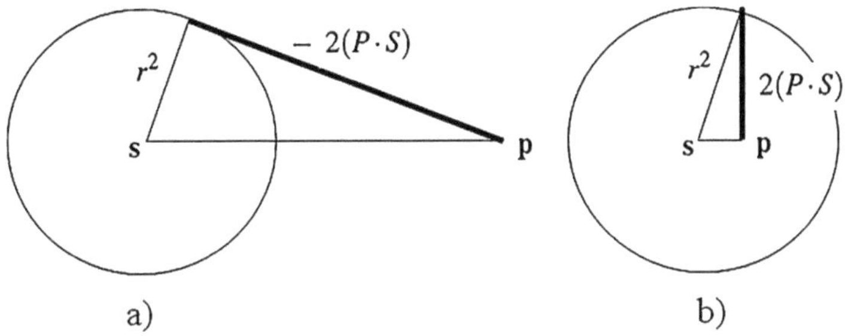

Fig. 3 The inner product of point and sphere [20], *the bold segments* describe the square root of the inner product depending on (**a**) $(\mathbf{s}-\mathbf{p})^2 = r^2 - 2(P\cdot S)$, the point **p** lies outside of the sphere; (**b**) $r^2 = 2(P\cdot S) + (\mathbf{s}-\mathbf{p})^2$, the point **p** lies inside of the sphere

For a vector P representing a point, we get

$$p_4 = \frac{1}{2}\mathbf{p}^2, \qquad p_5 = 1.$$

For a vector S representing a sphere, we get

$$s_4 = \frac{1}{2}\left(s_1^2 + s_2^2 + s_3^2 - r^2\right), \qquad s_5 = 1.$$

The inner product of point and sphere is according to (15)

$$\begin{aligned} P\cdot S &= \mathbf{p}\cdot\mathbf{s} - \frac{1}{2}\left(\mathbf{s}^2 - r^2\right) - \frac{1}{2}\mathbf{p}^2 \\ &= \mathbf{p}\cdot\mathbf{s} - \frac{1}{2}\mathbf{s}^2 + \frac{1}{2}r^2 - \frac{1}{2}\mathbf{p}^2 \\ &= \frac{1}{2}r^2 - \frac{1}{2}\left(\mathbf{s}^2 - 2\mathbf{p}\cdot\mathbf{s} - \mathbf{p}^2\right) \\ &= \frac{1}{2}r^2 - \frac{1}{2}(\mathbf{s}-\mathbf{p})^2. \end{aligned}$$

Finally, we get

$$2(P\cdot S) = r^2 - (\mathbf{s}-\mathbf{p})^2. \tag{19}$$

Twice the inner product $P\cdot S$ equals to the square of the radius minus the square of the distance between the point **p** and the center point **s** of the sphere. Figure 3 describes this relation geometrically. Equation (19) can be rearranged to

$$(\mathbf{s}-\mathbf{p})^2 = r^2 - 2(P\cdot S) \tag{20}$$

describing the relations of the right angle triangle in case (a) with the point **p** being outside of the sphere, while the equation

$$r^2 = 2(P\cdot S) + (\mathbf{s}-\mathbf{p})^2 \tag{21}$$

describes the relations of the right angle triangle in case (b) with the point $\mathbf{p}$ being inside of the sphere.

Please notice that, based on these observations, we can see that

$P \cdot S > 0$: $\mathbf{p}$ is inside of the sphere
$P \cdot S = 0$: $\mathbf{p}$ is on the sphere
$P \cdot S < 0$: $\mathbf{p}$ is outside of the sphere

4.3 Approximation of Points with the Help of Planes or Spheres

In this section, a point set $\mathbf{p_i} \in \mathbb{R}^3$, $i \in \{1, \ldots, n\}$, will be approximated with the help of the best fitting plane or sphere. Please find also an approach for the fitting of circles into point sets in [40].

Plane and sphere in conformal space are vectors of the form

$$S = s_1 e_1 + s_2 e_2 + s_3 e_3 + s_4 e_\infty + s_5 e_0, \tag{22}$$

while the points are specific vectors of the form

$$P_i = \mathbf{p}_i + \frac{1}{2}\mathbf{p}_i^2 e_\infty + e_0. \tag{23}$$

In order to solve the approximation problem, we

- Use the distance measure of the previous section between point and sphere/plane with the help of the inner product.
- Make a least squares approach to minimize the squares of the distances between the points and the sphere/plane.
- Solve the resulting eigenvalue problem.

4.3.1 Distance Measure

From Sect. 4.2.3 we already know that a distance measure between a point P_i and the sphere/plane S can be defined with the help of their inner product

$$P_i \cdot S = \left(\mathbf{p_i} + \frac{1}{2}\mathbf{p}_i^2 e_\infty + e_0\right) \cdot (\mathbf{s} + s_4 e_\infty + s_5 e_0). \tag{24}$$

According to (15), this results in

$$P_i \cdot S = \mathbf{p}_i \cdot \mathbf{s} - s_4 - \frac{1}{2} s_5 \mathbf{p}_i^2,$$

or equivalently

$$P_i \cdot S = \sum_{j=1}^{5} w_{i,j} s_j \tag{25}$$

with

$$w_{i,k} = \begin{cases} p_{i,k}, & k \in \{1,2,3\}, \\ -1, & k = 4, \\ -\frac{1}{2}\mathbf{p}_i^2, & k = 5. \end{cases}$$

4.3.2 Least Squares Approach

In the least-squares sense we consider the minimum of the sum of the squares of the distances (in terms of the inner product) between all the points and the plane/sphere

$$\min \sum_{i=1}^{n} (P_i \cdot S)^2. \tag{26}$$

In order to obtain the minimum, this can be rewritten in bilinear form to

$$\min\left(s^T B s\right) \tag{27}$$

with

$$s^T = (s_1, s_2, s_3, s_4, s_5)$$

and the 5×5 matrix

$$B = \begin{pmatrix} b_{1,1} & b_{1,2} & b_{1,3} & b_{1,4} & b_{1,5} \\ b_{2,1} & b_{2,2} & b_{2,3} & b_{2,4} & b_{2,5} \\ b_{3,1} & b_{3,2} & b_{3,3} & b_{3,4} & b_{3,5} \\ b_{4,1} & b_{4,2} & b_{4,3} & b_{4,4} & b_{4,5} \\ b_{5,1} & b_{5,2} & b_{5,3} & b_{5,4} & b_{5,5} \end{pmatrix}$$

with entries

$$b_{j,k} = \sum_{i=1}^{n} w_{i,j} w_{i,k}.$$

The matrix B is symmetric since $b_{j,k} = b_{k,j}$. We consider only normalized results $s^T s = 1$. A conventional approach to such a constrained optimization problem is introducing

$$\begin{aligned} L = s^T B s - 0 = s^T B s - \lambda\left(s^T s - 1\right), \\ s^T s = 1, \\ B^T = B. \end{aligned}$$

Necessary conditions for a minimum are

$$\begin{aligned} 0 &= \nabla L = 2 \cdot (Bs - \lambda s) = 0 \\ &\to Bs = \lambda s. \end{aligned}$$

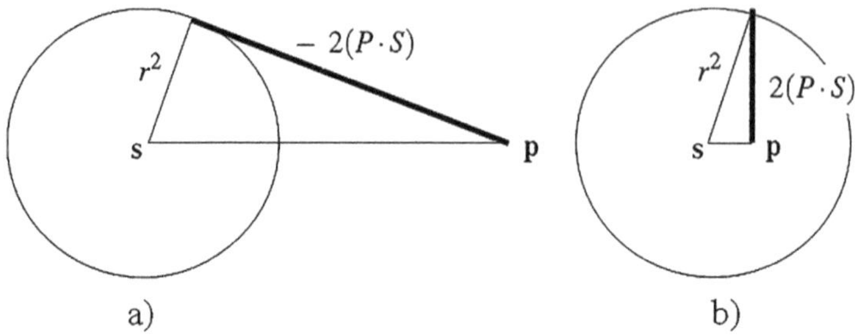

Fig. 4 The inner product of point and sphere, on one hand, describes already the square of a distance but, on the other hand, has to be squared again in the least squares sense since the inner product can be positive or negative depending on (**a**) the point **p** lies outside of the sphere and (**b**) the point **p** lies inside of the sphere

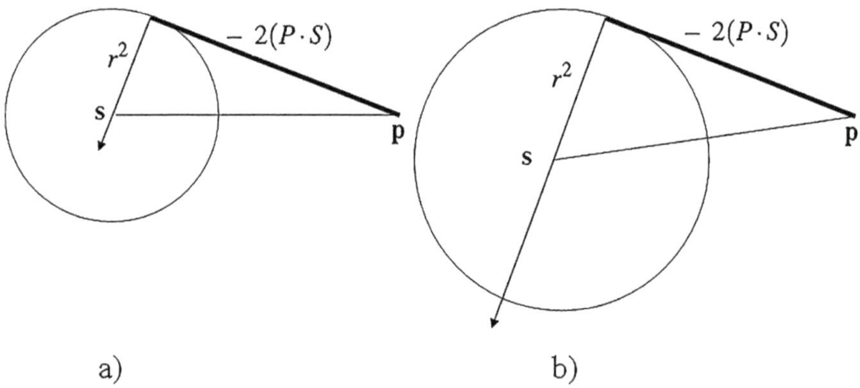

Fig. 5 The constraint $s^T s = 1$ leads implicitly to a scaling of the distance measure in order that it gets smaller with increasing radius, leading to a plane as a sphere with infinite radius

The solution of the minimization problem is given as the eigenvector of B that corresponds to the smallest eigenvalue.

Figures 4 and 5 discuss two properties of the distance measure of this approach dealing with the double squaring of the distance and with the limit process of the distance in the case of a plane as a sphere with infinite radius.

4.3.3 Example

Three distinct (not colinear) points are needed to describe a plane, while four distinct (not coplanar) points exactly describe a sphere. In this example we use five points in order to demonstrate that our approach is really able to fit the best fitting objects, whether it is a sphere or a plane. First, let us have a look on an example with the following five points with four of them being coplanar (see Table 2).

Table 2 Fitting test case with 5 ***not*** coplanar points

Point	x	y	z
$\mathbf{p}_1$	1	0	0
$\mathbf{p}_2$	1	1	0
$\mathbf{p}_3$	0	0	1
$\mathbf{p}_4$	0	1	1
$\mathbf{p}_5$	−1	0	1

Table 3 Fitting test case with 5 coplanar points

Point	x	y	z
$\mathbf{p}_1$	1	0	0
$\mathbf{p}_2$	1	1	0
$\mathbf{p}_3$	0	0	1
$\mathbf{p}_4$	0	1	1
$\mathbf{p}_5$	−1	0	2

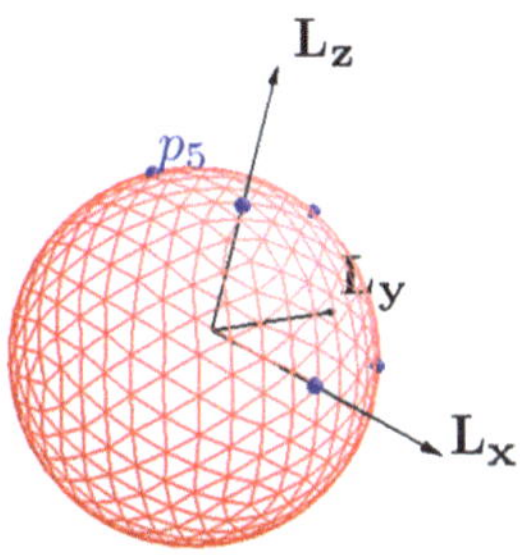

Fig. 6 Fitting a sphere into a set of 5 points

The least squares calculation results in

$$S = -0.301511e_1 + 0.301511e_2 - 0.301511e_3 - 0.603023e_\infty + 0.603023e_0.$$

Another scaled representation describing the same object is

$$S = -\frac{1}{2}e_1 + \frac{1}{2}e_2 - \frac{1}{2}e_3 - e_\infty + e_0.$$

This corresponds to a sphere with the center point $\mathbf{s} = (0.5, 0.5, -0.5)$ and the square of radius $r^2 = 2.75$ (see Fig. 6). Let us now change the fifth point in order that all the points are within one plane (see Table 3). Now, the result is

$$S = 0.57735e_1 + 0.57735e_3 + 0.57735e_\infty,$$

representing a plane according to Fig. 7.

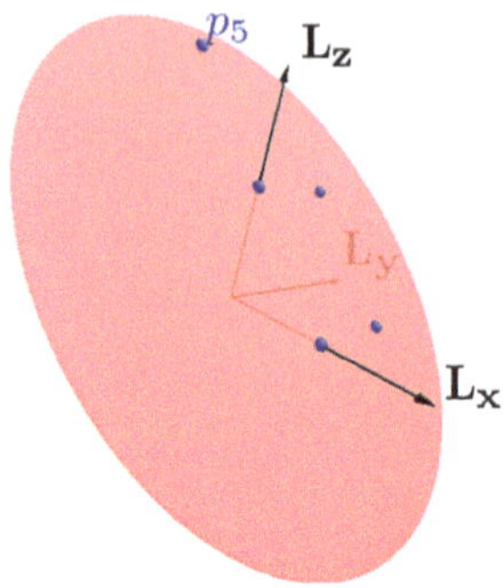

Fig. 7 Fitting a plane into a set of 5 points

5 Computational Efficiency of Geometric Algebra using Gaalop

Since many of the applications depend on an appropriate calculation platform for geometric algebra, it is worth investigating one approach in some detail. Gaalop [28, 30] uses a two-stage approach for the automatic optimization of geometric algebra algorithms. In a first step they optimize geometric algebra algorithms with the help of symbolic computing. This kind of optimization results in very basic algorithms leading to high efficient software implementations. These algorithms foster a high degree of parallelization and are then used for hardware optimizations in a second step.

They investigated performance issues with an inverse kinematics algorithm. Naively implemented, the first algorithm was slower than the conventional one. However, with the symbolic computation optimization approach the software implementation became three times faster [29] and with a hardware implementation about 300 times faster [30] (3 times by software optimization and 100 times by additional hardware optimization) than the conventional software implementation. This result served as a proof-of-concept for Gaalop [28].

Figure 8 shows an overview over the architecture of Gaalop. Its input is a geometric algebra algorithm written in CLUCalc (see [39]). Via symbolic simplification it is transformed into a generic intermediate representation (IR) that can be used for the generation of different output formats. Gaalop supports sequential platforms with the automatic generation of C and JAVA code, while its main focus is on supporting parallel platforms like reconfigurable hardware and modern accelerating GPUs. FPGAs (field programmable gate arrays) are currently supported as a structural hardware description, written in the Verilog language. Thanks to the lower prices of powerful GPUs, for instance, based on the CUDA technology [36] from NVIDIA or on the future Larrabee technology of INTEL, one can expect impressive results using the powerful language of geometric algebra.

One focus of Gaalop will lie on mixed solutions handling reasonable combinations of software and hardware implementations.

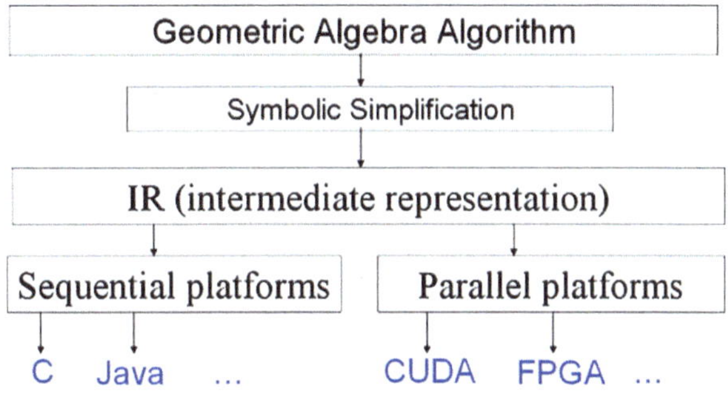

Fig. 8 Architecture of Gaalop

6 Conclusion

In this paper, we observed some properties of geometric algebra that have already proven helpful in computer graphics engineering applications. With these properties, together with the potential of being the base for highly efficient implementations using tools like Gaalop, we are convinced that geometric algebra will become more and more fruitful in a great variety of computational engineering applications. As a consequence, it is worth noting the benefits for students, researchers, and practitioners with geometric algebra. From the educational point of view, students do not have to learn the different mathematical systems and the translations between them, rather they learn one global mathematical system. Researchers gain new insights into their research area using geometric algebra. Practitioners in the field of computational engineering benefit from the easy development, testing, and maintenance of algorithms based on geometric algebra.

References

1. Bayro-Corrochano, E.: Geometric neural computing. IEEE Trans. Neural Netw. **12**(5), 968–986 (2001)
2. Bayro-Corrochano, E.: Robot perception and action using conformal geometry. In: Bayro-Corrochano, E. (ed.) The Handbook of Geometric Computing. Applications in Pattern Recognition, Computer Vision, Neurocomputing and Robotics, pp. 405–458. Springer, Heidelberg (2005). Chap. 13
3. Bayro-Corrochano, E., Banarer, V.: A geometric approach for the theory and applications of 3d projective invariants. J. Math. Imaging Vis. **16**, 131–154 (2001)
4. Bayro-Corrochano, E., Sobczyk, G. (eds.): Geometric Algebra with Applications in Science and Engineering. Birkhäuser, Basel (2001)
5. Bayro-Corrochano, E., Zamora-Esquivel, J.: Inverse kinematics, fixation and grasping using conformal geometric algebra. In: IROS 2004, September 2004, Sendai, Japan (2004)
6. Bayro-Corrochano, E., Zamora-Esquivel, J.: Kinematics and differential kinematics of binocular robot heads. In: Proceedings of ICRA Conference, Orlando, USA (2006)
7. Bayro-Corrochano, E., Daniilidis, K., Sommer, G.: Motor algebra for 3d kinematics: the case of the hand-eye calibration. J. Math. Imaging Vis. **13**, 79–99 (2000)
8. Bayro-Corrochano, E., Vallejo, R., Arana-Daniel, N.: Geometric preprocessing, geometric feedforward neural networks and Clifford support vector machines for visual learning. Neurocomputing **67**, 54–105 (2005). Special issue

9. Brendel, E., Kalbe, T., Hildenbrand, D., Schaefer, M.: Simulation of elastic rods using conformal geometric algebra. In: International Symposium on Frontiers of Computational Science, Nagoya, Japan (2008)
10. Buchholz, S., Hitzer, E.M.S., Tachibana, K.: Optimal learning rates for Clifford neurons. In: International Conference on Artificial Neural Networks, vol. 1, pp. 864–873, Porto, Portugal (2007)
11. Buchholz, S., Hitzer, E.M.S., Tachibana, K.: Coordinate independent update formulas for versor Clifford neurons. In: Proc. Joint 4th International Conference on Soft Computing and Intelligent Systems and 9th International Symposium on Advanced Intelligent Systems (SCIS and ISIS 2008), Nagoya, Japan (2008)
12. Cameron, J., Lasenby, J.: Oriented conformal geometric algebra. In: Proceedings of ICCA7 (2005)
13. Cibura, C., Hildenbrand, D.: Geometric algebra approach to fluid dynamics. In: AGACSE Conference Leipzig (2008)
14. Clifford, W.K.: Applications of Grassmann's extensive algebra. In: Tucker, R. (ed.) Mathematical Papers, pp. 266–276. Macmillian, London (1882)
15. Clifford, W.K.: On the classification of geometric algebras. In: Tucker, R. (ed.) Mathematical Papers, pp. 397–401. Macmillian, London (1882)
16. Dorst, L.: Honing geometric algebra for its use in the computer sciences. In: Sommer, G. (ed.) Geometric Computing with Clifford Algebra. Springer, Berlin (2001)
17. Dorst, L., Fontijne, D.: 3d Euclidean geometry through conformal geometric algebra (a gaviewer tutorial). Available from http://www.science.uva.nl/ga (2003)
18. Dorst, L., Mann, S.: Geometric algebra: a computational framework for geometrical applications (part i: algebra). Comput. Graph. Appl. **22**(3), 24–31 (2002)
19. Dorst, L., Doran, C., Lasenby, J. (eds.): Applications of Geometric Algebra in Computer Science and Engineering. Birkhäuser, Basel (2002)
20. Dorst, L., Fontijne, D., Mann, S.: Geometric Algebra for Computer Science, An Object-Oriented Approach to Geometry. Morgan Kaufman, San Mateo (2007)
21. Ebling, J.: Clifford Fourier transform on vector fields. IEEE Trans. Vis. Comput. Graph. **11**(4), 469–479 (2005). IEEE member Scheuermann, Gerik
22. Fontijne, D., Dorst, L.: Modeling 3D Euclidean geometry. IEEE Comput. Graph. Appl. **23**(2), 68–78 (2003)
23. Fontijne, D., Bouma, T., Dorst, L.: Gaigen: A geometric algebra implementation generator. Available at http://www.science.uva.nl/ga/gaigen (2005)
24. Hestenes, D.: New Foundations for Classical Mechanics. Springer, Dordrecht (1986)
25. Hestenes, D., Fasse, E.D.: Homogeneous rigid body mechanics with elastic coupling. In: Dorst, L., Doran, C., Lasenby, J. (eds.) Applications of Geometric Algebra in Computer Science and Engineering. Birkhäuser, Basel (2002)
26. Hestenes, D., Sobczyk, G.: Clifford Algebra to Geometric Calculus: A Unified Language for Mathematics and Physics. Springer, Dordrecht (1984)
27. Hildenbrand, D., Hitzer, E.M.S.: Analysis of point clouds using conformal geometric algebra. In: GRAPP Conference Madeira (2008)
28. Hildenbrand, D., Pitt, J.: The Gaalop home page. Available at http://www.gaalop.de (2008)
29. Hildenbrand, D., Fontijne, D., Wang, Y., Alexa, M., Dorst, L.: Competitive runtime performance for inverse kinematics algorithms using conformal geometric algebra. In: Eurographics Conference Vienna (2006)
30. Hildenbrand, D., Lange, H., Stock, F., Koch, A.: Efficient inverse kinematics algorithm based on conformal geometric algebra using reconfigurable hardware. In: GRAPP Conference Madeira (2008)
31. Lasenby, J., Bayro-Corrochano, E., Lasenby, A., Sommer, G.: A new methodology for computing invariants in computer vision. In: Proceedings of ICPR 96 (1996)
32. Lasenby, J., Fitzgerald, W.J., Lasenby, A., Doran, C.: New geometric methods for computer vision: an application to structure and motion estimation. Int. J. Comput. Vis. **3**(26), 191–213 (1998)

33. Mann, S., Dorst, L.: Geometric algebra: a computational framework for geometrical applications (part ii: applications). Comput. Graph. Appl. **22**(4), 58–67 (2002)
34. Mann, S., Dorst, L., Bouma, T.: The making of GABLE, a geometric algebra learning environment in matlab, pp. 491–511 (2001)
35. Naeve, A., Rockwood, A.: Course 53 geometric algebra. In: Siggraph Conference Los Angeles (2001)
36. NVIDIA: The CUDA home page. Available at http://www.nvidia.com/object/cuda_home.html (2009)
37. Perwass, C.: Applications of geometric algebra in computer vision. Ph.D. thesis, Cambridge University (2000)
38. Perwass, C.: Geometric Algebra with Applications in Engineering. Springer, Berlin (2009)
39. Perwass, C.: The CLU home page. Available at http://www.clucalc.info (2010)
40. Perwass, C., Förstner, W.: Uncertain geometry with circles, spheres and conics. In: Klette, R., Kozera, R., Noakes, L., Weickert, J. (eds.) Geometric Properties from Incomplete Data. Computational Imaging and Vision, vol. 31, pp. 23–41. Springer, Berlin (2006)
41. Perwass, C., Lasenby, J.: A geometric analysis of the trifocal tensor. In: Klette Reinhard, G.G.R.K. (ed.) Image and Vision Computing New Zealand, IVCNZ'98, Proceedings, pp. 157–162. The University of Auckland (1998)
42. Perwass, C., Lasenby, J.: A unified description of multiple view geometry. In: Sommer, G. (ed.) Geometric Computing with Clifford Algebra. Springer, Berlin (2001)
43. Perwass, C., Sommer, G.: The inversion camera model. In: 28. Symposium für Mustererkennung, DAGM 2006, Berlin, 12.–14.09.2006. Springer, Berlin (2006)
44. Perwass, C., Gebken, C., Sommer, G.: Geometry and kinematics with uncertain data. In: Leonardis, A., Bischof, H., Pinz, A. (eds.) 9th European Conference on Computer Vision. ECCV 2006, May 2006, Graz, Austria. LNCS, vol. 3951, pp. 225–237. Springer, Berlin (2006)
45. Petsche, H.J.: The Grassmann Bicentennial Conference home page. Available at http://www.uni-potsdam.de/u/philosophie/grassmann/Papers.htm (2009)
46. Pham, M.T., Tachibana, K., Hitzer, E.M.S., Yoshikawa, T., Furuhashi, T.: Classification and clustering of spatial patterns with geometric algebra. In: AGACSE Conference Leipzig (2008)
47. Reyes-Lozano, L., Medioni, G., Bayro-Corrochano, E.: Registration of 3d points using geometric algebra and tensor voting. J. Comput. Vis. **75**(3), 351–369 (2007)
48. Rosenhahn, B.: Pose estimation revisited. Ph.D. thesis, Inst. f. Informatik u. Prakt. Mathematik der Christian-Albrechts-Universität zu Kiel (2003)
49. Rosenhahn, B., Sommer, G.: Pose estimation in conformal geometric algebra. J. Math. Imaging Vis. **22**, 27–70 (2005)
50. Sommer, G. (ed.): Geometric Computing with Clifford Algebra. Springer, Berlin (2001)
51. Sommer, G.: Applications of geometric algebra in robot vision. In: Li, H., Olver, P.J., Sommer, G. (eds.) Computer Algebra and Geometric Algebra with Applications. LNCS, vol. 3519, pp. 258–277. Springer, Berlin (2005). 6th International Workshop IWMM 2004, Shanghai, China and International Workshop GIAE 2004, Xian, China
52. Sommer, G., Rosenhahn, B., Perwass, C.: The twist representation of free-form objects. In: Klette, R., Kozera, R., Noakes, L., Weickert, J. (eds.) Geometric Properties from Incomplete Data. Computational Imaging and Vision, vol. 31, pp. 3–22. Springer, Berlin (2006)
53. The homepage of geomerics ltd. Available at http://www.geomerics.com
54. Wareham, R., Lasenby, J.: Applications of conformal geometric algebra in computer vision and graphics. ACM Trans. Graph. (2004, submitted)
55. Wareham, R., Cameron, J., Lasenby, J.: Applications of conformal geometric algebra in computer vision and graphics. Lect. Notes Comput. Sci. **3519**, 329–349 (2005)
56. Zaharia, M.D., Dorst, L.: Modeling and visualization of 3d polygonal mesh surfaces using geometric algebra. Comput. Graph. **29**(5), 802–810 (2003)

Parameterization of 3D Conformal Transformations in Conformal Geometric Algebra

Hongbo Li

Abstract Conformal geometric algebra is a powerful mathematical language for describing and manipulating geometric configurations and their conformal transformations. By providing a 5D algebraic representation of 3D geometric configurations, conformal geometric algebra proves to be very helpful in pose estimation, motion design, and neuron-based machine learning (Bayro-Corrochano et al., J. Math. Imaging Vis. 24(1):55–81, 2006; Dorst et al., Geometric Algebra for Computer Science, Morgan Kaufmann, San Mateo, 2007; Hildenbrand, Comput. Graph. 29(5):795–803, 2005; Lasenby, Computer Algebra and Geometric Algebra with Applications, LNCS, vol. 3519, pp. 298–328, Springer, Berlin, 2005; Li et al., Geometric Computing with Clifford Algebras, pp. 27–60, Springer, Heidelberg, 2001; Mourrain and Stolfi, Invariant Methods in Discrete and Computational Geometry, pp. 107–139, Reidel, Dordrecht, 1995; Rosenhahn and Sommer, J. Math. Imaging Vis. 22:27–70, 2005; Sommer et al., Computer Algebra and Geometric Algebra with Applications, pp. 278–297, Springer, Berlin, 2005). In this chapter, we present some theoretical results on conformal geometric algebra which should prove to be useful in computer applications. The focus is on parameterizing 3D conformal transformations with either quaternionic Vahlen matrices or polynomial Cayley transform from the Lie algebra to the Lie group of conformal transformations in space.

1 Terminology and Notations

By embedding Euclidean space $\mathbb{R}^n$ into the set of null vectors in $\mathbb{R}^{n+1,1}$ in a nonlinear manner, we get the *conformal model* of nD Euclidean geometry [2, 5, 6, 11]. In the Minkowski space $\mathbb{R}^{n+1,1}$, a nonzero vector is said to be *null* if its inner product with itself is zero and is said to be *positive* if so is the inner product.

H. Li (✉)
Mathematics Mechanization Key Laboratory, Academy of Mathematics and Systems Science, Chinese Academy of Sciences, Beijing 100080, China
e-mail: hli@mmrc.iss.ac.cn

E. Bayro-Corrochano, G. Scheuermann (eds.), *Geometric Algebra Computing*,
DOI 10.1007/978-1-84996-108-0_4,

A point in $\mathbb{R}^n$ is represented by a null vector in $\mathbb{R}^{n+1,1}$, and the representation is unique up to scale. There is a unique-up-to-scale null vector $\mathbf{e} \in \mathbb{R}^{n+1,1}$ in the conformal model that does not represent any point in $\mathbb{R}^n$. We say that it represents the *conformal point at infinity*. A sphere or hyperplane in $\mathbb{R}^n$ is represented by a positive vector in $\mathbb{R}^{n+1,1}$, and the representation is unique up to scale. A positive vector represents a hyperplane if and only if its inner product with $\mathbf{e}$ equals zero.

The *Grassmann–Cayley algebra* [19] over $\mathbb{R}^{n+1,1}$, when equipped with the nD Euclidean geometric interpretations of the algebraic elements in this algebra, is called the *conformal Grassmann–Cayley algebra* of the nD space. The *Clifford algebra* over $\mathbb{R}^{n+1,1}$, when equipped with the nD conformal transformation interpretations of the algebraic elements in this algebra, is called the *conformal Clifford algebra* of the nD space. *Conformal geometric algebra* is an integration of conformal Grassmann–Cayley algebra and conformal Clifford algebra [9], the former representing geometric configurations, and the latter representing geometric transformations.

Terminology and notation:

1. *Tensor product*, denoted by "$\otimes$".
2. *Outer product*, denoted by "$\wedge$".
3. *Inner product*, denoted by "$\cdot$".
4. *Meet product*, denoted by "$\vee$".
5. *Geometric product*, denoted by juxtaposition of participating elements. The geometric product of r identical elements $\mathbf{A}$ is denoted by $\mathbf{A}^r$.
6. *Multivector*: any element in a Grassmann algebra (or Clifford algebra).
7. *Exponential* of a multivector $\mathbf{A}$: $\exp(\mathbf{A}) = e^{\mathbf{A}} = 1 + \mathbf{A} + \mathbf{A}^2/2! + \mathbf{A}^3/3! + \cdots$.
8. *Inverse* of a multivector $\mathbf{A}$, denoted by $\mathbf{A}^{-1}$.
9. *Blade*: a multivector which equals the outer product of several vectors.
10. *Grade*: the number of vector components in an outer product decomposition of a blade. A blade of grade r is called an r-blade.
11. *Homogeneous multivector*: a linear combination of blades of the same grade. The grade of a homogeneous multivector is that of any of the blade component. A homogeneous multivector of grade r is called an r-vector. A 2-vector is also called a bivector.
12. *r-graded part* of a multivector, denoted by "$\langle\ \rangle_r$".
13. *Scalar part* of a multivector: the 0-graded part, denoted by "$\langle\ \rangle$".
14. *Even (or odd) multivector*: a linear combination of homogeneous multivectors of even grades (or odd grades).
15. *Even-graded part* (or *odd-graded part*) of a multivector, denoted by "$\langle\ \rangle_+$" (or "$\langle\ \rangle_-$").
16. *Versor*: the geometric product of several invertible vectors.
17. *Rotor*: the geometric product of an even number of invertible vectors.
18. *Positive vector*: a vector whose inner product with itself is positive.
19. *Positive versor*: the geometric product of several positive vectors.
20. *Positive rotor*: the geometric product of an even number of positive vectors.
21. *Grassmann algebra over $\mathcal{V}^n$*, denoted by $\Lambda(\mathcal{V}^n)$.
22. *Clifford algebra over $\mathcal{V}^n$*, denoted by $C\ell(\mathcal{V}^n)$.

23. *Grassmann algebra* or *Clifford algebra* generated by a blade $\mathbf{I}_n$ and denoted by $\Lambda(\mathbf{I}_n)$ (or $C\ell(\mathbf{I}_n)$). The base vector space $\mathcal{V}^n$ of the Grassmann algebra (or Clifford algebra) is composed of all vectors whose outer product with $\mathbf{I}_n$ equals zero.
24. *Even Clifford subalgebra of* $C\ell(\mathcal{V}^n)$, composed of all even multivectors, denoted by $C\ell^+(\mathcal{V}^n)$.
25. *Odd vector subspace of* $C\ell(\mathcal{V}^n)$, composed of all odd multivectors, denoted by $C\ell^-(\mathcal{V}^n)$.
26. *Magnitude* of a multivector: denoted by "$|\ |$". The magnitude of a scalar is its absolute value. The magnitude of $\mathbf{A} \in C\ell(\mathcal{V}^n)$ is $|\mathbf{A}| = \sum_{i=0}^n \sqrt{|\langle \mathbf{A}\rangle_i \cdot \langle \mathbf{A}\rangle_i|}$.
27. *Pseudoscalar:* a blade of grade n in $\Lambda(\mathcal{V}^n)$ or $C\ell(\mathcal{V}^n)$.
28. *Dual operator* in a nondegenerate Clifford algebra, denoted by "$^\sim$". Fixing a pseudoscalar $\mathbf{I}_n$ of unit magnitude, for any multivector $\mathbf{A}$, $\mathbf{A}^\sim = \mathbf{A}\mathbf{I}_n^{-1}$.
29. *Reversion operator* in a Clifford algebra, denoted by "†". For vectors $\mathbf{a}_1, \mathbf{a}_2, \ldots, \mathbf{a}_r$, $(\mathbf{a}_1\mathbf{a}_2\cdots\mathbf{a}_r)^\dagger = \mathbf{a}_r\cdots\mathbf{a}_2\mathbf{a}_1$.
30. *Grade involution* in a Clifford algebra: denoted by overhat. For multivector $\mathbf{A}$, $\widehat{\mathbf{A}} = \langle \mathbf{A}\rangle_+ - \langle \mathbf{A}\rangle_-$.
31. *Conjugate operator* in a Clifford algebra: the composition of reversion and grade involution, denoted by overbar. For multivector $\mathbf{A}$, $\overline{\mathbf{A}} = \langle \mathbf{A}^\dagger\rangle_+ - \langle \mathbf{A}^\dagger\rangle_-$.
32. *Spin group over* $\mathcal{V}^n$, denoted by $\mathrm{Spin}(\mathcal{V}^n)$. It is composed of all rotors in $C\ell(\mathcal{V}^n)$ of unit magnitude together with the geometric product.

Example 1 In conformal Grassmann–Cayley algebra $C\ell(\mathbb{R}^{4,1})$, a circle passing through three points $\mathbf{1}, \mathbf{2}, \mathbf{3}$ in the space is represented by 3-blade $\mathbf{1} \wedge \mathbf{2} \wedge \mathbf{3}$, where $\mathbf{1}, \mathbf{2}, \mathbf{3}$ are null vectors of $\mathbb{R}^{4,1}$ representing points in $\mathbb{R}^3$. The blade is Minkowski, so its dual $(\mathbf{1} \wedge \mathbf{2} \wedge \mathbf{3})^\sim$ is a positive vector. Alternatively, the circle can be represented by a positive vector.

The set $\mathcal{N}$ of all null vectors in $\mathbb{R}^{n+1,1}$ has two connected components. In particular, null vectors $\pm\mathbf{a}$ are always in different connected components, as 0 is not a null vector. An orthogonal transformation in $\mathbb{R}^{n+1,1}$ keeping each component of $\mathcal{N}$ invariant is called a *positive orthogonal transformation*. All such transformations form a subgroup $O_+(n+1,1)$ of $O(n+1,1)$, called the *positive orthogonal group*. The orientation-preserving orthogonal transformations of $\mathbb{R}^{n+1,1}$ form another subgroup $SO(n+1,1)$ of $O(n+1,1)$, called the *special orthogonal group*. The intersection of the two subgroups, denoted by $SO_+(n+1,1)$, is called the *Lorentz group*, and its elements are called *Lorentz transformations*. Lorentz transformations are the linear isometries of $\mathbb{R}^{n+1,1}$ connected with the identity transformation $I_{\mathbb{R}^{n+1,1}}$.

In conformal Clifford algebra $C\ell(\mathbb{R}^{n+1,1})$, any positive rotor $\mathbf{U}$ induces a unique Lorentz transformation in $\mathbb{R}^{n+1,1}$ via the following *adjoint action*:

$$Ad_{\mathbf{U}}(\mathbf{x}) = \mathbf{U}\mathbf{x}\mathbf{U}^{-1} \quad \text{for all } \mathbf{x} \in \mathbb{R}^{n+1,1}. \tag{1}$$

Conversely, any Lorentz transformation in $\mathbb{R}^{n+1,1}$ is induced by a positive rotor that is unique up to scale.

In the conformal model, any Lorentz transformation in $\mathbb{R}^{n+1,1}$ induces a unique orientation-preserving conformal transformation in $\mathbb{R}^n$, and the converse is also true. Hence, any orientation-preserving conformal transformation in $\mathbb{R}^n$ is induced by a positive rotor in $C\ell(\mathbb{R}^{n+1,1})$ that is unique up to scale.

2 Exponential Map and Exterior Exponential Map

By a classical theorem of Riesz [14], any linear isometry of $\mathbb{R}^{n+1,1}$ connected with the identity is induced by a rotor of the exponential form. The group of rotors in $C\ell(\mathbb{R}^{n+1,1})$ differs from $\mathrm{Spin}(\mathbb{R}^{n+1,1})$ by a factor $\mathbb{R}-\{0\}$. Since the Lie algebra of the spin group is all bivectors in $\Lambda(\mathbb{R}^{n+1,1})$, any rotor connected with the identity can be expressed up to scale as the exponential $e^{\mathbf{B}_2}$ of a bivector $\mathbf{B}_2 \in \Lambda(\mathcal{V}^n)$. As a corollary, any positive rotor in $C\ell(\mathbb{R}^{n+1,1})$ is in the range of the exponential map.

Example 2 The Lie algebra representation of 3D rigid body motions via the exponential map.

Any rigid body motion in the space can be decomposed into a rotation followed by a translation. It can also be decomposed into a translation followed by a rotation. The decomposition is not unique without fixing the axis of rotation, which is a straight line in the space. However, there is a unique decomposition, in which the axis of rotation follows exactly the direction of translation. This is the *screw motion*, and the unique decomposition theorem is known as *Chasles' Theorem*.

In the conformal model of 3D Euclidean geometry, let $\mathbf{e}_1, \mathbf{e}_2, \mathbf{e}_3$ be an orthonormal basis of $\mathbb{R}^3$, and let $\mathbf{e}_0, \mathbf{e}$ be the pair of null vectors orthogonal to $\mathbb{R}^3$ in $\mathbb{R}^{4,1}$ and such that $\mathbf{e}_0 \cdot \mathbf{e} = -1$. The basis $(\mathbf{e}, \mathbf{e}_0, \mathbf{e}_1, \mathbf{e}_2, \mathbf{e}_3)$ is called a *Witt basis* of $\mathbb{R}^{4,1}$. The vector $\mathbf{e}_0$ represents the origin of $\mathbb{R}^3$ in the conformal model, while the vector $\mathbf{e}$ represents the conformal point at infinity.

In conformal geometric algebra $C\ell(\mathbb{R}^{4,1})$, the Lie algebra of the spin group of rigid body motions is $\Lambda^2(\mathbf{e}^{\sim})$, which is the bivector subspace of the Grassmann algebra generated by the vectors in $\mathbb{R}^{4,1}$ that are orthogonal to $\mathbf{e}$.

The vector space $\Lambda^2(\mathbf{e}^{\sim})$ has an orthonormal basis $\mathbf{e}_1\mathbf{e}_2, \mathbf{e}_2\mathbf{e}_3, \mathbf{e}_1\mathbf{e}_3, \mathbf{e}\mathbf{e}_1, \mathbf{e}\mathbf{e}_2, \mathbf{e}\mathbf{e}_3$. Any nonzero element in $\Lambda^2(\mathbf{e}^{\sim})$ can be written as

$$\mathbf{B}_2 = \frac{\mathbf{I}_2\theta + \mathbf{e}\mathbf{t}}{2}, \tag{2}$$

where 2-blade $\mathbf{I}_2 \in \Lambda(\mathbb{R}^3)$ is of unit magnitude, $\theta \in \mathbb{R}$, and $\mathbf{t} \in \mathbb{R}^3$.

If $\theta = 0$, then $e^{\mathbf{B}_2}$ induces the translation by the vector $\mathbf{t}$. If $\theta \neq 0$, then

$$\begin{aligned} e^{\mathbf{B}_2} &= e^{-\mathbf{e}(\mathbf{t}\cdot\mathbf{I}_2)/(2\theta)} e^{\mathbf{I}_2\theta/2} e^{\mathbf{e}(\mathbf{t}\cdot\mathbf{I}_2)/(2\theta)} e^{\mathbf{e}P^{\perp}_{\mathbf{I}_2}(\mathbf{t})/2} \\ &= \cos\frac{\theta}{2} + \mathbf{I}_2 \sin\frac{\theta}{2} + \frac{1}{\theta}\mathbf{e}P_{\mathbf{I}_2}(\mathbf{t}) \sin\frac{\theta}{2} + \frac{1}{2}\mathbf{e}P^{\perp}_{\mathbf{I}_2}(\mathbf{t}) \cos\frac{\theta}{2} \\ &\quad + \frac{1}{2}\mathbf{e}P^{\perp}_{\mathbf{I}_2}(\mathbf{t})\mathbf{I}_2 \sin\frac{\theta}{2}, \end{aligned} \tag{3}$$

where

$$
\begin{aligned}
P_{\mathbf{I}_2}(\mathbf{t}) &= (\mathbf{t} \cdot \mathbf{I}_2)\mathbf{I}_2^{-1}, \\
P_{\mathbf{I}_2}^{\perp}(\mathbf{t}) &= \mathbf{t} - P_{\mathbf{I}_2}(\mathbf{t}).
\end{aligned}
\tag{4}
$$

Equation (3) induces a screw motion with the vector of translation $P_{\mathbf{I}_2}^{\perp}(\mathbf{t})$, the axis of rotation passing through the point $-\mathbf{t} \cdot \mathbf{I}_2/\theta \in \mathbb{R}^3$, and the angle of rotation $-\theta$.

In parameterizing 3D conformal transformations, the 10D vector space $\Lambda^2(\mathbb{R}^{4,1})$ provides an ideal parametric space for the group of 3D conformal transformations. Since the exponential map from $\Lambda^2(\mathbb{R}^{4,1})$ to the group of positive rotors is surjective, any orientation-preserving 3D conformal transformation can be parameterized via the exponential map, although the parameterization is not unique.

The problem of parameterizing with the exponential map lies in evaluating the map and computing its inverse. While the map can be evaluated when restricted to some vector subspaces such as $\Lambda^2(\mathbf{e}^{\sim})$, the evaluation for the general case is still not available. Even when the evaluation exists, in many cases such as (3), it is computationally expensive because the map is transcendental instead of algebraic. Furthermore, the exponential map is not an isometry, and its tangent map preserves volume only at the origin of the Lie algebra taken as a vector space. The exponential map has infinitely many inverses in general.

The first alternative of exponential map is the following exterior exponential map:

Definition 1 Let $\mathcal{V}^n$ be a vector space over a field $\mathbb{K}$. The *exterior exponential* is the following map from $\Lambda^2(\mathcal{V}^n)$ to $\Lambda(\mathcal{V}^n)$:

$$
e^{\wedge \mathbf{B}_2} = 1 + \mathbf{B}_2 + \frac{\mathbf{B}_2 \wedge \mathbf{B}_2}{2!} + \cdots + \frac{\overbrace{\mathbf{B}_2 \wedge \mathbf{B}_2 \wedge \cdots \wedge \mathbf{B}_2}^{r}}{r!},
\tag{5}
$$

where r is the greatest integer such that $\overbrace{\mathbf{B}_2 \wedge \mathbf{B}_2 \wedge \cdots \wedge \mathbf{B}_2}^{r} \neq 0$.

The exterior exponential has two obvious properties: first, the scalar part of $e^{\wedge \mathbf{B}_2}$ is 1; second, the mapping is injective because the bivector part of $e^{\wedge \mathbf{B}_2}$ is $\mathbf{B}_2$.

Since any bivector has a *completely orthogonal decomposition*, i.e., for a bivector $\mathbf{B}_2$, there exist vectors $\mathbf{a}_1, \mathbf{b}_1, \mathbf{a}_2, \mathbf{b}_2, \ldots, \mathbf{a}_r, \mathbf{b}_r$ such that

$$
\mathbf{B}_2 = \lambda_1 \mathbf{a}_1 \wedge \mathbf{b}_1 + \lambda_2 \mathbf{a}_2 \wedge \mathbf{b}_2 + \cdots + \lambda_r \mathbf{a}_r \wedge \mathbf{b}_r,
\tag{6}
$$

where $\mathbf{a}_i \cdot \mathbf{b}_j = 0$ for any $1 \le i, j \le r$, and $\mathbf{a}_i \cdot \mathbf{a}_k = \mathbf{b}_i \cdot \mathbf{b}_k = 0$ for any $i \neq k$, we have

$$
e^{\wedge \mathbf{B}_2} = (1 + \lambda_1 \mathbf{a}_1 \wedge \mathbf{b}_1)(1 + \lambda_2 \mathbf{a}_2 \wedge \mathbf{b}_2) \cdots (1 + \lambda_r \mathbf{a}_r \wedge \mathbf{b}_r).
$$

So $e^{\wedge \mathbf{B}_2}$ is invertible if and only if each $\lambda_i \mathbf{a}_i \wedge \mathbf{b}_i$ is not a Minkowski blade of unit magnitude. When $\mathcal{V}^n$ is Minkowski or Euclidean, then if $e^{\wedge \mathbf{B}_2}$ is invertible, it must be a rotor connected with the identity, because so is each $1+\lambda_i \mathbf{a}_i \wedge \mathbf{b}_i$; furthermore,

$$\left(e^{\wedge \mathbf{B}_2}\right)^{-1} = \frac{e^{\wedge(-\mathbf{B}_2)}}{e^{\wedge \mathbf{B}_2} e^{\wedge(-\mathbf{B}_2)}}. \tag{7}$$

Below we assume that $\mathbf{B}_2$ is in the form of (6) and $e^{\wedge \mathbf{B}_2}$ is invertible, and analyze the range of the exterior exponential by restricting it to the conformal model $\mathbb{R}^{4,1}$ of 3D geometry.

If $\mathbf{B}_2$ is a nonzero blade, then $e^{\wedge \mathbf{B}_2} = 1 + \lambda_1 \mathbf{a}_1 \wedge \mathbf{b}_1$. When λ_1 varies, the range of $e^{\wedge \mathbf{B}_2}$ modulo scale contains all rotors in $\Lambda(\mathbf{a}_1 \wedge \mathbf{b}_1)$ whose 0-graded part and 2-graded part are both nonzero.

If $\mathbf{B}_2$ is not a blade, then

$$e^{\wedge \mathbf{B}_2} = 1 + \lambda_1 \mathbf{a}_1 \wedge \mathbf{b}_1 + \lambda_2 \mathbf{a}_2 \wedge \mathbf{b}_2 + \lambda_1 \lambda_2 \mathbf{a}_1 \wedge \mathbf{b}_1 \wedge \mathbf{a}_2 \wedge \mathbf{b}_2, \tag{8}$$

whose 0-graded part and 4-graded part are both nonzero. The range of $e^{\wedge \mathbf{B}_2}$ modulo scale is all rotors whose 0-graded part and 4-graded part are both nonzero.

Proposition 1 *When the range of the exterior exponential is restricted to rotors in $C\ell(\mathbb{R}^{4,1})$, the domain of definition is all bivectors in $\Lambda(\mathbb{R}^{4,1})$ satisfying*

$$(\mathbf{B}_2 \wedge \mathbf{B}_2)^2 \neq 4(\mathbf{B}_2 \cdot \mathbf{B}_2 - 1) \tag{9}$$

and is a set $\mathbb{R}^{10} - V^9$, where V^9 is a 9D algebraic variety in $\mathbb{R}^{10}$. The image space modulo scale is all rotors whose scalar parts are nonzero; topologically, it is the remainder of the positive orthogonal group $O_+(4,1)$, which is a 10D Lie group with two connected components, after removal of a 9D closed subset.

Proof By

$$e^{\wedge \mathbf{B}_2} = 1 + \mathbf{B}_2 + \frac{\mathbf{B}_2 \wedge \mathbf{B}_2}{2}, \tag{10}$$

we get $e^{\wedge \mathbf{B}_2} e^{\wedge(-\mathbf{B}_2)} = 1 - \mathbf{B}_2 \cdot \mathbf{B}_2 + (\mathbf{B}_2 \wedge \mathbf{B}_2)^2/4$, and (9) follows. □

Similar to the exponential map, the exterior exponential provides half-scaled bivector representations for rotations, translations, and dilations.

Given a rotor $\mathbf{A}$ in the image space of the exterior exponential, let $\mathbf{B}_2$ be a bivector whose exterior exponential equals $\mathbf{A}$ up to scale. Then

$$1 + \mathbf{B}_2 + \frac{\mathbf{B}_2 \wedge \mathbf{B}_2}{2} = \frac{\mathbf{A}}{\langle \mathbf{A} \rangle}, \tag{11}$$

so

$$\mathbf{B}_2 = \frac{\langle \mathbf{A} \rangle_2}{\langle \mathbf{A} \rangle}. \tag{12}$$

Example 3 Let $e^{\mathbf{I}_2\frac{\theta}{2}}$ be a rotor inducing the rotation in space with axis $\mathbf{I}_2^{\sim}$ and angle $-\theta \neq \pi \mod 2\pi$. Then up to scale,

$$e^{\mathbf{I}_2\frac{\theta}{2}} = e^{\wedge \mathbf{I}_2 \tan(\frac{\theta}{2})}. \tag{13}$$

Let $e^{\frac{1}{2}\mathbf{et}} = 1 + \mathbf{et}/2$ be a rotor inducing the translation by a vector $\mathbf{t}$. Then

$$e^{\frac{1}{2}\mathbf{et}} = e^{\wedge \frac{1}{2}\mathbf{et}}. \tag{14}$$

Let $e^{\frac{\theta}{2}\mathbf{I}_2}$ be a rotor inducing the dilation of scale $e^{-\theta}$ centering at point $\mathbf{I}_2$ (affine representation, cf. [9]). Then up to scale,

$$e^{\mathbf{I}_2\frac{\theta}{2}} = e^{\wedge \mathbf{I}_2 \tanh(\frac{\theta}{2})}. \tag{15}$$

Let $\mathbf{I}_2 e^{\frac{\theta}{2}\mathbf{I}_2}$ be a rotor inducing a dilation of scale $-e^{-\theta} \neq -1$. Then up to scale,

$$\mathbf{I}_2 e^{\mathbf{I}_2\frac{\theta}{2}} = e^{\wedge \mathbf{I}_2 \operatorname{ctanh}(\frac{\theta}{2})}. \tag{16}$$

On one hand, the exterior exponential is an injective quadratic map, which is superior to the exponential map algebraically. On the other hand, the exterior exponential has two severe drawbacks: first, the domain of definition is decomposed into several disconnected regions, which blocks the construction of large-scope bivector parameters in the design of continuous conformal transformations; second, the image space is also decomposed into several disconnected regions, making it impossible to represent rotors of large-scale continuous conformal transformations.

3 Twisted Vahlen Matrices and Quaternionic Vahlen Matrices

Recall that in complex analysis, any 2D conformal transformation can be represented by a fractional linear map from the Riemann sphere to itself, the sphere being the complex plane plus the complex point at infinity. In Clifford analysis, there is a similar representation for any nD conformal transformation. This is the so-called *Vahlen matrix representation* [1, 11, 12, 16].

In this section, we introduce the classical work of Vahlen (1902) on representing nD conformal transformations projectively by 2×2 matrices whose components are in $C\ell(\mathbb{R}^n)$, by means of introducing two alternatives of Vahlen's matrix representation:

- *Twisted Vahlen matrices*: the product of two 2×2 matrices of multivector components is no longer the usual matrix product, but a "twisted" one. The twisted matrix product is exactly the geometric product in the conformal geometric algebra $C\ell(\mathbb{R}^{n+1,1})$.
- *Quaternionic Vahlen matrices*: When $n = 3$, any twisted Vahlen matrix can be written as a 2×2 quaternionic matrix, and projectively, any 3D conformal transformation can be represented by such a quaternionic matrix. The composition of 3D conformal transformations is just the usual matrix product.

For 3D conformal transformations, twisted Vahlen matrices, Vahlen matrices, and quaternionic Vahlen matrices provide three equivalent transcendental parameterizations in the Lie algebra $\Lambda^2(\mathbb{R}^{4,1})$. The evaluation of each parameterization and the computing of the inverse are both very easy.

With respect to the Witt basis $(\mathbf{e}, \mathbf{e}_0, \mathbf{e}_1, \mathbf{e}_2, \ldots, \mathbf{e}_n)$ of $\mathbb{R}^{n+1,1}$, any vector has the decomposition $\mathbf{a}+\lambda\mathbf{e}+\mu\mathbf{e}_0$, where $\mathbf{a} \in \mathbb{R}^n$ is a linear combination of $\mathbf{e}_1, \mathbf{e}_2, \ldots, \mathbf{e}_n$. By $\mathbf{e}\mathbf{e}_0\mathbf{e} = -2\mathbf{e}$ and $\mathbf{e}_0\mathbf{e}\mathbf{e}_0 = -2\mathbf{e}_0$, any versor $\mathbf{M} = (\mathbf{a}_1 + \lambda_1\mathbf{e} + \mu_1\mathbf{e}_0)(\mathbf{a}_2 + \lambda_2\mathbf{e} + \mu_2\mathbf{e}_0)\cdots(\mathbf{a}_r + \lambda_r\mathbf{e} + \mu_r\mathbf{e}_0)$, where $\mathbf{a}_i \in \mathbb{R}^n$ and $\lambda_i, \mu_i \in \mathbb{R}$, after multilinear expansion, is changed into the following form:

$$\mathbf{M} = -\frac{\mathbf{A}}{2}\mathbf{e}\mathbf{e}_0 - \frac{\mathbf{B}}{2}\mathbf{e} + \mathbf{C}\mathbf{e}_0 - \frac{\mathbf{D}}{2}\mathbf{e}_0\mathbf{e}, \tag{17}$$

where $\mathbf{A}, \mathbf{B}, \mathbf{C}, \mathbf{D} \in C\ell(\mathbb{R}^n)$. More generally, by means of linearity any multivector $\mathbf{M} \in C\ell(\mathbb{R}^{n+1,1})$ has the unique decomposition (17).

Under the following correspondence of bases:

$$\begin{array}{lll} 1 = \begin{pmatrix} 1 & 0 \\ 0 & 1 \end{pmatrix}, & \mathbf{e} = \begin{pmatrix} 0 & -2 \\ 0 & 0 \end{pmatrix}, & \mathbf{e}_0 = \begin{pmatrix} 0 & 0 \\ 1 & 0 \end{pmatrix}, \\ \mathbf{e}\mathbf{e}_0 = \begin{pmatrix} -2 & 0 \\ 0 & 0 \end{pmatrix}, & \mathbf{e}_0\mathbf{e} = \begin{pmatrix} 0 & 0 \\ 0 & -2 \end{pmatrix}, & \mathbf{e} \wedge \mathbf{e}_0 = \begin{pmatrix} 1 & 0 \\ 0 & 1 \end{pmatrix}, \end{array} \tag{18}$$

(17) becomes

$$\mathbf{M} = -\frac{\mathbf{A}}{2}\mathbf{e}\mathbf{e}_0 - \frac{\mathbf{B}}{2}\mathbf{e} + \mathbf{C}\mathbf{e}_0 - \frac{\mathbf{D}}{2}\mathbf{e}_0\mathbf{e} = \begin{pmatrix} \mathbf{A} & \mathbf{B} \\ \mathbf{C} & \mathbf{D} \end{pmatrix}. \tag{19}$$

It is a classical result [11], which is also easy to verify, that (19) provides an algebraic isomorphism between $C\ell(\mathbb{R}^{n+1,1})$ and the following 2×2 twisted Clifford matrix algebra $\hat{M}_{2\times 2}(C\ell(\mathbb{R}^n))$:

Definition 2 The 2×2 *twisted Clifford matrix algebra* over $\mathbb{R}^n$ is the linear space of 2×2 matrices whose components are in $C\ell(\mathbb{R}^n)$, equipped with the *twisted multiplication* defined as follows: for any 2×2 matrices $\mathbf{M}_1, \mathbf{M}_2$ whose components are in $C\ell(\mathbb{R}^n)$,

$$\mathbf{M}_1\mathbf{M}_2 = \begin{pmatrix} \mathbf{A} & \mathbf{B} \\ \mathbf{C} & \mathbf{D} \end{pmatrix} \begin{pmatrix} \mathbf{A}' & \mathbf{B}' \\ \mathbf{C}' & \mathbf{D}' \end{pmatrix} := \begin{pmatrix} \mathbf{A}\mathbf{A}' + \mathbf{B}\hat{\mathbf{C}}' & \mathbf{A}\mathbf{B}' + \mathbf{B}\hat{\mathbf{D}}' \\ \mathbf{C}\hat{\mathbf{A}}' + \mathbf{D}\mathbf{C}' & \mathbf{C}\hat{\mathbf{B}}' + \mathbf{D}\mathbf{D}' \end{pmatrix}. \tag{20}$$

In each component on the right side of (20), the overhat (grade involution) is always added to the element of the second matrix that is not in the same row with the corresponding element of the first matrix multiplied with it. For example, in the first component $\mathbf{A}\mathbf{A}' + \mathbf{B}\hat{\mathbf{C}}'$, $\mathbf{A}, \mathbf{A}'$ are each in the first row of the corresponding matrix, while $\mathbf{B}, \mathbf{C}'$ are in different rows, so the overhat is added to $\mathbf{C}'$.

Let $\mathbf{I}_n$ be the unit pseudoscalar representing $\mathbb{R}^n$ with its positive orientation. The following formulas can be easily derived from (19):

$$\begin{pmatrix} \mathbf{A} & \mathbf{B} \\ \mathbf{C} & \mathbf{D} \end{pmatrix}^{\dagger} = \begin{pmatrix} \mathbf{D}^{\dagger} & \overline{\mathbf{B}} \\ \overline{\mathbf{C}} & \mathbf{A}^{\dagger} \end{pmatrix}, \quad \widehat{\begin{pmatrix} \mathbf{A} & \mathbf{B} \\ \mathbf{C} & \mathbf{D} \end{pmatrix}} = \begin{pmatrix} \hat{\mathbf{A}} & -\hat{\mathbf{B}} \\ -\hat{\mathbf{C}} & \hat{\mathbf{D}} \end{pmatrix},$$
$$\overline{\begin{pmatrix} \mathbf{A} & \mathbf{B} \\ \mathbf{C} & \mathbf{D} \end{pmatrix}} = \begin{pmatrix} \overline{\mathbf{D}} & -\mathbf{B}^{\dagger} \\ -\mathbf{C}^{\dagger} & \overline{\mathbf{A}} \end{pmatrix}, \quad \begin{pmatrix} \mathbf{A} & \mathbf{B} \\ \mathbf{C} & \mathbf{D} \end{pmatrix}^{\sim} = \begin{pmatrix} -\mathbf{A}\mathbf{I}_n^{-1} & \mathbf{B}\mathbf{I}_n^{-1} \\ -\mathbf{C}\mathbf{I}_n^{-1} & \mathbf{D}\mathbf{I}_n^{-1} \end{pmatrix}. \tag{21}$$

Under the correspondence (18), any vector $\mathbf{a} \in \mathbb{R}^{n+1,1}$ corresponds to the following matrix:

$$\begin{pmatrix} \mathbf{x} & \alpha \\ \beta & \mathbf{x} \end{pmatrix}, \tag{22}$$

where

$$\begin{aligned} \mathbf{x} &= P^{\perp}_{\mathbf{e}\wedge\mathbf{e}_0}(\mathbf{a}), \\ \alpha &= 2\mathbf{a}\cdot\mathbf{e}_0, \\ \beta &= -\mathbf{a}\cdot\mathbf{e}. \end{aligned} \tag{23}$$

In particular, the null vector $\mathbf{e}_0+\mathbf{x}+\mathbf{e}\mathbf{x}^2/2$, where $\mathbf{x} \in \mathbb{R}^n$, corresponds to the matrix

$$\begin{pmatrix} \mathbf{x} & -\mathbf{x}^2 \\ 1 & \mathbf{x} \end{pmatrix}. \tag{24}$$

Of particular interest are the matrices corresponding to versors in $C\ell(\mathbb{R}^{n+1,1})$. We first take a look at some examples. Let $\mathbf{I}_2 \in \Lambda^2(\mathbb{R}^n)$ and $\mathbf{t} \in \mathbb{R}^n$.

- The rotor of rotation $e^{\theta\mathbf{I}_2/2}$ corresponds to $\begin{pmatrix} e^{\theta\mathbf{I}_2/2} & 0 \\ 0 & e^{\theta\mathbf{I}_2/2} \end{pmatrix}$.
- The rotor of dilation $e^{\theta\mathbf{e}\wedge\mathbf{e}_0/2}$ corresponds to $\begin{pmatrix} e^{-\theta/2} & 0 \\ 0 & e^{\theta/2} \end{pmatrix}$.
- The rotor of dilation $(\mathbf{e}\wedge\mathbf{e}_0)e^{\theta\mathbf{e}\wedge\mathbf{e}_0/2}$ corresponds to $\begin{pmatrix} -e^{-\theta/2} & 0 \\ 0 & e^{\theta/2} \end{pmatrix}$.
- The rotor of translation $1+\mathbf{e}\mathbf{t}/2$ corresponds to $\begin{pmatrix} 1 & \mathbf{t} \\ 0 & 1 \end{pmatrix}$.
- The rotor of transversion $1-\mathbf{e}_0\mathbf{t}$ corresponds to $\begin{pmatrix} 1 & 0 \\ \mathbf{t} & 1 \end{pmatrix}$.

Definition 3 A 2×2 matrix $\mathbf{M} = \begin{pmatrix} \mathbf{A} & \mathbf{B} \\ \mathbf{C} & \mathbf{D} \end{pmatrix}$ over $C\ell(\mathbb{R}^n)$ is called a *twisted Vahlen matrix* if

1. $\mathbf{A}, \mathbf{B}, \mathbf{C}, \mathbf{D}$ are either versors or zero.
2. $\mathbf{A}\mathbf{B}^{\dagger}, \mathbf{B}\mathbf{D}^{\dagger}, \mathbf{D}\mathbf{C}^{\dagger}, \mathbf{C}\mathbf{A}^{\dagger}$ are vectors.
3. $\Delta = \mathbf{A}\mathbf{D}^{\dagger} + \mathbf{B}\mathbf{C}^{\dagger}$ is a nonzero scalar.

In the above definition, Condition 1 guarantees $\mathbf{B}^{\dagger}\mathbf{A} = \mathbf{A}^{-1}(\mathbf{A}\mathbf{B}^{\dagger})\mathbf{A} \in \mathbb{R}^n$ if $\mathbf{A}\mathbf{B}^{\dagger} \in \mathbb{R}^n$. So in Condition 2, $\mathbf{A}\mathbf{B}^{\dagger}$ can be replaced by any of $\mathbf{B}\mathbf{A}^{\dagger}$, $\mathbf{B}^{\dagger}\mathbf{A}$, $\mathbf{A}^{\dagger}\mathbf{B}$, and so for the other three elements in Condition 2. By Conditions 1 and 2,

$$\begin{pmatrix} \mathbf{A} & \mathbf{B} \\ \mathbf{C} & \mathbf{D} \end{pmatrix} \begin{pmatrix} \mathbf{A} & \mathbf{B} \\ \mathbf{C} & \mathbf{D} \end{pmatrix}^{\dagger} = \begin{pmatrix} \mathbf{A}\mathbf{D}^{\dagger} + \mathbf{B}\mathbf{C}^{\dagger} & \mathbf{A}\overline{\mathbf{B}} + \mathbf{B}\overline{\mathbf{A}} \\ \mathbf{C}\overline{\mathbf{D}} + \mathbf{D}\overline{\mathbf{C}} & \mathbf{C}\mathbf{B}^{\dagger} + \mathbf{D}\mathbf{A}^{\dagger} \end{pmatrix}$$

$$= \begin{pmatrix} \mathbf{A}\mathbf{D}^{\dagger} + \mathbf{B}\mathbf{C}^{\dagger} & 0 \\ 0 & (\mathbf{A}\mathbf{D}^{\dagger} + \mathbf{B}\mathbf{C}^{\dagger})^{\dagger} \end{pmatrix}, \tag{25}$$

so Condition 3 is equivalent to $\mathbf{M}\mathbf{M}^{\dagger}$ being a nonzero scalar.

Theorem 1 *In twisted Vahlen matrix* $\mathbf{M} = \left(\begin{smallmatrix} \mathbf{A} & \mathbf{B} \\ \mathbf{C} & \mathbf{D} \end{smallmatrix}\right)$, *when* $\mathbf{A} \neq 0$, *there exist* $\lambda \in \mathbb{R} - \{0\}$ *and* $\mathbf{b}, \mathbf{c} \in \mathbb{R}^n$ *such that*

$$\mathbf{M} = \mathbf{A} \begin{pmatrix} 1 & \mathbf{b} \\ \mathbf{c} & \lambda - \mathbf{c}\mathbf{b} \end{pmatrix}. \tag{26}$$

When $\mathbf{A} = 0$, *there exist* $\mu \in \mathbb{R} - \{0\}$ *and* $\mathbf{d} \in \mathbb{R}^n$ *such that*

$$\mathbf{M} = \mathbf{B} \begin{pmatrix} 0 & 1 \\ \mu & \mathbf{d} \end{pmatrix}. \tag{27}$$

Proof (i) If $\mathbf{A} \neq 0$ and $\mathbf{B} \neq 0$, by denoting

$$\mathbf{A}^{\dagger}\mathbf{B} = \mathbf{b}, \qquad \mathbf{A}^{\dagger}\mathbf{C} = \mathbf{c}, \qquad \mathbf{B}^{\dagger}\mathbf{D} = \mathbf{d}, \qquad \mathbf{A}^{\dagger}\mathbf{A} = \lambda^{-1}, \qquad \mathbf{B}^{\dagger}\mathbf{B} = \mu^{-1},$$

the matrix $\mathbf{M}$ can be written as

$$\begin{pmatrix} \mathbf{A} & \mathbf{B} \\ \mathbf{C} & \mathbf{D} \end{pmatrix} = \mathbf{A} \begin{pmatrix} 1 & \lambda\mathbf{b} \\ \lambda\mathbf{c} & \lambda\mu\mathbf{b}\mathbf{d} \end{pmatrix},$$

where $\mathbf{d}$ satisfies

$$\mathbf{d} = \mu^{-1}\Delta\mathbf{b}^{-1} - \mu^{-1}\mathbf{b}\mathbf{c}\mathbf{b}^{-1}. \tag{28}$$

By (28), $\lambda\mu\mathbf{b}\mathbf{d} = \lambda\Delta - \lambda^2\mathbf{c}\mathbf{b}$, so $\mathbf{M}$ can be written as

$$\begin{pmatrix} \mathbf{A} & \mathbf{B} \\ \mathbf{C} & \mathbf{D} \end{pmatrix} = \mathbf{A} \begin{pmatrix} 1 & \lambda\mathbf{b} \\ \lambda\mathbf{c} & \lambda\Delta - \lambda^2\mathbf{c}\mathbf{b} \end{pmatrix}.$$

(ii) If $\mathbf{A} \neq 0$ but $\mathbf{B} = 0$, then $\mathbf{M}$ can be written as

$$\begin{pmatrix} \mathbf{A} & \mathbf{B} \\ \mathbf{C} & \mathbf{D} \end{pmatrix} = \mathbf{A} \begin{pmatrix} 1 & 0 \\ \lambda\mathbf{c} & \lambda\Delta \end{pmatrix},$$

which is a special case of (i) where $\mathbf{b} = 0$.

(iii) If $\mathbf{A} = 0$, then $\mathbf{B} \neq 0$, and $\mathbf{M}$ can be written as

$$\begin{pmatrix} \mathbf{A} & \mathbf{B} \\ \mathbf{C} & \mathbf{D} \end{pmatrix} = \mathbf{B} \begin{pmatrix} 0 & 1 \\ \mu\Delta & \mu\mathbf{d} \end{pmatrix}.$$

□

Theorem 2 *Any versor in $C\ell(\mathbb{R}^{n+1,1})$ corresponds via* (19) *to a twisted Vahlen matrix.*

Proof Let $\mathbf{M}$ be a versor. When $\mathbf{M}$ is a vector, then it is of the form (22) and is a twisted Vahlen matrix if and only if it is neither zero nor null. To prove the theorem by induction, we need only prove that for any versor $\mathbf{M}$ and invertible vector $\mathbf{M}'$,

$$\mathbf{M}\mathbf{M}' = \begin{pmatrix} \mathbf{A} & \mathbf{B} \\ \mathbf{C} & \mathbf{D} \end{pmatrix} \begin{pmatrix} \mathbf{x} & \alpha \\ \beta & \mathbf{x} \end{pmatrix} = \begin{pmatrix} \mathbf{A}\mathbf{x} + \beta\mathbf{B} & \alpha\mathbf{A} - \mathbf{B}\mathbf{x} \\ -\mathbf{C}\mathbf{x} + \beta\mathbf{D} & \alpha\mathbf{C} + \mathbf{D}\mathbf{x} \end{pmatrix} \tag{29}$$

is a twisted Vahlen matrix.

By (26) and (27), we only need to consider two cases:

$$\text{(i)} \quad \mathbf{M} = \begin{pmatrix} 1 & \lambda\mathbf{b} \\ \lambda\mathbf{c} & \lambda\Delta - \lambda^2\mathbf{c}\mathbf{b} \end{pmatrix}, \qquad \text{(ii)} \quad \mathbf{M} = \begin{pmatrix} 0 & 1 \\ \mu\Delta & \mu\mathbf{d} \end{pmatrix}.$$

The corresponding matrix $\mathbf{M}\mathbf{M}'$ is respectively

$$\text{(i)} \quad \begin{pmatrix} \mathbf{x} + \lambda\beta\mathbf{b} & \alpha - \lambda\mathbf{b}\mathbf{x} \\ \lambda\Delta\beta - \lambda\mathbf{c}(\mathbf{x} + \lambda\beta\mathbf{b}) & \lambda\Delta\mathbf{x} + \lambda\mathbf{c}(\alpha - \lambda\mathbf{b}\mathbf{x}) \end{pmatrix},$$

$$\text{(ii)} \quad \begin{pmatrix} \beta & -\mathbf{x} \\ \mu(\beta\mathbf{d} - \Delta\mathbf{x}) & \mu(\Delta\alpha + \mathbf{d}\mathbf{x}) \end{pmatrix},$$

and it can be easily verified that each matrix is a twisted Vahlen matrix. □

Theorem 3 (Twisted version of Vahlen's Theorem) *Any twisted Vahlen matrix* $\mathbf{M}$ *generates the following conformal transformation in* $\mathbb{R}^n$:

$$\mathbf{x} \longmapsto \mathbf{M}(\mathbf{x}) = (\mathbf{A}\mathbf{x} + \mathbf{B})\big(\hat{\mathbf{C}}\mathbf{x} + \hat{\mathbf{D}}\big)^{-1} \quad \forall \mathbf{x} \in \mathbb{R}^n. \tag{30}$$

Conversely, any conformal transformation in $\mathbb{R}^n$ *has such a twisted fractional linear representation.*

Proof In the conformal model, a point $\mathbf{x} \in \mathbb{R}^n$ is represented by the null vector $\mathbf{e}_0 + \mathbf{x} + \mathbf{e}\mathbf{x}^2/2$ whose twisted Vahlen matrix representation is (24). The *graded adjoint action* of versor $\mathbf{M}$ on the null vector is, up to scale,

$$\begin{aligned} &\begin{pmatrix} \mathbf{A} & \mathbf{B} \\ \mathbf{C} & \mathbf{D} \end{pmatrix} \begin{pmatrix} \mathbf{x} & -\mathbf{x}^2 \\ 1 & \mathbf{x} \end{pmatrix} \overline{\begin{pmatrix} \mathbf{A} & \mathbf{B} \\ \mathbf{C} & \mathbf{D} \end{pmatrix}} \\ &= \begin{pmatrix} \mathbf{A}\mathbf{x}\mathbf{D}^\dagger + \mathbf{B}\mathbf{D}^\dagger + \mathbf{x}^2\mathbf{A}\overline{\mathbf{C}} + \mathbf{B}\mathbf{x}\overline{\mathbf{C}} & -\mathbf{A}\mathbf{x}\mathbf{B}^\dagger - \mathbf{B}\mathbf{B}^\dagger - \mathbf{x}^2\mathbf{A}\mathbf{A}^\dagger - \mathbf{B}\mathbf{x}\mathbf{A}^\dagger \\ -\mathbf{C}\mathbf{x}\overline{\mathbf{D}} + \mathbf{D}\overline{\mathbf{D}} + \mathbf{x}^2\mathbf{C}\mathbf{C}^\dagger - \mathbf{D}\mathbf{x}\mathbf{C}^\dagger & \mathbf{C}\mathbf{x}\overline{\mathbf{B}} - \mathbf{D}\overline{\mathbf{B}} - \mathbf{x}^2\mathbf{C}\overline{\mathbf{A}} + \mathbf{D}\mathbf{x}\overline{\mathbf{A}} \end{pmatrix} \\ &= \begin{pmatrix} (\mathbf{A}\mathbf{x} + \mathbf{B})(\mathbf{D}^\dagger + \mathbf{x}\overline{\mathbf{C}}) & -(\mathbf{A}\mathbf{x} + \mathbf{B})(\mathbf{B}^\dagger + \mathbf{x}\mathbf{A}^\dagger) \\ -(\mathbf{C}\mathbf{x} - \mathbf{D})(\overline{\mathbf{D}} - \mathbf{x}\mathbf{C}^\dagger) & (\mathbf{C}\mathbf{x} - \mathbf{D})(\overline{\mathbf{B}} - \mathbf{x}\overline{\mathbf{A}}) \end{pmatrix}. \end{aligned}$$

So in $\mathbb{R}^n$, $\mathbf{M}$ changes a vector $\mathbf{x}$ to the vector

$$-\frac{(\mathbf{Ax}+\mathbf{B})(\mathbf{D}^\dagger+\mathbf{x}\overline{\mathbf{C}})}{(\mathbf{Cx}-\mathbf{D})(\overline{\mathbf{D}}-\mathbf{x}\mathbf{C}^\dagger)} = -(\mathbf{Ax}+\mathbf{B})\big(\mathbf{D}^\dagger+\mathbf{x}\overline{\mathbf{C}}\big)\widehat{\big(\mathbf{D}^\dagger+\mathbf{x}\overline{\mathbf{C}}\big)}^{-1}(\mathbf{Cx}-\mathbf{D})^{-1}$$
$$= -(\mathbf{Ax}+\mathbf{B})\big(\widehat{\mathbf{Cx}-\mathbf{D}}\big)^{-1}$$
$$= (\mathbf{Ax}+\mathbf{B})\big(\hat{\mathbf{C}}\mathbf{x}+\hat{\mathbf{D}}\big)^{-1}. \qquad \square$$

When $C\ell(\mathbb{R}^n)$ is represented by a matrix algebra, the twisted matrix multiplication is very inconvenient and needs to be revised to usual matrix multiplication. The work was done by Vahlen in 1902.

Definition 4 The *algebra of* 2×2 *Clifford matrices* over $C\ell(\mathbb{R}^n)$, denoted by $M_{2\times 2}(C\ell(\mathbb{R}^n))$, is the linear space of matrices of the form $\mathbf{M}=\left(\begin{smallmatrix}\mathbf{A}&\mathbf{B}\\\mathbf{C}&\mathbf{D}\end{smallmatrix}\right)$, where $\mathbf{A},\mathbf{B},\mathbf{C},\mathbf{D}\in C\ell(\mathbb{R}^n)$, equipped with the usual matrix multiplication

$$\begin{pmatrix}\mathbf{A}&\mathbf{B}\\\mathbf{C}&\mathbf{D}\end{pmatrix}\begin{pmatrix}\mathbf{A}'&\mathbf{B}'\\\mathbf{C}'&\mathbf{D}'\end{pmatrix}=\begin{pmatrix}\mathbf{AA}'+\mathbf{BC}'&\mathbf{AB}'+\mathbf{BD}'\\\mathbf{CA}'+\mathbf{DC}'&\mathbf{CB}'+\mathbf{DD}'\end{pmatrix}. \tag{31}$$

Definition 5 2×2 matrix $\mathbf{M}=\left(\begin{smallmatrix}\mathbf{A}&\mathbf{B}\\\mathbf{C}&\mathbf{D}\end{smallmatrix}\right)$ over $C\ell(\mathbb{R}^n)$ is called a *Vahlen matrix* if

1. $\mathbf{A},\mathbf{B},\mathbf{C},\mathbf{D}$ are either versors or zero.
2. $\mathbf{AB}^\dagger,\mathbf{BD}^\dagger,\mathbf{DC}^\dagger,\mathbf{CA}^\dagger$ are vectors.
3. $\Delta=\mathbf{AD}^\dagger-\mathbf{BC}^\dagger$ is a nonzero scalar.

It is easy to verify that Condition 3 in the above definition is equivalent to matrix $\mathbf{M}$ being invertible.

Under the following correspondence, any twisted Clifford matrix corresponds to a unique Clifford matrix, and vice versa:

$$\text{twisted Clifford matrix }\begin{pmatrix}\mathbf{A}&\mathbf{B}\\\mathbf{C}&\mathbf{D}\end{pmatrix}\longleftrightarrow\text{ Clifford matrix }\begin{pmatrix}\mathbf{A}&\mathbf{B}\\\hat{\mathbf{C}}&\hat{\mathbf{D}}\end{pmatrix}. \tag{32}$$

The above correspondence is in fact an algebraic isomorphism. All the previous results presented in the form of twisted Clifford matrices can be translated easily into Clifford matrices. For example, the following is a translation of Theorem 3.

Theorem 4 (Vahlen's Theorem) *Any Vahlen matrix* $\mathbf{M}$ *generates the following conformal transformation in* $\mathbb{R}^n$:

$$\mathbf{x}\longmapsto\mathbf{M}(\mathbf{x})=(\mathbf{Ax}+\mathbf{B})(\mathbf{Cx}+\mathbf{D})^{-1}\quad\forall\mathbf{x}\in\mathbb{R}^n; \tag{33}$$

and any conformal transformation has such a fractional linear representation.

Consider the special case where $n=3$. Any 3D conformal transformation is induced by the adjoint action of a rotor in $C\ell(\mathbb{R}^{4,1})$, and the rotor is unique up to scale.

A rotor in $C\ell(\mathbb{R}^{4,1})$ corresponds to a twisted Vahlen matrix $\mathbf{M} = \left(\begin{smallmatrix} \mathbf{A} & \mathbf{B} \\ \mathbf{C} & \mathbf{D} \end{smallmatrix}\right)$, where $\mathbf{A}, \mathbf{D}$ are even and $\mathbf{B}, \mathbf{C}$ are odd. Such a matrix is called an *even twisted Vahlen matrix*.

Fix a Witt basis $(\mathbf{e}, \mathbf{e}_0, \mathbf{e}_1, \mathbf{e}_2, \mathbf{e}_3)$ of $\mathbb{R}^{4,1}$. Under the well-known correspondence

$$\begin{aligned} i &= \mathbf{e}_2 \wedge \mathbf{e}_3, \\ j &= \mathbf{e}_1 \wedge \mathbf{e}_3, \\ k &= \mathbf{e}_1 \wedge \mathbf{e}_2, \end{aligned} \tag{34}$$

the algebra of quaternions $\mathbb{Q}$ is isomorphic to the even subalgebra $C\ell^+(\mathbb{R}^3)$. Any nonzero element of $C\ell^+(\mathbb{R}^3)$ is a rotor, and by duality, any nonzero element of $C\ell^-(\mathbb{R}^3)$ is an odd versor.

Definition 6 A 2×2 quaternionic matrix $\mathbf{M} = \left(\begin{smallmatrix} \alpha & \beta \\ \gamma & \delta \end{smallmatrix}\right)$ is called a *quaternionic Vahlen matrix* if

1. $\alpha\overline{\beta},\ \beta\overline{\delta},\ \delta\overline{\gamma},\ \gamma\overline{\alpha}$ are all pure imaginary, or equivalently, $\langle\alpha\overline{\beta}\rangle = \langle\beta\overline{\delta}\rangle = \langle\delta\overline{\gamma}\rangle = \langle\gamma\overline{\alpha}\rangle = 0$;
2. The *determinant* $\Delta = \alpha\overline{\delta} + \beta\overline{\gamma} \neq 0$ is real.

It can be easily proved that a quaternionic Vahlen matrix is invertible if and only if its determinant is nonzero.

Theorem 5 *The following correspondence, together with* (34), *provides an algebraic isomorphism between the group of even twisted Vahlen matrices and the group of quaternionic Vahlen matrices*:

$$\mathbf{M} = \begin{pmatrix} \mathbf{A} & \mathbf{B} \\ \mathbf{C} & \mathbf{D} \end{pmatrix} \longmapsto \begin{pmatrix} \mathbf{A} & \mathbf{B}\mathbf{I}_3^{-1} \\ \mathbf{C}\mathbf{I}_3^{-1} & \mathbf{D} \end{pmatrix}. \tag{35}$$

Proof First, in $C\ell(\mathbf{I}_3)$, $\mathbf{A}\mathbf{B}^\dagger$ being a vector is equivalent to $\mathbf{A}(\overline{\mathbf{B}\mathbf{I}_3^{-1}})$ being a bivector. Second,

$$\Delta = \mathbf{A}\mathbf{D}^\dagger + \mathbf{B}\mathbf{C}^\dagger = \mathbf{A}\overline{\mathbf{D}} + (\mathbf{B}\mathbf{I}_3^{-1})(\mathbf{C}\mathbf{I}_3^{-1})^\dagger = \mathbf{A}\overline{\mathbf{D}} + (\mathbf{B}\mathbf{I}_3^{-1})(\overline{\mathbf{C}\mathbf{I}_3^{-1}}).$$

Third, by usual matrix multiplication,

$$\begin{aligned} \begin{pmatrix} \mathbf{A} & \mathbf{B}\mathbf{I}_3^{-1} \\ \mathbf{C}\mathbf{I}_3^{-1} & \mathbf{D} \end{pmatrix} \begin{pmatrix} \mathbf{A}' & \mathbf{B}'\mathbf{I}_3^{-1} \\ \mathbf{C}'\mathbf{I}_3^{-1} & \mathbf{D}' \end{pmatrix} &= \begin{pmatrix} \mathbf{A}\mathbf{A}' - \mathbf{B}\mathbf{C} & (\mathbf{A}\mathbf{B}' + \mathbf{B}\mathbf{D})\mathbf{I}_3^{-1} \\ (\mathbf{C}\mathbf{A}' + \mathbf{D}\mathbf{C}')\mathbf{I}_3^{-1} & \mathbf{D}\mathbf{D}' - \mathbf{C}\mathbf{B}' \end{pmatrix} \\ &= \begin{pmatrix} \mathbf{A}\mathbf{A}' + \mathbf{B}\hat{\mathbf{C}} & (\mathbf{A}\mathbf{B}' + \mathbf{B}\hat{\mathbf{D}})\mathbf{I}_3^{-1} \\ (\mathbf{C}\hat{\mathbf{A}}' + \mathbf{D}\mathbf{C}')\mathbf{I}_3^{-1} & \mathbf{D}\mathbf{D}' + \mathbf{C}\hat{\mathbf{B}}' \end{pmatrix}. \end{aligned}$$

□

A point $\mathbf{x} \in \mathbb{R}^3$ is represented by the pure imaginary quaternion $\mathbf{x}\mathbf{I}_3^{-1}$ under the correspondence (34), or in the 2D right-linear quaternionic vector space $\mathbb{Q}^2$ realizing the 1D projective space $\mathbb{QP}^1$, is represented by the vector $(\mathbf{x}\mathbf{I}_3^{-1}\ 1)^T$.

The matrix multiplication of a quaternionic Vahlen matrix $\mathbf{M} = \begin{pmatrix} \mathbf{A} & \mathbf{B}\mathbf{I}_3^{-1} \\ \mathbf{C}\mathbf{I}_3^{-1} & \mathbf{D} \end{pmatrix}$ with $(\mathbf{x}\mathbf{I}_3^{-1}\ 1)^T$ results in

$$\begin{pmatrix} \mathbf{A} & \mathbf{B}\mathbf{I}_3^{-1} \\ \mathbf{C}\mathbf{I}_3^{-1} & \mathbf{D} \end{pmatrix} \begin{pmatrix} \mathbf{x}\mathbf{I}_3^{-1} \\ 1 \end{pmatrix} = \begin{pmatrix} (\mathbf{A}\mathbf{x}+\mathbf{B})\mathbf{I}_3^{-1} \\ -\mathbf{C}\mathbf{x}+\mathbf{D} \end{pmatrix} = \begin{pmatrix} (\mathbf{A}\mathbf{x}+\mathbf{B})\mathbf{I}_3^{-1} \\ \hat{\mathbf{C}}\mathbf{x}+\hat{\mathbf{D}} \end{pmatrix}. \tag{36}$$

Combining the above result with (30), we get the following:

Theorem 6 (Vahlen's Theorem in quaternionic form) *Any quaternionic Vahlen matrix* $\mathbf{M} = \begin{pmatrix} \alpha & \beta \\ \gamma & \delta \end{pmatrix}$ *generates the following 3D conformal transformation: for any pure imaginary quaternion* υ *representing a point in space,*

$$\upsilon \longmapsto \mathbf{M}(\upsilon) = (\alpha\upsilon + \beta)(\gamma\upsilon + \delta)^{-1}; \tag{37}$$

or equivalently, in $\mathbb{QP}^1$ *where the point is represented homogeneously by* $(\upsilon : 1)$, *the conformal transformation is just the projectivity induced by the following invertible right-linear transformation over* $\mathbb{Q}$:

$$\begin{pmatrix} \upsilon \\ 1 \end{pmatrix} \longmapsto \begin{pmatrix} \alpha & \beta \\ \gamma & \delta \end{pmatrix} \begin{pmatrix} \upsilon \\ 1 \end{pmatrix} = \begin{pmatrix} \alpha\upsilon + \beta \\ \gamma\upsilon + \delta \end{pmatrix}. \tag{38}$$

Conversely, any 3D conformal transformation has such a quaternionic fractional linear representation.

Any bivector $\mathbf{B}_2 \in \Lambda^2(\mathbb{R}^{4,1})$ has the following decomposition:

$$\mathbf{B}_2 = \mathbf{A}_2 + \mathbf{b} \wedge \mathbf{e} + \mathbf{c} \wedge \mathbf{e}_0 + \lambda \mathbf{e} \wedge \mathbf{e}_0, \tag{39}$$

where $\mathbf{A}_2 \in \Lambda^2(\mathbb{R}^3)$, $\lambda \in \mathbb{R}$, and $\mathbf{b}, \mathbf{c} \in \mathbb{R}^3$. The following map from the Lie algebra $\Lambda^2(\mathbb{R}^{4,1})$ to the group of quaternionic Vahlen matrices under the correspondence (34) provides a transcendental parameterization of 3D conformal transformations, called *quaternionic Vahlen parameterization*:

$$\mathbf{B}_2 = \mathbf{A}_2 + \mathbf{b} \wedge \mathbf{e} + \mathbf{c} \wedge \mathbf{e}_0 + \lambda \mathbf{e} \wedge \mathbf{e}_0 \longmapsto e^{\mathbf{A}_2} \begin{pmatrix} 1 & \mathbf{b}\mathbf{I}_3^{-1} \\ \mathbf{c}\mathbf{I}_3^{-1} & \lambda - \mathbf{c}\mathbf{b} \end{pmatrix} \quad \text{if } \lambda \neq 0. \tag{40}$$

The Jacobian of the above parameterization is that of the exponential map $\mathbf{A}_2 \mapsto e^{\mathbf{A}_2}$ from $\Lambda^2(\mathbb{R}^3)$ to $\mathrm{Spin}(\mathbb{R}^3)$. It is always bounded. Those not in the range of the parameterization are conformal transformations induced by twisted Vahlen matrices of the form (27). The effect of such a transformation is

$$\mathbf{x} \in \mathbb{R}^3 \longmapsto Ad_{\mathbf{B}\mathbf{I}_3^{-1}}\left((\mu\mathbf{x} - \mathbf{d})^{-1}\right) \in \mathbb{R}^3, \tag{41}$$

which is the composition of the translation by vector $-\mathbf{d}/\mu$, the inversion with respect to the sphere centering at the origin and of radius $\mu^{-1/2}$, and a rotation whose axis passes through the origin of $\mathbb{R}^3$. The dimension of the conformal transformations outside the range of the parameterization is 7.

Collecting results from the above two paragraphs, we get the following:

Proposition 2 *The domain of definition of quaternionic Vahlen parameterization is a set* $\mathbb{R}^{10} - \mathbb{R}^9$ *parameterized by* $(\mathbf{A}_2, \mathbf{b}, \mathbf{c}, \lambda)$ *according to* (39), *where* $\lambda \neq 0$. *The image space is all* 3D *conformal transformations whose fractional linear representation* (30) *has the property that* $\mathbf{A} = 0$; *it is the remainder of* $O_+(4, 1)$ *after removal of a* 7D *closed subset and is topologically* $\mathbb{S}^3 \times (\mathbb{R} - \{0\}) \times \mathbb{R}^6$.

Compared with the exterior exponential, quaternionic Vahlen parameterization has the drawback that it is transcendental, and generally there are infinitely many inverses, but has the significant advantage that its domain of definition is simpler, and its image space is larger.

Example 4 Let there be a rotation in the space with fixed axis $\mathbf{I}_2^{\sim}$ and angle of rotation $\theta = \theta(t)$, where t is the time variable, and the range of θ is an interval of $\mathbb{R}$. The parameterization of the motion by outer exponential is $e^{\wedge \mathbf{I}_2 \tan(\frac{\theta(t)}{2})}$ and is invalid when $\theta(t) = \pi \mod 2\pi$.

In contrast, in the special case where the axis passes through the origin, the parameterization of the motion by quaternionic matrix is $e^{\mathbf{I}_2 \frac{\theta(t)}{2}} \begin{pmatrix} 1 & 0 \\ 0 & 1 \end{pmatrix}$. In the general case, let $\mathbf{I}_2^{\sim}$ represent the line passing through point $\mathbf{p} \in \mathbb{R}^3$ and following unit direction $\mathbf{n} \in \mathbb{R}^3$, i.e.,

$$\mathbf{I}_2^{\sim} = \mathbf{e} \wedge (\mathbf{e}_0 + \mathbf{p}) \wedge \mathbf{n}, \tag{42}$$

then the parameterization of the motion by quaternionic matrix is

$$\begin{aligned} &\begin{pmatrix} 1 & -\mathbf{p}\mathbf{I}_3^{-1} \\ 0 & 1 \end{pmatrix} e^{\mathbf{n}\mathbf{I}_3^{-1} \frac{\theta(t)}{2}} \begin{pmatrix} 1 & \mathbf{p}\mathbf{I}_3^{-1} \\ 0 & 1 \end{pmatrix} \\ &\quad = e^{\mathbf{n}\mathbf{I}_3^{-1} \frac{\theta(t)}{2}} \begin{pmatrix} 1 & (\mathbf{p} - e^{-\mathbf{n}\mathbf{I}_3^{-1} \frac{\theta(t)}{2}} \mathbf{p} e^{\mathbf{n}\mathbf{I}_3^{-1} \frac{\theta(t)}{2}})\mathbf{I}_3^{-1} \\ 0 & 1 \end{pmatrix}. \end{aligned} \tag{43}$$

It is valid for all $\theta(t) \in \mathbb{R}$.

4 Cayley Transform

In application, rational polynomial functions are much simpler than exponentials or trigonometric functions. For the special orthogonal group $SO(p, q)$, whose Lie algebra $so(p, q)$ is the set of antisymmetric linear transformations in $\mathbb{R}^{p,q}$, besides the exponential map, there is also a classical rational polynomial map from the Lie algebra to the Lie group, called *Cayley transform* [11]:

$$\begin{aligned} so(p, q) &\longrightarrow SO(p, q), \\ g &\longmapsto (I_{\mathbb{R}^{p,q}} + g)(I_{\mathbb{R}^{p,q}} - g)^{-1}, \quad \text{where } I_{\mathbb{R}^{p,q}} - g \text{ is invertible.} \end{aligned} \tag{44}$$

The map is injective but generally not surjective.

In the conformal model of 3D space, a natural idea is to consider simplifying the exponential map from $\Lambda^2(\mathbb{R}^{4,1})$ to the group of rotors by a fractional linear map similar to (44). The following mapping C:

$$\begin{aligned} \Lambda^2(\mathbb{R}^{4,1}) &\longrightarrow C\ell(\mathbb{R}^{4,1}), \\ \mathbf{B}_2 &\longmapsto (1+\mathbf{B}_2)(1-\mathbf{B}_2)^{-1}, \quad \text{where } 1-\mathbf{B}_2 \text{ is invertible,} \end{aligned} \tag{45}$$

is called the *Cayley transform* from Lie algebra $\Lambda^2(\mathbb{R}^{4,1})$ to the group of rotors in $C\ell(\mathbb{R}^{4,1})$.

The Cayley transform in terms of dual quaternions has been an important tool in describing and manipulating 3D rigid-body motions [17]. In this section, we enlarge the scope to 3D conformal transformations, explore the range and domain of definition of the Cayley transform, and present a degree-4 polynomial form of it, together with several neat formulas for the inverse of Cayley transform.

By computing the inverse $(1-\mathbf{B}_2)^{-1}$, we get that for any $\mathbf{B}_2 \in \Lambda^2(\mathbb{R}^{4,1})$ such that $\mathbf{B}_2^2 \neq 1$, the following equality holds up to scale:

$$C(\mathbf{B}_2) = (1+\mathbf{B}_2)^2(1-\mathbf{B}_2\cdot\mathbf{B}_2+\mathbf{B}_2\wedge\mathbf{B}_2). \tag{46}$$

If $C(\mathbf{B}_2)$ is required to be of unit magnitude, then

$$C(\mathbf{B}_2) = \frac{(1+\mathbf{B}_2)^2(1-\mathbf{B}_2\cdot\mathbf{B}_2+\mathbf{B}_2\wedge\mathbf{B}_2)}{(1-\mathbf{B}_2\cdot\mathbf{B}_2)^2-(\mathbf{B}_2\wedge\mathbf{B}_2)^2}. \tag{47}$$

Equation (46) can be used as an alternative definition of the Cayley transform. From this aspect, the Cayley transform is just a polynomial of degree 4 in $\mathbf{B}_2$, with values in the group of positive rotors of $C\ell(\mathbb{R}^{4,1})$; or equivalently, it is a rational polynomial of degree 4, with values in $\mathrm{Spin}_+(4, 1)$.

Theorem 7 [9] *The domain of definition of the Cayley transform C is all bivectors except the Minkowski blades of unit magnitude and is a set $\mathbb{R}^{10} - V^5$, where V^5 is a 5D algebraic variety in $\mathbb{R}^{10}$. The image space of C modulo scale is all positive rotors except those of the form $\mathbf{a}_1\mathbf{a}_2\mathbf{a}_3\mathbf{a}_4$, where the $\mathbf{a}_i$ are pairwise orthogonal positive vectors.*

Geometrically, the image space modulo scale is composed of positive rotors generating all orientation-preserving conformal transformations except the antipodal inversions, as shown in Fig. 1, *each of which is the composition of an inversion with respect to a sphere and the reflection with respect to the center of the sphere. Topologically, the image space modulo scale is the remainder of the Lorentz group of $\mathbb{R}^{4,1}$, which is a 10D connected Lie group, after removal of a 4D open disk.*

In the following, we present the "inverse" of the Cayley transform by finding all the preimages of a rotor in its range. Given a positive rotor $\mathbf{A}$ such that $\mathbf{A} \neq 1$ up to scale, let $\mathbf{B}_2$ be a bivector whose Cayley transform equals $\mathbf{A}$ up to scale.

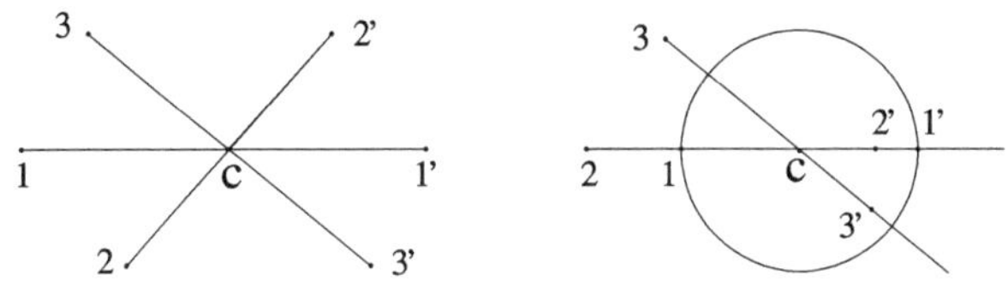

Fig. 1 Antipodal inversion: composition of a reflection with respect to a point and an inversion centering at the same point

A bivector is said to be *entangled*, or *coherent*, if in its completely orthogonal decomposition there are two components having equal square. It can be proved that for a bivector $\langle \mathbf{A} \rangle_2 \in \Lambda^2(\mathbb{R}^{4,1})$ to be entangled, it is necessary and sufficient that

$$\left(\langle \mathbf{A} \rangle_2 \cdot \langle \mathbf{A} \rangle_2\right)^2 = \left(\langle \mathbf{A} \rangle_2 \wedge \langle \mathbf{A} \rangle_2\right)^2. \tag{48}$$

Theorem 8 [9] *A positive rotor* $\mathbf{A}$ *in the range of the Cayley transform has exactly one bivector preimage if and only if either it is in* $\Lambda(\mathbf{C}_2)$, *where* $\mathbf{C}_2$ *is a 2-blade of degenerate signature, or its bivector part is entangled. The unique solution is*

$$\frac{\langle \mathbf{A} \rangle_2}{\langle \mathbf{A} \rangle_4 + 2\langle \mathbf{A} \rangle + |\langle \mathbf{A} \rangle \langle \mathbf{A} \rangle_4| / \langle \mathbf{A} \rangle}. \tag{49}$$

Any other positive rotor $\mathbf{A}$ *in the range of the Cayley transform has two bivector preimages, and they are inverse to each other:*

$$\frac{\langle \mathbf{A} \rangle_2}{\langle \mathbf{A} \rangle_4 + \langle \mathbf{A} \rangle \pm |\mathbf{A}|}. \tag{50}$$

Example 5 In $C\ell(\mathbb{R}^{4,1})$, let $\mathbf{A} = e^{\mathbf{I}_2 \frac{\theta}{2}}$ be a rotor inducing a 2D rotation, where $\mathbf{I}_2 \in \Lambda(\mathbf{e}^{\sim})$ is a Euclidean 2-blade of unit magnitude such that $\mathbf{I}_2^{\sim}$ is the axis of rotation, and $-\theta$ is the angle of rotation. Then

$$\mathbf{B}_2 = \frac{e^{\mathbf{I}_2 \frac{\theta}{2}} - e^{-\mathbf{I}_2 \frac{\theta}{2}}}{e^{\mathbf{I}_2 \frac{\theta}{2}} + e^{-\mathbf{I}_2 \frac{\theta}{2}} + 2} = \mathbf{I}_2 \tan \frac{\theta}{4}, \qquad \mathbf{B}_2^{-1} = -\mathbf{I}_2 / \tan \frac{\theta}{4}, \tag{51}$$

and both generate $\mathbf{A}$ by the Cayley transform.

While the bivector representation of a rotation via the exponential map is a half-angle representation, the bivector representation via the Cayley transform is a quarter-angle representation.

Example 6 In $C\ell(\mathbb{R}^{4,1})$, let $\mathbf{A} = 1 + \mathbf{et}/2$ be a rotor realizing a translation, where $\mathbf{t} \in \mathbf{e}^{\sim}$ is a positive vector. Then

$$\mathbf{B}_2 = \frac{\mathbf{et}}{4} \tag{52}$$

generates $\mathbf{A}$ by the Cayley transform. So the Cayley transform provides a quarter-distance bivector representation of the translation.

Example 7 Let $\mathbf{A} = e^{\frac{\theta}{2}\mathbf{e}\wedge\mathbf{a}}$ be a rotor realizing a dilation (or scaling), where $\theta \in \mathbb{R}$, $\mathbf{a}$ is a null vector representing a point, and $\mathbf{a}\cdot\mathbf{e} = -1$. Rotor $\mathbf{A}$ generates the dilation centering at point $\mathbf{a}$ and with scale $e^{-\theta}$. Denote $\mathbf{I}_2 = \mathbf{e}\wedge\mathbf{a}$. Then

$$\mathbf{B}_2 = \frac{e^{\mathbf{I}_2\frac{\theta}{2}} - e^{-\mathbf{I}_2\frac{\theta}{2}}}{e^{\mathbf{I}_2\frac{\theta}{2}} + e^{-\mathbf{I}_2\frac{\theta}{2}} + 2} = \mathbf{I}_2 \tanh\frac{\theta}{4}, \qquad \mathbf{B}_2^{-1} = \mathbf{I}_2/\tanh\frac{\theta}{4}, \tag{53}$$

and both generate $\mathbf{A}$ by the Cayley transform. So the Cayley transform provides a quarter-scale bivector representation of the dilation.

All orientation-preserving similarity transformations in $\mathbb{R}^3$ can be induced by bivectors in $\Lambda^2(\mathbb{R}^{4,1})$ through the Cayley transform and adjoint action. A translation is induced by the Cayley transform of a unique bivector. Any other orientation-preserving similarity transformation is induced by the Cayley transform of exactly two bivectors.

When choosing between the two bivector preimages $\mathbf{B}_2$ and $\mathbf{B}_2^{-1}$ of a rotor, since $|\mathbf{B}_2||\mathbf{B}_2^{-1}| = 1$, one of $|\mathbf{B}_2|$ and $|\mathbf{B}_2^{-1}|$ is greater than or equal to 1. By (50),

$$\left|\mathbf{B}_2 - \mathbf{B}_2^{-1}\right| = 2\frac{|\mathbf{A}|}{|\langle\mathbf{A}\rangle_2|} \geq 2. \tag{54}$$

So for two rotors that are close to each other, we can always choose their bivector preimages to be close to each other. If their magnitudes are greater than 1, we can choose their inverses so that the magnitudes become smaller than 1.

Then what is the use of having two bivector preimages for the same rotor? Take, as an example, the 2D rotation in Example 5. It is well known that $SO(2)$ is a circle which has the following rational parameterization induced by the stereographic projection from the north pole N:

$$e^{\mathbf{I}_2\theta} = \cos\theta + \mathbf{I}_2\sin\theta = \frac{2t}{1-t^2} + \mathbf{I}_2\frac{1+t^2}{1-t^2}, \tag{55}$$

where, as shown in Fig. 2, $t = \tan(\theta/2)$ is half the signed distance from the south pole S to point X in the horizontal direction, and point $e^{i\theta}$ in the complex plane is represented by the intersection R of line NX with the unit circle.

Since the map $\theta \mapsto e^{\mathbf{I}_2\theta}$ is an isometric immersion, the Jacobian of the rational parameterization (55) equals

$$\frac{d\theta}{dt} = 1\Big/\left(\frac{dt}{d\theta}\right) = 2\cos^2\frac{\theta}{2}. \tag{56}$$

When point R moves from the south pole S to point T, the Jacobian decreases from 2 to 1, while when point R continues to move from point T to the north pole N, the

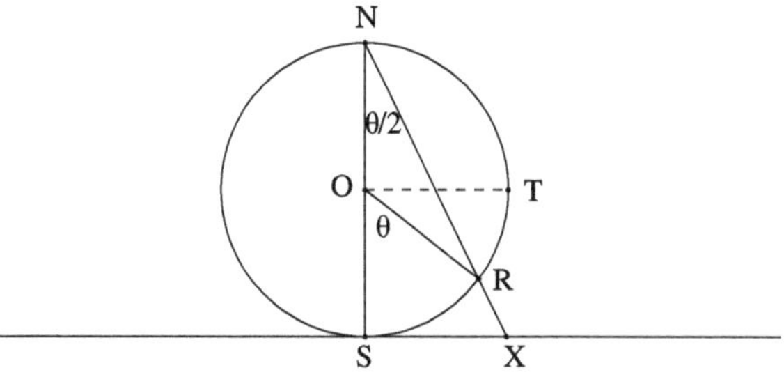

Fig. 2 Rational parametrization of a 2D rotation

Jacobian decreases from 1 to 0. The parameterization becomes practically useless nearby the north pole.

The two preimages in (51) are both mapped to the rotor $e^{\mathbf{I}_2\theta/2}$ by the Cayley transform. So the Cayley transform provides a generalization of the rational parameterization of a circle taken as the 2D rotation group. In fact, it serves as two rational parameterizations of the circle derived from two different stereographic projections: one from the north pole, and the other from the south pole.

Since the two maps

$$\theta \mapsto \mathbf{I}_2 \tan\frac{\theta}{4} \quad \text{and} \quad \theta \mapsto -\mathbf{I}_2 \operatorname{ctan}\frac{\theta}{4} \tag{57}$$

have Jacobians $1/(4\cos^2(\theta/4))$ and $1/(4\sin^2(\theta/4))$, respectively, the maps

$$\begin{aligned} t = \tan\frac{\theta}{4} &\overset{\text{Cayley}}{\longmapsto} e^{\mathbf{I}_2\theta/2}, \\ t = -\operatorname{ctan}\frac{\theta}{4} &\overset{\text{Cayley}}{\longmapsto} e^{\mathbf{I}_2\theta/2}, \end{aligned} \tag{58}$$

have Jacobians $J_1 = 2\cos^2\frac{\theta}{4}$ and $J_2 = 2\sin^2\frac{\theta}{4}$, respectively.

- When $(4k-1)\pi \leq \theta \leq (4k+1)\pi$, then $1 \leq J_1 \leq 2$ and $0 \leq J_2 \leq 1$.
- When $(4k+1)\pi \leq \theta \leq (4k+3)\pi$, then $1 \leq J_2 \leq 2$ and $0 \leq J_1 \leq 1$.

Hence the two maps in (58) cover different zones of the parameter $\theta \in \mathbb{R}$ for the Jacobians to be effective between 1 and 2. They serve as two different local coordinate charts whose union covers the whole Lie group.

The Cayley transform as a polynomial map of degree 4 is computationally superior; its inverse map has two branches and involves only one square-root computing. Its domain of definition and its image space, when restricted to orientation-preserving conformal transformations, are both larger than those of exterior differential and quaternionic Vahlen parameterization. Thus the Cayley transform and its inverse are an ideal tool for motion planning, interpolating, and fitting in the Lie algebra of 3D conformal transformations.

5 Conclusion

This chapter explores the issue of parameterizing 3D conformal transformations. Two new results are presented, one on quaternionic Vahlen parameterization, the other on the polynomial 3D Cayley transform. They provide compact representations of 3D conformal transformations and should prove to be useful in geometric applications.

References

1. Ahlfors, L.V.: Möbius transformations in $\mathbb{R}^n$ expressed through 2×2 matrices of Clifford numbers. Complex Var. **5**, 215–224 (1986)
2. Angles, P.: Conformal Groups in Geometry and Spin Structures. Birkhäuser, Basel (2008)
3. Bayro-Corrochano, E., Reyes-Lozano, L., Zamora-Esquivel, J.: Conformal geometric algebra for robotic vision. J. Math. Imaging Vis. **24**(1), 55–81 (2006)
4. Dorst, L., Fontijne, D., Mann, S.: Geometric Algebra for Computer Science. Morgan Kaufmann, San Mateo (2007)
5. Helmstetter, J., Micali, A.: Quadratic Mappings and Clifford Algebras. Birkhäuser, Basel (2000)
6. Hestenes, D., Sobczyk, G.: Clifford Algebra to Geometric Calculus. Kluwer, Dordrecht (1984)
7. Hildenbrand, D.: Geometric computing in computer graphics using conformal geometric algebra. Comput. Graph. **29**(5), 795–803 (2005)
8. Lasenby, A.: Recent applications of conformal geometric algebra. In: Li, H., et al. (eds.) Computer Algebra and Geometric Algebra with Applications. LNCS, vol. 3519, pp. 298–328. Springer, Berlin (2005)
9. Li, H.: Invariant Algebras and Geometric Reasoning. World Scientific, Singapore (2008)
10. Li, H., Hestenes, D., Rockwood, A.: Generalized homogeneous coordinates for computational geometry. In: Sommer, G. (ed.) Geometric Computing with Clifford Algebras, pp. 27–60. Springer, Heidelberg (2001)
11. Lounesto, P.: Clifford Algebras and Spinors. Cambridge University Press, Cambridge (1997)
12. Maks, J.: Clifford algebras and Möbius transformations. In: Micali, A., et al. (eds.) Clifford Algebras and Their Applications in Mathematical Physics, pp. 57–63. Kluwer, Dordrecht (1992)
13. Mourrain, B., Stolfi, N.: Computational symbolic geometry. In: White, N.L. (ed.) Invariant Methods in Discrete and Computational Geometry, pp. 107–139. Reidel, Dordrecht (1995)
14. Riesz, M.: Clifford Numbers and Spinors, 1958. Kluwer, Dordrecht (1993). From lecture notes made in 1957–1958, edited by Bolinder, E. and Lounesto, P.
15. Rosenhahn, B., Sommer, G.: Pose estimation in conformal geometric algebra I, II. J. Math. Imaging Vis. **22**, 27–70 (2005)
16. Ryan, J.: Conformal Clifford manifolds arising in Clifford analysis. Proc. R. Ir. Acad. A **85**, 1–23 (1985)
17. Selig, J.M.: Geometrical Methods in Robotics. Springer, New York (1996)
18. Sommer, G., Rosenhahn, B., Perwass, C.: Twists—an operational representation of shape. In: Li, H., et al. (eds.) Computer Algebra and Geometric Algebra with Applications. LNCS, vol. 3519, pp. 278–297. Springer, Berlin (2005)
19. White, N.: Grassmann–Cayley algebra and robotics applications. In: Bayro-Corrochano, E. (ed.) Handbook of Geometric Computing, pp. 629–656. Springer, Heidelberg (2005)

Part II
Clifford Fourier Transform

Two-Dimensional Clifford Windowed Fourier Transform

Mawardi Bahri, Eckhard M.S. Hitzer, and Sriwulan Adji

Abstract Recently several generalizations to higher dimension of the classical Fourier transform (FT) using Clifford geometric algebra have been introduced, including the two-dimensional (2D) Clifford–Fourier transform (CFT). Based on the 2D CFT, we establish the two-dimensional Clifford windowed Fourier transform (CWFT). Using the spectral representation of the CFT, we derive several important properties such as shift, modulation, a reproducing kernel, isometry, and an orthogonality relation. Finally, we discuss examples of the CWFT and compare the CFT and CWFT.

1 Introduction

One of the basic problems encountered in signal representations using the conventional Fourier transform (FT) is the ineffectiveness of the Fourier kernel to represent and compute location information. One method to overcome such a problem is the windowed Fourier transform (WFT). Recently, some authors [4, 7] have extensively studied the WFT and its properties from a mathematical point of view. In [6, 8] they applied the WFT as a tool of spatial-frequency analysis which is able to characterize the local frequency at any location in a fringe pattern.

On the other hand, Clifford geometric algebra leads to the consequent generalization of real and harmonic analysis to higher dimensions. Clifford algebra accurately treats geometric entities depending on their dimension as scalars, vectors, bivectors (oriented plane area elements), and trivectors (oriented volume elements), etc. Motivated by the above facts, we generalize the WFT in the framework of Clifford geometric algebra.

In the present paper we study the two-dimensional Clifford windowed Fourier transform (CWFT). A complementary motivation for studying this topic comes from

M. Bahri (✉)
School of Mathematical Sciences, Universiti Sains Malaysia, 11800 Penang, Malaysia
e-mail: mawardibahri@gmail.com

E. Bayro-Corrochano, G. Scheuermann (eds.), *Geometric Algebra Computing*,
DOI 10.1007/978-1-84996-108-0_5,

the understanding that the 2D CWFT is in fact intimately related with Clifford–Gabor filters [1] and quaternionic Gabor filters [2, 3]. This generalization also enables us to establish the two-dimensional Clifford–Gabor filters.

2 Real Clifford Algebra $\mathcal{G}_2$

Let us consider an orthonormal vector basis $\{\mathbf{e}_1, \mathbf{e}_2\}$ of the real 2D Euclidean vector space $\mathbb{R}^2 = \mathbb{R}^{2,0}$. The geometric algebra over $\mathbb{R}^2$ denoted by $\mathcal{G}_2$ then has the graded four-dimensional basis

$$\{1, \mathbf{e}_1, \mathbf{e}_2, \mathbf{e}_{12}\}, \tag{1}$$

where 1 is the real scalar identity element (grade 0), $\mathbf{e}_1, \mathbf{e}_2 \in \mathbb{R}^2$ are vectors (grade 1), and $\mathbf{e}_{12} = \mathbf{e}_1\mathbf{e}_2 = i_2$ defines the unit oriented pseudoscalar[1] (grade 2), i.e., the highest grade blade element in $\mathcal{G}_2$.

The associative geometric multiplication of the basis vectors obeys the following basic rules:

$$\mathbf{e}_1^2 = \mathbf{e}_2^2 = 1, \qquad \mathbf{e}_1\mathbf{e}_2 = -\mathbf{e}_2\mathbf{e}_1. \tag{2}$$

The general elements of a geometric algebra are called multivectors. Every multivector $f \in \mathcal{G}_2$ can be expressed as

$$f = \underbrace{\alpha_0}_{\text{scalar part}} + \underbrace{\alpha_1\mathbf{e}_1 + \alpha_2\mathbf{e}_2}_{\text{vector part}} + \underbrace{\alpha_{12}\mathbf{e}_{12}}_{\text{bivector part}} \qquad \forall \alpha_0, \alpha_1, \alpha_2, \alpha_{12} \in \mathbb{R}. \tag{3}$$

The grade selector is defined as $\langle f \rangle_k$ for the k-vector part of f. We often write $\langle \ldots \rangle = \langle \ldots \rangle_0$. Then (3) can be expressed as[2]

$$f = \langle f \rangle + \langle f \rangle_1 + \langle f \rangle_2. \tag{4}$$

The multivector f is called a parabivector if the vector part of (4) is zero, i.e.,

$$f = \alpha_0 + \alpha_{12}\mathbf{e}_{12}. \tag{5}$$

The reverse $\tilde{f}$ of a multivector $f \in \mathcal{G}_2$ is an anti-automorphism given by

$$\tilde{f} = \langle f \rangle + \langle f \rangle_1 - \langle f \rangle_2, \tag{6}$$

which fulfills $\widetilde{fg} = \tilde{g}\tilde{f}$ for every $f, g \in \mathcal{G}_2$. In particular, $\tilde{i}_2 = -i_2$.

The scalar product of two multivectors $f, \tilde{g}$ is defined as the scalar part of the geometric product $f\tilde{g}$,

$$f * \tilde{g} = \langle f\tilde{g} \rangle = \alpha_0\beta_0 + \alpha_1\beta_1 + \alpha_2\beta_2 + \alpha_{12}\beta_{12}, \tag{7}$$

---

[1] Other names in use are *bivector* or *oriented area element*.

[2] Note that (4) and (6) show grade selection and not component selection.

which leads to the cyclic product symmetry

$$\langle pqr\rangle = \langle qrp\rangle \quad \forall p,q,r \in \mathcal{G}_2. \tag{8}$$

For $f = g$ in (7), we obtain the modulus (or magnitude) $|f|$ of a multivector $f \in \mathcal{G}_2$ defined as

$$|f|^2 = f * \tilde{f} = \alpha_0^2 + \alpha_1^2 + \alpha_2^2 + \alpha_{12}^2. \tag{9}$$

It is convenient to introduce an inner product for two multivector-valued functions $f, g : \mathbb{R}^2 \to \mathcal{G}_2$ as follows:

$$(f, g)_{L^2(\mathbb{R}^2;\mathcal{G}_2)} = \int_{\mathbb{R}^2} f(\mathbf{x})\widetilde{g(\mathbf{x})}\, d^2\mathbf{x}. \tag{10}$$

One can check that this inner product satisfies the following rules:

$$\begin{aligned}
(f, g+h)_{L^2(\mathbb{R}^2;\mathcal{G}_2)} &= (f, g)_{L^2(\mathbb{R}^2;\mathcal{G}_2)} + (f, h)_{L^2(\mathbb{R}^2;\mathcal{G}_2)},\\
(f, \lambda g)_{L^2(\mathbb{R}^2;\mathcal{G}_2)} &= (f, g)_{L^2(\mathbb{R}^2;\mathcal{G}_2)}\tilde{\lambda},\\
(f\lambda, g)_{L^2(\mathbb{R}^2;\mathcal{G}_2)} &= (f, g\tilde{\lambda})_{L^2(\mathbb{R}^2;\mathcal{G}_2)},\\
(f, g)_{L^2(\mathbb{R}^2;\mathcal{G}_2)} &= \widetilde{(g, f)}_{L^2(\mathbb{R}^2;\mathcal{G}_2)},
\end{aligned} \tag{11}$$

where $f, g \in L^2(\mathbb{R}^2; \mathcal{G}_2)$, and $\lambda \in \mathcal{G}_2$ is a multivector constant. The scalar part of the inner product gives the L^2-norm

$$\|f\|^2_{L^2(\mathbb{R}^2;\mathcal{G}_2)} = \big\langle (f, f)_{L^2(\mathbb{R}^2;\mathcal{G}_2)}\big\rangle. \tag{12}$$

Definition 1 (Clifford module) Let $\mathcal{G}_2$ be the real Clifford algebra of 2D Euclidean space $\mathbb{R}^2$. The Clifford algebra module $L^2(\mathbb{R}^2; \mathcal{G}_2)$ is defined by

$$L^2(\mathbb{R}^2; \mathcal{G}_2) = \big\{f : \mathbb{R}^2 \longrightarrow \mathcal{G}_2 \mid \|f\|_{L^2(\mathbb{R}^2;\mathcal{G}_2)} < \infty\big\}. \tag{13}$$

3 Clifford Fourier Transform (CFT)

It is natural to extend the FT to the Clifford algebra $\mathcal{G}_2$. This extension is often called the Clifford–Fourier transform (CFT). For detailed discussions on the properties of the CFT and their proofs, see, e.g., [1, 5]. In the following we briefly review the 2D CFT.

Definition 2 The CFT of $f \in L^2(\mathbb{R}^2; \mathcal{G}_2) \cap L^1(\mathbb{R}^2; \mathcal{G}_2)$ is the function $\mathcal{F}\{f\}$: $\mathbb{R}^2 \to \mathcal{G}_2$ given by

$$\mathcal{F}\{f\}(\omega) = \int_{\mathbb{R}^2} f(\mathbf{x})e^{-i_2\omega\cdot\mathbf{x}}\, d^2\mathbf{x}, \tag{14}$$

where we can write $\omega = \omega_1 \mathbf{e}_1 + \omega_2 \mathbf{e}_2$ and $\mathbf{x} = x_1 \mathbf{e}_1 + x_2 \mathbf{e}_2$. Note that

$$d^2\mathbf{x} = \frac{d\mathbf{x}_1 \wedge d\mathbf{x}_2}{i_2} \tag{15}$$

is scalar valued ($d\mathbf{x}_k = dx_k \mathbf{e}_k$, $k = 1, 2$, no summation). Notice that the Clifford–Fourier kernel $e^{-i_2 \boldsymbol{\omega} \cdot \mathbf{x}}$ does not commute with every element of the Clifford algebra $\mathcal{G}_2$. Furthermore, the product has to be performed in a fixed order.

Theorem 1 *Suppose that $f \in L^2(\mathbb{R}^2; \mathcal{G}_2)$ and $\mathcal{F}\{f\} \in L^1(\mathbb{R}^2; \mathcal{G}_2)$. Then the CFT is an invertible transform, and its inverse is calculated by*

$$\mathcal{F}^{-1}\big[\mathcal{F}\{f\}(\boldsymbol{\omega})\big](\mathbf{x}) = f(\mathbf{x}) = \frac{1}{(2\pi)^2} \int_{\mathbb{R}^2} \mathcal{F}\{f\}(\boldsymbol{\omega})\, e^{i_2 \boldsymbol{\omega} \cdot \mathbf{x}}\, d^2\boldsymbol{\omega}. \tag{16}$$

4 2D Clifford Windowed Fourier Transform

In [1, 5] the 2D CFT has been introduced. This enables us to establish the 2D CWFT. We will see that several properties of the WFT can be established in the new construction with some modifications. We begin with the definition of the 2D CWFT.

4.1 Definition of the CWFT

Definition 3 A Clifford window function is a function $\phi \in L^2(\mathbb{R}^2; \mathcal{G}_2) \setminus \{0\}$ such that $|\mathbf{x}|^{1/2}\phi(\mathbf{x}) \in L^2(\mathbb{R}^2; \mathcal{G}_2)$.

$$\phi_{\omega,\mathbf{b}}(\mathbf{x}) = \frac{e^{i_2 \boldsymbol{\omega} \cdot \mathbf{x}} \phi(\mathbf{x} - \mathbf{b})}{(2\pi)^2} \tag{17}$$

denote the so-called Clifford window daughter functions.

Definition 4 (Clifford windowed Fourier transform) The Clifford windowed Fourier transform (CWFT) $G_\phi f$ of $f \in L^2(\mathbb{R}^2; \mathcal{G}_2)$ is defined by

$$\begin{aligned} f(\mathbf{x}) \longrightarrow G_\phi f(\omega, \mathbf{b}) &= (f, \phi_{\omega,\mathbf{b}})_{L^2(\mathbb{R}^2; \mathcal{G}_2)} \\ &= \frac{1}{(2\pi)^2} \int_{\mathbb{R}^2} f(\mathbf{x}) \left\{ e^{i_2 \boldsymbol{\omega} \cdot \mathbf{x}} \phi(\mathbf{x} - \mathbf{b}) \right\}^{\sim} d^2\mathbf{x} \\ &= \frac{1}{(2\pi)^2} \int_{\mathbb{R}^2} f(\mathbf{x})\, \widetilde{\phi(\mathbf{x} - \mathbf{b})}\, e^{-i_2 \boldsymbol{\omega} \cdot \mathbf{x}}\, d^2\mathbf{x}. \end{aligned} \tag{18}$$

This shows that the CWFT can be regarded as the CFT of the product of a Clifford-valued function f and a shifted and reversed Clifford window function ϕ, or as an inner product (10) of a Clifford-valued function f and the Clifford window daughter functions $\phi_{\boldsymbol{\omega},\mathbf{b}}$.

Taking the Gaussian function as the window function of (17) with fixed $\boldsymbol{\omega} = \boldsymbol{\omega}_0 = \omega_{0,1}\mathbf{e}_1 + \omega_{0,2}\mathbf{e}_2$, we obtain Clifford Gabor filters, i.e.,

$$g_c(\boldsymbol{x}, \sigma_1, \sigma_2) = \frac{1}{(2\pi)^2} e^{i_2 \boldsymbol{\omega}_0 \cdot \mathbf{x}} e^{-[(x_1/\sigma_1)^2 + (x_2/\sigma_2)^2]/2}, \tag{19}$$

where σ_1 and σ_2 are standard deviations of the Gaussian functions, and the translation parameters are $b_1 = b_2 = 0$.

In terms of the $\mathcal{G}_2$ Clifford–Fourier transform, (19) can be expressed as

$$\mathcal{F}\{g_c\}(\boldsymbol{\omega}) = \frac{1}{\pi \sigma_1 \sigma_2} e^{-\frac{1}{2}[(\sigma_1^2(\omega_1 - \omega_{0,1})^2 + \sigma_2^2(\omega_2 - \omega_{0,2})^2]}. \tag{20}$$

From (19) and (20) we see that Clifford–Gabor filters are well localized in the spatial and Clifford–Fourier domains.

The energy density is defined as the square modulus of the CWFT (18) given by

$$\left|G_\phi f(\boldsymbol{\omega}, \mathbf{b})\right|^2 = \frac{1}{(2\pi)^4} \left| \int_{\mathbb{R}^2} f(\mathbf{x}) \widetilde{\phi(\mathbf{x} - \mathbf{b})} e^{-i_2 \boldsymbol{\omega} \cdot \mathbf{x}} \, d^2\mathbf{x} \right|^2. \tag{21}$$

Equation (21) is often called a spectrogram which measures the energy of a Clifford-valued function f in the position–frequency neighborhood of $(\mathbf{b}, \boldsymbol{\omega})$.

In particular, when the Gaussian function (19) is chosen as the Clifford window function, the CWFT (18) is called the Clifford–Gabor transform.

4.2 Properties of the CWFT

We will discuss the properties of the CWFT. We find that many of the properties of the WFT are still valid for the CWFT, however, with certain modifications.

Theorem 2 (Left linearity) *Let $\phi \in L^2(\mathbb{R}^2; \mathcal{G}_2)$ be a Clifford window function. The CWFT of $f, g \in L^2(\mathbb{R}^2; \mathcal{G}_2)$ is a left linear operator,*[3] *which means that*

$$\left[G_\phi(\lambda f + \mu g)\right](\boldsymbol{\omega}, \mathbf{b}) = \lambda G_\phi f(\boldsymbol{\omega}, \mathbf{b}) + \mu G_\phi g(\boldsymbol{\omega}, \mathbf{b}) \tag{22}$$

with Clifford constants $\lambda, \mu \in \mathcal{G}_2$.

Proof Using the definition of the CWFT, the proof is obvious. □

Remark 1 Since the geometric multiplication is noncommutative, the right-linearity property of the CWFT does not hold in general.

---

[3]The CWFT of f is a *linear* operator for real constants $\mu, \lambda \in \mathbb{R}$.

Theorem 3 (Reversion) *Let $f \in L^2(\mathbb{R}^2; \mathcal{G}_2^+)$ be a parabivector-valued function. For a parabivector-valued window function ϕ, we have*

$$G_{\tilde{\phi}}\tilde{f}(\boldsymbol{\omega}, \mathbf{b}) = \{G_\phi f(-\boldsymbol{\omega}, \mathbf{b})\}^{\sim}. \tag{23}$$

Proof Application of Definition 4 to the left-hand side of (23) gives

$$\begin{aligned} G_{\tilde{\phi}}\tilde{f}(\boldsymbol{\omega}, \mathbf{b}) &= \frac{1}{(2\pi)^2}\int_{\mathbb{R}^2} \widetilde{f(\mathbf{x})}\phi(\mathbf{x}-\mathbf{b})\, e^{-i_2\boldsymbol{\omega}\cdot\mathbf{x}}\, d^2\mathbf{x} \\ &= \frac{1}{(2\pi)^2}\left\{\int_{\mathbb{R}^2} e^{i_2\boldsymbol{\omega}\cdot\mathbf{x}}\widetilde{\phi(\mathbf{x}-\mathbf{b})} f(\mathbf{x})\, d^2\mathbf{x}\right\}^{\sim} \\ &= \frac{1}{(2\pi)^2}\left\{\int_{\mathbb{R}^2} f(\mathbf{x})\, \widetilde{\phi(\mathbf{x}-\mathbf{b})}\, e^{i_2\boldsymbol{\omega}\cdot\mathbf{x}}\, d^2\mathbf{x}\right\}^{\sim}. \end{aligned} \tag{24}$$

This finishes the proof of the theorem. □

Theorem 4 (Switching) *If $|\mathbf{x}|^{1/2}f(\mathbf{x}) \in L^2(\mathbb{R}^2; \mathcal{G}_2)$ and $|\mathbf{x}|^{1/2}\phi(\mathbf{x}) \in L^2(\mathbb{R}^2; \mathcal{G}_2)$ are parabivector-valued functions, then we obtain*

$$G_\phi f(\boldsymbol{\omega}, \mathbf{b}) = e^{-i_2\boldsymbol{\omega}\cdot\mathbf{b}}\{G_f\phi(-\boldsymbol{\omega}, -\mathbf{b})\}^{\sim}. \tag{25}$$

Proof We have, by the CWFT definition,

$$\begin{aligned} G_\phi f(\boldsymbol{\omega}, \mathbf{b}) &= \frac{1}{(2\pi)^2}\int_{\mathbb{R}^2} f(\mathbf{x})\widetilde{\phi(\mathbf{x}-\mathbf{b})}\, e^{-i_2\boldsymbol{\omega}\cdot\mathbf{x}}\, d^2\mathbf{x} \\ &= \frac{1}{(2\pi)^2}\left\{\int_{\mathbb{R}^2} \phi(\mathbf{x}-\mathbf{b})\,\widetilde{f(\mathbf{x})}\, e^{i_2\boldsymbol{\omega}\cdot\mathbf{x}}\, d^2\mathbf{x}\right\}^{\sim}. \end{aligned} \tag{26}$$

The substitution $\mathbf{y} = \mathbf{x} - \mathbf{b}$ into the above expression gives

$$\begin{aligned} G_\phi f(\boldsymbol{\omega}, \mathbf{b}) &= \frac{1}{(2\pi)^2}\left\{\int_{\mathbb{R}^2} \phi(\mathbf{y})\,\widetilde{f(\mathbf{y}+\mathbf{b})}\, e^{i_2\boldsymbol{\omega}\cdot(\mathbf{y}+\mathbf{b})}\, d^2\mathbf{y}\right\}^{\sim} \\ &= \frac{1}{(2\pi)^2}e^{-i_2\boldsymbol{\omega}\cdot\mathbf{b}}\left\{\int_{\mathbb{R}^2} \phi(\mathbf{y})\,\widetilde{f(\mathbf{y}+\mathbf{b})}\, e^{i_2\boldsymbol{\omega}\cdot\mathbf{y}}\, d^2\mathbf{y}\right\}^{\sim} \\ &= \frac{1}{(2\pi)^2}e^{-i_2\boldsymbol{\omega}\cdot\mathbf{b}}\left\{\int_{\mathbb{R}^2} \phi(\mathbf{y})\,\widetilde{f(\mathbf{y}-(-\mathbf{b}))}\, e^{-i_2(-\boldsymbol{\omega})\cdot\mathbf{y}}\, d^2\mathbf{y}\right\}^{\sim}, \end{aligned} \tag{27}$$

which proves the theorem. □

Theorem 5 (Parity) *Let $\phi \in L^2(\mathbb{R}^2; \mathcal{G}_2)$ be a Clifford window function. If P is the parity operator defined as $P\phi(\mathbf{x}) = \phi(-\mathbf{x})$, then we have*

$$G_{P\phi}\{Pf\}(\boldsymbol{\omega}, \mathbf{b}) = G_\phi f(-\boldsymbol{\omega}, -\mathbf{b}). \tag{28}$$

Proof Direct calculations give, for every $f \in L^2(\mathbb{R}^2; \mathcal{G}_2)$,

$$\begin{aligned}
G_{P\phi}\{Pf\}(\boldsymbol{\omega}, \mathbf{b}) &= \frac{1}{(2\pi)^2} \int_{\mathbb{R}^2} f(-\mathbf{x}) \{\phi(-\mathbf{x}+\mathbf{b})\}^{\sim} e^{-i_2(-\boldsymbol{\omega})\cdot(-\mathbf{x})} \, d^2\mathbf{x} \\
&= \frac{1}{(2\pi)^2} \int_{\mathbb{R}^2} f(-\mathbf{x}) \{\phi(-\mathbf{x}-(-\mathbf{b}))\}^{\sim} e^{-i_2(-\boldsymbol{\omega})\cdot(-\mathbf{x})} \, d^2\mathbf{x} \\
&= \frac{1}{(2\pi)^2} \int_{\mathbb{R}^2} f(\mathbf{x}) \{\phi(\mathbf{x}-(-\mathbf{b}))\}^{\sim} e^{-i_2(-\boldsymbol{\omega})\cdot\mathbf{x}} \, d^2\mathbf{x}, \qquad (29)
\end{aligned}$$

which completes the proof. □

Theorem 6 (Shift in space domain, delay) *Let ϕ be a Clifford window function. Introducing the translation operator $T_{\mathbf{x}_0} f(\mathbf{x}) = f(\mathbf{x} - \mathbf{x}_0)$, we obtain*

$$G_\phi\{T_{\mathbf{x}_0} f\}(\boldsymbol{\omega}, \mathbf{b}) = \big(G_\phi f(\boldsymbol{\omega}, \mathbf{b} - \mathbf{x}_0)\big) \, e^{-i_2 \boldsymbol{\omega}\cdot\mathbf{x}_0}. \qquad (30)$$

Proof We have by using (18)

$$G_\phi\{T_{\mathbf{x}_0} f\}(\boldsymbol{\omega}, \mathbf{b}) = \frac{1}{(2\pi)^2} \int_{\mathbb{R}^2} f(\mathbf{x} - \mathbf{x}_0) \widetilde{\phi(\mathbf{x} - \mathbf{b})} \, e^{-i_2 \boldsymbol{\omega}\cdot\mathbf{x}} \, d^2\mathbf{x}. \qquad (31)$$

We substitute $\mathbf{t} = \mathbf{x} - \mathbf{x}_0$ into the above expression and get, with $d^2\mathbf{x} = d^2\mathbf{t}$,

$$\begin{aligned}
G_\phi\{T_{\mathbf{x}_0} f\}(\boldsymbol{\omega}, \mathbf{b}) &= \frac{1}{(2\pi)^2} \int_{\mathbb{R}^2} f(\mathbf{t}) \{\phi(\mathbf{t} - (\mathbf{b} - \mathbf{x}_0))\}^{\sim} e^{-i_2 \boldsymbol{\omega}\cdot(\mathbf{t}+\mathbf{x}_0)} \, d^2\mathbf{t} \\
&= \frac{1}{(2\pi)^2} \int_{\mathbb{R}^2} \big[f(\mathbf{t}) \{\phi(\mathbf{t} - (\mathbf{b} - \mathbf{x}_0))\}^{\sim} e^{-i_2 \boldsymbol{\omega}\cdot\mathbf{t}}\big] \, d^2\mathbf{t} \, e^{-i_2 \boldsymbol{\omega}\cdot\mathbf{x}_0}.
\end{aligned} \qquad (32)$$

This ends the proof of (30). □

Theorem 7 (Shift in frequency domain, modulation) *Let ϕ be a parabivector-valued Clifford window function. If $\boldsymbol{\omega}_0 \in \mathbb{R}^2$ and $f_0(\mathbf{x}) = f(\mathbf{x}) e^{i_2 \boldsymbol{\omega}_0\cdot\mathbf{x}}$, then*

$$G_\phi f_0(\boldsymbol{\omega}, \mathbf{b}) = G_\phi f(\boldsymbol{\omega} - \boldsymbol{\omega}_0, \mathbf{b}). \qquad (33)$$

Proof Using Definition 4 and simplifying it, we get

$$\begin{aligned}
G_\phi f_0(\boldsymbol{\omega}, \mathbf{b}) &= \frac{1}{(2\pi)^2} \int_{\mathbb{R}^2} f(\mathbf{x}) e^{i_2 \boldsymbol{\omega}_0\cdot\mathbf{x}} \widetilde{\phi(\mathbf{x} - \mathbf{b})} \, e^{-i_2 \boldsymbol{\omega}\cdot\mathbf{x}} \, d^2\mathbf{x} \\
&= \frac{1}{(2\pi)^2} \int_{\mathbb{R}^2} f(\mathbf{x}) \widetilde{\phi(\mathbf{x} - \mathbf{b})} \, e^{-i_2 (\boldsymbol{\omega} - \boldsymbol{\omega}_0)\cdot\mathbf{x}} \, d^2\mathbf{x}, \qquad (34)
\end{aligned}$$

which proves the theorem. □

Theorem 8 (Reconstruction formula) *Let ϕ be a Clifford window function. Then every 2D Clifford signal $f \in L^2(\mathbb{R}^2; \mathcal{G}_2)$ can be fully reconstructed by*

$$f(\mathbf{x}) = (2\pi)^2 \int_{\mathbb{R}^2}\int_{\mathbb{R}^2} G_\phi f(\boldsymbol{\omega}, \mathbf{b}) \phi_{\boldsymbol{\omega},\mathbf{b}}(\mathbf{x}) \big(\tilde{\phi}, \tilde{\phi}\big)^{-1}_{L^2(\mathbb{R}^2;\mathcal{G}_2)} \, d^2\mathbf{b}\, d^2\boldsymbol{\omega}. \tag{35}$$

Proof It follows from the CWFT defined by (18) that

$$G_\phi f(\boldsymbol{\omega}, \mathbf{b}) = \frac{1}{(2\pi)^2} \mathcal{F}\big\{ f(\mathbf{x}) \widetilde{\phi(\mathbf{x}-\mathbf{b})} \big\}(\boldsymbol{\omega}). \tag{36}$$

Taking the inverse CFT of both sides of (36), we obtain

$$\begin{aligned} f(\mathbf{x}) \widetilde{\phi(\mathbf{x}-\mathbf{b})} &= (2\pi)^2 \mathcal{F}^{-1}\big\{ G_\phi f(\boldsymbol{\omega}, \mathbf{b}) \big\}(\mathbf{x}) \\ &= \frac{(2\pi)^2}{(2\pi)^2} \int_{\mathbb{R}^2} G_\phi f(\boldsymbol{\omega}, \mathbf{b})\, e^{i_2 \boldsymbol{\omega}\cdot\mathbf{x}} \, d^2\boldsymbol{\omega}. \end{aligned} \tag{37}$$

Multiplying both sides of (37) by $\phi(\mathbf{x}-\mathbf{b})$ and then integrating with respect to $d^2\mathbf{b}$, we get

$$f(\mathbf{x}) \int_{\mathbb{R}^2} \widetilde{\phi(\mathbf{x}-\mathbf{b})} \phi(\mathbf{x}-\mathbf{b})\, d^2\mathbf{b} = \int_{\mathbb{R}^2}\int_{\mathbb{R}^2} G_\phi f(\boldsymbol{\omega}, \mathbf{b})\, e^{i_2 \boldsymbol{\omega}\cdot\mathbf{x}} \phi(\mathbf{x}-\mathbf{b})\, d^2\boldsymbol{\omega}\, d^2\mathbf{b} \tag{38}$$

or, equivalently,

$$f(\mathbf{x}) \big(\tilde{\phi}, \tilde{\phi}\big)_{L^2(\mathbb{R}^2;\mathcal{G}_2)} = (2\pi)^2 \int_{\mathbb{R}^2}\int_{\mathbb{R}^2} G_\phi f(\boldsymbol{\omega}, \mathbf{b})\, \phi_{\boldsymbol{\omega},\mathbf{b}}(\mathbf{x})\, d^2\boldsymbol{\omega}\, d^2\mathbf{b}, \tag{39}$$

which gives (35). □

It is worth noting here that if the Clifford window function is a parabivector-valued function, then the reconstruction formula (35) can be written in the form

$$f(\mathbf{x}) = \frac{(2\pi)^2}{\|\phi\|^2_{L^2(\mathbb{R}^2;\mathcal{G}_2)}} \int_{\mathbb{R}^2}\int_{\mathbb{R}^2} G_\phi f(\boldsymbol{\omega}, \mathbf{b}) \phi_{\boldsymbol{\omega},\mathbf{b}}(\mathbf{x})\, d^2\mathbf{b}\, d^2\boldsymbol{\omega}. \tag{40}$$

Theorem 9 (Orthogonality relation) *Assume that the Clifford window function ϕ is a parabivector-valued function. If two Clifford functions $f, g \in L^2(\mathbb{R}^2; \mathcal{G}_2)$, then we have*

$$\begin{aligned} &\int_{\mathbb{R}^2}\int_{\mathbb{R}^2} (f, \phi_{\boldsymbol{\omega},\mathbf{b}})_{L^2(\mathbb{R}^2;\mathcal{G}_2)} \widetilde{(g, \phi_{\boldsymbol{\omega},\mathbf{b}})_{L^2(\mathbb{R}^2;\mathcal{G}_2)}}\, d^2\boldsymbol{\omega}\, d^2\mathbf{b} \\ &\quad = \frac{\|\phi\|^2_{L^2(\mathbb{R}^2;\mathcal{G}_2)}}{(2\pi)^2} (f, g)_{L^2(\mathbb{R}^2;\mathcal{G}_2)}. \end{aligned} \tag{41}$$

Proof By inserting (18) into the left side of (41), we obtain

$$\int_{\mathbb{R}^2}\int_{\mathbb{R}^2}(f,\phi_{\boldsymbol{\omega},\mathbf{b}})_{L^2(\mathbb{R}^2;\mathcal{G}_2)}\widetilde{(g,\phi_{\boldsymbol{\omega},\mathbf{b}})}_{L^2(\mathbb{R}^2;\mathcal{G}_2)}\,d^2\boldsymbol{\omega}\,d^2\mathbf{b}$$

$$=\int_{\mathbb{R}^2}\int_{\mathbb{R}^2}(f,\phi_{\boldsymbol{\omega},\mathbf{b}})_{L^2(\mathbb{R}^2;\mathcal{G}_2)}\left(\int_{\mathbb{R}^2}\frac{1}{(2\pi)^2}e^{i_2\boldsymbol{\omega}\cdot\mathbf{x}}\phi(\mathbf{x}-\mathbf{b})\widetilde{g(\mathbf{x})}\,d^2\mathbf{x}\right)d^2\boldsymbol{\omega}\,d^2\mathbf{b}$$

$$=\int_{\mathbb{R}^2}\int_{\mathbb{R}^2}\left(\int_{\mathbb{R}^2}\int_{\mathbb{R}^2}\frac{1}{(2\pi)^4}f(\mathbf{x}')\,\widetilde{\phi(\mathbf{x}'-\mathbf{b})}\,e^{i_2\boldsymbol{\omega}\cdot(\mathbf{x}-\mathbf{x}')}d^2\boldsymbol{\omega}\,d^2\mathbf{x}'\right)$$
$$\times\,\phi(\mathbf{x}-\mathbf{b})\widetilde{g(\mathbf{x})}\,d^2\mathbf{x}\,d^2\mathbf{b}$$

$$=\frac{1}{(2\pi)^2}\int_{\mathbb{R}^2}\int_{\mathbb{R}^2}\left(\int_{\mathbb{R}^2}f(\mathbf{x}')\widetilde{\phi(\mathbf{x}'-\mathbf{b})}\delta(\mathbf{x}-\mathbf{x}')\phi(\mathbf{x}-\mathbf{b})\,d^2\mathbf{x}'\right)\widetilde{g(\mathbf{x})}\,d^2\mathbf{b}\,d^2\mathbf{x}$$

$$=\frac{1}{(2\pi)^2}\int_{\mathbb{R}^2}f(\mathbf{x})\int_{\mathbb{R}^2}\underbrace{\widetilde{\phi(\mathbf{x}-\mathbf{b})}\phi(\mathbf{x}-\mathbf{b})}_{\phi\text{ parabiv. funct.}}\,d^2\mathbf{b}\,\widetilde{g(\mathbf{x})}\,d^2\mathbf{x}$$

$$=\frac{1}{(2\pi)^2}\|\phi\|^2_{L^2(\mathbb{R}^2;\mathcal{G}_2)}\int_{\mathbb{R}^2}f(\mathbf{x})\widetilde{g(\mathbf{x})}\,d^2\mathbf{x},\tag{42}$$

which completes the proof of (41). □

Theorem 10 (Reproducing kernel) *For a parabivector-valued Clifford window function* $|\mathbf{x}|^{1/2}\phi\in L^2(\mathbb{R}^2;\mathcal{G}_2)$, *if*

$$\mathbb{K}_\phi\big(\boldsymbol{\omega},\mathbf{b};\boldsymbol{\omega}',\mathbf{b}'\big)=\frac{(2\pi)^2}{\|\phi\|^2_{L^2(\mathbb{R}^2;\mathcal{G}_2)}}(\phi_{\boldsymbol{\omega},\mathbf{b}},\phi_{\boldsymbol{\omega}',\mathbf{b}'})_{L^2(\mathbb{R}^2;\mathcal{G}_2)},\tag{43}$$

then $\mathbb{K}_\phi(\boldsymbol{\omega},\mathbf{b};\boldsymbol{\omega}',\mathbf{b}')$ *is a reproducing kernel, i.e.,*

$$G_\phi f\big(\boldsymbol{\omega}',\mathbf{b}'\big)=\int_{\mathbb{R}^2}\int_{\mathbb{R}^2}G_\phi f(\boldsymbol{\omega},\mathbf{b})\mathbb{K}_\phi\big(\boldsymbol{\omega},\mathbf{b};\boldsymbol{\omega}',\mathbf{b}'\big)\,d^2\boldsymbol{\omega}\,d^2\mathbf{b}.\tag{44}$$

Proof By inserting the inverse CWFT (40) into the definition of the CWFT (18) we easily obtain

$$G_\phi f\big(\boldsymbol{\omega}',\mathbf{b}'\big)$$
$$=\int_{\mathbb{R}^2}f(\mathbf{x})\widetilde{\phi_{\boldsymbol{\omega}',\mathbf{b}'}(\mathbf{x})}\,d^2\mathbf{x}$$
$$=\int_{\mathbb{R}^2}\left(\frac{(2\pi)^2}{\|\phi\|^2_{L^2(\mathbb{R}^2;\mathcal{G}_2)}}\int_{\mathbb{R}^2}\int_{\mathbb{R}^2}G_\phi f(\boldsymbol{\omega},\mathbf{b})\,\phi_{\boldsymbol{\omega},\mathbf{b}}(\mathbf{x})\,d^2\mathbf{b}\,d^2\boldsymbol{\omega}\right)\widetilde{\phi_{\boldsymbol{\omega}',\mathbf{b}'}(\mathbf{x})}\,d^2\mathbf{x}$$

$$= \int_{\mathbb{R}^2}\int_{\mathbb{R}^2} G_\phi f(\boldsymbol{\omega}, \mathbf{b}) \frac{(2\pi)^2}{\|\phi\|^2_{L^2(\mathbb{R}^2;\mathcal{G}_2)}} \left(\int_{\mathbb{R}^2} \phi_{\boldsymbol{\omega},\mathbf{b}}(\mathbf{x}) \widetilde{\phi_{\boldsymbol{\omega}',\mathbf{b}'}(\mathbf{x})}\, d^2\mathbf{x} \right) d^2\mathbf{b}\, d^2\boldsymbol{\omega}$$

$$= \int_{\mathbb{R}^2}\int_{\mathbb{R}^2} G_\phi f(\boldsymbol{\omega}, \mathbf{b}) \mathbb{K}_\phi\big(\boldsymbol{\omega}, \mathbf{b}; \boldsymbol{\omega}', \mathbf{b}'\big)\, d^2\mathbf{b}\, d^2\boldsymbol{\omega}, \tag{45}$$

which finishes the proof. □

Remark 2 Formulas (40), (41), and (43) also hold if the Clifford window function is a vector-valued function, i.e., $\phi(\mathbf{x}) = \phi_1(\mathbf{x})\mathbf{e}_1 + \phi_2(\mathbf{x})\mathbf{e}_2$.

The above properties of the CWFT are summarized in Table 1.

Table 1 Properties of the CWFT of $f, g \in L^2(\mathbb{R}^2; \mathcal{G}_2)$, $L^2 = L^2(\mathbb{R}^2; \mathcal{G}_2)$, where $\lambda, \mu \in \mathcal{G}_2$ are constants, $\omega_0 = \omega_{0,1}\mathbf{e}_1 + \omega_{0,2}\mathbf{e}_2 \in \mathbb{R}^2$, and $\mathbf{x}_0 = x_0\mathbf{e}_1 + y_0\mathbf{e}_2 \in \mathbb{R}^2$

Property	Clifford-valued function	2D CWFT
Left linearity	$\lambda f(x) + \mu g(x)$	$\lambda G_\phi f(\boldsymbol{\omega}, \mathbf{b}) + \mu G_\phi g(\boldsymbol{\omega}, \mathbf{b})$
Delay	$f(\mathbf{x} - \mathbf{x}_0)$	$(G_\phi f(\boldsymbol{\omega}, \mathbf{b} - \mathbf{x}_0))\, e^{-i_2\boldsymbol{\omega}\cdot\mathbf{x}_0}$
Modulation	$f(\mathbf{x})e^{i_2\boldsymbol{\omega}_0\cdot\mathbf{x}}$	$G_\phi f(\boldsymbol{\omega} - \boldsymbol{\omega}_0, \mathbf{b})$ if ϕ is a parabivector-valued function
Formulas		
Reversion	$G_{\tilde{\phi}} \tilde{f}(\boldsymbol{\omega}, \mathbf{b}) =$	$\{G_\phi f(-\boldsymbol{\omega}, \mathbf{b})\}^\sim$ if f and ϕ are parabivector-valued functions
Switching	$G_\phi f(\boldsymbol{\omega}, \mathbf{b}) =$	$e^{-i_2\boldsymbol{\omega}\cdot\mathbf{b}}\{G_f \phi(-\boldsymbol{\omega}, -\mathbf{b})\}^\sim$ if f and ϕ are parabivector-valued functions
Parity	$G_{P\phi}\{Pf\}(\boldsymbol{\omega}, \mathbf{b}) =$	$G_\phi f(-\boldsymbol{\omega}, -\mathbf{b})$
Orthogonality	$\frac{1}{(2\pi)^2}\|\phi\|^2_{L^2}(f, g)_{L^2} =$	$\int_{\mathbb{R}^2}\int_{\mathbb{R}^2} (f, \phi_{\boldsymbol{\omega},\mathbf{b}})_{L^2(\mathbb{R}^2;\mathcal{G}_2)} \times \widetilde{(g, \phi_{\boldsymbol{\omega},\mathbf{b}})}_{L^2(\mathbb{R}^2;\mathcal{G}_2)} d^2\boldsymbol{\omega}\, d^2\mathbf{b}$ if ϕ is a parabivector-valued function
Reconstruction	$f(\mathbf{x}) =$	$(2\pi)^2 \int_{\mathbb{R}^2}\int_{\mathbb{R}^2} G_\phi f(\boldsymbol{\omega}, \mathbf{b})\phi_{\boldsymbol{\omega},\mathbf{b}}(\mathbf{x}) \times (\tilde{\phi}, \tilde{\phi})^{-1}_{L^2(\mathbb{R}^2;\mathcal{G}_2)} d^2\mathbf{b}\, d^2\boldsymbol{\omega}$, $\frac{(2\pi)^2}{\|\phi\|^2_{L^2(\mathbb{R}^2;\mathcal{G}_2)}} \int_{\mathbb{R}^2}\int_{\mathbb{R}^2} G_\phi f(\boldsymbol{\omega}, \mathbf{b}) \times \phi_{\boldsymbol{\omega},\mathbf{b}}(\mathbf{x})\, d^2\mathbf{b} d^2\boldsymbol{\omega}$, if ϕ is a parabivector-valued function
Reproducing kernel	$G_\phi f(\boldsymbol{\omega}', \mathbf{b}') =$	$\int_{\mathbb{R}^2}\int_{\mathbb{R}^2} G_\phi f(\boldsymbol{\omega}, \mathbf{b})\mathbb{K}_\phi(\boldsymbol{\omega}, \mathbf{b}; \boldsymbol{\omega}', \mathbf{b}')\, d^2\boldsymbol{\omega}\, d^2\mathbf{b}$, $\mathbb{K}_\phi(\boldsymbol{\omega}, \mathbf{b}; \boldsymbol{\omega}', \mathbf{b}') = \frac{(2\pi)^2}{\|\phi\|^2_{L^2(\mathbb{R}^2;\mathcal{G}_2)}} (\phi_{\boldsymbol{\omega},\mathbf{b}}, \phi_{\boldsymbol{\omega}',\mathbf{b}'})_{L^2(\mathbb{R}^2;\mathcal{G}_2)}$, if ϕ is a parabivector-valued function

4.3 Examples of the CWFT

For illustrative purposes, we shall discuss examples of the CWFT. We then compute their energy densities.

Example 1 Consider the Clifford–Gabor filters (see Fig. 1) defined by ($\sigma_1 = \sigma_2 = 1/\sqrt{2}$)

$$f(\mathbf{x}) = \frac{1}{(2\pi)^2} e^{-\mathbf{x}^2 + i_2\omega_0\cdot\mathbf{x}}. \tag{46}$$

Obtain the CWFT of f with respect to the Gaussian window function $\phi(\mathbf{x}) = e^{-\mathbf{x}^2}$.

By the definition of the CWFT (18), we have

$$G_\phi f(\boldsymbol{\omega}, \mathbf{b}) = \frac{1}{(2\pi)^4} \int_{\mathbb{R}^2} e^{-\mathbf{x}^2 + i_2\omega_0\cdot\mathbf{x}} e^{-(\mathbf{x}-\mathbf{b})^2} e^{-i_2\boldsymbol{\omega}\cdot\mathbf{x}}\, d^2\mathbf{x}. \tag{47}$$

Substituting $\mathbf{x} = \mathbf{y} + \mathbf{b}/2$, we can rewrite (47) as

$$\begin{aligned}
G_\phi f(\boldsymbol{\omega}, \mathbf{b}) &= \frac{1}{(2\pi)^4} \int_{\mathbb{R}^2} e^{-(\mathbf{y}+\mathbf{b}/2)^2 + i_2\omega_0\cdot(\mathbf{y}+\mathbf{b}/2)} e^{-(\mathbf{y}-\mathbf{b}/2)^2} e^{-i_2\boldsymbol{\omega}\cdot(\mathbf{y}+\mathbf{b}/2)}\, d^2\mathbf{y} \\
&= \frac{e^{-\mathbf{b}^2/2}}{(2\pi)^4} \int_{\mathbb{R}^2} e^{-2\mathbf{y}^2} e^{-i_2\boldsymbol{\omega}\cdot\mathbf{y}} e^{i_2\omega_0\cdot\mathbf{y}}\, d^2\mathbf{y}\, e^{-i_2(\boldsymbol{\omega}-\omega_0)\cdot\mathbf{b}/2} \\
&= \frac{e^{-\mathbf{b}^2/2}}{(2\pi)^4} \int_{\mathbb{R}^2} e^{-2\mathbf{y}^2} e^{-i_2(\boldsymbol{\omega}-\omega_0)\cdot\mathbf{y}}\, d^2\mathbf{y}\, e^{-i_2(\boldsymbol{\omega}-\omega_0)\cdot\mathbf{b}/2} \\
&= \frac{e^{-\mathbf{b}^2/2}}{(2\pi)^4} \frac{\pi}{2} e^{-(\boldsymbol{\omega}-\omega_0)^2/8} e^{-i_2(\boldsymbol{\omega}-\omega_0)\cdot\mathbf{b}/2} \\
&= \frac{e^{-\mathbf{b}^2/2}}{32\pi^3} e^{-(\boldsymbol{\omega}-\omega_0)^2/8} e^{-i_2(\boldsymbol{\omega}-\omega_0)\cdot\mathbf{b}/2}.
\end{aligned} \tag{48}$$

The energy density is given by

$$\left|G_\phi f(\boldsymbol{\omega}, \mathbf{b})\right|^2 = \frac{e^{-\mathbf{b}^2}}{(32\pi^3)^2} e^{-(\boldsymbol{\omega}-\omega_0)^2/4}. \tag{49}$$

Example 2 Consider the first-order two-dimensional B-spline window function defined by

$$\phi(\mathbf{x}) = \begin{cases} 1 & \text{if } 0 \le x_1 \le 1 \text{ and } 0 \le x_2 \le 1, \\ 0 & \text{otherwise.} \end{cases} \tag{50}$$

Obtain the CWFT of the function defined as follows:

$$f(\mathbf{x}) = \begin{cases} \mathbf{x} & \text{if } 0 \le x_1 \le 1 \text{ and } 0 \le x_2 \le 1, \\ 0 & \text{otherwise.} \end{cases} \tag{51}$$

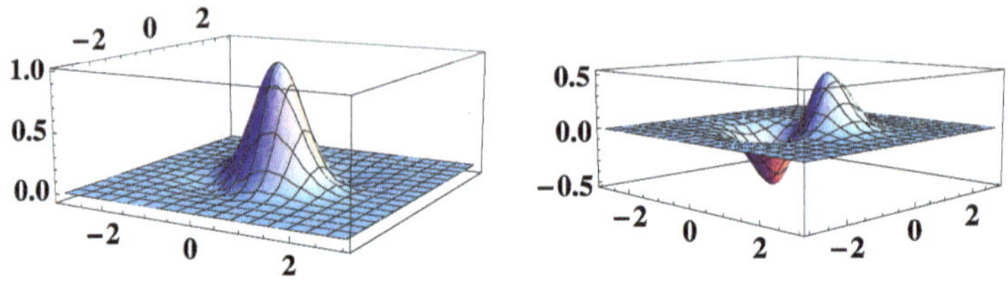

Fig. 1 The scalar part (*left*) and bivector part (*right*) of Clifford–Gabor filter for the parameters $\omega_{0,1} = \omega_{0,2} = 1, b_1 = b_2 = 0, \sigma_1 = \sigma_2 = 1/\sqrt{2}$ in the spatial domain using Mathematica 6.0

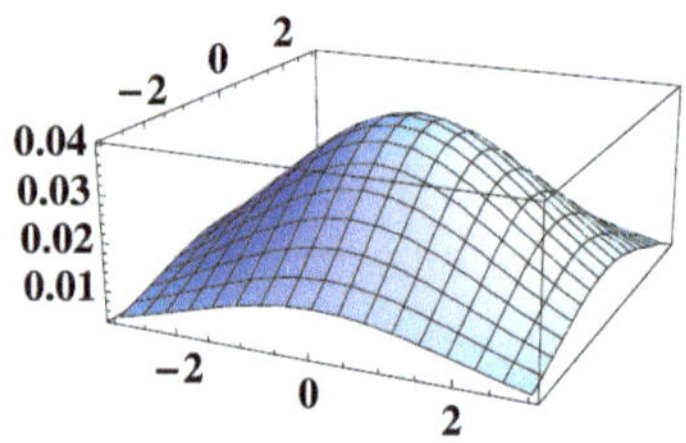

Fig. 2 Plot of the CWFT of Clifford–Gabor filter of Example 1 using Mathematica 6.0. Note that it is scalar valued for the parameters $b_1 = b_2 = 0$

Applying Definition 4 and simplifying it, we obtain

$$\begin{aligned}
&G_\phi f(\boldsymbol{\omega}, \mathbf{b}) \\
&= \frac{1}{(2\pi)^2} \int_{b_1}^{1+b_1} \int_{b_2}^{1+b_2} \mathbf{x}\, e^{-i_2 \boldsymbol{\omega}\cdot\mathbf{x}}\, dx_1\, dx_2 \\
&= \frac{1}{(2\pi)^2} \int_{b_1}^{1+b_1} \int_{b_2}^{1+b_2} (x_1\mathbf{e}_1 + x_2\mathbf{e}_2)\left(e^{-i_2\omega_1 x_1} e^{-i_2\omega_2 x_2}\right) dx_1\, dx_2 \\
&= \frac{1}{(2\pi)^2} \mathbf{e}_1 \int_{b_1}^{1+b_1} x_1 e^{-i_2\omega_1 x_1}\, dx_1 \int_{b_2}^{1+b_2} e^{-i_2\omega_2 x_2}\, dx_2 \\
&\quad + \frac{1}{(2\pi)^2} \mathbf{e}_2 \int_{b_1}^{1+b_1} e^{-i_2\omega_1 x_1} dx_1 \int_{b_2}^{1+b_2} x_2 e^{-i_2\omega_2 x_2}\, dx_2 \\
&= \big\{\mathbf{e}_2\omega_2\big[(1 + i_2\omega_1 b_1)\big(e^{-i_2\omega_1} - 1\big) + i_2\omega_1 e^{-i_2\omega_1}\big]\big(e^{-i_2\omega_2} - 1\big) \\
&\quad - \mathbf{e}_1\omega_1\big[(1 + i_2\omega_2 b_2)\big(e^{-i_2\omega_2} - 1\big) + i_2\omega_2 e^{-i_2\omega_2}\big]\big(e^{-i_2\omega_1} - 1\big)\big\} \\
&\quad \times \frac{e^{-i_2\boldsymbol{\omega}\cdot\mathbf{b}}}{(2\pi\omega_1\omega_2)^2}
\end{aligned} \tag{52}$$

with

$$\mathbf{b} = b_1\mathbf{e}_1 + b_2\mathbf{e}_2.$$

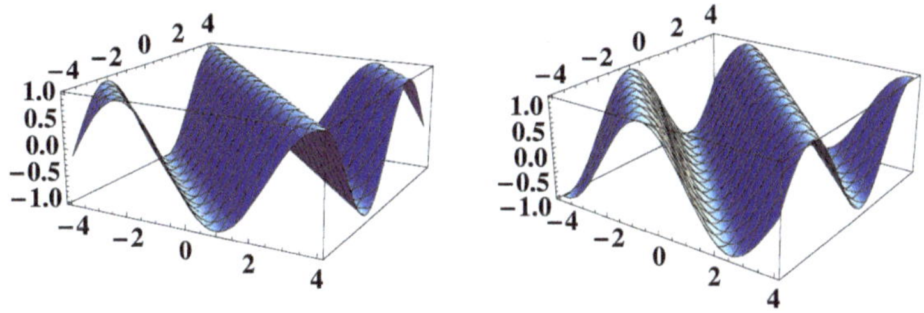

Fig. 3 Representation of the CFT basis for $\omega_1 = \omega_2 = 1$ with scalar part (*left*) and bivector part (*right*) using Mathematica 6.0

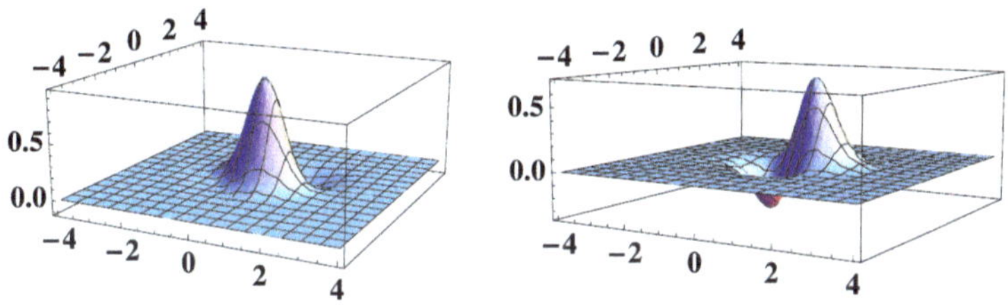

Fig. 4 Representation of the CWFT basis of a Gaussian window function for the parameters $\omega_{0,1} = \omega_{0,2} = 1, b_1 = b_2 = 0.2$ with scalar part (*left*) and bivector part (*right*) using Mathematica 6.0

5 Comparison of CFT and CWFT

Since the Clifford–Fourier kernel $e^{-i_2\boldsymbol{\omega}\cdot\mathbf{x}}$ is a global function, the CFT basis has an infinite spatial extension as shown in Fig. 3. In contrast, the CWFT basis $\phi(\mathbf{x}-\mathbf{b})\, e^{-i_2\boldsymbol{\omega}\cdot\mathbf{x}}$ has a limited spatial extension due to the local Clifford window function $\phi(\mathbf{x}-\mathbf{b})$ (see Fig. 4). This means that the CFT analysis cannot provide information about the signal with respect to position and frequency, so that we need the CWFT to fully describe the characteristics of the signal simultaneously in both spatial and frequency domains.

6 Conclusion

Using the basic concepts of Clifford geometric algebra and the CFT, we introduced the CWFT. Important properties of the CWFT were demonstrated. This generalization enables us to work with 2D Clifford–Gabor filters, which can extend the applications of the 2D complex Gabor filters.

Because the CWFT represents a signal in a joint space–frequency domain, it can be applied in many fields of science and engineering, such as image analysis and image compression, object and pattern recognition, computer vision, optics and filter banks.

Acknowledgements The authors would like to thank the referees for critically reading the manuscript. The authors further acknowledge R.U. Gobithaasan's assistance in producing the figures using Mathematica 6.0. This research was supported by the Malaysian Research Grant (Fundamental Research Grant Scheme) from the Universiti Sains Malaysia.

References

1. Brackx, F., De Schepper, N., Sommen, F.: The two-dimensional Clifford–Fourier transform. J. Math. Imaging Vis. **26**(1), 5–18 (2006)
2. Bülow, T.: Hypercomplex spectral signal representations for the processing and analysis of images. Ph.D. thesis, University of Kiel, Germany (1999)
3. Bülow, T., Felsberg, M., Sommer, G.: Non-commutative hypercomplex Fourier transforms of multidimensional signals. In: Sommer, G. (ed.) Geom. Comp. with Cliff. Alg., Theor. Found. and Appl. in Comp. Vision and Robotics, pp. 187–207. Springer, Berlin (2001)
4. Gröchenig, K., Zimmermann, G.: Hardy's Theorem and the short-time Fourier transform of Schwartz functions. J. Lond. Math. Soc. **2**(63), 205–214 (2001)
5. Hitzer, E., Mawardi, B.: Clifford Fourier transform on multivector fields and uncertainty principle for dimensions $n = 2$ (mod 4) and $n = 3$ (mod 4). Adv. Appl. Clifford Algebr. **18**(3–4), 715–736 (2008)
6. Kemao, Q.: Two-dimensional windowed Fourier transform for fringe pattern analysis: principles, applications, and implementations. Opt. Laser Eng. **45**, 304–317 (2007)
7. Weisz, F.: Multiplier theorems for the short-time Fourier transform. Integr. Equ. Oper. Theory **60**(1), 133–149 (2008)
8. Zhong, J., Zeng, H.: Multiscale windowed Fourier transform for phase extraction of fringe pattern. Appl. Opt. **46**(14), 2670–2675 (2007)

The Cylindrical Fourier Transform

Fred Brackx, Nele De Schepper, and Frank Sommen

Abstract The aim of this paper is to show the application potential of the cylindrical Fourier transform, which was recently devised as a new integral transform within the context of Clifford analysis. Next to the approximation approach where, using density arguments, the spectrum of various types of functions and distributions may be calculated starting from the cylindrical Fourier images of the L_2-basis functions in $\mathbb{R}^m$, direct computation methods are introduced for specific distributions supported on the unit sphere, and an illustrative example is worked out.

1 Introduction

The Fourier transform is by far the most important integral transform. Since its introduction by Fourier in the early 1800s, it has remained an indispensable and stimulating mathematical concept that is at the core of the highly evolved branch of mathematics called Fourier analysis.

The second subject of great relevance for the paper is Clifford analysis, an elegant and powerful higher-dimensional generalization of the theory of holomorphic functions, which is moreover closely related but complementary to harmonic analysis. Clifford analysis also offers the possibility to generalize one-dimensional mathematical analysis to higher dimension in a rather natural way by encompassing all dimensions at once, as opposed to the usual tensorial approaches.

It is precisely this last qualification which has been exploited in [2] and [3] to construct a genuine multidimensional Fourier transform within the context of Clifford analysis. This so-called Clifford–Fourier transform is briefly discussed in Sect. 3.

In [4] and [5] we devised and thoroughly studied the so-called cylindrical Fourier transform within the Clifford analysis setting. The idea is the following: for a fixed

N. De Schepper (✉)
Clifford Research Group, Department of Mathematical Analysis, Ghent University, Galglaan 2, 9000 Gent, Belgium
e-mail: nds@cage.ugent.be

E. Bayro-Corrochano, G. Scheuermann (eds.), *Geometric Algebra Computing*,
DOI 10.1007/978-1-84996-108-0_6,

vector in the image space, the level surfaces of the traditional Fourier kernel are planes perpendicular to that fixed vector. For this Fourier kernel, we now substitute a new Clifford–Fourier kernel such that, again for a fixed vector in the image space, its phase is constant on co-axial cylinders w.r.t. that fixed vector. The point is that, when restricting to dimension two, this new cylindrical Fourier transform coincides with the earlier introduced Clifford–Fourier transform. We are now faced with the following situation: in dimension greater than two we have a first Clifford–Fourier transform with elegant properties but no kernel in closed form, and a second cylindrical one with a kernel in closed form but more complicated calculation formulae. In dimension two both transforms coincide.

The aim of this paper is to show the application potential of the cylindrical Fourier transform.

To make the paper self-contained, we have also included an introductory section (Sect. 2) on Clifford analysis.

2 The Clifford Analysis Toolkit

Clifford analysis (see, e.g., [1]) offers a function theory which is a higher-dimensional analogue of the theory of the holomorphic functions of one complex variable.

The functions considered are defined in $\mathbb{R}^m$ $(m > 1)$ and take their values in the Clifford algebra $\mathbb{R}_{0,m}$ or its complexification $\mathbb{C}_m = \mathbb{R}_{0,m} \otimes \mathbb{C}$. If $(e_1, \ldots, e_m)$ is an orthonormal basis of $\mathbb{R}^m$, then a basis for the Clifford algebra $\mathbb{R}_{0,m}$ or $\mathbb{C}_m$ is given by all possible products of basis vectors $(e_A : A \subset \{1, \ldots, m\})$, where $e_\emptyset = 1$ is the identity element. The noncommutative multiplication in the Clifford algebra is governed by the rules $e_j e_k + e_k e_j = -2\delta_{j,k}$ $(j, k = 1, \ldots, m)$.

Conjugation is defined as the anti-involution for which $\overline{e_j} = -e_j$ $(j = 1, \ldots, m)$. In case of $\mathbb{C}_m$, the Hermitian conjugate of an element $\lambda = \sum_A \lambda_A e_A$ $(\lambda_A \in \mathbb{C})$ is defined by $\lambda^\dagger = \sum_A \lambda_A^c \, \overline{e_A}$, where λ_A^c denotes the complex conjugate of λ_A. This Hermitian conjugation leads to a Hermitian inner product and its associated norm on $\mathbb{C}_m$ given respectively by

$$(\lambda, \mu) = \left[\lambda^\dagger \mu\right]_0 \quad \text{and} \quad |\lambda|^2 = \left[\lambda^\dagger \lambda\right]_0 = \sum_A |\lambda_A|^2,$$

where $[\lambda]_0$ denotes the scalar part of the Clifford element λ.

The Euclidean space $\mathbb{R}^m$ is embedded in the Clifford algebras $\mathbb{R}_{0,m}$ and $\mathbb{C}_m$ by identifying the point $(x_1, \ldots, x_m)$ with the vector variable $\underline{x}$ given by $\underline{x} = \sum_{j=1}^m e_j x_j$. The product of two vectors splits up into a scalar part (the inner product up to a minus sign) and a so-called bivector part (the wedge product):

$$\underline{x}\,\underline{y} = \underline{x} . \underline{y} + \underline{x} \wedge \underline{y},$$

where

$$\underline{x}.\underline{y} = -\langle \underline{x}, \underline{y} \rangle = -\sum_{j=1}^{m} x_j y_j \quad \text{and} \quad \underline{x} \wedge \underline{y} = \sum_{i=1}^{m} \sum_{j=i+1}^{m} e_i e_j (x_i y_j - x_j y_i).$$

Note that the square of a vector variable $\underline{x}$ is scalar-valued and equals the norm squared up to a minus sign: $\underline{x}^2 = -\langle \underline{x}, \underline{x} \rangle = -|\underline{x}|^2$.

The central notion in Clifford analysis is the notion of monogenicity, a notion which is the multidimensional counterpart to that of holomorphy in the complex plane. A function $F(x_1, \ldots, x_m)$ defined and continuously differentiable in an open region of $\mathbb{R}^m$ and taking values in $\mathbb{R}_{0,m}$ or $\mathbb{C}_m$ is called left monogenic in that region if $\partial_{\underline{x}}[F] = 0$. Here $\partial_{\underline{x}}$ is the Dirac operator in $\mathbb{R}^m$: $\partial_{\underline{x}} = \sum_{j=1}^{m} e_j \partial_{x_j}$, an elliptic, rotation-invariant, vector differential operator of the first order, which may be looked upon as the "square root" of the Laplace operator in $\mathbb{R}^m$: $\Delta_m = -\partial_{\underline{x}}^2$. This factorization of the Laplace operator establishes a special relationship between Clifford analysis and harmonic analysis in that monogenic functions refine the properties of harmonic functions.

In the sequel the monogenic homogeneous polynomials will play an important role. A left-monogenic homogeneous polynomial P_k of degree k ($k \geq 0$) in $\mathbb{R}^m$ is called a left solid inner spherical monogenic of order k. The set of all left solid inner spherical monogenics of order k will be denoted by $M_\ell^+(k)$. The dimension of $M_\ell^+(k)$ is given by

$$\dim\bigl(M_\ell^+(k)\bigr) = \binom{m+k-2}{m-2} = \frac{(m+k-2)!}{(m-2)!k!}.$$

The set

$$\phi_{s,k,j}(\underline{x}) = \frac{2^{m/4}}{(\gamma_{s,k})^{1/2}} H_{s,k}\bigl(\sqrt{2}\underline{x}\bigr) P_k^{(j)}\bigl(\sqrt{2}\underline{x}\bigr) e^{(-|\underline{x}|^2/2)}, \tag{1}$$

$s, k \in \mathbb{N},\ j \leq \dim(M_\ell^+(k))$, constitutes an orthonormal basis for the space $L_2(\mathbb{R}^m)$ of square-integrable functions. Here $\{P_k^{(j)}(\underline{x});\ j \leq \dim(M_\ell^+(k))\}$ denotes an orthonormal basis of $M_\ell^+(k)$, and $\gamma_{s,k}$ a real constant depending on the parity of s. The polynomials $H_{s,k}(\underline{x})$ are the so-called generalized Clifford–Hermite polynomials introduced by Sommen; they are a multidimensional generalization to Clifford analysis of the classical Hermite polynomials on the real line. Note that $H_{s,k}(\underline{x})$ is a polynomial of degree s in the variable $\underline{x}$ with real coefficients depending on k. Furthermore, $H_{2s,k}(\underline{x})$ only contains even powers of $\underline{x}$ and is hence scalar valued, while $H_{2s+1,k}(\underline{x})$ only contains odd ones and is thus vector valued.

A result, which will be frequently used in Sect. 4.3, is the following generalization of the classical Funk–Hecke theorem.

Theorem 1 (Funk–Hecke theorem in space) *Let S_k be a spherical harmonic of degree k, and $\underline{\eta}$ a fixed point on the unit sphere S^{m-1} in $\mathbb{R}^m$. Denote $\langle \underline{\omega}, \underline{\eta} \rangle =$*

$\cos(\widehat{\underline{\omega}, \underline{\eta}}) = t_\eta$ *for* $\underline{\omega} \in S^{m-1}$. *Then*

$$\int_{\mathbb{R}^m} g(r) f(t_\eta) S_k(\underline{\omega})\, dV(\underline{x})$$

$$= A_{m-1}\left(\int_0^{+\infty} g(r) r^{m-1}\, dr\right)\left(\int_{-1}^{1} f(t)\left(1-t^2\right)^{(m-3)/2} P_{k,m}(t)\, dt\right) S_k(\underline{\eta}),$$

where $dV(\underline{x})$ *denotes the Lebesgue measure on* $\mathbb{R}^m$, $P_{k,m}(t)$ *the Legendre polynomial of degree* k *in the* m*-dimensional Euclidean space, and* $A_{m-1} = \frac{2\pi^{(m-1)/2}}{\Gamma(\frac{m-1}{2})}$ *the surface area of the unit sphere* S^{m-2} *in* $\mathbb{R}^{m-1}$.

As the Legendre polynomials are even or odd according to the parity of k, we can also state the following corollary.

Corollary 1 *Let* S_k *be a spherical harmonic of degree* k, *and* $\underline{\eta}$ *a fixed point on the unit sphere* S^{m-1}. *Denote* $\langle \underline{\omega}, \underline{\eta} \rangle = t_\eta$ *for* $\underline{\omega} \in S^{m-1}$. *Then the 3D-integral*

$$\int_{\mathbb{R}^m} g(r) f(t_\eta) S_k(\underline{\omega})\, dV(\underline{x})$$

is zero whenever

- f *is an odd function, and* k *is even;*
- f *is an even function, and* k *is odd.*

3 The Clifford–Fourier Transform

In [2] a new multidimensional Fourier transform in the framework of Clifford analysis, the so-called Clifford–Fourier transform, is introduced. The idea behind its definition originates from an alternative representation for the standard tensorial multidimensional Fourier transform given by

$$\mathcal{F}[f](\underline{\xi}) = \frac{1}{(2\pi)^{m/2}} \int_{\mathbb{R}^m} e^{(-i\langle \underline{x}, \underline{\xi} \rangle)} f(\underline{x})\, dV(\underline{x}).$$

It is indeed so that this classical Fourier transform can be seen as the operator exponential

$$\mathcal{F} = e^{(-i\pi/2\mathcal{H})} = \sum_{k=0}^{\infty} \frac{1}{k!}\left(-i\frac{\pi}{2}\right)^k \mathcal{H}^k,$$

where $\mathcal{H}$ is the scalar-valued differential operator $\mathcal{H} = \frac{1}{2}(-\Delta_m + r^2 - m)$. Note that, due to the scalar character of the standard Fourier kernel, the Fourier spectrum inherits its Clifford algebra character from the original signal, without any interaction with the Fourier kernel. So in order to genuinely introduce the Clifford analysis character in the Fourier transform, the idea occurred to us to replace the scalar-valued operator $\mathcal{H}$ in the operator exponential by a Clifford algebra-valued one. To that end, we aimed at factorizing the operator $\mathcal{H}$, making use of the factorization of the Laplace operator by the Dirac operator. Splitting $\mathcal{H}$ into a sum of Clifford algebra-valued second-order operators leads in a natural way to a *pair* of transforms $\mathcal{F}_{\mathcal{H}^\pm}$, the harmonic average of which is precisely the standard Fourier transform $\mathcal{F}$, i.e., $\mathcal{F}^2 = \mathcal{F}_{\mathcal{H}^+}\mathcal{F}_{\mathcal{H}^-}$.

The two-dimensional case of this Clifford–Fourier transform is special in that we are able to find a closed form for the kernel of the integral representation. Indeed, the two-dimensional Clifford–Fourier transform takes the form

$$\mathcal{F}_{\mathcal{H}^\pm}[f](\underline{\xi}) = \frac{1}{2\pi} \int_{\mathbb{R}^2} e^{(\pm(\underline{\xi} \wedge \underline{x}))} f(\underline{x})\, dV(\underline{x}).$$

This closed form enables us to generalize the well-known results for the standard Fourier transform both in the L_1- and L_2-contexts (see [3]). Note that we have not succeeded yet in obtaining such a closed form in arbitrary dimension. For a detailed account of the Clifford–Fourier transform, we refer the reader to the survey paper [5].

4 The Cylindrical Fourier Transform

4.1 Definition

The cylindrical Fourier transform is obtained by taking the multidimensional generalization of the two-dimensional Clifford–Fourier kernel.

Definition 1 The cylindrical Fourier transform of a function f is given by

$$\mathcal{F}_{\mathrm{cyl}}[f](\underline{\xi}) = \frac{1}{(2\pi)^{m/2}} \int_{\mathbb{R}^m} e^{(\underline{x} \wedge \underline{\xi})} f(\underline{x})\, dV(\underline{x})$$

with $e^{(\underline{x} \wedge \underline{\xi})} = \sum_{r=0}^{\infty} \frac{(\underline{x} \wedge \underline{\xi})^r}{r!}$.

The integral kernel of this cylindrical Fourier transform can be rewritten in terms of the cosine and the sinc function, which also reveals its form of a scalar plus a bivector, i.e., a so-called parabivector.

Proposition 1 *The kernel of the cylindrical Fourier transform can be rewritten as*

$$e^{(\underline{x} \wedge \underline{\xi})} = \cos\big(|\underline{x} \wedge \underline{\xi}|\big) + (\underline{x} \wedge \underline{\xi})\mathrm{sinc}\big(|\underline{x} \wedge \underline{\xi}|\big)$$

where $\mathrm{sinc}(x) := \frac{\sin(x)}{x}$ *is the unnormalized sinc function.*

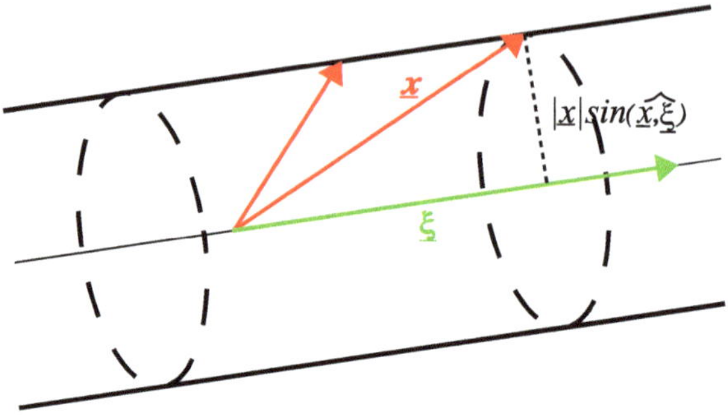

Fig. 1 In case of the cylindrical Fourier transform, for fixed $\underline{\xi}$, the phase $|\underline{x} \wedge \underline{\xi}|$ is constant on coaxial cylinders

Proof Splitting the defining series expansion of $e^{(\underline{x} \wedge \underline{\xi})}$ into its even and odd part and taking into account that $(\underline{x} \wedge \underline{\xi})^2 = -|\underline{x} \wedge \underline{\xi}|^2$ yields

$$\begin{aligned} e^{(\underline{x} \wedge \underline{\xi})} &= \sum_{\ell=0}^{\infty} (-1)^\ell \frac{|\underline{x} \wedge \underline{\xi}|^{2\ell}}{(2\ell)!} + (\underline{x} \wedge \underline{\xi}) \sum_{\ell=0}^{\infty} (-1)^\ell \frac{|\underline{x} \wedge \underline{\xi}|^{2\ell}}{(2\ell+1)!} \\ &= \cos\big(|\underline{x} \wedge \underline{\xi}|\big) + (\underline{x} \wedge \underline{\xi}) \operatorname{sinc}\big(|\underline{x} \wedge \underline{\xi}|\big). \end{aligned}$$ □

Let us now explain why we have chosen the name "cylindrical" for our new Fourier transform. From

$$|\underline{x} \wedge \underline{\xi}|^2 = |\underline{x}|^2 |\underline{\xi}|^2 - \big(\langle \underline{x}, \underline{\xi} \rangle\big)^2 = |\underline{x}|^2 |\underline{\xi}|^2 \big(1 - \cos(\widehat{\underline{x}, \underline{\xi}})^2\big) = |\underline{x}|^2 |\underline{\xi}|^2 \sin(\widehat{\underline{x}, \underline{\xi}})^2$$

it is clear that for $\underline{\xi}$ fixed, the "phase" $|\underline{x} \wedge \underline{\xi}|$ is constant if and only if $|\underline{x}| \sin(\widehat{\underline{x}, \underline{\xi}})$ is constant. In other words, for a fixed vector $\underline{\xi}$ in the image space, the phase $|\underline{x} \wedge \underline{\xi}|$ is constant on co-axial cylinders w.r.t. that fixed vector (see Fig. 1). For comparison, for a fixed vector $\underline{\xi}$ in the image space, the level surfaces of the traditional Fourier kernel are planes perpendicular to that fixed vector, since $\langle \underline{x}, \underline{\xi} \rangle = |\underline{x}||\underline{\xi}| \cos(\widehat{\underline{x}, \underline{\xi}})$ (see Fig. 2).

4.2 Properties

The cylindrical Fourier transform is well defined for each integrable function.

Theorem 2 *Let $f \in L_1(\mathbb{R}^m)$. Then $\mathcal{F}_{\text{cyl}}[f] \in L_\infty(\mathbb{R}^m) \cap C_0(\mathbb{R}^m)$, and, moreover,*

$$\big\|\mathcal{F}_{\text{cyl}}[f]\big\|_\infty \leq 2 \left(\frac{2}{\pi}\right)^{m/2} \|f\|_1.$$

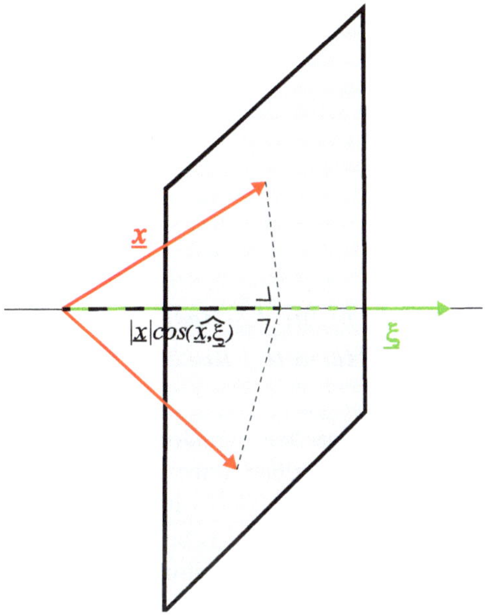

Fig. 2 In case of the classical Fourier transform, for fixed $\underline{\xi}$, the phase $\langle \underline{x}, \underline{\xi} \rangle$ is constant on planes perpendicular to $\underline{\xi}$

Proof Taking into account Proposition 1, we have that

$$\begin{aligned} \left| e^{(\underline{x} \wedge \underline{\xi})} \right| &= \left| \cos\left(|\underline{x} \wedge \underline{\xi}|\right) + \frac{(\underline{x} \wedge \underline{\xi})}{|\underline{x} \wedge \underline{\xi}|} \sin\left(|\underline{x} \wedge \underline{\xi}|\right) \right| \\ &\leq \left| \cos\left(|\underline{x} \wedge \underline{\xi}|\right) \right| + \left| \sin\left(|\underline{x} \wedge \underline{\xi}|\right) \right| \leq 2, \end{aligned}$$

which leads to the desired result. □

Although the cylindrical Fourier transform has a "simple" integral kernel, it satisfies calculation formulae that are more complicated than those of the multidimensional Clifford–Fourier transform (see [5]). For example, we state the differentiation and multiplication rules that nicely show that the two-dimensional case, in which the cylindrical Fourier transform and the Clifford–Fourier transform coincide, is special.

Proposition 2 (Differentiation and multiplication rule) *Let* $f, g \in L_1(\mathrm{R}^m)$. *The cylindrical Fourier transform satisfies*:

(i) *the differentiation rule*

$$\begin{aligned} \mathcal{F}_{\mathrm{cyl}}\left[\partial_{\underline{x}}\left[f(\underline{x})\right]\right](\underline{\xi}) &= -\underline{\xi}\,\mathcal{F}_{\mathrm{cyl}}\left[f(\underline{x})\right](-\underline{\xi}) + \frac{(2-m)}{(2\pi)^{m/2}}\underline{\xi} \\ &\quad \times \int_{\mathbb{R}^m} \operatorname{sinc}\left(|\underline{x} \wedge \underline{\xi}|\right) f(\underline{x})\, dV(\underline{x}) \end{aligned}$$

with $\operatorname{sinc}(x) := \frac{\sin(x)}{x}$ *the unnormalized sinc function*;

(ii) *the multiplication rule*

$$\mathcal{F}_{\text{cyl}}\big[\underline{x}\, f(\underline{x})\big](\underline{\xi}) = -\partial_{\underline{\xi}}\big[\mathcal{F}_{\text{cyl}}\big[f(\underline{x})\big](-\underline{\xi})\big] + \frac{(2-m)}{(2\pi)^{m/2}} \times \int_{\mathbb{R}^m} \text{sinc}\big(|\underline{x}\wedge\underline{\xi}|\big)\underline{x}\, f(\underline{x})\, dV(\underline{x}).$$

4.3 Spectrum of the L_2-Basis Consisting of Generalized Clifford–Hermite Functions

Let us now calculate the cylindrical Fourier spectrum of the L_2-basis (1). As these basis elements belong to the space of rapidly decreasing functions $\mathcal{S}(\mathbb{R}^m) \subset L_1(\mathbb{R}^m)$, their cylindrical Fourier image should be a bounded and continuous function. The calculation method is based on the Funk–Hecke theorem in space (see Theorem 1) and the following cylindrical Fourier kernel decomposition (see Proposition 1):

$$e^{(\underline{x}\wedge\underline{\xi})} = \cos\Big(r\rho\sqrt{1-t_{\underline{\eta}}^2}\Big) - r\rho t_{\underline{\eta}}\,\text{sinc}\Big(r\rho\sqrt{1-t_{\underline{\eta}}^2}\Big) - r\rho\underline{\eta}\underline{\omega}\,\text{sinc}\Big(r\rho\sqrt{1-t_{\underline{\eta}}^2}\Big), \quad (2)$$

where we have introduced the spherical coordinates

$$\underline{x} = r\underline{\omega}, \qquad \underline{\xi} = \rho\underline{\eta}, \qquad r = |\underline{x}|, \qquad \rho = |\underline{\xi}|, \qquad \underline{\omega}, \underline{\eta} \in S^{m-1}$$

and the notation $t_{\underline{\eta}} = \langle\underline{\omega}, \underline{\eta}\rangle$. For convenience, we denote the three terms in the decomposition (2) by A, B, and C.

As a first example, let us now calculate the cylindrical Fourier transform of the basis function $\phi_{0,k,j}(\underline{x})$ which is given, up to constants, by $P_k(\underline{x})e^{(-|\underline{x}|^2/2)}$ with P_k a left solid inner spherical monogenic of order k. By Corollary 1 it is obvious that we must make a distinction between k even and odd.

(A) *k even*

In the case where k is even, as a consequence of Corollary 1, the integrals containing the B- and C-terms of the kernel decomposition (2) reduce to zero. Furthermore, applying the Funk–Hecke theorem in space (see Theorem 1), we have that

$$\begin{aligned}\mathcal{F}_{\text{cyl}}\big[e^{(-|\underline{x}|^2/2)}\, P_k(\underline{x})\big](\underline{\xi}) &= \frac{1}{(2\pi)^{m/2}} \int_{\mathbb{R}^m} e^{(-r^2/2)} r^k \cos\Big(r\rho\sqrt{1-t_{\underline{\eta}}^2}\Big) P_k(\underline{\omega})\, dV(\underline{x}) \\ &= \frac{A_{m-1}}{(2\pi)^{m/2}} P_k(\underline{\eta}) \bigg(\int_0^{+\infty} e^{(-r^2/2)} r^{k+m-1}\, dr\bigg) \\ &\quad \times \bigg(\int_{-1}^{1} \cos\Big(r\rho\sqrt{1-t^2}\Big)(1-t^2)^{(m-3)/2} P_{k,m}(t)\, dt\bigg).\end{aligned}$$

Taking into account the series expansion of the cosine function, this result becomes

$$\begin{aligned}\mathcal{F}_{\mathrm{cyl}}\big[e^{(-|\underline{x}|^2/2)}P_k(\underline{x})\big](\underline{\xi}) &= \frac{k!(m-3)!}{(k+m-3)!}\frac{A_{m-1}}{(2\pi)^{m/2}}P_k(\underline{\eta})\sum_{\ell=0}^{\infty}\frac{(-1)^\ell}{(2\ell)!}\rho^{2\ell}\\ &\quad\times\left(\int_0^{+\infty}e^{(-r^2/2)}r^{2\ell+k+m-1}\,dr\right)\\ &\quad\times\left(\int_{-1}^{1}\big(1-t^2\big)^{(2\ell+m-3)/2}C_k^{(m-2)/2}(t)\,dt\right), \qquad (3)\end{aligned}$$

where we have also used the expression

$$P_{k,m}(t) = \frac{k!(m-3)!}{(k+m-3)!}C_k^{(m-2)/2}(t)$$

of the Legendre polynomials in $\mathbb{R}^m$ in terms of the Gegenbauer polynomials $C_k^\lambda(t)$. As these Gegenbauer polynomials C_k^λ are orthogonal on $]-1,1[$ w.r.t. the weight function $(1-t^2)^{\lambda-1/2}$ $(\lambda > -\frac{1}{2})$, it is easily seen that, for $\ell \leq \frac{k}{2}-1$,

$$\int_{-1}^{1}\big(1-t^2\big)^\ell\big(1-t^2\big)^{(m-3)/2}C_k^{(m-2)/2}(t)\,dt = 0.$$

Moreover, combining the integral formula (see [7], p. 826, formula (4) with $\alpha = \beta$)

$$\begin{aligned}&\int_{-1}^{1}\big(1-t^2\big)^\alpha C_k^\lambda(t)\,dt\\ &\quad= \frac{2^{2\alpha+1}(\Gamma(\alpha+1))^2\Gamma(k+2\lambda)}{k!\Gamma(2\lambda)\Gamma(2\alpha+2)}\,{}_3F_2\left(-k,k+2\lambda,\alpha+1;\lambda+\frac{1}{2},2\alpha+2;1\right),\end{aligned}$$

where $\mathrm{Re}(\alpha) > -1$, and ${}_3F_2(a,b,c;d,e;z)$ denotes the generalized hypergeometric series, with Watson's theorem (see, e.g., [6])

$${}_3F_2\left(a,b,c;\frac{a+b+1}{2},2c;1\right) = \frac{\sqrt{\pi}\,\Gamma(c+\frac{1}{2})\Gamma(\frac{a+b+1}{2})\Gamma(\frac{1-a-b+2c}{2})}{\Gamma(\frac{a+1}{2})\Gamma(\frac{b+1}{2})\Gamma(\frac{1-a+2c}{2})\,\Gamma(\frac{1-b+2c}{2})}$$

results into

$$\begin{aligned}&\int_{-1}^{1}(1-t^2)^\alpha\;C_k^\lambda(t)\,dt\\ &\quad= \frac{\sqrt{\pi}2^{2\alpha+1}(\Gamma(\alpha+1))^2\Gamma(k+2\lambda)\Gamma(\alpha+\frac{3}{2})\Gamma(\lambda+\frac{1}{2})\Gamma(\frac{2\alpha-2\lambda+3}{2})}{k!\Gamma(2\lambda)\Gamma(2\alpha+2)\Gamma(\frac{-k+1}{2})\Gamma(\frac{k+2\lambda+1}{2})\Gamma(\frac{2\alpha+3+k}{2})\Gamma(\frac{2\alpha-2\lambda+3-k}{2})}. \qquad (4)\end{aligned}$$

Applying the above result and taking into account that $\Gamma(2z) = \pi^{-1/2}2^{2z-1}\times\Gamma(z)\Gamma(z+1/2)$, (3) can be simplified to

$$\mathcal{F}_{\text{cyl}}\left[e^{(-|\underline{x}|^2/2)} P_k(\underline{x})\right](\underline{\xi})$$
$$= \frac{2^{k/2}\sqrt{\pi}}{\Gamma(\frac{-k+1}{2})\Gamma(\frac{k+m-1}{2})} P_k(\underline{\xi}) \sum_{\ell=k/2}^{\infty} \frac{(-1)^\ell 2^\ell \ell! \Gamma(\frac{2\ell+m-1}{2})}{(2\ell)!\, \Gamma(\frac{2\ell+2-k}{2})} |\underline{\xi}|^{2\ell-k}$$
$$= {}_1F_1\left(1-\frac{m}{2}; \frac{k+1}{2}; \frac{|\underline{\xi}|^2}{2}\right) e^{(-|\underline{\xi}|^2/2)} P_k(\underline{\xi})$$

with ${}_1F_1(a; c; z)$ Kummer's function, also called confluent hypergeometric function.

(B) *k odd*

For k odd, the integral containing the A-term of the kernel decomposition is zero, again as a consequence of Corollary 1. By means of the Funk–Hecke theorem in space we obtain

$$\mathcal{F}_{\text{cyl}}\left[e^{(-|\underline{x}|^2/2)} P_k(\underline{x})\right](\underline{\xi})$$
$$= \rho \frac{A_{m-1}}{(2\pi)^{m/2}} P_k(\underline{\eta}) \left(\int_0^{+\infty} e^{(-r^2/2)} r^{k+m}\, dr \right)$$
$$\times \left(\int_{-1}^{1} \operatorname{sinc}\left(r\rho\sqrt{1-t^2}\right)\left(1-t^2\right)^{(m-3)/2} \left(P_{k+1,m}(t) - t P_{k,m}(t)\right) dt \right).$$

Now, taking into account the Gegenbauer recurrence relation

$$(k+2\lambda) t C_k^\lambda(t) - (k+1) C_{k+1}^\lambda(t) = 2\lambda\left(1-t^2\right) C_{k-1}^{\lambda+1}(t),$$

we have that

$$P_{k+1,m}(t) - t P_{k,m}(t) = -\frac{k!(m-2)!}{(k+m-2)!}\left(1-t^2\right) C_{k-1}^{m/2}(t),$$

which in turn yields

$$\mathcal{F}_{\text{cyl}}\left[e^{(-|\underline{x}|^2/2)} P_k(\underline{x})\right](\underline{\xi}) = -\frac{k!(m-2)!}{(k+m-2)!} \frac{A_{m-1}}{(2\pi)^{m/2}} \rho P_k(\underline{\eta})$$
$$\times \left(\int_0^{+\infty} e^{(-r^2/2)} r^{k+m}\, dr \right)$$
$$\times \left(\int_{-1}^{1} \operatorname{sinc}\left(r\rho\sqrt{1-t^2}\right)\left(1-t^2\right)^{(m-1)/2} C_{k-1}^{m/2}(t)\, dt \right).$$

Next, applying consecutively the series expansion of the sinc function, the orthogonality of the Gegenbauer polynomials and expression (4), we find

$$\mathcal{F}_{\text{cyl}}\left[e^{(-|\underline{x}|^2/2)} P_k(\underline{x})\right](\underline{\xi}) = -{}_1F_1\left(1-\frac{m}{2}; \frac{k+2}{2}; \frac{|\underline{\xi}|^2}{2}\right) e^{(-|\underline{\xi}|^2/2)} P_k(\underline{\xi}).$$

So note that the cylindrical Fourier transform reproduces the Gaussian times the spherical monogenic up to a Kummer's function factor.

A second example is provided by the cylindrical Fourier transform of the basis function $\phi_{1,k,j}$ given, up to constants, by $e^{(-|\underline{x}|^2/2)}\underline{x}\, P_k(\underline{x})$. Its calculation runs along similar lines. Making again a distinction between k even and k odd, we find:

(A) *for k even,*

$$\begin{aligned}\mathcal{F}_{\mathrm{cyl}}\left[e^{(-|\underline{x}|^2/2)}\ \underline{x}\ P_k(\underline{x})\right](\underline{\xi})\\ = \frac{(k+m-1)}{(k+1)}\ {}_1F_1\left(1-\frac{m}{2};\frac{k+3}{2};\frac{|\underline{\xi}|^2}{2}\right)e^{(-|\underline{\xi}|^2/2)}\underline{\xi}\ P_k(\underline{\xi});\end{aligned}$$

(B) *for k odd,*

$$\mathcal{F}_{\mathrm{cyl}}\left[e^{(-|\underline{x}|^2/2)}\ \underline{x}\ P_k(\underline{x})\right](\underline{\xi}) = {}_1F_1\left(1-\frac{m}{2};\frac{k+2}{2};\frac{|\underline{\xi}|^2}{2}\right)e^{(-|\underline{\xi}|^2/2)}\underline{\xi}\ P_k(\underline{\xi}),$$

showing again the reproducing property up to a Kummer's function factor.

For the calculation of the cylindrical Fourier spectrum of a general basis element $\phi_{s,k,j}$, we refer the reader to [4] and the survey paper [5].

5 Application Potential of the Cylindrical Fourier Transform

In the foregoing section we established the image under the cylindrical Fourier transform of an L_2-basis for the space of all L_2-functions in $\mathbb{R}^m$. Using density arguments, these results may be used to approximate the cylindrical Fourier image of various types of functions and distributions in $\mathbb{R}^m$. However, for certain types of functions or distributions, direct calculation methods are available on top of this approximation approach. A typical example is provided by the case of distributions concentrated on the unit sphere, which are of the form

$$F(\underline{x}) = \delta(r-1)f(\underline{\omega}), \quad \underline{x} = r\underline{\omega},\ r = |\underline{x}| \in [0,\infty[\ ,\ \underline{\omega} \in S^{m-1}.$$

The corresponding cylindrical Fourier transform is given by

$$\mathcal{F}_{\mathrm{cyl}}[F](\underline{\xi}) = \mathcal{F}_{\mathrm{cyl}}[f](\underline{\xi}) = \frac{1}{(2\pi)^{m/2}}\int_{S^{m-1}} \exp(\underline{\omega}\wedge\underline{\xi})f(\underline{\omega})\,dS(\underline{\omega}).$$

Hereby $\underline{\xi}$ still belongs to the whole space $\mathbb{R}^m$, while the data $f(\underline{\omega})$ are defined on the unit sphere, a codimension one surface of $\mathbb{R}^m$. It is hence expected that the data $f(\underline{\omega})$ are already determined by the cylindrical Fourier image restricted to a suitable codimension one surface as well, typical examples being:

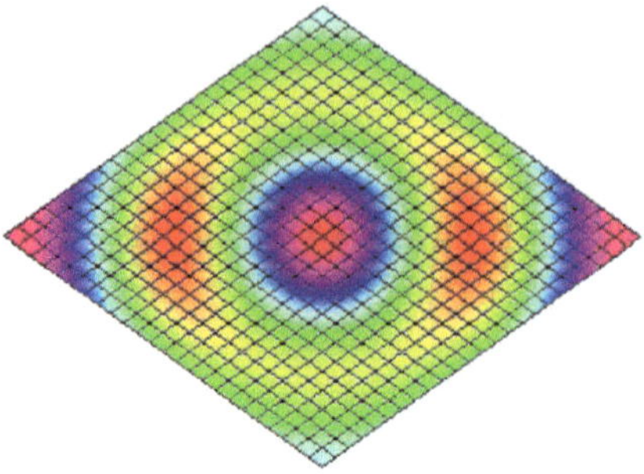

Fig. 3 The real part of the cylindrical Fourier spectrum of the characteristic function of a geodesic triangle on S^2

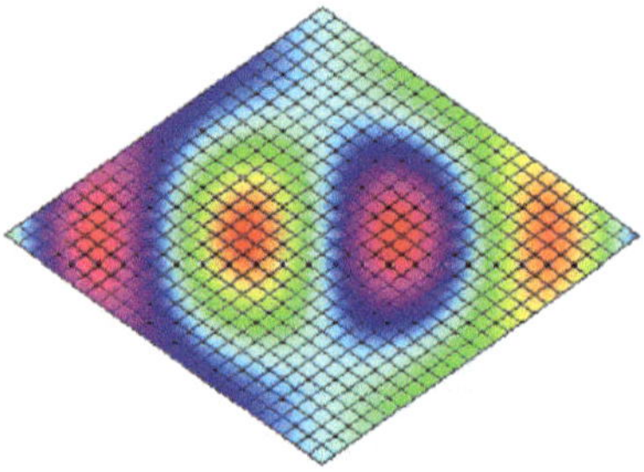

Fig. 4 The e_1e_2-component of the cylindrical Fourier spectrum of the characteristic function of a geodesic triangle on S^2

(i) $\underline{\xi} = \underline{\eta} \in S^{m-1}$, leading to an integral transform from S^{m-1} to S^{m-1},
(ii) $\underline{\xi} = \xi_1 e_1 + \cdots + \xi_{m-1}e_{m-1}$, i.e., $\underline{\xi}$ belongs to the affine subspace given by $\xi_m = 0$.

To evaluate the cylindrical Fourier transform explicitly, it suffices in both cases to express the function $f(\underline{\omega})$ as a series of spherical monogenics and to apply a Funk–Hecke argument on the spherical monogenics. This may lead to correspondences between function spaces on S^{m-1} and isomorphisms between them including inversion methods. The establishment of direct inversion formulae remains an independent and interesting problem for future research.

As an example (see Figs. 3, 4, 5, and 6), we have computed directly the cylindrical Fourier image of the characteristic function of a geodesic triangle on the two sphere S^2 that may be expressed in spherical coordinates by the integral

$$\frac{1}{(2\pi)^{3/2}} \int_0^{\pi/2} \int_0^{\pi/2} \exp(\underline{\omega} \wedge \underline{\xi}) \sin(\theta)\, d\theta\, d\phi$$

with $\underline{\omega} = \sin(\theta)\cos(\phi)e_1 + \sin(\theta)\sin(\phi)e_2 + \cos(\theta)e_3$ and $\underline{\xi} = ae_1 + be_2$.

6 Conclusion

There is a recent increasing interest in integral transforms and in particular Fourier transforms which take advantage of the algebraic structure inherent in hypercomplex function theories, especially quaternionic and Clifford analysis. In this pa-

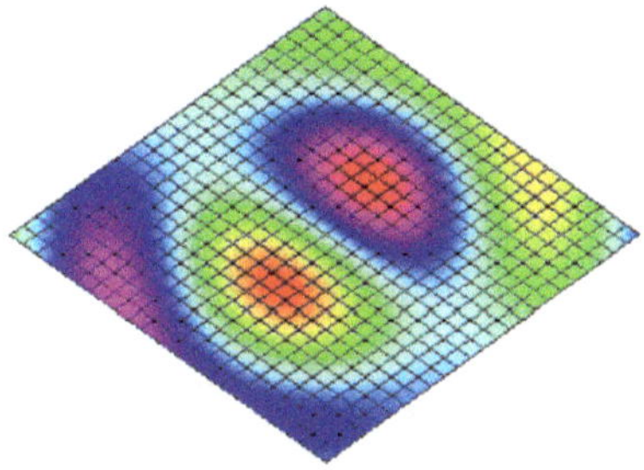

Fig. 5 The e_1e_3-component of the cylindrical Fourier spectrum of the characteristic function of a geodesic triangle on S^2

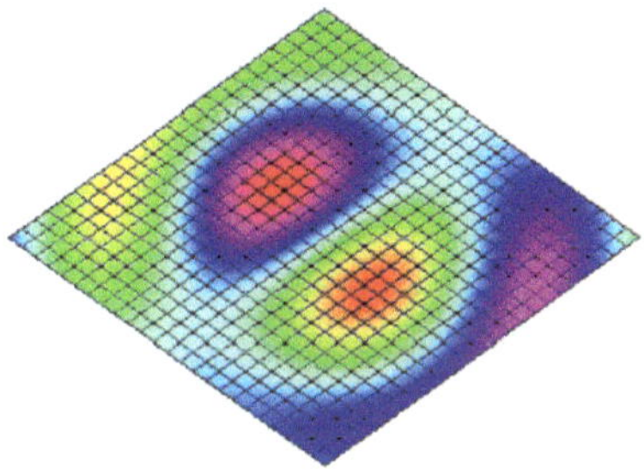

Fig. 6 The e_2e_3-component of the cylindrical Fourier spectrum of the characteristic function of a geodesic triangle on S^2

per we have shown that the recently developed cylindrical Fourier transform of Clifford analysis in Euclidean space of arbitrary dimension is a promising higher-dimensional integral transform with application potential. We have introduced a few methods for the practical computation of the corresponding spectra and illustrated one of these methods by working out an explicit example. For the theory underlying the cylindrical Fourier transform and similar integral transforms in Clifford analysis, we refer the reader to, e.g., [2–4] and in particular to the survey paper [5], and the references contained therein.

References

1. Brackx, F., Delanghe, R., Sommen, F.: Clifford Analysis. Pitman Publishers, London (1982)
2. Brackx, F., De Schepper, N., Sommen, F.: The Clifford–Fourier transform. J. Fourier Anal. Appl. (2005). doi:10.1007/s00041-005-4079-9
3. Brackx, F., De Schepper, N., Sommen, F.: The two-dimensional Clifford–Fourier transform. J. Math. Imaging Vis. (2006). doi:10.1007/s10851-006-3605-y
4. Brackx, F., De Schepper, N., Sommen, F.: The cylindrical Fourier spectrum of an L_2-basis consisting of generalized Clifford–Hermite functions. In: Simos, T.E., Psihoyios, G., Tsitouras, Ch. (eds.) Numerical Analysis and Applied Mathematics, AIP Conference Proceedings, Kos, Greece, pp. 686–690 (2008)
5. Brackx, F., De Schepper, N., Sommen, F.: The Fourier transform in Clifford analysis. Adv. Imaging Electron Phys. (2009). doi:10.1016/S1076-5670(08)01402-x
6. Erdélyi, A., Magnus, W., Oberhettinger, F., Tricomi, F.G.: Higher Transcendental Functions. vol. 1. McGraw-Hill, New York (1953)
7. Gradshteyn, I.S., Ryzhik, I.M.: Table of Integrals, Series, and Products. Academic Press, San Diego (1980)

Analyzing Real Vector Fields with Clifford Convolution and Clifford–Fourier Transform

Wieland Reich and Gerik Scheuermann

Abstract Postprocessing in computational fluid dynamics and processing of fluid flow measurements need robust methods that can deal with scalar and vector fields. While image processing of scalar data is a well-established discipline, there is a lack of similar methods for vector data. This paper surveys a particular approach defining convolution operators on vector fields using geometric algebra. This includes a corresponding Clifford–Fourier transform including a convolution theorem. Finally, a comparison is tried with related approaches for a Fourier transform of spatial vector or multivector data. In particular, we analyze the Fourier series based on quaternion holomorphic functions of Gürlebeck et al. (Funktionentheorie in der Ebene und im Raum, Birkhäuser, Basel, 2006), the quaternion Fourier transform of Hitzer (Proceedings of Function Theories in Higher Dimensions, 2006) and the biquaternion Fourier transform of Sangwine et al. (IEEE Trans. Signal Process. 56(4),1522–1531, 2007).

1 Fluid Flow Analysis

Fluid flow, especially of air and water, is usually modeled by the Navier–Stokes equations or simplifications like the Euler equations [1]. The physical fields in this model include pressure, density, velocity, and internal energy [22]. These variables depend on space and often also on time. While there are mainly scalar fields, velocity is a vector field and of high importance for any analysis of numerical or physical experiments. Some numerical simulations use a discretization of the spatial domain and calculate the variables at a finite number of positions on a regular lattice (finite difference methods). Other methods split space into volume elements and assume a polynomial solution of a certain degree in each volume element (finite element methods or finite volume methods). These numerical methods create a large amount

W. Reich (✉)
Institut fuer Informatik, Universität Leipzig, 04009 Leipzig, Germany
e-mail: reich@informatik.uni-leipzig.de

E. Bayro-Corrochano, G. Scheuermann (eds.), *Geometric Algebra Computing*,
DOI 10.1007/978-1-84996-108-0_7,

of data and its analysis, i.e., postprocessing, usually uses computer graphics to create images. Research about this process is an important part of scientific visualization [21].

Besides numerical simulations, modern measurement techniques like particle image velocimetry (PIV) [16] create velocity vector field measurements on a regular lattice using laser sheets and image processing. Therefore, simulations and experiments in fluid mechanics create discretized vector fields as part of their output. The analysis of these fields is an important postprocessing task. In the following, we will assume that there is a way to continuously integrate the data to simplify the notation. Of course, one can also discretize the integrals conversely.

Flow visualization knows a lot of direct methods like hedgehogs, streamlines, or streamsurfaces [21] and also techniques based on mathematical data analysis like topology or feature detection methods [10, 17, 20]. Since these methods are often not very robust, a transfer of image processing to vector data is an attractive approach. A look into any image processing book, e.g., by Jähne [15], reveals the importance of shift-invariant linear filters based on convolution. Of course, there is vector data processing in image processing, but the usual techniques for optical flow, which is the velocity field warping one image into another, do not really help as they concentrate mainly on noncontinuities in the field which is not a typical event in fluid flow velocity fields.

2 Geometric Algebra

In classical linear algebra, there are several multiplications involving vectors. Some multiplications have led to approaches for a convolution on vector fields. Scalar multiplication can be easily applied and can be seen as a special case of component-wise multiplication of two vectors which has been used by Granlund and Knutsson [8]. The scalar product has been used by Heiberg et al. [11] as a convolution operator. Obviously, one would like a unified convolution operator that incorporates these approaches and can be applied several times, in contrast to the scalar product version that creates a scalar field using two vector fields. Furthermore, one looks for all the nice theorems like convolution theorem with a suitable generalized Fourier transform, derivation theorem, shift theorem, and Parseval's theorem. Geometric algebra allows such a convolution [5, 6].

Let $\mathbb{R}^d$, $d = 2, 3$, be the Euclidean space with the orthonormal basis

$$\{e_1, e_2\} \text{ resp. } \{e_1, e_2, e_3\}. \tag{1}$$

We use the 2^d-dimensional real geometric algebras $G_d, d = 2, 3$, which have a associative, bilinear geometric product satisfying

$$1e_j = e_j, \quad j = 1, \ldots, d, \tag{2}$$

$$e_j^2 = 1, \quad j = 1, \ldots, d, \tag{3}$$

$$e_j e_k = -e_k e_j, \quad j, k = 1, \ldots, d, \ j \neq k. \tag{4}$$

Their basis is given by

$$\{1, e_1, e_2, i_2 := e_1 e_2\} \text{ resp. } \{1, e_1, e_2, e_3, e_1 e_2, e_2 e_3, e_3 e_1, i_3 := e_1 e_2 e_3\}. \quad (5)$$

It can be verified quickly that the squares of i_2, i_3 and those of the bivectors $e_1 e_2$, $e_2 e_3$, and $e_3 e_1$ are -1. General elements in G_d are called multivectors, while elements of the form

$$v = \sum_{j=1}^{d} \alpha_j e_j \quad (6)$$

are called vectors, i.e., $v \in \mathbb{R}^d \subset G_d$. The geometric product of two vectors $a, b \in \mathbb{R}^d$ results in

$$ab = a \cdot b + a \wedge b, \quad (7)$$

where $\cdot$ is the inner product, and $\wedge$ is the outer product. We have

$$a \cdot b = |a||b| \cos(\alpha), \quad (8)$$

$$|a \wedge b| = |a||b| \sin(\alpha) \quad (9)$$

with the usual norm for vectors and the angle α from a to b.

Let F be a multivector-valued function (field) of a vector variable x defined on some region G of the Euclidean space $\mathbb{R}^d$; compare (19) for $d = 3$ and (23) for $d = 2$. We define the Riemann integral of F by

$$\int_G F(x)|dx| = \lim_{|\Delta x| \to 0, n \to \infty} \sum_{j=1}^{n} F(x_j)|\Delta x_j|. \quad (10)$$

We define $\Delta x = dx_1 \wedge dx_2 i_2^{-1}$ $(d = 2)$ resp. $\Delta x = dx_1 \wedge dx_2 \wedge dx_3 i_3^{-1}$ $(d = 3)$ as the dual oriented scalar magnitude. The quantity $|\Delta x|$ is used here to make the integral grade preserving since dx is a vector within geometric algebra in general.

The directional derivative of F in direction r is

$$F_r(x) = \lim_{h \to 0} \frac{F(x + hr) - F(x)}{h} \quad (11)$$

with $r \in \mathbb{R}^3$, $h \in \mathbb{R}$. With the vector derivative

$$\nabla = \sum_{j=1}^{d} e_j \frac{\partial}{\partial e_j} \quad (12)$$

(vector valued), the complete (left) derivative of F is defined as

$$\nabla F(x) = \sum_{j=1}^{d} e_j F_{e_j}(x). \quad (13)$$

Similarly, we define the complete right derivative as

$$F(x)\nabla = \sum_{j=1}^{d} F_{e_j}(x)e_j. \tag{14}$$

The curl and divergence of a vector field f can be computed as scalar and bivector parts of (12):

$$\operatorname{curl} f = \nabla \wedge f = \frac{1}{2}(\nabla f - f\nabla), \qquad \operatorname{div} f = \nabla \cdot f = \frac{1}{2}(\nabla f + f\nabla). \tag{15}$$

The readers interested in the basics and more applications of geometric algebra are also referred to [12] and [4]. That fluid flow dynamics is accessible by geometric algebra methods and is also discussed in [3].

3 Clifford Convolution

Let $V, H : \mathbb{R}^d \to G_d$ be two multivector fields. We define the *Clifford convolution* as

$$(H * V)(r) := \int_{\mathbb{R}^d} H(\xi)V(r - \xi)|d\xi|. \tag{16}$$

If both fields are scalar fields, this is the usual convolution in image processing. If H is a scalar field, e.g., a Gaussian kernel, and V is a vector field, we get a scalar multiplication and can model smoothing of a vector field. If H is a vector field, and V a is multivector field, the simple relation

$$H(\xi)V(r - \xi) = H(\xi) \cdot V(r - \xi) + H(\xi) \wedge V(r - \xi) \tag{17}$$

shows that the scalar part of the result is Heiberg's convolution, while the bivector part contains additional information. General multivector fields allow a closed operation in G_d, so that several convolutions can be combined. We have shown [5, 7] that this convolution can be used for the analysis of velocity vector fields from computational fluid dynamics (CFD) simulations and PIV measurements.

4 Clifford–Fourier Transform

Our group has found a generalization of the Fourier transform of complex signals to multivector fields [6] that allows the generalization of the well-known theorems like the convolution theorem for the convolution defined in the previous section. There are different approaches of transforming multivector valued data, e.g., in [2, 14], and [18]. In Sect. 5 we discuss the relation to ours.

Let $F : \mathbb{R}^d \to G_3$ be a multivector field. We define

$$\mathcal{F}\{F\}(u) := \int_{\mathbb{R}^d} F(x) \exp(-2\pi i_d x \cdot u)|dx| \tag{18}$$

as the *Clifford–Fourier transform* (*CFT*) with the inverse

$$\mathcal{F}^{-1}\{F\}(x) := \int_{\mathbb{R}^d} F(u) \exp(2\pi i_d x \cdot u)|du|. \tag{19}$$

In three dimensions, the CFT is a linear combination of four classical complex Fourier transforms as can be seen by looking at the real components. Since

$$\begin{aligned} F(x) &= F_0(x) + F_1(x)e_1 + F_2(x)e_2 + F_3(x)e_3 \\ &\quad + F_{12}(x)e_1e_2 + F_{23}(x)e_2e_3 + F_{31}(x)e_3e_1 + F_{123}(x)e_1e_2e_3 &(20)\\ &= F_0(x) + F_1(x)e_1 + F_2(x)e_2 + F_3(x)e_3 + F_{12}(x)i_3e_3 \\ &\quad + F_{23}(x)i_3e_1 + F_{31}(x)i_3e_2 + F_{123}(x)i_3 &(21)\\ &= \big(F_0(x) + F_{123}(x)i_3\big)1 + \big(F_1(x) + F_{23}(x)i_3\big)e_1 \\ &\quad + \big(F_2(x) + F_{31}(x)i_3\big)e_2 + \big(F_3(x) + F_{12}(x)i_3\big)e_3, &(22) \end{aligned}$$

we get

$$\begin{aligned} \mathcal{F}\{F\}(u) = {}& \big[\mathcal{F}\{F_0 + F_{123}i_3\}(u)\big]1 + \big[\mathcal{F}\{F_1 + F_{23}i_3\}(u)\big]e_1 \\ & + \big[\mathcal{F}\{F_2 + F_{31}i_3\}(u)\big]e_2 + \big[\mathcal{F}\{F_3 + F_{12}i_3\}(u)\big]e_3. \end{aligned} \tag{23}$$

We have proven earlier [6] that the convolution, derivative, shift, and Parseval theorem hold. For vector fields, we can see that the CFT treats each component as a real signal that is transformed independently from the other components. Various examples of 3D patterns and their CFT are shown in Fig. 1.

In two dimensions, the CFT is a linear combination of two classical complex Fourier transforms. We have

$$\begin{aligned} F(x) &= F_0(x) + F_1(x)e_1 + F_2(x)e_2 + F_{12}(x)e_1e_2 &(24)\\ &= F_0(x) + F_1(x)e_1 + F_2(x)e_1i_2 + F_{12}(x)i_2 &(25)\\ &= 1\big[F_0(x) + F_{12}(x)i_2\big] + e_1\big[F_1(x) + F_2(x)i_2\big] &(26) \end{aligned}$$

and obtain

$$\mathcal{F}\{F\}(u) = 1\big[\mathcal{F}\{F_0 + F_{12}i_2\}(u)\big] + e_1\big[\mathcal{F}\{F_1 + F_2i_2\}(u)\big]. \tag{27}$$

Regarding the convolution theorem, one has to separate the vector and spinor parts since i_2 does not commute with all algebra elements. With this restriction, the theorem holds again [6]. Figure 2 shows a turbulent fluid in 2D, in Figs. 3 and 4 its discrete CFT is visualized.

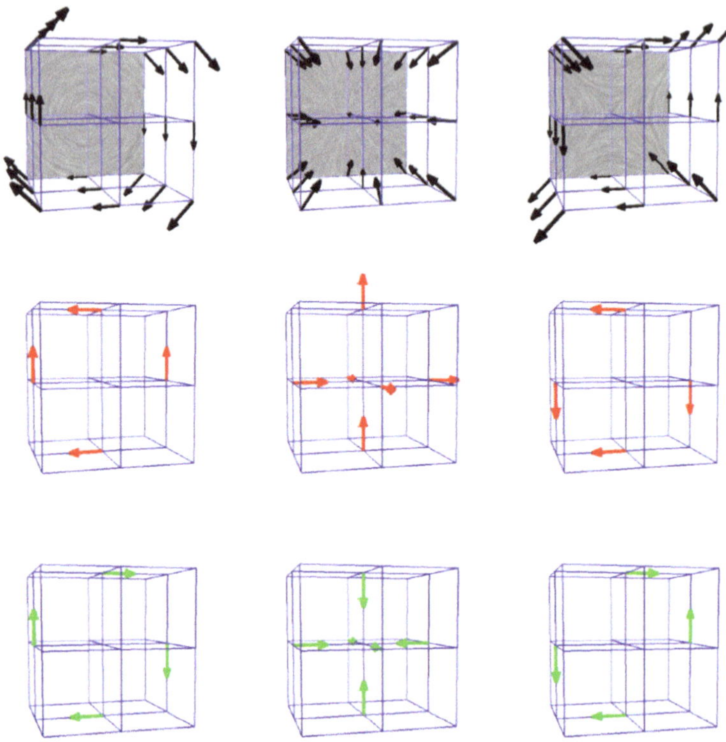

Fig. 1 *Top:* Various 3D patterns. *Middle*: The vector part of their DCFT. *Bottom:* The bivector part of their DCFT, displayed as normal vector of the plane. *Left*: $3 \times 3 \times 3$ rotation in one coordinate plane. *Middle*: $3 \times 3 \times 3$ convergence. *Right*: $3 \times 3 \times 3$ saddle line. The mean value of the discrete Clifford–Fourier transform (DCFT) is situated in the center of the field. In 3D, the waves forming the patterns can be easily seen in the frequency domain. The magnitude of the bivectors of the DCFT is only half the magnitude of the corresponding vectors, though both are displayed with same length

5 Relation to Other Fourier Transforms

In recent years, other definitions of a Fourier transform have appeared in the literature that can be applied to vector fields. In this section, we compare our approach with the Fourier series based on quaternion analysis by Gürlebeck, Habetha, and Sprößig [9], the biquaternion Fourier transform by Sangwine et al. [19], and the quaternion Fourier transform by Hitzer in [13]. Therefore we use three important isomorphisms. First of all, quaternions are isomorphic to the even subalgebra G_3^+ of G_3 by

$$i \mapsto e_1e_2, \qquad j \mapsto e_2e_3, \qquad k \mapsto e_1e_3, \tag{28}$$

biquaternions are isomorphic to G_3 by additionally

$$I \mapsto e_1e_2e_3 = i_3. \tag{29}$$

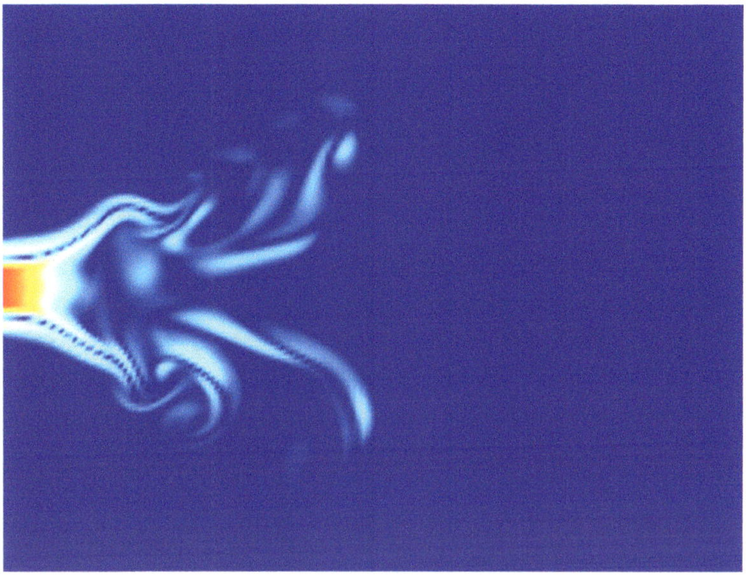

Fig. 2 A 2D slice of a turbulent swirling jet entering a fluid at rest. The image shows color coding of the absolute magnitude of the vectors. The colors are scaled from zero (*blue*) to the maximal magnitude (*red*)

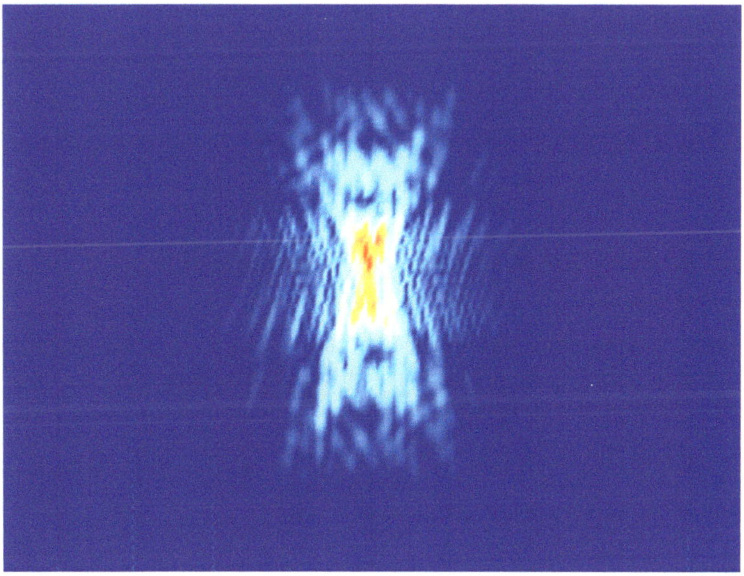

Fig. 3 This image shows a (fast) discrete Clifford–Fourier transform of the data set. Zero frequency is located in *the middle* of the image. Vectors are treated as rotors when using Clifford algebra in the frequency domain, thus color coding is based on the magnitude of the transformed rotor. The scaling of the colors is the same as the last image. We zoomed in to get more information

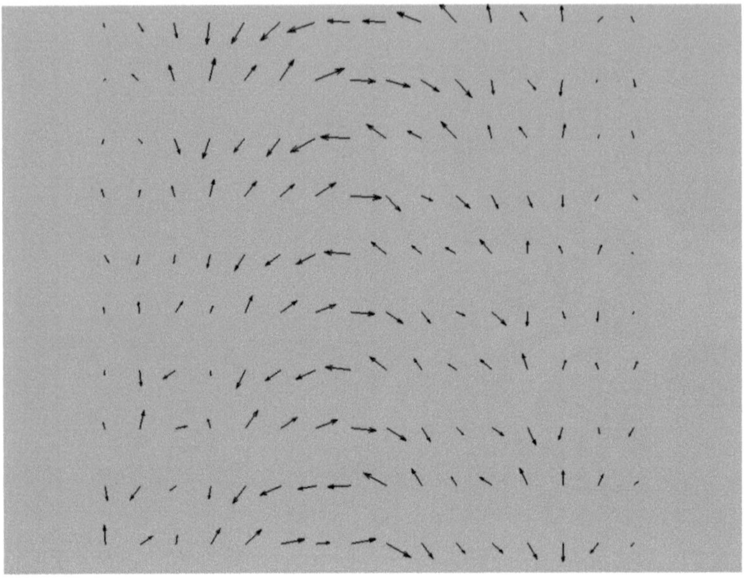

Fig. 4 In that image we zoom in further and take a look at the direction of the "vectors" in a neighborhood of zero frequency

Further the quaternions are isomorphic to the Clifford Algebra $Cl_{0,2}$ of the Anti-Euclidean Space $\mathbb{R}^{0,2}$ by

$$i \mapsto e_1, \qquad i \mapsto e_2, \qquad k \mapsto e_1 e_2, \tag{30}$$

which we will only use in for the definition of holomorphicity in Sect. 5.1.

5.1 $\mathbb{H}$-Holomorphic Functions and Fourier Series

We follow the definitions by Gürlebeck et al. [9] and identify the quaternions as in (29). Let $f : \mathbb{H} \to \mathbb{H}$ be a function with real partial derivatives $\partial_k := \frac{\partial}{\partial q_k}$. We define the complete real differential as

$$df = \sum_{k=0}^{3} \partial_k f \, dq_k \tag{31}$$

and set

$$dq = dq_0 + \sum_{k=1}^{3} e_k \, dq_k, \qquad d\bar{q} = dq_0 - \sum_{k=1}^{3} e_k \, dq_k. \tag{32}$$

This leads to

$$df = \frac{1}{2}\left(\sum_{k=0}^{3} \partial_k f\, e_k\right) d\bar{q} + \frac{1}{2}\left(\partial_0 f\, dq - \sum_{k=1}^{3} \partial_k f\, dq\, e_k\right). \tag{33}$$

A real C^1-function f is right $\mathbb{H}$-*holomorphic* in $G \subset \mathbb{H}$ if, for every $q \in G$ and $h \to 0$, there exist $a_k(q) \in \mathbb{H}$ with

$$f(q+h) = f(q) + \sum_{k=1}^{3} a_k(q)\,(h_k - h_0 e_k) + o(h) \tag{34}$$

and *left* $\mathbb{H}$-*holomorphic* if

$$f(q+h) = f(q) + \sum_{k=1}^{3} (h_k - h_0 e_k) a_k(q) + o(h) \tag{35}$$

with Landau symbol $o(h)$. The h_0, h_k are the 4D coordinates of $h \in \mathbb{H}$. With the operator

$$\bar{\partial} := \frac{\partial}{\partial q_0} + \sum_{k=1}^{3} \frac{\partial}{\partial q_k} e_k, \tag{36}$$

we have

$$f \text{ is left } \mathbb{H}\text{-holomorphic} \iff \bar{\partial} f = 0, \tag{37}$$

$$f \text{ is right } \mathbb{H}\text{-holomorphic} \iff f\bar{\partial} = 0. \tag{38}$$

Let $\mathbb{B}_3 := \{q \in \mathbb{H} | |q| = 1\}$ be the unit sphere in $\mathbb{H}$. Let $L^2(\mathbb{B}_3)$ be the functions on $\mathbb{B}_3$ with existing integral of the squared function. Then one can write

$$L^2(\mathbb{B}_3) \cap \ker \bar{\partial} = \bigoplus_{k=0}^{\infty} H_k^+, \tag{39}$$

where H_k^+ are homogenous $\mathbb{H}$-holomorphic polynomials. There is an orthogonal basis for this space that allows a Fourier series approximation [9].

If we look at a vector field

$$v : \mathbb{R}^3 \to \mathbb{R}^3 \subset G_3, \tag{40}$$

we have to find a related $\mathbb{H}$-holomorphic function $f : \mathbb{B}_3 \to \mathbb{H} \subset G_3$ to apply the construction above. We tried

$$v(x) = f(x) e_1 \overline{f(x)} \tag{41}$$

and

$$v(x) = f(x) e_1, \qquad v(x) = e_1 f(x), \tag{42}$$

as well as

$$f(x) = v_1(x)e_1e_2 + v_2(x)e_2e_3 + v_3(x)e_1e_3 \quad (43)$$

with $e_i \in G_3$.

In all examined cases, general linear vector fields v do not generate a $\mathbb{H}$-holomorphic function f, which makes applying the Fourier series expansion to our purpose inconvenient.

5.2 Biquaternion Fourier Transform

Let $\mathbb{H}_\mathbb{C}$ be the biquaternions, i.e.,

$$\mathbb{H}_\mathbb{C} = \{q_0 + q_1 i + q_2 j + q_3 k \mid q_k = \Re(q_k) + I\Im(q_k) \in \mathbb{C}\} \quad (44)$$

with the algebra isomorphism $\mathbb{H}_\mathbb{C} \to G_3$ as in (28)–(29). Sangwine et al. [19] choose a $\mu \in G_3$ with $\mu^2 = -1$ and define the *right biquaternion Fourier transform* (*BiQFT*) for a signal $F : \mathbb{R}^3 \to G_3$ by

$$\mathcal{F}_r^\mu\{F\}(u) = \int_{\mathbb{R}^3} F(x)\exp(-2\pi\mu x \cdot u)|dx| \quad (45)$$

and the *left biquaternion Fourier transform* by

$$\mathcal{F}_l^\mu\{F\}(u) = \int_{\mathbb{R}^3} \exp(-2\pi\mu x \cdot u)F(x)|dx|. \quad (46)$$

For $\mu = i_3$, this is the 3D-CFT. But for a pure bi-quarternion, i.e., a bivector, one can choose an orthogonal basis $\mu, \nu, \xi = \mu\nu$, with $\{\mu, \nu, \xi\}$ being quaternionic roots of -1 such that any $q \in G_3$ can be written as

$$\begin{aligned} q &= q_0 + q_1e_1e_2 + q_2e_2e_3 + q_3e_1e_3 \\ &= q_0 + \tilde{q}_1\mu + \tilde{q}_2\nu + \tilde{q}_3\xi \\ &= (q_0 + \tilde{q}_1\mu) + (\tilde{q}_2 + \tilde{q}_3\mu)\nu \\ &= Simp(q) + Perp(q)\nu \end{aligned} \quad (47)$$

with $Simp(q)$ and $Perp(q)$ denoting the so-called simplex and perplex of q. For a pure bi-quarternion μ, the corresponding BiQFT fulfills

$$\mathcal{F}^{e_1e_2} = T^{-1}\mathcal{F}^\mu T \quad (48)$$

with the linear operator $T(1) = 1, T(e_1e_2) = \mu, T(e_2e_3) = \nu, T(e_1e_3) = \xi$, so any two BiQFTs with pure bi-quarternion μ differ just by an orthogonal transformation. The BiQFT splits like the CFT in four independent classical complex Fourier transforms:

$$\begin{aligned}
\mathcal{F}_r^{\mu}\{F\}(u) &= \int_{\mathbb{R}^3} F(x)\exp(-2\pi\mu x\cdot u)|dx| \\
&= \int_{\mathbb{R}^3} \big(f_0(x)+\tilde{f}_1(x)\mu\big)\exp\big(-2\pi\mu x\cdot u\big)|dx| \\
&\quad + \int_{\mathbb{R}^3} \big(\tilde{f}_2(x)+\tilde{f}_3(x)\mu\big)\nu\exp(-2\pi\mu x\cdot u)|dx| \\
&= \int_{\mathbb{R}^3} \big(\Re\big(f_0(x)\big)+\Re\big(\tilde{f}_1(x)\big)\mu\big)\exp\big(-2\pi\mu x\cdot u\big)|dx| \\
&\quad + \int_{\mathbb{R}^3} \big(\Im\big(f_0(x)\big)+\Im\big(\tilde{f}_1(x)\big)\mu\big)\exp\big(-2\pi\mu x\cdot u\big)|dx|i_3 \\
&\quad + \int_{\mathbb{R}^3} \big(\Re\big(\tilde{f}_2(x)\big)+\Re\big(\tilde{f}_3(x)\big)\mu\big)\nu\exp\big(-2\pi\mu x\cdot u\big)|dx| \\
&\quad + \int_{\mathbb{R}^3} \big(\Im\big(\tilde{f}_2(x)\big)+\Im\big(\tilde{f}_3(x)\big)\mu\big)\nu\exp\big(-2\pi\mu x\cdot u\big)|dx|i_3. \qquad (49)
\end{aligned}$$

For a real vector field

$$v:\mathbb{R}^3\to\mathbb{R}^3\subset G_3, \qquad x\mapsto\sum_{k=1}^{3} v_k(x)e_k, \qquad (50)$$

we have

$$\begin{aligned}
v(x) &= \big(-v_3(x)e_1e_2 - v_1(x)e_2e_3 + v_2(x)e_1e_3\big)i_3 \qquad (51)\\
&= \big(\big(-v_3(x)e_1e_2\big)+\big(-v_1(x)+v_2(x)e_1e_2\big)e_2e_3\big)i_3. \qquad (52)
\end{aligned}$$

We get for $\mu = e_1e_2$:

$$\begin{aligned}
\mathcal{F}_r^{i}\{v\}(u) &= \int_{\mathbb{R}^3} \big(-v_3(x)e_1e_2\big)\exp\big(-2\pi e_1e_2x\cdot u\big)|dx|i_3 \\
&\quad + \int_{\mathbb{R}^3} \big(-v_1(x)+v_2(x)e_1e_2\big)\exp(-2\pi e_1e_2x\cdot u)|dx|i_3, \qquad (53)
\end{aligned}$$

which means that the vector field is split in a purely complex signal, $-v_3(x)e_1e_2$ and a complex signal $-v_1(x)+v_2(x)e_1e_2$, which are transformed independently by two classical Fourier transforms.

In essence, the BiQFT means choosing a planar direction μ in $\mathbb{R}^3$, transforming the planar part of the vector field with a 2D-CFT in each plane parallel to μ, and transforming the scalar part orthogonal to μ as an independent real signal. One can say that the BiQFT is the direct generalization of the 2D-CFT to three dimensions, while the 3D-CFT looks at a vector field as three independent real signals.

5.3 Two-Sided Quaternion Fourier Transform

Another approach for transforming a function F including the main QFT Theorems can be found in [2], and its generalization in [13]. Hitzer also stated a Plancherel Theorem for the QFT and, together with Mawardi, extended the theory to higher-dimensional Clifford algebras in [14]. Let $F : \mathbb{R}^2 \to G_3^+$; then the QFT is defined as

$$\mathcal{F}\{F\}(u) = \int_{\mathbb{R}^2} e^{-2\pi e_1 e_2 x_1 u_1} F(x_1, x_2) e^{-2\pi e_2 e_3 x_2 u_2} |dx|. \tag{54}$$

The inverse QFT is given by

$$\mathcal{F}^{-1}\{F\}(x) = \int_{\mathbb{R}^2} e^{2\pi e_1 e_2 x_1 u_1} F(u_1, u_2) e^{2\pi e_2 e_3 x_2 u_2} |du|. \tag{55}$$

Though the usual decomposition of F into four real-valued resp. two complex-valued signals is possible via

$$F = F_0 + e_1 e_2 F_1 + (F_2 + e_1 e_2 F_3) e_2 e_3, \tag{56}$$

and there are several options to embed two real variables in a quaternion for applying the transform to a real vector field, the Two-Sided QFT is different from the 2D-CFT. Not only the multiplication from two sides and using two distinct axes of transformation at once lead to different numerical results; even if the Fourier kernel is all right-sided, we cannot merge the two exponentials, because the functional equation does not hold for arbitrary quaternions.

Investigating the precise relationship of both transforms is left for future work.

6 Conclusion

It has been shown that a convolution of vector fields is a nice asset for the analysis of fluid flow simulations or physical velocity measurements. Geometric algebra allows a formulation of a suitable convolution as closed operation. Furthermore, one can define a Clifford–Fourier transform in two- and three- dimensional Euclidean spaces that allows the well-known theorems like convolution theorem, derivative theorem, and Parseval's theorem. Looking into the two CFT transforms reveals that they look at the vector field in a totally different manner, i.e., the 2D-CFT transforms the vector field as one complex signal, while the 3D-CFT transforms the vector field as three independent real signals. This mismatch can be interpreted by the BiQFT of Sangwine et al. which needs an element $\mu \in G_3$ with $\mu^2 = -1$. For a pure bivector, this means choosing a planar direction in which the vector field is transformed as complex signal. The perpendicular part of the vector field is independently transformed as real signal. For $\mu = i_3$, one gets the 3D-CFT. We have also shown that several constructions do not allow a use of the Fourier series approach based on [9].

Acknowledgements We want to thank our former colleague Julia Ebling for her work. We also thank Heinz Krüger, TU Kaiserslautern, for many valuable hints and comments.

References

1. Batchelor, G.K.: An Introduction to Fluid Dynamics. Cambridge University Press, Cambridge (1967)
2. Bülow, T., Felsberg, M., Sommer, G.: Non-commutative hypercomplex Fourier transforms of multidimensional signals. In: Geometric Computing with Clifford Algebra, pp. 187–207. Springer, Berlin (2001)
3. Cibura, C.: Geometric algebra approach to fluid dynamics. Presentation at AGACSE 2008
4. Dorst, L., Fontijne, D., Mann, S.: Geometric Algebra for Computer Science. Morgan Kaufmann, San Mateo (2007)
5. Ebling, J., Scheuermann, G.: Clifford convolution and pattern matching on vector fields. In: IEEE Visualization 2003 Proceedings, pp. 193–200. IEEE Comput. Soc., Los Alamitos (2003)
6. Ebling, J., Scheuermann, G.: Clifford Fourier transform on vector fields. IEEE Trans. Vis. Comput. Graph. **11**(4), 469–479 (2005)
7. Ebling, J., Scheuermann, G., van der Wall, B.G.: Analysis and visualization of 3-c piv images from hart ii using image processing methods. In: Brodlie, K.W., Duke, D.J., Joy, K.I. (eds.) EUROGRAPHICS—IEEE VGTC Symposium on Visualization 2005 Proceedings, pp. 161–168. IEEE Comput. Soc., Los Alamitos (2005)
8. Granlund, G.H., Knutsson, H.: Signal Processing for Computer Vision. Kluwer Academic, Dordrecht (1995)
9. Gürlebeck, K., Habetha, K., Sprößig, W.: Funktionentheorie in der Ebene und im Raum. Birkhäuser, Basel (2006)
10. Hauser, H., Hagen, H., Theisel, H.: Topology-Based Methods in Visualization. Springer, Berlin (2007)
11. Heiberg, E.B., Ebbers, T., Wigström, L., Karlsson, M.: Three dimensional flow characterization using vector pattern matching. IEEE Trans. Vis. Comput. Graph. **9**(3), 313–319 (2003)
12. Hestenes, D.: New Foundations for Classical Mechanics. Kluwer Academic, Dordrecht (1986)
13. Hitzer, E.: Quaternion Fourier transform on quaternion fields and generalizations. In: Proceedings of Function Theories in Higher Dimensions (2006)
14. Hitzer, E., Mawardi, B.: Clifford Fourier transform on multivector fields and uncertainty principles for dimensions $n = 2$ (mod 4) and $n = 3$ (mod 4). In: Proceedings of 7th Int. Conf. on Clifford Algebras and their Applications (2005)
15. Jähne, B.: Digitale Bildverarbeitung, 5th edn. Springer, Berlin (2000)
16. Raffel, M., et al.: Recording and evaluation methods of piv investigations on a helicopter rotor model. Exp. Fluids **36**, 146–156 (2004)
17. Roth, M.: Automatic extraction of vortex core lines and other line-type features for scientific visualization. Ph.D. thesis, ETH Zürich (2000)
18. Sangwine, S.J., Ell, T.: Hypercomplex Fourier transforms of color images. IEEE Trans. Image Process. **16**, 22–35 (2007)
19. Sangwine, S.J., Le Bihan, N., Said, S.: Fast complexified Fourier transform. IEEE Trans. Signal Process. **56**(4), 1522–1531 (2007)
20. Scheuermann, G., Tricoche, X.: Topological methods in flow visualization. In: The Visualization Handbook, pp. 341–358. Elsevier, Amsterdam (2005)
21. Schroeder, W., Martin, K.W., Lorensen, B.: The Visualization Toolkit, 2nd edn. Prentice-Hall, New York (1998)
22. Versteeg, H.K., Malalasekera: An Introduction to Computational Fluid Dynamics—The Finite Volume Method. Pearson Education, Upper Saddle River (2007)

Clifford–Fourier Transform for Color Image Processing

Thomas Batard, Michel Berthier, and Christophe Saint-Jean

Abstract The aim of this paper is to define a Clifford–Fourier transform that is suitable for color image spectral analysis. There have been many attempts to define such a transformation using quaternions or Clifford algebras. We focus here on a geometric approach using group actions. The idea is to generalize the usual definition based on the characters of abelian groups by considering group morphisms from $\mathbb{R}^2$ to spinor groups Spin(3) and Spin(4). The transformation we propose is parameterized by a bivector and a quadratic form, the choice of which is related to the application to be treated. A general definition for 4D signal defined on the plane is also given; for particular choices of spinors, it coincides with the definitions of S. Sangwine and T. Bülow.

1 Introduction

During the last years several attempts have been made to generalize the classical approach of scalar signal processing with the Fourier transform to higher-dimensional signals. The reader will find a detailed overview of the related works at the beginning of [1]. We only mention in this introduction some of the approaches investigated by several authors.

Motivated by the spectral analysis of color images, S. Sangwine and T. Ell have proposed in [13] and [5] a generalization based on the use of quaternions: a color corresponds to an imaginary quaternion, and the imaginary complex i is replaced by the unit quaternion μ coding the grey axis. A quaternionic definition is also given by T. Bülow and G. Sommer in the context of analytic signals, for signals defined on the plane and with values in the algebra $\mathbb{H}$ of quaternions [3]. Concerning analytic signals, M. Felsberg makes use of the Clifford algebras $\mathbb{R}_{2,0}$ and $\mathbb{R}_{3,0}$ to define an appropriate Clifford–Fourier transform [6].

T. Batard (✉)
Lab. MIA, La Rochelle University, La Rochelle, France
e-mail: tbatar01@univ-lr.fr

E. Bayro-Corrochano, G. Scheuermann (eds.), *Geometric Algebra Computing*,
DOI 10.1007/978-1-84996-108-0_8,

A generalization in the Clifford algebras context appears also in J. Ebling and G. Scheuermann [4]. The authors mainly use their transformation to analyze frequencies of vector fields. Using the same Fourier kernel, B. Mawardi and E. Hitzer obtain an uncertainty principle for $\mathbb{R}_{3,0}$ multivector functions [11]. The reader may find in [1] definitions of Clifford–Fourier transform and Clifford–Gabor filters based on the Dirac operator and Clifford analysis.

One could ask the reason why to propose a new generalization. An important thing when dealing with Fourier transform is its link with group representations. We then recall in Sect. 2 the usual definition of the Fourier transform of a function defined on an abelian Lie group by means of the characters of the group. The definition we propose in Sect. 3 relies mainly on the generalization of the notion of characters; that is why we study the group morphisms from $\mathbb{R}^2$ to Spin(3) and Spin(4). These morphisms help to understand the behavior of the Fourier transform with respect to well chosen spinors. We treat in Sect. 4 three applications corresponding to specific bivectors of $\mathbb{R}_{4,0}$. They consist in filtering frequencies according to color, hue, and chrominance part of a given color. In Sect. 5, we show that for particular choices of group morphisms and under well-chosen identification with quaternions, the Clifford–Fourier transform we propose coincides with the definitions of S. Sangwine [5] and T. Bülow [2].

2 Fourier Transform and Group Actions

Let us recall briefly some basic ideas related to the group approach of the definition of the Fourier transform. Details can be found in the Appendix; see also [15] for examples of applications to Fourier descriptors.

Let G be a Lie group. The Pontryagin dual of G, denoted $\widehat{G}$, is the set of equivalence classes of unitary irreducible representations of G. It appears that if G is abelian, every irreducible unitary representation of G is of dimension 1, i.e., is a continuous group morphism from G to S^1. This is precisely the definition of a character. It is well known that the characters of $\mathbb{R}^m$ are given by

$$(x_1, \dots, x_m) \longmapsto e^{i(u_1 x_1 + \dots + u_m x_m)}$$

with real $u_1, u_2, \dots, u_m$. This shows that $\widehat{\mathbb{R}^m} = \mathbb{R}^m$. The characters of SO(2) are the group morphisms

$$\theta \longmapsto e^{in\theta}$$

for $n \in \mathbb{Z}$, and the corresponding Pontryagin dual is $\mathbb{Z}$. The characters of $\mathbb{Z}/n\mathbb{Z}$ are the group morphisms

$$u \longmapsto e^{i\frac{2\pi k u}{n}}$$

for $k \in \mathbb{Z}/n\mathbb{Z}$, from which we deduce that $\mathbb{Z}/n\mathbb{Z}$ is its own Pontryagin dual.

In the general case (provided that G is unimodular), the Fourier transform of a function $f \in L^2(G; \mathbb{C})$ is defined on $\widehat{G}$ by

$$\widehat{f}(\varphi) = \int_G f(x)\varphi(x^{-1})\,d\nu(x)$$

(for ν a well-chosen invariant measure on G). Applying this formula to the case $G = \mathbb{R}^m$, resp. $G =$SO(2), resp. $G = \mathbb{Z}/n\mathbb{Z}$ leads to the usual definition of the Fourier transform, resp. Fourier coefficients, resp. discrete Fourier transform.

Traditionally, the Fourier transform in $L^2(\mathbb{R}^m, (\mathbb{R}^n, \|\|_2))$ is defined by n standard Fourier transforms in $L^2(\mathbb{R}^m, \mathbb{R})$ on each one of the components, embedding $\mathbb{R}$ into $\mathbb{C}$. Using group representations theory, we are able to define Fourier transforms that treat jointly the different components.

From now on, we deal with the abelian group $G = (\mathbb{R}^2, +)$ since this paper is devoted to image processing applications.

Let us make some crucial remarks about the case $n = 2$.

Let f be a real- or complex-valued function defined on $\mathbb{R}^2$. Its Fourier transform is given by

$$\widehat{f}(a, b) = \int_{\mathbb{R}^2} f(x, y)e^{-i(ax+by)}\,dx\,dy.$$

Identifying $\mathbb{C}$ with $(\mathbb{R}^2, \|\|_2)$, we have $S^1 = \mathrm{SO}(2)$, and the action of S^1 on $\mathbb{C}$, given by the complex multiplication, corresponds to the action of the group SO(2) on $(\mathbb{R}^2, \|\|_2)$. Hence, we can define a Fourier transform in $L^2(\mathbb{R}^2, (\mathbb{R}^2, \|\|_2))$ using the action of group morphisms from $\mathbb{R}^2$ to SO(2) on $(\mathbb{R}^2, \|\|_2)$. These ones are real unitary representations of the group $\mathbb{R}^2$ of dimension 2.

The Fourier transform of $f \in L^2(\mathbb{R}^2, (\mathbb{R}^2, \|\;\|_2))$ defined above can be written in the Clifford algebra language. Indeed, from the embedding of $(\mathbb{R}^2, \|\|_2)$ into $\mathbb{R}_{2,0}$, f may be viewed as an $\mathbb{R}^1_{2,0}$-valued function

$$f(x, y) = f_1(x, y)e_1 + f_2(x, y)e_2,$$

where $e_1^2 = e_2^2 = 1$ and $e_1e_2 = -e_2e_1$. From this point of view, the Fourier transform of f is given by

$$\begin{aligned}\widehat{f}(a, b) = \int_{\mathbb{R}^2} &\big[\cos\big((ax + by)/2\big)1 + \sin\big((ax + by)/2\big)e_1e_2\big] \\ &\times \big(f_1(x, y)e_1 + f_2(x, y)e_2\big) \\ &\times \big[\cos\big(-(ax + by)/2\big)1 + \sin\big(-(ax + by)/2\big)e_1e_2\big]\,dx\,dy\end{aligned}$$

using the fact that the action of Spin(2) on $\mathbb{R}^1_{2,0}$ corresponds to the action of SO(2) on $(\mathbb{R}^2, \|\;\|_2)$ (see Appendix). We can write this last formula in the following form:

$$\widehat{f}(a, b) = \int_{\mathbb{R}^2} \big(f_1(x, y)e_1 + f_2(x, y)e_2\big) \perp \varphi_{a,b}(-x, -y)\,dx\,dy,$$

where $\varphi_{a,b}$ is the morphism from $\mathbb{R}^2$ to Spin(2) that sends (x, y) to $\exp[((ax + by)/2)(e_1e_2)]$, and $\perp$ denotes the action $v \perp s = s^{-1}vs$ of Spin(2) on $\mathbb{R}^1_{2,0}$ and, more generally, the action of Spin(n) on $\mathbb{R}^1_{n,0}$.

Note that group morphisms from $\mathbb{R}^2$ to Spin(2) followed by the action on $\mathbb{R}^1_{2,0}$ correspond to the action of group morphisms from $\mathbb{R}^2$ to SO(2) on $(\mathbb{R}^2, \| \|_2)$. In other words, they are real unitary representations of $\mathbb{R}^2$ of dimension 2 too.

Remark 1 As in the standard case, where the Fourier transform of a real-valued function is defined by embedding $\mathbb{R}$ into $\mathbb{C}$, we define here the Fourier transform of a real-valued function by embedding $\mathbb{R}$ into $\mathbb{R}^2$.

Starting from these elementary observations, we now proceed to generalize this construction for $\mathbb{R}^n$-valued functions defined in $\mathbb{R}^2$. In other words, we are looking for a generalization of the action of group morphisms to SO(2) on the values of an $(\mathbb{R}^2, \| \|_2)$-valued function.

3 Clifford–Fourier Transform in $L^2(\mathbb{R}^2, (\mathbb{R}^n, Q))$

Let $f \in L^2(\mathbb{R}^2, (\mathbb{R}^n, Q))$ where Q is a positive definite quadratic form. We propose to associate the Fourier transform of f with the action of the following group morphisms on the values of f, depending on the parity of n.

If n is even, then we consider the morphisms

$$\varphi : \mathbb{R}^2 \longrightarrow \mathrm{SO}(Q).$$

If n is odd, then we embed $(\mathbb{R}^n, Q)$ into $(\mathbb{R}^{n+1}, Q \oplus 1)$ and consider the morphisms

$$\varphi : \mathbb{R}^2 \longrightarrow \mathrm{SO}(Q \oplus 1).$$

Thus the generalization we propose is based on the computation of real unitary representations of dimension n or $n+1$ of the abelian group $\mathbb{R}^2$. The main fact is that we no more consider equivalent classes of representations. This means in particular that the Fourier transform we define depends on the positive definite quadratic form of $\mathbb{R}^n$.

Remark 2 Recall that up to a change of the basis, a positive definite quadratic form is given by the identity matrix. Thus, f may always be viewed as an $(\mathbb{R}^p, \| \|_2)$-valued function (p denotes n if n is even and $n + 1$ if n is odd). As a consequence of the change of the basis, SO(Q) become SO(p) and group morphisms from $\mathbb{R}^2$ to SO(Q) become group morphisms from $\mathbb{R}^2$ to SO(p).

As for the case of $\mathbb{R}^2$-valued functions, we can rewrite the Fourier transform in the Clifford algebra language, using the fact that the action of Spin(p) on $\mathbb{R}^1_{p,0}$ corresponds to the action of SO(p) on $\mathbb{R}^p$. Moreover, it appears to be much more easier to compute group morphisms to Spin(p) rather than group morphisms to the matrix group SO(p).

If n is even, then from the embedding of $\mathbb{R}^n$ into $\mathbb{R}_{n,0}$, f may be viewed as an $\mathbb{R}^1_{n,0}$-valued function:

$$f(x,y) = f_1(x,y)e_1 + f_2(x,y)e_2 + \cdots + f_n(x,y)e_n,$$

where $e_i^2 = 1$ and $e_i e_j = -e_j e_i$. Denoting by φ a group morphism from $\mathbb{R}^2$ to Spin(n), we define the Clifford–Fourier transform of f by

$$\widehat{f}(\varphi) = \int_{\mathbb{R}^2} \varphi(x,y) f(x,y) \varphi(-x,-y)\, dx\, dy = \int_{\mathbb{R}^2} f(x,y) \perp \varphi(-x,-y)\, dx\, dy.$$

If n is odd, we first embed $\mathbb{R}^n$ into $\mathbb{R}^{n+1}$. Then, from the embedding of $\mathbb{R}^{n+1}$ into $\mathbb{R}_{n+1,0}$, f may be viewed as an $\mathbb{R}^1_{n+1,0}$-valued function:

$$f(x,y) = f_1(x,y)e_1 + f_2(x,y)e_2 + \cdots + f_n(x,y)e_n + 0e_{n+1},$$

where $e_i^2 = 1$ and $e_i e_j = -e_j e_i$. Denoting by φ a group morphism from $\mathbb{R}^2$ to Spin$(n+1)$, we define the Clifford–Fourier transform of f by

$$\widehat{f}(\varphi) = \int_{\mathbb{R}^2} \varphi(x,y) f(x,y) \varphi(-x,-y)\, dx\, dy = \int_{\mathbb{R}^2} f(x,y) \perp \varphi(-x,-y)\, dx\, dy.$$

Remark 3 If n is even, the Clifford–Fourier transform of f is an $\mathbb{R}^1_{n,0}$-valued function. If n is odd, the Clifford–Fourier transform of f is an $\mathbb{R}^1_{n+1,0}$-valued function.

Remark 4 For $Q = 1$ on $\mathbb{R}$, the Fourier transforms we define correspond to the standard Fourier transforms of $\mathbb{R}$-valued functions.

From now on, we deal with the case $n = 3$ since this paper is devoted to color image processing. However, we have seen above that we treat the cases $n = 3$ and $n = 4$ in the same manner, by computing group morphisms from $\mathbb{R}^2$ to Spin(4).

3.1 *The Cases n* = 3, 4*: Group Morphisms from* $\mathbb{R}^2$ *to* Spin(4)

This part is devoted to the computation of group morphisms from $\mathbb{R}^2$ to Spin(4).

Using the fact that the group Spin(4) is isomorphic to the group Spin(3) × Spin(3), we first compute group morphisms from $\mathbb{R}^2$ to Spin(3).

One can verify that Spin(3) is the group

$$\text{Spin}(3) = \left\{a1 + be_1e_2 + ce_2e_3 + de_3e_1,\ a^2 + b^2 + c^2 + d^2 = 1\right\}$$

and is isomorphic to the group of unit quaternions.

Proposition 1 *The group morphisms from* $\mathbb{R}^2$ *to* Spin(3) *are given by*

$$(x,y) \longmapsto e^{\frac{1}{2}(ux+vy)B},$$

where B *belongs to* $\mathbb{S}^2_{3,0}$, *the set of unit bivectors in* $\mathbb{R}_{3,0}$ *(see Appendix), and* u *and* v *are real.*

Proof We have to determine the abelian subalgebras of the Lie algebra $\mathfrak{spin}(3) = \mathbb{R}^2_{3,0}$ of the Lie group Spin(3). More precisely, as the exponential map of $\mathbb{R}^2$ is onto, group morphisms from $\mathbb{R}^2$ to Spin(3) are given by Lie algebra morphisms from the abelian Lie algebra $\mathfrak{R}^2$ of $\mathbb{R}^2$ to $\mathfrak{spin}(3)$. Taking two generators (f_1, f_2) of $\mathfrak{R}^2$, any morphism φ from $\mathfrak{R}^2$ to $\mathfrak{spin}(3)$ satisfies

$$\varphi(f_1) \times \varphi(f_2) = 0.$$

We deduce that $\mathrm{Im}(\varphi)$ is an abelian subalgebra of $\mathbb{R}^2_{3,0}$ whose dimension is inferior or equal to 2. If $a = a_1e_1e_2 + a_2e_3e_1 + a_3e_2e_3$ and $b = b_1e_1e_2 + b_2e_3e_1 + b_3e_2e_3$ satisfy $a \times b = 0$, then the structure relations of $\mathbb{R}^2_{3,0}$, i.e.,

$$e_1e_2 \times e_3e_1 = e_2e_3, \qquad e_3e_1 \times e_2e_3 = e_1e_2, \qquad e_2e_3 \times e_1e_2 = e_3e_1,$$

imply

$$(a_1b_2 - a_2b_1)e_2e_3 - (a_1b_3 - a_3b_1)e_3e_1 + (a_2b_3 - a_3b_2)e_1e_2 = 0.$$

This shows that two commuting elements of $\mathbb{R}^2_{3,0}$ are colinear and that the abelian subalgebras of $\mathbb{R}^2_{3,0}$ are of dimension 1. If we write $\varphi(f_1) = \frac{1}{2}uB$ and $\varphi(f_2) = \frac{1}{2}vB$ for some $u, v \in \mathbb{R}$ and $B \in \mathbb{S}^2_{3,0}$, we see that the morphisms from $\mathfrak{R}^2$ to $\mathbb{R}^2_{3,0}$ are parameterized by two real numbers and one unit bivector and are given by

$$\varphi_{u,v,B} : (x,y) \mapsto \frac{1}{2}(ux+vy)B.$$

Consequently, the group morphisms from $\mathbb{R}^2$ to Spin(3) are the morphisms $\widetilde{\varphi}_{u,v,B}$ with

$$\widetilde{\varphi}_{u,v,B} : (x,y) \mapsto e^{\frac{1}{2}(ux+vy)B}.$$

□

Let us recall what group is Spin(4). Every τ in Spin(4) is of the form

$$\begin{aligned}\tau &= u + Iv \\ &= (a1 + be_1e_2 + ce_2e_3 + de_3e_1) + I\big(a'1 + b'e_1e_2 + c'e_2e_3 + d'e_3e_1\big),\end{aligned}$$

where I denotes the pseudoscalar of $\mathbb{R}_{4,0}$, and the following relations hold:

$$u\overline{u} + v\overline{v} = 1, \qquad u\overline{v} + v\overline{u} = 0.$$

The morphism $\chi : \mathrm{Spin}(4) \longrightarrow \mathrm{Spin}(3) \times \mathrm{Spin}(3)$ with

$$\chi(u + Iv) = (u + v, u - v)$$

is an isomorphism. An alternative description of Spin(4) relies on the following fact: the morphism $\psi : \mathbb{H}_1 \times \mathbb{H}_1 \longrightarrow \mathrm{SO}(4)$ defined by

$$(\tau, \rho) \longmapsto (v \longmapsto \tau v \overline{\rho})$$

(where v is a vector of $\mathbb{R}^4$ considered as a quaternion) is a universal covering of SO(4) (see [12]). This means that Spin(4) is isomorphic to $\mathbb{H}_1 \times \mathbb{H}_1$. We will use this remark later on to compare our transform to Sangwine's and Bülow's ones.

Proposition 2 *The group morphisms from $\mathbb{R}^2$ to Spin(4) are the morphisms $\widetilde{\phi}_{u,v,B,w,z,C}$ that send (x, y) to*

$$e^{\frac{1}{8}[x(u+w)+y(v+z)][B+C+I(B-C)]}\, e^{\frac{1}{8}[x(u-w)+y(v-z)][B-C+I(B+C)]}$$

with u, v, w, z real and B, C two elements of $\mathbb{S}^2_{3,0}$.

Proof The group law of $\mathrm{Spin}(3) \times \mathrm{Spin}(3)$ being

$$\big((a, b), (c, d)\big) \rightarrow (ac, bd),$$

the group morphisms from $\mathbb{R}^2$ to $\mathrm{Spin}(3) \times \mathrm{Spin}(3)$ are the morphisms $\widetilde{\varphi}_{u,v,B,w,z,C}$ defined by

$$\widetilde{\varphi}_{u,v,B,w,z,C} : (x, y) \mapsto \big(e^{\frac{1}{2}(ux+vy)B}, e^{\frac{1}{2}(wx+zy)C}\big)$$

with u, v, w, z real and B, C two elements of $\mathbb{S}^2_{3,0}$.

By χ^{-1}, the group morphisms from $\mathbb{R}^2$ to Spin(4) are the $\widetilde{\phi}_{u,v,B,w,z,C}$ that send (x, y) to

$$\frac{e^{\frac{1}{2}(ux+vy)B} + e^{\frac{1}{2}(wx+zy)C}}{2} + I\frac{e^{\frac{1}{2}(ux+vy)B} - e^{\frac{1}{2}(wx+zy)C}}{2}.$$

However, this writing is not convenient to determine group morphisms to SO(4) since it does not provide explicitly the rotations in $\mathbb{R}^4$ that $\widetilde{\phi}_{u,v,B,w,z,C}$ generates by its action on $\mathbb{R}^1_{4,0}$. The solution comes from an "orthogonalization" of the corresponding Lie algebras morphism from $\mathfrak{R}^2$ to $\mathbb{R}^2_{4,0}$, namely the linear map

$$\phi_{u,v,B,w,z,C}(X, Y) = T_{(0,0)}\, \widetilde{\phi}_{u,v,B,w,z,C}(X, Y),$$

where T denotes the linear tangent map. By definition,

$$\phi_{u,v,B,w,z,C}(X, Y) = \frac{d}{dt}\big(\widetilde{\phi}_{u,v,B,w,z,C}\big(\exp\big(t(X, Y)\big)\big)\big)\big|_{t=0}.$$

The exponential map of $\mathbb{R}^2$ being the identity map, we get

$$\begin{aligned}\phi_{u,v,B,w,z,C}(X,Y) &= \frac{d}{dt}\big(\widetilde{\phi}_{u,v,B,w,z,C}\big(t(X,Y)\big)\big)\big|_{t=0}\\ &= \frac{d}{dt}\left(\frac{e^{\frac{1}{2}t(uX+vY)B}+e^{\frac{1}{2}t(wX+zY)C}}{2}\right.\\ &\quad \left.+I\frac{e^{\frac{1}{2}t(uX+vY)B}-e^{\frac{1}{2}t(wX+zY)C}}{2}\right)\Bigg|_{t=0}\\ &= \frac{(uX+vY)B+(wX+zY)C}{4}\\ &\quad +I\frac{(uX+vY)B-(wX+zY)C}{4}.\end{aligned}$$

The orthogonalization of the morphism $\phi_{u,v,B,w,z,C}$ consists in decomposing the bivector $\phi_{u,v,B,w,z,C}(X,Y)$ for each X,Y into commuting bivectors whose squares are real. The corresponding spinor is written as a product of commuting spinors of the form e^{F_i} with $F_i^2<0$. These ones represent rotations of angle $-F_i^2$ in the oriented planes given by the F_i's. In our case, the bivector $\phi_{u,v,B,w,z,C}(X,Y)$ is decomposed into F_1+F_2 where

$$\begin{aligned}F_1 &= \frac{1}{8}\big[\big(X(u+w)+Y(v+z)\big)\big(B+C+I(B-C)\big)\big],\\ F_2 &= \frac{1}{8}\big[\big(X(u-w)+Y(v-z)\big)\big(B-C+I(B+C)\big)\big]\end{aligned}$$

(see the Appendix for details). The group morphisms $\widetilde{\phi}_{u,v,B,w,z,C}$ from $\mathbb{R}^2$ to Spin(4) can then be written as

$$\begin{aligned}\widetilde{\phi}_{u,v,B,w,z,C}(x,y) &= e^{[\frac{(ux+vy)B+(wx+zy)C}{4}+I\,\frac{(ux+vy)B-(wx+zy)C}{4}]}\\ &= e^{\frac{1}{8}[(x(u+w)+y(v+z))(B+C+I(B-C))]}\\ &\quad \times e^{\frac{1}{8}[(x(u-w)+y(v-z))(B-C+I(B+C))]}.\end{aligned}$$ □

This is a convenient form to describe group morphisms from $\mathbb{R}^2$ to SO(4).

To conclude this part, let us remark that the expression of the morphisms $\widetilde{\phi}_{u,v,B,w,z,C}$ may be simplified. Indeed, when B and C describe $\mathbb{S}^2_{3,0}\subset\mathbb{R}_{4,0}$, the unit bivectors

$$D=\frac{1}{4}\big(B+C+I(B-C)\big)\quad\text{and}\quad ID=\frac{1}{4}\big(B-C+I(B+C)\big)$$

describe $\mathbb{S}^2_{4,0}$, the set of unit bivectors in $\mathbb{R}_{4,0}$.

Therefore, the morphisms $\widetilde{\phi}_{u,v,B,w,z,C}$ are parameterized by four real numbers and one unit bivector $D \in \mathbb{S}^2_{4,0}$ and may be written

$$\widetilde{\Phi}_{u,v,w,z,D}(x,y) = e^{\frac{1}{2}[(x(u+w)+y(v+z))D]} e^{\frac{1}{2}[(x(u-w)+y(v-z))ID]}.$$

3.2 The Cases $n = 3, 4$: The Clifford–Fourier Transform

From the computation of group morphisms from $\mathbb{R}^2$ to Spin(4), we give an explicit formula of the Clifford–Fourier transform $\widehat{f}$ of $f \in L^2(\mathbb{R}^2, (\mathbb{R}^3, Q))$ or $L^2(\mathbb{R}^2, (\mathbb{R}^4, Q))$.

Definition 1 Let $f \in L^2(\mathbb{R}^2, (\mathbb{R}^3, Q))$ resp. $L^2(\mathbb{R}^2, (\mathbb{R}^4, Q))$ and denote by f the embedding of f into the Clifford algebra $Cl(\mathbb{R}^4, Q \oplus 1)$ resp. $Cl(\mathbb{R}^4, Q)$. The Clifford–Fourier transform of f is given by

$$\begin{aligned}\widehat{f}(u,v,w,z,D) &= \int_{\mathbb{R}^2} f(x,y) \perp \widetilde{\Phi}_{u,v,w,z,D}(-x,-y)\,dx\,dy \\ &= \int_{\mathbb{R}^2} e^{\frac{1}{2}[(x(u+w)+y(v+z))D]} e^{\frac{1}{2}[(x(u-w)+y(v-z))ID]} f(x,y) \\ &\quad \times e^{-\frac{1}{2}[(x(u+w)+y(v+z))D]} e^{-\frac{1}{2}[(x(u-w)+y(v-z))ID]}\,dx\,dy.\end{aligned}$$

Decomposing f as $f_{||} + f_{\perp}$ with respect to the plane generated by the bivector D, we get

$$\begin{aligned}\widehat{f}(u,v,w,z,D) &= \int_{\mathbb{R}^2} f_{||}(x,y) e^{[-(x(u+w)+y(v+z))D]}\,dx\,dy \\ &\quad + \int_{\mathbb{R}^2} f_{\perp}(x,y) e^{[-(x(u-w)+y(v-z))ID]}\,dx\,dy.\end{aligned}$$

Indeed, the plane generated by ID represents the orthogonal of the plane generated by D in $\mathbb{R}^4$.

Proposition 3 *The Clifford–Fourier transform is left-invertible. Its inverse is the map* $\check{}$ *given by*

$$\check{g}(a,b) = \int_{\mathbb{R}^4 \times \mathbb{S}^2_{4,0}} g(u,v,w,z,D) \perp \widetilde{\Phi}_{u,v,w,z,D}(a,b)\,du\,dv\,dw\,dz\,d\nu,$$

where ν *is a unit measure on* $\mathbb{S}^2_{4,0}$.

Proof We have to verify that $\check{} \circ \hat{}(f)(\lambda, \mu) = f(\lambda, \mu)$ for all $(\lambda, \mu) \in \mathbb{R}^2$.

$$\begin{aligned}\check{} \circ \hat{}(f)(\lambda, \mu) = &\int_{\mathbb{R}^4 \times \mathbb{S}^2_{4,0}} \left[\int_{\mathbb{R}^2} f_{||}(x, y) e^{[-(x(u+w)+y(v+z))D]} dx\, dy\right] \\ &\times e^{[(\lambda(u+w)+\mu(v+z))D]} du\, dv\, dw\, dz\, d\nu \qquad (1) \\ &+ \int_{\mathbb{R}^4 \times \mathbb{S}^2_{4,0}} \left[\int_{\mathbb{R}^2} f_{\perp}(x, y) e^{[-(x(u-w)+y(v-z))ID]} dx\, dy\right] \\ &\times e^{[(\lambda(u-w)+\mu(v-z))ID]} du\, dv\, dw\, dz\, d\nu. \qquad (2)\end{aligned}$$

It is sufficient to prove that $(1) = f_{||}(\lambda, \mu)$.

$$\begin{aligned}(1) &= \int_{\mathbb{R}^4 \times \mathbb{S}^2_{4,0}} \int_{\mathbb{R}^2} f_{||}(x, y) e^{[(\lambda-x)(u+w)+(\mu-y)(v+z)]D} dx\, dy\, du\, dv\, dw\, dz\, d\nu \\ &= \int_{\mathbb{R}^4 \times \mathbb{S}^2_{4,0}} \int_{\mathbb{R}^2} f_{||}(x, y) e^{u(\lambda-x)D} e^{w(\lambda-x)D} e^{v(\mu-y)D} \\ &\quad \times e^{z(\mu-y)D} dx\, dy\, du\, dv\, dw\, dz\, d\nu \\ &= \int_{\mathbb{R}^2} \int_{\mathbb{R}^3 \times \mathbb{S}^2_{4,0}} f_{||}(x, y) \left(\int_{\mathbb{R}} e^{u(\lambda-x)D} du\right) e^{w(\lambda-x)D} e^{v(\mu-y)D} \\ &\quad \times e^{z(\mu-y)D} dw\, dv\, dz\, d\nu\, dx\, dy \\ &= \int_{\mathbb{R}^2} \int_{\mathbb{R}^2 \times \mathbb{S}^2_{4,0}} f_{||}(x, y) \delta_{\lambda,x} \left(\int_{\mathbb{R}} e^{w(\lambda-x)D} dw\right) e^{v(\mu-y)D} \\ &\quad \times e^{z(\mu-y)D} dv\, dz\, d\nu\, dx\, dy \\ &= \int_{\mathbb{R}^2} \int_{\mathbb{R} \times \mathbb{S}^2_{4,0}} f_{||}(x, y) \delta_{\lambda,x} \delta_{\lambda,x} \left(\int_{\mathbb{R}} e^{v(\mu-y)D} dv\right) dz\, d\nu\, dx\, dy \\ &= \int_{\mathbb{R}^2} \int_{\mathbb{S}^2_{4,0}} f_{||}(x, y) \delta_{\lambda,x} \delta_{\lambda,x} \delta_{\mu,y} \left(\int_{\mathbb{R}} e^{z(\mu-y)D} dz\right) d\nu\, dx\, dy \\ &= \int_{\mathbb{R}^2} \int_{\mathbb{S}^2_{4,0}} f_{||}(x, y) \delta_{\lambda,x} \delta_{\lambda,x} \delta_{\mu,y} \delta_{\mu,y}\, d\nu\, dx\, dy \\ &= \int_{\mathbb{R}^2} f_{||}(x, y) \delta_{\lambda,x} \delta_{\lambda,x} \delta_{\mu,y} \delta_{\mu,y}\, dx\, dy = f_{||}(\lambda, \mu).\end{aligned}$$

□

4 Application to Color Image Filtering

4.1 Clifford–Fourier Transform of Color Images

For the applications we have in mind to color image filtering, we define a partial Clifford–Fourier transform, i.e., we deal with a subset of the set of unitary group representations of $\mathbb{R}^2$ of dimension 4. The subset we consider will depend of the colors we aim at filtering.

More precisely, we restrict Definition 1 to the set of group morphisms $\widetilde{\Phi}_{u,v,0,0,D}$ where the bivector D is fixed.

Definition 2 (Clifford–Fourier transform with respect to a bivector) Let $f \in L^2(\mathbb{R}^2, (\mathbb{R}^3, Q))$ resp. $L^2(\mathbb{R}^2, (\mathbb{R}^4, Q))$ and denote by f the embedding of f into the Clifford algebra $Cl(\mathbb{R}^4, Q \oplus 1)$ resp. $Cl(\mathbb{R}^4, Q)$. The Clifford–Fourier transform of f with respect to the bivector D is defined by

$$\begin{aligned}\widehat{f}_D(u,v) &= \int_{\mathbb{R}^2} f(x,y) \perp \widetilde{\Phi}_{u,v,0,0,D}(-x,-y)\, dx\, dy \\ &= \int_{\mathbb{R}^2} e^{\frac{1}{2}(xu+yv)ID} e^{\frac{1}{2}(xu+yv)D} f(x,y) e^{-\frac{1}{2}(xu+yv)D} e^{-\frac{1}{2}(xu+yv)ID}\, dx\, dy.\end{aligned}$$

It follows the definition of the Clifford–Fourier transform of a color image.

Definition 3 (Clifford–Fourier transform of a color image) Let I be a color image. We associate to I a function $f \in L^2(\mathbb{R}^2, (\mathbb{R}^3, Q))$ defined by

$$f(x,y) = r(x,y)e_1 + g(x,y)e_2 + b(x,y)e_3 + 0e_4,$$

where r, g, and b correspond to the red, green, and blue levels.

The Clifford–Fourier transform of I with respect to Q and D is the $Cl(\mathbb{R}^4, Q \oplus 1)$-valued function $\widehat{I}_{Q,D}$ defined by

$$\widehat{I}_{Q,D}(u,v) = \widehat{f}_D(u,v) = \int_{\mathbb{R}^2} f(x,y) \perp \widetilde{\Phi}_{u,v,0,0,D}(-x,-y)\, dx\, dy.$$

Thus, given a color image, we define a set of associated Clifford–Fourier transforms parameterized by the set of positive definite quadratic forms on $\mathbb{R}^3$ and unit bivectors in $\mathbb{R}_{4,0}$.

As the Clifford–Fourier transform in $L^2(\mathbb{R}^3, Q)$ and $L^2(\mathbb{R}^4, Q)$, we can show that the Clifford–Fourier transform of a color image is invertible.

Proposition 4 *Let $f \in L^2(\mathbb{R}^2, (\mathbb{R}^3, Q))$, and D be a unit bivector in $Cl(\mathbb{R}^4, Q \oplus 1)$. Then, the Clifford–Fourier transform of f with respect to D is invertible. Its inverse is the map* $\check{}$ *defined by*

$$\check{g}(x,y) = \int_{\mathbb{R}^2} g(u,v) \perp \widetilde{\Phi}_{u,v,0,0,D}(x,y)\, du\, dv.$$

Proof Decomposing f with respect to the plane generated by D as $f = f_{||} + f_{\perp}$, we have

$$\widehat{f_D}(u,v) = \int_{\mathbb{R}^2} \big(f_{||}(x,y) + f_{\perp}(x,y)\big) \perp \widetilde{\Phi}_{u,v,0,0,D}(-x,-y)\,dx\,dy.$$

This can be written

$$\widehat{f_D}(u,v) = \widehat{f_{D_{||}}}(u,v) + \widehat{f_{D_{\perp}}}(u,v),$$

where

$$\begin{aligned}\widehat{f_{D_{||}}}(u,v) &= \int_{\mathbb{R}^2} f_{||}(x,y) \perp \widetilde{\Phi}_{u,v,0,0,D}(-x,-y)\,dx\,dy \\ &= \int_{\mathbb{R}^2} f_{||}(x,y) e^{-(ux+vy)D}\,dx\,dy\end{aligned}$$

and

$$\begin{aligned}\widehat{f_{D_{\perp}}}(u,v) &= \int_{\mathbb{R}^2} f_{\perp}(x,y) \perp \widetilde{\Phi}_{u,v,0,0,D}(-x,-y)\,dx\,dy \\ &= \int_{\mathbb{R}^2} f_{\perp}(x,y) e^{-(ux+vy)ID}\,dx\,dy.\end{aligned}$$

Let us remark that each of the two integrals may be identified with the Fourier transform of a function from $\mathbb{R}^2$ to $\mathbb{C}$. Then, we deduce that there exists an inversion formula (left and right) for the Clifford–Fourier transform $\widehat{f_D}$ given by

$$f(x,y) = \int_{\mathbb{R}^2} \widehat{f_D}(u,v) \perp \widetilde{\Phi}_{u,v,0,0,D}(x,y)\,du\,dv.$$

Indeed, the right term equals

$$\begin{aligned}&\int_{\mathbb{R}^2} \big(\widehat{f_{D_{||}}}(u,v) + \widehat{f_{D_{\perp}}}(u,v)\big) \perp \widetilde{\Phi}_{u,v,0,0,D}(x,y)\,du\,dv \\ &= \int_{\mathbb{R}^2} \widehat{f_{D_{||}}}(u,v) e^{(ux+vy)D}\,du\,dv + \int_{\mathbb{R}^2} \widehat{f_{D_{\perp}}}(u,v) e^{(ux+vy)ID}\,du\,dv. \qquad (3)\end{aligned}$$

Each of these integrals may be identified with the inversion formula of the Fourier transform of a function from $\mathbb{R}^2$ to $\mathbb{C}$; hence,

$$(3) = f_{||}(x,y) + f_{\perp}(x,y) = f(x,y).$$ □

The following proposition is useful for applications and in particular for applications to the frequencies filtering developed in the next section. It gives an integral representation of any 3D-valued signal defined on the plane by 3D-valued cosinusoidal signals. This representation is obtained from the Clifford–Fourier transform with respect to some bivector. In this proposition we show that the representation is invariant with respect to the choice of the bivector. In the discrete case, we obtain a decomposition of the signal as a sum of cosinusoidal signals.

Proposition 5 *Using the previous notation, if B and D are elements of $\mathbb{S}^2_{4,0}$, we have*

$$\begin{aligned}&\widehat{f_B}(u,v)\perp\widetilde{\Phi}_{u,v,0,0,B}(x,y)+\widehat{f_B}(-u,-v)\perp\widetilde{\Phi}_{-u,-v,0,0,B}(x,y)\\&\quad=\widehat{f_D}(u,v)\perp\widetilde{\Phi}_{u,v,0,0,D}(x,y)+\widehat{f_D}(-u,-v)\perp\widetilde{\Phi}_{-u,-v,0,0,D}(x,y).\end{aligned}$$

Moreover, the e_4 component of this expression is null.

Proof Simple computations show that

$$\begin{aligned}&\widehat{f_B}(u,v)\perp\widetilde{\Phi}_{u,v,0,0,B}(x,y)+\widehat{f_B}(-u,-v)\perp\widetilde{\Phi}_{-u,-v,0,0,B}(x,y)\\&=\int_{\mathbb{R}^2}e^{-\frac{xu+yv}{2}(B+IB)}e^{\frac{\lambda u+\mu v}{2}(B+IB)}f(\lambda,\mu)e^{-\frac{\lambda u+\mu v}{2}(B+IB)}e^{\frac{xu+yv}{2}(B+IB)}\,d\lambda\,d\mu\\&\quad+\int_{\mathbb{R}^2}e^{\frac{xu+yv}{2}(B+IB)}e^{-\frac{\lambda u+\mu v}{2}(B+IB)}f(\lambda,\mu)e^{\frac{\lambda u+\mu v}{2}(B+IB)}e^{-\frac{xu+yv}{2}(B+IB)}\,d\lambda\,d\mu\\&=\int_{\mathbb{R}^2}2\cos\bigl(u(x-\lambda)+v(y-\mu)\bigr)f_{\|}(\lambda,\mu)\,d\lambda\,d\mu\\&\quad+\int_{\mathbb{R}^2}2\cos\bigl(u(x-\lambda)+v(y-\mu)\bigr)f_{\perp}(\lambda,\mu)\,d\lambda\,d\mu.\end{aligned}\tag{4}$$

Hence,

$$(4)=\int_{\mathbb{R}^2}2\cos\bigl(u(x-\lambda)+v(y-\mu)\bigr)f(\lambda,\mu)\,d\lambda\,d\mu.\qquad\square$$

This proposition justifies the fact that these filters are symmetric with respect to the transformation $(u,v)\mapsto(-u,-v)$.

4.2 Color Image Filtering

We now present applications to color image filtering. The use of the Fourier transform is motivated by the well-known fact that nontrivial filters in the spatial domain may be implemented efficiently with masks in the Fourier domain. Although it seems natural to believe that the results on grey level images may be generalized, there are not so many works dedicated to the specific case of color images. Let us mention [14], where an attempt is made through the use of an ad hoc quaternionic transform. The mathematical construction we propose appears to be well founded since it explains the fundamental role of bivectors and scalar products in terms of group actions. As explained before, the possibility to choose the bivector D and the quadratic form Q is an asset allowing a wider range of applications. Indeed, Sangwine et al. proposal can be written in our formalism by considering appropriate D and Q.

The applications proposed in this paper are based on the following fact:

$$(\widehat{f_D})_{\|} = \widehat{(f_{\|})_D} \quad \text{and} \quad (\widehat{f_D})_{\perp} = \widehat{(f_{\perp})_D}.$$

In other words, the part of the Clifford–Fourier transform of f that is parallel to D corresponds to the standard Fourier transform of the part of f that is parallel to D. The same principle holds for the orthogonal part.

We use low pass, high pass, and directional filters on the D-parallel part resp. D-orthogonal part, leaving the D-parallel part resp. D-orthogonal unmodified. The choice of the bivector D and the quadratic form Q (that determines the D-orthogonal part) will depend on the colors we aim at filtering. Then, we show the action of such filters using the inversion formula of the Clifford–Fourier transform.

There is another way to decompose a color $\alpha = (r, g, b)$, that is, with respect to its luminance and chrominance parts, respectively denoted by l_α and v_α. Embedding the color space RGB into the Clifford algebra $\mathbb{R}_{4,0}$ by

$$i_\alpha = r\, e_1 + g\, e_2 + b\, e_3 + 0\, e_4,$$

the former corresponds to the projection of i_α on the axis generated by the unit vector $(e_1 + e_2 + e_3)/\sqrt{3}$; the latter its projection on the orthogonal plane in $e_1e_2e_3$, called the chrominance plane, represented by the unit bivector $(e_1e_2 - e_1e_3 + e_2e_3)/3$. In what follows we make use of the following fact too: every hue can be represented as an equivalence class of bivectors of $\mathbb{R}_{4,0}$. More precisely, we have the following result.

Proposition 6 *Let T be the set of bivectors*

$$T = \big\{(e_1 + e_2 + e_3) \wedge i_\alpha,\ \alpha \in \text{RGB}\big\}$$

with the following equivalence relation:

$$B \simeq C \quad \Longleftrightarrow \quad B = \lambda C \quad \text{for } \lambda > 0.$$

Then, there is a bijection between $T/\simeq$ and the set of hues.

Proof We have

$$(e_1 + e_2 + e_3) \wedge i_\alpha = (e_1 + e_2 + e_3) v_\alpha.$$

Then, there is a bijection between $T/\simeq$ and the set $(e_1 + e_2 + e_3)\, v$ for v a unit vector in the chrominance plane. This latter being in bijection with the set of different hues, we conclude that there exists a bijection between $T/\simeq$ and the set of hues. □

(a) Fourier house

(b) Flowers

Fig. 1 Original images

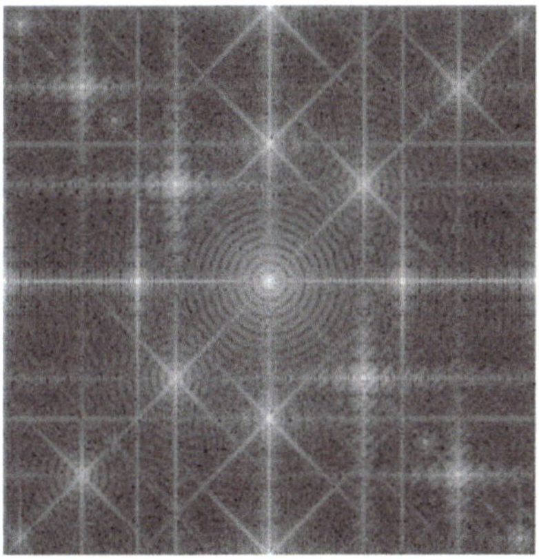

(a) Log-modulus of the parallel part

(b) Directional cut filtering for the red color

Fig. 2 The Clifford–Fourier transform $\widehat{H}_{Q1,e_1e_4}$

Figure 1 shows the original images used for these experiments.[1] Figure 1(a) H is a modified color version of the Fourier house containing red, desatured red, green, cyan stripes in various directions, a uniform red circle and a red square with lower luminance. Figure 1(b) F is a natural image taken from the Berkeley image segmentation database [10].

Figure 2(a) is the centered log-modulus of the D-parallel part of $\widehat{H}_{Q_1,D}$, where Q_1 is the quadratic form such that $Q_1 \oplus 1$ is given by the identity matrix I_4 in the basis (e_1, e_2, e_3, e_4), and D is the bivector e_1e_4. Figure 2(b) is the result of a directional cut filter around $\pi/2$ which removes of vertical frequencies. Let us point

[1] Available at http://mia.univ-larochelle.fr/ → Production → Démos.

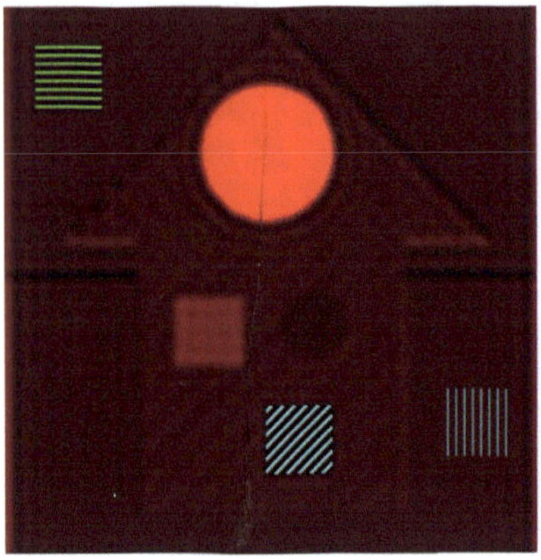

(a) red color

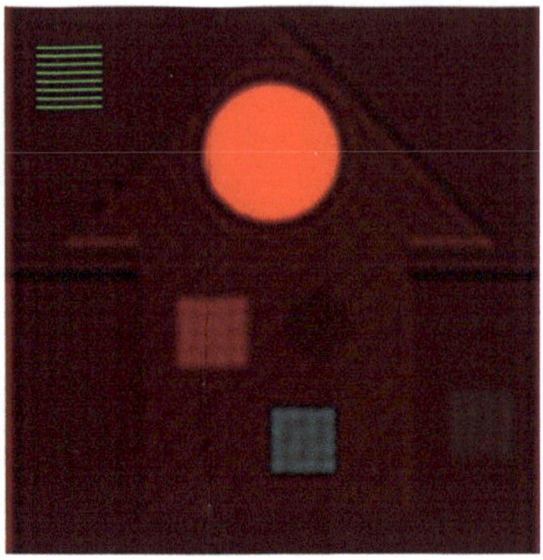

(b) red hue

Fig. 3 Low pass filtering

out that horizontal green stripes are not altered since green color 255 e_2 belongs to $ID = e_2e_3$.

Figure 3 shows the difference between a low pass filter in the D-parallel part of $\widehat{H}_{Q_1,e_1e_4}$ (Fig. 3(a)) and the D-parallel part $\widehat{H}_{Q_1,\frac{1}{\sqrt{2}}(e_2+e_3)e_1}$ (Fig. 3(b)). The first one consists in removing high frequencies of the red components of the image, whereas the second one consists in removing high frequencies of the red hue part of the image.

In Fig. 3(a), we can see that both green and cyan stripes are not modified. As in the previous case, this comes from the fact that both green color and cyan color $255\, e_2 + 255\, e_3$ belong to ID. The result is different in Fig. 3(b). The unit bivector $\frac{1}{\sqrt{2}}(e_2 + e_3)e_1 = \frac{1}{\sqrt{2}}(e_1 + e_2 + e_3) \wedge e_1$ represents the red hue, involving that the cyan stripes are blurred. Indeed, unit bivectors representing cyan and red hues are opposite, and therefore they generate the same plane. Green stripes are no more invariant to the low pass filter since the green axis e_2 is not orthogonal to the bivector $\frac{1}{\sqrt{2}}(e_2 + e_3)e_1$.

In Fig. 4, the color α has been chosen to match with the color of the background green leaves. As the low pass filter (Fig. 4(a)) removes green high frequencies, the center of flowers containing yellow high frequencies turns red. In Fig. 4(b), background pixels corresponding to green low frequencies appear almost grey.

To conclude this part, we propose to compare the results of two low pass filters on the D-orthogonal part with respect to the same bivector $D = e_1e_4$ but changing the quadratic form. As a consequence, the bivector ID differs in the two cases. For the first one (Fig. 5(a)), we take Q_1, whereas for the second one (Fig. 5(b)), we construct the quadratic form Q_2 such that Q_2 is given by I_4 in the basis $(e_1, \frac{1}{\sqrt{2}}(e_1 + e_2), \frac{i_\alpha}{\|i_\alpha\|}, e_4)$. In other words, we orthogonalize the red, the yellow, and the color of leaves which are the main colors in the image.

In Fig. 5(a), the unit bivector ID is e_2e_3. Hence, the low pass filter removes green and blue high frequencies but preserves red high frequencies. This explains

(a) Low pass filter in the parallel part (b) High pass filter in the parallel part

Fig. 4 The Clifford–Fourier transform $\widehat{F}_{Q_1, \frac{v_\alpha \wedge e_4}{\|v_\alpha \wedge e_4\|}}$ with $\alpha = (96, 109, 65)$

(a) The Clifford-Fourier transform $\widehat{F}_{Q_1, e_1 e_4}$ (b) The Clifford-Fourier transform $\widehat{F}_{Q_2, e_1 e_4}$

Fig. 5 Low pass filters in the ID part

why the image turns red. In Fig. 5(b), the unit bivector ID is $\frac{(e_1+e_2)}{\sqrt{2}} \frac{i_\alpha}{\|i_\alpha\|}$; it contains the colors of the background and inside the flowers. Therefore, the low pass filter removes all the high frequencies in the image except the ones of the red petals.

For some specific applications, a fine tuning of the quadratic form Q should give better results.

5 Related Works

To conclude this paper, we show how to recover the hypercomplex Fourier transform of S. Sangwine and the quaternionic Fourier transform of T. Bülow in the Clifford algebras context, using the appropriate morphism from $\mathbb{R}^2$ to Spin(4). First of all, let us recall the definitions of these Fourier transforms.

5.1 The Hypercomplex Fourier Transform of Sangwine et al.

In [5], the authors define the discrete hypercomplex Fourier transform. It can be extended to $\mathbb{R}^2$ as follows. Let $f : \mathbb{R}^2 \to \mathbb{H}$; then its hypercomplex Fourier transform is given by

$$F(u, v) = \int_{\mathbb{R}^2} e^{-\mu(xu+yv)} f(x, y)\, dx\, dy,$$

where $\mu \in \mathbb{H}_0 \cap \mathbb{H}_1$.

There is a freedom in the choice of μ in the hypercomplex Fourier transform as we have a freedom in the choice of the bivector D in the Clifford–Fourier transform for color images. In fact, they have the same role, i.e., they decompose the four-dimensional space $\mathbb{R}^4$ into two orthogonal two-dimensional subspaces and decompose the Fourier transform into two standard Fourier transforms.

This is shown in the following proposition.

Proposition 7 *Let $\mu = \mu_1 i + \mu_2 j + \mu_3 k$ be a unit quaternion. Let $f \in L^2(\mathbb{R}^2, (\mathbb{R}^4, Q))$ where Q is the quadratic form represented by I_4 in the basis (e_1, e_2, e_3, e_4), and let C be the unit bivector $e_4 \wedge (\mu_1 e_1 + \mu_2 e_2 + \mu_3 e_3)$. Then, $\widehat{f_C}$ given by*

$$\begin{aligned}\widehat{f_C}(u, v) &= \int_{\mathbb{R}^2} f(x, y) \rfloor \widetilde{\Phi}_{u,v,0,0,C}(-x, -y)\, dx\, dy \\ &= \int_{\mathbb{R}^2} e^{\frac{1}{2}(xu+yv)IC} e^{\frac{1}{2}(xu+yv)C} f(x, y) e^{-\frac{1}{2}(xu+yv)C} e^{-\frac{1}{2}(xu+yv)IC}\, dx\, dy\end{aligned}$$

corresponds to the hypercomplex Fourier transform of f seen as an $\mathbb{H}$-valued function under the identification[2]

$$c_1 \leftrightarrow i, \qquad e_2 \leftrightarrow j, \qquad e_3 \leftrightarrow k, \qquad e_4 \leftrightarrow 1.$$

Proof We have to determine the four-dimensional rotation that is generated by the action of the unit quaternion $e^{\mu\phi}$ on $\mathbb{H}$ given by

$$q \longmapsto e^{\mu\phi} q.$$

It is explained in [5] that this rotation may be decomposed as the sum of two two-dimensional rotations of angle $-\phi$ in the planes generated by $(1, \mu)$ and its orthogonal (with respect to the euclidean quadratic form).

---

[2]The product law needs not to be respected since we just use an isomorphism of vector spaces.

Therefore, we can identify this rotation with the action of the spinor

$$e^{-\frac{\phi}{2}(C+IC)}$$

on the four-dimensional space $\mathbb{R}^1_{4,0}$. As a consequence, the action of group morphisms $(x, y) \longmapsto e^{\mu(xu+yv)}$ from $\mathbb{R}^2$ to $\mathbb{H}_1$ on $\mathbb{H}$ corresponds to the action of group morphisms $(x, y) \longmapsto e^{-\frac{1}{2}(xu+yv)(C+IC)}$ from $\mathbb{R}^2$ to Spin(4) on $\mathbb{R}^1_{4,0}$. □

Remark 5 To the best of our knowledge, the authors restrict for their applications to μ taken as the grey axis, i.e.,

$$\mu = \frac{1}{\sqrt{3}}(i + j + k).$$

In other words, the Fourier transform they propose is decomposed as a standard Fourier transform of the luminance part and a standard Fourier transform of the chrominance part.

5.2 *The Quaternionic Fourier Transform of Bülow*

The quaternionic Fourier transform [2] of a function $f : \mathbb{R}^2 \to \mathbb{R}$ is the quaternion-valued function $\mathcal{F}(f)$ defined by

$$\mathcal{F}(f)(y_1, y_2) = \int_{\mathbb{R}^2} \exp(-2\pi i y_1 x_1) f(x_1, x_2) \exp(-2\pi j y_2 x_2)\, dx_1\, dx_2.$$

The link between this Fourier transform and the one proposed here is given by the next result.

Proposition 8 *Let* $f \in L^2(\mathbb{R}^2; \mathbb{R}e_4)$ *where* (e_1, e_2, e_3, e_4) *is the basis of* $\mathbb{R}^4$ *that generates* $\mathbb{R}_{4,0}$. *The Clifford–Fourier transform of* f *defined by*

$$\begin{aligned}&\widehat{f}_C(2\pi y_1, 0, 0, 2\pi y_2)\\ &\quad = \int_{\mathbb{R}^2} f(x_1, x_2) \perp \widetilde{\Phi}_{2\pi y_1, 0, 0, 2\pi y_2, C}(-x_1, -x_2)\, dx_1\, dx_2,\end{aligned}$$

where C *is the bivector*

$$-\frac{1}{4}(e_1 + e_2)(e_3 - e_4),$$

corresponds to the quaternionic Fourier transform of f *seen as an* $\mathbb{H}$*-valued function under the following identification:*[3]

$$e_1 \leftrightarrow i, \qquad e_2 \leftrightarrow j, \qquad e_3 \leftrightarrow k, \qquad e_4 \leftrightarrow 1.$$

[3]The product law needs not to be respected since we just use an isomorphism of vector spaces.

Proof We have to determine one of the two elements of Spin(3) × Spin(3) that generate the following rotation in $\mathbb{H}$:

$$f(x_1, x_2) \mapsto \exp(-2\pi i y_1 x_1) f(x_1, x_2) \exp(-2\pi j y_2 x_2).$$

Simple computations show that the rotation

$$f(x_1, x_2) \mapsto \exp(-2\pi i y_1 x_1) f(x_1, x_2)$$

can be written in $\mathbb{R}^1_{4,0}$ as

$$f(x_1, x_2) \mapsto e^{-\pi x_1 y_1 (e_4 e_1 + e_2 e_3)} f(x_1, x_2) e^{\pi x_1 y_1 (e_4 e_1 + e_2 e_3)}.$$

In the same way,

$$f(x_1, x_2) \mapsto f(x_1, x_2) \exp(-2\pi j y_2 x_2)$$

corresponds to

$$f(x_1, x_2) \mapsto e^{-\pi x_2 y_2 (e_4 e_2 + e_1 e_3)} f(x_1, x_2)\, e^{\pi x_2 y_2 (e_4 e_2 + e_1 e_3)}.$$

By associativity, this shows that

$$\exp(-2\pi i y_1 x_1)\, f(x_1, x_2) \exp(-2\pi j y_2 x_2) = e^{-\tau} e^{-\rho} f(x_1, x_2) e^{\rho} e^{\tau},$$

where

$$\tau = \pi x_2 y_2 (e_4 e_2 + e_1 e_3)$$

and

$$\rho = \pi x_1 y_1 (e_4 e_1 + e_2 e_3).$$

By definition,

$$\chi\left(e^{\rho} e^{\tau}\right) = \chi\left(e^{\pi x_1 y_1 e_4 e_1}\right) \chi\left(e^{\pi x_1 y_1 e_2 e_3}\right) \chi\left(e^{\pi x_2 y_2 e_4 e_2}\right) \chi\left(e^{\pi x_2 y_2 e_1 e_3}\right).$$

By simple computations we get

$$\chi\left(e^{\rho} e^{\tau}\right) = \left(e^{2\pi x_1 y_1 e_2 e_3}, e^{2\pi x_2 y_2 e_1 e_3}\right)$$

and conclude therefore that

$$(x_1, x_2) \mapsto \left(e^{2\pi x_1 y_1 e_2 e_3}, e^{2\pi x_2 y_2 e_1 e_3}\right)$$

is the morphism $\widetilde{\phi}_{2\pi y_1, 0, e_2 e_3, 0, 2\pi y_2, e_1 e_3}$.

From Sect. 3, this latter may be rewritten $\widetilde{\Phi}_{2\pi y_1, 0, 0, 2\pi y_2, \frac{1}{4}(e_1 + e_2)(e_3 - e_4)}$.

Indeed, we have

$$\begin{aligned}\frac{1}{4}\big(e_2e_3+e_1e_3+I(e_2e_3-e_1e_3)\big)&=\frac{1}{4}(e_2e_3+e_1e_3-e_1e_4-e_2e_4)\\&=\frac{1}{4}\big(e_1(e_3-e_4)+e_2(e_3-e_4)\big)\\&=\frac{1}{4}\big((e_1+e_2)(e_3-e_4)\big).\end{aligned}$$ □

6 Conclusion

We proposed in this paper a definition of Clifford–Fourier transform that is motivated by group actions considerations. We defined a Clifford–Fourier transform that is associated with the action of all the group morphisms $\widetilde{\Phi}_{u,v,w,z,D}$ from $\mathbb{R}^2$ to Spin(4), parameterized by four real numbers and one unit bivector. This transform has the property of being left invertible. For the particular case of a color image, we associate the Clifford–Fourier transform with the action of group morphisms $\widetilde{\Phi}_{u,v,0,0,D}$, specified by only two real numbers (the frequencies) and where the bivector D is fixed. This transform is parameterized by a quadratic form on $\mathbb{R}^4$ and a unit bivector in the corresponding Clifford algebra. Some previous Fourier transforms based on quaternions are proved to be particular settings of ours. We have treated in this context an application to color image filtering. Future works will be devoted to find applications of the general transform that should easily deal with relations between colors in the image. Applications to multispectral images such as color/infrared images will be also investigated.

Acknowledgement This work is partially supported by the "Communauté d'agglomération de La Rochelle."

Appendix

A.1 Lie Groups Representations and Fourier Transforms

From the group theory approach, the basic structure we need to define Fourier transforms is locally compact unimodular groups. Let us start by the definition of the dual of a topological group G, that is, the set of the equivalence classes of its unitary irreducible representations, denoted by $\widehat{G}$. We refer to [16] for details.

Definition 4 (Group representation) Let G be a topological group, and V be a topological vector space over $\mathbb{R}$ or $\mathbb{C}$.

A continuous linear representation (φ, V) from G to V is a group morphism

$$\varphi : g \mapsto \varphi(g)$$

from G to $GL(V)$ such that the map

$$(a, g) \mapsto \varphi(g)(a)$$

from $V \times G$ to V is continuous.

In general, V is a Hilbert space. If V is finite-dimensional, then the representation is said to be finite, and the dimension of V is called the degree of the representation.

Definition 5 (Irreducible representation) A subspace W of V is said to be invariant by φ if $\varphi(g)(W) \subset W \ \forall g \in G$.

Then, the representation φ is said to be irreducible if W, and $\{0\}$ are the only subspaces of V that are invariant by φ.

Definition 6 (Equivalent representations) Let (φ_1, V_1) and (φ_2, V_2) be two linear representations of the same group G. We say that they are equivalent if there exists an isomorphism $\gamma : V_1 \to V_2$ such that

$$\gamma \circ \varphi_1(g) = \varphi_2(g) \circ \gamma \quad \forall g \in G.$$

From now on, V is a $\mathbb{C}$-vector space equipped with a hermitian form $\langle\, , \rangle$.

Definition 7 (Unitary representation) The representation φ is unitary with respect to $\langle\, , \rangle$ if

$$\big\langle \varphi(g)(a), \varphi(g)(b) \big\rangle = \langle a, b \rangle \quad \forall a, b \in V,\ \forall g \in G.$$

We now restrict to locally compact unimodular groups. On such groups, we can construct a measure that is invariant with respect to both left and right translations. It a called a Haar measure. From a Haar measure a Haar integral of the group is defined.

Proposition 9 *Let G be a locally compact unimodular group, and let ν denote a Haar measure. Then, for $f \in L^2(G; \mathbb{C})$ and $h \in G$, we have*

$$\int_G f(g)\, d\nu(g) = \int_G f(gh)\, d\nu(g) = \int_G f(hg)\, d\nu(g).$$

Remark 6 Locally compact abelian groups and compact groups are unimodular.

Definition 8 (Fourier transform on locally compact unimodular groups) Let G be a locally compact unimodular group with Haar measure ν. The Fourier transform of $f \in L^2(G; \mathbb{C})$ is the map $\hat{f}$ defined on $\widehat{G}$ by

$$\hat{f}(\varphi) = \int_G f(g)\varphi\big(g^{-1}\big)\, d\nu(g).$$

Theorem 1 (Inversion formula of the Fourier transform) *$\hat{f}(\varphi)$ is a Hilbert–Schmidt operator over the space of the representation φ. There is a measure over $\widehat{G}$ denoted by $\widehat{\nu}$ such that $\hat{f} \in L^2(\widehat{G}; \mathbb{C})$ and $f \mapsto \hat{f}$ is an isometry. Moreover, the following inverse formula holds*:

$$f(g) = \int_{\widehat{G}} \mathrm{Trace}\big(\hat{f}(\varphi)\varphi(g)\big)\, d\widehat{\nu}(\varphi).$$

Let us now have a closer look on Lie groups. We refer to [7] for an introduction to differential geometry.

Definition 9 (Lie group and Lie algebra) A real C^∞ Lie group is a topological group endowed with a structure of real C^∞-manifold. The Lie algebra of G is (isomorphic to) the tangent space of G at the neutral element e: T_eG. It is usually denoted by $\mathfrak{g}$. It can be made into an algebra over $\mathbb{R}$ by considering the Lie bracket $[\,,\,]$ that satisfies: $(X, Y) \mapsto [X, Y]$ from $\mathfrak{g} \times \mathfrak{g}$ to $\mathfrak{g}$ is $\mathbb{R}$-bilinear. Moreover, it satisfies

$$[X, X] = 0 \quad \forall X \in \mathfrak{g}$$

and

$$\big[X, [Y, Z]\big] + \big[Y, [Z, X]\big] + \big[Z, [X, Y]\big] = 0 \quad \forall X, Y, Z \in \mathfrak{g}.$$

Definition 10 (Exponential map) Let G be a C^∞ Lie group. The exponential map of G is the map from $\mathfrak{g}$ to G

$$\exp : X \longmapsto f(1),$$

where $f : \mathbb{R} \to G$ satisfies

$$f(t+s) = f(t)f(s) \quad \forall t, s \in \mathbb{R}$$

and

$$f'(0) = X.$$

f is called a one-parameter subgroup.

To compute group morphisms from $\mathbb{R}^2$ to Spin(3) and Spin(4), we use the following result on Lie groups morphisms.

Proposition 10 *Let G and H be two C^∞ Lie groups, and $\exp_G$, $\exp_H$ be the corresponding exponential maps. Let $\phi : G \to H$ be a Lie group morphism. The linear tangent map of ϕ at g, denoted by $T_g\phi$, is the linear map from T_gG to $T_{\phi(g)}H$ given by*

$$T_g\phi(X) = \frac{d}{dt}\phi\big(g \exp_G(tX)\big)|_{t=0}.$$

Then, if we note e the neutral element of G, we have

$$\phi\big(\exp_G(X)\big) = \exp_H\big(T_e\phi(X)\big). \tag{5}$$

The map $T_e\phi$ is a Lie algebra morphism, i.e., it satisfies

$$T_e\phi\big([X,Y]\big) = \big[T_e\phi(X), T_e\phi(Y)\big] \quad \forall X, Y \in \mathfrak{g}.$$

From (5) we deduce that if the group G is connected and the exponential map of G is onto, then the Lie group morphisms from G to H are determined by Lie algebras morphisms from $\mathfrak{g}$ to $\mathfrak{h}$.

A.2 Clifford Algebras

Let V be a vector space of finite dimension n over $\mathbb{R}$ equipped with a quadratic form Q. Formally speaking, the Clifford algebra $Cl(V,Q)$ is the solution of the following universal problem. We search a couple $(Cl(V,Q), i_Q)$ where $Cl(V,Q)$ is an $\mathbb{R}$-algebra and $i_Q : V \longrightarrow Cl(V,Q)$ is $\mathbb{R}$-linear satisfying

$$\big(i_Q(v)\big)^2 = Q(v).1$$

for all v in V (1 denotes the unit of $Cl(V,Q)$) such that, for each $\mathbb{R}$-algebra A and each $\mathbb{R}$-linear map $f : V \longrightarrow A$ with

$$\big(f(v)\big)^2 = Q(v).1$$

for all v in V (1 denotes the unit of A), then there exists a unique morphism

$$g : Cl(V,Q) \longrightarrow A$$

of $\mathbb{R}$-algebras such that $f = g \circ i_Q$.

The solution is unique up to isomorphisms and is given as the (noncommutative) quotient

$$T(V)/\big(v \otimes v - Q(v).1\big)$$

of the tensor algebra of V by the ideal generated by $v \otimes v - Q(v).1$, where v belongs to V (see [12] for a proof).

It is well known that there exists a unique anti-automorphism t on $Cl(V,Q)$ such that

$$t\big(i_Q(v)\big) = i_Q(v)$$

for all v in V. It is called reversion and usually denoted by $x \longmapsto x^\dagger$, x in $Cl(V,Q)$. In the same way there exists a unique automorphism α on $Cl(V,Q)$ such that

$$\alpha\big(i_Q(v)\big) = -i_Q(v)$$

for all v in V. In this paper we write v for $i_Q(v)$ (according to the fact that i_Q embeds V in $Cl(V,Q)$).

As a vector space, $Cl(V,Q)$ is of dimension 2^n on $\mathbb{R}$ and a basis is given by the set

$$\left\{e_{i_1}e_{i_2}\cdots e_{i_k},\ i_1 < i_2 < \cdots < i_k,\ k \in \{1,\ldots,n\}\right\}$$

and the unit 1. An element of degree k

$$\sum_{i_1<\cdots<i_k} \alpha_{i_1\ldots i_k} e_{i_1}e_{i_2}\cdots e_{i_k}$$

is called a k-vector. A 0-vector is a scalar, and $e_1e_2\cdots e_n$ is called the pseudoscalar. We denote $\langle x\rangle_k$ the component of degree k of an element x of $Cl(V,Q)$.

The inner product of x_r of degree r and y_s of degree s is defined by

$$x_r \cdot y_s = \langle x_r y_s\rangle_{|r-s|}$$

if r and s are positive and by

$$x_r \cdot y_s = 0$$

otherwise.

The outer product of x_r of degree r and y_s of degree s is defined by

$$x_r \wedge y_s = \langle x_r y_s\rangle_{r+s}.$$

These products extend by linearity on $Cl(V,Q)$. Clearly, if a and b are vectors of V, then the inner product of a and b coincides with the scalar product defined by Q. When it is defined (for example, when x is a versor and Q is positive), we denote

$$\|x\| = \sqrt{xx^\dagger}$$

and say that x is a unit if $xx^\dagger = \pm 1$.

In this paper, we deal in particular with the Clifford algebra of the Euclidean $\mathbb{R}^n$ denoted by $\mathbb{R}_{n,0}$. $\mathbb{R}^k_{n,0}$ is the subspace of elements of degree k, and $\mathbb{R}^*_{n,0}$ is the group of elements that admit an inverse in $\mathbb{R}_{n,0}$. We denote by $\mathbb{S}^2_{n,0}$ the set of elements of $\mathbb{R}^2_{n,0}$ of norm 1.

Let a be a vector in $\mathbb{R}_{n,0}$, and B be the k-vector $a_1 \wedge a_2 \wedge \cdots \wedge a_k$. Then the orthogonal projection of a on the k-plane generated by the a_i's is the vector

$$P_B(a) = (a \cdot B)B^{-1}.$$

The vector

$$a - (a \cdot B)B^{-1} = (a \wedge B)B^{-1}$$

is called the rejection of a on B.

A.3 *The Spinor Group* Spin(n)

It is defined by

$$\mathrm{Spin}(n) = \left\{ \prod_{i=1}^{2k} a_i,\ a_i \in \mathbb{R}^1_{n,0},\ \|a_i\| = 1 \right\}$$

or equivalently

$$\mathrm{Spin}(n) = \left\{ x \in \mathbb{R}_{n,0},\ \alpha(x) = x,\ xx^\dagger = 1, xvx^{-1} \in \mathbb{R}^1_{n,0}\ \forall v \in \mathbb{R}^1_{n,0} \right\}.$$

It is well known that Spin(n) is a connected compact Lie group that universally covers SO(n) ($n \geq 3$). One can verify that Spin(3) is the group

$$\left\{ a1 + be_1e_2 + ce_2e_3 + de_3e_1,\ a^2 + b^2 + c^2 + d^2 = 1 \right\}$$

and is isomorphic to the group $\mathbb{H}^1$ of unit quaternions. It is also a classical result that Spin(4) is isomorphic to Spin(3) $\times$ Spin(3) (see [9] for more information on spinors in $\mathbb{R}^3$ and $\mathbb{R}^4$).

The Lie algebra of Spin(n) is $\mathbb{R}^2_{n,0}$ with Lie bracket

$$A \times B = AB - BA.$$

As the exponential map from its Lie algebra to Spin(n) is onto (see [7] for a proof), every spinor can be written as

$$S = \sum_{i=0}^{\infty} \frac{1}{i!} A^i$$

for some bivector A.

From Hestenes and Sobczyk [8] we know that every A in $\mathbb{R}^2_{n,0}$ can be written as

$$A = A_1 + A_2 + \cdots + A_m,$$

where $m \leq n/2$, and

$$A_j = \|A_j\| a_j b_j, \quad j \in \{1, \ldots, m\}$$

with

$$\{a_1, \ldots, a_m, b_1, \ldots, b_m\}$$

a set of orthonormal vectors. Thus,

$$A_j A_k = A_k A_j = A_k \wedge A_j$$

whenever $j \neq k$ and

$$A_k^2 = -\|A_k\|^2 < 0.$$

This means that the planes encoded by A_k and A_j are orthogonal and implies that

$$e^{A_1+A_2+\cdots+A_m} = e^{A_{\sigma(1)}} e^{A_{\sigma(2)}} \dots e^{A_{\sigma(m)}}$$

for all σ in the permutation group $\mathfrak{S}(m)$. Actually, as A_k^2 is negative, we have

$$e^{A_i} = \cos\big(\|A_i\|\big) + \sin\big(\|A_i\|\big)\frac{A_i}{\|A_i\|}.$$

The corresponding rotation

$$R_i : x \longmapsto e^{-A_i} x e^{A_i}$$

acts in the oriented plane defined by A_i as a plane rotation of angle $2\|A_i\|$. The vectors orthogonal to A_i are invariant under R_i.

It then appears that any element R of SO(n) is a composition of commuting simple rotations, in the sense that they have only one invariant plane. The vectors left invariant by R are those of the orthogonal subspace to A. If $m = n/2$, this latter is trivial. The previous decomposition is not unique if $\|A_k\| = \|A_j\|$ for some j and k with $j \neq k$. In this case infinitely many planes are left invariant by R.

References

1. Brackx, F., De Schepper, N., Sommen, F.: The two-dimensional Clifford–Fourier transform. J. Math. Imaging Vis. **26**, 5–18 (2006)
2. Bülow, T.: Hypercomplex spectral signal representations for the processing and analysis of images. Ph.D. thesis, Kiel (1999)
3. Bülow, T., Sommer, G.: The hypercomplex signal—a novel approach to the multidimensional analytic signal. IEEE Trans. Signal Process. **49**(11), 2844–2852 (2001)
4. Ebling, J., Scheuermann, G.: Clifford Fourier transform on vector fields. IEEE Trans. Vis. Comput. Graph. **11**(4), 2844–2852 (2005)
5. Ell, T.A., Sangwine, S.: Hypercomplex Fourier transforms of color Images. IEEE Trans. Image Process. **16**(1), 5–18 (2007)
6. Felsberg, M.: Low-level image processing with the structure multivector. Ph.D. thesis, Kiel (2002)
7. Helgason, S.: Differential Geometry, Lie Groups and Symmetric Spaces. Academic Press, San Diego (1978)
8. Hestenes, D., Sobczyk, G.: Clifford Algebra to Geometric Calculus. Reidel, Dordrecht (1984)
9. Lounesto, P.: Clifford Algebras and Spinors. London Mathematical Society Lecture Notes Series. Cambridge University Press, Cambridge (1997)
10. Martin, D., Fowlkes, C., Tal, D., Malik, J.: A database of human segmented natural images and its application to evaluating segmentation algorithms and measuring ecological statistics. In: Proc. 8th Intl. Conf. Computer Vision, vol. 2, pp. 416–423 (2001)
11. Mawardi, B., Hitzer, E.: Clifford Fourier transformation and uncertainty principle for the Clifford algebra $cl_{3,0}$. Adv. App. Clifford Algebr. **16**(1), 41–61 (2006)
12. Postnikov, M.: Leçons de géométrie. Groupes et algèbres de Lie. Mir (1982)
13. Sangwine, S., Ell, T.A.: Hypercomplex Fourier transforms of color images. In: IEEE International Conference on Image Processing (ICIP), vol. 1, pp. 137–140. Thessaloniki, Greece (2001)

14. Sangwine, S., Ell, T.A., Gatsheni, B.: Colour-dependent linear vector image filtering. In: Twelfth European Signal Processing Conference, EUSIPCO 2004, pp. 585–588. Vienna, Austria (2004)
15. Smach, F., Lemaître, C., Gauthier, J.P., Miteran, J., Atri, M.: Generalized Fourier descriptors with applications to objects recognition in SVM context. J. Math. Imaging Vis. **30**(1), 43–71 (2008)
16. Vilenkin, N.J.: Special Functions and the Theory of Group Representations. Translations of Mathematical Monographs, vol. 22. Am. Math. Soc., Providence (1968)

Hilbert Transforms in Clifford Analysis

**Fred Brackx, Bram De Knock,
and Hennie De Schepper**

Abstract The Hilbert transform on the real line has applications in many fields. In particular in one-dimensional signal processing, the Hilbert operator is used to extract global and instantaneous characteristics, such as frequency, amplitude, and phase, from real signals. The multidimensional approach to the Hilbert transform usually is a tensorial one, considering the so-called Riesz transforms in each of the cartesian variables separately. In this paper we give an overview of generalized Hilbert transforms in Euclidean space developed within the framework of Clifford analysis. Roughly speaking, this is a function theory of higher-dimensional holomorphic functions particularly suited for a treatment of multidimensional phenomena since all dimensions are encompassed at once as an intrinsic feature.

1 Introduction: The Hilbert Transform on the Real Line

The Hilbert transform is named after D. Hilbert, who, in his studies of integral equations, was the first to observe what is nowadays known as the Hilbert transform pair. However, the Hilbert transform theory was developed mainly by E.C. Titchmarsh and G.H. Hardy. It was Hardy who named it after Hilbert. The Hilbert transform is applied in the theoretical description of many devices and has become an indispensable tool for both global and local descriptions of a signal. It has been directly implemented in the form of Hilbert analogue or digital filters which allow one to distinguish different frequency components and therefore locally refine the structure analysis. Those filters are essentially based on the notion of *analytic signal*, which consists of the linear combination of a bandpass filter, selecting a small part of the spectral information and its Hilbert transform, the latter basically being the result of a phase shift by $\frac{\pi}{2}$ on the original filter (see, e.g., [33]).

F. Brackx (✉)
Clifford Research Group, Faculty of Engineering, Ghent University, Galglaan 2, 9000 Gent, Belgium
e-mail: Freddy.Brackx@UGent.be

E. Bayro-Corrochano, G. Scheuermann (eds.), *Geometric Algebra Computing*,
DOI 10.1007/978-1-84996-108-0_9,

For a real one-dimensional finite energy signal f, i.e., $f \in L_2(\mathbb{R})$, its Hilbert transform on the real line is given by

$$\mathcal{H}[f](x) = \frac{1}{\pi} \operatorname{Pv} \int_{-\infty}^{+\infty} \frac{f(t)}{x-t}\, dt, \tag{1}$$

where Pv denotes the Cauchy principal value, meaning that in the integral the singularity at $t = x$ is approached in a symmetrical way. Infinite energy signals, such as (piecewise) constant functions and sines and cosines, should be interpreted as tempered distributions for which the Hilbert transform is defined as the convolution

$$\mathcal{H}[f](x) = \frac{1}{\pi}\left(\operatorname{Pv} \frac{1}{t} * f(t)\right)(x), \tag{2}$$

where $\operatorname{Pv} \frac{1}{t}$ is the Principal Value distribution satisfying, in the distributional sense,

$$\frac{d}{dt} \ln|t| = \operatorname{Pv} \frac{1}{t} \quad \text{and} \quad t \operatorname{Pv} \frac{1}{t} = 1.$$

In order to recall the fundamental properties of the Hilbert transform on the real line, we introduce the Cauchy integral of a function $f \in L_2(\mathbb{R})$:

$$\mathcal{C}[f](x,y) = -\frac{1}{2\pi i} \int_{-\infty}^{+\infty} \frac{f(t)}{(x-t)+iy}\, dt, \quad y \neq 0. \tag{3}$$

This Cauchy integral is, as a function of the complex variable $z = x + iy$, holomorphic in the upper and lower halves of the complex plane and decays to zero for $y \to \pm\infty$. In other words, for $f \in L_2(\mathbb{R})$, its Cauchy integral $\mathcal{C}[f](x,y)$ belongs to the Hardy spaces $H_2(\mathbb{C}^{\pm})$, respectively defined by

$$H_2\left(\mathbb{C}^{\pm}\right) = \left\{F \text{ holomorphic in } \mathbb{C}^{\pm} \text{ such that } \sup_{y \gtrless 0} \int_{-\infty}^{+\infty} \left|F(z)\right|^2 dx < +\infty\right\}. \tag{4}$$

Proposition 1 *The Hilbert operator $\mathcal{H} : L_2(\mathbb{R}) \to L_2(\mathbb{R})$,* (1), *enjoys the following properties*:

P(1) *$\mathcal{H}$ is translation invariant, i.e.,*

$$\tau_a\left[\mathcal{H}[f]\right] = \mathcal{H}\left[\tau_a[f]\right]$$

with $\tau_a[f](t) = f(t-a)$.

P(2) *$\mathcal{H}$ is dilation invariant, i.e.,*

$$d_a\left[\mathcal{H}[f]\right] = \operatorname{sgn}(a) \mathcal{H}\left[d_a[f]\right]$$

with $d_a[f](t) = f(t/a)/\sqrt{|a|}$.

P(3) *$\mathcal{H}$ is a linear, bounded, and norm-preserving operator.*

P(4) *$\mathcal{H}$ is invertible with $\mathcal{H}^{-1} = -\mathcal{H}$, and thus $\mathcal{H}^2 = -\mathbf{1}$.*

P(5) *$\mathcal{S}$ is unitary, i.e., $\mathcal{S}^*\mathcal{S} = \mathcal{S}\mathcal{S}^* = \mathbf{1}$.*
P(6) *$\mathcal{S}$ commutes with differentiation, i.e.,*

$$\frac{d}{dt}\big(\mathcal{S}[f](t)\big) = \mathcal{S}\left[\frac{d}{dt} f(t)\right].$$

P(7) *$\mathcal{S}$ arises in a natural way by considering the nontangential boundary limits (in L_2 sense) of the Cauchy integral* (3), *i.e.,*

$$\mathcal{C}^{\pm}[f](x) = \lim_{\substack{y \to 0\pm \\ \text{NT}}} \mathcal{C}[f](x, y) = \pm\frac{1}{2} f(x) + \frac{1}{2} i\, \mathcal{S}[f](x), \quad x \in \mathbb{R}. \tag{5}$$

The operators $\mathcal{C}^{\pm}$ are usually called the Cauchy transforms, and formulae (5) and P(7) are the Plemelj–Sokhotzki formulae in Clifford analysis.

Thus putting $\mathcal{H} = i\mathcal{S}$ we obtain an involutive, norm-preserving, and bounded linear operator $\mathcal{H} : L_2(\mathbb{R}) \to L_2(\mathbb{R})$, which may be used to define the Hardy space $H_2(\mathbb{R})$ as the closed subspace of $L_2(\mathbb{R})$ consisting of functions g for which $\mathcal{H}[g] = g$. We call those functions $g \in H_2(\mathbb{R})$ *analytic signals*, inspired by the fact that the nontangential boundary limit $\mathcal{C}^{+}[f]$ of the holomorphic (or analytic) Cauchy integral $\mathcal{C}[f]$, (3), indeed belongs to the Hardy space $H_2(\mathbb{R})$. The real and imaginary parts $u = \mathbb{R}e[g]$ and $v = \mathbb{I}m[g]$ of an analytic signal g satisfy the *Hilbert formulae*

$$\mathcal{H}[u] = iv \quad \text{and} \quad \mathcal{H}[iv] = u. \tag{6}$$

It follows that

$$g = (\mathbf{1} + \mathcal{H})[u] \quad \text{and} \quad g = (\mathbf{1} + \mathcal{H})[iv], \tag{7}$$

showing that an analytic signal contains redundant information since it can be recovered from its real (or its imaginary) part solely. Note that the Hardy spaces $H_2(\mathbb{R})$ and $H_2(\mathbb{C}^+)$, (4), are isomorphic, since the nontangential boundary limit for $y \to 0+$ of $F(z) \in H_2(\mathbb{C}^+)$ exists a.e. and belongs to $H_2(\mathbb{R})$, and the Cauchy integral in $\mathbb{C}^+$ of $F(x + i0)$ precisely is $F(z)$.

In the frequency space the Hilbert transform, which is convolutional in nature, takes the form of a multiplication operator. Denoting by $\mathcal{F}$ the usual Fourier transform, for a function $f \in L_2(\mathbb{R})$, we have

$$\begin{aligned} &\mathcal{F}\big[\mathcal{H}[f]\big](\omega) = \operatorname{sgn}\omega \mathcal{F}[f](\omega) \quad \text{and} \\ &\mathcal{H}\big[\mathcal{F}[f]\big](\omega) = -\mathcal{F}\big[\operatorname{sgn} t f(t)\big](\omega). \end{aligned} \tag{8}$$

In particular the Fourier spectrum of an analytic signal $g \in H_2(\mathbb{R})$ is a causal function with only positive frequencies, and conversely; more explicitly, it reads:

$$\begin{aligned} \mathcal{F}[g](\omega) &= \mathcal{F}(1 + \mathcal{H})[u](\omega) = (1 + \operatorname{sgn}\omega)\mathcal{F}[u](\omega) \\ &= \begin{cases} 2\mathcal{F}[u](\omega), & \omega > 0, \\ 0, & \omega < 0. \end{cases} \end{aligned} \tag{9}$$

2 Hilbert Transforms in Euclidean Space

The Hilbert transform was first generalized to m-dimensional Euclidean space by means of the Riesz transforms R_j in each of the cartesian coordinates x_j, $j = 1, \ldots, m$, given by

$$R_j[f](\underline{x}) = \lim_{\varepsilon \to 0+} \frac{2}{a_{m+1}} \int_{\mathbb{R}^m \setminus B(\underline{x},\varepsilon)} \frac{x_j - y_j}{|\underline{x} - \underline{y}|^{m+1}} f(\underline{y})\, dV(\underline{y}), \tag{10}$$

where $a_{m+1} = \frac{2\pi^{(m+1)/2}}{\Gamma((m+1)/2)}$ denotes the area of the unit sphere S^m in $\mathbb{R}^{m+1}$. It was Horváth who, already in his 1953 paper [29], introduced the Clifford vector-valued Hilbert operator

$$\mathcal{S} = \sum_{j=1}^{m} e_j R_j. \tag{11}$$

The multidimensional Hilbert transform was taken up again in the 1980s and further studied in, e.g., [21, 22, 27, 32, 36] in the Clifford analysis setting.

Clifford analysis is a function theory which offers an elegant and powerful generalization to higher dimension of the theory of holomorphic functions in the complex plane. In its most simple but still useful setting, flat m-dimensional Euclidean space, Clifford analysis focusses on so-called monogenic functions, i.e., null solutions of the Clifford vector-valued Dirac operator

$$\partial_{\underline{x}} = \sum_{j=1}^{m} e_j\, \partial_{x_j}, \tag{12}$$

where $(e_1, \ldots, e_m)$ forms an orthonormal basis for the quadratic space $\mathbb{R}^m$ underlying the construction of the Clifford algebra $\mathbb{R}_{0,m}$, and where the basis vectors satisfy the multiplication rules

$$e_j e_k + e_k e_j = -2\delta_{j,k}, \quad j,k = 1, \ldots, m. \tag{13}$$

Monogenic functions have a special relationship with harmonic functions of several variables: they are refining their properties, since the Dirac operator factorizes the m-dimensional Laplacian $\partial_{\underline{x}}^2 = -\Delta_m$. Euclidean space $\mathbb{R}^m$ is embedded in the Clifford algebra $\mathbb{R}_{0,m}$ by identifying the point $(x_1, \ldots, x_m) \in \mathbb{R}^m$ with the vector variable

$$\underline{x} = \sum_{j=1}^{m} e_j x_j. \tag{14}$$

For more details on Clifford analysis and its applications, we refer to, e.g., [2, 20, 23].

2.1 Definition and Properties

In the framework of Euclidean Clifford analysis, the (Clifford–)Hilbert transform for a suitable function or distribution f is given by

$$\begin{aligned}\mathcal{H}[f](\underline{x}) &= \frac{2}{a_{m+1}}\,\overline{e_0}\,\mathrm{Pv}\int_{\mathbb{R}^m}\frac{\overline{\underline{x}}-\overline{\underline{y}}}{|\underline{x}-\underline{y}|^{m+1}}\,f(\underline{y})\,dV(\underline{y})\\ &= \frac{2}{a_{m+1}}\,\overline{e_0}\,\lim_{\varepsilon\to 0+}\int_{|\underline{x}-\underline{y}|>\varepsilon}\frac{\overline{\underline{x}}-\overline{\underline{y}}}{|\underline{x}-\underline{y}|^{m+1}}\,f(\underline{y})\,dV(\underline{y}).\end{aligned} \tag{15}$$

In the above expression, e_0 is an additional basis vector for which also

$$e_0^2 = -1 \quad \text{and} \quad e_0\,e_j + e_j\,e_0 = -2\,\delta_{0,j}, \quad j = 1,\ldots,m. \tag{16}$$

Furthermore, $\bar{\cdot}$ stands for the usual conjugation in the Clifford algebra $\mathbb{R}_{0,m+1}$, i.e., the main anti-involution for which $\overline{e_j} = -e_j$, $j = 0,\ldots,m$. As in the one-dimensional case, there is a strong relationship between the Hilbert transform and the Cauchy integral of a function $f \in L_2(\mathbb{R}^m)$. The functions considered here take their values in the Clifford algebra $\mathbb{R}_{0,m+1}$. The space $L_2(\mathbb{R}^m)$ is equipped with the $\mathbb{R}_{0,m+1}$-valued inner product and corresponding squared norm:

$$\langle f, g\rangle = \int_{\mathbb{R}^m}\overline{f(\underline{x})}\,g(\underline{x})\,dV(\underline{x}), \quad \|f\|^2 = \big[\langle f, f\rangle\big]_0, \tag{17}$$

where $[\lambda]_0$ denotes the scalar part of the Clifford number λ. The Cauchy integral of $f \in L_2(\mathbb{R}^m)$ is defined by

$$\mathcal{C}[f](x) = \mathcal{C}[f](x_0,\underline{x}) = \frac{1}{a_{m+1}}\int_{\mathbb{R}^m}\frac{x_0 + e_0(\underline{x}-\underline{y})}{|x_0+\underline{x}-\underline{y}|^{m+1}}\,f(\underline{y})\,dV(\underline{y}), \quad x_0 \neq 0. \tag{18}$$

Observe the formal similarity with the Cauchy integral (3) of $f \in L_2(\mathbb{R})$, x_0 taking the role of y, and the vector $\underline{y}$ taking the role of t. It is a (left-)monogenic function in the upper and lower half spaces $\mathbb{R}^{m+1}_{\pm} = \{x_0 e_0 + \underline{x} : \underline{x} \in \mathbb{R}^m,\ x_0 \gtrless 0\}$. By a monogenic function in $\mathbb{R}^{m+1}$ is meant a function annihilated by the Cauchy–Riemann operator,

$$D_x = \overline{e_0}\,\partial_x = \overline{e_0}(e_0\partial_{x_0} + \partial_{\underline{x}}) = \partial_{x_0} + \overline{e_0}\,\partial_{\underline{x}}, \tag{19}$$

which decomposes the Laplace operator in $\mathbb{R}^{m+1}$, $D_x\overline{D_x} = \Delta_{m+1}$.

Moreover the Cauchy integral decays to zero as $x_0 \to \pm\infty$. Summarizing, for a function $f \in L_2(\mathbb{R}^m)$, its Cauchy integral $\mathcal{C}[f](x_0,\underline{x})$, (18), belongs to the Hardy spaces $H_2(\mathbb{R}^{m+1}_{\pm})$, respectively defined by

$$H_2\big(\mathbb{R}^{m+1}_{\pm}\big) = \left\{D_x F = 0 \text{ in } \mathbb{R}^{m+1}_{\pm} \text{ such that } \sup_{x_0 \gtrless 0}\int_{-\infty}^{+\infty}\big|F(x_0+\underline{x})\big|^2\,dx < +\infty\right\}. \tag{20}$$

The properties of the multidimensional Hilbert transform are summarized in the following proposition; they show a remarkable similarity with those of the one-dimensional Hilbert transform listed in Proposition 1.

Proposition 2 *The Hilbert transform $\mathcal{H} : L_2(\mathbb{R}^m) \to L_2(\mathbb{R}^m)$ enjoys the following properties*:

P(1) *$\mathcal{H}$ is translation invariant, i.e.,*

$$\tau_{\underline{b}}\big[\mathcal{H}[f]\big] = \mathcal{H}\big[\tau_{\underline{b}}[f]\big]$$

with $\tau_{\underline{b}}[f](\underline{x}) = f(\underline{x} - \underline{b}), \underline{b} \in \mathbb{R}^m$.

P(2) *$\mathcal{H}$ is dilation invariant, i.e.,*

$$d_a\big[\mathcal{H}[f]\big] = \mathcal{H}\big[d_a[f]\big]$$

with $d_a[f](\underline{x}) = \frac{1}{a^{m/2}} f(\underline{x}/a), a > 0$.

P(3) *$\mathcal{H}$ is a norm-preserving, bounded, and linear operator.*

P(4) *$\mathcal{H}$ is an involution and thus invertible with $\mathcal{H}^{-1} = \mathcal{H}$.*

P(5) *$\mathcal{H}$ is unitary with $\mathcal{H}^* = \mathcal{H}$.*

P(6) *$\mathcal{H}$ anticommutes with the Dirac operator* (12).

P(7) *$\mathcal{H}$ arises in a natural way by considering the nontangential boundary limits in L_2 sense of the Cauchy integral* (18):

$$\mathcal{C}^{\pm}[f](\underline{x}) = \lim_{\substack{x_0 \to 0\pm \\ \mathrm{NT}}} \mathcal{C}[f](x_0, \underline{x}) = \pm\frac{1}{2} f(\underline{x}) + \frac{1}{2}\mathcal{H}[f](\underline{x}), \quad \underline{x} \in \mathbb{R}^m. \tag{21}$$

In the distributional sense, this boundary behavior is explicited by

$$E(0\pm, \underline{x}) = \lim_{x_0 \to 0\pm} E(x_0, \underline{x}) = \pm\frac{1}{2}\delta(\underline{x}) + \frac{1}{2}\mathcal{K}(\underline{x}), \tag{22}$$

where $E(x_0, \underline{x})$ is the fundamental solution of the Cauchy–Riemann operator D_x, (19):

$$D_x E(x_0, \underline{x}) - D_x\left(\frac{1}{a_{m+1}} \frac{x_0 - \overline{e_0}\underline{x}}{|x_0 + e_0\underline{x}|^{m+1}}\right) = \delta(x_0, \underline{x}), \tag{23}$$

and $\mathcal{K}$ is the Hilbert convolution kernel:

$$\mathcal{H}[f] = \mathcal{K} * f = \frac{2}{a_{m+1}}\overline{e_0}\,\mathrm{Pv}\frac{\overline{\underline{x}}}{|\underline{x}|^{m+1}} * f. \tag{24}$$

As each function in the Hardy space $H_2(\mathbb{R}^{m+1}_{\pm})$, (20), possesses a nontangential L_2 boundary limit as $x_0 \to 0\pm$, one is lead to the introduction of the Hardy space $H_2(\mathbb{R}^m)$ as the closure in $L_2(\mathbb{R}^m)$ of the nontangential boundary limits $F(\underline{x} + 0)$ as $x_0 \to 0+$ of the functions $F(x_0, \underline{x})$ in $H_2(\mathbb{R}^{m+1}_+)$. As moreover the Cauchy integral of $F(\underline{x} + 0)$ is precisely $F(x_0, \underline{x})$, we may conclude that the Hardy spaces $H_2(\mathbb{R}^{m+1}_+)$ and $H_2(\mathbb{R}^m)$ are isomorphic.

As the Hardy space $H_2(\mathbb{R}^m)$ is, by definition, a closed subspace of the space $L_2(\mathbb{R}^m)$, the latter space may be decomposed as the orthogonal direct sum

$$L_2(\mathbb{R}^m) = H_2(\mathbb{R}^m) \oplus H_2(\mathbb{R}^m)^{\perp}. \tag{25}$$

The corresponding projection operators are precisely the Cauchy transforms $\pm\mathcal{C}^{\pm}$ since it can be directly verified that

$$\begin{aligned} f &= \mathcal{C}^+[f] - \mathcal{C}^-[f]; \\ \mathcal{C}^+\big[\mathcal{C}^+[f]\big] &= \mathcal{C}^+[f]; \\ (-\mathcal{C}^-)\big[-\mathcal{C}^-[f]\big] &= (-\mathcal{C}^-)[f]; \\ \mathcal{C}^+\big[\mathcal{C}^-[f]\big] &= \mathcal{C}^-\big[\mathcal{C}^+[f]\big] = 0; \\ \big\langle \mathcal{C}^+[f], \mathcal{C}^-[f]\big\rangle_{L_2} &= 0. \end{aligned}$$

The analytic signal $\mathcal{C}^+[f] \in H_2(\mathbb{R}^m)$ and the anti-analytic signal $(-\mathcal{C}^-[f]) \in H_2(\mathbb{R}^m)^{\perp}$ thus possess a monogenic extension to $H_2(\mathbb{R}^{m+1}_{\pm})$, respectively. Note that the Hardy space $H_2(\mathbb{R}^m)$ and its orthogonal complement $H_2(\mathbb{R}^m)^{\perp}$ are nicely characterized by means of the Hilbert and Cauchy transforms:

Lemma 1 *A function $g \in L_2(\mathbb{R}^m)$ belongs to $H_2(\mathbb{R}^m)$ if and only if $\mathcal{H}[g] = g$, or $\mathcal{C}^+[g] = g$, or $\mathcal{C}^-[g] = 0$.*

Lemma 2 *A function $h \in L_2(\mathbb{R}^m)$ belongs to $H_2(\mathbb{R}^m)^{\perp}$ if and only if $\mathcal{H}[g] = -g$, or $\mathcal{C}^+[h] = 0$, or $\mathcal{C}^-[h] = -h$.*

2.2 Analytic Signals

Because of the properties mentioned in the preceding subsection, the functions in $H_2(\mathbb{R}^m)$ already deserve to be called *analytic signals* in $\mathbb{R}^m$. But then their frequency contents should show a property similar to one-dimensional causality (9), thus involving a multidimensional analogue of the Heaviside step function. As in the one-dimensional case, the Hilbert transform (15) in frequency space takes the form of a multiplication operator; for a function $f \in L_2(\mathbb{R}^m)$, there holds

$$\mathcal{F}\big[\mathcal{H}[f]\big](\underline{y}) = \overline{e_0} i \underline{\xi}\, \mathcal{F}[f](\underline{y}) \quad \text{and} \quad \mathcal{H}\big[\mathcal{F}[f]\big](\underline{y}) = e_0\, \mathcal{F}\big[i\, \underline{\omega} f(\underline{x})\big](\underline{y}), \tag{26}$$

where $\mathcal{F}$ denotes the standard Fourier transform in $\mathbb{R}^m$ given by

$$\mathcal{F}\big[f(\underline{x})\big](\underline{y}) = \int_{\mathbb{R}^m} f(\underline{x}) \exp\big(-i\langle \underline{x}, \underline{y}\rangle\big)\, dV(\underline{x}), \tag{27}$$

and $\underline{\omega} = \underline{x}/|\underline{x}|$ and $\underline{\xi} = \underline{y}/|\underline{y}|$ may be interpreted as the multidimensional analogues of the signum-function $\operatorname{sgn}(x) = x/|x|$ on the real line. As an aside, these formulae allow the practical computation of the Hilbert transform by means of the Fourier transform:

$$\mathcal{H}[f](\underline{x}) = \mathcal{F}^{-1}\big[\overline{e_0}\, i\, \underline{\xi}\, \mathcal{F}[f](\underline{y})\big]. \tag{28}$$

The Fourier spectrum of the Cauchy transforms $\mathcal{C}^{\pm}[f]$ (21) of a function $f \in L_2(\mathbb{R}^m)$ then read

$$\begin{aligned}\mathcal{F}\big[\mathcal{C}^{\pm}[f]\big] &= \pm\frac{1}{2}\,\mathcal{F}[f] + \frac{1}{2}\,\overline{e_0}\, i\, \underline{\xi}\; \mathcal{F}[f]\\ &= \pm\, \psi_{\pm}\, \mathcal{F}[f],\end{aligned} \tag{29}$$

where we have introduced the mutually annihilating idempotents

$$\psi_{+} = \frac{1}{2}(1 + \overline{e_0}\, i\, \underline{\omega}) \quad \text{and} \quad \psi_{-} = \frac{1}{2}\,(1 - \overline{e_0}\, i\, \underline{\omega}) \tag{30}$$

satisfying the following properties:

(i) $\psi_{\pm}^2 = \psi_{\pm}$
(ii) $\psi_{+}\psi_{-} = \psi_{-}\psi_{+} = 0$
(iii) $\psi_{+} + \psi_{-} = 1$
(iv) $\psi_{+} - \psi_{-} = \overline{e_0}\, i\, \underline{\omega}$
(v) $i e_0 \underline{\omega} \psi_{\pm} = \pm \psi_{\pm}$

The functions $\psi_{\pm}$, (30), thus are the Clifford algebra-valued multidimensional analogues to the Heaviside step function. They were introduced independently by Sommen [35] and McIntosh [32]. They allow for the practical computation of the Cauchy transforms of a function $f \in L_2(\mathbb{R}^m)$ through

$$\mathcal{C}^{\pm}[f] = \mathcal{F}^{-1}\big[\pm\, \psi_{\pm}\, \mathcal{F}[f]\big], \tag{31}$$

which will be used in the next subsection. Now take an analytic signal $g \in H_2(\mathbb{R}^m)$; then, in accordance with Lemma 1, $g = \mathcal{H}[g]$ or $g = \frac{1}{2}\,(g + \mathcal{H}[g]) = \mathcal{C}^{+}[g]$ and $\mathcal{C}^{-}[g] = 0$, from which it follows that

$$\mathcal{F}[g] = \mathcal{F}\big[\mathcal{C}^{+}[g]\big] = \psi_{+}\mathcal{F}[g], \tag{32}$$

whereas, trivially,

$$\mathcal{F}\big[\mathcal{C}^{-}[g]\big] = -\psi_{-}\mathcal{F}[g] = 0, \tag{33}$$

which is the multidimensional counterpart to the "vanishing negative frequencies" in one dimension.

We now show that, similarly to the splitting of a complex signal into its real and imaginary parts, see (6), a Clifford algebra-valued analytic signal can be split into two components satisfying multidimensional *Hilbert formulae*. To that end, we observe that, by the introduction of the additional basis vector e_0, the Clifford algebra $\mathbb{R}_{0,m+1}$ may be decomposed, using two copies of the Clifford algebra $\mathbb{R}_{0,m}$, as follows:

$$\mathbb{R}_{0,m+1} = \mathbb{R}_{0,m} \oplus \overline{e_0}\, \mathbb{R}_{0,m}. \tag{34}$$

Thus, if g is an $\mathbb{R}_{0,m+1}$-valued analytic signal, it can be written as $g = u + \overline{e_0}\, v$, where u and v are $\mathbb{R}_{0,m}$-valued functions satisfying, in view of Lemma 1,

$$\mathcal{H}[u] = \overline{e_0}\, v \quad \text{and} \quad \mathcal{H}[\overline{e_0}\, v] = u. \tag{35}$$

This means that an analytic signal g is completely determined by one of its components u or v:

$$g = (\mathbf{1} + \mathcal{H})[u] = (\mathbf{1} + \mathcal{H})[\overline{e_0}\, v], \tag{36}$$

and moreover shows a Fourier spectrum only containing ψ_+-frequencies and doubling those of u or v:

$$\begin{aligned} \mathcal{F}[g] &= (1 + \overline{e_0}\, i\, \underline{\xi})\; \mathcal{F}[u] = (1 + \overline{e_0}\, i\, \underline{\xi})\; \mathcal{F}[\overline{e_0}\, v] \\ &= 2\, \psi_+\, \mathcal{F}[u] = 2\, \psi_+\, \mathcal{F}[\overline{e_0}\, v]. \end{aligned} \tag{37}$$

Similar considerations hold for anti-analytic signals in $H_2(\mathbb{R}^m)^{\perp}$.

2.3 Monogenic Extensions of Analytic Signals

For any $f \in L_2(\mathbb{R}^m)$, the Cauchy transforms $\pm\mathcal{C}^{\pm}[f]$, (21), are (anti-)analytic signals, thus showing monogenic extensions to $\mathbb{R}^m_{\pm}$. A first possibility to construct these monogenic extensions is by using the Cauchy integral, leading to the monogenic functions

$$\mathcal{C}\big[\mathcal{C}^+[f]\big] = \begin{cases} \mathcal{C}[f] & \text{in } \mathbb{R}^m_+, \\ 0 & \text{in } \mathbb{R}^m_-, \end{cases} \tag{38}$$

$$\mathcal{C}\big[-\mathcal{C}^-[f]\big] = \begin{cases} 0 & \text{in } \mathbb{R}^m_+, \\ -\mathcal{C}[f] & \text{in } \mathbb{R}^m_-, \end{cases} \tag{39}$$

which moreover tend to zero as $x_0 \to \pm\infty$. However there is also another way to construct monogenic extensions to $\mathbb{R}^{m+1}$ of functions in $\mathbb{R}^m$, albeit that they have to be real-analytic. This method, the so-called Cauchy–Kowalewska extension principle, is a typical construct of Clifford analysis; for a given real-analytic function ϕ in $\mathbb{R}^m$, a monogenic extension in an open neighborhood in $\mathbb{R}^{m+1}$ of $\mathbb{R}^m$ is given by

$$CK[\phi] = \exp\left(-x_0 \overline{e_0} \partial_{\underline{x}}\right)[\phi] = \sum_{j=0}^{+\infty} \frac{(-1)^j}{j!}\, x_0^j\, (\overline{e_0} \partial_{\underline{x}})^j [\phi]. \tag{40}$$

In particular the scalar-valued and real-analytic Fourier kernel $\exp(i\langle \underline{x}, \underline{y}\rangle)$ in $\mathbb{R}^m$ is monogenically extended to the whole of $\mathbb{R}^{m+1}$ by

$$CK\big[\exp\big(i\langle \underline{x}, \underline{y}\rangle\big)\big] = \sum_{j=0}^{+\infty} \frac{(-1)^j}{j!} x_0^j (\overline{e_0} i \underline{y})^j \big[\exp\big(i\langle \underline{x}, \underline{y}\rangle\big)\big]$$
$$= \exp(-i x_0 \overline{e_0} \underline{y}) \exp\big(i\langle \underline{x}, \underline{y}\rangle\big), \tag{41}$$

which takes its values in $\operatorname{span}_{\mathbb{C}}(\overline{e_0} e_1, \ldots, \overline{e_0} e_m)$.

In view of (31), i.e.,

$$\mathcal{C}^+[f] = \mathcal{F}^{-1}\big[\psi_+ \, \mathcal{F}[f]\big], \tag{42}$$

we thus obtain, following an idea of [34] and [30], as a monogenic extension of $\mathcal{C}^+[f]$:

$$CK\big[\mathcal{C}^+[f]\big](x_0, \underline{x})$$
$$= (2\pi)^{-m} \int_{\mathbb{R}^m} \exp\big(i\langle \underline{x}, \underline{y}\rangle\big) \exp\big(-i x_0 \overline{e_0} \underline{y}\big) \psi_+ \mathcal{F}[f](\underline{y}) \, dV(\underline{y}). \tag{43}$$

A direct computation yields

$$CK\big[\mathcal{C}^+[f]\big](x_0, \underline{x})$$
$$= (2\pi)^{-m} \int_{\mathbb{R}^m} \exp\big(i\langle \underline{x}, \underline{y}\rangle\big) \exp(-x_0 \rho) \psi_+ \mathcal{F}[f](\underline{y}) \, dV(\underline{y})$$
$$= (2\pi)^{-m} \int_{S^{m-1}} \psi_+ \, dS(\underline{\xi}) \int_0^{+\infty} \exp\big(\big(i\langle \underline{x}, \underline{\xi}\rangle - x_0\big)\rho\big) \rho^{m-1} \mathcal{F}[f](\rho \underline{\xi}) \, d\rho, \tag{44}$$

since

$$\exp(-i x_0 \overline{e_0} \underline{y}) \psi_+ = \exp(-x_0 \rho) \psi_+, \tag{45}$$

where we have once more used spherical coordinates with $\underline{y} = \rho \underline{\xi}$. This further leads to

$$CK\big[\mathcal{C}^+[f]\big](x_0, \underline{x})$$
$$= (2\pi)^{-m} \int_{S^{m-1}} \psi_+ \, dS(\underline{\xi}) \mathcal{L}\big[\rho^{m-1} \mathcal{F}[f](\rho \underline{\xi})\big]\big(x_0 - i\langle \underline{x}, \underline{\xi}\rangle\big), \tag{46}$$

where $\mathcal{L}$ denotes the Laplace transform. It is clear that this monogenic extension tends to zero only as $x_0 \to +\infty$. Thus, with restriction to $\mathbb{R}^m_+$, we obtain

$$\mathcal{C}[f](x_0, \underline{x}) = (2\pi)^{-m} \int_{S^{m-1}} \psi_+ \, dS(\underline{\xi}) \; \mathcal{L}\big[\rho^{m-1} \, \mathcal{F}[f](\rho \underline{\xi})\big]$$
$$\times \big(x_0 - i\langle \underline{x}, \underline{\xi}\rangle\big), \quad x_0 > 0. \tag{47}$$

In a similar way, we obtain in $\mathbb{R}^m_-$

$$\mathcal{C}[f](x_0, \underline{x}) = (2\pi)^{-m} \int_{S^{m-1}} \psi_- \, dS(\underline{\xi}) \, \mathcal{L}\big[\rho^{m-1} \, \mathcal{F}[f](\rho\underline{\xi})\big] \times \big(-x_0 - i\langle \underline{x}, \underline{\xi}\rangle\big), \quad x_0 < 0. \tag{48}$$

2.4 Example 1

The direct sum decomposition of finite-energy signals goes through for tempered distributions and even more so for compactly supported distributions. Let us illustrate this by the case of the delta- or Dirac-distribution $\delta(\underline{x})$ in $\mathbb{R}^m$. Its Cauchy integral is given by

$$\mathcal{C}[\delta](x_0, \underline{x}) = E(x_0, \underline{x}) * \delta(\underline{x}) = E(x_0, \underline{x}) = \frac{1}{a_{m+1}} \frac{x_0 - \overline{e_0}\, \underline{x}}{|x_0 e_0 + \underline{x}|^{m+1}}, \tag{49}$$

which is monogenic in $\mathbb{R}^{m+1}_\pm$ and even in $\mathbb{R}^{m+1} \backslash \{0\}$ w.r.t. the Cauchy–Riemann operator D_x (19). This implies that as long as $\underline{x} \neq 0$, there is a continuous transition of this Cauchy integral over $\mathbb{R}^m$ as the common boundary of $\mathbb{R}^{m+1}_+$ and $\mathbb{R}^{m+1}_-$. Thus the "jump" over $\mathbb{R}^m$ of $\mathcal{C}[\delta](x_0, \underline{x})$ will occur at $\underline{x} = 0$, and indeed

$$\mathcal{C}^\pm[\delta](\underline{x}) = \pm \frac{1}{2} \, \delta(\underline{x}) + \frac{1}{2} \, \mathcal{K}(\underline{x}) \tag{50}$$

with $\mathcal{K}$ the Hilbert kernel (24), since

$$\mathcal{H}[\delta](\underline{x}) = \mathcal{K} * \delta(\underline{x}) = \mathcal{K}(\underline{x}) = \frac{2}{a_{m+1}} \overline{e_0} \, \mathrm{Pv} \, \frac{\overline{\underline{x}}}{|\underline{x}|^{m+1}}. \tag{51}$$

The direct sum decomposition of the Dirac-distribution $\delta(\underline{x})$ now follows readily:

$$\delta(\underline{x}) = \left(\frac{1}{2}\delta(\underline{x}) + \frac{1}{2}\mathcal{K}(\underline{x})\right) + \left(\frac{1}{2}\delta(\underline{x}) - \frac{1}{2}\mathcal{K}(\underline{x})\right). \tag{52}$$

The Cauchy integral of the first component is given by

$$\mathcal{C}\big[\mathcal{C}^+[\delta]\big] = \begin{cases} \mathcal{C}[\delta] = E(x_0, \underline{x}) & \text{in } \mathbb{R}^{m+1}_+, \\ 0 & \text{in } \mathbb{R}^{m+1}_-, \end{cases} \tag{53}$$

while the Cauchy integral of the second component is given by

$$\mathcal{C}\big[-\mathcal{C}^-[\delta]\big] = \begin{cases} 0 & \text{in } \mathbb{R}^{m+1}_+, \\ -\mathcal{C}[\delta] = -E(x_0, \underline{x}) & \text{in } \mathbb{R}^{m+1}_-. \end{cases} \tag{54}$$

As the Hilbert transform is involutive, we obtain for the transform of the Hilbert kernel itself:

$$\mathcal{H}[\mathcal{K}](\underline{x}) = \mathcal{H}^2[\delta](\underline{x}) = \delta(\underline{x}), \tag{55}$$

which is confirmed by the convolution

$$\mathcal{H}[\mathcal{K}] = \mathcal{K} * \mathcal{K} = \frac{4}{a_{m+1}^2} \operatorname{Pv} \frac{\overline{\underline{x}}}{|\underline{x}|^{m+1}} * \operatorname{Pv} \frac{\overline{\underline{x}}}{|\underline{x}|^{m+1}} = \delta(\underline{x}). \tag{56}$$

This leads to the direct sum decomposition of the Hilbert kernel $\mathcal{K}(\underline{x})$:

$$\mathcal{K}(\underline{x}) = \left(\frac{1}{2}\mathcal{K}(\underline{x}) + \frac{1}{2}\delta(\underline{x})\right) + \left(\frac{1}{2}\mathcal{K}(\underline{x}) - \frac{1}{2}\delta(\underline{x})\right), \tag{57}$$

where both components may be monogenically extended through their Cauchy integral to respectively $\mathbb{R}^{m+1}_{\pm}$ by the functions $\pm E(x_0, \underline{x})$. Note in this connection that $(\pm\mathcal{C}^{\pm})[\delta] = \mathcal{C}^{\pm}[\mathcal{K}]$.

As the delta-distribution $\delta(\underline{x})$ is $\mathbb{R}_{0,m}$ valued—in fact real valued—and its Hilbert transform $\mathcal{K}(\underline{x})$ is $\overline{e_0}\,\mathbb{R}_{0,m}$ valued, they sum up to an $\mathbb{R}_{0,m+1}$-valued analytic signal $\delta(\underline{x}) + \mathcal{K}(\underline{x})$ which has its frequencies supported by ψ_+ and doubling those of $\delta(\underline{x})$. This is confirmed by the following results in frequency space. For the standard Fourier transform (27), we have $\mathcal{F}[\delta] = 1$; thus,

$$\mathcal{F}[\mathcal{K}] = \mathcal{F}\left[\frac{2}{a_{m+1}} \operatorname{Pv} \frac{\overline{\underline{x}}}{|\underline{x}|^{m+1}}\right] = \overline{e_0}\, i\, \underline{\xi}, \tag{58}$$

and thus also

$$\mathcal{F}\big[\delta(\underline{x}) + \mathcal{K}(\underline{x})\big] = 1 + \overline{e_0}\, i\, \underline{\xi} = 2\,\psi_+. \tag{59}$$

As already mentioned, the (anti-)analytic signals $\pm\mathcal{C}^{\pm}[\delta](\underline{x}) = \frac{1}{2}\,\delta(\underline{x}) \pm \frac{1}{2}\,\mathcal{K}(\underline{x}) = \mathcal{C}^{\pm}[\mathcal{K}](\underline{x})$ may be monogenically extended to $\mathbb{R}^{m+1}_{\pm}$ by the functions $\pm\, E(x_0, \underline{x}) \in H_2(\mathbb{R}^{m+1}_{\pm})$, respectively, defined in (23). Alternatively the Cauchy–Kowalewska technique (40) leads to

$$CK\big[\mathcal{C}^+[\delta]\big] = (2\pi)^{-m} \int_{S^{m-1}} \psi_+ \, dS(\underline{\xi}) \mathcal{L}\big[\rho^{m-1}\big]\big(x_0 - i\langle \underline{x}, \underline{\xi}\rangle\big), \quad x_0 > 0, \tag{60}$$

and

$$CK\big[-\mathcal{C}^-[\delta]\big] = (2\pi)^{-m} \int_{S^{m-1}} \psi_- \, dS(\underline{\xi}) \mathcal{L}\big[\rho^{m-1}\big]\big(-x_0 - i\langle \underline{x}, \underline{\xi}\rangle\big), \quad x_0 < 0. \tag{61}$$

As $\mathcal{L}[\rho^{m-1}] = \frac{\Gamma(m)}{z^m}$ for $\mathbb{R}e(z) > 0$, we arrive at

$$CK\big[\mathcal{C}^+[\delta]\big] = \frac{(m-1)!}{(2\pi)^m} \int_{S^{m-1}} \frac{\psi_+}{(x_0 - i\langle \underline{x}, \underline{\xi}\rangle)^m} \, dS(\underline{\xi}), \quad x_0 > 0, \tag{62}$$

and

$$CK\big[-\mathcal{C}^{-}[\delta]\big] = \frac{(m-1)!}{(2\pi)^m} \int_{S^{m-1}} \frac{\psi_-}{(-x_0 - i\langle \underline{x}, \underline{\xi}\rangle)^m}\, dS(\underline{\xi}), \quad x_0 < 0. \tag{63}$$

But $i\psi_+ = -\overline{e_0}\underline{\xi}\,\psi_+$ and $i\psi_- = \overline{e_0}\underline{\xi}\,\psi_-$, from which it follows that

$$\frac{\psi_+}{(x_0 - i\langle \underline{x}, \underline{\xi}\rangle)^m} = \frac{\psi_+}{(x_0 + \langle \underline{x}, \underline{\xi}\rangle ole_0 \underline{\xi})^m} \tag{64}$$

and

$$\frac{\psi_-}{(x_0 + i\langle \underline{x}, \underline{\xi}\rangle)^m} = \frac{\psi_-}{(x_0 + \langle \underline{x}, \underline{\xi}\rangle \overline{e_0}\underline{\xi})^m}. \tag{65}$$

Moreover the CK-extensions under consideration $CK[\mathcal{C}^{+}[\delta]] = E(x_0, \underline{x})$, $x_0 > 0$, and $CK[-\mathcal{C}^{-}[\delta]] = -E(x_0, \underline{x})$, $x_0 < 0$, both are $\mathbb{R}_{0,m+1}$ valued, so their complex-imaginary parts should vanish, which implies that

$$\int_{S^{m-1}} \frac{\underline{\xi}}{(x_0 + \langle \underline{x}, \underline{\xi}\rangle \overline{e_0}\underline{\xi})^m}\, dS(\underline{\xi}) = 0, \tag{66}$$

finally leading to

$$E(x_0, \underline{x}) = \frac{1}{2}\frac{(m-1)!}{(2\pi)^m} \int_{S^{m-1}} \frac{1}{(x_0 + \langle \underline{x}, \underline{\xi}\rangle \overline{e_0}\underline{\xi})^m}\, dS(\underline{\xi}), \quad x_0 > 0, \tag{67}$$

and

$$E(x_0, \underline{x}) = \frac{(-1)^{m-1}}{2}\frac{(m-1)!}{(2\pi)^m} \int_{S^{m-1}} \frac{1}{(x_0 + \langle \underline{x}, \underline{\xi}\rangle \overline{e_0}\underline{\xi})^m}\, dS(\underline{\xi}), \quad x_0 < 0. \tag{68}$$

2.5 Example 2

Again we start with a scalar-valued tempered distribution

$$u(\underline{x}) = \exp\big(i\langle \underline{a}, \underline{x}\rangle\big) = \cos\langle \underline{a}, \underline{x}\rangle + i \sin\langle \underline{a}, \underline{x}\rangle, \tag{69}$$

with a nonzero constant Clifford vector $\underline{a}$, for which we put $\underline{\alpha} = \underline{a}/|\underline{a}|$.

From one-dimensional theory it is known that the Hilbert transform $\mathcal{H} = i\mathcal{S}$ acts as a *rotator*, mapping $\cos ax$ and $\sin ax$ to $i\,\mathrm{sgn}(a)\sin ax$ and $-i\,\mathrm{sgn}(a)\cos ax$, respectively. It is now shown that the Clifford–Hilbert transform (15) enjoys a similar property in higher dimension. We have successively

$$\mathcal{F}\big[u(\underline{x})\big](\underline{y}) = (2\pi)^m \delta(\underline{y} - \underline{a}), \tag{70}$$

and thus

$$\mathcal{F}\big[\mathcal{H}\big[u(\underline{x})\big]\big](\underline{y}) = \overline{e_0} i \underline{\xi} \mathcal{F}\big[u(\underline{x})\big](\underline{y}) = (2\pi)^m \overline{e_0} i \underline{\xi} \delta(\underline{y} - \underline{a})$$
$$= (2\pi)^m i \overline{e_0} \underline{\alpha} \delta(\underline{y} - \underline{a}), \tag{71}$$

from which it follows that

$$\mathcal{H}[u](\underline{x}) = i\, \overline{e_0}\, \underline{\alpha} \exp\big(i\, \langle \underline{a}, \underline{x} \rangle\big) = \overline{e_0}\, \underline{\alpha} \big(-\sin\langle \underline{a}, \underline{x} \rangle + i \cos\langle \underline{a}, \underline{x} \rangle\big), \tag{72}$$

and thus

$$\mathcal{H}\big[\cos\langle \underline{a}, \underline{x} \rangle\big] = -(\overline{e_0}\, \underline{\alpha}) \sin\langle \underline{a}, \underline{x} \rangle \tag{73}$$

and

$$\mathcal{H}\big[\sin\langle \underline{a}, \underline{x} \rangle\big] = (\overline{e_0}\, \underline{\alpha}) \cos\langle \underline{a}, \underline{x} \rangle. \tag{74}$$

Note that $\underline{\alpha} = \underline{a}/|\underline{a}|$ is the multidimensional counterpart to the one-dimensional $\mathrm{sgn}(a)$ and that $(\overline{e_0}\underline{\alpha})^2 = -1$.

We also obtain the following analytic signals:

(i) $\cos\langle \underline{a}, \underline{x} \rangle - (\overline{e_0}\, \underline{\alpha}) \sin\langle \underline{a}, \underline{x} \rangle = \exp(-(\overline{e_0}\, \underline{\alpha}) \langle \underline{a}, \underline{x} \rangle)$
(ii) $\sin\langle \underline{a}, \underline{x} \rangle + (\overline{e_0}\underline{\alpha}) \cos\langle \underline{a}, \underline{x} \rangle = (\overline{e_0}\underline{\alpha}) \exp(-(\overline{e_0}\underline{\alpha}) \langle \underline{a}, \underline{x} \rangle)$
(iii) $(1 + i(\overline{e_0}\, \underline{\alpha})) \exp(i \langle \underline{a}, \underline{x} \rangle) = (1 + i(\overline{e_0}\, \underline{\alpha})) \exp(-(\overline{e_0}\underline{\alpha}) \langle \underline{a}, \underline{x} \rangle)$

3 Generalized Hilbert Transforms in Euclidean Space

In the early 2000s, four broad families $T_{\lambda,p}$, $U_{\lambda,p}$, $V_{\lambda,p}$, and $W_{\lambda,p}$, with $\lambda \in \mathbb{C}$ and $p \in \mathbb{N}_0$, of specific distributions in Clifford analysis were introduced and studied by Brackx, Delanghe, and Sommen (see [3, 5]), and it was shown that the Hilbert kernel $\mathcal{K}$, introduced in the preceding section, is one of those distributions acting as a convolution operator (see, e.g., [1]). Later on, those distributions were normalized and thoroughly discussed in a series of papers [4, 6, 7, 9–11, 14]. We recall the definitions of those normalized distributions, where $l \in \mathbb{N}_0$:

$$\begin{cases} T^*_{\lambda,p} = \pi^{\frac{\lambda+m}{2}+p} \dfrac{T_{\lambda,p}}{\Gamma(\frac{\lambda+m}{2}+p)}, & \lambda \neq -m-2p-2l, \\ T^*_{-m-2p-2l,p} = \dfrac{(-1)^p l! \pi^{\frac{m}{2}-l}}{2^{2p+2l}(p+l)!\Gamma(\frac{m}{2}+p+l)} P_p(\underline{x}) \partial_{\underline{x}}{}^{2p+2l} \delta(\underline{x}), \end{cases} \tag{75}$$

$$\begin{cases} U^*_{\lambda,p} = \pi^{\frac{\lambda+m+1}{2}+p} \dfrac{U_{\lambda,p}}{\Gamma(\frac{\lambda+m+1}{2}+p)}, & \lambda \neq -m-2p-2l-1, \\ U^*_{-m-2p-2l-1,p} = \dfrac{(-1)^{p+1} l! \pi^{\frac{m}{2}-l}}{2^{2p+2l+1}(p+l)!\Gamma(\frac{m}{2}+p+l+1)} \big(\partial_{\underline{x}}{}^{2p+2l+1} \delta(\underline{x})\big) P_p(\underline{x}), \end{cases} \tag{76}$$

$$\begin{cases} V^*_{\lambda,p} = \pi^{\frac{\lambda+m+1}{2}+p} \dfrac{V_{\lambda,p}}{\Gamma(\frac{\lambda+m+1}{2}+p)}, & \lambda \neq -m-2p-2l-1, \\ V^*_{-m-2p-2l-1,p} = \dfrac{(-1)^{p+1} l! \pi^{\frac{m}{2}-l}}{2^{2p+2l+1}(p+l)!\Gamma(\frac{m}{2}+p+l+1)} P_p(\underline{x}) \big(\partial_{\underline{x}}{}^{2p+2l+1} \delta(\underline{x})\big), \end{cases} \tag{77}$$

$$\begin{cases} W^*_{\lambda,p} = \pi^{\frac{\lambda+m}{2}+p} \dfrac{W_{\lambda,p}}{\Gamma(\frac{\lambda+m}{2}+p)}, & \lambda \neq -m-2p-2l, \\ W^*_{-m-2p-2l,p} = \dfrac{(-1)^{p+1} l! \pi^{\frac{m}{2}-l}}{2^{2p+2l+2}(p+l+1)!\Gamma(\frac{m}{2}+p+l+1)} \underline{x} P_p(\underline{x}) \underline{x}\, \partial_{\underline{x}}{}^{2p+2l+2} \delta(\underline{x}), \end{cases} \tag{78}$$

the action of the original distributions $T_{\lambda,p}$, $U_{\lambda,p}$, $V_{\lambda,p}$, and $W_{\lambda,p}$ on a testing function ϕ being given by

$$\langle T_{\lambda,p}, \phi \rangle = a_m \big\langle \mathrm{Fp}\, r_+^{\mu+p_e}, \Sigma_p^{(0)}[\phi] \big\rangle, \tag{79}$$

$$\langle U_{\lambda,p}, \phi \rangle = a_m \big\langle \mathrm{Fp}\, r_+^{\mu+p_e}, \Sigma_p^{(1)}[\phi] \big\rangle, \tag{80}$$

$$\langle V_{\lambda,p}, \phi \rangle = a_m \big\langle \mathrm{Fp}\, r_+^{\mu+p_e}, \Sigma_p^{(3)}[\phi] \big\rangle, \tag{81}$$

$$\langle W_{\lambda,p}, \phi \rangle = a_m \big\langle \mathrm{Fp}\, r_+^{\mu+p_e}, \Sigma_p^{(2)}[\phi] \big\rangle. \tag{82}$$

We explain the notation in the above expressions. First, the symbol Fp stands for the well-known distribution "finite parts" on the real line; furthermore, $\mu = \lambda + m - 1$, and p_e denotes the "even part of p," defined by $p_e = p$ if p is even and $p_e = p - 1$ if p is odd. Finally, $\Sigma_p^{(0)}$, $\Sigma_p^{(1)}$, $\Sigma_p^{(2)}$, and $\Sigma_p^{(3)}$ are the generalized spherical mean operators defined on scalar-valued testing functions ϕ by

$$\Sigma_p^{(0)}[\phi] = r^{p-p_e} \Sigma^{(0)}\big[P_p(\underline{\omega})\, \phi(\underline{x})\big] = \frac{r^{p-p_e}}{a_m} \int_{S^{m-1}} P_p(\underline{\omega}) \phi(\underline{x})\, dS(\underline{\omega}), \tag{83}$$

$$\Sigma_p^{(1)}[\phi] = r^{p-p_e} \Sigma^{(0)}\big[\underline{\omega}\, P_p(\underline{\omega})\, \phi(\underline{x})\big] = \frac{r^{p-p_e}}{a_m} \int_{S^{m-1}} \underline{\omega}\, P_p(\underline{\omega})\, \phi(\underline{x})\, dS(\underline{\omega}), \tag{84}$$

$$\begin{aligned} \Sigma_p^{(2)}[\phi] &= r^{p-p_e} \Sigma^{(0)}\big[\underline{\omega}\, P_p(\underline{\omega})\, \underline{\omega}\, \phi(\underline{x})\big] \\ &= \frac{r^{p-p_e}}{a_m} \int_{S^{m-1}} \underline{\omega}\, P_p(\underline{\omega})\, \underline{\omega}\, \phi(\underline{x})\, dS(\underline{\omega}), \end{aligned} \tag{85}$$

$$\Sigma_p^{(3)}[\phi] = r^{p-p_e} \Sigma^{(0)}\big[P_p(\underline{\omega})\, \underline{\omega}\, \phi(\underline{x})\big] = \frac{r^{p-p_e}}{a_m} \int_{S^{m-1}} P_p(\underline{\omega})\, \underline{\omega}\, \phi(\underline{x})\, dS(\underline{\omega}), \tag{86}$$

where $P_p(\underline{\omega})$ is an inner spherical monogenic of degree p, i.e., a restriction to the unit sphere S^{m-1} of a monogenic homogeneous polynomial in $\mathbb{R}^m$.

Making use of those Clifford distributions, we have then constructed two possible generalizations of the Hilbert transform (15), aiming at preserving as much as possible of its traditional properties P(1)–P(7) listed in Proposition 2 (see also [6, 9]). It is shown that in each case some of the properties—different ones—are inevitably lost. Nevertheless we will obtain in Sect. 3.1 a bounded singular integral operator on $L_2(\mathbb{R}^m)$ and in Sect. 3.2 a bounded singular integral operator on appropriate Sobolev spaces.

3.1 First generalization

In the first approach the Hilbert transform is generalized by using other kernels for the convolution, stemming from the families of distributions mentioned above. They constitute a refinement of the generalized Hilbert kernels introduced by Horváth in [28], who considered convolution kernels, homogeneous of degree $(-m)$, of the form

$$\mathrm{Pv}\,\frac{S(\underline{\omega})}{r^m}, \qquad \underline{x}=r\underline{\omega}, \qquad r=|\underline{x}|, \qquad \underline{\omega}\in S^{m-1}, \tag{87}$$

where $S(\underline{\omega})$ is a surface spherical harmonic. We investigate generalized Hilbert convolution kernels which are homogeneous of degree $(-m)$ as well, however involving inner and outer spherical monogenics. We already have mentioned that an inner spherical monogenic is the restriction to the unit sphere S^{m-1} of a monogenic homogeneous polynomial in $\mathbb{R}^m$. An outer spherical monogenic is the restriction to the unit sphere S^{m-1} of a monogenic homogeneous function in the complement of the origin; an example of an outer spherical monogenic is the "signum" function $\underline{\omega}$ since it is the restriction to S^{m-1} of the function $\underline{x}/|\underline{x}|^{m+1}$, which is monogenic in $\mathbb{R}^m\backslash\{\underline{0}\}$.

We consider the following specific distributions:

$$\begin{aligned}
T_{-m-p,p} &= \mathrm{Fp}\,\frac{1}{r^m}P_p(\underline{\omega}) = \mathrm{Pv}\,\frac{P_p(\underline{\omega})}{r^m},\\
U_{-m-p,p} &= \mathrm{Fp}\,\frac{1}{r^m}\underline{\omega}P_p(\underline{\omega}) = \mathrm{Pv}\,\frac{\underline{\omega}P_p(\underline{\omega})}{r^m},\\
V_{-m-p,p} &= \mathrm{Fp}\,\frac{1}{r^m}P_p(\underline{\omega})\underline{\omega} = \mathrm{Pv}\,\frac{P_p(\underline{\omega})\underline{\omega}}{r^m},\\
W_{-m-p,p} &= \mathrm{Fp}\,\frac{1}{r^m}\underline{\omega}P_p(\underline{\omega})\underline{\omega} = \mathrm{Pv}\,\frac{\underline{\omega}P_p(\underline{\omega})\underline{\omega}}{r^m},\\
\mathrm{Pv}\,\frac{S_{p+1}(\underline{\omega})}{r^m} &= -\frac{1}{2(p+1)}(U_{-m-p,p}+V_{-m-p,p}),\\
\mathrm{Pv}\,\frac{\underline{\omega}S_{p+1}(\underline{\omega})}{r^m} &= -\frac{1}{2(p+1)}(W_{-m-p,p}-T_{-m-p,p}),
\end{aligned} \tag{88}$$

where $P_p(\underline{x})=\partial_{\underline{x}}S_{p+1}(\underline{x})$, $S_{p+1}(\underline{x})$ being a scalar-valued solid spherical harmonic and hence $P_p(\underline{x})$ being a vector-valued solid spherical monogenic. These distributions are homogeneous of degree $(-m)$, and the functions occurring in the numerator satisfy the cancellation condition

$$\int_{S^{m-1}} \Omega(\underline{\omega})\,d\underline{\omega}=0, \tag{89}$$

$\Omega(\underline{\omega})$ being either of $P_p(\underline{\omega})$, $\underline{\omega}P_p(\underline{\omega})$, $P_p(\underline{\omega})\underline{\omega}$, or $\underline{\omega}P_p(\underline{\omega})\underline{\omega}$.

Their Fourier symbols, given by (see [14])

$$\mathcal{F}[T_{-m-p}] = i^{-p} \pi^{\frac{m}{2}} \frac{\Gamma(\frac{p}{2})}{\Gamma(\frac{m+p}{2})} P_p(\underline{\omega}), \tag{90}$$

$$\mathcal{F}[U_{-m-p}] = i^{-p-1} \pi^{\frac{m}{2}} \frac{\Gamma(\frac{p+1}{2})}{\Gamma(\frac{m+p+1}{2})} \underline{\omega}\, P_p(\underline{\omega}), \tag{91}$$

$$\mathcal{F}[V_{-m-p}] = i^{-p-1} \pi^{\frac{m}{2}} \frac{\Gamma(\frac{p+1}{2})}{\Gamma(\frac{m+p+1}{2})} P_p(\underline{\omega})\underline{\omega}, \tag{92}$$

$$\mathcal{F}[W_{-m-p}] = i^{-p-2} \pi^{\frac{m}{2}} \frac{p\Gamma(\frac{p}{2})}{(m+p)\Gamma(\frac{m+p}{2})} \left(\underline{\omega} P_p(\underline{\omega})\underline{\omega} - \frac{m-2}{p} P_p(\underline{\omega})\right), \tag{93}$$

are homogeneous of degree 0 and moreover are bounded functions, whence

$$T_{-m-p,p} * f, \qquad U_{-m-p,p} * f, \qquad V_{-m-p,p} * f, \qquad W_{-m-p,p} * f \tag{94}$$

are bounded Singular Integral Operators on $L_2(\mathbb{R}^m)$, which are direct generalizations of the Hilbert transform $\mathcal{H}$ (15), preserving (properly adapted analogues of the) properties P(1)–P(3).

We now investigate whether the new operators (94) will fulfil some appropriate analogues of the remaining properties P(4)–P(7) as well. To this end, we closely examine the kernel $T_{-m-p,p}$. First, we observe that

$$T_{-m-p,p} * T_{-m-p,p} = \frac{(-1)^p}{2^p} \pi^{\frac{m}{2}} \frac{\Gamma(\frac{m}{2})}{\Gamma(p)} \left[\frac{\Gamma(\frac{p}{2})}{\Gamma(\frac{m+p}{2})}\right]^2 T_{-m,p} P_p(\partial_{\underline{x}}), \tag{95}$$

which directly implies that the generalized Hilbert transform $T_{-m-p,p} * f$ does not satisfy an analogue of property P(4). Next, as it can easily be shown that the considered operator coincides with its adjoint—up to a minus sign when p is even—we may also conclude, in view of (95), that it will not be unitary, neither does it commute with the Dirac operator.

Finally, the most important drawback of this first generalization is undoubtedly the fact that we cannot establish an analogue of property P(7), since it turned out impossible to find a generalized Cauchy kernel in $\mathbb{R}^{m+1} \setminus \{0\}$, for which a part of the boundary values is precisely the generalized Hilbert kernel $T_{-m-p,p}$. Similar conclusions hold for the other generalized kernels used in (88).

3.2 Second Generalization

Subsequent to the observations made in the previous subsection, we now want to find a type of generalized Hilbert kernel which actually preserves property P(7). To that end, we define the function

$$E_p(x) = E_p(x_0, \underline{x}) = \frac{1}{a_{m+1,p}} \frac{\bar{x} e_0}{|x|^{m+1+2p}} P_p(\underline{x})$$
$$= \frac{1}{a_{m+1,p}} \frac{x_0 + e_0 \underline{x}}{|x_0 e_0 + \underline{x}|^{m+1+2p}} P_p(\underline{x}), \tag{96}$$

where

$$a_{m+1,p} = \frac{(-1)^p}{2^p} \frac{2\pi^{\frac{m+1}{2}}}{\Gamma(\frac{m+1}{2} + p)}, \tag{97}$$

involving a homogeneous polynomial $P_p(\underline{x})$ of degree p which we take once more to be vector valued and monogenic. It is stated in the next proposition that these functions E_p are good candidates for generalized Cauchy kernels. Note that for $p = 1$ and taking $P_0(\underline{x}) = 1$, the standard Cauchy kernel, i.e., the fundamental solution of the Cauchy–Riemann operator D_x in $\mathbb{R}^{m+1}$ is reobtained.

For the proofs of all results mentioned in the remainder of this section, we refer the reader to [9].

Proposition 3 *The function* E_p, (96), *satisfies the following properties*:

(i) $E_p \in L_1^{\text{loc}}(\mathbb{R}^{m+1})$ *and* $\lim_{|x| \to \infty} E_p(x) = 0 \ \forall p \in \mathbb{N}$
(ii) $D_x \, E_p(x) = P_p(\partial_{\underline{x}})\delta(x)$ *in distributional sense* $\forall p \in \mathbb{N}$

Here, $L_1^{\text{loc}}(\mathbb{R}^{m+1})$ *stands for the locally integrable functions on* $\mathbb{R}^{m+1}$.

Next, we calculate their nontangential distributional boundary values as $x_0 \to 0\pm$. To that end, we first formulate an interesting auxiliary result in the following lemma.

Lemma 3 *For* $p \in \mathbb{N}_0$, *one has*

$$\lim_{x_0 \to 0+} \frac{x_0}{|x_0 + \underline{x}|^{m+1+2p}} = \frac{1}{2^{p+1} p!} a_{m+1,p} \, \partial_{\underline{x}}^{2p} \delta(\underline{x}), \tag{98}$$

$a_{m+1,p}$ *being given by* (97).

Proposition 4 *For each* $p \in \mathbb{N}_0$, *one has*

$$E_p(0+, \underline{x}) = \lim_{x_0 \to 0+} E_p(x_0, \underline{x}) = \frac{1}{2} P_p(\partial_{\underline{x}})\delta(\underline{x}) + \frac{1}{2} \mathcal{K}_p(\underline{x}), \tag{99}$$

$$E_p(0-, \underline{x}) = \lim_{x_0 \to 0-} E_p(x_0, \underline{x}) = -\frac{1}{2} P_p(\partial_{\underline{x}})\delta(\underline{x}) + \frac{1}{2} \mathcal{K}_p(\underline{x}), \tag{100}$$

where

$$\mathcal{K}_p(\underline{x}) = \frac{2}{a_{m+1,p}} \overline{e_0} \, \text{Fp} \, \frac{\bar{\underline{\omega}} P_p(\underline{\omega})}{r^{m+p}} = -\frac{2}{a_{m+1,p}} \overline{e_0} U^*_{-m-2p,p}. \tag{101}$$

The distribution $\mathcal{K}_p$ (101) arising in the previous proposition allows for the definition of a generalized Hilbert transform $\mathcal{H}_p$ given by

$$\mathcal{H}_p[f] = \mathcal{K}_p * f. \tag{102}$$

Because the Fourier symbol

$$\mathcal{F}[\mathcal{K}_p] = -\frac{2}{a_{m+1,p}} \overline{e_0} i^{-p-1} U^*_{0,p} \tag{103}$$

of the kernel $\mathcal{K}_p$ is not a bounded function, the operator $\mathcal{H}_p$, (102), will also not be bounded on $L_2(\mathbb{R}^m)$. However, the Fourier symbol (103) is a polynomial of degree p, implying that $\mathcal{H}_p$ is a bounded operator between the Sobolev spaces $W_2^n(\mathbb{R}^m) \to W_2^{n-p}(\mathbb{R}^m)$ for $n \geq p$. It can indeed be proved that:

Proposition 5 *The generalized Cauchy integral $\mathfrak{C}_p$ given by $\mathfrak{C}_p[f] = E_p * f$ maps the Sobolev space $W_2^n(\mathbb{R}^m)$ into the Hardy space $H^2(\mathbb{R}^{m+1}_+)$ for each natural number $n \geq p$.*

Corollary 1 *The generalized Hilbert transform $\mathcal{H}_p$, (102), is a bounded linear operator between the Sobolev spaces $W_2^n(\mathbb{R}^m)$ and $W_2^{n-p}(\mathbb{R}^m)$ for each natural number $n \geq p$.*

Comparing further the properties of $\mathcal{H}_p$ with those of the standard Hilbert transform $\mathcal{H}$ in Clifford analysis shows that the main objective for this second generalization is fulfilled on account of Proposition 4: the kernel $\mathcal{K}_p$ arises as a part of the boundary values of a generalized Cauchy kernel E_p, which constitutes an analogue of the "classical" property P(7). However, the kernel $\mathcal{K}_p$ is a homogeneous distribution of degree $(-m - p)$, meaning that $\mathcal{H}_p$ is not dilation invariant.

4 The Anisotropic Hilbert Transform

The (generalized) multidimensional Hilbert transforms on $\mathbb{R}^m$ considered in Sects. 2 and 3 might be characterized as isotropic, since the metric in the underlying space is the standard Euclidean one. In this section we adopt the idea of an *anisotropic* (also called *metric-dependent* or *metrodynamical*) Clifford setting, which offers the possibility of adjusting the coordinate system to preferential, not necessarily mutually orthogonal, directions intrinsically present in the m-dimensional structures or signals to be analyzed. In this new area of Clifford analysis (see, e.g., [13, 19]), we have constructed the so-called anisotropic (Clifford–)Hilbert transform (see [15, 17]), a special case of which was already introduced and used for two-dimensional image processing in [26].

4.1 Definition of the Anisotropic Hilbert Transform

For the basic language of anisotropic Clifford analysis, we first present the notion of metric tensor, namely a real, symmetric, and positive definite tensor $\widetilde{G} = (g_{kl})_{k,l=0,\ldots,m}$ of order $(m+1)$, which gives rise to two bases in $\mathbb{R}^{m+1}$: a covariant basis $(e_0, \ldots, e_m)$ and a contravariant basis $(e^0, \ldots, e^m)$ corresponding to each other through the metric tensor, viz

$$e_k = \sum_{l=0}^{m} g_{kl} e^l \quad \text{and} \quad e^l = \sum_{k=0}^{m} g^{lk} e_k \quad \text{with } \widetilde{G}^{-1} = \left(g^{kl}\right)_{k,l=0,\ldots,m}. \tag{104}$$

Then, a Clifford algebra is constructed, depending on the metric tensor involved, and all necessary definitions and results of Euclidean Clifford analysis are adapted to this metric-dependent setting. We mention, e.g., that the classical scalar product is replaced by the symmetric bilinear form

$$\langle x, y \rangle_{\widetilde{G}} = \sum_{k=0}^{m} \sum_{l=0}^{m} g_{kl} x^k y^l. \tag{105}$$

The anisotropic Dirac and Cauchy–Riemann operators in $\mathbb{R}^{m+1}$ take the forms

$$\partial_{\widetilde{G}} = \sum_{k=0}^{m} e^k \, \partial_{x^k} \tag{106}$$

and

$$D_{\widetilde{G}} = \overline{e}_0 \partial_{\widetilde{G}} = \partial_{x^0} + \overline{e}_0 \underline{\partial}_G, \tag{107}$$

where $G = (g_{kl})_{k,l=1,\ldots,m}$ in $\mathbb{R}^{m\times m}$ is the subtensor of the metric tensor $\widetilde{G}$ in $\mathbb{R}^{(m+1)\times(m+1)}$. The fundamental solution of the latter operator,

$$E_{\widetilde{G}}(x) = \frac{1}{a_{m+1}} \frac{\overline{x} e^0}{(\langle x, x \rangle_{\widetilde{G}})^{(m+1)/2}}, \tag{108}$$

is now used as the kernel in the definition of the metrodynamical Cauchy integral given, for a function $f \in L_2(\mathbb{R}^m)$ or a tempered distribution, by

$$\mathcal{C}_{\widetilde{G}}[f] = E_{\widetilde{G}} * f, \tag{109}$$

which is monogenic in $\mathbb{R}^{m+1}_+$ (and in $\mathbb{R}^{m+1}_-$). Taking limits in L_2 or distributional sense as $x^0 \to 0+$ gives, through careful calculation (see [15]),

$$\lim_{x^0 \to 0+} \mathcal{C}_{\widetilde{G}}[f] = \frac{1}{\sqrt{\det \widetilde{G}}} \left(\frac{1}{2} f + \frac{1}{2} \overline{e}^0 H_{\text{ani}} * f \right) \tag{110}$$

with

$$H_{\mathrm{ani}}(\underline{x}) = \sqrt{\det \widetilde{G}}\left(\frac{2}{a_{m+1}}\,\mathrm{Pv}\,\frac{\overline{\underline{x}}}{(\langle \underline{x}, \underline{x}\rangle_G)^{(m+1)/2}}\right). \tag{111}$$

Similarly, as $x^0 \to 0-$, we obtain

$$\lim_{x^0 \to 0-} \mathcal{C}_{\widetilde{G}}[f] = \frac{1}{\sqrt{\det \widetilde{G}}}\left(-\frac{1}{2}f + \frac{1}{2}\overline{e}^0 H_{\mathrm{ani}} * f\right). \tag{112}$$

The above results are the anisotropic Plemelj–Sokhotzki formulae, and they give rise to the definition of the anisotropic Hilbert transform:

$$\mathcal{H}_{\mathrm{ani}}[f] = \overline{e}^0 H_{\mathrm{ani}} * f. \tag{113}$$

As already mentioned in the introduction of this section, for $m = 2$, such an anisotropic Hilbert transform was considered in [26], however, for the special case where the e_0-direction in $\mathbb{R}^3$ is chosen perpendicular to the $\mathbb{R}^2$-plane spanned by (e_1, e_2). This corresponds to a $\widetilde{G}$-matrix of order 3 in which $g_{01} = g_{02} = 0$.

4.2 Properties of the Anisotropic Hilbert Transform

In order to study the properties of the linear operator $\mathcal{H}_{\mathrm{ani}}$, (113), we will also have to pass to frequency space, so we need to introduce a proper definition for the anisotropic Fourier transform on $\mathbb{R}^m$ in the present metric-dependent setting:

$$\begin{aligned}\mathcal{F}_G[f](\underline{x}) &= \int_{\mathbb{R}^m} \exp\bigl(-2\pi i \langle \underline{x}, \underline{y}\rangle_G\bigr) f(\underline{y})\, dV(\underline{y}) \\ &= \int_{\mathbb{R}^m} \exp\bigl(-2\pi i \underline{x}^T G \underline{y}\bigr) f(\underline{y})\, dV(\underline{y}).\end{aligned} \tag{114}$$

Due to the assumed symmetric character of the tensor G, it is found that

$$\mathcal{F}_G[f](\underline{x}) = \mathcal{F}[f](G\underline{x}). \tag{115}$$

The following properties of $\mathcal{H}_{\mathrm{ani}}$ may then be proved (see [15]):

(P1) $\mathcal{H}_{\mathrm{ani}}$ is translation invariant.
(P2) $\mathcal{H}_{\mathrm{ani}}$ is dilation invariant, which is equivalent to its kernel H_{ani}, (111), being a homogeneous distribution of degree $-m$.
(P3) $\mathcal{H}_{\mathrm{ani}}$ is a bounded operator on $L_2(\mathbb{R}^m)$, which is equivalent to its Fourier symbol

$$\mathcal{F}_G[H_{\mathrm{ani}}](\underline{x}) = \sqrt{\frac{\det \widetilde{G}}{\det G}}\, i\, \frac{\underline{x}}{\langle \underline{x}, \underline{x}\rangle_G} \tag{116}$$

being a bounded function.

(P4) Up to a metric related constant, $\mathcal{H}_{\text{ani}}$ squares to unity or, more explicitly,

$$(\mathcal{H}_{\text{ani}})^2 = \frac{\det \widetilde{G}}{\det G}\mathbf{1}. \tag{117}$$

(P5) $\mathcal{H}_{\text{ani}}$ is selfadjoint.
(P6) $\mathcal{H}_{\text{ani}}$ arises in a natural way by considering nontangential boundary values of the Cauchy integral $\mathcal{C}_{\widetilde{G}}$, (109), in $\mathbb{R}^{m+1}$.

Note that the anisotropic Hilbert transform shows the influence of the underlying metric in two different ways: (1) the determinant of the "mother" metric $\widetilde{G}$ on $\mathbb{R}^{m+1}$ arises as an explicit factor in the expression for the kernel, and (2) the induced metric G on $\mathbb{R}^m$ comes into play explicitly through the denominator of the kernel and also implicitly through its numerator since the vector $\overline{\underline{x}}$ contains the (skew) basis vectors $(e_k)_{k=1}^m$.

The particularity of this metric dependence may also be seen in frequency space, where the metric G not only arises in the Fourier symbol (116) of $\mathcal{H}_{\text{ani}}$ but is also hidden in the definition of the anisotropic Fourier transform itself, while the "mother" metric $\widetilde{G}$ again only is seen to arise through its determinant.

The above observations do raise the question whether there exists a one-to-one correspondence between a given Hilbert transform on $(\mathbb{R}^m, G)$ and the associated Cauchy integral on $(\mathbb{R}^{m+1}, \widetilde{G})$ from which it originates, or in other words: does the Hilbert transform contain enough geometrical information to completely determine the "mother" metric $\widetilde{G}$? The answer is negative. It turns out that, given a Hilbert kernel

$$H_{\text{ani}} = c\left(\frac{2}{a_{m+1}} \operatorname{Pv} \frac{\overline{\underline{x}}}{(\langle \underline{x}, \underline{x}\rangle)^{(m+1)/2}}\right) \tag{118}$$

being dependent on the m-dimensional metric G and on the strictly positive constant c, it is part of the boundary value of a Cauchy kernel in $(\mathbb{R}^{m+1}, \widetilde{G})$ with

$$\widetilde{G} = \begin{pmatrix} g_{00} & \underline{u}^T \\ \underline{u} & G \end{pmatrix}, \tag{119}$$

where $(g_{00}, \underline{u}^T)$ are characterized, but not uniquely determined, by the equation

$$g_{00} - \underline{u}^T G^{-1} \underline{u} = \frac{c}{\det G}. \tag{120}$$

4.3 Example

It is interesting to demonstrate the difference between the Clifford–Hilbert transform of Sect. 2 and its anisotropic counterpart. So consider in $\mathbb{R}^m$ again the scalar-valued tempered distribution $f(\underline{x}) = \exp(i\langle \underline{a}, \underline{x}\rangle)$, where $\underline{a}$ is a constant, nonzero Clifford vector.

In the isotropic case we find (see Sect. 2.4)

$$\mathcal{H}[f](\underline{x}) = i\overline{e_0}\frac{\underline{a}}{|\underline{a}|}\exp\big(i\langle\underline{a},\underline{x}\rangle\big). \tag{121}$$

In the anisotropic case we successively obtain

$$\mathcal{F}_G[f](\underline{y}) = \mathcal{F}[f](G\underline{y}) = \delta(G\underline{y}-\underline{a}), \tag{122}$$

and thus

$$\mathcal{F}_G\big[\mathcal{H}_{G,c}[f]\big](\underline{y}) = \overline{e^0}\, i\sqrt{\frac{\det(\widetilde{G})}{\det(G)}}\frac{G^{-1}\underline{a}}{|G^{-1}\underline{a}|_G}\delta(G\underline{y}-\underline{a}) \tag{123}$$

with

$$\left|G^{-1}\underline{a}\right|_G = \left[(G^{-1}\underline{a})^T G(G^{-1}\underline{a})\right]^{\frac{1}{2}} = \left[\underline{a}^T G^{-1}\underline{a}\right]^{\frac{1}{2}}. \tag{124}$$

Subsequent calculations reveal that

$$\begin{aligned}
\mathcal{F}_G^{-1}\left[\delta(G\underline{y}-\underline{a})\right](\underline{x}) &= \int_{\mathbb{R}^m}\exp\big(i\,\underline{x}^T G\underline{y}\big)\delta(G\underline{y}-\underline{a})\,dV(\underline{y}) \\
&= \frac{1}{\det(G)}\int_{\mathbb{R}^m}\exp\big(i\underline{x}^T\underline{y}'\big)\delta\big(\underline{y}'-\underline{a}\big)\,dV\big(\underline{y}'\big) \\
&= \frac{1}{\det(G)}\exp\big(i\langle\underline{a},\underline{x}\rangle\big).
\end{aligned} \tag{125}$$

Hence,

$$\mathcal{H}_{\text{ani}}[f](\underline{x}) = i\overline{e^0}\sqrt{\frac{\det(\widetilde{G})}{(\det(G))^3}}\frac{G^{-1}\underline{a}}{|G^{-1}\underline{a}|}\exp\big(i\langle\underline{a},\underline{x}\,\rangle\big). \tag{126}$$

5 Conclusion

The concept of *analytic signal* on the real time axis is fundamental in signal processing. It contains the original signal and its Hilbert transform and allows for the decomposition of a finite-energy signal into its analytic and anti-analytic components. In mathematical terms, this is rephrased as the direct sum decomposition of $L_2(\mathbb{R})$ into the Hardy space $H_2(\mathbb{R})$ and its orthogonal complement, and the analytic signals are precisely the functions in $H_2(\mathbb{R})$. In this paper we have presented several generalizations of the Hilbert transform and the corresponding analytic signal to Euclidean space of arbitrary dimension, and we have indicated the properties which are characteristic in the one-dimensional case and persist in each of those generalizations. It becomes apparent, also from the given examples, that the Clifford analysis framework is most appropriate to develop these multidimensional Hilbert transforms. That Clifford analysis could be a powerful tool in multidimensional signal analysis became already clear during the last decade from the several constructions of multidimensional Fourier transforms with quaternionic or Clifford algebra-valued kernels with direct applications in signal analysis and pattern recognition, see [8, 12, 18, 24–26, 31] and also the review paper [16], wherein the relations between the different approaches are established. In view of the fact that in the underlying paper the interaction of the Clifford–Hilbert transforms with only the standard Fourier transform was considered, their interplay with the various Clifford–Fourier transforms remains an intriguing and promising topic for further research.

References

1. Brackx, F., De Schepper, H.: Hilbert–Dirac operators in Clifford analysis. Chin. Ann. Math. Ser. B **26**(1), 1–14 (2005)
2. Brackx, F., Delanghe, R., Sommen, F.: Clifford Analysis. Research Notes in Mathematics, vol. 76. Pitman, London (1982)
3. Brackx, F., Delanghe, R., Sommen, F.: Spherical means, distributions and convolution operators in Clifford analysis. Chin. Ann. Math. Ser. B **24**(2), 133–146 (2003)
4. Brackx, F., De Knock, B., De Schepper, H.: A specific family of Clifford distributions. In: Son, L.H., Tutschke, W., Jain, S. (eds.) Methods of Complex and Clifford Analysis, Proceedings of ICAM, Hanoi, August 2004, pp. 215–228. SAS International Publication, Delhi (2004)
5. Brackx, F., Delanghe, R., Sommen, F.: Spherical means and distributions in Clifford analysis. In: Qian, T., Hempfling, T., McIntosch, A., Sommen, F. (eds.) Advances in Analysis and Geometry: New Developments Using Clifford Algebra. Trends in Mathematics. pp. 65–96. Birkhäuser, Basel (2004)
6. Brackx, F., De Knock, B., De Schepper, H.: Generalized multi-dimensional Hilbert transforms involving spherical monogenics in the framework of Clifford analysis. In: Simos, T.E., Psihoyios, G., Tsitouras, Ch. (eds.) ICNAAM 2005, Official Conference of the European Society of Computational Methods in Sciences and Engineering, pp. 911–914. Wiley, VCH, Weinheim, New York (2005)
7. Brackx, F., De Knock, B., De Schepper, H.: Multi-vector spherical monogenics, spherical means and distributions in Clifford analysis. Acta Math. Sin. (Engl. Ser.) **21**(5), 1197–1208 (2005)
8. Brackx, F., De Schepper, N., Sommen, F.: The Clifford–Fourier transform. J. Fourier Anal. Appl. **11**(6), 669–681 (2005)
9. Brackx, F., De Knock, B., De Schepper, H.: Generalized multidimensional Hilbert transforms in Clifford analysis. Int. J. Math. Math. Sci. **2006**, 1–19 (2006)
10. Brackx, F., De Knock, B., De Schepper, H.: On the Fourier spectra of distributions in Clifford analysis. Chin. Ann. Math. Ser. B **27**(5), 495–506 (2006)
11. Brackx, F., De Knock, B., De Schepper, H., Sommen, F.: Distributions in Clifford analysis: an overview. In: Eriksson, S.-L. (ed.) Clifford Analysis and Applications (Proceedings of the Summer School, Tampere, August 2004), pp. 59–73. Tampere University of Technology, Institute of Mathematics, Research Report, vol. 82 (2006)
12. Brackx, F., De Schepper, N., Sommen, F.: The two-dimensional Clifford–Fourier transform. J. Mat: Imaging Vis. **26**(1–2), 5–18 (2006)
13. Brackx, F., De Schepper, N., Sommen, F.: Metric dependent Clifford analysis with applications to wavelet analysis. In: Alpay, D. (ed.) Wavelets, Multiscale Systems and Hypercomplex Analysis. Operator Theory: Advances and Applications, vol. 167, pp. 17–67. Birkhäuser, Basel (2006)
14. Brackx, F., De Knock, B., De Schepper, H.: On generalized Hilbert transforms and their interaction with the Radon transform in Clifford analysis. Math. Methods Appl. Sci. **30**(9), 1071–1092 (2007)

15. Brackx, F., De Knock, B., De Schepper, H.: A metric dependent Hilbert transform in Clifford analysis. Bull. Belg. Math. Soc. Simon Stevin **14**(3), 445–453 (2007)
16. Brackx, F., De Schepper, N., Sommen, F.: The Fourier transform in Clifford analysis. Adv. Imaging Electron Phys. **156**(6), 55–201 (2008)
17. Brackx, F., De Knock, B., De Schepper, H.: A multidimensional Hilbert transform in anisotropic Clifford analysis. In: Gürlebeck, K., Könke, C. (eds.) Proceedings of the 17th International Conference on the Application of Computer Science and Mathematics in Architecture and Civil Engineering. Bauhaus Universität Weimar, Germany, 12–14 July 2006 [cd-rom proceedings]
18. Bülow, T., Sommer, G.: The hypercomplex signal—a novel approach to the multidimensional signal. IEEE Trans. Signal Process. **49**(11), 2844–2852 (2001)
19. De Schepper, N.: Multi-dimensional continuous wavelet transforms and generalized Fourier transforms in Clifford analysis. Ph.D. thesis, Ghent University, Ghent (2006)
20. Delanghe, R.: Clifford analysis: history and perspective. Comput. Methods Funct. Theory **1**(1), 107–153 (2001)
21. Delanghe, R.: Some remarks on the principal value kernel in $\mathbb{R}^m$. Complex Var. Theory Appl. **47**(8), 653–662 (2002)
22. Delanghe, R.: On some properties of the Hilbert transform in Euclidean space. Bull. Belg. Math. Soc. Simon Stevin **11**, 163–180 (2004)
23. Delanghe, R., Sommen, F., Souček, V.: Clifford algebra and spinor-valued functions, a function theory for the Dirac operator. Mathematics and its Applications, vol. 53. Kluwer Academic, Dordrecht (1992)
24. Ebling, J., Scheuermann, G.: Clifford convolution and pattern matching on vector fields. In: Proceedings of IEEE Visualization'03, pp. 193–200. Comput. Soc., Los Alamitos (2003)
25. Ebling, J., Scheuermann, G.: Clifford Fourier transform on vector fields. IEEE Trans. Vis. Comput. Graph. **11**(4), 469–479 (2005)
26. Felsberg, M.: Low-level image processing with the structure multivector. Ph.D. thesis. Christian-Albrechts-Universität, Kiel (2002)
27. Gilbert, J.E., Murray, M.A.M.: Clifford Algebra and Dirac Operators in Harmonic Analysis. Cambridge Studies in Advanced Mathematics, vol. 26. Cambridge University Press, Cambridge (1991)
28. Horváth, J.: Singular integral operators and spherical harmonics. Trans. Am. Math. Soc. **82**, 52–63 (1950)
29. Horváth, J.: Sur les fonctions conjuguées à plusieurs variables. Nederl. Akad. Wetensch. Proc. Ser. A 56. Indag. Math. **15**, 17–29 (1953) (in French)
30. Li, C., McIntosh, A., Qian, T.: Clifford algebras, Fourier transforms, and singular convolution operators on Lipschitz surfaces. Rev. Math. Iberoam. **10**, 665–721 (1994)
31. Mawardi, B., Hitzer, E.: Clifford fourier transformation and uncertainty principle for the clifford geometric algebra $Cl_{3,0}$. Adv. Appl. Clifford Algebr. **16**(1), 41–61 (2006)
32. McIntosh, A.: Clifford algebras, Fourier theory, singular integrals and harmonic functions on Lipschitz domains. In: Ryan, J. (ed.) Clifford Algebras in Analysis and Related Topics, Fayetteville, 1993. Studies in Advanced Mathematics. pp. 33–87. CRC Press, Boca Raton (1996)
33. Poularikas, A.D. (ed.): The Transforms and Applications Handbook. CRC Press, Boca Raton (1996)
34. Ryan, J.: Clifford analysis. In: Ablamowicz, R., Sobczyk, G. (eds.) Lectures on Clifford (Geometric) Algebras and Applications, pp. 53–89. Birkhäuser, Basel (2004)
35. Sommen, F.: Some connections between Clifford analysis and complex analysis. Complex Var. Theory Appl. **1**(1), 97–118 (1982)
36. Sommen, F.: Hypercomplex Fourier and Laplace transforms II. Complex Var. Theory Appl. **1**(2–3), 209–238 (1982/1983)

Part III
Image Processing, Wavelets and Neurocomputing

Geometric Neural Computing for 2D Contour and 3D Surface Reconstruction

Jorge Rivera-Rovelo, Eduardo Bayro-Corrochano, and Ruediger Dillmann

Abstract In this work we present an algorithm to approximate the surface of 2D or 3D objects combining concepts from geometric algebra and artificial neural networks. Our approach is based on the self-organized neural network called Growing Neural Gas (GNG), incorporating versors of the geometric algebra in its neural units; such versors are the transformations that will be determined during the training stage and then applied to a point to approximate the surface of the object. We also incorporate the information given by the generalized gradient vector flow to select automatically the input patterns, and also in the learning stage in order to improve the performance of the net. Several examples using medical images are presented, as well as images of automatic visual inspection. We compared the results obtained using snakes against the GSOM incorporating the gradient information and using versors. Such results confirm that our approach is very promising. As a second application, a kind of morphing or registration procedure is shown; namely the algorithm can be used when transforming one model at time t_1 into another at time t_2. We include also examples applying the same procedure, now extended to models based on spheres.

1 Introduction

To approximate the contour or surface of an object is a task that can be carried out by different methods. If we want to preserve the topology of the data, a good choice is the use of self-organizing neural networks [1]. In this work we propose a

J. Rivera-Rovelo (✉)
Universidad Anahuac Mayab, Carr. Merida-Progreso Km. 15.5, Merida, Mexico
e-mail: jorge.rivera@unimayab.edu.mx

E. Bayro-Corrochano (✉)
Dept. Electrical Eng. & Computer Science, CINVESTAV, Unidad Guadalajara, Av. Científica 1145, 45015 Colonia el Bajío, Zapopan, JAL, Mexico
e-mail: edb@gdl.cinvestav.mx

E. Bayro-Corrochano, G. Scheuermann (eds.), *Geometric Algebra Computing*,
DOI 10.1007/978-1-84996-108-0_10,

very advanced algorithm using early vision preprocessing and self-organizing neural computing in terms of geometric algebra techniques. Other similar approaches can be found in [3, 7]. We believe that the early vision preprocessing, together with self-organizing neurocomputing, resembles in certain manner the geometric visual processing in biological creatures.

The proposed approach uses the *Generalized Gradient Vector Flow (GGVF)* [10] to guide the automatic selection of the input patterns and to guide the learning process of the self-organized neural network GNG [4], to obtain a set of transformations M expressed in the conformal geometric algebra framework. In this framework rigid body transformations of geometric entities (like points, lines, planes, circles, spheres) are expressed in compact form as operators called *versors* that are applied in a multiplicative way to any entity of the conformal geometric algebra. Thus, training the network, we obtain the transformation that can be applied to entities resulting in the definition of the object contour or shape. The experimental results show applications in medical image processing and visual inspection tasks, confirming that our approach is very promising.

2 Geometric Algebra

The Geometric Algebra [2, 6, 9] $G_{p,q,r}$ is constructed over the vector space $\mathcal{V}^{p,q,r}$, where p, q, r denote the signature of the algebra; if $p \neq 0$ and $q = r = 0$, the metric is Euclidean; if only $r = 0$, the metric is pseudo-Euclidean; if $p \neq 0$, $q \neq 0$, and $r \neq 0$, the metric is degenerate. In this algebra, we have the *geometric product* which is defined as in (1) for two vectors a, b and has two parts: the inner product $a \cdot b$ is the symmetric part, while the wedge product $a \wedge b$ is the antisymmetric part:

$$ab = a \cdot b + a \wedge b. \tag{1}$$

The dimension of $G_{n=p,q,r}$ is 2^n, and G_n is constructed by the application of the geometric product over the vector basis e_i,

$$e_i e_j = \begin{cases} 1 & \text{for } i = j \in 1, \ldots, p, \\ -1 & \text{for } i = j \in p+1, \ldots, p+q, \\ 0 & \text{for } i = j \in p+q+1, \ldots, p+q+r, \\ e_i \wedge e_j & \text{for } i \neq j. \end{cases}$$

This leads to a basis for the entire algebra: $\{1\}, \{e_i\}, \{e_i \wedge e_j\}, \{e_i \wedge e_j \wedge e_k\}, \ldots, \{e_1 \wedge e_2 \wedge \cdots \wedge e_n\}$. Any multivector can be expressed in terms of this basis. In the nD space there are multivectors of grade 0 (scalars), grade 1 (vectors), grade 2 (bivectors), grade 3 (trivectors), ..., up to grade n. This results in a basis for G_n containing elements of different grade called *blades* (e.g., scalars, vectors, bivectors, trivectors, etc.): $1, e_1, \ldots, e_{12}, \ldots, e_{123}, \ldots, I$, which are called *basis blades*, where the element of maximum grade is the pseudoscalar $I = e_1 \wedge e_2 \cdots \wedge e_n$. A linear combination of basis blades, all of the same grade k, is called k-vector. The linear combination of such k-vectors is called *multivector*, and multivectors with certain characteristics represent different geometric objects (as points, lines, planes, circles, spheres, etc.), depending on the GA where we are working on. Given a multivector M, if we are interested in extracting only the blades of a given grade, we write $\langle M\rangle_r$, where r is the grade of the blades we want to extract (obtaining a homogeneous multivector M' or an r-vector).

The reverse of a multivector is given by

$$\left\langle M^{\dagger}\right\rangle_i = (-1)^{\frac{i(i-1)}{2}} \langle M\rangle_i. \tag{2}$$

That is, the reversion is a linear mapping that inverts the geometric product's order and stays in the same space.

2.1 The OPNS and IPNS

In Geometric Algebra, the blades have geometric meaning based on their interpretation as linear subspaces. For example, suppose that you have a vector $a \in \mathbb{R}^n$; we define the function $\mathcal{O}_a$ as

$$\mathcal{O}_a : x \in \mathbb{R}^n \rightarrow x \wedge a \in G_n,$$

where G_n is the Clifford Algebra over the space $\mathbb{R}^n$. The *kernel* of this function is the set of vectors in $\mathbb{R}^n$ such that $\mathcal{O}_a$ maps to zero. This *kernel* is named *Outer Product Null Space (OPNS)* of a and is denoted by $\mathcal{N}\mathcal{O}(a)$. From the wedge product definition we know that $x \wedge a = 0$ given that x and a are linearly dependent. Thus, $\mathcal{N}\mathcal{O}(a)$ can be expressed in terms of a as

$$\mathcal{N}\mathcal{O}(a) = \{\alpha a : \alpha \in \mathbb{R}\}.$$

In general, the OPNS of a k-blade $\langle A\rangle_k \in G_n$ is a k-dimensional linear subspace of $\mathbb{R}^n$,

$$\mathcal{N}\mathcal{O}\big(\langle A\rangle_k\big) := \big\{x \in \mathbb{R}^n : x \wedge \langle A\rangle_k = 0\big\}. \tag{3}$$

The *Inner Product Null Space (IPNS)* of a blade $\langle A\rangle_k \in G_n$, denoted as $\mathcal{N}\mathcal{I}(\langle A\rangle_k)$, is the *kernel* of the function $\mathcal{I}_{\langle A\rangle_k}$ defined as

$$\mathcal{I}_{\langle A\rangle_k} : x \in \mathbb{R}^n \rightarrow x \cdot \mathcal{I}_{\langle A\rangle_k} \in G_n. \tag{4}$$

Thus, the *kernel* is

$$\mathcal{N}\mathcal{I}\big(\langle A\rangle_k\big) := \big\{x \in \mathbb{R}^n : \mathcal{I}_{\langle A\rangle_k}(x) = 0 \in G_n\big\}. \tag{5}$$

For example, consider a vector $a \in \mathbb{R}^3$; then $\mathcal{N}\mathcal{I}(a)$ is given by

$$\mathcal{N}\mathcal{I}(a) := \big\{x \in \mathbb{R}^3 : x \cdot a = 0\big\}.$$

That is, all vectors that are perpendicular to the vector a belong to its IPNS. The representation of an entity expressed in the IPNS can be obtained from its corresponding OPNS representation by multiplying the latter by the pseudo-scalar of the GA we are working on. Sometimes the OPNS representation is called *dual representation*.

2.2 Conformal Geometric Algebra

To work in Conformal Geometric Algebra (CGA) $G_{4,1,0}$ means to embed the Euclidean space in a higher-dimensional space with two extra basis vectors which have particular meaning; in this way, we represent particular objects of the Euclidean space with subspaces of the conformal space. The vectors we add are e_+ and e_-, which square to 1 and -1, respectively. With these two vectors, we define the null vectors

$$e_0 = \frac{1}{2}(e_- - e_+), \qquad e_\infty = (e_- + e_+), \tag{6}$$

interpreted as the origin and the point at infinity, respectively. From now and in the rest of the paper, points in the 3D-Euclidean space will be denoted in lowercase letters, while conformal points in uppercase letters; also the conformal entities will be expressed in the *Inner Product Null Space* (IPNS, Sect. 2.1) and not in the *Outer Product Null Space* (OPNS, Sect. 2.1) unless it is specified explicitly. To map a point $x \in \mathcal{V}^3$ to the conformal space in $G_{4,1}$, we use

$$X = x + \frac{1}{2}x^2 e_\infty + e_0. \tag{7}$$

As mentioned before, we can use CGA to represent particular objects of the 3D-Euclidean space; the spheres are specially interesting because they are the basic entities in CGA from which other entities are derived. Spheres with center c and radius ρ are represented as

$$S = c + \frac{1}{2}(c^2 - \rho^2)e_\infty + e_0. \tag{8}$$

In fact, we can think in conformal points X as degenerated spheres of radius $\rho = 0$. Let X_1, X_2 be two conformal points. If we subtract X_2 from X_1, we obtain

$$X_1 - X_2 = (x_1 - x_2) + \frac{1}{2}(x_1^2 - x_2^2)e_\infty + e_0, \tag{9}$$

and if we square this result, we obtain

$$(X_1 - X_2)^2 = (x_1 - x_2)^2. \tag{10}$$

So, if we want a measure of the Euclidean distance between the two points, we can apply (10). To obtain the 3D magnitude of a vector, we simply multiply its conformal representation X as follows:

$$|x| = \sqrt{-2(X \cdot e_0)}. \tag{11}$$

The dot product between two conformal vectors X_1 and X_2 results in $X_1 \cdot X_2 = (x_1 \cdot x_2) - \frac{1}{2}x_1^2 - \frac{1}{2}x_2^2$; therefore,

$$x_1 \cdot x_2 = X_1 \cdot X_2 + \frac{1}{2}x_1^2 + \frac{1}{2}x_2^2. \tag{12}$$

Then, the angle θ between two vectors X_1 and X_2 in their conformal representation can be computed using (12) and (11):

$$\theta = \arccos\left(\frac{X_1 \cdot X_2 + \frac{1}{2}x_1^2 + \frac{1}{2}x_2^2}{|x_1||x_2|}\right). \tag{13}$$

On the other hand, to compute the perpendicular vector $x_3 \in \mathcal{E}^3$ to the vectors X_1 and X_2, we first compute the wedge product $A = X_1 \wedge X_2$, and then we take the coefficients of the components e_{12}, e_{23}, e_{31}, represented as $\langle A\rangle_{e_{12}}, \langle A\rangle_{e_{23}}, \langle A\rangle_{e_{31}}$, which define the plane **b** dual to the vector, to finally obtain the perpendicular vector multiplying by $I_E = e_1 \wedge e_2 \wedge e_3$:

$$x3 = I_E(\mathbf{b}) = I_E\big(\langle A\rangle_{e_{12}} \wedge e_{12} + \langle A\rangle_{e_{23}} \wedge e_{23} + \langle A\rangle_{e_{31}} \wedge e_{31}\big). \tag{14}$$

Note that the bivector $\mathbf{b} = \langle A\rangle_{e_{12}} \wedge e_{12} + \langle A\rangle_{e_{23}} \wedge e_{23} + \langle A\rangle_{e_{31}} \wedge e_{31}$ can be used to build a rotor as explained further. Reader is encouraged to see the CGA representation of other entities consulting [9]. All such entities and its transformations can be managed easily using the rigid body motion operators described further.

2.3 Rigid Body Motion

In GA there exist specific operators named *versors* to model rotations, translations, and dilations and are called rotors, translators, and dilators, respectively. In general, a versor G is a multivector which can be expressed as the geometric product of nonsingular vectors,

$$G = \pm a_1 a_2 \ldots a_k. \tag{15}$$

In CGA, such operators are defined by (16), (17), and (18), R being the rotor, T the translator, and D_λ the dilator:

$$R = e^{\frac{1}{2}\theta\mathbf{b}}, \tag{16}$$

$$T = e^{-\frac{te_\infty}{2}}, \tag{17}$$

$$D_\lambda = e^{\frac{-\log(\lambda)\wedge E}{2}}, \tag{18}$$

where **b** is the bivector dual to the rotation axis, θ is the rotation angle, $t \in \mathcal{E}^3$ is the translation vector, λ is the factor of dilation, and $E = e_\infty \wedge e_0$.

Such operators are applied to any entity of any dimension by multiplying the entity by the operator from the left and by the reverse of the operator (Sect. 2) from the right. Let X_i be any entity in CGA; then to rotate it, we do $X_1' = RX_1\tilde{R}$, while to translate it, we do $X_2' = TX_2\tilde{T}$, and to dilate it, we use $X_3' = D_\lambda X_3 \tilde{D}_\lambda$.

3 Determining the Shape of an Object

To determine the shape of an object, we can use a topographic mapping which uses selected points of interest along the contour of the object to fit a low-dimensional map to the high-dimensional manifold of such contour. This mapping is commonly achieved by using self-organized neural networks as Kohonen's Maps (SOM) or Neural Gas (NG) [5]; however, if we desire a better topology preservation, we should not specify the number of neurons of the network a priori (as specified for neurons in SOM or NG, together with its neighborhood relations) but allow the network to grow using an incremental training algorithm, as in the case of the Growing Neural Gas (GNG) [4]. In this work we follow the idea of growing neural networks and present an approach based on the GNG algorithm to determine the shape of objects by means of applying versors of the CGA, resulting in a model easy to handle in post processing stages; a scheme of our approach is shown in Fig. 1. The neural network has versors associated to its neurons, and its learning algorithm determines their parameters that best fit the input patterns, allowing us to get every point on the contour by interpolation of such versors.

Additionally, we modify the acquisition of input patterns by adding a preprocessing stage which determines the inputs to the net; this is done by computing the Generalized Gradient Vector Flow (GGVF) and analyzing the *streamlines* followed by particles (points) placed on the vertices of small squares defined by dividing the 2D/3D space in such squares/cubes. The streamline or the path followed by a particle that is placed on $\mathbf{x} = (x, y, z)$ coordinates will be denoted as $S(\mathbf{x})$. The

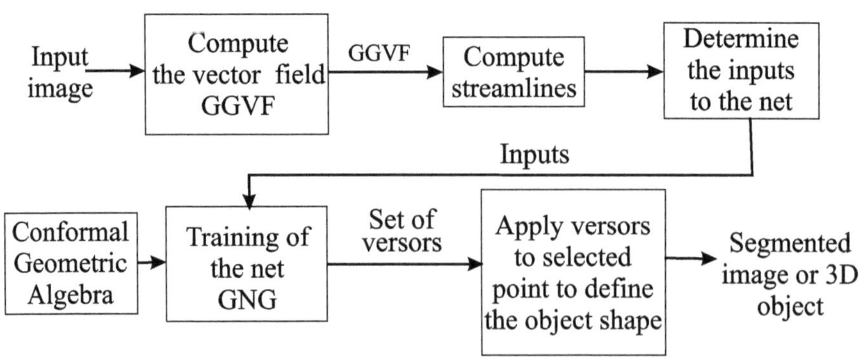

Fig. 1 A block diagram of our approach

information obtained with GGVF is also used in the learning stage as explained further.

3.1 Automatic Samples Selection Using GGVF

In order to select automatically the input patterns, we use the GGVF [10], which is a *dense vector field* derived from the volumetric data by minimizing a certain energy functional in a variational framework. The minimization is achieved by solving linear partial differential equations, which diffuses the gradient vectors computed from the volumetric data. To define the GGVF, the edge map is defined at first as

$$f(\mathbf{x}) : \Omega \to \mathcal{R}. \tag{19}$$

For the 2D image, it is defined as $f(x, y) = -|\nabla G(x, y) * I(x, y)|^2$, where $I(x, y)$ is the gray level of the image on pixel (x, y), $G(x, y)$ is a 2D Gaussian function (for robustness in presence of noise), and ∇ is the gradient operator. With this edge map, the GGVF is defined as to be the vector field $\mathbf{v}(x, y, z) = [u(x, y, z), v(x, y, z), w(x, y, z)]$ that minimizes the energy functional

$$\mathcal{E} = \int \int g(|\nabla f|)\nabla^2 \mathbf{v} - h(|\nabla f|)(\mathbf{v} - \nabla f), \tag{20}$$

where

$$g(|\nabla f|) = e^{-\frac{|\nabla f|}{\mu}} \quad \text{and} \quad h(|\nabla f|) = 1 - g(|\nabla f|), \tag{21}$$

and μ is a coefficient. An example of such dense vector field obtained in a 2D image is shown in Fig. 2(a), while an example of the vector field for a volumetric data is shown in Fig. 2(b). Observe the large range of capture of the forces in the image. Due to this large capture range, if we put particles (points) on any place over the image, they can be guided to the contour of the object. The automatic selection of input patterns is done by analyzing the *streamlines* of points on a 3D grid topology defined over the volumetric data. This means that the algorithm follows the streamlines of each point of the grid, which will guide the point to the more evident contour of the object; then the algorithm selects the point where the streamline finds a peak in the edge map and gets its conformal representation X as in (7) to make the input pattern set. Additionally to the X (conformal position of the point), the inputs have the vector $\mathbf{v}_\zeta = [u,\ v,\ w]$ which is the value of the GGVF in such pixel and will be used in the training stage as a parameter determining the amount of energy the input has to attract neurons. This information will be used in the training stage together with the position $\mathbf{x}$ for learning the topology of the data. Summarizing, the input set $\mathbf{I}$ will be

$$\mathbf{I} = \{\zeta_k = X_{\zeta_k}, \mathbf{v}_{\zeta_\mathbf{k}} | \mathbf{x}_\zeta \in S(\mathbf{x}') \text{ and } f(\mathbf{x}_\zeta) = 1\}, \tag{22}$$

where X_ζ is the conformal representation of $\mathbf{x}_\zeta$; $\mathbf{x}_\zeta \in S(\mathbf{x}')$ means that $\mathbf{x}_\zeta$ is on the path followed by a particle placed in $\mathbf{x}'$, and $f(\mathbf{x}_\zeta)$ is the value of the edge map

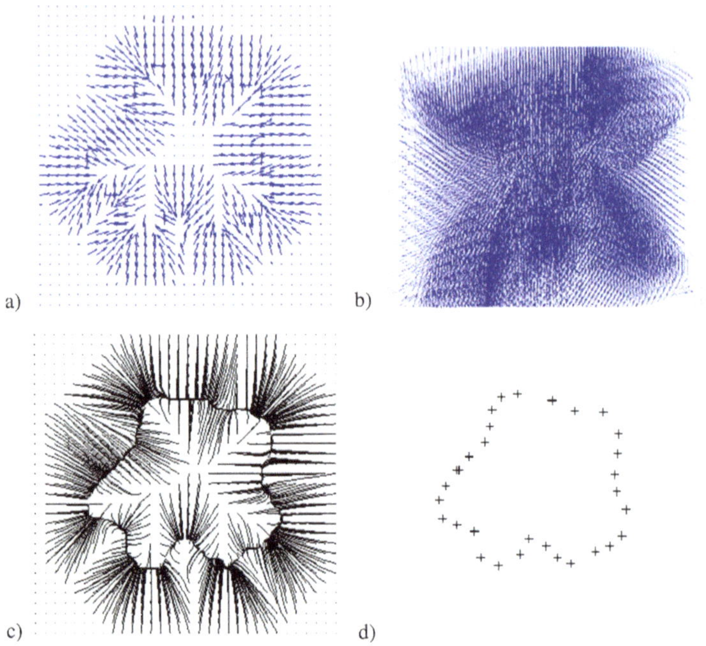

Fig. 2 Example of the dense vector field called GGVF (not the whole vector field is shown, but only representative samples of a grid). (**a**) Samples of the vector field for a 2D image; (**b**) Samples of the vector field for volumetric data; (**c**) Example of streamlines for particles arranged in a 32×32 grid according to the vector field shown in (**a**); (**d**) Points selected as input patterns according to (22)

in position $\mathbf{x}_\zeta$ (assuming that it is binarized). As some streamlines can carry to the same point or very close points, we can add constraints to avoid very close samples; one very simple restriction is that the candidate to be included in the input set must be at least at a fixed distance d_{thresh} of any other input.

Figure 2(c) shows the streamlines according to the vector field shown in Fig. 2(a) and the input patterns selected as described before are shown in Fig. 2.

3.2 Learning the Shape Using Versors

It is important to note that although we will explain the algorithm using points, the versors can be applied to any entity in GA that we had selected to model the object. The network starts with a minimum number of versors (neural units), and new units are inserted successively. The network is specified by

- A set of units (neurons) named N, where each $\mathbf{n}_l \in N$ has its associated versor $M_{\mathbf{n}_l}$; each versor is the transformation that must be applied to a point to place it in the contour of the object. The set of transformations will ultimately describe the shape of the object.

- A set of connections between neurons defining the topological structure.

Also, take into account that

- There are two learning parameters, e_w and e_n, for the winner neuron and for the direct neighbors of it; such parameters remain constant during all the process.
- Each neuron $\mathbf{n}_l$ will be composed by its versor $M_{\mathbf{n}_l}$, the *signal counter* sc_l, and the *relative signal frequency* rsf_l. The signal counter sc_l is incremented for the neuron $\mathbf{n}_l$ every time it is the winner neuron. The relative signal frequency rsf_l is defined as

$$rsf_l = \frac{sc_l}{\sum_{\forall \mathbf{n}_j} sc_j}. \tag{23}$$

  This parameter will act as an indicator to insert new neural units.

With these elements, we define the learning algorithm of the GNG to find the versors that will define the contour as follows:

1. Let P_0 be a fixed initial point over which the transformations will be applied. This point corresponds to the conformal representation of p_0, which can be a random point or the centroid defined by the inputs. The initial transformations will be expressed as $M = e^{-\frac{t}{2}e_\infty}$ in the conformal geometric algebra. The vector t initially is a random displacement.
2. Start with the minimal number of neurons, which have associated random motors M and a vector $\mathbf{v}_l = [u_l,\ v_l,\ w_l]$, whose magnitude is interpreted as the capacity of learning for such neuron.
3. Select one input ζ from the inputs set $\mathbf{I}$ and find the winner neuron; this means finding the neuron n_l having the versor M_l which moves the point P_0 closer to such input:

$$M_{\text{win}} = \min_{\forall M} \sqrt{\left(X_\zeta - M P_0 \tilde{M}\right)^2}. \tag{24}$$

4. Modify M_{win} and all others versors of neighboring neurons M_l in such a way that the modified M will represent a transformation moving the point P_0 nearer the input. Note that each motor is composed by a rotation ΔR and a translation ΔT. The rotation is computed as in (16), θ being the angle between the actual position a and $a' = a + \mathbf{v}$ ($\mathbf{v}$ is the GGVF vector value in such position); the bivector dual to the rotation axis is computed as in (14); the rotors and translators are defined as

$$\Delta R_{\text{win}} = e^{\frac{e_w \eta(\mathbf{v}_\zeta, \mathbf{v}_{\text{win}})}{2}\theta \mathbf{b}}, \tag{25}$$

$$\Delta T = e^{-\frac{\Delta t_{\text{win}}}{2} e_\infty}, \tag{26}$$

$$\Delta t_{\text{win}} = e_w \eta(\mathbf{v}_\zeta, \mathbf{v}_{\text{win}})(\mathbf{x}_\zeta - p_0) \tag{27}$$

   for the winner neuron, and

$$\Delta R_n = e^{\frac{e_n \eta(\mathbf{v}_\zeta, \mathbf{v}_n)}{2}\theta \mathrm{b}}, \tag{28}$$

$$\Delta T = e^{-\frac{\Delta t_n}{2} e_\infty}, \tag{29}$$

$$\Delta t_n = e_n \eta(\mathbf{v}_\zeta, \mathbf{v}_n)(\mathbf{x}_\zeta - p_0) \tag{30}$$

for its direct neighbors, to obtain $\Delta M = \Delta T \Delta R$. Finally, the new motor is

$$M_{l_{\text{new}}} = \Delta M M_{l_{\text{old}}}, \tag{31}$$

ϕ is a function defining the amount a neuron can learn according to its distance to the winner one (defined as in (32)), and $\eta(\mathbf{v}_\zeta, \mathbf{v}_l)$ is defined as in (33).

$$\phi = e^{-\frac{(M_{\text{win}} P_0 \tilde{M}_{\text{win}} - M_l P_0 \tilde{M}_l)^2}{2\sigma}}, \tag{32}$$

$$\eta(\mathbf{v}_\zeta, \mathbf{v}_l) = \|\mathbf{v}_\zeta - \mathbf{v}_l\|^2, \tag{33}$$

which is a function defining a quantity of learning depending on the strength to teach of the input ζ and the capacity to learn of the neuron, given in $\mathbf{v}_\zeta$ and $\mathbf{v}_l$, respectively. Also update

$$\mathbf{v}_{\text{win}}^{\text{new}} = \mathbf{v}_{\text{win}} + e_w \mathbf{v}_{\text{win}}, \tag{34}$$

$$\mathbf{v}_n^{\text{new}} = \mathbf{v}_n + e_n \mathbf{v}_n. \tag{35}$$

5. Every certain number λ of iterations, determine the neuron with the rsf_l with highest value. Then, if any of the direct neighbors of that neuron is at a distance larger than $c_{\max}$, do

   - Determine neighboring neurons $\mathbf{n}_i$ and $\mathbf{n}_j$.
   - Create a new neuron n_{new} between $\mathbf{n}_i$ and $\mathbf{n}_j$ whose associated M and $\mathbf{v}_l$ will be

   $$M_{n_{\text{new}}} = \frac{M_i + M_j}{2}, \qquad \mathbf{v}_{l\_\text{new}} = \frac{\mathbf{v}_i + \mathbf{v}_j}{2}; \tag{36}$$

   the new units will have the values $sc_{\text{new}} = 0$ and $rsf_{\text{new}} = 0$.
   - Delete the old edge connecting $\mathbf{n}_i$ and $\mathbf{n}_j$ and create two new edges connecting n_{new} with $\mathbf{n}_i$ and $\mathbf{n}_j$.

6. Repeat Steps 3 to 5 if the stopping criterion is not achieved. The stop criterion is when a maximum number of neurons is reached or when the learning capacity of neurons approaches to zero (is less than a threshold $c_{\min}$), and the first that happens stops the learning process.

Training the network, we find the set of M defining positions on a trajectory; such positions minimize the error measured as the average distance between X_ζ and the result of $M_\zeta P_0 \tilde{M}_\zeta$:

$$\chi = \frac{\sum_{\forall \zeta} \sqrt{(M_\zeta P_0 \tilde{M}_\zeta - X_\zeta)^2}}{N}, \tag{37}$$

where M_ζ moves P_0 closer to input X_ζ, and N is the number of inputs.

4 Experiments

Figure 3 shows the result when the algorithm is applied to a magnetic resonance image (MRI); the goal is to obtain the shape of the ventricle. Figure 3(a) shows the original brain image and the region of interest (ROI); Fig. 3(b) shows the computed vector field for the ROI; Fig. 3(c) shows the streamlines in the ROI defined for particles placed on the vertices of a 32×32 grid; Fig. 3(d) shows the initial shape as defined for the two initial random motors M_a, M_b; Fig. 3(e) shows the final shape obtained; and finally Fig. 3(f) shows the original image with the segmented object.

Figure 4 presents an image showing that our approach can also be used for automated visual inspection tasks; reader can observe that such image contains a very blurred object. That image is for the inspection of hard disk head sliders. Figure 4(a) shows the original image and the region of interest (ROI); Fig. 4(b) shows the computed vector field of the ROI; Fig. 4(c) shows the streamlines defined for particles placed on the vertices of a 32×32 grid; Fig. 4(d) shows the inputs selected according the streamlines and the initial shape as defined for the two initial random motors M_a and M_b; and Fig. 4(e) shows the final shape obtained overlapped with the original image, showing that the algorithm gives good results if it is used for segmentation.

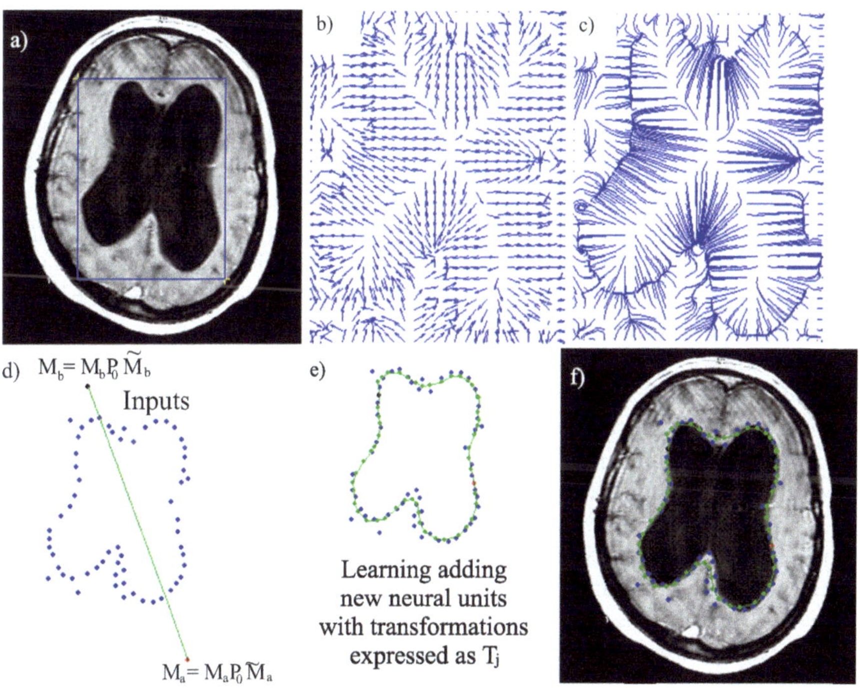

Fig. 3 (**a**) Original image and region of interest (ROI); (**b**) Zoom of the dense vector field of the ROI; (**c**) Zoom of the streamlines in ROI; (**d**) Inputs and initial shape; (**e**) Final shape defined according the 54 estimated motors; (**f**) Image segmented according the results

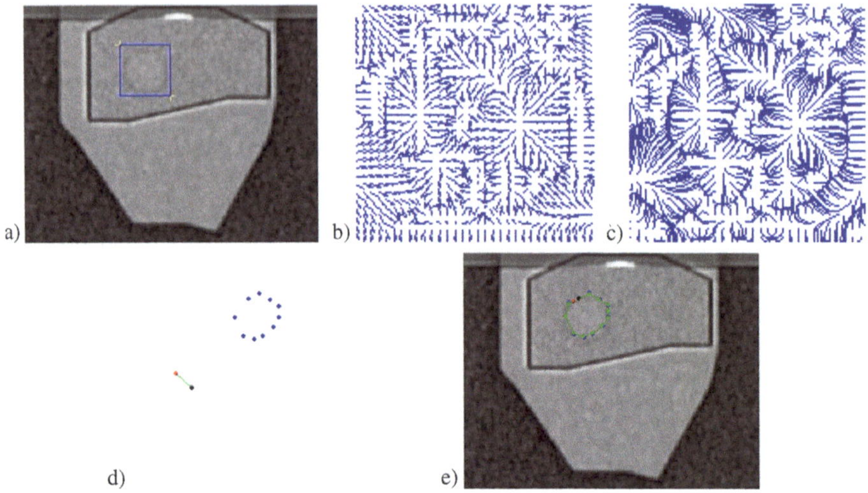

Fig. 4 Application in visual inspection tasks: (**a**) Original image and the region of interest (ROI); (**b**) Zoom of the dense vector field of the ROI; (**c**) Zoom of the streamlines in ROI; (**d**) Inputs and initial shape according the two initial random transformations M_a and M_b; (**e**) Final shape defined according the 15 estimated motors (original image with the segmented object)

Figure 5 shows the application of the ggvf-snakes algorithm in the same problem. It is important to note that although the approaches of Figs. 4 and 5 use GGVF information to find the shape of an object, the estimated final shape using the neural approach is better than the one using active contours; the second approach (see Fig. 5) fails to segment the object no matter if the initialization of the snake is given inside, outside, or over the contour we are interested in. Additionally, the fact of expressing such shape as a set of motors allows us to have a model best suited to be used in further applications which can require the deformation of the model, especially if such model is not based on points but on the other GA entities, because we do not need to change the motors (recall that they are applied in the same way to any other entity).

The proposed algorithm was applied to different sets of medical images. Figure 6 shows some images of such sets. The first row of each figure shows the original image and the region of interest, while the second row shows the result of the proposed approach. Table 1 shows the errors obtained with our approach using and not using the GGVF information. We can observe that the inclusion of GGVF information improves the approximation of the surface.

To compare our algorithm, we use the GNG with and without GGVF information, as well as a growing version of SOM, also using and not using the GGVF information. These algorithms were applied to a set of 2D medical images (some obtained with computer tomography (CT) and some with magnetic resonance (MR)). Figure 7(a) shows the average errors when GSOM stops for different examples: segmenting a ventricle, a blurred object, a free form curve, and a column disk. Note that using the GGVF information the error is reduced. This means that using the GGVF

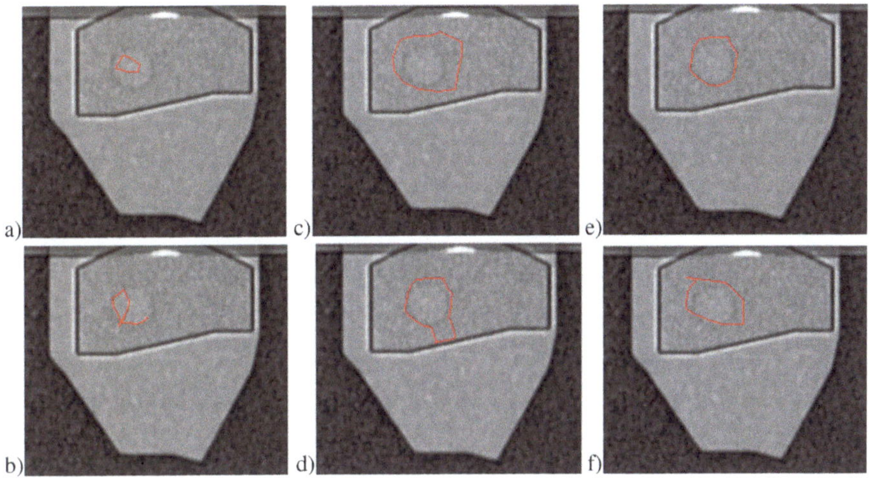

Fig. 5 Result obtained when using the active contour approach to segment the object in the same image as in Fig. 4. (**a**) Initialization of snake inside the object; (**b**) Final result obtained with initialization showed in (**a**); (**c**) Initialization of snake outside the object; (**d**) Final result obtained with initialization showed in (**c**); (**e**) Initialization of snake over the contour; (**f**) Final result obtained with initialization showed in (**e**)

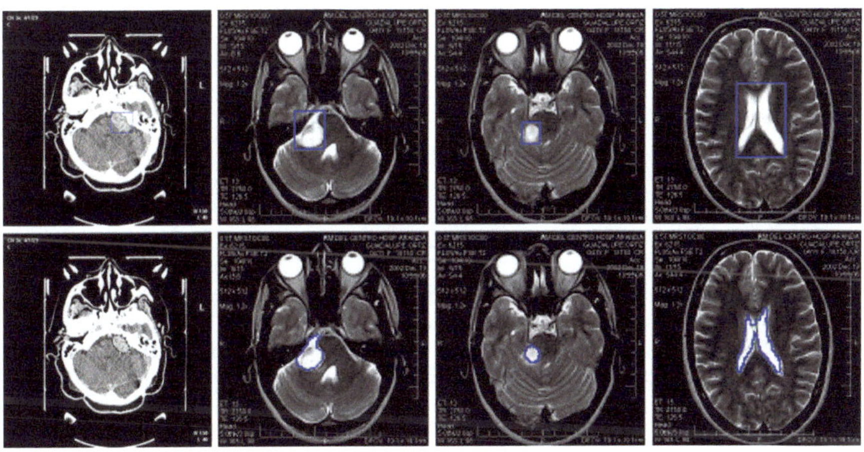

Fig. 6 *First row* (*upper row*): original image and the region of interest. *Second row*: Result of segmentation

information, as we propose, a better approximation of the object shape can be obtained. Figure 7(b) shows the average errors obtained for several examples but using the GNG with and without the GGVF information. Note that again, the GGVF contributes to obtain a better approximation to the object surface. Also note that the average errors obtained with the GNG algorithm are smaller than the errors ob-

Table 1 Errors obtained by the algorithm with and without the GGVF information. ε_1: error without GGVF; ε_2: error with GGVF

Example	ε_1	ε_2	Example	ε_1	ε_2
Ventricle 1	3.29	2.51	Eye 1	7.63	6.8
Eye 2	3.43	2.98	Column disk 1	4.65	4.1
Tumor 1	3.41	2.85	Tumor 2	2.95	2.41
Free form curve	2.84	1.97	Column disk 2	2.9	2.5

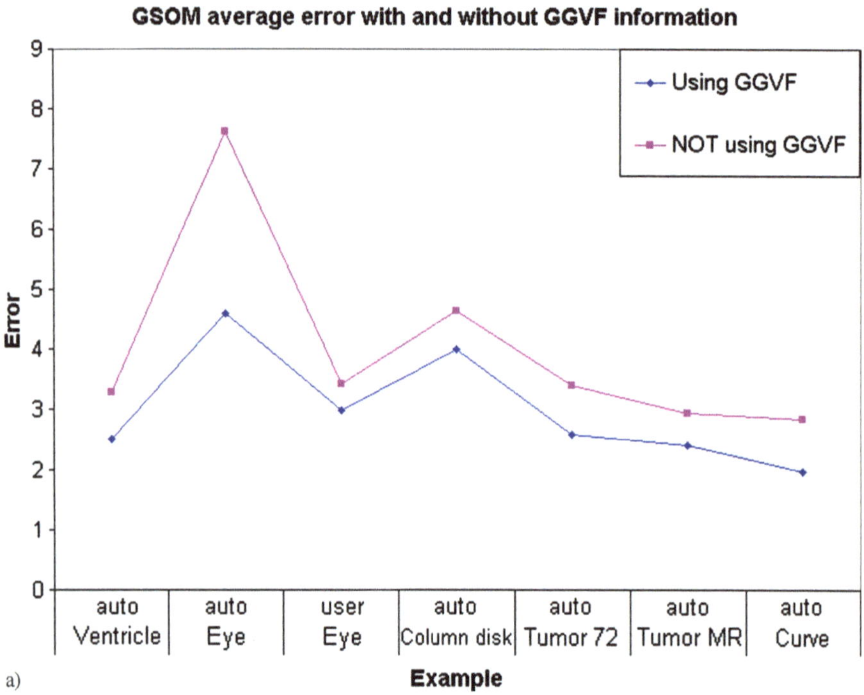

a)

Fig. 7 (**a**) Average errors for different examples using the GSOM algorithm with and without GGVF information; (**b**) Average errors for different examples using the GNG algorithm with and without GGVF information; (**c**) Comparison between the errors obtained with GSOM and GNG using and without using GGVF information. Note that both are improved with GGVF information, although GNG gives better results

tained with the GSOM, as can be seen in Fig. 7(c) and that both are improved with GGVF information, although GNG gives better results.

It is necessary to mention that the whole process is quick enough; in fact, the computational time required for all the images showed in this work took only few seconds. The computation of the GGVF is the most time-consuming task in the algorithm, but it only takes about 3 seconds for 64×64 images, 20 seconds for 256×256 images, and 110 seconds for 512×512 images. This is the reason why we decide not to compute it for the whole image but only for selected region of interest. The same criterion was applied to 3D examples.

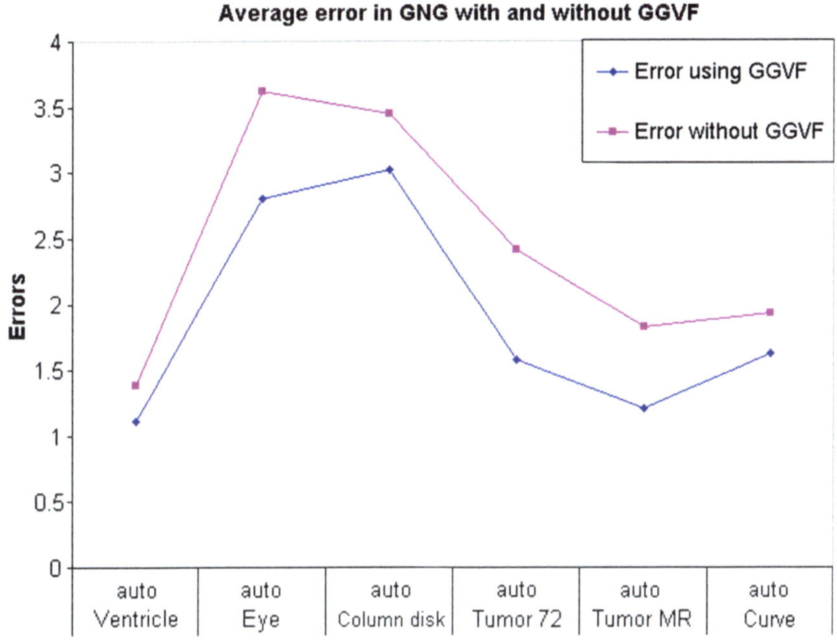

b)

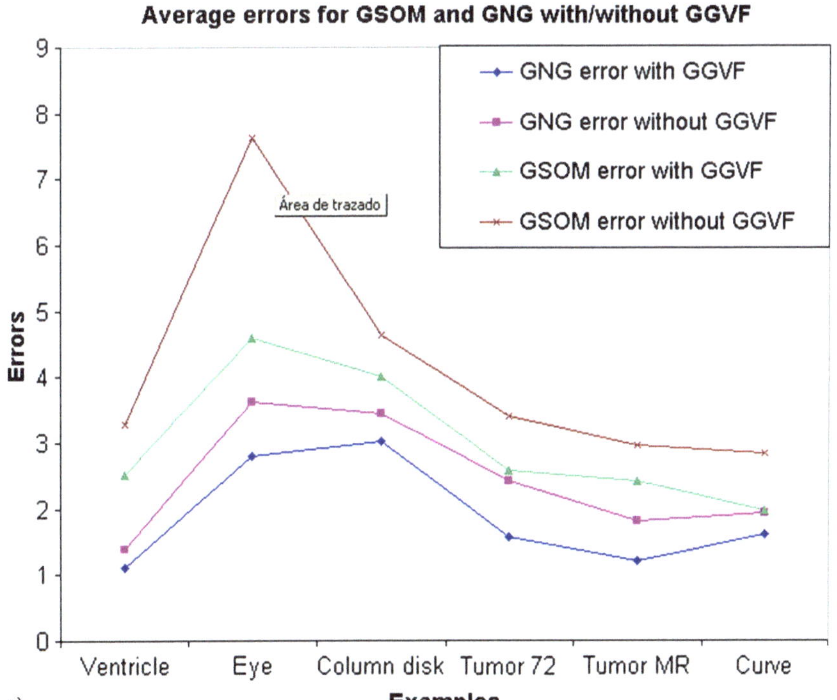

c)

Fig. 7 (Continued)

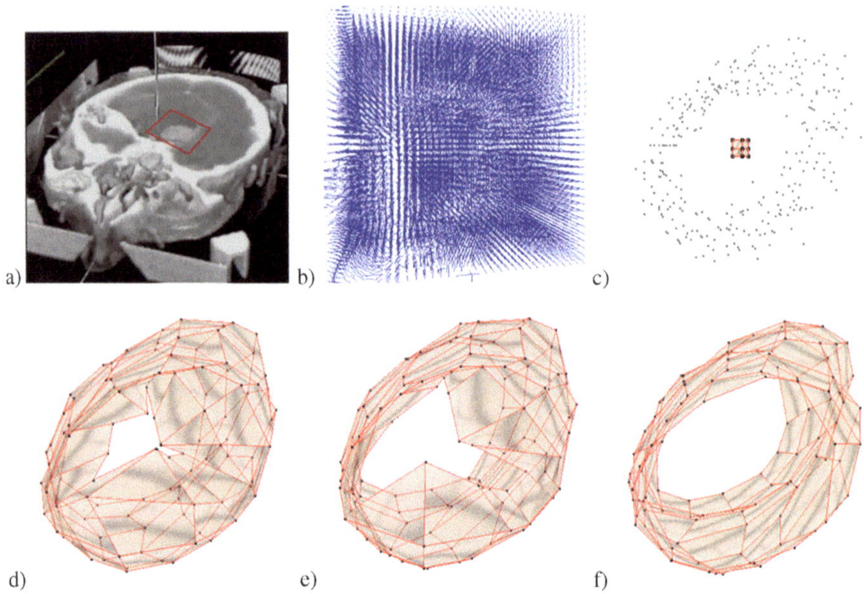

Fig. 8 The algorithm for 3D object's shape determination. (**a**) 3D model of the patient's head containing a tumor in the marked region; (**b**) Vectors of the dense GGVF on a 3D grid arrangement of $32 \times 32 \times 16$; (**c**) Inputs determined by GGVF and edge map and the initialization of the net GNG; (**d**)–(**e**) Two stages during the learning; (**f**) Final shape after training has finished with a total of 170 versors M (associated with 170 neural units)

Figure 8(a) shows the patient head with the tumor which surface we need to approximate; Fig. 8(b) shows the vectors of the dense GGVF on a 3D grid arrangement of size $32 \times 32 \times 16$; Fig. 8(c) shows the inputs determined by GGVF and edge map and also shows the initialization of the net GNG; and Figs. 8(d)–(f) show some stages of the adaptation process, while the net is determining the set of transformations M (Fig. 8(f) is the final shape after training has finished with a total of 170 versors M (associated with 170 neural units)).

Figure 9 shows another 3D example, corresponding to a pear, and the surface is well approximated. Figure 9(b) shows the inputs and the initialization of the net with nine neural units (the topology of the net is defined as a sort of pyramid around the centroid of input points); while Fig. 9(c) shows the result after the net has been reached the maximum number of neurons, which was fixed to 300; finally, Fig. 9(d) shows the minimization of the error according to (37).

Another useful application of the algorithm using the gradient information of the GGVF during the training of the GNG neural net in the geometric algebra framework is the transformation of one model obtained at time t_1 into another obtained at time t_2 (a kind of morphing of 3D surfaces).

Figure 10(a) shows the initial shape which will be transformed into the one showed in Fig. 10(b); Figs. 10(c)–(f) show some stages during the process. Note that Fig. 10(f) looks like Fig. 10(b), as expected.

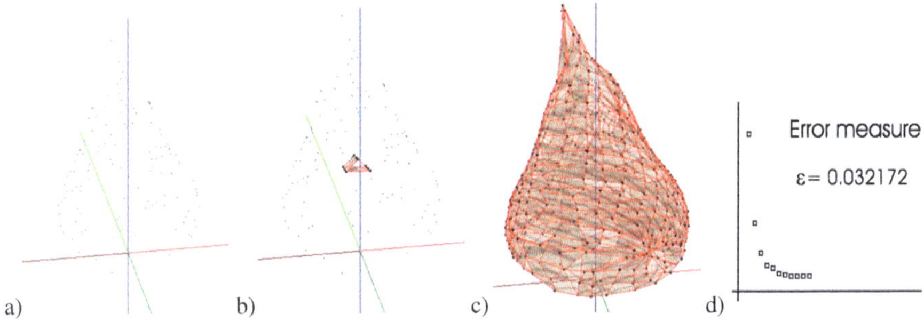

Fig. 9 3D object shape definition for the case of a pear. (**a**) Inputs to the net selected using GGVF and streamlines; (**b**) Inputs and the initialization of the net with nine neural units; (**c**) Result after the net has been reached the maximum number of neurons (300 neurons); (**d**) Error measurement using (37)

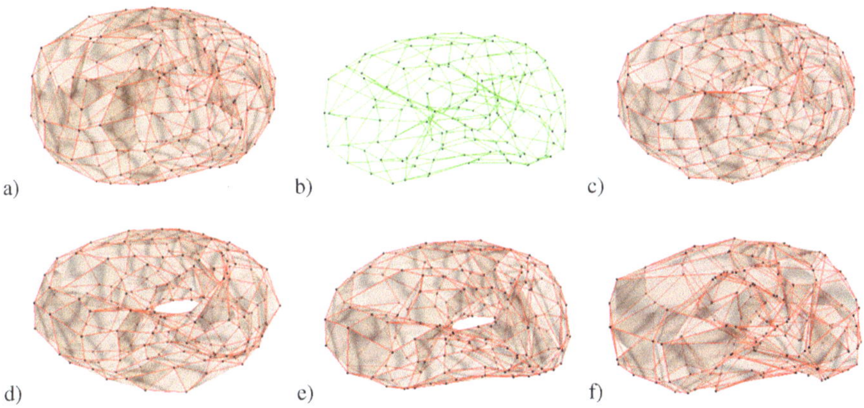

Fig. 10 The 3D surface shown in (**a**) is transformed into the one shown in (**b**). Different stages during the evolution are shown in (**c**) to (**f**), where (**f**) is the final shape (that is, the final shape of (**a**)) after finishing the evolution of the net, which should look like (**b**)

In the case shown in Fig. 11, we have onc 3D model with an irregular shape which will be transformed to take a shape similar to a pear; Fig. 11(a) shows the initial shape which will be transformed into the one showed in Fig. 11(b); Figs. 11(c)–(f) show some stages during the process. Again, the resulting volume looks like the one expected (Fig. 11(b)).

To illustrate the application of the presented algorithm in cases having models based on entities different to the points, we show in Fig. 12 models based on spheres [8]. The goal is the same: morphing the model shown in Figs. 12(a) and 12(d) to the one showed in Figs. 12(b) and 12(e), respectively. The results are shown in Figs. 12(c) and 12(f).

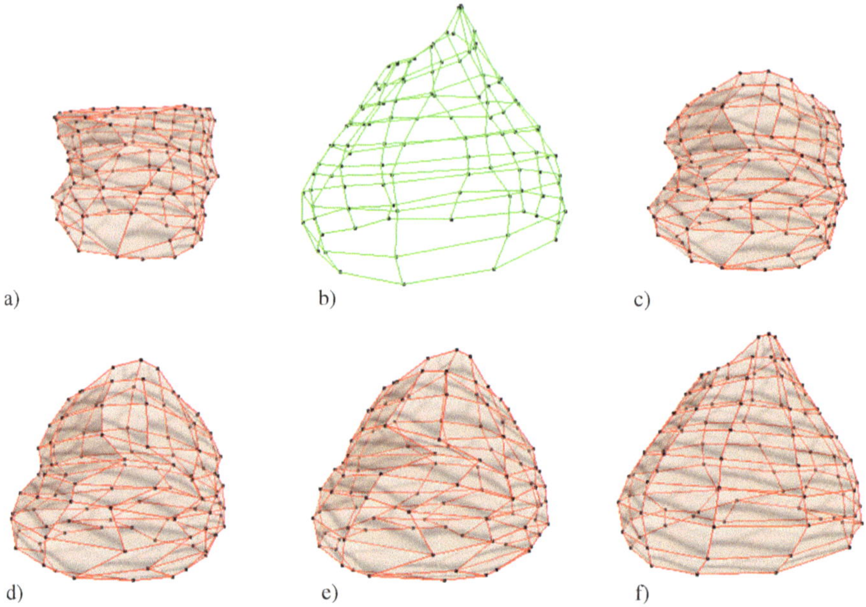

Fig. 11 The 3D surface shown in (**a**) is transformed into the one shown in (**b**). Different stages during the evolution are shown in (**c**) to (**f**), where (**f**) is the final shape (that is, the final shape of (**a**)) after finishing the evolution of the net, which should look like (**b**)

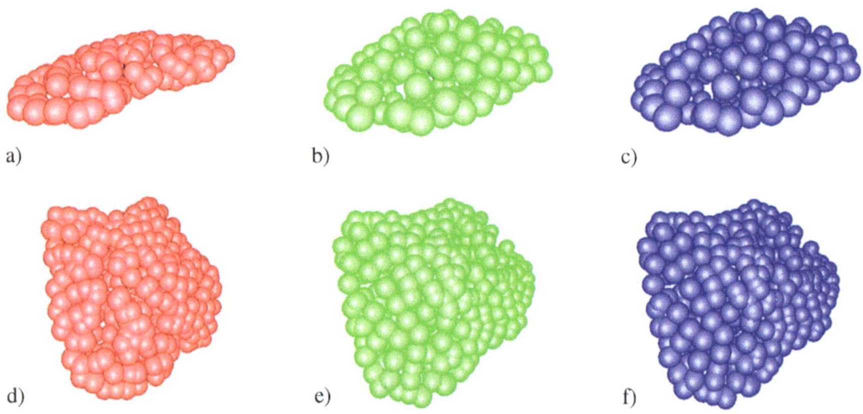

Fig. 12 The 3D models based on spheres shown in (**a**) and (**d**) are transformed into the one shown in (**b**) and (**e**), respectively, resulting in the models showed in (**c**) and (**f**), respectively

5 Conclusion

In this work the authors show how to incorporate geometric algebra techniques in an artificial neural network approach to approximate 2D contours or 3D surfaces. In addition, they show the use of the dense vector field named Generalized Gradient

Vector Flow (GGVF) not only to select the inputs to the neural network GNG but also as a parameter guiding its learning process. This network was used to find a set of transformations expressed in the conformal geometric algebra framework, which move a point by means of a versor along the contour of an object, defining by this way the shape of the object. This has the advantage that versors of the conformal geometric algebra can be used to transform any entity exactly in the same way: multiplying the entity from the left by M and from the right by $\tilde{M}$.

There were presented some experiments showing the application of the proposed method in medical image processing and for visual inspection tasks. The results obtained show that by incorporating the GGVF information we can get automatically the set of inputs to the net, and also we improve its performance. Some comparisons between the results obtained with this algorithm, against the results obtained by a modified version of the GSOM net and also against the ggvf-snakes, showed that our proposal is better. When dealing with the 3D case, we presented two different applications: surface approximation and the transformation of a model at time t_1 into another at time t_2, obtaining good results even using models based on spheres.

References

1. Angelopoulou, A., Psarrou, A., García Rodríguez, J., Revett, K.: Automatic landmarking of 2D medical shapes using the growing neural gas network. In: Proceedings of the International Conference on Computer Vision, ICCV 2005, October 13–21, Beijing, China, pp. 210–219
2. Bayro-Corrochano, E.: Robot perception and action using conformal geometry. In: Handbook of Geometric Computing. Applications in Pattern Recognition, Computer Vision, Neurocomputing and Robotics, pp. 405–458. Springer, Heidelberg (2005). Chap. 13. Bayro-Corrochano, E. (ed.)
3. Buchholz, S., Hitzer, E., Tachibana, K.: Coordinate independent update formulas for versor Clifford neurons. In: Joint 4th International Conference on Soft Computing and Intelligent Systems and 9th International Symposium on Advanced Intelligent Systems (SCIS&ISIS2008), Nagoya, Japan, 17–21 Sep. 2008, pp. 814–819
4. Fritzke, B.: A Growing Neural Gas Network Learns Topologies. Advances in Neural Information Processing Systems, vol. 7. MIT Press, Cambridge (1995)
5. Mehrotra, K., Mohan, C., Ranka, S.: Unsupervised learning. In: Elements of Artificial Neural Networks, Chap. 5, pp. 157–213
6. Perwass, C., Hildenbrand, D.: Aspects of geometric algebra in Euclidean, projective and conformal space. Christian-Albrechts-University of Kiel, Technical Report No. 0310 (2003)
7. Pham, M.T., Tachibana, K., Hitzer, E.M.S., Buchholz, S., Yoshikawa, T., Furuhashi, T.: Feature extractions with geometric algebra for classification of objects. In: Proceedings of the International Joint Conference on Neural Networks, IJCNN 2008, Part of the IEEE World Congress on Computational Intelligence, WCCI 2008, Hong Kong, China, 1–6 June 2008
8. Ranjan, V., Fournier, A.: Union of spheres (UoS) model for volumetric data. In: Proceedings of the Eleventh Annual Symposium on Computational Geometry, Vancouver, Canada, C2–C3, pp. 402–403 (1995)
9. Rosenhahn, B., Sommer, G.: Pose estimation in conformal geometric algebra. Christian-Albrechts-University of Kiel, Technical Report No. 0206, pp. 13–36 (2002)
10. Xu, Ch.: Deformable models with applications to human cerebral cortex reconstruction from magnetic resonance images. Ph.D. thesis, Johns Hopkins University, pp. 14–63 (1999)

Geometric Associative Memories and Their Applications to Pattern Classification

Benjamin Cruz, Ricardo Barron, and Humberto Sossa

Abstract Associative memories (AMs) were proposed as tools usually used in the restoration and classification of distorted patterns. Many interesting models have emerged in the last years with this aim. In this chapter a novel associative memory model (Geometric Associative Memory, GAM) based on Conformal Geometric Algebra (CGA) principles is described. At a low level, CGA provides a new coordinate-free framework for numeric processing in problem solving. The proposed model makes use of CGA and quadratic programming to store associations among patterns and their respective class. To classify an unknown pattern, an inner product is applied between it and the obtained GAM. Numerical and real examples to test the proposal are given. Formal conditions are also provided that assure the correct functioning of the proposal.

1 Introduction

Two main problems in pattern recognition are pattern classification and pattern restoration. One approach usually used to restore or classify desired patterns is by means of an associative memory. Lots of models of associative memories have emerged in the last 40 years, starting with the Lernmatrix of Steinbouch [25], then the Linear Associator of Anderson [1] and Kohonen [19], and the well-known model proposed by Hopfield, the Hopfield Memory [18]. For their operation, all of these models use the same kind of algebraic operations. Later there appeared the so-called *Morphological Associative Memories* (MAMs) [23] that are based on the mathematical morphology paradigm.

An associative memory **M** is a device whose main function is associating input patterns with output patterns. The notation for a pattern association between two

R. Barron (✉)
Center of Computing Research, Juan de Dios Batiz s/n Col. Nueva Industrial Vallejo Gustavo A. Madero, C.P. 07738, Mexico DF, Mexico
e-mail: rbarron@cic.ipn.mx

E. Bayro-Corrochano, G. Scheuermann (eds.), *Geometric Algebra Computing*,
DOI 10.1007/978-1-84996-108-0_11,

vectors x and y can be seen as an ordered pair (x, y). The whole set of all associations that form the associative memory is called *fundamental patterns set* or simply *fundamental set* (FS). Patterns belonging to the FS are called *fundamental patterns*.

Associations are completely stored in a weighted matrix. This matrix can be used to generate output patterns using the associated input patterns. This weighted matrix is the associative memory **M**. The process by which **M** is built is called *learning* or *training phase,* and the process through which an output pattern is generated using an input pattern is called *restoration* or *classification phase*.

When by means of an associative memory **M** a specific fundamental pattern is correctly classified, then **M** presents a *perfect recall* for that pattern [5]. An associative memory that presents perfect recall for all patterns of the FS is called a *perfect recalling memory*. When an associative memory **M** recovers or classifies patterns affected with noise correctly, it is said that **M** presents a *robust recall* or *robust classification*.

1.1 Classic Associative Memory Models

In 1961, a first work of associative memories was developed by Karl Steinbouch, the so-called *Die Lernmatrix*. This memory was proposed in 1961 and is capable of both pattern classification and pattern association [25]. In 1972, two papers by James A. Anderson [1] and Teuvo Kohonen [19] proposed the same model of associative memory, the so-called *Linear Associator* model for associative memories.

In the same year, a new associative memory device was presented by Kaouru Nakano, *the Associatron* [22]. This device was able to store entities represented by bit-patterns in a distributed form. It was able to restore complete patterns using a portion of them. Ten years later, John J. Hopfield introduced the so-called *Hopfield Memory* [18]. Hopfield considers this model as a physical system described by x that has locally stable points.

Almost 20 years after the Hopfield Memory, a new set of lattice algebra-based associative memories appeared, the *Morphological Associative Memories* (MAMs) [23]. Minima or maxima of sums are used for their operation, in contrast to the sums of products used in previous models. A variant of these MAMs are the *Alpha–Beta* Associative Memories ($\alpha\beta$) [27]. For these memories, two new operators are defined: α (alpha) and β (beta). These are detailed and discussed in [5].

There are two types of MAMs and $\alpha\beta$, the *min memories* that can cope with patterns altered with subtractive noise and the *max memories* that can cope with patterns altered with additive noise. However, contrary to what one might think, the performance of these models in the presence of patterns altered with mixed noise (most common in real situations) is too deficient [26].

In the literature, three ways to face the problem of mixed noise by means of an associative memory can be highlighted. In [26], a way to solve this problem by means of the so-called *kernels* is given. A kernel is a reduced version of an original pattern; the basic idea is to use two associative memories, one for recalling the kernel using the distorted pattern and the other one for recalling the original pattern using the obtained kernel. Kernels for MAMs are however difficult to find, and if new patterns have to be added to the fundamental set, the kernels need to be computed again [11].

Another approximation is by means of the so-called *median memories* [24]. These memories use the well-known *median* operator widely used in signal processing instead of the maximum or minimum operators. Due the characteristics of the median operator, these memories can cope with mixed noise directly. However, the conditions for obtaining a perfect recall are difficult to achieve.

Finally in [11], it is shown how to solve the problem of the mixed noise by decomposing a pattern into parts (*sub-patterns*). This method is feasible due to the locality of the noise. Some parts of the pattern are less affected by noise than the other ones. However this method can consume a lot of computing time.

2 Basics of Conformal Geometric Algebra

In the XIX century many scientists worked on the development of algebraic systems. Among these, William K. Clifford (1845–1879) introduced *Geometric Algebras* (GA) called *Clifford Algebras* by mathematicians. They were completely described in his paper *Applications of Grassmann's Extensive Algebra* [10].

A geometric algebra is a priori coordinate-free [14]. In GA, geometric objects and operators over these objects are treated in a single algebra [13]. A special characteristic of GA is its geometric intuition. Another important feature is that the expressions in GA usually have low symbolic complexity [17].

The Conformal Geometric Algebra (CGA) is a (3,2)-dimensional coordinate-free theory and provides a conformal representation for 3D objects. Spheres and circles are both algebraic objects with a geometric meaning. In CGA, points, spheres, and planes are easily represented as *multivectors* [15]. A multivector is the outer product of various vectors [20].

CGA provides a great variety of basic geometric entities to compute with [17]. Intersections between lines, circles, planes, and spheres are directly generated. The creation of such elementary geometric objects simply occurs by algebraically joining a minimal number of points in the object subspace. The resulting multivector expressions completely encode in their components positions, orientations, and radii [13].

The main products of Geometric Algebra are the *geometric* or *Clifford product*, the *outer product* and the *inner product*. The inner product is used for the computation of angles and distances.

For notation purposes, Euclidean vectors will be noted by lowercase italic letters $(p, q, s, \ldots)$, with the exception of the letters i, j, k, l, m, n that will be used to refer to indices. The corresponding conformal points will be noted by italic capital letters $(P, Q, S, \ldots)$. A Euclidean matrix will be noted by bold capital letters (**M**). To denote that an element belongs to an object (vector), a subscript will be used. To refer that an object belongs to a set of objects of the same type, a superscript will be used.

For example, if S is a sphere, then S_k is the kth component of it, and S^k is the kth sphere of a set of spheres. To denote scalars Greek letters will be used.

In particular, an original Euclidean point $p \in \mathbb{R}^n$ is extended to an $(n+2)$-dimensional conformal space [16] as

$$P = p + \frac{1}{2}(p)^2 e_\infty + e_0, \tag{1}$$

where p is a linear combination of the Euclidean base vectors. e_0 and e_∞ represent the Euclidean origin and the point at infinity, respectively, such that $e_0^2 = e_\infty = 0$ and $e_0 \cdot e_\infty = -1$ [13].

Equation (1) expresses a homogeneous relationship between both Euclidean and conformal domains since, given a scalar α and a conformal point P, αP and P both represent the same Euclidean point p. When the coefficient of e_0 is equal to 1, then P has a canonic representation. In this section, the algebra works in the conformal domain, while the geometric semantics lies in the Euclidean domain. In the same way, the sphere has the canonical form

$$S = C - \frac{1}{2}(\gamma)^2 e_\infty = c + \frac{1}{2}\big((c)^2 - (\gamma)^2\big)e_\infty + e_0, \tag{2}$$

where C is the central point in conformal form as defined in (2), where γ is the radius of the sphere. Also, a sphere can be easily obtained by four points that lie on it [13], as follows:

$$S = P^1 \wedge P^2 \wedge P^3 \wedge P^4. \tag{3}$$

In this case, it is said that (3) is a dual representation of (2). In the same way, a plane can be defined by three points that lie on it and the point at infinity [13] as follows:

$$T = P^1 \wedge P^2 \wedge P^3 \wedge e_\infty. \tag{4}$$

From (3) and (4) we can see that a plane is a sphere that passes through the point at infinity [15].

A distance measure between two conformal points P and Q can be defined with the help of the inner product [16] as follows:

$$P \cdot Q = p \cdot q - \frac{1}{2}(p)^2 - \frac{1}{2}(q)^2 = -\frac{1}{2}(p-q)^2 \quad \Longleftrightarrow \quad (p-q)^2 = -2(P \cdot Q), \tag{5}$$

resulting in the square of the Euclidean distance. In the same way, a distance measure between one conformal point P and a sphere S can be defined with the help of the inner product [16] as

$$P \cdot S = p \cdot s - \frac{1}{2}\left((s)^2 - \frac{1}{2}(\gamma)^2\right) - \frac{1}{2}(p)^2 = \frac{1}{2}\big((\gamma)^2 - (s-p)^2\big) \tag{6}$$

or in a simplified form as

$$2(P \cdot S) = (\gamma)^2 - (s-p)^2. \tag{7}$$

Based on (7), if $P \cdot S > 0$, then p is inside of the sphere; if $P \cdot S < 0$, then p is outside of the sphere; and if $P \cdot S = 0$, then p is on the sphere. In pattern classification, therefore, if a CGA spherical neighborhood is used, the inner product makes it possible to know when a pattern is inside or outside of the neighborhood.

3 Geometric Algebra Classification Models

While classic models all use the same kind of algebraic operations, MAMs make use of the mathematical morphological paradigm (min and max operations). Next, a description of how geometric algebra operations can be used to store the association among a subset of patterns and their corresponding index classes is shown. It is worth mentioning that the idea of using geometric algebra in classification is not new.

In [2], a *Quaternionic Multilayer Perceptron* (QMLP) in *Quaternion Algebra* is developed. A QMLP is a Multilayer Perceptron (MLP) in which both the weights of connections and the biases are *quaternions*, as well as input and output signals. With the help of the QMLP, the number of parameters of an MLP needed to perform a multidimensional series prediction decreases [2].

A new set of *Geometric Algebra Neural Networks* was introduced in [7]. Real, complex, and quaternionic neural networks can be further generalized in the geometric algebra framework [7]. The weights, the activation functions, and the outputs are represented by multivectors. The geometric product is used to operate these multivectors.

Bayro and Vallejo extended the McCulloch–Pitts neuron [21] to the *geometric neuron* by substituting the scalar product with the Clifford or geometric product. A *feed-forward geometric neural network* is then built, where the inner vector product is extended to the geometric product, and the activation functions are a generalization of the function proposed in [2].

In [7], a new approach is also proposed, the *Support Multivector Machines*. The basic idea is generating neural networks using *Support Vector Machines* (SVM) for the processing of multivectors in geometric algebra. The use of geometric algebra in SVMs offers both new tools and new understanding of SVMs for multidimensional learning [7].

In [4], a special higher-order neuron, the so-called *Hyper-sphere neuron* was introduced. A hypersphere neuron may be implemented as a *perceptron* with two bias inputs. In that work, a perceptron based on conformal geometric algebra principles was described. An iterative hypersphere neuron was also proposed. The decision surface of the perceptron presented is not a hyperplane but a hypersphere. An advantage of this representation is that only a standard scalar product needs to be evaluated in order to decide whether an input vector is inside or outside a hypersphere.

So-called *Clifford Neurons* are introduced in [8], the weights and the threshold of a classical neuron are replaced by multivectors, and the real multiplication is replaced by the Clifford product. Two types of Clifford Neurons are described, the

Basic Clifford Neuron, which can be viewed as a *Linear Associator*, and the *Spinnor Clifford Neuron*. Both types of neurons can be the starting point to fast second-order training methods for Clifford and Spinnor MLPs in the future [9].

In the following section, a new idea that has never been used before to develop an associative memory model based on the conformal geometric algebra principles will be explained.

4 Geometric Associative Memories

Definition 1 When two sets of points in $\mathbb{R}^n$ can be completely separated by a hyperplane, they are said to be *linearly separable*.

Linear separation is important for pattern classification; that hyperplane works as a *decision surface*; it can be used for deciding to which class an unclassified will be assigned by finding which side of the hyperplane the pattern is located. Many classification models (i.e., *neural networks*) have better results when the patterns are linearly separable.

In the same way, the next definition can be enunciated.

Definition 2 When two sets of points in $\mathbb{R}^n$ can be completely separated by a hypersphere, they are said to be *spherically separable*.

In this case the decision is made by finding if the pattern is located *inside* or *outside* of the sphere. Thus, the following theorem can be established:

Theorem 1 *Any two sets of linearly separable points in* $\mathbb{R}^n$ *are spherically separable too.*

Proof Consider any two sets of linearly separable points in $\mathbb{R}^n$. From (3) and (4) the hyper-plane that separate them is a sphere that passes through the point at infinity. Then, there is a sphere that separates them. Therefore, the two sets are spherically separable. □

It is worth mentioning that Theorem 1 does not guarantee that two sets of spherically separable points are linearly separable.

Spherical neighborhoods are usually difficult to handle, but in the context of geometric algebra, this is not a problem. In [4], a method for building a hypersphere neuron in an iterative way is described. In the following, three one-shot methods to build a sphere are explained.

4.1 Creating Spheres

The goal of a *Geometric Associative Memory* (GAM) is to classify a pattern as belonging to a specific class if and only if the pattern is inside of the support region (hypersphere) of that class. Building spherical neighborhoods implies to find the center of each sphere and then a suitable radius. Some procedures to achieve this with CGA have been reported in the literature. Three of them are described in the following.

In [12], a *one-shot* method is described, where given a set of points $\{p^i, i = 1, \ldots, m\}$, a spherical neighborhood is constructed. The center is computed as

$$c = \sum_{i=1}^{n} p^i / m. \tag{8}$$

In other words, the center is the average among all the patterns of each class. To compute the radius, the following expression is used:

$$\gamma = \min\left[\left(C \cdot P^i\right), i = 1, \ldots, m\right], \tag{9}$$

where C and P^i are the conformal representations of c and p^i, respectively. This procedure guarantees that all the patterns in the respective class will be inside of a sphere of class. A disadvantage of this procedure is its high computational cost.

In [17], a second approach is presented: planes or spheres are fitted into point sets by using a least squares approach. The algorithm uses the distance measure between points and spheres with the help of the inner product. It performs a least squares approach to minimize the square of the distances between a point and a sphere.

With the help of this approach, spherical neighborhoods that fit a set of patterns can be created. In this case, the spheres work as attractors with their corresponding class patterns as centers. The drawback for this method is that generally some points might appear located outside the resulting sphere.

In [17], bounding a sphere of cloud points is presented. The case for one and two points is described, and the case of expanding an existing bounding sphere when adding more points (or spheres) is presented. Using both cases, bounding a sphere of a set of points can be easily performed.

In [6], a method to construct a smallest enclosing hypersphere using quadratic programming and conformal geometric algebra was presented. The method combines characteristics of the first two methods, i.e., fit an optimal sphere that contains all the points.

The above methods can be used to build spherical neighborhoods for a specific class by using the points (patterns) of that class. But they do not take into account the patterns of other classes or the separation between classes.

In the following a new approach will be presented. It is inspired by ideas from [12]. The proposal can be used to find an optimal spherical neighborhood taking into account the patterns of the class that the sphere covers and the patterns of the other classes.

Let $P = \{P^i \cup P^j \mid i = 1, \ldots, l,\ j = l + 1, \ldots, m\}$ be a set of spherically separable points in $\mathbb{R}^n$, where $\{P^i \mid i = 1, \ldots, l\}$ are points belonging to one class, and $\{P^j \mid j = l + 1, \ldots, m\}$, are points belong to the other class. The problem is to find an optimal sphere S with the least square error, such that P^i are inside S and P^j are outside of it or, in other words, to solve

$$\min_{S} \sum_{i=1}^{m} (P^i \cdot S), \tag{10}$$

subject to (11) for points inside of the sphere and (12) for points outside of it

$$P^i \cdot S \geq 0, \quad i = 1, \ldots, l, \tag{11}$$

$$P^j \cdot S < 0, \quad j = l+1, \ldots, l. \tag{12}$$

In order to find an optimal solution, a quadratic programming algorithm must be applied. Therefore this problem must be changed into a Euclidean problem of optimization, starting from (10) and considering that all the spheres are in canonical form such that the term S_{n+2} of them can be omitted:

$$\begin{aligned}
\sum_{i=1}^{m} (P^i \cdot S)^2 &= \sum_{i=1}^{m} \left(p^i \cdot s - S_{n+1} - \frac{1}{2}(p^i)^2 \right)^2 \\
&= \sum_{i=1}^{m} \left([p^i \cdot s - S_{n+1}] - \frac{1}{2}(p^i)^2 \right)^2 \\
&= \sum_{i=1}^{m} \left([p^i \cdot s - S_{n+1}]^2 - [p^i \cdot s - S_{n+1}](p^i)^2 + \frac{1}{4}(p^i)^4 \right),
\end{aligned} \tag{13}$$

where $S_{n+1} = \frac{1}{2}(p^i - s)^2$. Thus,

$$\sum_{i=1}^{m} (P^i \cdot S)^2 = \sum_{i=1}^{m} (p^i \cdot s - S_{n+1})^2 + \sum_{i=1}^{m} (-p^i \cdot s + S_{n+1})(p^i)^2 + \frac{1}{4} \sum_{i=1}^{m} (p^i)^4. \tag{14}$$

Here, the third term is irrelevant because it does not depend on parameter S, and thus it can be omitted. Without losing generality it can be rewritten in Euclidean notation as in (15), where **W** and **F** are matrices whose components are (16) and (17) respectively, and $x = [S_1, \ldots, S_{n+1}]$.

$$\sum_{i=1}^{m} (P^i \cdot S)^2 = \sum_{i=1}^{m} \left(\sum_{k=1}^{n+1} \mathbf{W}_{i,k} x_k \right) + \sum_{i=1}^{m} \left(\sum_{k=1}^{n+1} \mathbf{F}_{i,k} x_k \right), \tag{15}$$

$$\mathbf{W}_{i,k} = \begin{cases} p_k^i & \text{for } k = 1, \ldots, n, \\ -1 & \text{otherwise,} \end{cases} \tag{16}$$

$$\mathbf{F}_{i,k} = \begin{cases} -(p_k^i)(p^i)^2 & \text{for } k = 1, \ldots, n, \\ (p^i)^2 & \text{otherwise.} \end{cases} \tag{17}$$

Let $w_i = [\mathbf{W}_{i,1}, \ldots, \mathbf{W}_{i,n+1}]$; then for the first term of the right side of expression (15) and considering that w_i^t is the transpose of the vector w_i, we have

$$\begin{aligned} \sum_{i=1}^{m}\left(\sum_{k=1}^{n+1}\mathbf{W}_i^t S_k\right)^2 &= \sum_{i=1}^{m}(w_i^t x)^2 \\ &= \left(w_1^t x + \cdots + w_m^t x\right)^2 \\ &= w_1^t x w_1^t x + \cdots + w_m^t x w_m^t x \\ &= \left(w_1^t x w_1^t + \cdots + w_m^t x w_m^t\right)x \\ &= \left(\left[x^t w_1 + \cdots + x^t w_m\right]\mathbf{W}\right)x \\ &= x^t \mathbf{W}^t \mathbf{W} x. \end{aligned} \tag{18}$$

By considering that $\mathbf{H} = \mathbf{W}^t\mathbf{W}$ we have

$$\sum_{i=1}^{m}\left(\sum_{k=1}^{n+1}\mathbf{W}_i^t S_k\right)^2 = x^t\mathbf{H}x. \tag{19}$$

For the second term of the right side of the expression (15), let $y_k = \sum_{i=1}^{m}\mathbf{F}_{i,k}$:

$$\begin{aligned} \sum_{i=1}^{m}\left(\sum_{k=1}^{n+1}\mathbf{F}_{i,k}x_k\right) &= \sum_{k=1}^{n+1}\left(\sum_{i=1}^{m}\mathbf{F}_{i,k}x_k\right) \\ &= \sum_{k=1}^{n+1}(y_k x_k). \end{aligned} \tag{20}$$

Let $y = [y_1, \ldots, y_{n+1}]$; then

$$\sum_{i=1}^{m}\left(\sum_{k=1}^{n+1}\mathbf{F}_{i,k}S_k\right) = y^t x. \tag{21}$$

With the help of (19) and (21), expression (10) can be converted into Euclidean matrix notation as follows:

$$x^t\mathbf{H}x + y^t x. \tag{22}$$

The constraint (11) for points inside of the sphere will change into

$$\begin{aligned} P^i \cdot S &\geq 0, \\ p^i \cdot s - S_{n+1} - \frac{1}{2}(p^i)^2 &\geq 0, \\ p^i \cdot s - S_{n+1} &\geq \frac{1}{2}(p^i)^2, \\ \sum_{k=1}^{n+1}(-\mathbf{W}_{k,i})S_k &\leq -\frac{1}{2}(p^i)^2, \end{aligned} \tag{23}$$

where $\mathbf{W}_{i,k}$ was defined in (16). Equation (23) can be rewritten as

$$-\mathbf{W}x \leq -\frac{1}{2}p_i^2, \tag{24}$$

where $x = [S_1, \dots, S_{n+1}]$, and the constraint (12) for points outside of the sphere will be

$$\begin{aligned} P^i \cdot S &< 0, \\ p^i \cdot s - S_{n+1} - \frac{1}{2}(p^i)^2 &< 0, \\ p^i \cdot s - S_{n+1} &< \frac{1}{2}(p^i)^2, \\ \sum_{k=1}^{n+1}(-\mathbf{W}_{i,k})S_k &< \frac{1}{2}(p^i)^2, \end{aligned} \tag{25}$$

$$\mathbf{W}x < \frac{1}{2}p_i^2. \tag{26}$$

Let $\mathbf{A} = [-\mathbf{W}, \mathbf{W}]$, and b be a vector whose ith component is $-\frac{1}{2}(p^i)^2$ for $i = 1, \dots, l$ and $\frac{1}{2}(p^i)^2 - \varepsilon$ for $i = l + m, \dots, m$. The ε is a smallest positive quantity used to change the "$<$" of (26) to be the "$\leq$". Then the constraints (11) and (12) can be converted to a Euclidean matrix notation,

$$\mathbf{A}x \leq b. \tag{27}$$

Finally, the problem of solving (10) has changed to a classical optimization problem with constraints

$$\begin{aligned} &\min_x \left(x^t \mathbf{W} x + y^t x\right), \\ &\text{s.t.} \quad \mathbf{A}x \leq b. \end{aligned} \tag{28}$$

Thus, the optimal sphere S is given by solving (28), where $S_k = x_k$ for $k = 1, \dots, n+1$ and $S_{n+2} = 1$. It is clear that by including in the restrictions all the points that stay out of the sphere, the solution S results in a separation surface that allows differentiating between two classes (i.e., inner and outer points).

The procedure works perfectly for two spherically separable classes. In a multiclass situation, the procedure is similar. In this case, the subset $\{P^i, i = 1, \ldots, l\}$ will be all patterns for class k, and $\{P^j,\ j = l+1, \ldots, m\}$ will be all patterns for the other classes. The kth sphere S_k is then found by solving (28). The same procedure must be applied for all the other classes.

4.2 Pattern Learning and Classification

The learning phase of an associative memory consists on storing associations among input patterns and output patterns. In the case of Geometric Associative Memories (GAMs), the learning phase consists on creating the spherical neighborhoods for each class. A GAM $\mathbf{M}$ is thus a matrix of size $m \times (n+2)$ (m is the number of classes, and n is the dimension of the space). The kth row are the components of the kth sphere, as it can be seen in (29). C^k and γ^k are the center and radius of the kth sphere, respectively.

$$\mathbf{M} = \begin{bmatrix} S^1 \\ S^2 \\ \vdots \\ S^m \end{bmatrix}. \tag{29}$$

Classification of a pattern can be performed by using the idea from [4], an inner product between the conformal notation of an unclassified pattern $x \in \mathbb{R}^n$ and $\mathbf{M}$ must be applied to getting, as a result, a vector u. This vector will contain all the inner products between the unclassified pattern and the spheres of class. This vector is given as

$$u_k = \mathbf{M}_k \cdot X = S^k \cdot X. \tag{30}$$

When X is inside a sphere, (30) returns a positive number (or zero) and a negative number otherwise. Note that the classification phase is independent of the training phase.

In some cases (mainly noisy patterns), X could be inside two or more spheres or could be outside of all spheres. To decide to which sphere a given pattern belongs, the following remapping must be used:

$$v_k = \begin{cases} -\infty & \text{if } u_k < 0, \\ u_k - (r^k)^2 & \text{otherwise.} \end{cases} \tag{31}$$

This must be done for $k = 1, \ldots, m$. The class identifier j can be obtained as follows:

$$j = \arg\max_k [v_k \mid k = 1, \ldots, m]. \tag{32}$$

As it can be seen, when x is outside of the k-sphere, expression (31) returns $-\infty$, and when x is inside of the k-sphere, the same expression returns the distance (with minus sign) between x and c^k. By doing this, x will be classified by a class sphere covering its conformal representation; with the help of expression (32), x will be classified by the sphere with center closest to it.

Note that, in some cases, $v_k = -\infty$ for $k = 1, \dots, m$, that is, x is outside of all the spheres. Then, when expression (32) is applied, it cannot return a value. At this point, two choices can be taken. First, x does not belong to any class. Second, using expression (32) directly on $u_k - (r^k)^2$. In this case, the GAM works as a minimum distance classifier, but the use of neighborhoods is relegated.

The classification phase is independent of the training phase. The proposed method works perfectly when the classes are spherically separable.

4.3 Conditions for Perfect Classification

In associative memories, when an associative memory **M** recovers or classifies the fundamental set correctly, it is said that **M** presents perfect recall or perfect classification. Let **M** be a trained GAM, as it was presented in the previous section.

Theorem 2 *Assume m sets of spherically separable classes in $\mathbb{R}^n$, and let $\mathbf{M}$ be a trained GAM for those classes. Then $\mathbf{M}$ presents perfect classification.*

Proof Let k be an index class whose sphere S^k is the kth component in **M**, and let p be a fundamental pattern of class k, and let j be an index $j = 1, \dots, m$ such that $j \neq k$, S^k having being obtained using expression (10). Then according to condition (11), $P \cdot S^k \geq 0$ because it is a pattern of class k and $P \cdot S^j \geq 0$ for some $j \neq k$.

When (31) is applied to P, v has a positive number or zero in position k and $-\infty$ in the other positions. Therefore, (32) returns k. This covers all patterns in all classes. □

4.4 Conditions for Robust Classification

In associative memories, when an associative memory **M** recovers or classifies correctly patterns affected with noise, it is said that **M** presents *robust recall* or *robust classification*. The robustness in a GAM depends on the size of its radius; the GAM can classify any noise pattern as belonging to its class when that pattern is located inside of it. Patterns located outside of a specific sphere (i.e., some noise patterns) will not be classified as belonging to that class sphere.

The quantity of noise that can admit a fundamental pattern depends on the position of it with respect to the center and the border of the sphere. Patterns nearest to the center can admit more quantity of noise than patterns located near of the border.

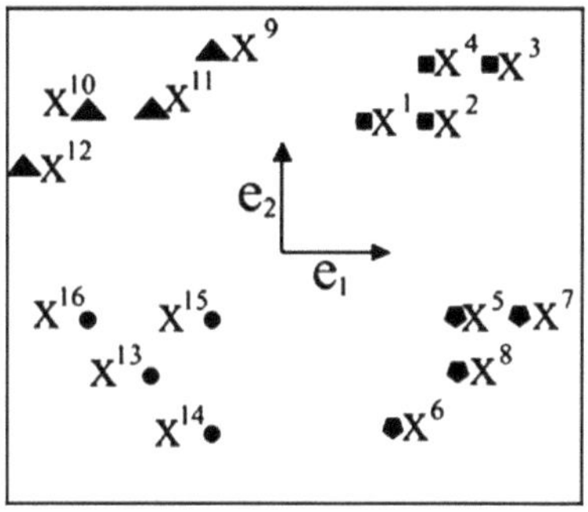

Fig. 1 Example 1, sets of patterns. *Square*, *pentagon*, *triangle*, and *circle* shapes are patterns belonging to classes 1, 2, 3, and 4, respectively

5 Numerical Examples

In this section, two illustrative examples for the problem of classifying sets of patterns are presented. For simplicity, in order to clarify the results, a 2D and 3D Euclidean space for the geometric problem are used.

In both cases, a function of the Optimization Toolbox of MatLab was used to solve the minimization problem. Function *quadprog* solves quadratic programming problems. It finds an initial feasible solution by first solving a linear programming problem.

Example 1 The following are linearly separable patterns set in $\mathbb{R}^n$:

$$\begin{array}{lllll}
\text{Class 1} & x^1=[1\quad 1], & x^2=[2\quad 1], & x^3=[3\quad 2], & x^4=[2\quad 2],\\
\text{Class 2} & x^5=[2\quad -1], & x^6=[1\quad -3], & x^7=[3\quad -1], & \\
 & x^8=[2\quad -2], & & & \\
\text{Class 3} & x^9=[-1\quad 3], & x^{10}=[-3], & x^{11}=[-2\quad 2], & \\
 & x^{12}=[-4\quad 1], & & & \\
\text{Class 4} & x^{13}=[-2\quad -2], & x^{14}=[-1\quad -3], & x^{15}=[-1\quad -1], & \\
 & x^{16}=[-3\quad -1]. & & &
\end{array} \tag{33}$$

Figure 1 shows a graphical representation of these patterns. By using (10), the corresponding spheres (in this case, circles) are obtained. Their respective centers and radii are

$$\begin{aligned}
c^1 &= [0.65\quad 1.11], & \gamma^1 &= 2.51,\\
c^2 &= [1\quad -1], & \gamma^2 &= 2,\\
c^3 &= [-1.5\quad 1.5], & \gamma^3 &= 2.55,\\
c^4 &= [-0.8\quad -0.6], & \gamma^4 &= 2.41.
\end{aligned} \tag{34}$$

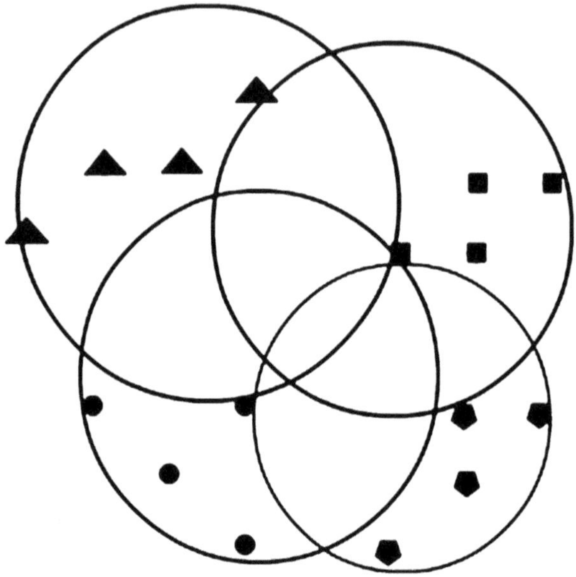

Fig. 2 Circles obtained using the method of Sect. 4.2. They function as separation surfaces

The value used for ε in this example was 0.0001. Finally, the GAM **M** is

$$\mathbf{M} = \begin{bmatrix} S^1 = C^1 - (\frac{1}{2})(\gamma^1)^2 e_\infty \\ S^2 = C^2 - (\frac{1}{2})(\gamma^1)^2 e_\infty \\ S^3 = C^3 - (\frac{1}{2})(\gamma^1)^3 e_\infty \\ S^4 = C^4 - (\frac{1}{2})(\gamma^1)^4 e_\infty \end{bmatrix}$$

$$= \begin{bmatrix} S^1 = 0.65e_1 + 1.11e_2 - 2.31e_\infty + e_0 \\ S^2 = e_1 - e_2 - e_\infty + e_0 \\ S^3 = -1.5e_1 + 1.5e_2 - e_\infty + e_0 \\ S^4 = -0.8e_1 - 0.6e_2 - 2.4e_\infty + e_0 \end{bmatrix}. \tag{35}$$

In Fig. 2, the corresponding circles are presented. Note that the circle of class 1 is optimal, because if it grows a bit more, then X^{11} could fall inside of it, and if it decreases a bit more, X^3 could fall outside of it. The same happens with the other circles. Now, let the following set of noisy patterns to be classified:

$$\begin{aligned} \tilde{x}^1 &= x^1 + [0 \quad 2] = [1 \quad 3], \\ \tilde{x}^8 &= x^8 + [-2 \quad 0] = [0 \quad -2], \\ \tilde{x}^9 &= x^9 + [-1 \quad 1] = [-2 \quad 4], \\ \tilde{x}^{15} &= x^{15} + [0 \quad -1] = [-1 \quad -2]; \end{aligned} \tag{36}$$

they have been affected with noise. If (30) is applied, the result is

$$u^1 = \begin{bmatrix} 1.31 \\ -6 \\ -1 \\ -5.2 \end{bmatrix}, \quad u^8 = \begin{bmatrix} -1.92 \\ 1 \\ -4 \\ 1.6 \end{bmatrix},$$
$$u^9 = \begin{bmatrix} -4.53 \\ -15 \\ 0 \\ -8.4 \end{bmatrix}, \quad u^{15} = \begin{bmatrix} 1.31 \\ -6 \\ -1 \\ -5.2 \end{bmatrix}. \tag{37}$$

The next step is to apply expression (31). By doing this, the following expressions are obtained:

$$v^1 = \begin{bmatrix} -1.83 \\ -\infty \\ -\infty \\ -\infty \end{bmatrix}, \quad v^8 = \begin{bmatrix} -\infty \\ -1 \\ -\infty \\ -\infty \end{bmatrix},$$
$$v^9 = \begin{bmatrix} -\infty \\ -\infty \\ -3.25 \\ -\infty \end{bmatrix}, \quad v^{15} = \begin{bmatrix} -\infty \\ -\infty \\ -\infty \\ -1 \end{bmatrix}. \tag{38}$$

The class index is then obtained by means of (32) for $\tilde{x}^1$, $\tilde{x}^8$, $\tilde{x}^9$, and $\tilde{x}^{15}$, $j = 1, 2, 3, 4$, respectively. Note that in these cases, classification is correct even when they are affected with noise and although $\tilde{x}^8$ falls inside of two spheres. However, consider the following pattern:

$$\tilde{x}^3 = x^3 + \begin{bmatrix} 0.05 & 0 \end{bmatrix} = \begin{bmatrix} 3.05 & 2 \end{bmatrix}. \tag{39}$$

Note that, in this case, the noise is minimum, but when expressions (30) and (31) are applied,

$$u^3 = \begin{bmatrix} -0.12 \\ -4.6 \\ -7.23 \\ -7.89 \end{bmatrix}, \quad v^3 = \begin{bmatrix} -\infty \\ -\infty \\ -\infty \\ -\infty \end{bmatrix}, \tag{40}$$

and in this case, expression (32) cannot classify it, due to that it is located outside of all spheres.

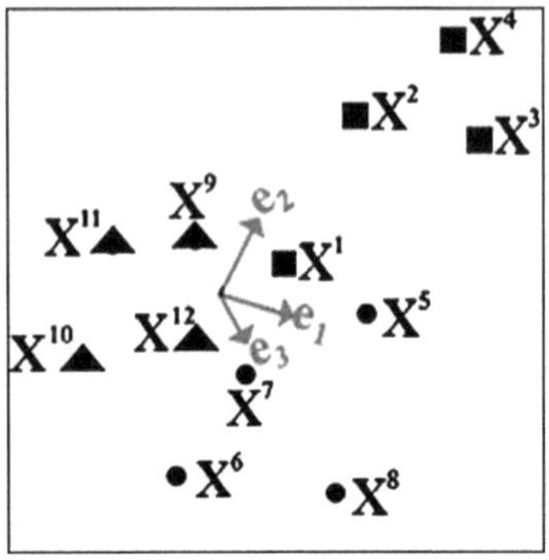

Fig. 3 Example 2, sets of patterns in 3D. *Square*, *circle*, and *triangle* shapes are patterns belonging to classes 1, 2, and 3 respectively

Example 2 Consider the following set of nonlinearly separable patterns in $\mathbb{R}^3$:

$$\begin{aligned} &\text{Class 1} \quad x^1 = [0.5 \quad 0.5 \quad 0.5], \qquad x^2 = [-0.5 \quad 2.5 \quad -1], \\ &\qquad\qquad x^3 = [2.5 \quad 2 \quad -1], \qquad x^4 = [1 \quad 3 \quad 0], \\ &\text{Class 2} \quad x^5 = [2 \quad 0 \quad -0.5], \qquad x^6 = [0.5 \quad -2 \quad 0], \\ &\qquad\qquad x^7 = [2 \quad -1.5 \quad 0.5], \qquad x^8 = [1 \quad -1 \quad -0.5], \\ &\text{Class 3} \quad x^9 = [-0.5 \quad 0.5 \quad 0], \qquad x^{10} = [-1 \quad -1 \quad -0.5], \\ &\qquad\qquad x^{11} = [-1 \quad 0 \quad -0.5], \qquad x^{12} = [0 \quad -0.5 \quad 0]. \end{aligned} \tag{41}$$

Figure 3 shows a graphical representation of these patterns.

By using (10), the corresponding class spheres were obtained. Their respective centers and radii are

$$\begin{aligned} c^1 &= [1.19 \quad 1.60 \quad 0.61], \qquad \gamma^1 = 2.11, \\ c^2 &= [1.75 \quad -0.75 \quad 1.72], \qquad \gamma^2 = 2.47, \\ c^3 &= [-0.29 \quad -0.20 \quad 0.05], \qquad \gamma^3 = 1.06; \end{aligned} \tag{42}$$

the value used for ε in this example was 0.0001. Finally, the GAM $\mathbf{M}$ is

$$\mathbf{M} = \begin{bmatrix} S^1 = 1.19e_1 + 1.6e_2 + 0.61e_3 - 0.04e_\infty + e_0 \\ S^2 = 1.75e_1 - 0.75e_2 + 1.72e_3 + 0.25e_\infty + e_0 \\ S^3 = -0.29e_1 - 0.2e_2 + 0.05e_3 - 0.5e_\infty + e_0 \end{bmatrix}. \tag{43}$$

In Fig. 4, the corresponding spheres are presented. As in the previous example, the spheres are optimal.

Now, let the following set of noisy patterns to be classified:

$$\begin{aligned} \tilde{x}^1 &= x^1 + [0.5 \quad -0.5 \quad -1] = [0.5 \quad 0.5 \quad 0], \\ \tilde{x}^7 &= x^7 + [-1 \quad 0.5 \quad -0.5] = [1 \quad -1 \quad 0], \end{aligned} \tag{44}$$

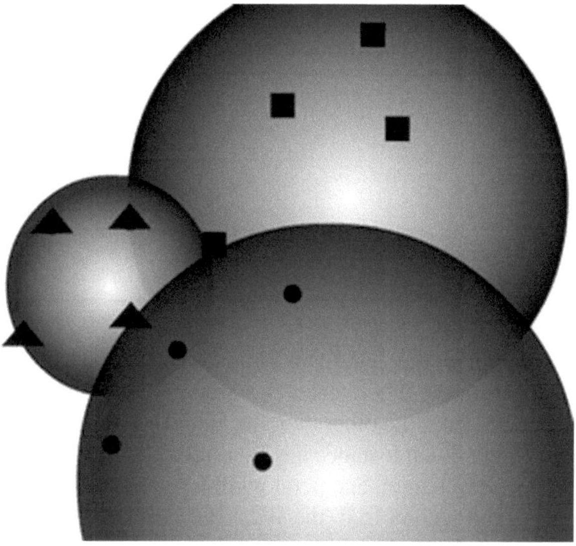

Fig. 4 Spheres obtained using the method of Sect. 4.2. They function as separation surfaces

$$\tilde{x}^{10} = x^{10} + \begin{bmatrix} 1 & 1 & 0 \end{bmatrix} = \begin{bmatrix} 0 & 0 & 0 \end{bmatrix}.$$

If (30) is applied, the result is

$$u^1 = \begin{bmatrix} 1.25 \\ -0.27 \\ -0.65 \end{bmatrix}, \qquad u^7 = \begin{bmatrix} -1.36 \\ 1.25 \\ -0.59 \end{bmatrix}, \qquad u^{10} = \begin{bmatrix} 0.04 \\ -0.25 \\ 0.5 \end{bmatrix}. \tag{45}$$

The next step is to apply expression (31). By doing this, the following expressions are obtained:

$$v^1 = \begin{bmatrix} -0.97 \\ -\infty \\ -\infty \end{bmatrix}, \qquad v^7 = \begin{bmatrix} -\infty \\ -1.79 \\ -\infty \end{bmatrix}, \qquad v^{10} = \begin{bmatrix} -2.19 \\ -\infty \\ -0.06 \end{bmatrix}. \tag{46}$$

The class index is then obtained by means of (32) for $\tilde{x}^1$, $\tilde{x}^7$, and $\tilde{x}^{10}$, $j = 1, 2, 3$, respectively. As in Example 1, the classification is correct for some patterns altered with noise.

As can be observed, although GAMs can classify some patterns altered with noise, they are very sensitive in some other cases, mainly in the case of patterns located at the border of the sphere. This problem might be fixed by adding a positive value to restriction (24) and a negative value to restriction (26).

6 Real Examples

To test the potential of the proposal, two trials with real data were performed. In the first trial, the best known database used by the pattern recognition community

Table 1 Results of the classification phase

Data Set	Patterns used in FS	Classification of FS	Classification of TS
Iris Plant	15	100%	90%
Iris Plant	25	100%	93.3%
Iris Plant	30	100%	94.6%
Wines	15	90.6%	71.9%
Wines	25	91.1%	80.9%
Wines	30	97.7%	84.8%

was adopted, the *Iris Plant Data Base*. This data set contains three classes of 50 instances each, where each class refers to a type of iris plant. One class is linearly separable from the other two; the latter two classes are NOT linearly separable from each other. Each instance has four numeric, predictive attributes (sepal length, sepal width, petal length, and petal width).

For the second trial, the *Wine Recognition Data Base* was used. These data are the results of a chemical analysis of wines grown in the same region in Italy but derived from three different cultivations. The analysis determined the quantities of 13 constituents found in each of the three types of wines. This data set contains three classes of 59, 48, and 71 instances, respectively, where each class refers to a type of wine. Each instance has 13 continuous attributes.

Both data bases were obtained from [3]. As in the previous section, the function *quadprog* of the optimization toolbox of Matlab was used in order to solve the quadratic programming problem.

15, 25, and 30 instances, respectively, were used to form the Fundamental Set (FS) for each type of problem. All of the instances were used to form the Test Set (TS). In the training phase, patterns of each FS were used to build the GAMs. In the classification phase, patterns of the TS were classified; the results are shown in Table 1. The first column shows the number of patterns used for the learning phase, and the second and third columns show the percentages of the patterns correctly classified using the FS and the TS, respectively. In the case of the GAM from Iris Plant data set, perfect recall was obtained because the patterns used for the FS are spherically separable. In the case of the GAM from the Wine data set, perfect recall is not obtained because the patterns are not spherically separable. Thus some patterns of a specific class may fall outside their respective class sphere.

It can be observed that classification rate of the patterns of the TS (for both data sets) increases when the number of patterns used in the FS increases. When the FS has more patterns, the corresponding spheres grow, and then more patterns of the TS could fall inside of the class spheres.

7 Conclusions and Future Work

Geometric Algebra allows one to model situations and to formulate problems in terms of high-level symbolic expressions. Nevertheless, it is possible to achieve an implementation working in an elementary coordinate system. Of course, in some cases, it is possible to find a solution by purely handling symbolic expressions, but this is rare for realistic problems.

In this work, a new associative memory model based on Conformal Geometric Algebra has been described, the Geometric Associative Memory (GAM). The training phase is done by finding an optimal sphere with quadratic programming. GAMs can perfectly operate when the classes are spherically separable.

For classification purposes, an inner product between the unclassified pattern and the GAM was applied. Then a minimum function is used to obtain the index class.

Numerical and real examples were given to show the potential of the proposal. As shown, the method can operate both with linearly and nonlinearly separable patterns. The proposed model can also cope with distorted patterns.

Patterns located on the border of the sphere might not be well classified. At this moment, a way to extend the radius of the sphere is been developing. The basic idea is to change the restrictions of the optimization problem.

Formal conditions under which the proposed model can work were also given and proven. In particular, the case of the perfect classification was presented. A brief explanation about the functioning of the GAMs against the noise was presented; the GAMs can cope with noise patterns when the noise version falls inside of the class sphere.

Nowadays, we are also interested to test our method in more realistic situations and in comparison (in computing time and performance) between the proposed model and other geometric classification models. We are working too in GAMs that work with separation surfaces other than spheres, like ellipses, squares, or other irregular shapes; then, the GAMs can work with nonspherically separable classes.

Acknowledgements The authors thank the National Polytechnic Institute of Mexico (SIP-IPN) under grants 20090620 and 20091421. Authors also thank the European Union, the European Commission, and CONACyT for the economical support. This paper has been prepared for economical support of the European Commission under grant FONCICYT 93829. The content of this paper is an exclusive responsibility of the CIC-IPN, and it cannot be considered that it reflects the position of the European Union. We thank also the reviewers for their comments for the improvement of this paper.

References

1. Anderson, J.: A simple neural network generating an interactive memory. Math. Biosci. **14**, 197–220 (1972)
2. Arena, P., Baglio, S., Fortuna, L., Xibilia, M.: Chaotic time series prediction via quaternionic multilayer perceptrons. Syst., Man and Cybern., Intell. Syst. for the 21st Century, IEEE Int. Conf., vol. 2, pp. 1790–1794 (1995)
3. Asuncion, A., Newman, D.: UCI machine learning repository (2007). http://www.ics.uci.edu/mlearn/MLRepository.html
4. Banarer, V., Perwass, C., Sommer, G.: The hypersphere neuron. 11th Eur. Symp. on Artif. Neural Netw. Evere, Belgium: d-side publ., pp. 469–474 (2003)
5. Barron, R.: Memorias asociativas y redes neuronales morfologicas para la recuperacion de patrones. Ph.D, thesis, Mexico, DF: National Institute Politechnic—Center of Computing Research (2006)

6. Barron, R., Cruz, B., Sossa, H., Laguna, G.: Conformal geometric algebra for spherical convex hull optimization. In: Proc. 3rd Internat. Conf. on Appl. of Geom. Algebras in Comput. Sci. and Eng., AGACSE 2008 (2008)
7. Bayro, E., Vallejo, R.: Geometric feedforward neural networks and support vector machines. In: Bayro-Corrochano, E., Sobczyk, G. (eds.) Geometric Algebra with Applications in Science and Engineering, pp. 309–325. Birkhäuser, Basel (2001)
8. Buchholz, S.: A theory of neural computation with Clifford algebras. Thesis, Kiel: Christian-Albrechts-Universitat (2005)
9. Buchholz, S., Tachibana, K., Hitzer, E.: Optimal learning rates for Clifford neurons. In: Proc. of ICANN 2007, Part I. LNCS, vol. 4668, pp. 864–873. Springer, Berlin (2007)
10. Clifford, W.: Applications of Grassmann's extensive algebra. Am. J. Math. **1**(4), 350–358 (1878)
11. Cruz, B., Sossa, H., Barron, R.: A new two level associative memory for efficient pattern restoration. Neural Process. Lett. **25**, 1–16 (2007)
12. Cruz, B., Barron, R., Sossa, H.: Geometric associative memory model with application to pattern classification. In: Proc. 3rd Internat. Conf. on Appl. of Geom. Algebras in Comput. Sci. and Eng., AGACSE 2008 (2008)
13. Hitzer, E.: Euclidean geometric objects in the Clifford geometric algebra of origin, 3-space, infinity. Bull. Belg. Math. Soc. **11**(5), 653–662 (2004)
14. Hestenes, D., Sobczyk, G.: Clifford Algebra to Geometric Calculus. Springer, Berlin (1984)
15. Hestenes, D.: Old wine in new bottles. In: Bayro-Corrochano, E., Sobczyk, G. (eds.) Geometric Algebra: A Geometric Approach to Computer Vision, Quantum and Neural Computing, Robotics, and Engineering, pp. 498–520. Birkhäuser, Basel (2001)
16. Hestenes, D., Li, H., Rockwood, A.: New algebraic tools for classical geometry. In: Sommer, G. (ed.) Geometric Computing with Clifford Algebras. vol. 40, pp. 3–23. Springer, Heidelberg (2001)
17. Hildebrand, D.: Geometric computing in computer graphics using conformal geometric algebra. Tutorial, TU Darmstadt, Germany: Interact. Graph. Syst. Group (2005)
18. Hopfield, J.: Neural networks and physical systems with emergent collective computational abilities. Proc. Natl. Acad. Sci. **79**, 2554–2558 (1982)
19. Kohonen, T.: Correlation matrix memories. IEEE Trans. Comput. **C-21**(4), 353–359 (1972)
20. Li, H., Hestenes, D., Rockwood, A.: Generalized homogeneous coordinates for computational geometry. In: Sommer, G. (ed.) Geometric Computing with Clifford Algebras. vol. 40, pp. 27–52. Springer, Heidelberg (2001)
21. McCulloch, W., Pitts, W.: A logical calculus of the ideas immanent in nervous activity. Bull. Math. Biophys. **5**, 115–133 (1943)
22. Nakano, K.: Associatron a model or associative memory. IEEE Trans. Syst., Man Cybern **12**, 380–388 (1972)
23. Ritter, G., Sussner, P., Diaz-de-Leon, J.: Morphological associative memories. IEEE Trans. Neural Netw. **9**(2), 281–293 (1998)
24. Sossa, H., Barron, R.: New associative model for pattern recall in the presence of mixed noise. IASTED Fifth Int. Conf. on Signal and Image Process. (SIP 2003), pp. 485–490 (2003)
25. Steinbouch, K.: Die Lernmatrix. Kybernetik **1**(1), 26–45 (1961)
26. Sussner, P.: Observations on morphological associative memories and the kernel method. Neurocomputing **31**, 167–183 (2003)
27. Yañez, C., Diaz-de-Leon, J.: Introducción a las memorias asociativas. Mexico: Res. in Comp. Sci. (2003)

Classification and Clustering of Spatial Patterns with Geometric Algebra

**Minh Tuan Pham, Kanta Tachibana,
Eckhard M.S. Hitzer, Tomohiro Yoshikawa,
and Takeshi Furuhashi**

Abstract In fields of classification and clustering of patterns most conventional methods of feature extraction do not pay much attention to the geometric properties of data, even in cases where the data have spatial features. This paper proposes to use geometric algebra to systematically extract geometric features from data given in a vector space. We show the results of classification of handwritten digits and those of clustering of consumers' impression with the proposed method.

1 Introduction

Nowadays classification and clustering of patterns are of central importance for discovery of information from enormous amounts of data available in various practical fields. An appropriate method to extract features from patterns is needed for good classification and clustering. But so far most conventional methods of feature extraction ignore the geometric properties of data even in the case where the data have spatial features. For example, when m vectors are measured from an object in three-dimensional space, conventional methods represent the object by $x \in \mathbf{R}^{3m}$ which is the vector made by arranging m groups of three coordinates of each vector in a row. However, using only these coordinate values fails to capture geometric relationships among m vectors, e.g., the coordinate values depend on the definition of the coordinate system, and inference or classification becomes remarkably bad when objects are measured in a coordinate system different from the one used for learning. Some conventional methods may extract coordinate-free features, but whether such features arise and are adopted depends on experience of the model builder. For example, some image recognition methods used moment vectors [1, 2] as feature vectors for learning. The moment vectors are the generalization moments of inertia of each spatial vector, and it is expressed by a linear sum of $\{x_{i,1}^a x_{i,2}^b \mid i = 1, \ldots, m; a, b \in N\}$,

M.T. Pham (✉)
Nagoya University, Furou 1-1, Chikusa, Nagoya, Japan
e-mail: minhtuan@cmplx.cse.nagoya-u.ac.jp

E. Bayro-Corrochano, G. Scheuermann (eds.), *Geometric Algebra Computing*,
DOI 10.1007/978-1-84996-108-0_12,

where $(x_{i,1}, x_{i,2})$ is the coordinate of vector i. However, the relations between different spatial vectors are not considered in the moment, i.e., the components of $\{x_{i,1}^a x_{j,2}^b \mid i \neq j\}$ are not considered at all. So, it is hard to classify data that contain relations between spatial vectors.

In this study, we use geometric algebra (GA) [3–5] to systematically undertake various kinds of feature extractions and to improve precision and robustness in classification and clustering problems. There are already many successful examples of its use in, e.g., colored image processing or multidimensional time-series signal processing with low-dimensional GAs [6–12]. In addition, GA-valued neural network learning methods for learning input–output relationships [13] are well studied. In our proposed method, geometric features extracted with GA can also be used for learning a distribution and for semi-supervised clustering.

We use geometric features to learn a Gaussian mixture model (GMM) with the expectation maximization (EM) algorithm [14]. Because each feature extraction derived by the proposed method has its own advantages and disadvantages, we apply a plural mixture of GMMs for a classification problem. As an example of multiclass classification of geometric data, we use a handwritten digit dataset. When classifying new handwritten digits in practice with the learning model, it is natural to expect that the coordinate system in a real new environment differs from the one used for obtaining the learning dataset. Therefore, in this paper, we evaluate the classification performance for randomly rotated test data.

As a second application, we analyze a dataset of questionnaire for a newly developed product. Characteristics of this dataset are:

1. The same m questions are asked for n different objects (usage scenes of the product).
2. Each respondent answers his/her willingness to buy for either of three different prices and does not answer for the other prices.

Considering the first characteristic, we regard a pattern of answering to the questions by a respondent as a tuple of m points in an n-dimensional space. This aims to extract features of n-dimensional *shape* formed by the m vectors with GA. For the second characteristic, we utilize harmonic functions [17] for the semi-supervised learning. In our proposed method, geometric features extracted with GA can be used for defining a weighted graph over unlabeled and labeled data where the weights are given in terms of a similarity between respondents. To evaluate the effect of features extracted with GA, we examine kernel matrices induced from the geometric features using kernel alignment [18] between them. This paper reports a result of semi-supervised clustering of respondents taking geometric properties of questionnaire into consideration.

2 Method

This section describes our proposal to extract geometric features from spatial patterns for classification and clustering. Our general scheme is based on a description of *shape* formed by m-tuple of n-dimensional vectors by GA.

2.1 Feature Extraction for Geometric Data

An orthonormal basis $\{\mathbf{e}_1, \mathbf{e}_2, \ldots, \mathbf{e}_n\}$ can be chosen for a real vector space $\mathbf{R}^n$. The GA of $\mathbf{R}^n$, denoted by $\mathcal{G}_n$, is constructed by an associative and bilinear product of vectors, the geometric product, which is defined by

$$\mathbf{e}_i\mathbf{e}_j = \begin{cases} 1 & (i = j), \\ -\mathbf{e}_j\mathbf{e}_i & (i \neq j). \end{cases} \tag{1}$$

GAs are also defined for negative squares $\mathbf{e}_i^2 = -1$ of some or all basis vectors. Such GAs have many applications in computer graphics, robotics, virtual reality, etc. [5]. However, for our purposes, definition (1) will be sufficient.

The geometric product of linearly independent vectors $\mathbf{a}_1, \ldots, \mathbf{a}_k$ $(k \leq n)$ has its maximum grade term as the k-blade $\mathbf{a}_1 \wedge \cdots \wedge \mathbf{a}_k$. Linear combinations of k-blades are called k-vectors, represented by $\sum_{I \in \mathcal{I}_k} w_I \mathbf{e}_I$, where $\mathcal{I}_k = \{i_1 \ldots i_k \mid 1 \leq i_1 < \cdots < i_k \leq n\}$. For $\mathcal{G}_n$, $\bigwedge^k \mathbf{R}^n$ denotes the set of all k-blades, and $\mathcal{G}_n^k$ denotes set of k-vectors. The geometric product of k vectors yields

$$\mathbf{a}_1 \ldots \mathbf{a}_k \in \begin{cases} \mathcal{G}_n^1 \oplus \cdots \oplus \mathcal{G}_n^{k-2} \oplus \bigwedge^k \mathbf{R}^n & (\text{odd } k), \\ \mathcal{G}_n^0 \oplus \cdots \oplus \mathcal{G}_n^{k-2} \oplus \bigwedge^k \mathbf{R}^n & (\text{even } k). \end{cases} \tag{2}$$

Now we propose a systematic derivation of feature extractions from a series or a set of spatial vectors $\xi = \{\mathbf{p}_l \in \mathbf{R}^n, l = 1, \ldots, m\}$. Our method is to extract the scalar part of products of k-vector data that encode the different features.

First, assuming that ξ is a series of n-dimensional vectors, $n' + 1$ feature extractions are derived, where $n' = \min\{n, m\}$. For $k = 1, \ldots, n'$,

$$f_k(\xi) = \left\{\left\langle \mathbf{p}_l \ldots \mathbf{p}_{l+k-1}\mathbf{e}_I^{-1}\right\rangle, I \in \mathcal{I}_k, l = 1, \ldots, m-k+1\right\} \in \mathbf{R}^{(m-k+1)|\mathcal{I}_k|}, \tag{3}$$

where $\langle\cdot\rangle$ denotes the operator that selects the scalar part, $|\mathcal{I}_k|$ is the number of combinations of k elements from n elements, and $\mathbf{e}_I^{-1}$ is the inverse of $\mathbf{e}_I$. For $I = i_1 \ldots i_k$, $\mathbf{e}_I^{-1} = \mathbf{e}_{i_k} \ldots \mathbf{e}_{i_2}\mathbf{e}_{i_1}$. When $k > n'$, this means that $\forall l \in \{1, \ldots, m-k+1\}$, $\langle \mathbf{p}_l \ldots \mathbf{p}_{l+k-1}\mathbf{e}_I^{-1}\rangle = 0$. This feature is not possible in the case of classification problem or clustering. We further define

$$f_0(\xi) = \left\{\langle \mathbf{p}_l\mathbf{p}_{l+1}\rangle, l = 1, \ldots, m-1\right\} \in \mathbf{R}^{m-1}. \tag{4}$$

Next, assuming that ξ is a set of vectors, $n' + 1$ feature extractions can also be derived in the same way:

$$f_k(\xi) = \left\{\left\langle \mathbf{p}_{l_1} \cdots \mathbf{p}_{l_k}\mathbf{e}_I^{-1}\right\rangle, I \in \mathcal{I}_k\right\} \in \mathbf{R}^{({}_mC_k)|\mathcal{I}_k|}, \tag{5}$$

$$f_0(\xi) = \left\{\langle \mathbf{p}_{l_1}\mathbf{p}_{l_2}\rangle\right\} \in \mathbf{R}^{({}_mC_2+m)}. \tag{6}$$

The dimension of the feature space becomes different from the case where ξ is a series.

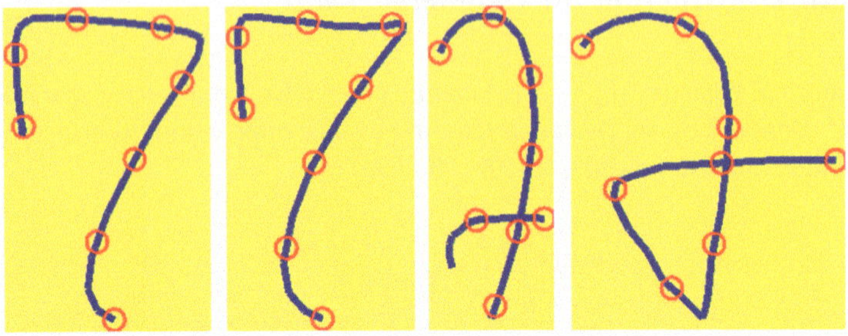

Fig. 1 Examples of handwritten digit "7"

We denote by f_k a feature vector extracted by a feature extraction f_k. f_0 is the scalar part in the geometric product of two vectors chosen from n vectors. f_k consists of the coefficient of k-blade in the geometric product of k vectors chosen from n vectors. f_0 and f_n do not depend on the measurement coordinate system.

Below we show several feature extractions for the case of handwritten digit data of the UCI Machine Learning Repository [15] and the case of questionnaire data.

Each of the digit data is given by eight points $\xi = \{\mathbf{p}_1, \dots, \mathbf{p}_8\}$, measured by dividing the handwritten curves in equally long curve segments. A two-dimensional point is given by $\mathbf{p}_l = x_l\mathbf{e}_1 + y_l\mathbf{e}_2$ with $\sum_{l=1}^{8} x_l = \sum_{l=1}^{8} y_l = 0$. Using GA, various kinds of feature extraction can be undertaken systematically. Figure 1 shows some examples of the handwritten digit "7". They are shown with straight line segments, different from the real curved trajectories of a pen.

The simplest feature extraction f_1, which is also used in conventional methods, is

$$\begin{aligned}\mathbf{f}_1(\xi) &= \left[\langle \mathbf{p}_1\mathbf{e}_1^{-1}\rangle, \langle \mathbf{p}_1\mathbf{e}_2^{-1}\rangle, \dots, \langle \mathbf{p}_8\mathbf{e}_1^{-1}\rangle, \langle \mathbf{p}_8\mathbf{e}_2^{-1}\rangle\right] \\ &= [\mathbf{p}_1 \cdot \mathbf{e}_1, \mathbf{p}_1 \cdot \mathbf{e}_2, \dots, \mathbf{p}_8 \cdot \mathbf{e}_1, \mathbf{p}_8 \cdot \mathbf{e}_2] \\ &= [x_1, y_1, \dots, x_8, y_8] \in \mathbf{R}^{16}. \end{aligned} \tag{7}$$

A second feature extraction f_2 uses the directed magnitudes of outer products of consecutive points:

$$\begin{aligned}\mathbf{f}_2(\xi) &= \left[\langle \mathbf{p}_1\mathbf{p}_2\mathbf{e}_{12}^{-1}\rangle, \dots, \langle \mathbf{p}_7\mathbf{p}_8\mathbf{e}_{12}^{-1}\rangle\right] \\ &= [x_1y_2 - x_2y_1, \dots, x_7y_8 - x_8y_7] \in \mathbf{R}^7. \end{aligned} \tag{8}$$

A third feature extraction f_0 uses the inner product of consecutive points:

$$\begin{aligned}\mathbf{f}_0(\xi) &= \left[\langle \mathbf{p}_1\mathbf{p}_2\rangle, \dots, \langle \mathbf{p}_7\mathbf{p}_8\rangle\right] \\ &= [x_1x_2 + y_1y_2, \dots, x_7x_8 + y_7y_8] \in \mathbf{R}^7. \end{aligned} \tag{9}$$

Each respondent gave evaluation values to the same ten questions for six objects. We therefore regard a filled out questionnaire as $m(=10)$ points in an $n(=6)$-

dimensional space: $\xi = \{\mathbf{p}_1, \ldots, \mathbf{p}_{10}\}$, $\mathbf{p}_l = \sum_i^6 x_{l,i}\mathbf{e}_i$ with $x_{l,i} \in \{-2, -1, 0, 1, 2\}$. Using GA, various kinds of feature extractions can be undertaken systematically. In this paper, we use three kinds of features extracted from ξ with GA.

The simplest feature extraction f_1, which is coordinate value, also used in conventional methods, is

$$\begin{aligned} \mathbf{f}_1(\xi) &= \left[\langle \mathbf{p}_1\mathbf{e}_1^{-1}\rangle, \ldots, \langle \mathbf{p}_1\mathbf{e}_6^{-1}\rangle, \ldots, \langle \mathbf{p}_{10}\mathbf{e}_1^{-1}\rangle, \ldots, \langle \mathbf{p}_{10}\mathbf{e}_6^{-1}\rangle\right] \\ &= [x_{1,1}, x_{1,2}, \ldots, x_{10,5}, x_{10,6}] \in \mathbf{R}^{60}. \end{aligned} \tag{10}$$

A second feature extraction f_0 uses the inner product of two points:

$$\mathbf{f}_0(\xi) = \left[\langle \mathbf{p}_1\mathbf{p}_1\rangle, \langle \mathbf{p}_1\mathbf{p}_2\rangle, \ldots, \langle \mathbf{p}_{10}\mathbf{p}_{10}\rangle\right] \in \mathbf{R}^{55}. \tag{11}$$

If two questions are correlated for six objects, then the corresponding element of $\mathbf{f}_0$ becomes large.

Finally, a third feature extraction f_2 uses outer product of two points:

$$\begin{aligned} \mathbf{f}_2(\xi) = \big[&\langle \mathbf{p}_1\mathbf{p}_2\mathbf{e}_{12}^{-1}\rangle, \ldots, \langle \mathbf{p}_1\mathbf{p}_2\mathbf{e}_{56}^{-1}\rangle, \ldots, \\ &\langle \mathbf{p}_9\mathbf{p}_{10}\mathbf{e}_{12}^{-1}\rangle, \ldots, \langle \mathbf{p}_9\mathbf{p}_{10}\mathbf{e}_{56}^{-1}\rangle\big] \in \mathbf{R}^{675}. \end{aligned} \tag{12}$$

Each $|I_2|$ elements of $\mathbf{f}_2$ express independence of two questions and the direction of hyper-plane spanned by the two questions in the six-dimensional space. If two questions are uncorrelated, then the corresponding element becomes large.

2.2 Distribution Learning and Its Mixture for Classification

A GMM is useful to approximate a data distribution in a data space. A GMM is characterized by parameters $\Theta = \{\beta_j, \mu_j, \Sigma_j\}$, where β_j, μ_j, and Σ_j are the mixture ratio, the mean vector, and the variance covariance matrix of the jth Gaussian, respectively. The output is

$$p(\xi \mid \Theta) = \sum_{j=1}^{M} \beta_j \mathcal{N}_d\big(f(\xi) - \mu_j; \Sigma_j\big), \tag{13}$$

where $\mathcal{N}_d(\cdot;\cdot)$ is the d-dimensional Gaussian distribution function with center fixed at the origin.

To train M Gaussians with given incomplete data $X = \{x_i = f(\xi_i) \mid 1 \le i \le N\}$, the EM algorithm [14] is often utilized. The algorithm identifies both parameters Θ and latent variables $Z = \{z_{ij} \in \{0, 1\} \mid 1 \le j \le M\}$. The z_{ij} are random variables with $z_{ij} = 1$ indicating that the individual datum x_i belongs to the jth of M Gaussian distributions. Thus $\sum_{j=1}^{M} P(z_{ij}) = 1$. The EM algorithm repeats the E-step and

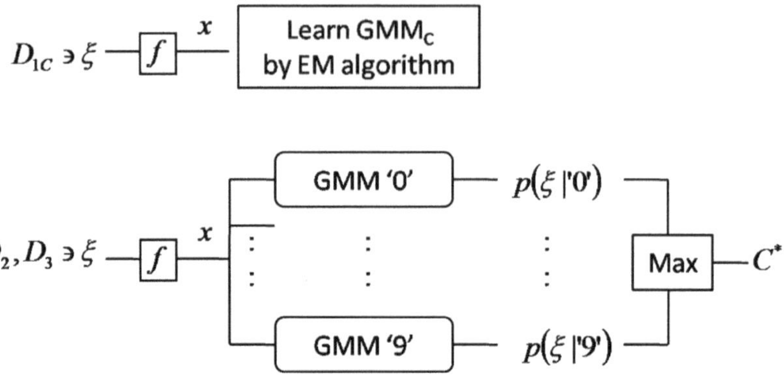

Fig. 2 Flow of multiclass classification. *The top diagram* shows the training of the GMM for class $C \in \{\text{'0'}, \ldots, \text{'9'}\}$. The D_{1C} denotes a subset of training samples with label C. The $f : \xi \mapsto x$ shows feature extraction. Either of $\{f_1, f_2, f_0\}$ is chosen as f. *The bottom diagram* shows estimation by the learned GMMs. The same f chosen for training is used here. The GMM_C outputs $p(\xi \mid C)$. The final estimation is $C^* = \arg\max_C \; p(\xi \mid C)P(C)$, where $P(C) = \frac{1}{10}$ is the prior distribution. The set D_3 consists of independent test data

the M-step until $P(Z)$ and Θ converge. The E-step updates the probabilities of Z according to Bayes' theorem $P(Z \mid X, \Theta) \propto p(X \mid Z, \Theta)$. The M-step updates the parameters Θ of the Gaussians to maximize the likelihood $l(\Theta, X, Z) = p(X \mid Z, \Theta)$.

A major drawback of the GMM is its large number of free parameters whose order is $O(Md^2)$. Moreover, when the correlation between any pair of features is close to 1, the calculation of the inverse matrix is numerically unstable. So, the GMM becomes unable to compute the correct probability distribution. To remedy this, Tipping and Bishop [16] proposed to use only the eigencomponents with the largest q eigenvalues, where q is a preset value of a Gaussian distribution. For further better approximation of a probability distribution, Meinicke and Ritter [19] proposed to use only the eigencomponents of Gaussian distributions that are larger than a certain cutoff. The cutoff eigenvalue is set at $\lambda_- = \alpha\lambda_{\max}$, where $\alpha \in (0, 1]$ is a hyperparameter, and $\lambda_{\max}$ is the largest eigenvalue of the variance covariance matrix of incomplete data X. This means that we develop (13) as follows:

$$\begin{aligned} p(x \mid \Theta) &= \sum_{j=1}^{M} \beta_j \prod_{k=1}^{d} \mathcal{N}_1\big((x - \mu_j) \cdot v_k; \lambda_k\big) \\ &\approx \sum_{j=1}^{M} \beta_j \left\{ \prod_{k=1}^{q_j} \mathcal{N}_1\big((x - \mu_j) \cdot v_k; \lambda_k\big) \right\} \mathcal{N}_1\big((x - \mu_j)_-; \lambda_-\big)^{d-q_j}, \end{aligned} \tag{14}$$

where λ_k is the kth largest eigenvalue of the jth Gaussian distribution, and v_k is the corresponding eigenvector. Because $k \le q_j$ is equivalent to $\lambda_k > \lambda_-$, the bracket $(x - \mu_j)_- = (x - \mu_j) \cdot v_{q_j}$ is the length of the component which is perpendicular to all of the q_j eigenvectors with the largest eigenvalues.

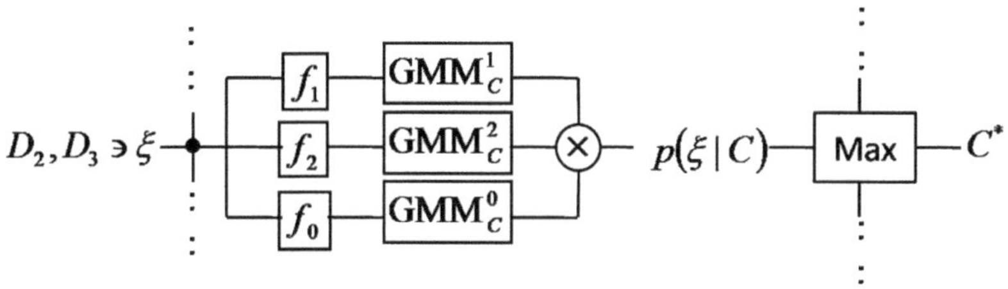

Fig. 3 Mixture of GMMs. Three GMMs via different feature extractions are mixed to yield output $p(\xi \mid C)$

The flow of training and estimation of handwritten digit classification, as an example of multiclass classification, is shown in Fig. 2. The M and α for each GMM are decided by validation with dataset D_2.

Each feature extraction derived with GA has advantages and disadvantages. A big merit of adopting learning of distributions rather than learning of input–output relations is that the learned distributions allow us to obtain reliable inference by mixing plural weak learners. In this study, we therefore use a mixture of GMMs. Inferences are mixed as

$$p(\xi \mid C) = \prod_{k=0}^{2} p\bigl(f_k(\xi) \mid C\bigr). \tag{15}$$

Figure 3 shows the mixture of different GMMs. Inferences via different feature extractions are mixed to produce the output $p(\xi \mid C)$.

2.3 GA Kernel and Alignment and Semi-Supervised Learning for Clustering

For feature extractions $f_k,\ k = 0, 1, 2$, with GA, we define a similarity between two instances $i, j \in \{1, \dots, p\}$ as

$$w_{ij;k} = \exp\biggl(-\frac{\|\mathbf{f}_k(\xi_i) - \mathbf{f}_k(\xi_j)\|^2}{\sigma_k^2}\biggr), \tag{16}$$

where $\|\cdot\|$ is the Euclidean distance in the feature space, and a parameter σ_k is decided by Zhu et al. [17] as described below. The kernel matrix $W_k = [w_{ij;k}]$ is a symmetric matrix with p rows and p columns.

In this study, we combine three kinds of feature extractions to cluster instances. The effect of combining two feature extractions becomes small if their kernel matrices are aligned. The alignment [18] between two kernel matrices is defined as

$$A(W_k, W_l) = \sum_{i,j} \widetilde{w_{ij;k}} \widetilde{w_{ij;l}} \in (0, 1], \tag{17}$$

where, $\widetilde{w_{ij;k}} = w_{ij;k}(\sum_{\tilde{i},\tilde{j}} w_{\tilde{i}\tilde{j};k}^2)^{-\frac{1}{2}}$ is the element of the matrix $\widetilde{W_k}$ which is the matrix W_k normalized so that the squared sum of all its elements becomes 1.

In addition, cluster structure embedded in the data distribution is evaluated as alignment with identity matrix E,

$$A(W_k, E) = \frac{1}{\sqrt{p}} \sum_i \widetilde{w_{ii;k}} \in (0, 1]. \tag{18}$$

The alignment with E becomes 1 when similarity between any two different instances is 0, i.e., no cluster structure is embedded.

On the other hand, when binary label $y_i \in \{-1, 1\}$ is given for each instance, if instances with the same label are allocated near and those with the different label are allocated far, then the clustering result of this feature agrees with the labels. This is evaluated by the alignment between W_k and $Y = [Y_{ij} = y_i y_j]$,

$$A(W_k, Y) = \frac{1}{p}\left(\sum_{i,j|y_i=y_j} \widetilde{w_{ij;k}} - \sum_{i,j|y_i\neq y_j} \widetilde{w_{ij;k}}\right) \in [-1, 1]. \tag{19}$$

A good combination of two feature extractions has a low $A(W_k, W_l)$ value. And, if $A(W_k, E)$ is low and $A(W_k, Y)$ is high, f_k induces a feature space with rich cluster structure and agreement with the given labels.

Semi-supervised learning is a problem to infer labels for unlabeled data $U = \{l+1, \ldots, l+u = p\}$ or unknown unlabeled data when the label $y_i \in \{-1, 1\}$ of a part $L = \{1, \ldots, l\}$ of the instance set is known. In this paper, we solve a problem in the case of labeling U. The goal is to find a binary function $\gamma : U \to \{-1, 1\}$ such that similar points have the same label.

Referring Zhu et al. [17], we decide a parameter σ of the kernel as follows. During making a minimum spanning tree over all data points with Kruskal's Algorithm, we label temporarily an unlabeled data $a \in U$ to the same label as the labeled datum which connects with a for the first time. Then we find the median distance of tree edges that connect two instances with different labels. We regard this distance d_o as a heuristic to the median distance between class regions. We arbitrarily set $\sigma = \frac{d_0}{3}$ following the 3σ rule of Normal distribution, so that the weight of this edge is close to 0.

Then, we construct a $p \times p$ symmetric weight matrix $W = [w_{ij}]$. The weight matrix can be separated as

$$W = \begin{bmatrix} W_{LL} & W_{LU} \\ W_{UL} & W_{UU} \end{bmatrix} \tag{20}$$

at the lth row and lth column. For this purpose, Zhu et al. proposed to first compute a real-valued function $g: U \to [0, 1]$ which minimizes the energy

$$E(g) = \sum_{i,j} w_{ij}\big(g(i) - g(j)\big)^2. \tag{21}$$

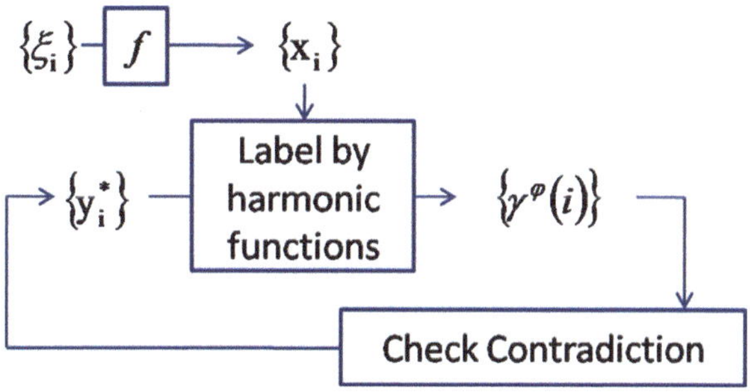

Fig. 4 Algorithm to find latent willingness to buy. From questionnaire data ξ_i, $f \in \{f_0, f_1, f_2\}$ extracts geometric features x_i. Label y_i^* is initially set to y_i. After calculating $\gamma^{\varphi}(i)$, labels of contradicted respondents are excluded from $\{y_i^*\}$. The algorithm ends when no more contradictions occur

Restricting

$$g(i) = g_L(i) \equiv \begin{cases} 0 & (y_i = -1), \\ 1 & (y_i = 1) \end{cases} \tag{22}$$

for the labeled data, g for the unlabeled data can be calculated by

$$g_U = (D_{UU} - W_{UU})^{-1} W_{UL}\, g_L, \tag{23}$$

where $D_{UU} = \mathrm{diag}(d_i)$ is the diagonal matrix with entries $d_i = \sum_j w_{ij}$ for the unlabeled data. Then $\gamma(i)$ is decided using the class mass normalization proposed by Zhu et al.:

$$\gamma(i) = \begin{cases} 1 & \left(|\{j \mid y_j = 1\}| \frac{g_U(i)}{\sum_i g_U(i)} > |\{j \mid y_j = -1\}| \frac{1-g_U(i)}{\sum_i (1-g_U(i))}\right), \\ -1 & \text{(otherwise)}. \end{cases} \tag{24}$$

Because either of three prices $\{\varphi_1, \varphi_2, \varphi_3 \mid \varphi_1 < \varphi_2 < \varphi_3\}$ is indicated to a respondent when he/she indicates willingness to buy, we subdivide the respondents into three groups according to the indicated price. Then, we calculate γ^{φ_k} for each $k \in \{1, 2, 3\}$ regarding one group of respondents as labeled and the other groups as unlabeled. After that, we check the consistency of respondent i, i.e., whether the willingness decreases weakly monotonously with the price.

$$\gamma^{\varphi_1}(i) \geq \gamma^{\varphi_2}(i) \geq \gamma^{\varphi_3}(i). \tag{25}$$

If respondent i contradicts to this condition then we clear the label y_i and repeat the semi-supervised learning regarding such respondents as unlabeled from this time on. As shown in Fig. 4, we repeat this procedure until contradictions do not occur any more.

3 Experimental Results and Discussion

This section shows experimental results of the proposed methods. Sects. 3.1 and 3.2 show applications of the multiclass classification method proposed in Sect. 2.2 and the semi-supervised learning method proposed in Sect. 2.3.

3.1 Classification of Handwritten Digits

3.1.1 Handwritten Digits

We used the Pen-Based Recognition of Handwritten Digits dataset of the UCI Repository [15] as an example application for multiclass classification because digits have two-dimensional spatial features. The dataset consists of 10992 samples written by 44 people. Among these samples, 7494 samples were written by 30 people divided into learning data D_1 and validation data D_2. The 3498 remaining samples were written by 14 other people and are used as test data D_3. Eight points $\{\mathbf{r}_l\}$ dividing the orbit of the pen point into seven equally long segments were chosen. In this study, we carry out the feature extraction with GA, after computing $\mathbf{p}_l = \mathbf{r}_l - \bar{\mathbf{r}}$, i.e., setting the origin $\bar{\mathbf{r}}$ at the center of the digit.

3.1.2 Classification Result

For each feature extraction $f \in \{f_1, f_2, f_0\}$, the GMM learned from D_1 with four values of $\alpha \in \{0.1, 0.01, 0.001, 0.0001\}$, and we then evaluated the correct classification rate for D_2. Table 1 shows the results. In Table 1, the maximal classification rate is underlined. $\alpha = 0.01$ was therefore chosen for all feature extractions $\{f_1, f_2, f_0\}$.

Table 2 shows the results of the identified models M with $\bar{q} = \sum_j q_j/M$. The numbers of misclassifications via coordinates f_1, via outer products f_2, and via inner products f_0 were 100, 201, and 494, respectively. The correct classification rates were 97.14%, 94.25%, and 85.88%, respectively.

Tables 2 clearly show the differences in the misclassifications by different feature extractions. Because the feature extraction f_1 keeps the coordinate information of each point, it results in a smaller number of misclassifications. However, it has other disadvantages. For example, because the first point $\mathbf{p}_1$ of most learning data of '0' is in the top left area, some test data whose $\mathbf{p}_1$ is in the top right area are misclassified. The feature extractions f_2 and f_0, on the other hand, loose the coordinate information of each point. However, they extract partial shape features not affected by the position of the part of the digit under concern. Therefore, f_2 and f_0 correctly classified the '0' data with curves beginning in the top right area. Therefore, a mixture of different f_1, f_2, f_0 expert feature extractions works best. The total number

Table 1 Correct classification rate for D_2 for various α

α	f_1	f_2	f_0
0.1	93.22%	95.09%	84.08%
0.01	<u>99.36%</u>	<u>96.53%</u>	<u>90.13%</u>
0.001	99.28%	96.43%	89.44%
0.0001	98.88%	96.45%	89.78%

Table 2 Identified models for different features f_1, f_2, f_0

		'0'	'1'	'2'	'3'	'4'	'5'	'6'	'7'	'8'	'9'
M	f_1	6	7	5	5	5	8	7	6	8	9
	f_2	2	7	3	3	3	4	3	6	7	8
	f_0	2	8	9	4	4	8	6	6	5	5
$\bar{q}$	f_1	6	6.1	8.4	9.4	8.4	3	6.4	7.8	7.5	8.7
	f_2	7	4.2	5.3	6	7	4.7	5.3	5.5	5.1	5.5
	f_0	6.5	4.9	4.8	6.3	6	5.1	5.3	4.8	6.6	6.8

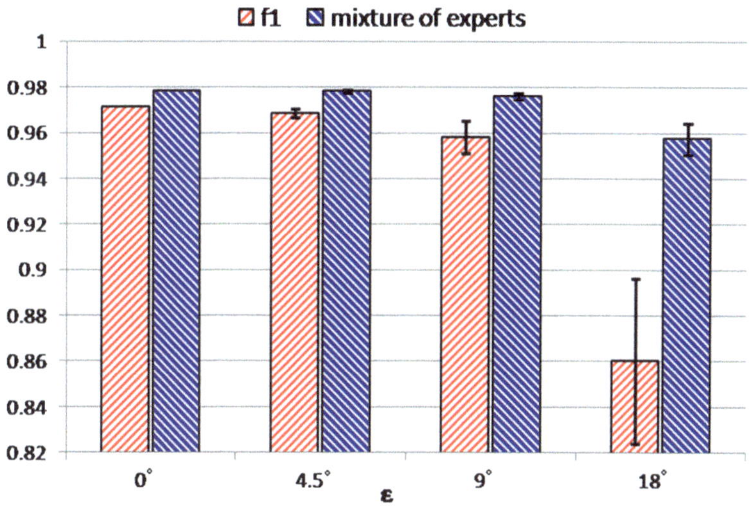

Fig. 5 Correct classification rate with f_1 and mixture of experts

of misclassifications using a mixture of experts was 75, and the correct classification rate was 97.86%, better than the result of only using f_1, f_2, or f_0.

We assumed cases of different measurement environments from the one in which both D_1 and D_2 have been measured. We generated a dataset $D'_3 = \{R(\xi, \varphi), \xi \in D_3, \varphi \sim \mathcal{U}(\varepsilon)\}$ that randomly rotated each digit of the test data D_3. $\mathcal{U}(\varepsilon)$ is the uniform distribution on $[-\varepsilon, \varepsilon]$, and φ is a random variable. We generated 20 different sets D'_3 from the test dataset D_3 for each $\varepsilon \in \{\pi/40, \pi/20, \pi/10\}$ and classified them. $R(\xi, \theta)$ rotates all the points of ξ by θ. Figure 5 shows the average and the standard deviation of the correct classification rate when using the feature extraction f_1 and the mixture of experts.

The classification precision using only feature extraction f_1 decreased remarkably with increasing ε. On the other hand, the classification precision using the mixture of expert did not decrease that much. The rotations had no influence in the cases of f_2 and f_0. Their classification success rates were 94.25% and 85.88%, respectively.

3.2 Kernel Alignment and Web Questionnaire Analysis Results

3.2.1 Web Questionnaire Data

We analyzed a web questionnaire data for a new product by extracting features with GA and finding the latent willingness to buy. Subjects answered ten questions about each of six objects, i.e., scenes in which the product was used. Subjects were asked to give an evaluation value to each question in five levels $\{1, 2, 3, 4, 5\}$, where "5" means "I agree very much" and "1" means "I disagree very much." Subjects answered willingness to buy the product for a price randomly selected from three prices.

We carried out the feature extractions with GA after subtracting 3 from all evaluation values so that $x_{l,i} = \{-2, -1, 0, 1, 2\}$. For simplicity, the five level willingness values were binarized: "5", "4"$\mapsto$ 1 and "3", "2", "1"$\mapsto$ -1.

3.2.2 Analysis Results

We calculated alignment of kernel matrix W_k derived by feature $\mathbf{f}_k$, $k = 0, 1, 2$, which extracted with GA. Table 3 shows kernel alignment with other feature kernels, identity matrix E, and the latent willingness matrix Y. The binary label $y_i = 1$ when the respondent had latent willingness as a result of three analyses via different feature and $y_i = -1$ otherwise, i.e., when the respondent was judged not having latent willingness at least one analysis.

With kernel matrix W_1, the kernel matrix W_2 had a smaller alignment (0.51) than W_0 had (0.92). Therefore, the effect of combination with feature extraction f_2 was better than with f_0. Also, the result showed that because both feature extractions f_0 and f_2 had alignment with E lower than f_1, they have more abundant structure of cluster between respondents, and they contribute the total inference because alignment with Y is higher than f_1.

Table 4 shows the result without introducing GA to find latent willingness to buy. In the table, "C" shows the percentage of respondents whose γ contradicted to condition (25) after the algorithm ended, and thus we ignore those respondents. "F–F" shows the percentage of respondents whose $y_i = -1$, where, for simplicity, we do not mind what price was indicated to the respondent, and $\gamma^{\varphi_1}(i) = -1$, i.e., the respondent did not have either apparent or latent willingness even if the price was the

Table 3 Kernel alignment evaluation

	W_0	W_1	W_2
W_0	1	0.92	0.63
W_1	0.92	1	0.51
W_2	0.63	0.51	1
E	0.71	0.82	0.36
Y	0.048	0.036	0.103

Table 4 Result without GA

	C	0.1%
f_1	F–F	39.9%
	F–T	22.0%
	T–F	6.7%
	T–T	31.3%

lowest. "F–T" shows the percentage of respondents whose $y_i = -1$ but $\gamma^{\varphi_1}(i) = 1$, i.e., the respondent answered not to have willingness, but he/she had latent willingness at least for the lowest price. "T–F" shows the percentage of respondents whose $y_i = 1$ but $\gamma^{\varphi_1}(i) = -1$, i.e., the respondent showed apparent willingness, but from the similarity of answering patterns he/she was not willing to buy. "T–T" shows the percentage of respondents whose $y_i = 1$ and $\gamma^{\varphi_1}(i) = 1$, i.e., the respondent had both apparent and latent willingnesses. The analysis made the following clear:

- Out of 38.0% of respondents ("T–F" or "T–T") who answered positively to the direct question of willingness, 31.3% of all respondents were detected as "truly" willing to buy ("T–T").
- Out of 61.9% of respondents ("F–F" or "F–T") who answered negatively to the direct willingness question, 22.0% of all respondents were detected as latently willing to buy ("F–T").

As a conclusion, 53.3% of respondents had a latent willingness to buy ("F–T" or "T–T"), and the other 46.7% respondents did not.

Next, we conducted a more detailed analysis introducing GA to define two more feature spaces which are based on f_0, f_2, respectively. Respondents were subdivided into the five groups of "C", "F–F", "F–T", "T–F", and "T–T" for each feature aspect. Thus Table 5 shows four subtables each of which further divides the corresponding respondents divided by f_1 to a matrix of judgements based on f_0 and f_2. Though 52.1% of respondents had the same result by all analyses based on f_0, f_1, f_2 (total of diagonal cells), the other 47.9% of respondents had different result. Especially,

- The second table shows that out of 22.0% of respondents who did not have apparent but had latent willingness according to analysis based on f_1 alone, only 1.3% of all respondents were judged as so according both to analyses based on f_0 and f_2.
- The bottom table shows that out of 31.3% of respondents who were judged as "truly" willing to buy in the analysis based on f_1 alone, 17.6% of all respondents were judged differently in at least one aspect of his/her answering pattern. On the other hand, the remaining 13.7% of all respondents can be judged as willing to buy with more confidence supported by the judgements based on f_0 and f_2.

As a conclusion, 15.0% of respondents were found to have latent willingness to buy ("F–T" or "T–T" by all analyses) with more confidence than in analysis without introducing GA.

Table 5 Detailed labeling result

f_1 (F–F) 39.9%		f_2 F–F	F–T
f_0	F–F	32.6%	0.2%
	F–T	7.0%	0.1%

f_1 (F–T) 22.0%		f_2 F–F	F–T
f_0	F–F	6.0%	0.3%
	F–T	14.4%	1.3%

f_1 (T -F) 6.7%		f_2 T–F	T–T
f_0	T–F	5.5%	0.1%
	T–T	0.9%	0.2%

f_1 (T–T) 31.3%		f_2 T–F	T–T
f_0	T–F	3.0%	0.9%
	T–T	13.7%	13.7%

Finally, we utilize principal component analysis (PCA) to visualize the data given by f_1. Figure 6 shows apparent and latent willingnesses.

The top figure shows apparent willingness to buy. In top figure, blue '$\times$' shows a respondent who did not have willingness even price φ_1, and red '$\circ$' shows a willing respondent in the case of price φ_1. The gray marks show apparent willingness in other case of prices.

The bottom figure show latent willingness to buy. In the bottom figure, blue '$\times$' and red '$\circ$' show the result with feature extractions f_1. Green '$\diamond$' show respondents who were judged as willing to buy with all feature extractions f_0, f_1, f_2. From Fig. 6 we can find that green '$\diamond$' in the bottom figure resembles distribution of blue '$\times$' in the top figure. This means that the proposed method can find respondents who have *strong* latent willingness.

4 Conclusions

In this study, we proposed systematic feature extraction methods by using GA. Based on the extracted features, we solved two machine learning problems, i.e.,

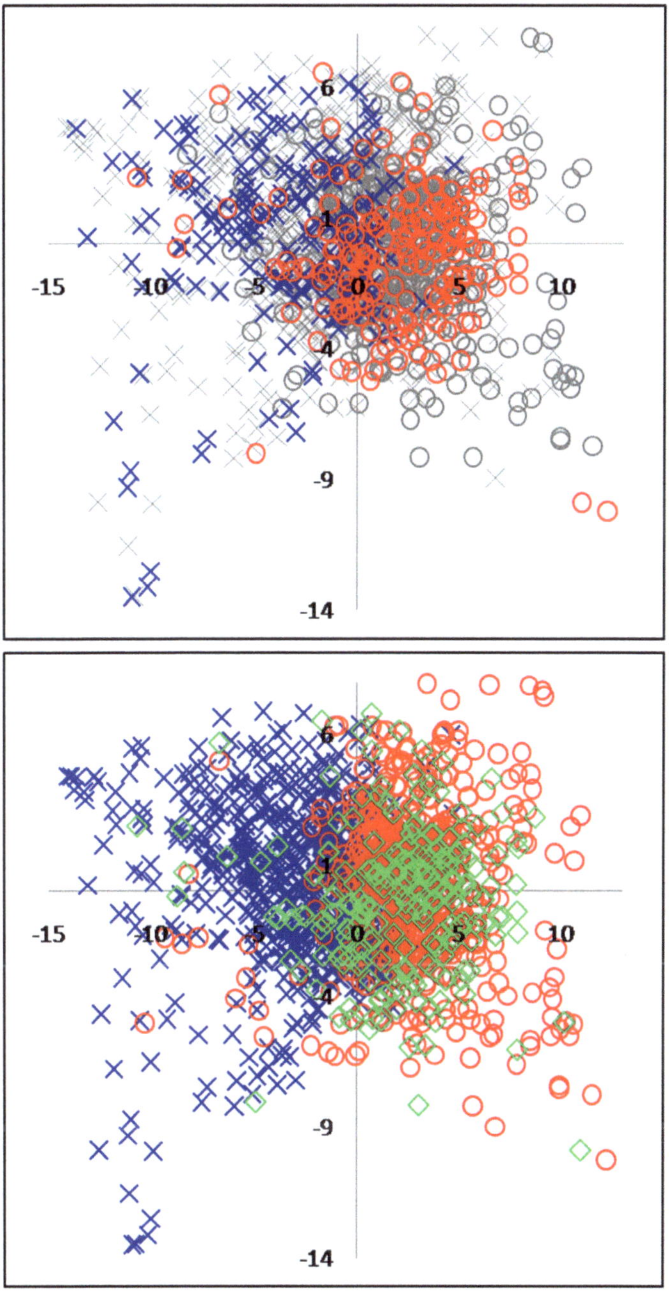

Fig. 6 The visualization of latent willingness to buy the product by PCA. *The top figure* shows apparent willingness, and *the bottom figure* shows latent willingness to buy. A respondent who did not have willingness is shown by '×'. A willing respondent is shown by '○'. *Green* '◇' in *the bottom figure* show respondents who were judged as willing to buy with all feature extractions f_0, f_1, f_2

the classification of the handwritten digits by using GMMs and the discovery of latent willingness to buy using semi-supervised learning.

We applied the proposed method to two-dimensional objects of handwritten digits deriving three ways of feature extraction, via coordinates, outer products, and inner products. When we assumed cases of different measurement environments, the classification success rate by pure coordinate value feature extraction dropped substantially for rotated test data. In contrast to this, with the mixture of experts, the classification success rate was not only higher in the case of a constant measurement environment; it was also much more stable in the case of large rotations of the test data. Therefore, we can confirm that the strategy to mix different GA feature extractions is superior in both classification precision and robustness when compared with pure coordinate value features, which is the most often used conventional method.

We also applied the proposed method of feature extraction to clustering of answering patterns for a web questionnaire. We proposed the clustering algorithm by using the result based on the similarity in each feature space. Then, we applied the proposed method to the clustering of answering patterns for a web questionnaire, deriving three kinds of feature extraction i.e., coordinates, outer products, and inner products. The result showed that feature extractions based on outer product and inner product, respectively, had more abundant structure of cluster between respondents and higher alignment with latent willingness to buy than in $(m \times n)$-dimensional vector space. Based on the extracted features, we found latent willingness to buy from the questionnaire data. The results showed that semi-supervised learning based on coordinates may detect respondents who had latent willingness to buy and that introducing GA to the analysis may further find respondents who have *strong* latent willingness.

Acknowledgements This work was supported by Grant-in-Aid for the 21st century COE program "Frontiers of Computational Science" (Nagoya University) and Grant-in-Aid for Young Scientists (B) #19700218.

References

1. Hu, M.K.: Visual pattern recognition by moment invariants. IRE Trans. Inf. Theory **8**(2), 179–187 (1962)
2. Mukundan, R., Ramakrishman, K.R.: Moment Functions in Image Analysis, Theory and Application. World Scientific, Singapore (1998)
3. Doran, C., Lasenby, A.: Geometric Algebra for Physicists. Cambridge University Press, Cambridge (2003)
4. Hestenes, D.: New Foundations for Classical Mechanics. Springer, Dordrecht (1986)
5. Dorst, L., Fontijne, D., Mann, S.: Geometric Algebra for Computer Science: An Object-oriented Approach to Geometry. Morgan Kaufmann Series in Computer Graphics. Morgan Kaufmann, San Mateo (2007)
6. Sekita, I., Kurita, T., Otsu, N.: Complex autoregressive model for shape recognition. IEEE Trans. Pattern Anal. Mach. Intell. **14**(4), 489–496 (1992)
7. Hirose, A.: Complex-Valued Neural Networks: Theories and Applications. Series on Innovative Intelligence, vol. 5. World Scientific, Singapore (2006)

8. Matsui, N., Isokawa, T., Kusamichi, H., Peper, F., Nishimura, H.: Quaternion neural network with geometrical operators. J. Intell. Fuzzy Syst. **15**(3–4), 149–164 (2004)
9. Buchholz, S., Le Bihan, N.: Optimal separation of polarized signals by quaternionic neural networks. In: 14th European Signal Processing Conference, EUSIPCO 2006, September 4–8, Florence, Italy (2006)
10. Nitta, T.: An extension of the back-propagation algorithm to complex numbers. Neural Netw. **10**(8), 1391–1415 (1997)
11. Hildenbrand, D., Hitzer, E.: Analysis of point clouds using conformal geometric algebra. In: 3rd International Conference on Computer Graphics Theory and Applications, Funchal, Madeira, Portugal (2008)
12. Hitzer, E.: Quaternion Fourier transform on quaternion fields and generalizations. Adv. Appl. Clifford Algebr. **17**(3), 497–517 (2007)
13. Sommer, G.: Geometric Computing with Clifford Algebras. Springer, Berlin (2001)
14. Dempster, A., Laird, N., Rubin, D.: Maximum likelihood from incomplete data via the EM algorithm. J. R. Stat. Soc. Ser. B **39**(1), 1–38 (1977)
15. Asuncion, A., Newman, D.J.: UCI Machine Learning Repository. University of California, School of Information and Computer Science, Irvine (2007)
16. Tipping, M.E., Bishop, C.M.: Mixtures of probabilistic principal component analysers. Neural Comput. **11**, 443–482 (1999)
17. Zhu, X., Lafferty, J., Ghahramani, Z.: Combining active learning and semi-supervised learning using Gaussian fields and harmonic functions. In: ICML 2003 Workshop on the Continuum from Labeled to Unlabeled Data in Machine Learning and Data Mining (2003)
18. Cristianini, N., Kandola, J., Elisseeff, A., Shawe-Taylor, J.: On kernel target alignment. J. Mach. Learn. Res. (2002)
19. Meinicke, P., Ritter, H.: Resolution-based complexity control for Gaussian mixture models. Neural Comput. **13**(2), 453–475 (2001)

QWT: Retrospective and New Applications

Yi Xu, Xiaokang Yang, Li Song,
Leonardo Traversoni, and Wei Lu

Abstract Quaternion wavelet transform (QWT) achieves much attention in recent years as a new image analysis tool. In most cases, it is an extension of the real wavelet transform and complex wavelet transform (CWT) by using the quaternion algebra and the 2D Hilbert transform of filter theory, where analytic signal representation is desirable to retrieve phase-magnitude description of intrinsically 2D geometric structures in a grayscale image. In the context of color image processing, however, it is adapted to analyze the image pattern and color information as a whole unit by mapping sequential color pixels to a quaternion-valued vector signal. This paper provides a retrospective of QWT and investigates its potential use in the domain of image registration, image fusion, and color image recognition. It is indicated that it is important for QWT to induce the mechanism of adaptive scale representation of geometric features, which is further clarified through two application instances of uncalibrated stereo matching and optical flow estimation. Moreover, quaternionic phase congruency model is defined based on analytic signal representation so as to operate as an invariant feature detector for image registration. To achieve better localization of edges and textures in image fusion task, we incorporate directional filter bank (DFB) into the quaternion wavelet decomposition scheme to greatly enhance the direction selectivity and anisotropy of QWT. Finally, the strong potential use of QWT in color image recognition is materialized in a chromatic face recognition system by establishing invariant color features. Extensive experimental results are presented to highlight the exciting properties of QWT.

1 Introduction

Quaternion wavelet transform (QWT) attracts increasing research interests recently as a new analysis tool for various image processing tasks. It is usually formulated as

L. Traversoni (✉)
Ciencias Básicas e Ingenieria, Univ. Autonoma Met. (Iztapalapa), 09340 Mexico, Mexico
e-mail: ltd@xanum.uam.mx

E. Bayro-Corrochano, G. Scheuermann (eds.), *Geometric Algebra Computing*,
DOI 10.1007/978-1-84996-108-0_13,

an extension of the real wavelet transform and complex wavelet transform (CWT) by using the quaternion algebra and the 2D Hilbert transform of filter theory. Typically, analytic signal representation is desirable to retrieve phase-magnitude description of intrinsically 2D geometric structures in a scalar image. The amplitude component reveals the energy of the filter response and therefore serves for the detection of events, while the phase component uncovers the type of the detected event and encodes the relative location of image structures [1]. In the applications involved in comparison of series of images, such as optical flow [1, 2] and stereo matching [3, 4], the phases of quaternion wavelet transformed image are very important to provide essential features and inherent 2D shift clues at corresponding points. In addition, the confidence map of the movement measurement can be built according to the complementary amplitude spectrum [4].

Meanwhile, some preliminary research works provide an insight of the use of QWT in vector signal representation, such as the spectrum analysis for quaternion-valued color image [5] and color pattern estimation [6, 7]. Compared with the traditional color image filtering techniques, which are commonly based on separate processing of the color components, quaternion filters can depict a color pixel as a whole unit, namely pure quaternion, and naturally compute transformation in three-dimensional color/vector space. This operation would make full use of inter-channel color information and efficiently suppress the artifacts. The pioneer work of Ell utilized quaternion Fourier transform (QFT) to treat color as a single entity and achieve higher color information accuracy [5]. To improve the strength of local quaternion filtering in color space, some face recognition systems defined a family of quaternion Gabors to extract local color features for high face recognition accuracy [6, 7]. Philippe Carré and Patrice Denis built a color quaternionic filter bank called the color Shannon wavelet based on a windowing process in the quaternionic Fourier space and established joint spatio-frequential representation of color images [8].

Aforementioned discussion demonstrates that QWT is a very useful image analysis tool and could be applied in extensive scalar/vector image processing tasks. This paper is motivated to give a suggestive reference for the use of QWT. It attempts to summarize the lessons from the QWT development experience and explores the potential applications of QWT. The remainder parts of this paper are structured as follows. Section 2 surveys the evolution of QWT and presents the basic principles of quaternion wavelet construction for analytic signal analysis. Section 3 indicates that the mechanism of adaptive scale representation of geometric features is important for image analysis, which is testified in two application instances of uncalibrated stereo matching and optical flow estimation. Sections 4 and 5 switch the focus to the potential use of QWT in two new applications, namely image registration and image fusion. As for image registration application, the quaternionic phase congruency model is defined to give an invariant feature detector in scale space. The accordingly extracted features are matched to robustly estimate the image affine transformation in registration task. With regard to image fusion application, we incorporate directional filter bank (DFB) into the quaternion wavelet decomposition scheme to enhance the direction selectivity and anisotropy of QWT. Consequently, the modified QWT scheme could provide a better representation of edges and textures. Section 6 materializes the strong potential use of QWT in color image recognition by establishing invariant color features in a chromatic face recognition system. Section 7 is devoted to conclusions and future work.

2 Evolution of Qwt and Principles of Quaternion Wavelet Construction

The notion of quaternion was introduced by Hamilton in 1843 [9]. Recently, a new research branch called quaternion Fourier transform (QFT) has been developed based on quaternion algebra and is important for quaternion linear time-invariant system analysis [10, 11]. In analogue to complex Fourier space, the earliest definition of QFT is the two-side form as follows [1]:

$$F^q(u,v) = \int_{-\infty}^{\infty}\int_{-\infty}^{\infty} e^{-iux} f(x,y) e^{-jvy}\,dx\,dy. \tag{1}$$

In fact, there are many variants of QFT, e.g., the right-side QFT

$$F^q(u,v) = \int_{-\infty}^{\infty}\int_{-\infty}^{\infty} f(x,y) e^{-\mu(ux+vy)}\,dx\,dy \tag{2}$$

where μ is a unit pure quaternion. Similar to 1D wavelets, the existing quaternion wavelets have been constructed as a family of functions based on windowing process in the quaternionic Fourier space [1–4, 8].

2.1 Evolution of QWT

Unlike Fourier basis functions, the locality of quaternion wavelet basis leads to sparse representation of singularity-rich signals by compacting the signal energy into a small number of coefficients. The generated spatial features associated with a given scale and spatial support form the foundation for the analysis of linear time-varying system, commonly the various procedures of local image analysis and estimation.

One of the first applications of quaternion wavelet is Bülow's work [1]. He defined the concept of quaternion Gabor, i.e., a Gaussian windowed basis function of the QFT, for use with scalar images. Nowadays, quaternion Gabors are the most commonly used quaternion wavelets and can be obtained from tensor product extension of complex Gabors in quaternion domain:

Quaternion Gabor:

$$G^q(\boldsymbol{x},\boldsymbol{u},m) = \frac{uv}{2\pi m^2\sigma_f^2} e^{-0.5(\frac{xu}{m\sigma_f})^2} e^{-0.5(\frac{yv}{m\sigma_f})^2} e^{-i2\pi ux} e^{-j2\pi vy}, \tag{3}$$

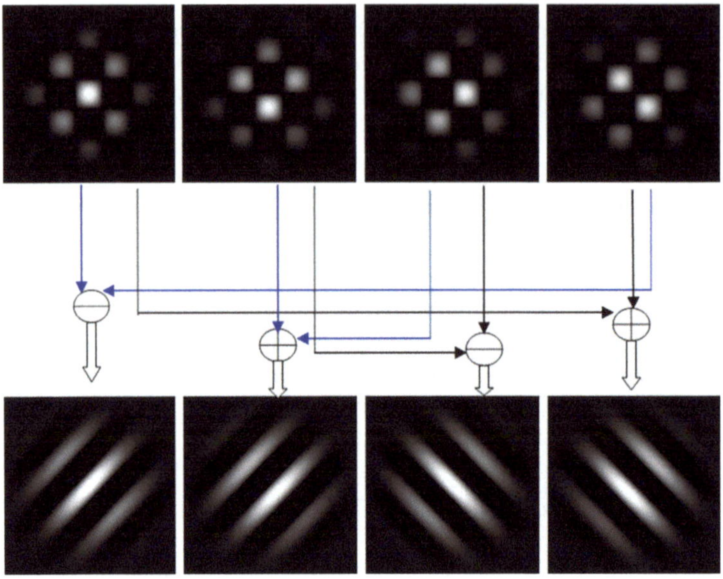

Fig. 1 Two complex Gabors (*bottom*) contained in a quaternion Gabor (*top*)

Complex Gabor:

$$G(\boldsymbol{x}, \boldsymbol{u}, m) = \frac{uv}{2\pi m^2 \sigma_f^2} e^{-0.5(\frac{xu}{m\sigma_f})^2} e^{-0.5(\frac{yv}{m\sigma_f})^2} e^{-i2\pi(ux+vy)} \tag{4}$$

where $\boldsymbol{u} = (u, v)^T$ is the radial center frequency of the filter, and the Gaussian envelope is truncated by the window $x \in [-\frac{m}{2u}, \frac{m}{2u}]$, $y \in [-\frac{m}{2v}, \frac{m}{2v}]$. Parameter m is the number of wavelength included in the window, and σ_f denotes the fraction of the window size that corresponds to one standard deviation of the Gaussian envelope along x, y directions. Since 99.7% of the envelope is located within three standard deviations from the origin, usually we select $\sigma_f = 1/6$. Fixing parameter and varying radial center frequency $\boldsymbol{u}$, we get a family of constant-octave quaternion Gabors. From (3) and (4) it is noted that two quaternion bases i and j ($i^2 = j^2 = k^2 = -1$ and $i \cdot j = k$) replace the single complex root i in the complex Gabor filter. Therefore the quaternion Gabor in (3) contains the complex Gabor in (4) and can generate another complex Gabor, which is the complementary part to represent a real 2D signal in the complex Fourier frequency domain, as shown in Fig. 1. The use of quaternion Gabors in biometrics is justified since the profiles of cortical receptive were found to strongly resemble the impulse responses of complex Gabors. Meanwhile, one can deduce that quaternion Gabors share with their complex counterparts the property of being jointly optimally localized in the spatial and frequency domains [1].

Similar to complex Gabor, a quaternion Gabor has side-slopes in negative frequency quadrants. In addition, a typical quaternion Gabor image analysis is noninvertible and expensive to compute due to the Fourier kernel. Some researchers built quaternion wavelets from tensor products of Kinsbury's q-shift Dual-Tree complex

(a)
(b)
(c)

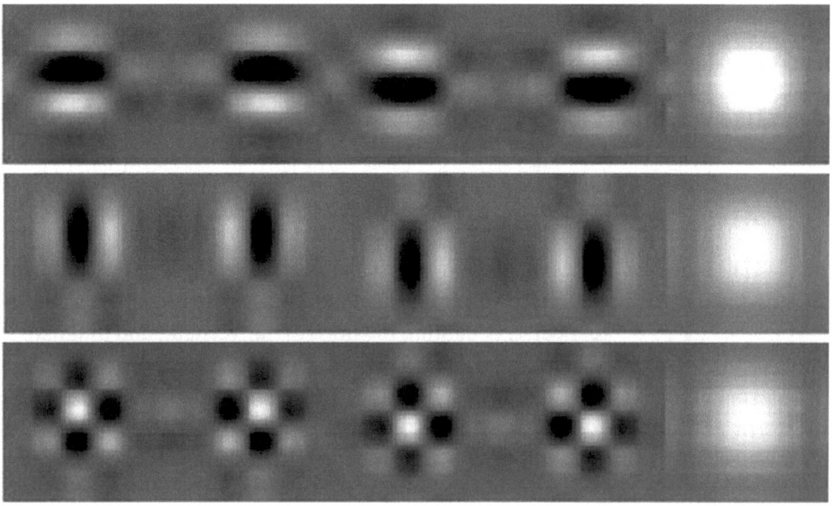

Fig. 2 Three quaternion wavelets built from Dual-Tree complex wavelet [12], capturing horizontal, vertical, and diagonal subbands respectively from row (**a**) to row (**c**) [3]. (From *left* to *right* at each row: the real part, three imaginary parts, and the quaternion magnitude)

(a)
(b)
(c)

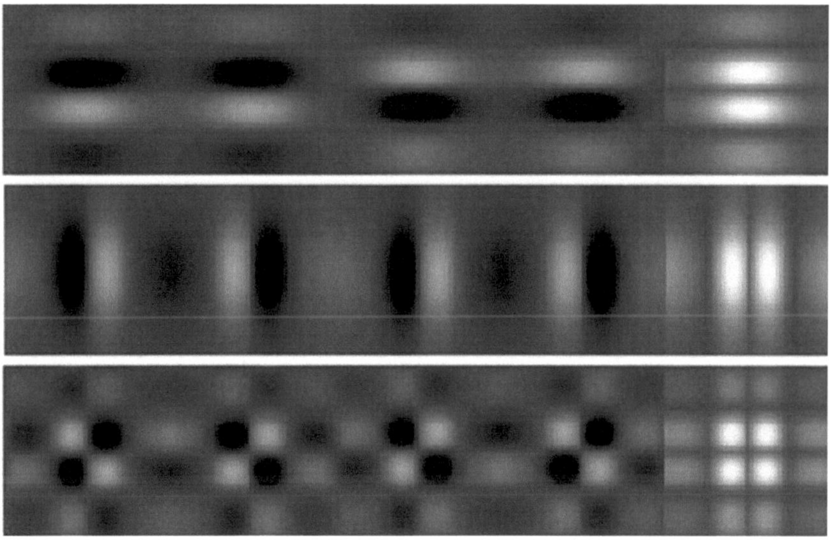

Fig. 3 Three quaternion wavelets built from biorthogonal wavelet basis [13] based on quaternion algebra and Hilbert transform, capturing horizontal, vertical, and diagonal subbands respectively from row (**a**) to row (**c**). (From *left* to *right* at each row: the real part, three imaginary parts, and the quaternion magnitude) [4]

wavelets [3] or from biorthogonal wavelet basis based on the quaternion algebra and 2D Hilbert transform [4]. These quaternion wavelets have been constructed through finite impulse response (FIR) filter banks and thus have a fast invertible implementation.[1] Moreover, these quaternion wavelets have the general formulation as a Hilbert quadruple:

$$\Psi_1^q(x,y) = \phi(x)\psi(y) + i\phi_H(x)\psi(y) + j\phi(x)\psi_H(y) + k\phi_H(x)\psi_H(y), \quad (5)$$

$$\Psi_2^q(x,y) = \psi(x)\phi(y) + i\psi_H(x)\phi(y) + j\psi(x)\phi_H(y) + k\psi_H(x)\phi_H(y), \quad (6)$$

$$\Psi_3^q(x,y) = \psi(x)\psi(y) + i\psi_H(x)\psi(y) + j\psi(x)\psi_H(y) + k\psi_H(x)\psi_H(y). \quad (7)$$

In (5)–(7), ϕ and ψ represent the low-pass and high-pass filter pair. Subscript H denotes the 1D Hilbert transform along the given axis. Therefore the quaternion wavelets Ψ_1^q, Ψ_2^q, Ψ_3^q respectively capture the horizontal, vertical, and diagonal subbands of the input 2D scalar signal.

In terms of analytic signal construction, QWT is an extension of the real wavelet transform and complex wavelet transform (CWT) by using the quaternion algebra and the 2D Hilbert transform of filter theory. The concept of the analytic signal is important in signal theory and introduced in 1946 by Gabor [14]. It makes the instantaneous amplitude and phase of local signal directly accessible. As the strengthening extension of CWT, QWT preserves the properties of CWT and adds new features by extending local signal phase from 1D complex phase to 2D quaternionic phase. QWT can realize the analysis of intrinsically 2D features (corner-like). In contrast, CWT provides a powerful tool in intrinsically 1D features (edge-like) analysis, while real wavelet transform cannot be exactly analytic. There were four fundamental local structures which could be distinguished from the 1D local phase, while we can find sixteen such structures using the 2D local phase, as compared in Figs. 4 and 5, where the phase value is determined by the signal and the filter.[2]

2.2 *Principles of Quaternion Wavelet Construction*

In the existing works, researchers have paid much attention to the particular use of quaternion wavelets and the comparison of QWT with DWT and CWT [2, 3]. The principles of quaternion wavelet construction are somewhat fragmentary to present in these works. To seek for such a guideline, we would investigate the desirable properties of the quaternion wavelets in this section.

Linear-Phase and Shift-Invariance Property

It is noted that the linear-phase property of quaternion wavelets is important for analytic signal representation, and thus no phase compensation is needed in multiscale signal decomposition. Besides the most common quaternion Gabors, the tensor

[1] Invertible quaternion wavelet transform (QWT) for quaternion-valued color images is still an open problem. Section 2 discusses the invertible QWT for scalar images.

[2] The filter's shape should match the local variation of the image structures.

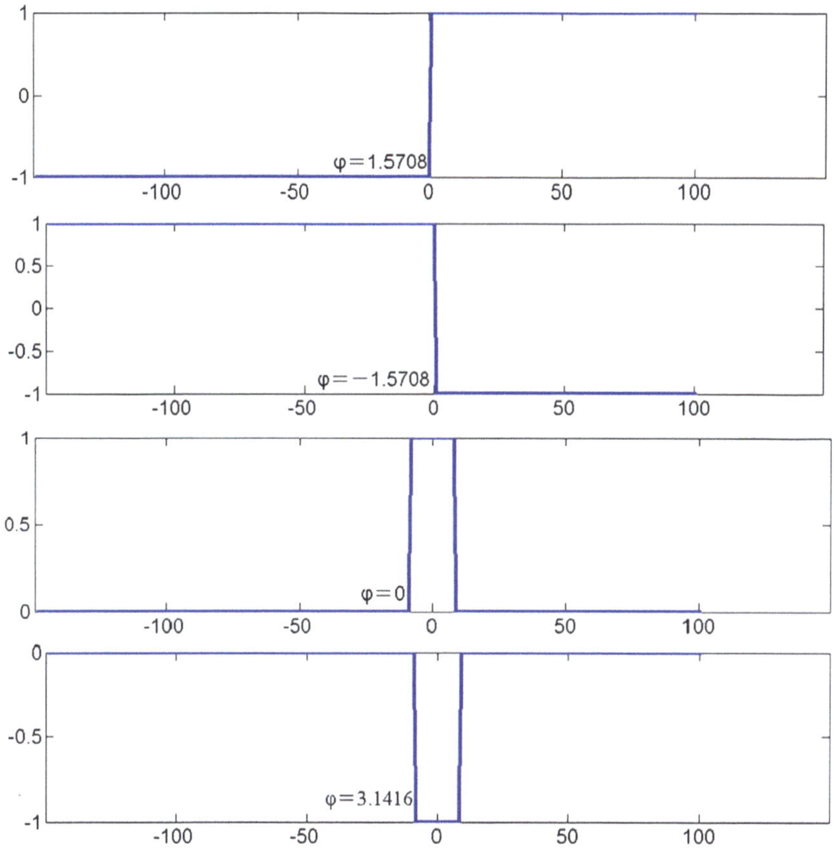

Fig. 4 Four fundamental 1D local structures and the local complex phases ϕ evaluated at the central points

products of linear-phase complex wavelets are usually exploited to build quaternion wavelets [3, 4]. To capture geometric image features nonoscillatorily, one important property of the filter is that a shift in the time domain should cause no change in the magnitude spectrum. As an instance to demonstrate the importance of shift-invariance property, Fig. 6 shows two examples of 1D shifted step responses. It is noted in (b) that the magnitude of DWT varies significantly across time-scale domain. The smoothly varied response in (a) indicates the near shift-invariance of Dual-tree CWT. Similarly, this is also a fundamental assumption for supporting QWT to resolve those problems involved in comparison of time-variant signals.

Hilbert Quadruple with no DC Response

Four real 2D functions $f_\eta,\ \eta \in \{1,2,3,4\}$, are called a Hilbert quadruple if

$$I\big[(f_k)_A^q\big] = f_\lambda, \qquad J\big[(f_k)_A^q\big] = f_\mu, \qquad K\big[(f_k)_A^q\big] = f_\nu, \tag{8}$$

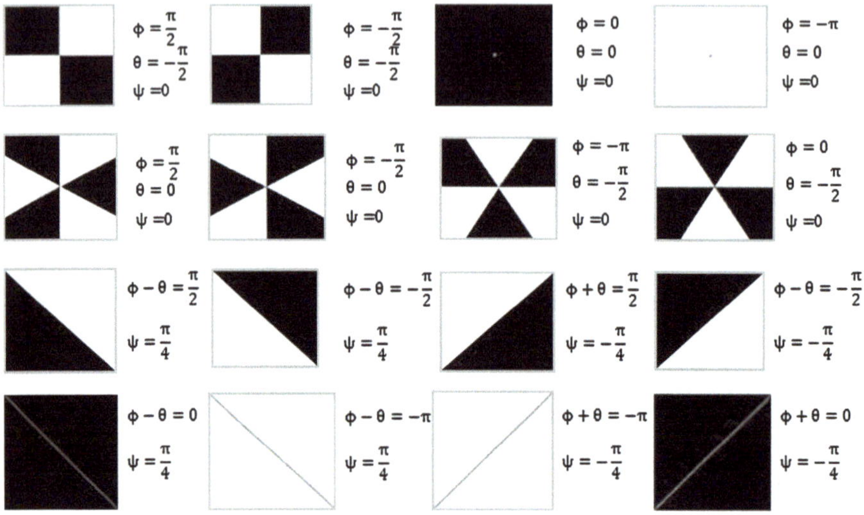

Fig. 5 Sixteen fundamental 2D structures and the local quaternionic phases $(\phi, \theta, \psi)^T$ evaluated at the central points

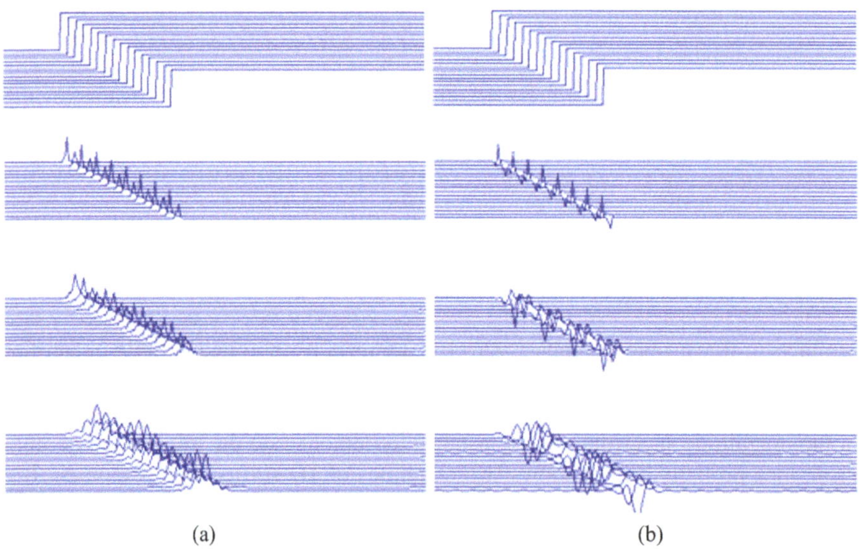

Fig. 6 (**a**) Magnitude of 1D shifted step response of Dual-Tree CWT at three successive scales. (**b**) Magnitude of 1D shifted step response of the real part of Dual-Tree CWT at three successive scales

for some permutation of pairwise different $k, \lambda, \mu, \upsilon \in \{1, 2, 3, 4\}$, where

$$f_A^q = f + i f_{Hi_x} + j f_{Hi_y} + k f_{Hi_{xy}}. \tag{9}$$

Operators $I(\cdot)$, $J(\cdot)$, $K(\cdot)$ in (8) respectively extract the three imaginary parts of the quaternion analytic signal $(f_k)_A^q$ that are in turn related to quaternion units i, j, k. Subscripts Hi_x, Hi_y in (9) respectively represent the partial Hilbert transform along x-axis and y-axis, while Hi_{xy} denotes the total Hilbert transform in $x-y$ coordinates. To constitute a bandpass analytic signal, we could use the following configuration:

$$f_A^q = \Psi_A^q \otimes f = (\Psi + i\Psi_{Hi_x} + j\Psi_{Hi_y} + k\Psi_{Hi_{xy}}) \otimes f. \quad (10)$$

Due to the commutability of convolution operation $\otimes$ and Hilbert transform, we can first build a quaternion wavelet from real-valued filters through Hilbert transform and then construct an analytic signal using convolution operator. The resultant analytic signal is supported only on the upper right quadrant ($u \geq 0, v \geq 0$). In addition, no DC response is expected in the filtering output. Because of the substantial power in natural signals at low frequencies, this DC sensitivity often introduces a positive bias in the real part of the response. This is the main reason why most of authors select the Gabors with small bandwidth (usually less than one octave) in 1D analytic signal construction.

Short-Length Filters with Good Localization in Space–Frequency Domain

It is generally accepted that the measurement of geometric image structures should require only local support in time–frequency domain. Quaternion Gabor filters are appropriate when one is interested in local spectral properties of a signal since they fulfill the uncertainty relation as an equality [1]. However, they are irreversible and usually require heavy computations, especially for the calculation of the quaternion Fourier transform. As a substitute for the quaternion Gabors, the methods in [3, 4] formed pixel-wise quaternion wavelets by imposing Hilbert transform respectively on q-shift orthonormal wavelets and biorthogonal wavelets. These relatively short filters would accelerate the measurement of geometric image structures. In the following experiments, this kind of quaternion wavelets is used to extract multiscale 2D phase structures.

3 The Mechanism of Adaptive Scale Representation in QWT

Current works have proved that the local quaternion phase is equivariant with the 2D spatial position in the scalar image [1, 2]. Thus it is possible to estimate the 2D disparity or optical flow field based on the local quaternionic phases [3, 4]. Usually, the phase-difference model is utilized to implement such tasks [1–3]. The main merit of the phase use is that the measurement is insensitive to illumination variations and geometric distortions [15]. However, most of existing phase-based matching methods have not dealt well with the issue of the inherent phase singularity. At the points where the amplitude falls to zero, phases are undefined, and thus disparity/optical flow estimation is unreliable. In the following content, we would point out that the mechanism of adaptive scale representation of geometric features is important to alleviate the phase singularity problem.

It is known that objects in the world appear in different ways depending on the scale of observation [16]. Here we use a set of quaternion Gabor filters as kernel functions to give the complete scale representation of the original signal. During the adaptation of the local scales of quaternion Gabor filtering, as shown in Fig. 7, the response at the given pattern reaches the maximum value at its characteristic scale and descends at neighboring scales. When the scale of filtering is rather far from the characteristic scale of the local pattern, the amplitude is expected to be very small, and thus the singularity problem arises. Therefore it is important to select local appropriate scales for further analysis of unknown image pattern.

To robustly model the correspondence of point sets in an image pair, which are captured at different time or views, our previous work proposed a phase-based data measure for assignment $\boldsymbol{a} = (p, q)$ using adaptive-scale quaternion wavelet kernel representation as below [4]:

$$\begin{aligned} M\big(\boldsymbol{a} = (p,q)\big) = \sum_{\sigma \in S} \sum_{p' \in \Omega_p, q' \in \Omega_q} & \rho_1(p';\sigma)\rho_2(q';\sigma)\big(\big|\big[\phi_1(p';\sigma) - \phi_2(q';\sigma)\big]_{2\pi}\big| \\ & + \big|\big[\theta_1(p';\sigma) - \theta_2(q';\sigma)\big]_{\pi}\big| \\ & + \big|\big[\psi_1(p';\sigma) - \psi_2(q';\sigma)\big]_{\pi/2}\big|\big), \end{aligned} \tag{11}$$

where S is the scale-space of the given quaternion filter set, Ω_p and Ω_q are the neighborhoods of the points p and q, respectively, ρ_1 and ρ_2 are the output amplitude spectrums of two images, which either are the left and right views in the stereo matching problem or the sample images captured at different time in the optical flow problem. The operator $[\Phi]_A$ extracts the principal value of Φ within the range of $[-A/2, A/2)$. It is noted that the phase pattern $(\phi, \theta, \psi)^T$ in a local neighborhood is directly used to set up the data measure and the absolute distance metric of two phase patterns is weighted by the amplitude. As a result, only the phase values extracted by the quaternion wavelet kernels that well match the characteristic scale of the given pattern with strong responses contribute to the data measure. Therefore the mechanism of adaptive-scale kernel representation efficiently alleviates the negative effects of phase singularity.

To further emphasize the importance of the mechanism of the adaptive scale kernel representation, we give two comparison experiments to demonstrate the improvements introduced by this mechanism. As shown in Figs. 8 and 9, we compare our computational model with Chan's phase-difference model [3] for 2D shift estimation between images, where the latter one is typical of the existing phase-based method [1, 2].

Chan [3] used the phase-difference model to compute the disparity/optical flow field. The phase singularities are detected and removed by threshold constraints of magnitude and local frequency. Then the shift estimation would be given up if the local structure cannot be captured under the given scale of the filtering. In his pyramid decomposition structure, these holes would outspread from coarse scales to fine

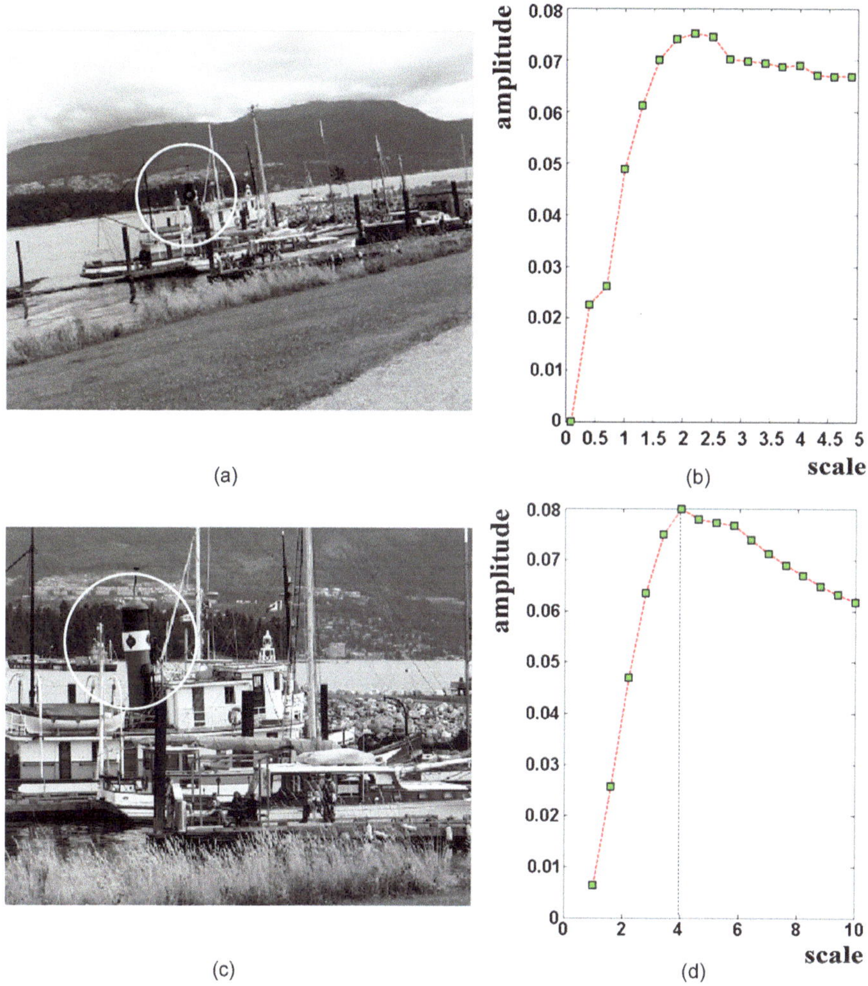

Fig. 7 Characteristic scales of two image patterns at the circle centers. (**a**), (**c**) Two images taken with different zoom. (**b**), (**d**) The responses of the quaternion Gabors in scale space at two circle centers

scales. In our model, we introduce the mechanism of the adaptive-scale kernel representation and select two multiscale decomposition schemes to testify its validity. For the disparity estimation, the pyramid decomposition structure is used to get the multiscale phase structures. The phase singularities have chances to revival once the disparity computation reaches their characteristic scales. In addition, the active points with high amplitude serve an important role of propagating disparities within their neighborhood Ω. Thus the singularities are still assigned a good match with high probability as long as enough active points are distributed around. From coarse scales to fine scales, the inactive points/singularities and the active points might exchange their roles according to the approximation of the current scale to their

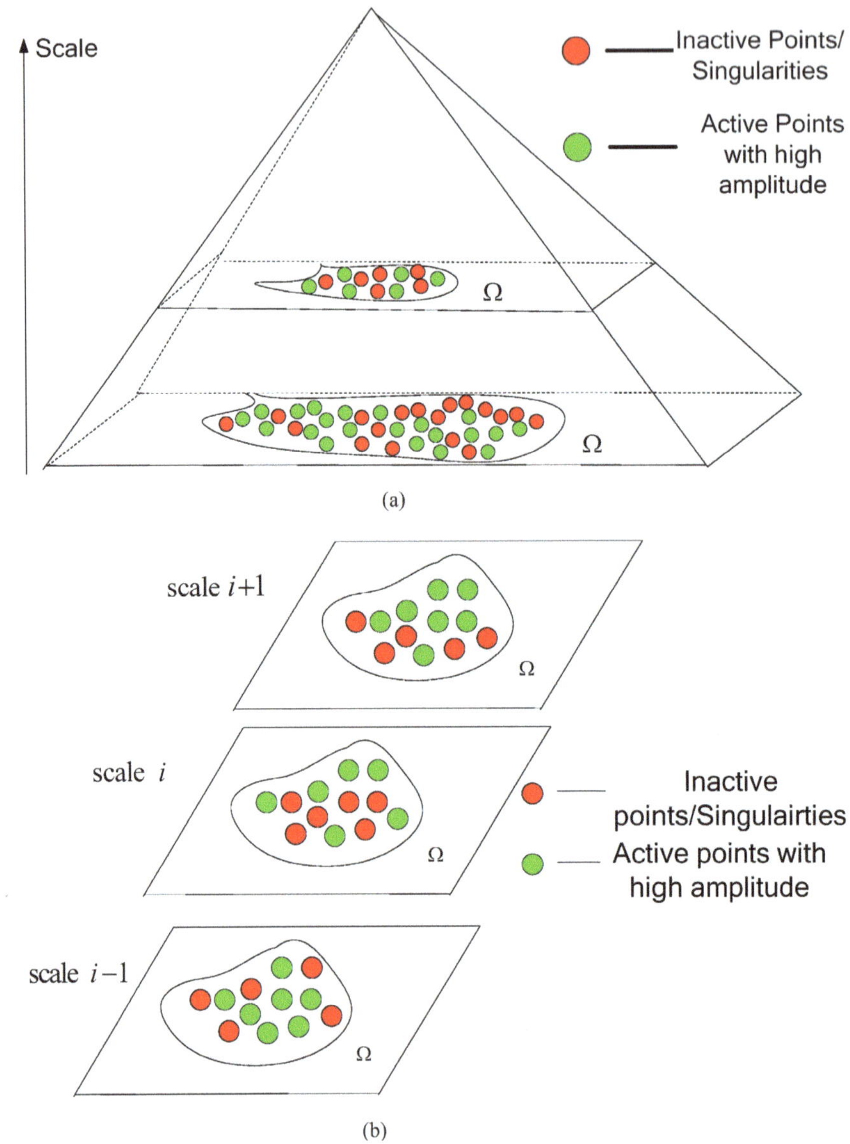

Fig. 8 The transit of active points under the adaptive-scale kernel representation

characteristic scales (as shown in Fig. 8(a)). For the optical flow computation, we select a set of quaternion Gabors to successively cover the characteristic scales of image structures. At each point, the phase structures are extracted by adaptive-scale kernels, resulting in low density of phase singularities (as shown in Fig. 8(b)).

As shown in Fig. 9, the correct depth clues are obtained in our model, while it is difficult to distinguish objects of varied depth in Chan's results under large disparity

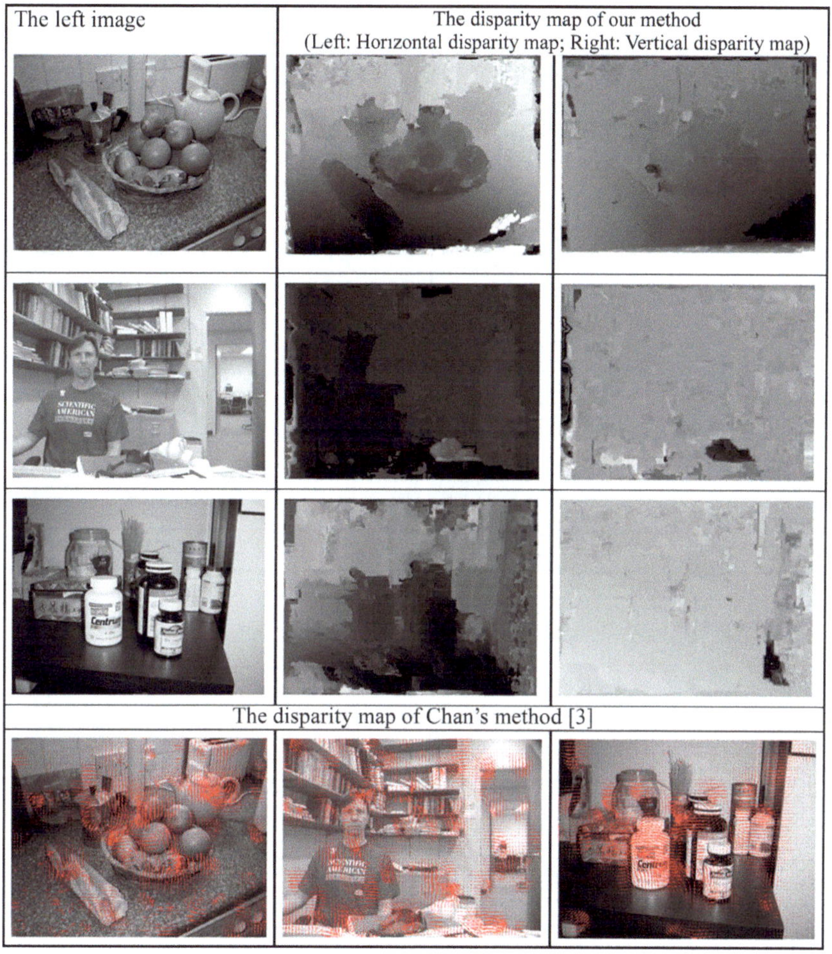

Fig. 9 Comparison of two computational models for disparity estimation

range. Less depth clues can be observed from vertical disparity map due to the uniform distribution. According to the movement distribution in ground truth image shown in Fig. 10, distinct improvement can also be observed in our optical flow field as a comparison with Chan's results in the "Yosemite" sequence. We only use two Gabors in the optical flow estimation to demonstrate the improvement introduced by the adaptive-scale kernel representation. Better performance can be gained when more successive-scale Gabors are used.

4 The Potential Use of QWT in Image Registration

It has been shown in Sect. 3 that quaternionic phases were a reliable tool to settle the correspondence problem. Similar to most of existing matching methods, there is

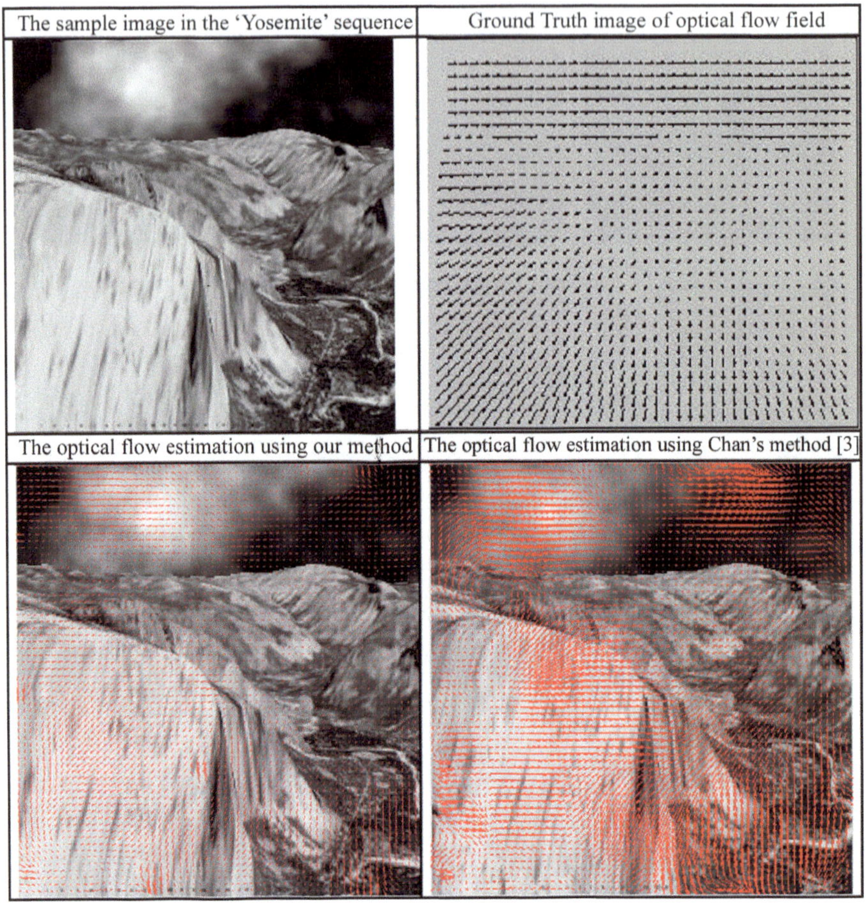

Fig. 10 Comparison of two computational models for optical flow estimation

an implicit assumption of weak foreshortening between the images. Thus the phase structures can be compared at the similar scale in two views. However, this assumption would be violated in wide baseline stereo, 3D model alignment, and creation of panoramic views due to the two cameras far apart or with a large vergence angle. The issue of image registration should be considered as one of these problems. Currently, the concept of affine invariant features is proposed to deal with the distinct affine deformations under changing viewpoint [17]. In this section, we establish the quaternionic phase congruency model to detect affine-invariant features. Combining it with the scale invariant descriptor, it provides encouraging matching results in image registration task.

The measurement of significance is important for feature detectors. We always expect only those highly distinguished features to be used in image registration. Phase congruency model in complex wavelet domain has already been proved as a dimensionless measure of feature significance and demonstrates its superior behavior in noise and illumination invariance [18]. We extend this model to quaternion wavelet domain and reveal the tighter constraint implied in the quaternionic phase congruency model.

Here we first analyze the phase congruency model in the quaternionic Fourier domain; then the conclusions can be directly extended to the quaternion wavelet domain since it can be considered as the windowed version of the former one. Tracing back to (1), we give the definition of phase congruency function in quaternionic Fourier domain as the extension of Morrone and Owens's work [19]

$$F^q(u,v) = \int_{-\infty}^{\infty}\int_{-\infty}^{\infty} e^{-iux} f(x,y) e^{-jvy}\,dx\,dy. \tag{1}$$

As for the real 2D signal $f(x,y)$, (1) can be written as

$$F^q(u,v) = \int_{-\infty}^{\infty}\int_{-\infty}^{\infty} f(x,y) e^{-i\phi(x,y)} e^{-j\theta(x,y)}\,dx\,dy, \tag{12}$$

where $\phi(x,y) = 2\pi ux$ and $\theta(x,y) = 2\pi vy$. We denote $\boldsymbol{\Phi}$ as $(\phi,\theta)^T$; then the definition of quaternionic phase congruency function at point $\boldsymbol{x} = (x,y)^T$ is based on the function Φ:

$$PC(\boldsymbol{x}) = \frac{\sum_n A_n \cos((\Phi_n(\boldsymbol{x}) - \bar{\Phi}(\boldsymbol{x})))}{\sum_n A_n}, \tag{13}$$

where A_n and Φ_n respectively represent the amplitude and the phase vector of the nth scale component in the quaternion space. The function $\bar{\Phi}(\boldsymbol{x})$ denotes the phase vector of the composite vector formed from all the scale components. As follows, we reveal the tighter constraint implied in the quaternionic phase congruency model as compared with the one in CWT domain. An arbitrary 2D signal can be decomposed into its symmetrical components as

$$\begin{aligned}
f_{ee}(x,y) &= \frac{1}{4}\big(f(x,y) + f(-x,y) + f(x,-y) + f(-x,-y)\big),\\
f_{oe}(x,y) &= \frac{1}{4}\big(f(x,y) - f(-x,y) + f(x,-y) - f(-x,-y)\big),\\
f_{eo}(x,y) &= \frac{1}{4}\big(f(x,y) + f(-x,y) - f(x,-y) - f(-x,-y)\big),\\
f_{oo}(x,y) &= \frac{1}{4}\big(f(x,y) - f(-x,y) - f(x,-y) + f(-x,-y)\big).
\end{aligned} \tag{14}$$

As we know, the complex Fourier transform of a real 2D signal is formulated as

$$F(u,v) = \int_{-\infty}^{\infty}\int_{-\infty}^{\infty} f(x,y) e^{-i(ux+vy)}\,dx\,dy. \tag{15}$$

According to the symmetry components in QFT and CFT, it is noted that the quaternion Fourier kernel can be expanded into four individual oscillation terms:

(a) Original ‘Mandrill’ image (b) CWT phase congruency (c) QWT phase congruency

Fig. 11 The comparison of quaternionic phase congruency image and complex phase congruency image

$\cos\phi\cos\theta$, $-\sin\phi\cos\theta$, $-\cos\phi\sin\theta$, and $\sin\theta\sin\phi$, which are combined into two terms $\cos(\phi+\theta)$ and $\sin(\phi+\theta)$ in the complex Fourier kernel. As a result, we can get four symmetrical components from QFT while two symmetrical components from CFT of a 2D real signal. Therefore our quaternionic phase congruency function is defined based on four symmetrical components, where the extreme value is acquired only when both ϕ and θ are congruent across scales. However, complex phase congruency is defined based on two symmetry components which are combined from the former four components, where the extreme value is acquired as long as $\phi+\theta$ are congruent across scales. So a tighter constraint is contained in the quaternionic phase congruency model in contrast to the complex phase congruency model. For good localization, we introduce the model of quaternionic phase congruency to QWT, and, as an example, we compute the quaternionic phase congruency image using a set of quaternion Gabors which are tuned to four scales and six orientations. As shown in Fig. 11, only the features with higher local energy are emphasized in the quaternionic phase congruency. By the way, we adopt Kovesi's idea [18] to impress the ill-effects of noises in quaternion domain.

Based on the quaternionic phase congruency model shown in (13), we propose a wide-baseline stereo matching method as depicted in Fig. 12. In terms of the existing affine-invariant feature matching methods [17], the performance is affected dominantly by the robustness of the extreme point/region detectors in the joint time–scale space. Fortunately, the quaternionic phase congruency provides important clues for such points/regions. According to Morron's work, the extreme points of complex phase congruency are consistent with the extreme points of local energy [19]. This conclusion can be directly extended to quaternion Fourier domain due to the similar formulations in (12) and (15). The congruency of the quaternion phase vectors also imply the reliability of the local structure in current successive scales. Hence, we select the extreme points in the congruency image as the candidate features. To capture more features at varied scales, we subsample the original image pair into pyramidal structures. On each layer of the pyramid, we employ a set of quaternion Gabors to obtain the phase congruency image, where the extreme-valued points are assigned with the median scale of the given Gabor filter set as their characteristic

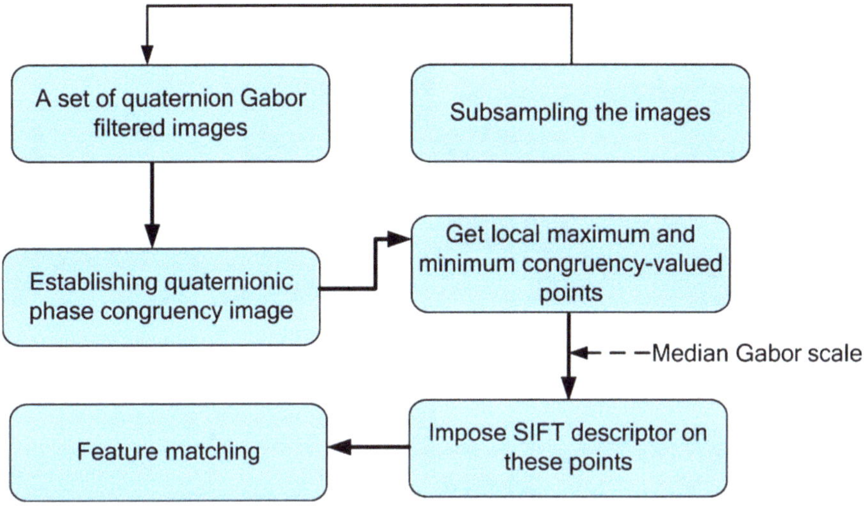

Fig. 12 The wide-baseline matching scheme of invariant features obtained from quaternionic phase congruency model

scales. The ratio of the maximum scale to the minimum scale is fixed at 2 for each Gabor set. Referring to the state-of-the-art of wide-baseline stereo matching methods, we use the best ranked SIFT descriptor [20] to match these features. In the matching experiments, we even select the minimum scale of the given Gabor group as the characteristic scales of the extreme congruency-valued points and find that the matching results are still encouraging under distinct affine distortions, as shown in Fig. 13(a). Here we only show nine point pairs for visual evaluation. In fact, much more good matches can be obtained. To further testify our matching scheme, we compute the fundamental matrix using the robust estimator in [21] based on the matching results and demonstrate three corresponding epipolar lines, as shown in Fig. 13(b).

A further issue which has to be considered is the affine invariance of these feature matching results. Since it is impossible to completely avoid outliers, here we exploit the popular LMeds method [21] to remove them and reliably estimate the homography matrix between the corresponding points. Then we construct the mosaic images based on homography matrices to do a visual inspection. Two image mosaicing tests have been performed, and the related homography matrices are listed in Fig. 14. It can be noticed that these homography matrices take into account projective distortions (the nonzero values in the last entry).

5 The Potential Use of QWT in Image Fusion

As the tensor product of two separable shift-invariant CWTs, the quaternion wavelet transform constructed in Sect. 2 is nearly shift-invariant. However, the separability

(a) The corresponding points.

(b) The corresponding epipolar lines.

Fig. 13 The related matching results obtained by SIFT descriptor of the extreme phase-congruency-valued points in QWT domain

of QWT leads to bad directionality as shown in Figs. 1–3. In this section, we introduce the directional filter bank (DFB) into QWT to get a better representation of edges and textures.

It is known that contourlet transform (CT) provides features with high directionality and anisotropy [22]. The architecture of CT is shown in Fig. 15, where the main components are a Laplacian pyramid (LP) and a directional filter bank (DFB). The LP decomposition at each step generates a sampled lowpass version of the original and the difference between the original and the prediction, resulting in a bandpass image. The process can be iterated on the coarse version. The DFB is designed to capture the high-frequency components (representing directionality) of images. Those bandpass images obtained in LP can be fed into DFB so that directional information can be captured efficiently. The architecture of CT allows for a different number of directions at each scale. Thus, we can obtain a better representation of edges and textures.

Since edges and textures are an intrinsic information in image representation, it is important for QWT to enhance its directionality and anisotropy. Since QWT is able to get multiscale bandpass images, we only introduce DFB in CT scheme into QWT and observe what would happen then. To testify whether the hybrid transformation scheme "QWT + DFB" better localizes the high-frequency components, we propose a new fusion method based on this hybrid transformation scheme, as shown

Image pairs	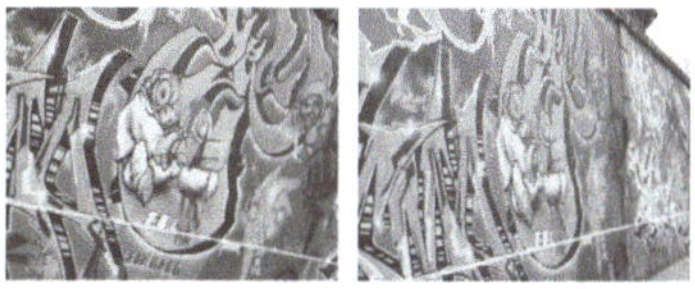	
Mosaicing results		
Homography Matrix	$\begin{pmatrix} 0.807 & -0.190 & -27.425 \\ 0.189 & 0.484 & -36.267 \\ 0 & 0 & 0.616 \end{pmatrix}$	$\begin{pmatrix} 0.460 & 0.004 & -125.42 \\ 0.054 & -0.574 & -3.960 \\ 0 & 0 & -0.617 \end{pmatrix}$

Fig. 14 Two image mosaicing tests and corresponding homography matrices

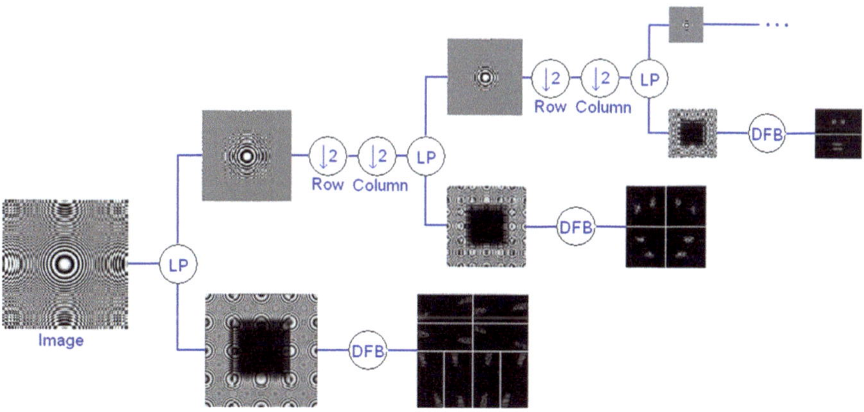

Fig. 15 The architecture of CT

in Fig. 16, where IDFB and IQWT respectively denote the inverse DFB scheme and the inverse QWT scheme.

A simple energy-based fusion method is adopted in the fusion tests, where the larger coefficients are preferred for generating the fusion result across scales except that at the coarsest scale the coefficients are averaged. The test images consist of parts with different resolutions. The corresponding fusion results are illustrated in Fig. 17. Due to the invertibility and near shift-invariance of QWT and DFB, the proposed hybrid transformation scheme runs well in the image fusion task. It is distinct that the fusion results of "QWT + DFB" scheme represent the high-frequency

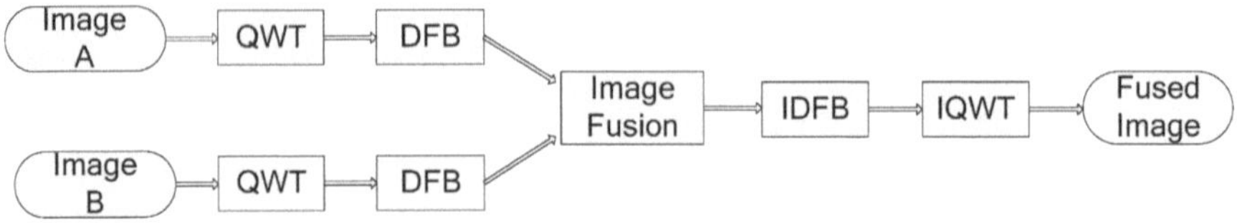

Fig. 16 The scheme of the proposed image fusion method

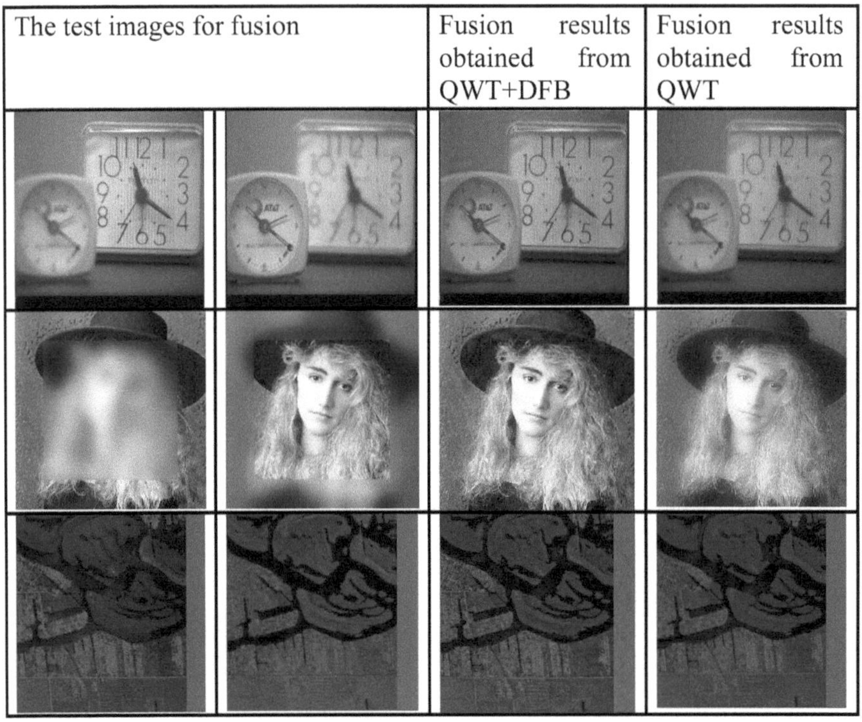

Fig. 17 Comparison of image fusion results

components of image pairs much better in contrast to those of QWT scheme. The potential of QWT might be further enhanced by selecting proper fusion methods. Here we mainly focus on the practicability of QWT in image fusion task.

6 The Potential Use of QWT in Color Image Recognition

A major motivation for the study of quaternion Fourier transforms (QFT) is that they handle color image pixels as vectors and thus offer scope to process color images holistically, rather than by separating luminance and chrominance or color space components. Nowadays QFT finds its extensive use in color image processing, such as hypercomplex spectrum filtering [5], color image registration [23], and color watermarking [24]. Similar to complex Fourier transform, QFT looses the notion of chronology in the frequency domain. The researchers hold high hopes on developing QWT into a useful tool for local color image analysis. Until now the invertible QWT for multiscale vector-valued color image analysis is still an open issue [8]. The most commonly used quaternion wavelets in color image processing are quaternion Gabors, which can achieve joint spatial-frequency representation along a given direction in color space [6, 7]. In this section, we show how to use quaternion Gabors to establish invariant color features, which have been successfully applied in our color face recognition system [7].

There are many variants for the definition of quaternionic Gabors. In Jones's work [6], it is formulated as

$$F_h(\vec{x}) = \mu \frac{\|\vec{k}_h\|^2}{\sigma^2} e^{-\frac{\|\vec{k}_h\|^2 \vec{x}^2}{2\sigma^2}} \left[\cos(\vec{k}_h \vec{x}) - e^{-\frac{\sigma^2}{2}}\right] \tag{16}$$

so that it maps to pure quaternionic values, conceptually pointing in a particular direction in three-channel color space. In (16), $\vec{x}$ is the 2D spatial location vector, $\vec{k}_h$ indicates the direction of the sinusoidal oscillation of the filter, and μ is a unit pure quaternion which represents a specific direction in RGB color space. To extract multiscale and multioriented features, we define a quaternion Gabor set with constant octave band,

$$F_{u,v}(x, y) = \mu \frac{f_m^2}{2^u \sigma^2} e^{-\frac{f_m^2}{2^{u+1}\sigma^2}(x^2+y^2)} \left[e^{\frac{f_m}{\sqrt{2^u}}(x \cos\frac{v}{8} + y \sin\frac{v}{8})} - e^{-\frac{\sigma^2}{2}}\right], \tag{17}$$

where parameters σ and f_m respectively denote the standard deviation of the Gaussian envelope and the center frequency at the finest scale. Here we select $u = 0, 1, \ldots, 4$ and $v = 0, 1, \ldots, 7$ to uniformly partition the scale and the orientation scope, whereby it results in image decomposition at five scales and eight orientations. Because of the substantial power in natural signals at low frequencies, this DC sensitivity is eliminated by the term in the square bracket to avoid a positive bias of the response. In order to find an adaptive selection of μ and thus realize homogeneous projection in the color vector space, the computation of μ depends on quaternion principal component analysis (QPCA) and is summarized as follows:

Step 1. Provide the covariance matrix S for the three color planes in the original image,

$$S = \frac{1}{M} \sum_m \begin{bmatrix} \boldsymbol{P}_r \\ \boldsymbol{P}_g \\ \boldsymbol{P}_b \end{bmatrix} \begin{bmatrix} \boldsymbol{P}_r \\ \boldsymbol{P}_g \\ \boldsymbol{P}_b \end{bmatrix}^r - \frac{1}{M^2} \begin{bmatrix} \sum_m \boldsymbol{P}_r \\ \sum_m \boldsymbol{P}_g \\ \sum_m \boldsymbol{P}_b \end{bmatrix} \begin{bmatrix} \sum_m \boldsymbol{P}_r \\ \sum_m \boldsymbol{P}_g \\ \sum_m \boldsymbol{P}_b \end{bmatrix}^r, \tag{18}$$

where M is the number of color pixels; $\boldsymbol{P}_r$, $\boldsymbol{P}_g$, and $\boldsymbol{P}_b$ are respectively the normalized pixel value matrix of red, green, and blue channel.

Step 2. Impose QPCA on the covariance matrix S and determine the pure quaternion μ as the principal eigenvector of S. Once the color axis μ is determined, quaternion Gabor Features are extracted by the convolution of the original color image $I^q(x, y)$ and the given quaternionic Gabor kernel $F_{u,v}(x, y)$. Here color pixel values are represented as pure quaternion $I^q(x, y) = I_r(x, y)i + I_g(x, y)j + I_b(x, y)k$,

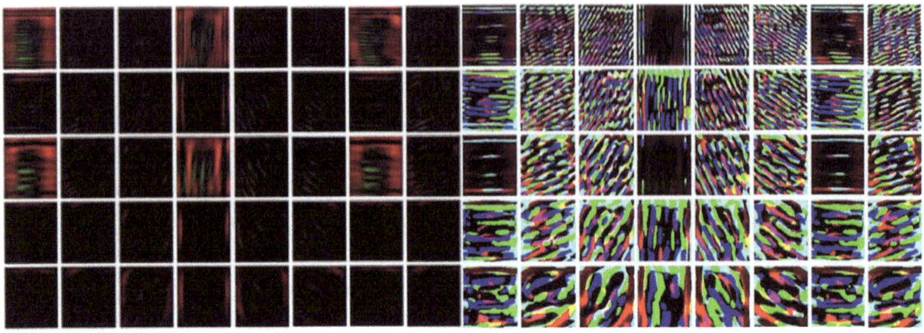

Fig. 18 Quaternion Gabor Feature images: (**a**) magnitude; (**b**) argument

where the r, g, and b subscripts denote the red, green, and blue color channels, respectively. No matter what transformation happens in the RGB vector space between two corresponding color face images, we always depict the convolution output $O^q(x, y) = Iq(x, y) \otimes F_{u,v}(x, y) = a + b \cdot i + c \cdot j + d \cdot k$ as one scalar part and one vector part, i.e., the real part $\|O^q_{\|\mu}(x, y)\|$ and the imaginary part $\|O^q_{\perp\mu}(x, y)\|$, where the former one is the negative projection of $O^q(x, y)$ to the principal color component μ of the original image, while the latter one is orthogonal to color axis μ.

Two kinds of quaternion Gabor feature images are established according to the convolution result. The magnitude color image is the one-to-one mapping from imaginary coefficients b, c, and d to red, green, and blue channels. Another color image is set up from the arguments between the imaginary parts and the real part, that is,

$$\varphi_1 = \arctan(b/a), \tag{19}$$

$$\varphi_2 = \arctan(c/a), \tag{20}$$

$$\varphi_3 = \arctan(d/a), \tag{21}$$

where $\arctan(\cdot) \in [-\pi, \pi)$. One example of quaternion Gabor feature images is illustrated in Fig. 18. There are eight orientations along each row and five scales along each column for two kinds of images.

It can be deduced that argument image is robust to the illumination contrast variations. Since nonparametric histogram method could enhance the robustness of features against variations of pose, illumination, and expressions in face dataset, we use Local Binary Pattern (LBP) histogram [25] to establish local invariant descriptors for quaternion Gabor features. These descriptors are named here as local quaternionic Gabor binary pattern (LQGBP) descriptors that encode the relationship between the quaternion Gabor features. We unfold the construction procedure of LQGBP descriptors as follows:

Step 1. Generate quaternion Gabor feature images at five scales and eight orientations, as shown in Fig. 18.

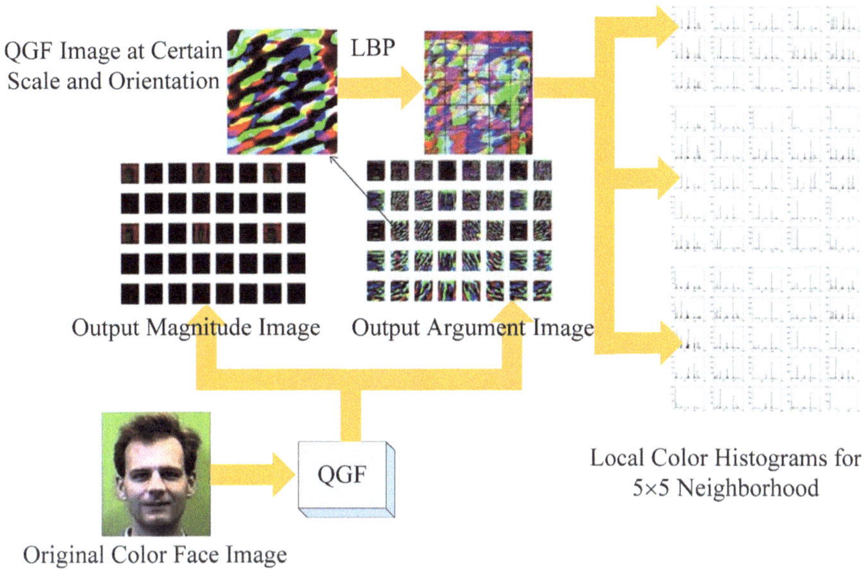

Fig. 19 The configuration of LQGBP-based color face recognition system

Step 2. Each quaternion Gabor feature image $f(c)$ is further mapped to $f'(c)$, where $f'(c)$ is obtained by the following rule:

$$A(c_1) = \begin{cases} 1 & \text{if } f(c_1) \geq f(c) \text{ and } c_1 \subset \Omega_c, \\ 0 & \text{otherwise,} \end{cases} \tag{22}$$

$$f'(c) = \sum A(c_1) \times 2^i, \tag{23}$$

where c indicates the position of image pixel, and Ω denotes a neighborhood system.

Step 3. The output quaternion Gabor feature images after local binary mapping are divided into 8×8 blocks and represented by the stacks of the local histograms of these subregions. These local histograms contribute to a global invariant descriptor of quaternion Gabor features. For two magnitude/argument quaternion Gabor feature images, the similarity measure E is defined as follows:

$$E = \sum_{s,o,c} (H_1 \cap H_2)/N, \tag{24}$$

$$N = 6N_p \times N_s \times N_o, \tag{25}$$

where the subscript of H is used to indicate the local histogram from different face images; symbols s, o, and c further point out at which scale, orientation, and position histogram H is computed; N_p is the number of pixels in the original image; N_s and N_o respectively compute the number of scales and orientations; the operator $\cap$ retains the smaller individual of the two matrices. The flowchart shown in

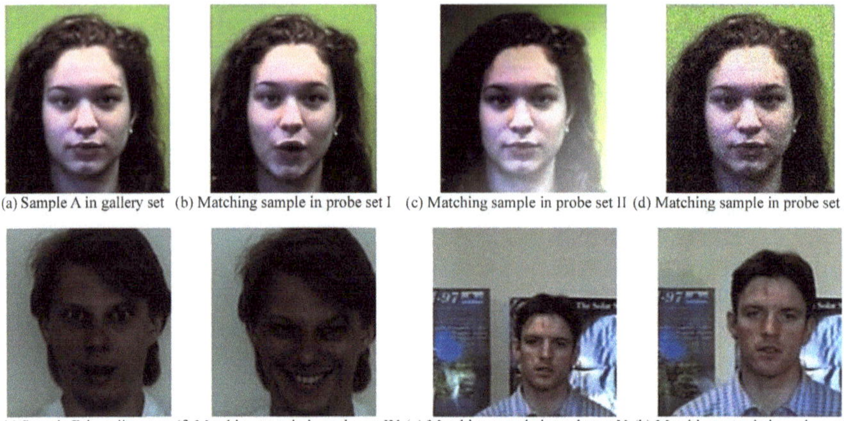

(a) Sample A in gallery set (b) Matching sample in probe set I (c) Matching sample in probe set II (d) Matching sample in probe set III

(e) Sample B in gallery set (f) Matching sample in probe set IV (g) Matching sample in probe set V (h) Matching sample in probe set VI

Fig. 20 Color face recognition results

Fig. 19 delineates the proposed face recognition method, which is formulated as an automatic face feature extraction module and a feature matching procedure. It makes full use of the interrelationship among different color channels and achieves promising performance against noises, variations of lighting, and expressions in the open face dataset downloaded from http://cswww.wssex.ac.uk/mv/allfaces/index.html, as shown in Fig. 20. Detailed quantitative analysis can be referred to [7].

7 Conclusion

In this paper, we give an overview of the development of QWT and investigate its potential applications in scalar image processing and vector image processing. QWT is natural to deal with the operation in color/vector space, but nowadays the construction of invertible multiscale QWT for vector-valued color images is still a far-reaching issue. Several important problems remain open: constraints of quaternion wavelet needed to reconstruct a color signal, the definition of sparse quaternion wavelet transform in color space, the selection of color space, and so on.

Acknowledgements This work was supported by Research Fund for the Doctoral Program of Higher Education of China (200802481006), National Natural Science Foundation of China (60702044, 60902073), NCET-06-0409, and the Cultivation Fund of the Key Scientific and Technical Innovation Project, Ministry of Education of China (706022).

References

1. Bülow, T.: Hypercomplex spectral signal representations for the processing and analysis of images. Dissertation, Kiel: Institut für Informatik und Praktische Mathematik der Christian-Albrechts-Universitat zu Kiel (1999)

2. Corrochano, E.B.: The theory and use of the quaternion wavelet transform. J. Math. Imaging Vis. **24**, 19–35 (2006)
3. Chan, W.L., Choi, H., Baraniuk, R.: Quaternion wavelets for image analysis and processing. Proc. IEEE Int. Conf. Image Process. **5**, 3057–3060 (2004)
4. Xu, Y., Zhou, J., Yang, X.: Quaternion wavelet phase based stereo matching for uncalibrated images. Pattern Recogn. Lett. **28**(12), 1509–1522 (2007)
5. Sangwine, S., Ell, T.: Hypercomplex Fourier transforms of color images. IEEE Trans. Image Process. **16**(1), 22–35 (2007)
6. Jones, C., Abbott, A.: Color face recognition by hypercomplex Gabor analysis. In: 7th International Conference on Automatic Face and Gesture Recognition, pp. 126–131 (2006)
7. Lu, W., Xu, X.Y.Y., Song, L.: Local quaternionic Gabor binary patterns for color face recognition. In: International Conference on Acoustics, Speech, and Signal Processing, pp. 741–744 (2008)
8. Carré, P., Denis, P.: Quaternionic wavelet transform for colour images. Proc. SPIE **6383**, 638 301/1–638 301/15 (2006). Invited paper
9. Hamilton, W.: Elements of Quaternions. Longman, Harlow (1866)
10. Felsberg, M.: Optimized fast algorithms for the quaternionic Fourier transform. In: Proc. 8th International Conference on Computer Analysis of Images and Patterns, vol. 1689, pp. 209–216 (1999)
11. Pei, S., Ding, J., Chang, J.: Efficient implementation of quaternion Fourier transform, convolution, and correlation by 2-D complex FFT. IEEE Trans. Signal Process. **49**(11), 2783–2797 (2001)
12. Kingsbury, N.: Complex wavelets for shift invariant analysis and filtering of signals. Appl. Comput. Harmon. **10**(3), 234–253 (2001)
13. da Silva, E.A.B., Ghanbari, M.: On the performance of linear phase wavelet transform in low bitrate coding. IEEE Trans. Image Process. **5**(5), 689–704 (1996)
14. Gabor, D.: Theory of communication. J. IEE **93**, 429–457 (1946)
15. Fleet, D., Jepson, A.: Stability of phase information. IEEE Trans. Pattern Anal. Mach. Intell. **15**(12), 1253–1268 (1993)
16. Lindeberg, T.: Feature detection with automatic scale selection. Int. J. Comput. Vis. **30**(2), 79–116 (1998)
17. Mikolajczyk, K., Tuytelaars, T.: A comparison of affine region detectors. Int. J. Comput. Vis. **65**(1/2), 43–72 (2005)
18. Kovesi, P.: Invariant feature measures of image features from phase information. Ph.D. Dissertation, University of Western Australia (1996)
19. Morrone, M., Owens, R.: Feature detection from local energy. Pattern Recogn. Lett. **6**, 303–313 (1987)
20. Mikolajczyk, K., Schmid, C.: A performance evaluation of local descriptors. IEEE Trans. Pattern Anal. Mach. Intell. **27**(10), 1615–1630 (2005)
21. Zhang, Z., Deriche, R., Faugeras, O., Luong, Q.: A robust technique for matching two uncalibrated images through the recovery of the unknown epipolar geometry. Artif. Intell. J. **78**, 87–119 (1995)
22. Do, M., Vetterli, M.: Contourlets. Beyond wavelets, pp. 1–27 (2001)
23. Lu, Z., Xu, Y., Yang, X., Song, L., Traversoni, L.: 2D quaternion Fourier transform: the spectrum properties and its application in color image registration. In: International Conference on Multimedia and Expo, pp. 1715–1718 (2007)
24. Ma, X., Xu, Y., Song, L., Yang, X., Burkhardt, H.: Color image watermarking using local quaternion Fourier spectral analysis. In: International Conference on Multimedia and Expo, pp. 233–236 (2008)
25. Ahonen, T., Hadid, A., Pietikainen, M.: Face description with local binary patterns: application to face recognition. IEEE Trans. Pattern Anal. Mach. Intell. **28**(12), 2037–2041 (2006)

Part IV
Computer Vision

Image Sensor Model Using Geometric Algebra: From Calibration to Motion Estimation

Thibaud Debaecker, Ryad Benosman, and Sio H. Ieng

Abstract In computer vision image sensors have universally been defined as the nonparametric association of projection rays in the 3D world to pixels in the images. If the pixels' physical topology can be often neglected in the case of perspective cameras, this approximation is no longer valid in the case of variant scale sensors, which are now widely used in robotics. Neglecting the nonnull pixel area and then the pixel volumic field of view implies that geometric reconstruction problems are solved by minimizing a cost function that combines the reprojection errors in the 2D images. This paper provides a complete and realistic cone-pixel camera model that equally fits constant or variant scale resolution together with a protocol to calibrate such a sensor. The proposed model involves a new characterization of pixel correspondences with 3D-cone intersections computed using convex hull and twists in Conformal Geometric Algebra. Simulated experiments show that standard methods and especially Bundle Adjustment are sometimes unable to reach the correct motion, because of their ray-pixel approach and the choice of reprojection error as a cost function which does not particularly fit the physical reality. This problem can be solved using a nonprojective cone intersection cost function as introduced below.

1 Introduction

A large amount of work has been carried out on perspective cameras introducing the pinhole model and the use of projective geometry. This model turns out to be very efficient in most cases, and it is still widely used within the computer vision community. Several computation improvements have been introduced [5]; nevertheless, this model has limitations, notably that it can only be applied to projective sensors. The pixel sampling and slight fluctuations in the calibration process lead to the fact

T. Debaecker (✉)
Institut des Systèmes Intelligents et de Robotique (ISIR), 4 place Jussieu, Paris, France
e-mail: thibaud.debaecker@isir.jussieu.fr

E. Bayro-Corrochano, G. Scheuermann (eds.), *Geometric Algebra Computing*,
DOI 10.1007/978-1-84996-108-0_14, © Springer-Verlag London Limited 2010

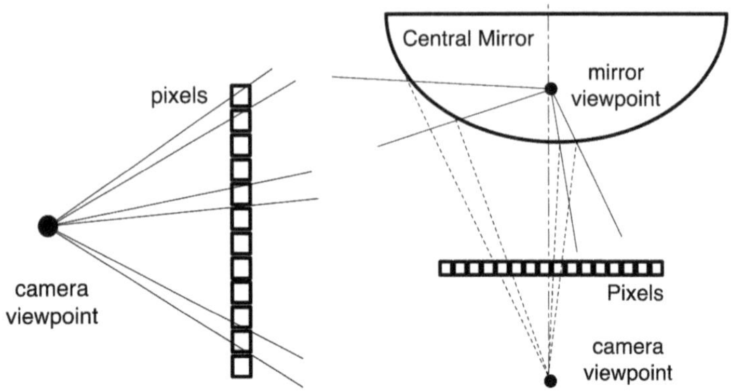

(a) Pixel-cones of a perspective camera. (b) Catadioptric pixel-cones after reflection on a central mirror

Fig. 1 Pixel-cones in the case of perspective cameras and variant scale sensors (here a central catadioptric sensor). All cones are almost the same in (**a**), whereas in (**b**) the pixel-cones vary drastically according to the position of the pixel within the perspective camera observing the mirror

that two rays of view of pixels representing the same point in the scene never exactly intersect across the same 3D point they should represent. Finally, this model fails to introduce the reality of the sensor, since the approximation of the field of view of the pixel is restricted to a ray instead of a small surface in the image plane.

The limitations pointed out become drastically problematic with the appearance of new kinds of nonlinear visual sensors like foveolar retinas [3] or panoramic sensors (see [2] for an overview). As shown in Fig. 1(a), in the case of perspective cameras, all pixels produce a similar cone of view. Cones being barely the same, it is easily understandable why cones can be approximated using lines in this case, but as an illustration of the consequences on a pinhole case, let us consider two identical 640×480 cameras with a 30-centimeter space between each of them, and watching a scene from a five-meter distance. Following the formulation we will describe in the following sections, for a pair of matched points, it is possible to compute in 3D space the real volume which represents the solution set of the triangulation problem. It appears that the mere pixel sampling leads to a 12-cm^3 volume of solutions. It becomes obvious that this approximation using rays will lead to significant imprecision for a catadioptric sensor (combination of a perspective camera and a hyperboloid mirror), especially in the computation of intersections, as shown in Fig. 1(b). Cones then become an absolute necessity.

Different methods have been developed to address the motion estimation problem despite the issue of nonperfect intersection of rays. It is important to notice two particular points in these methods. First, a cost function is chosen to evaluate the accuracy of a solution, and then a seeking strategy is chosen to reach the optimal solution according to this cost function. If the panel of seeking strategies is wide (Branch and Bound Algorithm [11], Levenberg-Marquardt and other least-squared minimization used in the classic Bundle Adjustment (BA) [18], Second Order Cone Programming [10]), the cost function is in most cases the reprojection error. Because of the 5 DOF resolution of the motion estimation problem, this cost function is logically chosen instead of distances of rays in 3D space.

The aim of this paper is to show that introducing cones gives an opportunity to be closer to the real physics of pixel correspondences and thus generates more accurate situations. Euclidean spaces do not allow an easy-manipulating cone expression, especially if intersections of such cones have to be computed. Conformal Geometric Algebra is introduced to enable a simple formulation of cones using twists [17]. A simple line is used as the twist axis to rotate a second line used as the cone directrix. A wide variety of shapes can be generated with twists combination, constructing cones with different kinds of basis. Motion estimation has been chosen here as an application of this cone-pixel camera model to show its reliability. The use of this model enables us to introduce a new cost function as the intersection of cones in space. We show here through experiments that BA is unable to estimate the correct motion because the solution does not correspond to a minimum of its cost function. However it can be found with the cone intersection criterion using a stochastic optimization method like Simulated Annealing [12].

Recently, cones have been introduced to modelize the uncertainties of ray directions rather than the pixel field of view. The most related work have been done by Perwass et al. [14] by showing how the uncertainty of all elements of the Geometric Algebra of conformal space can be appropriately described by covariance matrices. Giving an uncertain expression of the projection point, this approach can modelize noncentral sensors. In [11], cone aperture is set as an arbitrary error parameter of matched points. Other approaches which do not use non-least-square minimization methods have been used to correct these problems in multiple view geometry problematics. All these methods still remain mathematical instead of physical approaches [8, 9] and consider that nonintersections reflect necessarily an imprecise calibration result despite that it corresponds to the real geometry.

This chapter is structured as follows. After introducing the basis element of Conformal Geometric Algebra in Sect. 2, we describe in Sect. 3 the mathematical formulation of the general pixel-cones model using twists [17]. An experimental protocol to find the pixel cones of light is presented in Sect. 4, and results obtained from this protocol are applied on a pinhole camera and a catadioptric sensor. In Sect. 5, we introduce a cone intersection score function to address the motion estimation problem and present results in simulation experiments. Conclusions and future works are included in Sect. 6.

2 Introduction to Conformal Geometric Algebra

This section has been widely inspired by [15], which presents a good understanding and more detailed introduction to Geometric Algebra. The reader unfamiliar with CGA should refer to [6, 7] for an overview, and examples of its use in computer vision can be found in [13, 16].

2.1 Geometric Algebras

Geometric Algebra can be seen as a Clifford Algebra with its focus on a suited geometric interpretation. A Geometric Algebra $\mathcal{G}_{p,q,r}$ is a nonlinear space of dimension $2^n, n = p+q+r$, with a space structure, called blades, to represent so-called multivectors as higher-grade algebraic entities in comparison to vectors of a vector space as first grade entities. A geometric Algebra $\mathcal{G}_{p,q,r}$ is constructed from a vector space $\mathbb{R}^{p,q,r}$, endowed with the signature (p,q,r), by application of a *geometric product*. This means that the generating vector space is always an element of its generated geometric algebra, and therefore the vectors of the vector space can be found as elements in each geometric algebra.

The product defining a geometric algebra is called *geometric product* and is denoted by juxtaposition, e.g., **AB** for two algebraic elements **A** and **B** called multivectors. The geometric product of vectors consists of an outer ($\wedge$) product and an inner (.) product. Their effect is to increase or to decrease the grade of the algebraic entities, respectively. Let $\mathbf{e}_i, \mathbf{e}_j \in \mathbb{R}^{p,q,r}$ be two orthonormal basis vectors of the vector space. Then the geometric product for these vectors of the geometric algebra $\mathcal{G}_{p,q,r}$ is defined as

$$\mathbf{e}_i\mathbf{e}_j := \begin{cases} 1 & \text{for } i = j \in \{1,\ldots,p\}, \\ -1 & \text{for } i = j \in \{p+1,\ldots,p+q\}, \\ 0 & \text{for } i = j \in \{p+q+1,\ldots,n\}, \\ \mathbf{e}_{ij} = \mathbf{e}_i \wedge \mathbf{e}_j = -\mathbf{e}_j \wedge \mathbf{e}_i & \text{for } i \neq j. \end{cases}$$

The geometric product of the same two basis vectors leads to a scalar, whereas the geometric product of two different basis vectors leads to a new entity, which is called a *bivector*. Geometric algebras can be expressed on the basis of graded elements. Scalars are of grade zero, vectors of grade one, bivectors of grade two, etc. A linear combination of elements of different grades is called a multivector **M**, which can be expressed as

$$\mathbf{M} = \sum_{i=0}^{n} \langle \mathbf{M} \rangle_i,$$

where the operator $\langle . \rangle$ denotes the projection of a general mutlivector to the entities of grade s. A multivector **A** of grade i can be written as $\mathbf{A}_{\langle i \rangle}$.

The inner (.) and outer ($\wedge$) products of two vectors $\mathbf{u}, \mathbf{v} \in \mathbb{R}^{4,1}$ are defined as

$$\mathbf{u}.\mathbf{v} := \frac{1}{2}(\mathbf{uv} + \mathbf{vu}), \tag{1}$$

$$\mathbf{u} \wedge \mathbf{v} := \frac{1}{2}(\mathbf{uv} - \mathbf{vu}). \tag{2}$$

Two blades of highest grade are called pseudoscalars and noted as $\pm\mathbf{I}$. The *dual* $\mathbf{X}^*$ of a blade **X** is defined by

$$\mathbf{X}^* := \mathbf{X}\mathbf{I}^{-1}.$$

It follows that the dual of an r-blade is an $(n-r)$-blade. The reverse $\tilde{\mathbf{A}}_{\langle S\rangle}$ of an s-blade $\mathbf{A}_{\langle S\rangle} = \mathbf{a}_1 \wedge \cdots \wedge \mathbf{a}_s$ is defined as the reverse outer product of the vectors $\mathbf{a}_i$,

$$\tilde{\mathbf{A}}_{\langle S\rangle} = (\mathbf{a}_1 \wedge \mathbf{a}_2 \wedge \cdots \wedge \mathbf{a}_{s-1} \wedge \mathbf{a}_s)^{\tilde{}},$$

$$\tilde{\mathbf{A}}_{\langle S\rangle} = (\mathbf{a}_s \wedge \mathbf{a}_{s-1} \wedge \cdots \wedge \mathbf{a}_2 \wedge \mathbf{a}_1).$$

The *join* $\mathbf{A}\dot{\wedge}\mathbf{B}$ is the pseudoscalar of the space given by the sum of spaces spanned by $\mathbf{A}$ and $\mathbf{B}$. For blades $\mathbf{A}$ and $\mathbf{B}$, the *dual shuffle* product $\mathbf{A} \vee \mathbf{B}$ is defined by the DeMorgan rule

$$(\mathbf{A} \vee \mathbf{B})^* := \mathbf{A}^* \dot{\wedge} \mathbf{B}^*.$$

For blades $\mathbf{A}$ and $\mathbf{B}$, it is possible to use the join to express *meet* operations:

$$\mathbf{A} \vee \mathbf{B} := \left(\mathbf{A}J^{-1} \wedge \mathbf{B}J^{-1}\right)J \tag{3}$$

with $J = \mathbf{A}\dot{\wedge}\mathbf{B}$.

2.2 Conformal Geometric Algebra (CGA)

A *Minkowski* plane is used to introduce CGA. Its vector space $\mathbb{R}^{1,1}$ has the orthonormal basis $\{\mathbf{e}_+, \mathbf{e}_-\}$ defined by the properties

$$\mathbf{e}_+^2 = 1, \qquad \mathbf{e}_-^2 = -1, \qquad \mathbf{e}_+ . \mathbf{e}_- = 0.$$

In addition, a *null basis* can now be introduced by the vectors

$$\mathbf{e}_0 = \frac{1}{2}(\mathbf{e}_- - \mathbf{e}_+) \quad \text{and} \quad \mathbf{e} = \mathbf{e}_- + \mathbf{e}_+.$$

These vectors can be interpreted as the origin $\mathbf{e}_0$ of the coordinate system and the point at infinity $\mathbf{e}$, respectively. Furthermore, $\mathbf{E}$ is defined as $\mathbf{E} := \mathbf{e} \wedge \mathbf{e}_0 = \mathbf{e}_+ \wedge \mathbf{e}_-$.

The role of the Minkowski plane is to generate null vectors, and so to extend an Euclidean vector space $\mathbb{R}^n$ to $\mathbb{R}^{n+1,1} = \mathbb{R}^n \oplus \mathbb{R}^{1,1}$. The conformal vector space derived from $\mathbb{R}^3$ is thus denoted as $\mathbb{R}^{4,1}$, and a basis is given by $\{\mathbf{e}_1, \mathbf{e}_2, \mathbf{e}_3, \mathbf{e}_+, \mathbf{e}_-\}$. The conformal unit pseudoscalar is denoted as

$$\mathbf{I}_C = \mathbf{e}_{+-123} = \mathbf{E}\mathbf{I}_E,$$

where $\mathbf{I}_E$ is the unit pseudoscalar in the Euclidean Geometric Algebra, i.e., $\mathbf{I}_E = \mathbf{e}_{123}$. The points in CGA are related to those of Euclidean space by

$$\underline{\mathbf{x}} = \mathbf{x} + \frac{1}{2}\mathbf{x}^2\mathbf{e} + \mathbf{e}_0.$$

In CGA, spheres can be interpreted as the basis entities from which the other entities are derived. A sphere with center $\mathbf{p} \in \mathcal{G}_3$ and radius $\rho \in \mathbb{R}^3$ can be written as

$$(\mathbf{x} - \mathbf{p})^2 = \rho^2. \tag{4}$$

It turns out that a point $\mathbf{x}$ is nothing more than a degenerate sphere with radius $\rho = 0$. Equation (4) can therefore be represented more compactly as

$$(\mathbf{x} - \mathbf{p})^2 = \rho^2 \quad \Longleftrightarrow \quad \underline{\mathbf{x}}.\underline{\mathbf{s}} = 0.$$

A sphere can be defined with its dual form which can be calculated directly from at least four points on it,

$$\underline{\mathbf{s}}^* = \underline{\mathbf{a}} \wedge \underline{\mathbf{b}} \wedge \underline{\mathbf{c}} \wedge \underline{\mathbf{d}},$$

and a point $\underline{\mathbf{x}}$ is on this sphere if and only if

$$\underline{\mathbf{x}}.\underline{\mathbf{s}} = 0 \quad \Longleftrightarrow \quad \underline{\mathbf{x}} \wedge \underline{\mathbf{s}}^* = 0.$$

Geometrically, a circle $\underline{\mathbf{z}}$ can be described by the intersection $\underline{\mathbf{z}} = \underline{\mathbf{s}_1} \wedge \underline{\mathbf{s}_2}$ of two linearly independent spheres $\underline{\mathbf{s}_1}$ and $\underline{\mathbf{s}_1}$. This means that

$$\underline{\mathbf{x}} \in \underline{\mathbf{z}} \quad \Longleftrightarrow \quad \underline{\mathbf{x}}.\underline{\mathbf{z}} = 0.$$

In the same way, the dual form of a circle is geometrically defined by three points on it,

$$\underline{\mathbf{z}}^* = \underline{\mathbf{a}} \wedge \underline{\mathbf{b}} \wedge \underline{\mathbf{c}},$$

and a line is a degenerate case of a circle passing through a point at infinity:

$$\underline{\mathbf{L}}^* = \underline{\mathbf{e}} \wedge \underline{\mathbf{a}} \wedge \underline{\mathbf{b}}.$$

The bivectors of the geometric algebra can be used to represent rotations of points in the 3D space. A *rotor* $\mathbf{R}$ is an even grade element of the algebra which satisfies $\mathbf{R}\tilde{\mathbf{R}} = 1$. By using the Euler representation of a rotor, we have

$$\mathbf{R} = \exp\left(-\frac{\theta}{2}n\right).$$

The rotation of a point represented by its vector $\mathbf{x}$ can be carried out by multiplying the rotor $\mathbf{R}$ from the left and its reverse from the right to the point such as

$$\mathbf{x}' = \mathbf{R}\mathbf{x}\tilde{\mathbf{R}}.$$

CGA enables one a multiplicative expression of translation $\mathbf{t}$ as a special rotation acting at infinity by using the null vector $\mathbf{e}$:

$$\underline{\mathbf{x}}' = \mathbf{T}\underline{\mathbf{x}}\tilde{\mathbf{T}} \quad \text{with } \mathbf{T} = \exp\left(-\frac{\mathbf{et}}{2}\right).$$

It is possible in GA to generate kinematic shapes which result from the orbit effect of points under the action of a set of coupled operators. The nice idea is that the operators are what describes the curve (or shape). To model a rotation of a point $\underline{\mathbf{X}}$ around an arbitrary line $\underline{\mathbf{L}}$ in the space, the general idea is to translate the point $\underline{\mathbf{X}}$ with the distance vector between the line $\underline{\mathbf{L}}$ and the origin, to perform a rotation and to translate the transformed point back. So a motor $\mathcal{M}$ describing a general rotation has the form

$$\mathcal{M} = \mathbf{TR}\tilde{\mathbf{T}}.$$

Screw motions can be used to describe rigid motions by combining a rotation around an axis with a translation parallel to that axis $\mathbf{T}_{d\mathbf{n}}$. The resulting motor is

$$\mathbf{M} = \mathbf{T}_{d\mathbf{n}}\mathbf{TR}\tilde{\mathbf{T}}.$$

As introduced in [17], these operators are the motors which are the representation of $SE(3)$ in $\mathcal{R}_{4,1}$. The use of twists gives a compact representation of cones and brings the heavy computation of the intersection of two general cones to a simple intersection of lines.

3 General Model of a Cone-Pixels Camera

This section will present a general model of a camera using cone-pixels. There are several possible ways to write the equation of a cone. A single-sided cone with vertex V, axis ray with origin at V, unit-length direction A, and cone angle $\alpha \in (0, \pi/2)$ is defined by the set of points X such that vector $X - V$ forms an angle α with A. The algebraic condition is $A \cdot (X - V) = |X - V| \cos(\alpha)$. The solid cone is the cone plus the region it bounds, specified as $A \cdot (X - V) \geq |X - V| \cos(\alpha)$. It is somewhat painful to compute the intersection of two cones, and this can become even more complicated integrating rigid motion parameters between cones. CGA is used to enable us a simple formulation of cones using twists, as introduced in Sect. 2.

3.1 Geometric Settings

As shown in Fig. 3 in the case of a perspective camera, the image plane here represented by I contains several rectangular pixels $p(i, j)$, where i, j corresponds to the position of the pixel. Considering $p(i, j)$, its surface is represented by a rectangle defined by points $A_0 \ldots A_8$, with A_0 corresponding to the center of the rectangle.

Given a line $\underline{\mathbf{l}}$ (with unit direction) in space, the corresponding motor describing a general rotation around this line is given by $\mathcal{M}(\theta, \underline{\mathbf{l}}) = \exp(-\frac{\theta}{2}\underline{\mathbf{l}})$. The general rotation of a point $\underline{\mathbf{x}}$ around any arbitrary line $\underline{\mathbf{l}}$ is

$$\underline{\mathbf{x}}' = \mathcal{M}(\theta, \underline{\mathbf{l}})\underline{\mathbf{x}}\tilde{\mathcal{M}}(\theta, \underline{\mathbf{l}}). \tag{5}$$

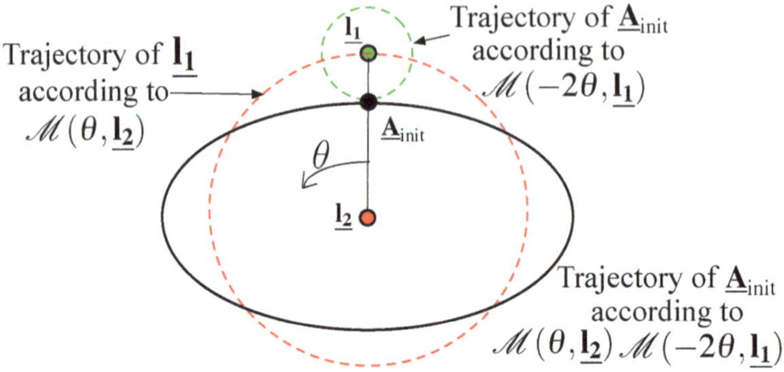

Fig. 2 Generating an ellipse with a *2twist* combination

The general form of the *2twist* generated curve is the set of points $\underline{\mathbf{x}}'$ defined as

$$\underline{\mathbf{x}}' = \mathcal{M}(\lambda_2\theta, \underline{\mathbf{l}_2})\mathcal{M}(\lambda_1\theta, \underline{\mathbf{l}_1})\underline{\mathbf{x}}\tilde{\mathcal{M}}(\lambda_1\theta, \underline{\mathbf{l}_1})\tilde{\mathcal{M}}(\lambda_2\theta, \underline{\mathbf{l}_2}). \tag{6}$$

In the following, we are interested in generating ellipses to approximate the form of the pixel rather than squares. Ellipses are indeed expressible mathematically in such compact form as it will be shown in what follows, while using square instead will involve discontinuities in the expression of the cones. In the previous equation, an ellipse corresponds to the values $\lambda_1 = -2$ and $\lambda_2 = 1$. $\underline{\mathbf{l}_1}$ and $\underline{\mathbf{l}_2}$ are the two rotation axes needed to define the ellipses [17]. An ellipse generated with twists is illustrated in Fig. 2.

Considering a single pixel $p_{i,j}$ (see Fig. 3), its surface can be approximated by the ellipse generated by a point A that rotates around point A_0 with a rotation axis corresponding to $\mathbf{e}_3$ normal to the plane I. The ellipse $\mathcal{E}^{i,j}$ generated corresponding to the pixel $p_{i,j}$ is the set of all the positions of $A(\theta)$:

$$\forall\theta \in [0,\ldots,2\pi],$$
$$\mathcal{E}^{i,j} = \big\{\underline{\mathbf{A}}(\theta) = \mathcal{M}(\theta, \underline{\mathbf{l}_2})\mathcal{M}(-2\theta, \underline{\mathbf{l}_1})\underline{\mathbf{A}}_{\text{init}}\tilde{\mathcal{M}}(-2\theta, \underline{\mathbf{l}_1})\tilde{\mathcal{M}}(\theta, \underline{\mathbf{l}_2}) \mid\big\}.$$

The initial position of $\underline{\mathbf{A}}$ is $\underline{\mathbf{A}}_{\text{init}}$. The elliptic curve is generated by setting the two connected twists in order to obtain an ellipse with principal axes $(\overline{A_8A_0}, \overline{A_6A_0})$ in order to fit the rectangular surface of the projection of the pixel as shown in Fig. 3. It is now possible to generate the cone corresponding to the field of view of the pixel. To set the different lines, we use the *dual* expression of geometric objects in CGA. We set the line $\underline{\mathbf{l}^*_{0_{i,j}}}$, the cone axis corresponding to $p_{i,j}$, as

$$\underline{\mathbf{l}^*_{0_{i,j}}} = e \wedge \underline{\mathbf{O}} \wedge \underline{\mathbf{A}}_0.$$

The generatrix of the cone is the line joining O to A. Since the position of A depends on the cone aperture α, the generatrix is noted as $\underline{\mathbf{l}_{\mathbf{OA}(\alpha)}}^{i,j}$.

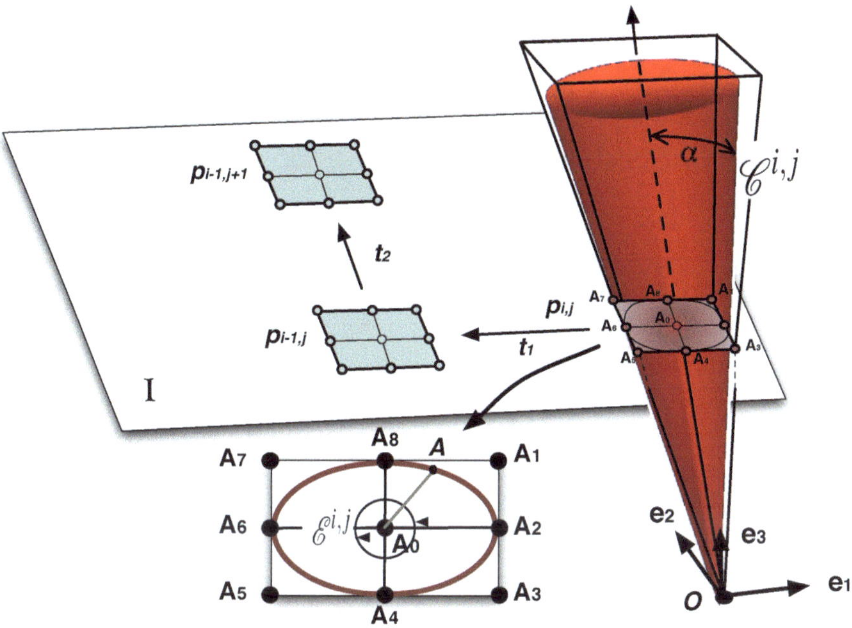

Fig. 3 Cones Geometric settings

A priori, one can think that the pixel-cone of view of $p_{i,j}$ is the cone $\mathcal{C}^{i,j}(\alpha)$ defined by the projection point and the surface of the pixel,

$$\mathcal{C}^{i,j}(\alpha) = \mathcal{M}(\theta, \underline{\mathbf{l}_0}_{i,j}) \underline{\mathbf{l}_{\mathbf{OA}(\alpha)}}^{i,j} \tilde{\mathcal{M}}(\theta, \underline{\mathbf{l}_0}_{i,j}), \tag{7}$$

with $\mathcal{M}(\theta, \underline{\mathbf{l}_0}_{i,j}) = \exp(-\frac{\theta}{2}\underline{\mathbf{l}_0}_{i,j})$, but it will appear that this first intuition is not true because of the optical refraction combined with proper camera design. However, this cone has to be computed and used as a preliminary guess to find the real cone of view which can only be wider. In the following, this overlapping will be pointed out experimentally, and overlapping cones of view of neighboring pixels are shown in Fig. 9. Two criterions will be introduced taking into account an eventual overlapping.

Note the equivalent expression of the cone using the outer product generating a line after having generated an ellipse from the point $A(\alpha)$:

$$\mathcal{C}^{i,j}(\alpha) = e \wedge \underline{\mathbf{O}} \wedge \big(\mathcal{M}(\theta, \underline{\mathbf{l}_0}_{i,j}) \underline{\mathbf{A}(\alpha)}^{i,j} \tilde{\mathcal{M}}(\theta, \underline{\mathbf{l}_0}_{i,j})\big). \tag{8}$$

The same process is to be applied again after translating the pixel $p_{i,j}$ using $\mathbf{t}_1$ and $\mathbf{t}_2$, which corresponds to the translation to switch from one pixel to the other. The projection $p_{i,j}$ of a pixel is moved to a next pixel:

$$p_{i+1,j+1} = \mathcal{T}^2(\mathbf{t}_2)\mathcal{T}^1(\mathbf{t}_1) p_{i,j} \tilde{\mathcal{T}}^1(\mathbf{t}_1)\tilde{\mathcal{T}}^2(\mathbf{t}_2),$$

where $\mathcal{T}(\mathbf{t})$ corresponds to a translation operator in CGA.

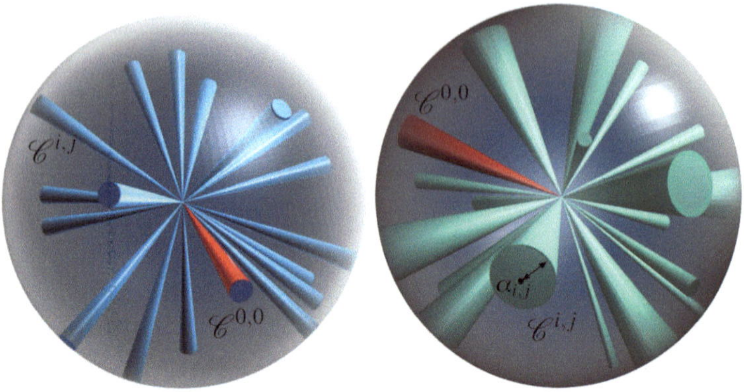

(a) A bundle of pixel-cones of a central linear sensor. (b) A bundle of pixel-cones of a central variant linear sensor

Fig. 4 Different configurations of pixel-cones in the case of linear and variant scale sensors. In *red* the principal cone according which every other is located

3.2 The General Model of a Central Cone-Pixel Camera

The general form of a central sensor, whether it has linear resolution (cones vary slightly) or variant resolution, is the expression of a bundle of cones. All cones $\mathfrak{C}^{i,j}$ will be located using spherical coordinates and located according to an origin set as the cone $\mathfrak{C}^{0,0}(\alpha)$ that has $\mathbf{e}_3$ as a principal axis. The general form of a central linear scale camera (Fig. 4(a)) is then simply given by

$$\mathfrak{C}^{i,j}_{\phi,\psi}(\alpha) = \mathcal{M}(\psi, \mathbf{e}_{23})\mathcal{M}(\phi, \mathbf{e}_{13})\mathfrak{C}^{0,0}_{0,0}(\alpha)\tilde{\mathcal{M}}(\phi, \mathbf{e}_{13})\tilde{\mathcal{M}}(\psi, \mathbf{e}_{23}), \tag{9}$$

where ψ, ϕ denote the spherical coordinates of the cone, and α is the constant aperture for an uniform resolution.

The general form of a central variant scale sensor is slightly different. Each cone having a different aperture α, cones need to be defined according to their position. The general form becomes

$$\mathfrak{C}^{i,j}_{\phi,\psi}(\alpha_{i,j}) = \mathcal{M}(\psi, \mathbf{e}_{23})\mathcal{M}(\phi, \mathbf{e}_{13})\mathfrak{C}^{0,0}_{0,0}(\alpha_{i,j})\tilde{\mathcal{M}}(\phi, \mathbf{e}_{13})\tilde{\mathcal{M}}(\psi, \mathbf{e}_{23}). \tag{10}$$

3.3 Intersection of Cones

The previous formulation of cones established in (7) is a parameterized line bundle and cannot be considered as a GA entity and cannot be used as easier with all GA tools. Then to express the intersection of two cones $\mathfrak{C}^{i_m,j_m}_{\phi_m,\psi_m}$ and $\mathfrak{C}^{i_n,j_n}_{\phi_n,\psi_n}$, we used the usual $\cap$ operator instead of using a GA meet operator defined in (3) which should have been the correct expression of the intersection of classic objects in GA.

The result is a set of points of intersection $P_{m,n}$ between the generatrix lines of the cones,

$$P_{m,n} = \{\mathfrak{C}^{i_m,j_m}_{\phi_m,\psi_m} \cap \mathfrak{C}^{i_n,j_n}_{\phi_n,\psi_n}\}. \tag{11}$$

If the intersection exists, $P_{m,n}$ is not empty, and the set of points then forms a convex hull the volume of which can be computed using [1].

4 General Cone-Pixel Camera Calibration

4.1 Experimental Protocol

Cones being at the heart of the model, we will now give an experimental setup of the calibration procedure to provide an estimation of the cone of view of each pixel of a camera. The method is not restricted to a specific camera geometry; it relies on the use of multiple planes calibration [4, 20]. As shown in Fig. 5, the camera to be calibrated is observing a calibration plane (in our case, a computer screen), the aim is to estimate the cone of view of a pixel $p_{i,j}$ by computing for each position of the screen its projection surface $SP^k(i,j)$, k being the index of the calibration plane. The metric is provided using a reference high-resolution calibrated camera (RC)[1] observing the calibration planes, whose positions and metrics can then be known in the RC coordinates. The impact surfaces $SP^k(i,j)$, once determined on each screen, normally lead (as shown in Fig. 5) to the determination of all pixel-cones parameters.

Figure 6 shows the experimental setup carried out for the experiments. The key-point of the calibration protocol therefore relies on the determination of pixels' impact $SP^k(i,j)$.

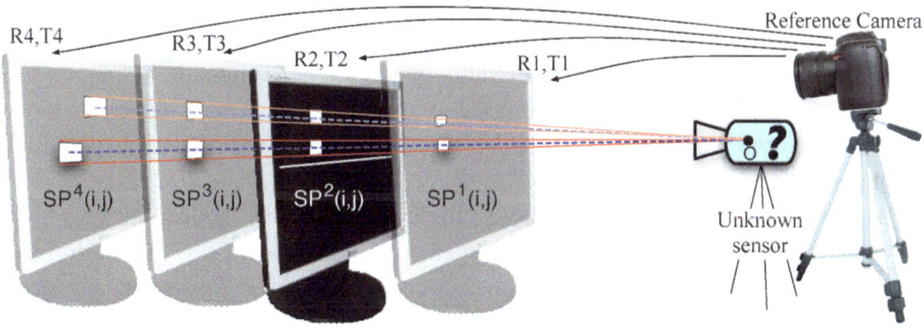

Fig. 5 Experimental protocol: Cone construction and determination of the sensor projection center The R_i, T_i represent the rigid motion between the reference camera and the calibration planes coordinate systems

[1] 6 Megapixel digital single-lens Nikon D70 reflex camera fitted with 18–70-mm Nikkor micro lens. The micro lens and the focus are fixed during the whole experiment.

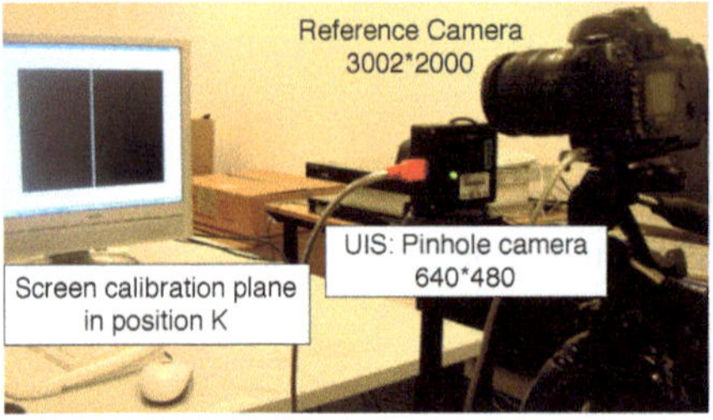

(a) Case of a pinhole camera as UIS.

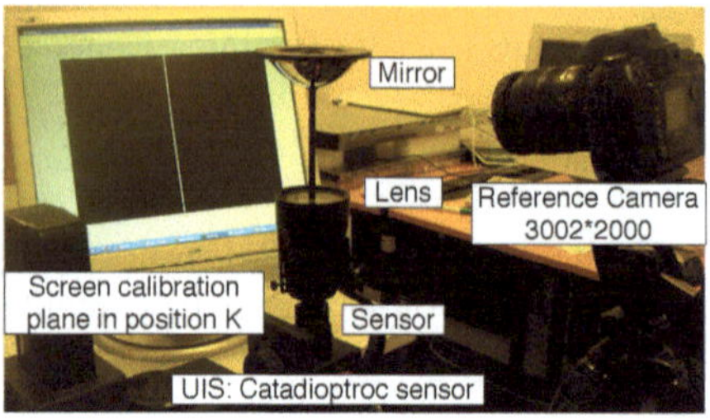

(b) Case of a catadioptric sensor as UIS.

Fig. 6 Experimental protocol for two kinds of unknown image sensor

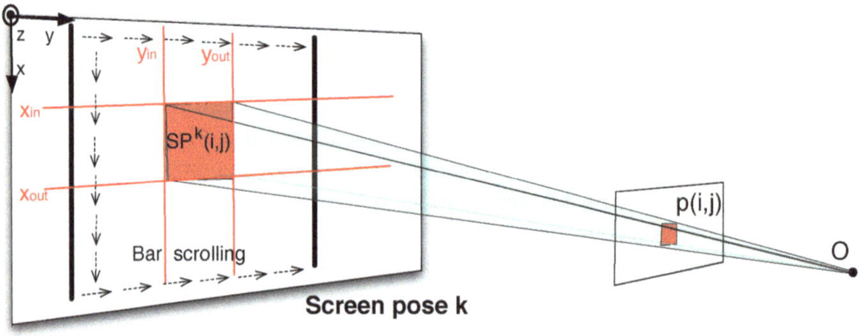

Fig. 7 Intersection surface $SP^k(i, j)$ between the calibration plane k and the cone $\mathcal{C}(i, j)$

The activity of each pixel $p(i, j)$ is then tracked while RC observes the screen calibration planes (see Fig. 7). At each position the screen displays a white bar scrolling on a uniform black background (see Fig. 6). The bar will cause a change in the grey level values of pixels when it is in their cone of vision. The pixels' gray level increases from a minimum value (when the bar is outside $SP^k(i, j)$) to a maximum value (when the bar is completely inside $SP^k(i, j)$) and decreases down to zero when the bar is again outside $SP^k(i, j)$. Figure 8 gives a visual explanation of the process.

A sensitivity threshold can be chosen to determine pixels' activation. Using the reference camera calibration results, it is then possible, once $SP^k(i, j)$ is determined, to compute its edges as the positions of y_{in} and y_{out} in the RC coordinate system. The bar is scrolled in two orthogonal directions producing two other edges x_{in} and x_{out} (Fig. 7). At this stage, the edges of $SP^k(i, j)$ are then completely known. The location and size of pixel-cones can then in a second stage be estimated once all $SP^k(i, j)$ are known. Cones are computed using the center of $SP^k(i, j)$ providing the rotation axis, the cone envelope is given by computing rays that pass through all the intersection points of the vertex of each $SP^k(i, j)$ corresponding to each pixel (Fig. 5).

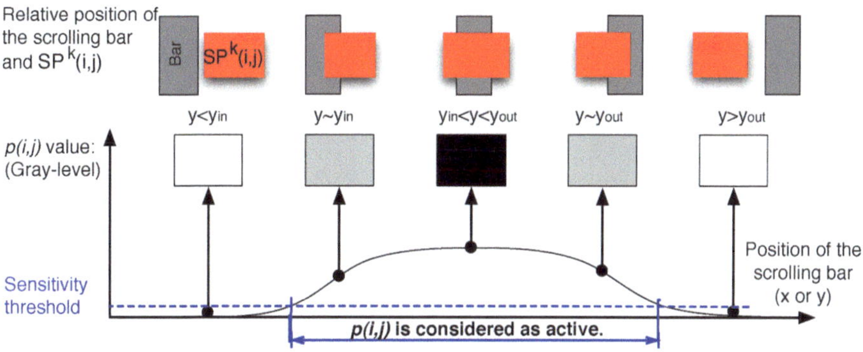

Fig. 8 Pixel response according to the scroll bar position

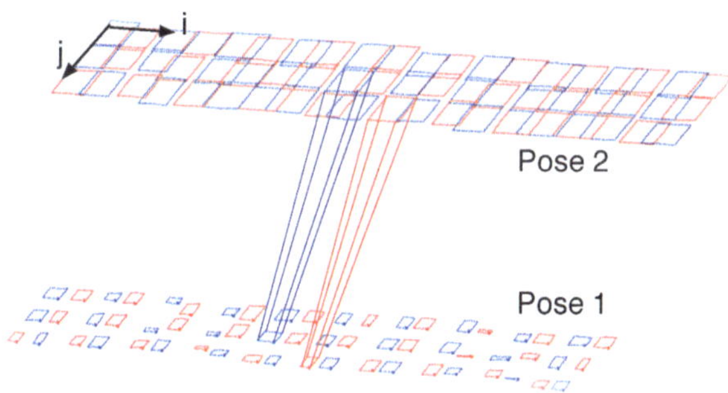

Fig. 9 Experimental results: Cones of view in the case of a pinhole camera

4.2 Calibration Experimental Results

The following experiments were carried out using PointGrey DragonFly®2, with a 640×480 resolution, and a parabolic catadioptric sensor with a telecentric lens; both are central sensors (Fig. 6(a)). Figure 9 shows cones reconstruction on $SP^1(i, j)$ and $SP^2(i, j)$ in the case of a pinhole camera. For clarity, only two cones were drawn. As expected, the results show repetitive pattern as corresponding pixel's impact. We can see few bad measurements, especially in Pose 1. Therefore, we use four planes to avoid this problem, using the redundancy of the information. This figure shows that the spatial sensitivity of pixels overlaps and does not correspond to the dimension of the pixels as the informed reader could imagine.

With a catadioptric camera, it is a geometric truth that the aperture angle of each cone increases as pixels are set off the optical axis of the camera. This phenomenon is experimentally shown in Fig. 10, which represents the evolution of the solid angle of pixel-cones. In principle, the solid angle should not vary according to the position of the calibration plane that was used to compute it. The curves are logically very close even if a small bias appears for very large cone-pixels at the periphery of the mirror (where the uncertainties of the measure on the surface due to the nonlinearity of the mirror are the highest).

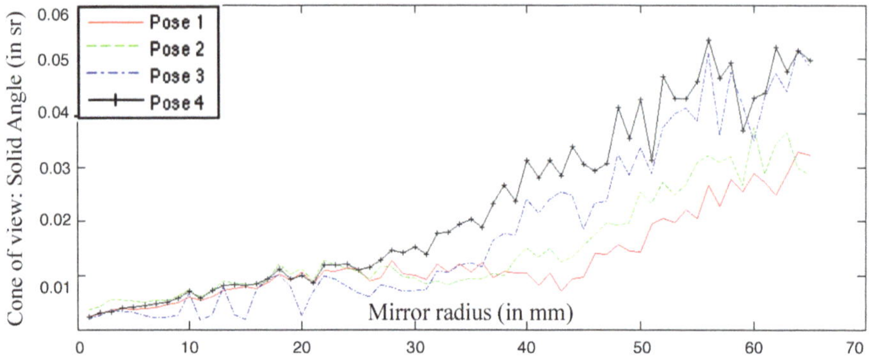

Fig. 10 Solid angle of view according to the mirror radius

Table 1 Central projection point estimation coordinates: case of a pinhole camera

	Ground truth	Axis estimation	Error	Apex estimation	Error
x	−78.33	−78.97	0.65	−78.97	0.65
y	45.36	44.08	1.28	44.07	1.29
z	45.74	57.89	12.15	57.90	12.16

The method allows us the estimation of the central point position of the calibrated sensor. In the case of a pinhole camera the calibration screens were located between 800–1050 mm from the reference camera, whilst the camera to be calibrated was set a few centimeters away (see Fig. 6). In order to obtain a ground truth data, the pinhole camera was calibrated using the classic ray method [19]. Three positions of the optic center are then computed for comparison. The first is given by the classic calibration, the second by the intersection of the rotation axis of estimated cones, and the last by the intersection of all rays (traced as shown in Fig. 9) representing estimated cones. The results are shown in Table 1.

A different single viewpoint is found in each case. There are slight variations in the position of the center in the third coordinate. This can be explained by the fact that the calibrated portion of the sensor used to estimate cones is limited (55×60 pixels located around the center of the image). In this configuration, the depth estimation is obviously less accurate. Concerning the catadioptric sensor, the results show that the cones intersect at a single point. The combination of a parabola and a telecentric lens can only produce a central sensor, which proves the method to be efficient. The estimation of the position of the viewpoint using the principal axis of the estimated cones and all the rays that form the estimated cones produce similar results (in mm: $x = -23.38$, $y = 147.55$, $z = 384.79$ and $x = -23.32$, $y = 147.33$, $z = 385.57$). The mean distance between the rotation axis and their estimated single point is 3.71 mm. The mean distance between the apex and their estimated single point is 2.97 mm.

5 Motion Estimation

5.1 Problem Formulation

Estimating relative camera motion from two calibrated views is a classic problem in computer vision. Till now, this problem has been expressed as minimizing the geometric distances between the measured image features and the reprojected ones with the new motion parameters [10, 11, 18]. In this paper we would like to propose a new geometrical criteria defined with cone intersections which better fits the physical reality while providing more accurate results.

Consider two views I_1 and I_2 of a scene acquired by the same full calibrated projective camera moved from a first to a second location. We assume that this calibration step has provided the intrinsic parameters in the two following ways: Cone-Pixel model for the test presented here and pinhole camera model to ensure a reliable comparison with existing techniques like Bundle Adjustment. We assume that the camera is successively located at $[\mathbf{I}|\mathbf{0}]$ and $[\mathbf{R}|\mathbf{t}]$, where $\mathbf{R}$ is the rotation, and $\mathbf{t}$ is the translation of the motion. Let now X be a set of N 3D feature points and $x_1 \leftrightarrow x_2$ the sets of image points observed respectively in the first and the second image. The problem statement is therefore the following: knowing the measurement x_1 and x_2, how to retrieve the motion $[\mathbf{R}|\mathbf{t}]$.

From a physical point of view and as described above, both rays never strongly intersect (see Fig. 11(a)). The cone defined by the surface of the pixel encompasses all the rays of view of the pixel (Fig. 11(b)), and each ray of view corresponds to the directrix of each cone. A reliable pixel correspondence involves that these two pixel cones of view have a nonnull intersection Fig. 11(c). Noise measurement is not required to encounter this problem, since due to the pixel sampling, it arises in every case of numerical cameras. However, noise will be taken into account in the following section.

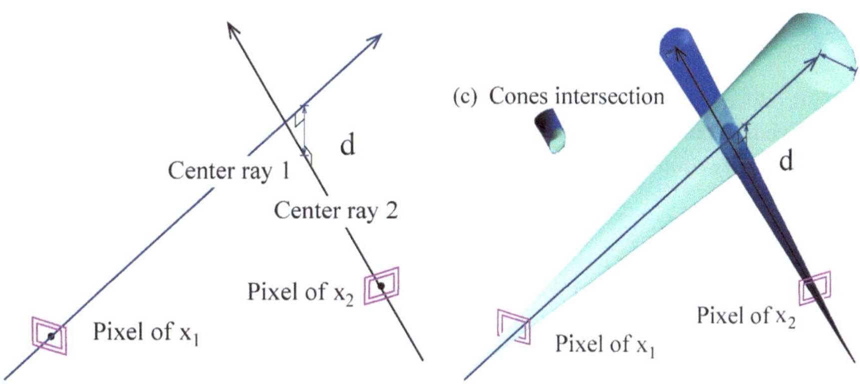

(a) Two rays of light in a 3D space. (b) Cones of light.

Fig. 11 Difference between rays and cones of light intersections

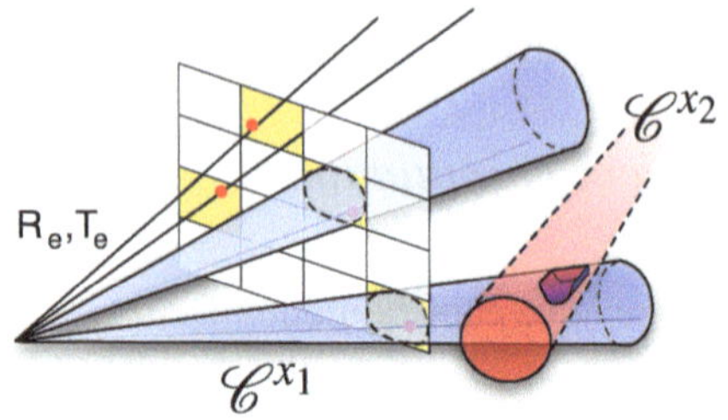

Fig. 12 Motion estimated with cones intersections criterion

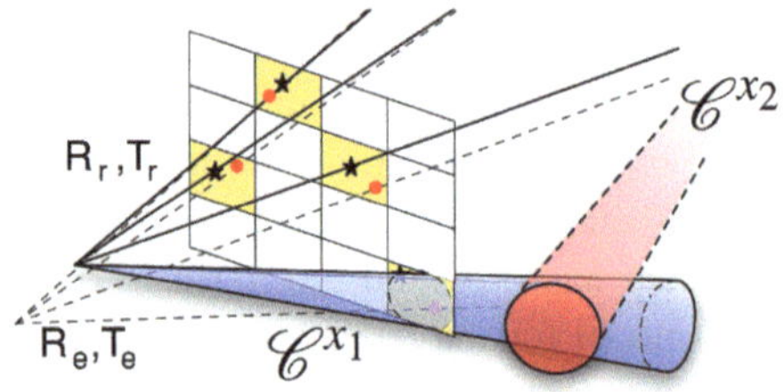

Fig. 13 Motion estimated with reprojection error criterion

It comes then that finding a discrete motion minimizing these quantities for all points could lead to a solution $[\mathbf{R}_r|\mathbf{t}_r]$ which does not provide nonnull cone intersections for all reliable correspondences. An illustration of this major drawback of reprojection error minimization is illustrated in Figs. 12 and 13.

Figure 12 shows an example of four 3D point projections assuming the exact motion $[\mathbf{R}_e, \mathbf{t}_e]$ (for clarity purposes, only a few cones of view are drawn). In this example, most of projected points are located on the pixels' periphery. It follows that using a minimization of the reprojection error as a cost function to estimate a new motion $[\mathbf{R}_r, \mathbf{t}_r]$ could involve the set out of Fig. 13, where the reprojection error is lower (see the black stars representing the reprojected points) even if it eliminates a correspondence of cones which no longer intersect.

5.2 Cone Intersection Score Functions

We introduce here an initial score function S_1 computed according to the following steps: we first compute, for each correspondence, the minimal cone aperture α such that the intersection

$$P_{x_1,x_2} = \left\{\mathfrak{C}^{x_1}(\alpha) \vee \mathfrak{C}^{x_2}(\alpha)\right\} \tag{12}$$

exists (see the (7) in Sect. 3.3 and Fig. 14).

A binary score s_1 (Fig. 15(a)) is provided for each match according to α by the expression

$$s_1(\alpha) = u(\alpha) - u(\alpha - \rho), \tag{13}$$

where $u(\alpha)$ is the classic Heaviside step function ($u(\alpha) = 0$ if $\alpha < 0$ and $u(\alpha) = 1$ otherwise), and ρ is the nominal pixel aperture, which is known from the full cone-

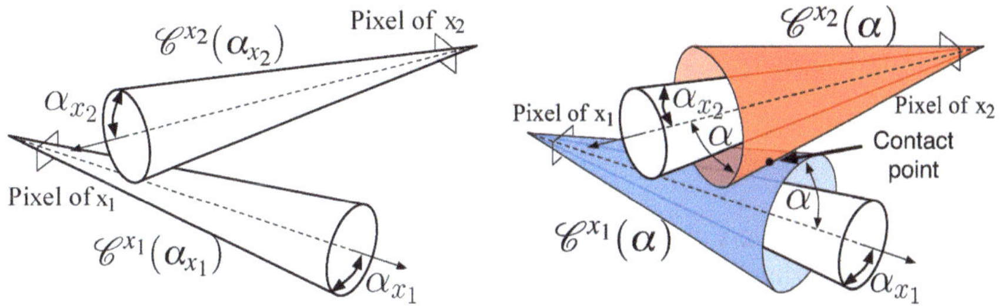

(a) Cones of nominal aperture ρ do not intersect.
(b) Finding the aperture α such as the intersection exists.

Fig. 14 Case of nonintersecting nominal cones

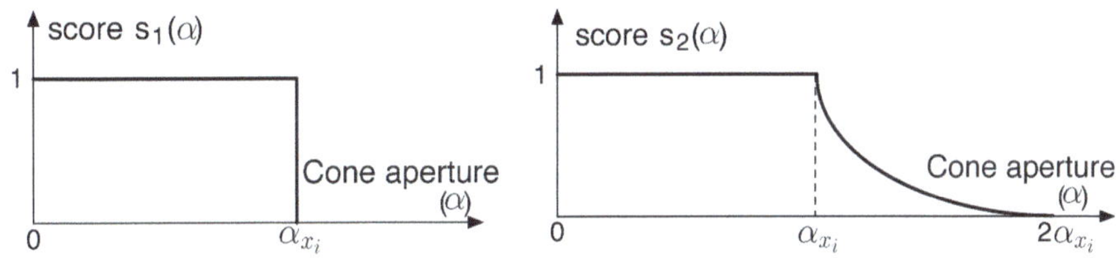

(a) Score function s_1.
(b) Score function s_2.

Fig. 15 Score functions expressions

pixel calibration step. The total score corresponding to the motion estimation is given by

$$S_1(\mathbf{R},\mathbf{t}) = \sum_{n=1}^{N} s_1(\alpha_k). \tag{14}$$

This score function statement is *the number of correspondences with strongly intersecting cones, given a motion estimation*. It comes from this expression that

$$0 \le S_1(\mathbf{R},\mathbf{t}) \le N \quad \text{and} \quad S_1(\mathbf{R},\mathbf{t}) \in \mathbb{N}. \tag{15}$$

Assuming that there is no error in the pixel matching, a solution can be found such that $S_1(\mathbf{R},\mathbf{t}) = N$.

This constraint fits physical reality better than classic reprojection error but is difficult to use because of its discrete formulation. To ease its use in seeking strategies, a second continuous criterion inspired by S_1 is introduced. A score for each correspondence is computed (see Fig. 15(b)) according to

$$s_2(\alpha) = s_1(\alpha) + \left(2 - \frac{\alpha}{\rho}\right)^2 \big(u(\alpha-\rho) - u(\alpha-2\rho)\big); \tag{16}$$

the global score function S_2 is then defined as

$$S_2(\mathbf{R}, \mathbf{t}) = \sum_{n=1}^{N} s_2(\alpha_k). \tag{17}$$

In addition to being very close to the physical reality (it modelizes the small blurring between neighboring pixels pointed out in the calibration experiments, see Fig. 8), this score function provides good results, as will be shown in the following section.

Because of the convex hull computation, this cost function is not analytic. It will not be possible to use classic minimization (gradient descent, nonlinear programming, SOCP) as seeking strategies. The optimal motion solution according to the cone intersection cost function will be found using stochastic optimization. The method chosen here is Simulated Annealing (SA) [12] because of its simplicity and its convergence properties.

5.3 *Simulation Experiments for Motion Estimation Using Cone Intersection Criterion*

The validation of this method has been carried out in simulation to control the whole parameter set. It did not depend on possible error measurement or any noise. Let X be a set of $N = 592$ 3D points generated randomly in the field of view of two identical full calibrated (intrinsic and extrinsic parameters) 640×480 views I_1 and I_2. The optimal solution is then exactly known and will be noted as $[\mathbf{R}_e, |\mathbf{t}_e]$ (index e is chosen for "exact"). Many motion estimation algorithms for this problem have been developed. In this section we will review two of these according to cone intersection score functions. A first motion estimation is provided from the fundamental matrix [5] and is then used as the initial step for a BA [18]. Results provided by BA are finally modified with Simulated Annealing. This sequential protocol enables a step-by-step study of each method with the different cost and score functions discussed here. Results are shown in Table 2. It can be noticed that the score functions S_1 and S_2 can be computed to evaluate and compare the different guesses $(\mathbf{R}, \mathbf{T})$. Therefore, the corresponding columns are not empty even for the Bundle Adjustment and for the Fundamental Matrix. Two particular facts can be drawn from these results:

(i) The reprojection error is greater for the perfect motion than for the BA solution, entailing that in this case BA is not able to reach the correct motion, since it does not correspond to the minimum of the cost function.
(ii) We considered two images of the same scene and 592 pixel correspondences between both images. There is a set of motion (R, T) between both frame capture such that every couple of cones intersect. Then we can be sure that the correct motion is in this set, and consequently, every motion which does not verify this property cannot be a correct motion. Both columns S_1 and S_2 show consequently that BA provides a nonrealistic solution.

Table 2 Motion estimation results: case of a simple pixel sampling

Method	Rep. error	S_1	S_2	Translation error (%)	Rotation error (%)	Iterations
Fundamental matrix	1.4×10^4	538	577.95	0.073	0.13	–
Bundle adjustment	13.80	587	588.57	0.17	0.054	304
Cone approach with SA	28.59	592	592	0.04	0.068	421
Exact motion $[\mathbf{R}_e, \mathbf{t}_e]$	13.84	592	592	0	0	–

Table 3 Motion estimation results: case of a Gaussian noise ($\sigma = 0.5$)

Method	Rep. error	S_1	S_2	Translation error (%)	Rotation error (%)	Iterations
Fundamental matrix	1.32×10^4	309	388.9	0.426	0.58	–
Bundle adjustment	27.97	471	516.52	0.428	0.084	288
Cone approach with SA	41.27	491	584.3	0.64	0.086	506
Exact motion $[\mathbf{R}_e, \mathbf{t}_e]$	28.05	490	534.92	0	0	–

Algorithm 1 Motion Estimation using cone intersection criterion

Require: Initial estimation $[\mathbf{R}_i, \mathbf{t}_i]$, nominal aperture ρ_n, S

while $S \leq S_{ok}$ **do**
  Generate a Guess $[\mathbf{R}, \mathbf{t}]$: $\mathbf{R} = \mathbf{R}_i + \lambda_R$, $\mathbf{t} = \mathbf{t}_i + \lambda_t$
  for $n = 1, n \leq N$ **do**
    find α such as P_{x_1,x_2} exists (12)
    $s_n = s_2(\alpha)$ (16)
  end for
  $S2 = \sum(s_n)$ (17)
  if $S2 > S$ **then**
    $S = S2$
    $\mathbf{R}_i = \mathbf{R}$ and $\mathbf{t}_i = \mathbf{t}$
    $\lambda_R = \frac{\lambda_R}{2}$ and $\lambda_t = \frac{\lambda_t}{2}$
  end if
end while

Rotations provided by both BA and Simulated Annealing are very close (difference less than 1×10^{-4}). However, a significant accuracy have been gained in translation estimation.

A second test is carried out by applying an additive Gaussian noise (standard deviation $\sigma = 0.5$) to the same data x_1 and x_2. Similarly, results provided by the different methods are shown in Table 3. It can be noticed that the results provided by BA are a little closer to the exact motion, compared to those provided by the cone intersection criterion and simulated annealing, even if they remain very similar. The notion of "better results" could not be used strictly speaking, because precisely of the noisy aspect. The reader should keep in mind that the Gaussian noise model used there can obviously be easier handled by a quadratic error minimization instead of a cone intersection criterion. Moreover, the score obtained by this presented method exceed those obtained for the exact motion. We can consider that this solution is satisfying and physically more reliable.

6 Conclusion and Future Works

This paper presented a general method to modelize cameras introducing the use of cones to give a better approximation of the pixels' field of view (rather than the usual use of lines). We also introduced an experimental protocol to estimate cones that is not restricted to any geometry of cameras. The model used Conformal Geometric Algebra that allowed us to handle cones in a simple manner using twists. This formulation enabled the introduction of a new pixel matching characterization as a nonnull intersection of cones of view. On this basis, it was possible to successfully address the motion estimation problem using this characterization as a new score function, with better results than classic ray approach. Simulated Annealing was chosen as a seeking strategy. A large panel of others methods could be used instead. The aim of this paper was not to discuss these strategies but to prove that cone intersection score criterion is closer to the physics and better to address computer vision problems such as motion estimation. Current work is focusing on these seeking strategies and the computation of a direct Cone Adjustment algorithm.

References

1. Barber, C.B., Dobkin, D.P., Huhdanpaa, H.: The quickhull algorithm for convex hulls. ACM Trans. Math. Softw. **22**(4), 469–483 (1996)
2. Benosman, R., Kang, S.: Panoramic Vision: Sensors, Theory, Applications. Springer, Berlin (2001)
3. Debaecker, T., Benosman, R.: Bio-inspired model of visual information codification for localization: from retina to the lateral geniculate nucleus. J. Integr. Neurosci. **6**(3), 1–33 (2007)
4. Grossberg, M.D., Nayar, S.K.: A general imaging model and a method for finding its parameters. In: ICCV, pp. 108–115 (2001)
5. Hartley, R., Zisserman, A.: Multiple View Geometry in Computer Vision. Cambridge University Press, Cambridge (2003)
6. Hestenes, D.: The design of linear algebra and geometry. Acta Appl. Math.: Int. Surv. J. Appl. Math. Math. Appl. **23**, 65–93 (1991)
7. Hestenes, D., Sobczyk, G.: Clifford Algebra to Geometric Calculus. Reidel, Dordrecht (1984)
8. Kahl, F., Hartley, R.: Multiple view geometry under the l-infinity norm. In: PAMI (2008)
9. Ke, Q., Kanade, T.: Quasiconvex optimization for robust geometric reconstruction. In: ICCV (2005)
10. Kim, J.-H., Hartley, R.I., Frahm, J.-M., Pollefeys, M.: Visual odometry for non-overlapping views using second-order cone programming. In: ACCV (2), pp. 353–362 (2007)
11. Kim, J.-H., Li, H., Hartley, R.: Motion estimation for multi-camera systems using global optimization (2008)

12. Kirkpatrick, S., Gelatt, C.D., Vecchi, M.P.: Optimization by simulated annealing. Science **220**, 671–680 (1983)
13. Perwass, C.: Applications of geometric algebra in computer vision. Ph.D. thesis, University of Cambridge (2000)
14. Perwass, C., Gebken, C., Sommer, G.: Geometry and kinematics with uncertain data. In: ECCV (1), pp. 225–237 (2006)
15. Rosenhahn, B.: Pose estimation revisited. Ph.D. thesis, Christian-Albrechts-Universitat zu Kiel, Institut für Informatik und Praktische Mathematik (2003)
16. Rosenhahn, B., Perwass, C., Sommer, G.: Free-form pose estimation by using twist representations. Algorithmica **38**(1), 91–113 (2003)
17. Sommer, G., Rosenhahn, B., Perwass, C.: Twists—an operational representation of shape. In: IWMM GIAE, pp. 278–297 (2004)
18. Triggs, B., McLauchlan, P., Hartley, R., Fitzgibbon, A.: Bundle adjustment—a modern synthesis, pp. 298–375 (2000)
19. Zhang, Z.: Flexible camera calibration by viewing a plane from unknown orientations, vol. 1, pp. 666–673 (1999)
20. Zhang, Z.: A flexible new technique for camera calibration. IEEE Trans. Pattern Anal. Mach. Intell. **22**(11), 1330–1334 (2000)

Model-Based Visual Self-localization Using Gaussian Spheres

David Gonzalez-Aguirre, Tamim Asfour, Eduardo Bayro-Corrochano, and Ruediger Dillmann

Abstract A novel model-based approach for global self-localization using active stereo vision and density Gaussian spheres is presented. The proposed object recognition components deliver noisy percept subgraphs, which are filtered and fused into an ego-centered reference frame. In subsequent stages, the required vision-to-model associations are extracted by selecting ego-percept subsets in order to prune and match the corresponding world-model subgraph. Ideally, these coupled subgraphs hold necessary information to obtain the model-to-world transformation, i.e., the pose of the robot. However, the estimation of the pose is not robust due to the uncertainties introduced when recovering Euclidean metric from images and during the mapping from the camera to the ego-center. The approach models the uncertainty of the percepts with a radial normal distribution. This formulation allows a closed-form solution which not only derives the maximal density position depicting the optimal ego-center but also ensures the solution even in situations where pure geometric spheres might not intersect.

1 Motivation

Autonomous systems require the fundamental capability of self-localization in order to properly process, associate, and interpret the incoming environmental sensor signals and properly act in the environment. Remarkable examples of such systems are humanoid robots operating in *human-centered* environments [1], see Fig. 1(a).

A formal representation of the elements composing the surroundings and their interrelationships is needed to enable the robot to perform complex tasks through the composition of multimodal skills accomplished through a perception–action cycle.

D. Gonzalez-Aguirre (✉)
Humanoids and Intelligence Systems Lab, Karlsruhe Institute of Technology, Adenauerring 2, 76131 Karlsruhe, Germany
e-mail: gonzalez@ira.uka.de

E. Bayro-Corrochano, G. Scheuermann (eds.), *Geometric Algebra Computing*, DOI 10.1007/978-1-84996-108-0_15,

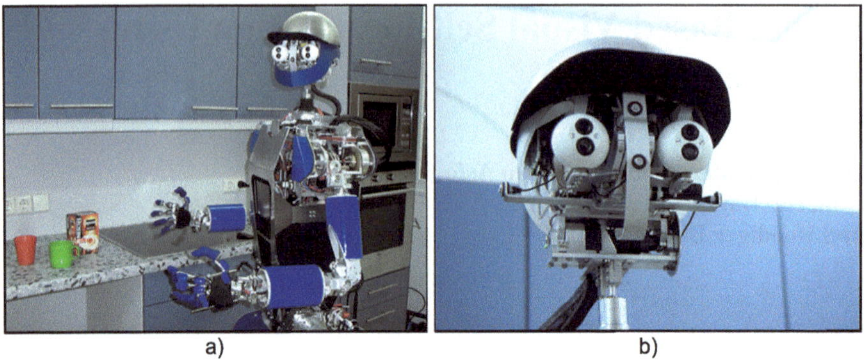

Fig. 1 (**a**) The humanoid robot ARMAR-IIIa and its kitchen environment, see [2]. (**b**) The active vision Karlsruhe humanoid head, equipped with seven DoF and two cameras per eye, see [3]. The wide-angle lens are used for peripheral vision, while the narrow-angle lens are applied for foveated vision

An effective mechanism to achieve the self-localization in these environments ought to profit from the intrinsic topological and geometric structure of the world by either constraining the search within a tailored feature space or by extracting invariant properties of the world elements. This mechanism has to sagaciously face many diminishing factors that complicate the self-localizing task, i.e., the granularity of the model, the nature of the sensors, and the uncertainty of the perception–recognition cycle.

This chapter presents a novel geometric and statistical approach for model-based global self-localization using an active-vision sensing paradigm for humanoid robots. The global localization concerns about the position and orientation (6D-*pose*) of the robot during the initialization.

The natural and inherent usage of conformal geometric algebra [5] arises from the fundamental key idea of using conjuncted restriction subspaces in order to constraint and find the location of the robot. In this manner, the formulation profits from those interesting features of this powerful mathematical framework [6]. For instance, the generalized intersection operator of geometric entities such as planes, lines, spheres, circles, point pairs, and points is an ideal instrument to attain the generation and validation of the ego-center location candidates of the robot.

This proper treatment of subspaces helps to reduce the complexity of the percept-to-model matching by a computationally efficient, conceptually clear, and consistent apparatus for expressing the intersection among the geometric primitives.

In contrast to standard methods in linear algebra, where usually a case-based procedure is applied to determine the intersection subspaces, the conformal geometric algebra provides a generalized mechanism, the meet operator [5, 6].

2 Outline of Visual Self-localization

The upper bar of the Fig. 2 shows the three strata comprising the self-localization. First, the *physical space* encloses the real world where the robot is located. The *visual space* refers to the stratum where the image information from the world is contained.

Finally, the *world-model space* is a graph-based representation of the surroundings consisting of two sublayers, the geometric-level with the 3D vertices and their composition information and the topological-level describing the interrelation of object components.

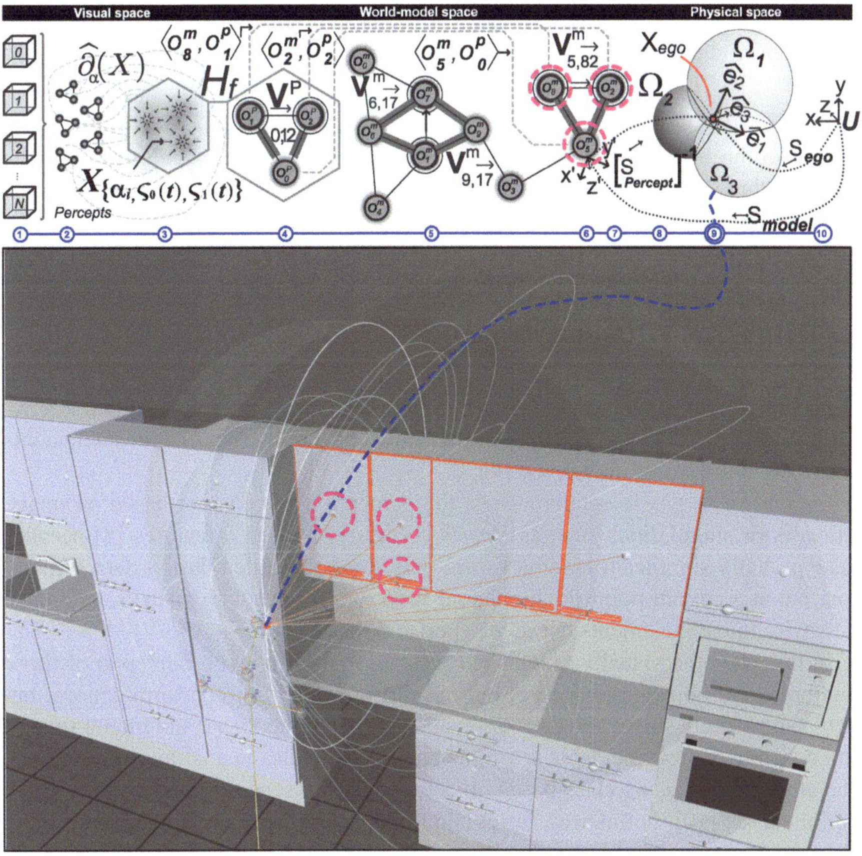

Fig. 2 Model-based visual self-localization approach (see [9]). (**1**) Appearance-based object recognition components. (**2**) Extracted percepts mapped into the ego-frame. (**3**) Multitrial percepts fusion. (**4**) Fused ego-percepts with their corresponding world-model associations. (**5**) Proximity filtering for pruning purposes upon world model. (**6**) Orientation filtering. (**7**) Hypotheses generation. (**8**) Hypotheses validation. (**9**) Geometric and statistical pose-estimation optimization. (**10**) Resulting pose

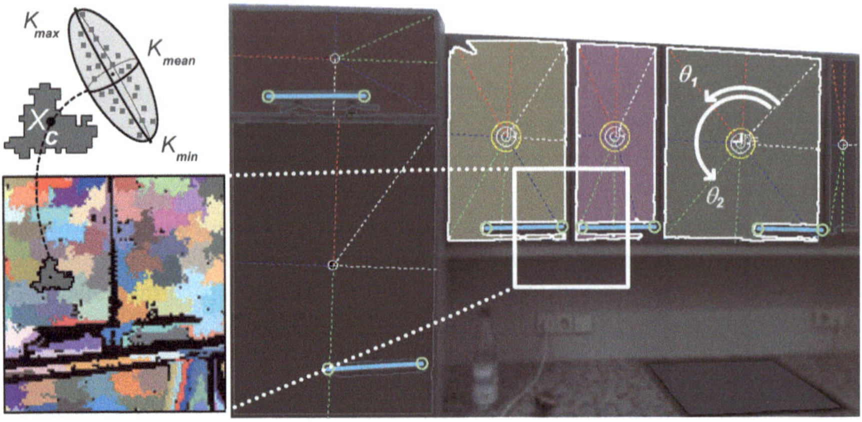

Fig. 3 Results of the class specific object recognition algorithms for door and door-handle, for a detailed description, see [9]

Due to the model-based nature of the problem, the global localization can be split into three sequential phases: visual acquisition of landmarks, data association for model matching, and optimization of pose estimation.

2.1 *Visual Acquisition of Landmarks*

The active-vision perception and recognition components[1] are responsible for delivering the 3D position and orientation (6D-pose) of the instances of those elements described in the world model, see Fig. 2.

In contrast to previous approaches, the perception layer is not based on image saliences or singularities such as *Harris* corners [7] or *SIFT* features [8] because these partially significant landmarks not only imply a burden during data association, but at a certain point the humanoid robot utterly needs to visually recognize the environmental elements in order to perform tasks.

In this way, the visually recognized instances (from now on *Percepts*) of those environmental objects provide not only useful information to perform actions, but they also partially solve[2] the data association between the visual and model spaces. In a concrete context, percepts are doors and door-handles in a building, see Fig. 3. The advantage of using class-based object recognition schema has been previously exploited, see [10]. In this way, faster and more robust methods can be applied.

In contrast to general feature approaches, like in [11], they lack of feature model association, besides offering poor reliability compared to those approaches designed

[1]These are class-specific object recognition modules that were implemented as stated in the authors' previous publication [9].

[2]Up to the class instance association level.

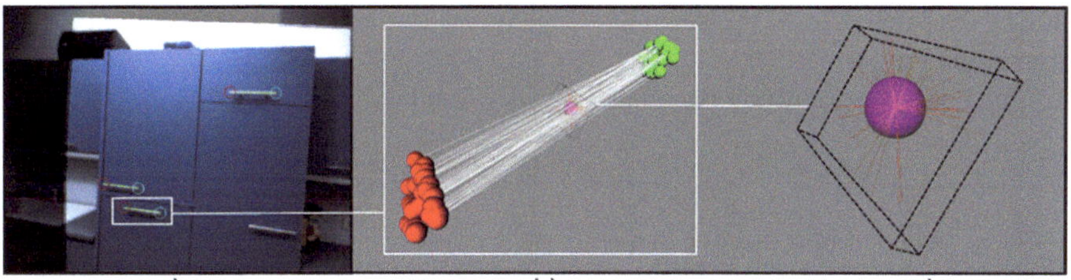

Fig. 4 (**a**) Door-handle percepts recognized during scanning. (**b**) Multiple percepts corresponding to the same element in the world. (**c**) Percepts fused percepts into a stationary point $X_{\{\alpha_i, \varsigma_0(t), \varsigma_1(t)\}}$ of the underlying multimodal density function $\widehat{\partial_\alpha}(x)$, delineation set, and its bounding box

for specific domains. In this implementation, doors and door-handles were robustly recognized by means of Gaussian classification over characteristic feature spaces extracted from class specific descriptors[3] of the eigenvectors[4] from color-segmented regions in stereo images, i.e., 2D recognition. For a detailed description of the methods, the reader is referred to [9]. Many specific recognition components may be added to improve the performance of the system at graph filtering by increasing the partition of the graph, i.e., reinforcing constraints and increasing pruning.

2.2 Data Association for Model Matching

There are two fundamental questions to be answered in order to properly solve the data association:

- How to fuse multiple percepts corresponding to the same world element arising from multiple vantage points, see Fig. 4(b).
- How to match these fused percepts against the world model in order to compose the kinematic chain linking the selected perceptions to the world model, i.e., the backwards transformation from the world to the robot, see (5).

Percepts Fusion

Initially, a reference ego-space frame is defined; it is attached to a references element of the humanoid robot, i.e., a kinematic frame of the robot which remains stationary during the visual scanning phase. Then, the time-varying kinematic chain of transformations coupling the stereo vision system with the ego-frame is taken into account for the registration of the percepts. Subsequently, the percepts acquired

[3]Specific tailored feature vector.

[4]From the covariance matrix of the clustered binary regions.

during discrete steps of the scanning trajectory are mapped into the reference ego-frame, see Fig. 4.

The underlying multimodal spatial density function

$$\widehat{\partial_\alpha}(x) : \mathbb{R}^3 \mapsto \mathbb{R}$$

of the α-type percepts implies that stationary points

$$X_{\{\alpha_i, \varsigma_0(t), \varsigma_1(t)\}}$$

are the high-density locations (α-modes) of elements of α type, i.e., door, window, etc. These points describe the fused locations of the α-elements.[5] Percepts converging to $X_{\{\alpha_i, \varsigma_0(t), \varsigma_1(t)\}}$ constitute the fusion set, i.e., the cluster delineation in [13]. This is the key to properly fuse the multiple view percepts, see Fig. 2(3).

These ideas are commonly used in the nonparametric density estimation techniques as *Parzen* Windows [12] and Mean Shift [13]. The problem of estimating the bandwidth matrix and kernel type is coherently solved by using the geometric class-description of the percept, i.e., the inverse covariance matrix obtained from the 3D vertices of the geometric model.

The *Epanechnikov* [13] kernel was chosen over the Gaussian kernel because of its faster convergence producing only negligible differences in the resulting delineation set compared with the results when using the Gaussian kernel. By exploiting these ideas, the multiple view perceptions are efficiently fused into a common reference space constituting the fused percepts set H_f, see Fig. 2(4).

Fused Percepts Matching

Previously merged landmarks are matched with the model by simultaneously trimming and coupling the elements of the world and those fused percepts, see Fig. 2(5–6).

In order to achieve this mechanism, a graph-based representation of the world was implemented, whereas the fused percepts are arranged into a set of subgraphs according to their spatial distribution.

This coupling process requires to adequately incorporate the previous noisy fused-percept subgraphs as proper constraints to trim the model graph. In this way, the elements in the model which correspond to the selected acquired percepts remain active in the model space. The elements that cannot satisfy the constraints are dismissed.

A selected percept subset could be partially matched against the model by using relative distances and orientations among them, i.e., removing elements which have no relative incidence within the perceived range of relative distances and orientations. These are the key ideas of the *proximity* and *orientation* filtering.

---

[5]In Figs. 2–3 the α elements are the door-handles acquired in multiple views; in this case the α label refers to the class door-handle.

For these purposes, the world has been computationally modeled with two levels of abstraction. The first one describes the geometric composition of the elements and their relative pose. This is basically a CAD[6] structure. On this level the entities are data arrangements with information concerning 3D vertices and their composition describing geometric primitives. In the second level, the latter structures compose instances of *object-model*[7] O_i^m with attributes, e.g., identifier, type, size, and pose.

The collection of object-model instances constitutes the *node set* ν, whereas the *link set*

$$\Lambda \subset \left\{ O_i^m \times O_j^m : O_i^m, O_j^m \in \nu,\ i > j,\ \|X_i - X_j\| < \zeta \right\}$$

depicts the connections $\lambda_{i,j}$ formed by all object model instances with the relative distance[8] falling below $\zeta \in \mathbb{R}$.

Proximity Filtering

When filtering links in the world-model graph, noise is taken into account in the form of deviation parameter ϵ_i of the distance between the perceived-recognized objects[9] O_i^{pf}:

$$\epsilon_i = \frac{1}{\zeta}\left(\|X_i^f - C_L\|\right)^2 \tag{1}$$

with location X_i^f and center of the left camera C_L [15]. The result of the proximity filter is the set of links

$$\psi_{\{\alpha,\beta,\phi,\tau\}} \subset \Lambda$$

connecting nodes of type α to nodes of type β, e.g., door to door-handle, which are separated by a distance ϕ with error-tolerance

$$\tau = \max_{k \in \Theta}(\epsilon_k),$$

where Θ denotes the subset of recognized objects of both types:

$$\psi_{\{\alpha,\beta,\phi,\tau\}} \subset \left\{ O_{(i,\alpha)}^m \times O_{(j,\beta)}^m : \left|\left(\phi - \|X_i - Xj\|\right)\right| < \tau \right\}.$$

The *active link set* ψ_{act} consists of nodes from the intersection of those q proximity filtering partial results

$$\psi_{\text{act}} := \bigcap_{i=1}^{q} \psi_{\{\alpha_i,\beta_i,\phi_i,\tau_i\}}.$$

---

[6]Coin3D: www.coin3d.org.

[7]Note that the superscript "m" emphasizes the model object instance.

[8]The magnitude of the threshold ζ corresponds to the maximal length of the 3D-FOV, see [14].

[9]Note that the superscript "f" emphasizes the fused-percept instance.

Each filtering stage performs a strong reduction of the cardinality of the active link set, because those remaining nodes are tightly constrained, i.e., nodes should have neighbors with restricted types at constrained distance ranges. Fast performance was achieved by using a distance lookup table and filtering only previously selected nodes.

Orientation Filtering

A more powerful, but computationally expensive, technique to reduce the nodes within active link set is attained by accepting only the nodes with incidences having a certain relative pose. In this sense, the definition of the frame transformation has to be consistent while considering the noisy nature of the percept as follows:

First, three noncollinear elements are selected,

$$O_i^{Pf},\ O_j^{Pf},\ \text{and}\ O_k^{Pf} \in H_f;$$

then a frame is specified

$$S_{\text{Percept}}^{i,j,k} = \left[R_{\text{Percept}}^{i,j,k}, X_i^f\right]$$

relative to the ego-perception frame[10]

$$\widehat{\delta_1} = \frac{X_j^f - X_i^f}{\|X_j^f - X_i^f\|}, \qquad \widehat{\delta_2} = \frac{[\widehat{\delta_1} \wedge (X_k^f - X_i^f)]^*}{\|[\widehat{\delta_1} \wedge (X_k^f - X_i^f)]^*\|}, \qquad \text{and} \quad \widehat{\delta_3} = \frac{[\widehat{\delta_1} \wedge \widehat{\delta_2}]^*}{\|[\widehat{\delta_1} \wedge \widehat{\delta_2}]^*\|},$$

which leads to

$$R_{\text{Percept}}^{i,j,k} = \left[\widehat{\delta^n} \cdot \widehat{e_n}\right]_{n=1\ldots3}.$$

Note that these computations take place in $\mathbf{G}_{(3,0)}$, and thus the dual of the wedge product of two vectors corresponds to the cross product in vector calculus.

Next, the relative displacement from O_j^{Pf} to O_k^{Pf} expressed on the frame of perception is computed:

$$V_{i,\mathbf{jk}}^{Pf} = S_{\text{Percept}}^{i,j,k}\left(X_j^{Pf} - X_k^{Pf}\right).$$

Such a vector merges the relative orientations of the three percepts in a signature-like consistent manner. Therefore, it is possible to reject nodes which do not have a "*similar*" displacement vector among two of the neighbors with corresponding type and proximity. This *noisy similarity* is quantified by the length and angle discrepancies μ and κ between the perception signature $V_{i,\mathbf{jk}}^{P}$ and the model signature $V_{u,\mathbf{uw}}^{m}$ vectors, expressed on the world model $S_{\text{model}}^{u,w,v}$.

---

[10]With orthonormal basis vectors $\{\widehat{e_1}, \widehat{e_2}, \widehat{e_3}\}$.

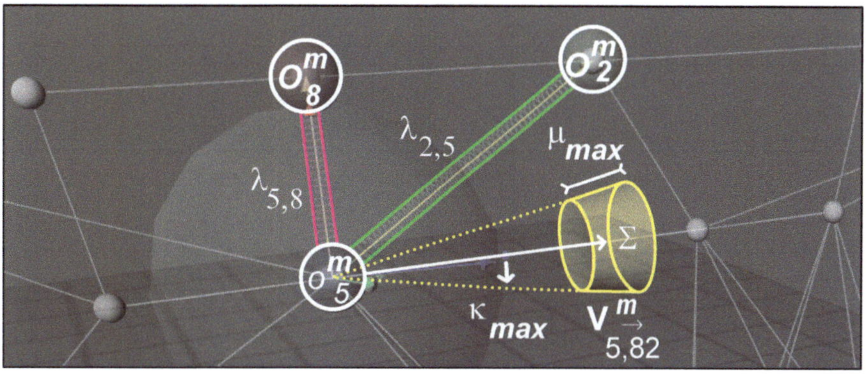

Fig. 5 World-model graph at pruning by means of proximity and orientation filtering. Example of accepted node O_5^m with vector $V^m_{5,\mathbf{8},\mathbf{2}}$ inside Σ. Notice that the subspace Σ corresponds to the boolean subtraction of two spherical cones [4]. The aperture of the implicit cone depicts the noise parametric tolerance of the orientation filtering, see (3). The radii of both implicit spheres differ by $\mu_{\max}$, i.e., the proximity filtering noise parametrical tolerance, see (2)

Figure 5 shows the subspace Σ bounded by

$$\left\| V^p_{i,\mathbf{jk}} - V^m_{u,\mathbf{vw}} \right\| < \mu_{\max}, \tag{2}$$

$$\arccos(\widehat{V^p_{i,\mathbf{jk}}} \cdot \widehat{V^m_{u,\mathbf{vw}}}) < \kappa_{\max}. \tag{3}$$

When filtering a node, the combinational explosion is avoided by computing only the subgraphs with link lengths falling into the range

$$\left(\left\| V^p_{i,\mathbf{jk}} \right\| - \mu_{\max}\right) < \left\| \left[O_j^{p^f}, O_k^{p^f} \right] \right\| < \left(\left\| V^p_{i,\mathbf{jk}} \right\| + \mu_{\max}\right).$$

2.3 Pose-Estimation Optimization

Previously extracted model subgraphs that simultaneously match the typed incidences and relative pose of those acquired *percepts subgraphs*, embody the association coupling the *visual space*, *world model*, and *physical world*.

They simultaneously impose restraints which are the *geometric-compelling keys* to deduct the pose of the robot. Each association

$$\left\langle O_i^{p_f}, O_j^m \right\rangle$$

constraints the position of the robot to the subspace of all points that are $\|X_i^{p_f}\|$ units away from X_j^m. This subspace is actually the surface on a sphere, i.e.,

$$\underbrace{\Omega\left\langle O_i^{p_f}, O_j^m \right\rangle}_{\text{Restriction Subspace}} := X_j^m + \frac{1}{2} \underbrace{\left(\|X_j^m\| - \|X_i^{p_f}\|\right)}_{\text{Perception-Model Matching}} e_\infty + e_0 \in PK^3 \tag{4}$$

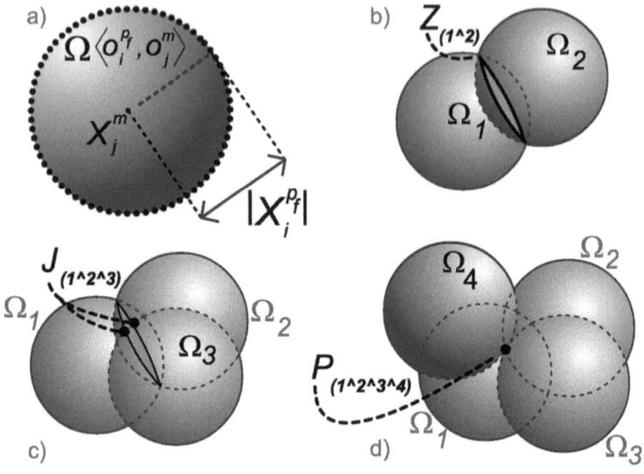

Fig. 6 (**a**) Constrained-subspace embodies the surface of the sphere. (**b**) Cooccurring constrained-subspaces depicting a circle. (**c**) Three constrained-subspaces acting in conjunction yielding to a point pair. (**d**) Four constrained-subspaces yielding to a *simultaneity* point, i.e., the point within the intersection of these four constrained-subspaces

centered at X_j^m with radius $\|X_i^{Pf}\|$, see Fig. 6(a).

Note that the sphere in (4) is an element of the conformal geometric space PK^3, which has the Clifford algebra signature $\mathbf{G}_{(4,1)}$, see [5].

For a single percept, this idea provides no benefit, but on second thought, when observing the same concept with two different percepts, it turns out to be a very profitable formulation because the ego-center should reside in both constrained subspaces, meaning that it has to be on the surface of both spheres at the same time.

Consider two *restriction spheres* simultaneously constraining the position of the robot,

$$\Omega_1\langle O_i^{Pf}, O_j^m\rangle \quad \text{and} \quad \Omega_2\langle O_k^{Pf}, O_l^m\rangle;$$

they implicate that the position of the robot belongs to both subspaces. Thus, the restricted subspace is a circle, i.e., an intersection of spheres, see Fig. 6(b),

$$Z_{(1\wedge 2)} = \Omega_1\langle O_i^{Pf}, O_j^m\rangle \wedge \Omega_2\langle O_k^{Pf}, O_l^m\rangle.$$

Following the same pattern, a third sphere Ω_3 enforces the restriction to a point pair

$$J_{(1\wedge 2\wedge 3)} = Z_{(1\wedge 2)} \wedge \Omega_3\langle O_r^{Pf}, O_s^m\rangle,$$

i.e., circle–sphere intersection, see Fig. 6(c). Finally, a fourth sphere Ω_4 determines the position of the robot, i.e., the intersection point from the latter point pair, see Fig. 6(d),

$$P_{(1\wedge 2\wedge 3\wedge 4)} = J_{(1\wedge 2\wedge 3)} \wedge \Omega_4\langle O_t^{Pf}, O_h^m\rangle.$$

Latter concepts outline a technique which uses the previously partially matched elements of the world model and process them by a geometric apparatus for generating the ego-center candidates. This apparatus uses the centers of the spheres within the model space and the radii from the fused-percepts, see Fig. 2(6–9). The formulation and treatment of the uncertainty acquired during perception is presented in Sect. 3.

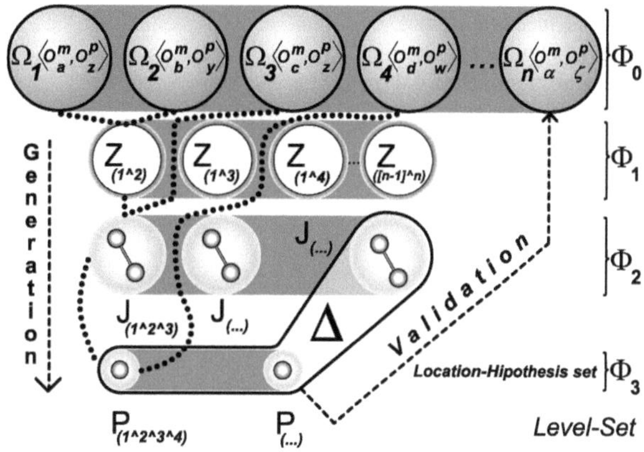

Fig. 7 Location hypotheses generation–validation mechanism systematically manages the location hypotheses

The computational complexity of this *location hypotheses* management process is upper bounded by $\mathbf{O}(n^4)$, where n is the cardinality of the subset of percept-spheres. The amount of spheres n is never greater than 6 while generating candidates; besides, in rare cases the internal partial result is that the intersection stages are densely populated. This could be easily seen when intersecting two spheres. The resulting circle occupies a smaller subspace which in successive stages meets only fewer remaining spheres. One important factor why there are less operations in this combinational computation is because the child primitives that result from the intersection of parent spheres should not be combined with their relatives avoiding useless computation effort and memory usage.

Hypotheses Generation

Each percept subgraph is used to produce the *zero-level set*, composed of spheres, see Fig. 7,

$$\Phi_0 = \left\{\Omega_\zeta\langle O_i^m, O_j^p\rangle\right\}_{\zeta=1\ldots n}.$$

These spheres are then intersected by means of the *wedge* operator $\wedge$ in an *upper triangular* fashion producing the *first-level set* Φ_1 containing circles.

The *second-level set* Φ_2 is computed by intersecting the circles with spheres from Φ_0 excluding those directly above. Then the latter resulting point-pairs are intersected in the same way creating the highest possible level (*third-level set*) Φ_3; here the points resulting of the intersection of four spheres are contained.

Finally, elements of Φ_2 that have no descendants in Φ_3 and all elements on Φ_3 represent the location hypotheses

$$\Delta := \bigwedge_{\xi} \Omega_\xi\langle O_i^m, O_j^p\rangle.$$

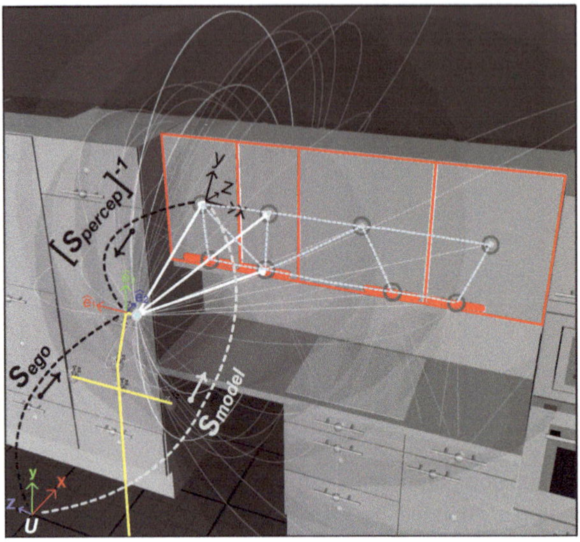

Fig. 8 Kinematic frames involved in the ideal visual self-localization. Notice the directions of the coupling transformations in order to reveal the frame S_{ego}

Hypotheses Validation

Hypotheses are checked by selecting associations, see Figs. 2–8,

$$\langle O_i^{p_f}, O_j^m \rangle$$

that were not considered in the generation of the current validating hypothesis. In case there is more than one prevailing hypothesis, which rarely happens in nonsymmetric repetitive environments, an active validation needs to take place selecting objects from the model and then localizing them in the visual space. The criterion to select the discriminator percept $D_{i,j}^m$ (*priming instance*) is the maximal pose difference between hypotheses pairs.

Ideal Pose Estimation

Once the location hypothesis has revealed, the position of the robot X_{ego} (see Fig. 8) and the orientation S_{ego} are expressed as

$$\underbrace{S_{\text{ego}}}_{\text{Self-Localization}} = \underbrace{S_{\text{model}}^{u,w,v}}_{\text{Model-Matching}} \underbrace{\left[S_{\text{Percept}}^{i,j,k}\right]^{-1}}_{\text{Visual-Perception}}, \tag{5}$$

which is actually the transformation from the kinematic chain that couples the world-model frame S_{model} (*forwards*) and the perception frame $[S_{\text{Percept}}^{i,j,k}]^{-1}$ (*backwards*), see Fig. 8.

There are situations where a variety of diminishing effects alter the depth calculations of the percepts in a way that the ideal pose calculation may not be robust or could not be assessed. The subsequent sections describe the sources and nature of the uncertainties, which are modeled and optimized by the proposed technique to determine the optimal location of the robot, i.e., the maximal probabilistic position.

3 Uncertainty

The critical role of the uncertainty cannot only strongly diminish the precision of the estimated pose, but it can also prevent the existence of it by drawing away the intersection of the restriction subspaces, i.e., the spheres might not intersect due to numerical instability and errors introduced by the perception layer.

In order to sagaciously manage these conditions and other derived side effects, it is crucial to reflect upon the nature of the acquired uncertainties regarding this localization approach. There are two remarkable categorical sources of uncertainty, *image-to-space* and *space-to-ego* uncertainties.

3.1 Image-to-Space Uncertainty

Image-to-space uncertainty is obtained from the appearance-based vision recognition process. It begins with the pixel precision limitations, e.g., noise, discretization, quantization, etc., and ends with the error limitations of the camera model and its calibration, e.g., radial–tangential distortion and intrinsic parameters [16]. This uncertainty could be modeled, according to the central limit theorem [17], as a normal distribution where the standard deviation σ_i is strongly related to the perception depth ρ_i:

$$\sigma_i \cong \frac{1}{\zeta}\rho_i^2, \tag{6}$$

where $\zeta > 1 \in \mathbb{R}$ is an empirical scalar factor depending on the resolution of the images and the vergence angle of the stereo rig, whereas the perception depth

$$\rho_i = (x_i - C_L) \cdot \hat{e}_d \tag{7}$$

depicts the distance between camera center C_L and point in space x_i along the stereo rig normal vector $\hat{e}_d$, see Fig. 9. This deviation model arises from the following superposed facts: first, considering only the monocular influence in each camera of the stereo rig.

The surface patch A_i on the plane perpendicular to the optical axis of the camera imaged into a single pixel P_A grows as function of the distance ρ_i:

$$A_i = \rho_i^2 \tan\left(\frac{\theta_h}{h}\right) \tan\left(\frac{\theta_v}{v}\right),$$

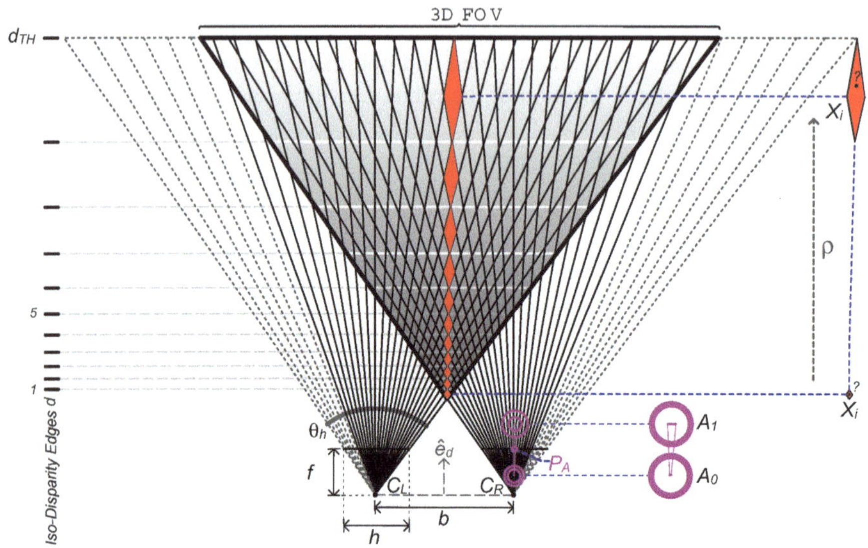

Fig. 9 The image-to-space uncertainty factors in a front-parallel configuration

where θ_h and θ_v are the horizontal and vertical angular apertures of the field of view, whereas h and v depict the width and height resolutions of the image, see Fig. 9.

Consequently, the stereo triangulation has an additional effect during the estimation of the 3D position $\mathbf{M}_{\text{stereo}}(X_i)$ of a matched point pair. The distance ρ_i affects the magnitude of the disparity d_i. Therefore, the precision of the pixel computations plays a decisive role, i.e., the 3D space points which are closer to the base line have wider disparities along the epipolar lines, meanwhile the points located after distance $\rho_{Th} > fb$ have a very narrow disparity, falling in the subpixel domain $d < 1$, which results in inaccurate depth calculations.

This situation also produces a sparse distribution of the iso-disparity surfaces [15], meaning that the subspace contained between this surface-strata grows as

$$d_i = \frac{fb}{\rho_i}, \tag{8}$$

where the focal distance f and the base line size b play relevant roles in the measurement precision

$$b = \|C_L - C_R\|.$$

Figure 9 shows the ideal front parallel case iso-disparity edges delineating the subspaces contained between two discrete steps in the disparity relation of (8).

In this manner, points contained within one of these subspaces produce the same discrete disparity when matching corresponding pixels. Hence, the location uncertainty should be proportional to the distance contained between iso-disparity surfaces. These two applied factors produce an uncertainty growing in an attenuated

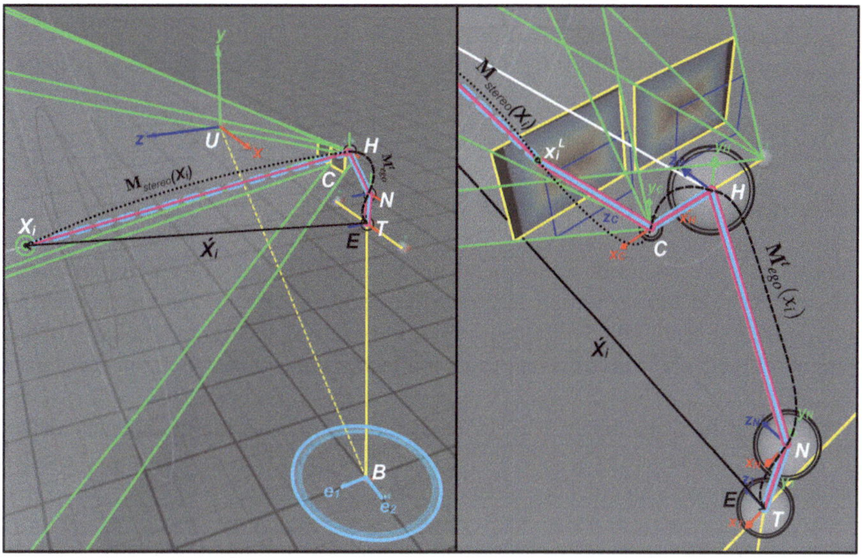

Fig. 10 The space-to-ego uncertainty acquisition process produced by the mapping of percepts from camera coordinates to the ego-frame. (**a**) The whole transformation $\acute{X}_i = \mathbf{M}^t_{\text{ego}}(\mathbf{M}_{\text{stereo}}(X_i))$. (**b**) The transformation $\mathbf{M}^t_{\text{ego}} = [T_{(t)} N_{(t)} H C_L]^{-1}$

quadratic fashion, which is reflected in the model as a deviation spreading in the same pattern reflected upon (7).

3.2 Space-to-Ego Uncertainty

The space-to-ego uncertainty is caused while relating the pose of the percepts from the left camera frame to the ego-frame, i.e., head-base frame of the humanoid robot, see Fig. 10(a).

It is caused by the physical and measurement inaccuracies, which are substantially magnified by projective effects, i.e., the almost negligible errors in the encoders and mechanical joints of the active head of the humanoid robot are amplified proportionally to the distance ρ_i between the ego-center and the location of the percept.

Figure 10(b) shows the kinematic chain starting at x_i^L, the left camera coordinates of the space point X_i. Subsequently, the transformation from the left camera frame C_L to the shoulders base $T(t)$ passing through the eyes base H and neck frame $N(t)$ is given by

$$\acute{X}_i = \mathbf{M}^t_{\text{ego}}(x_i), \tag{9}$$

$$\mathbf{M}^t_{\text{ego}} = [T_{(t)} N_{(t)} H C_L]^{-1}, \tag{10}$$

where $\mathbf{M}^t_{\text{ego}}$ is the ego-mapping at time t. Here, the transformations $T_{(t)}$ and $N_{(t)}$ are time-dependent because they are active during the execution of the scanning strategy, see Fig. 10(b).

4 Geometry and Uncertainty Model

Once the visual recognition components have provided all classified percepts within a discrete step of the scanning trajectory, these percepts are mapped into the reference ego-frame using (9). This ego-frame is fixed during the scanning phase. In this fashion all percepts from different trials are located in a static common frame, see Fig. 10(b).

The unification-blending process done by the fusion phase simultaneously allows the rejection of the percepts that are far from being properly clustered and creates the delineation set which is later melted into a fused percept.

Next, the geometric and statistical phase for determining the position of the robot based on intersection of spheres is properly formulated by introducing the Gaussian sphere and its apparatus for intersection-optimization.

4.1 Gaussian Spheres

The considered restriction spheres Ω_i are endowed with a soft density function

$$\widehat{f}(\Omega_i, x), \quad \Omega_i \in PK^3, \ x \in \mathbb{R}^3 \mapsto (0, 1] \in \mathbb{R}.$$

The density value decreases exponentially as a function of the distance from an arbitrary point x to the surface of the sphere Ω_i:

$$S(x, X_i, r_i) = \left|\left(\|x - X_i\| - r_i\right)\right|, \tag{11}$$

$$\widehat{f}(\Omega_i, x) = e^{\frac{-S(x, X_i, r_i)^2}{2\sigma_i^2}}. \tag{12}$$

The latter function depicts the nonnormalized[11] radial normal distribution

$$\check{N}\left(\mu := \left\{x \mid \ker\left(S(x, X_i, r_i)\right)\right\}, \sigma_i^2\right)$$

for x to be in the surface of Ω_i, i.e., the null space of $S(x, X_i, r_i)$. Note that here the standard deviation σ_i refers to (6).

The density of a point x in relation with a sphere Ω_i represents the nonnormalized probability for the point x to belong to the surface of the sphere Ω_i. Obviously the maximal density is on the surface of the sphere itself.

[11] By the factor $\frac{1}{\sigma\sqrt{2\pi}}$.

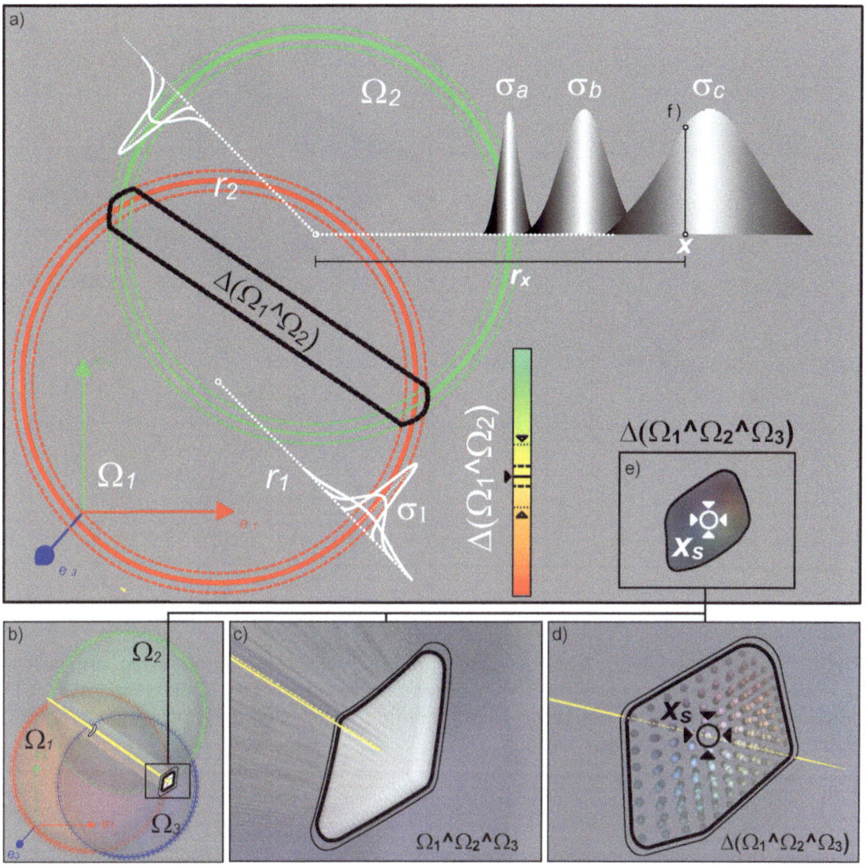

Fig. 11 Gaussian spheres meeting. (**a**) Two Gaussian spheres meeting $\Omega_1 \wedge \Omega_2$ describing a density-subspace $\Delta(\Omega_1 \wedge \Omega_2)$. (**b**) Three Gaussian spheres $\Omega_{i=1,2,3}$ meeting in two regions depicting a subspace $\Omega_1 \wedge \Omega_2 \wedge \Omega_3$. (**c**) Detailed view of one of the previous subspaces. (**d**) Discrete approximation of the maximal density location x_s. (**e**) Details of the implicit density-space $\Delta(\Omega_1 \wedge \Omega_2 \wedge \Omega_3)$. (**f**) (*Upper-right*) Implicit radius r_x when estimating the density at position x

It is necessary to propose an effective mechanism which applies intersections of restriction spherical subspaces as the essential idea for determining the robot position. The nature of the applied intersection has to consider the endowed spatial density of the involved Gaussian spheres.

In the following sections, the restriction spheres and their conjuncted composition properly model both uncertainties, allowing the meeting of spheres by finding the subspace where the maximal density is located, see Fig. 11.

This could be interpreted as an isotropic dilatation or contraction of each sphere in order to meet at maximal density of the total density function, see Figs. 12 and 13.

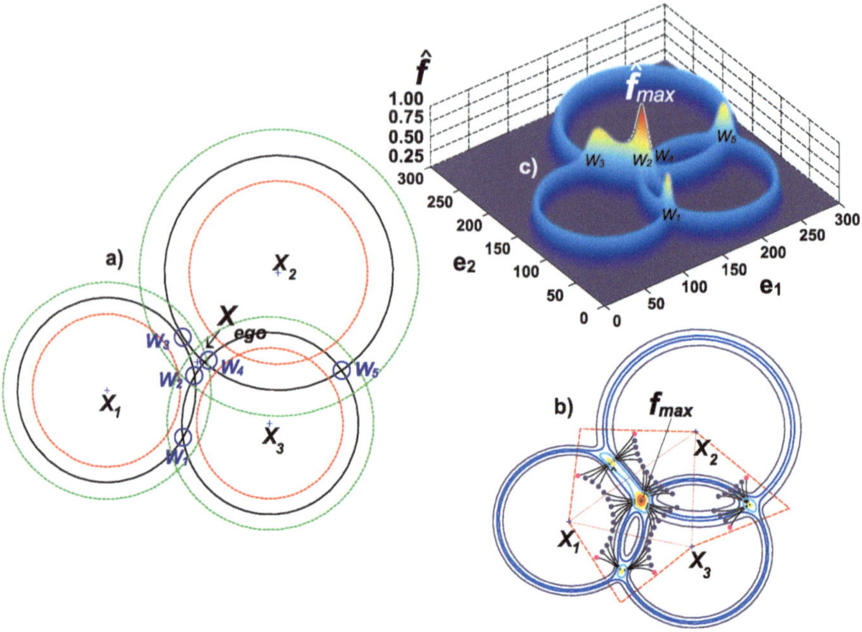

Fig. 12 Gaussian circles, i.e., 2D Gaussian spheres. (**a**) Three Gaussian circles setup. (**b**) The total accumulative density $\widehat{f}_c(x) = \sum_i^n \widehat{f}(\Omega_i, x)$ allows a better visualization of the composition of its product counterpart $\widehat{f}_t(x)$, see also Fig. 13. (**c**) Density contours with seeds and their convergence by means of gradient ascendant methods

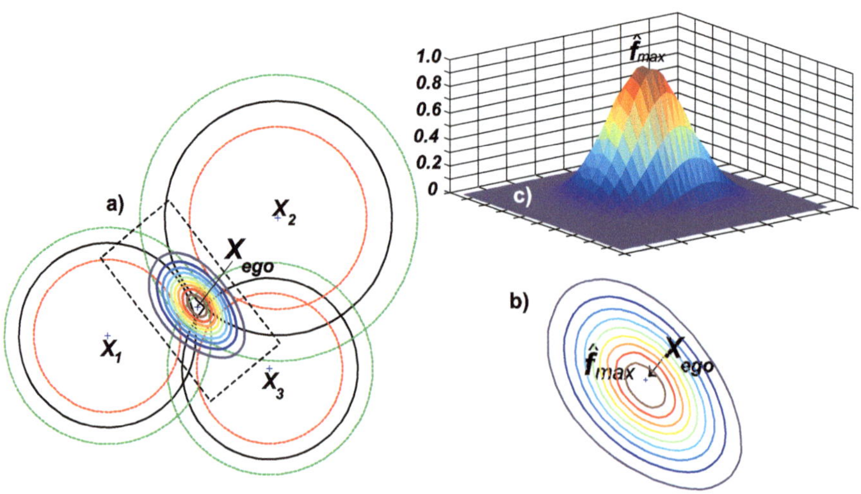

Fig. 13 The Gaussian circles, i.e., 2D Gaussian spheres. (**a**) Three Gaussian circles setup. (**b**) The total density $\widehat{f}_t(x) = \prod_i^n \widehat{f}(\Omega_i, x)$. (**c**) Density contours and ego-center X_{ego}; notice that the resulting distribution is not Gaussian

$$\widehat{f_t}(x) \longrightarrow (0, 1] \in \mathbb{R}, \quad x \in \mathbb{R}^3, \tag{13}$$

$$\widehat{f_t}(x) = \prod_i^n \widehat{f}(\Omega_i, x). \tag{14}$$

Due to the geometric structure composed by n spheres, it is possible to foresee the amount of peaks and the regions $\mathbf{W}_s$ where the density peaks are located, see Fig. 12(c). Therefore, it is feasible to use state-of-the-art gradient ascendant methods [18] to converge to the modes using multiple seeds. These should be strategically located based on the spheres centers and intersection zones, see Fig. 12(b).

Finally, the seed with maximal density represents the solution position x_s,

$$x_s = \arg\max \widehat{f_t}(x). \tag{15}$$

However, there are many issues of this shortcoming solution. The iterative solution has a precision limited by the parameter used to stop the shifting of the seeds. In addition, the location and spreading of the seeds could have a tendency to produce undesired oscillation phenomena, under- or oversampling and all other disadvantages that iterative methods present.

The optimization expressed by (15) could be properly solved in a convenient closed form. In order to address the solution x_s, it is necessary to observe the configuration within a more propitious space, which simultaneously allows an advantageous representation of the geometrical constraint and empowers an efficient treatment of the density, i.e., incorporating the measurements according to their uncertainty and relevance while avoiding density decay.

4.2 Radial Space

The key to attain a suitable representation of the latter optimization resides in the exponent of (12). There, the directed distance from a point x to the closest point on the surface of the sphere is expressed by (11). When considering the total density function, see (14), it unfolds the complexity by expressing the total density as a tensor product.

The inherent nature of the problem lies in the radial domain, i.e., the expression $S(x, X_i, r_i)^2$ is actually the square magnitude of the difference between the radius r_i and the implicit defined radius r_x between the center of the spheres X_i and the point x, see Fig. 11(f). Hence, the optimization configuration can be better expressed in radial terms, and the geometrical constraints restricting the relative positions of the spheres are properly and naturally clarified in the following sections.

4.3 Restriction Lines

Consider the case of two spheres Ω_1 and Ω_2, see Fig. 14(a). Here, the radii of both spheres and the distance between their centers

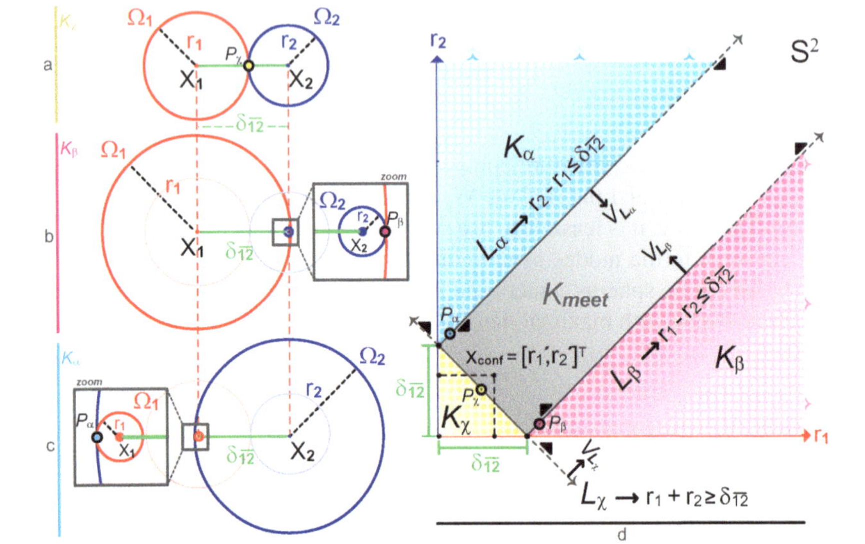

Fig. 14 The spheres's intersection restriction lines derivation in the radial space $\mathbf{S}^2$. (**a**) The line L_χ is the first restriction for ensuring nonempty intersection of spheres. (**b**) The derivation of reminding right side empty intersection restriction line L_β. (**c**) The left side symmetric case, generating the third restriction Line L_α

$$\delta_{\overline{1,2}} = \|X_1 - X_2\| = \sqrt{-2(\Omega_1 \cdot \Omega_2)}$$

allow the formulation of the geometric restrictions which ensure the intersection of the spheres in at least a single point P_χ.

These restrictions are expressed by the inequality line L_χ which describes the radial configuration subspace represented by pairs of the form

$$P_\chi = [r_1, r_2]^T \in \mathbf{S}^2,$$

the intersection of spheres $\Omega_1 \wedge \Omega_2$, i.e., a circle with zero radius, where the $\mathbf{S}^2$ refers to the radial configuration space of two spheres.

Note that in Fig. 14(d), the inequality line divides the configuration space into two regions. The half space holding the restriction imposed by the inequality line L_χ still contains configurations which produce no intersection of spheres, in fact any configuration holding

$$r_2 \geq \delta_{\overline{1,2}} + r_1.$$

In order to prevent these degenerated configurations two additional restriction inequality lines arise, unveiled by following similar pattern.

In the same fashion, Fig. 14(b) shows the case where the minimal contact point P_β occurs, subject to

$$r_1 \geq \delta_{\overline{1,2}} + r_2.$$

In this configuration subspace, the sphere Ω_1 fully contains the sphere Ω_2, and their surfaces intersect solely at P_β. Once again, in order to ensure at least this contact point, the fluctuation of the radii of both spheres is restricted by a linear relation expressed by the inequality line L_β. The latter restriction actually happens in a symmetric manner by interchanging the roles of Ω_1 and Ω_2, resulting in a third restriction, i.e., the inequality line L_α, see Fig. 14(c–d).

As a result, the space is divided in four regions K_α, K_β, K_χ, and K_{meet}, all open except K_χ. Only those configurations within the subspace K_{meet} represent nonempty intersections of spheres, e.g., the point x_{conf} in Fig. 14(d) with

$$x_{\text{conf}} = \left[r_1', r_2'\right]^T \in K_{\text{meet}}.$$

The edge surface separating K_{meet} from the other regions depict single-point intersections of spheres, whereas elements within K_{meet} represent intersection depicting a circle with nonzero radius.

Latter conceptualization soundly compounds the distance among centers of the spheres with their radii. It produces a robust and general criterion to establish intersection guarantee, see Fig. 14(d).

4.4 Restriction Hyperplanes

The previous derivation of the restriction lines was achieved by considering only the case involving two spheres; however, it is possible to extend these restrictions to n spheres.

Formally, this affirmation is theoretically supported by representing the n sphere radial configuration space $\mathbf{S}^n$ as the *Hilbert* space $\mathbf{C}^n$, where each dimension depicts the radius of one sphere. An element $x_{\text{conf}} \in \mathbf{S}^n$ of the n-dimensional radial configuration space can be uniquely specified by its coordinates with respect to orthonormal basis vectors

$$\hat{e}_i \in \mathbf{S}^n \mid i \in \{1, \dots, n\} \subset \mathbb{Z},$$

which are, as expected in a *Hilbert* space, perpendicular to each other, because the radius of each sphere is independent from the others. In this manner, the previous restriction lines could be perpendicularly extruded in $n-2$ dimensions creating the restriction hyperplanes $\Phi_\alpha^{(i,j)}$.

Here again, each hyperplane divides the space in two subspaces. Configurations within the region opposite to the normal vector V_{L_α} (back of the hyperplane) represent nonintersecting spheres, see Fig. 15.

Even more, the set of hyperplanes expressed in their Hesse normal form could be used to compose a matrix inequality

$$\mathbf{A}x \leq b, \tag{16}$$

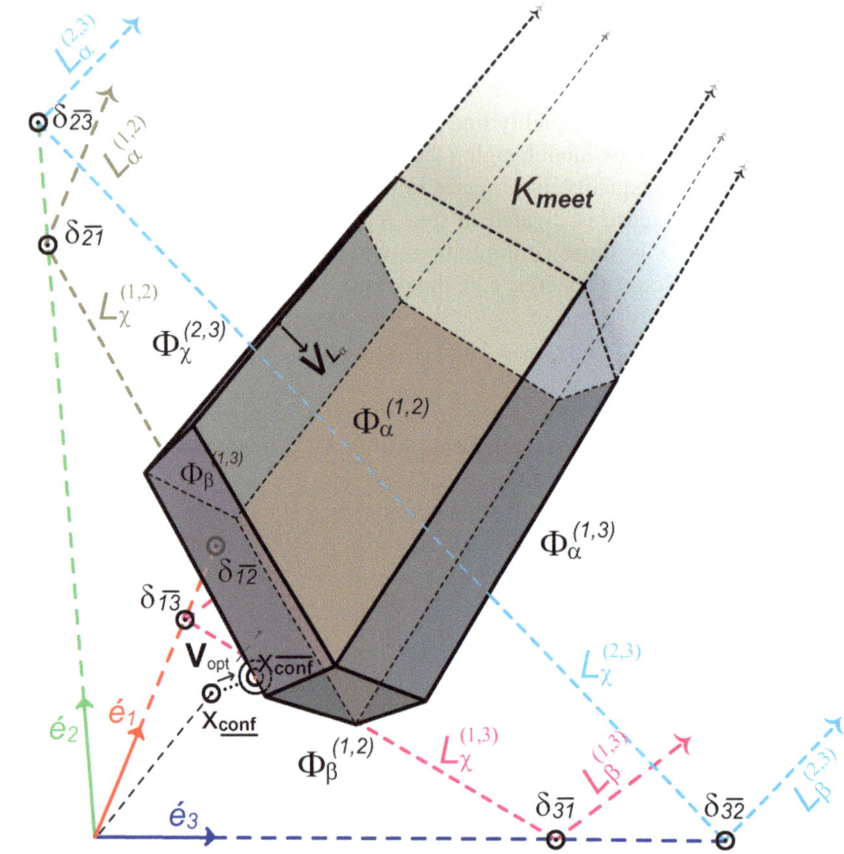

Fig. 15 The radial density space $\mathbf{Sb}^3$ containing the open polytope which delineates the subspace K_{meet}. The transformation-optimization vector V_{opt} implies an isotropic variation in the underlying density domain while creating a general dilatation within the implicit radial domain

where $\mathbf{A}$ is an $m \times n$ matrix with m bounding half-spaces (normal vectors of the hyperplanes), and b represents an $m \times 1$ column vector formed by stacking the Hesse distances of the hyperplanes, i.e., an open polytope, see Fig. 15.

Consider the case where $n = 3$. Three spheres implying an open polyhedron, within the radial space each line

$$L_\alpha^{(i,j)},\ L_\beta^{(i,j)},\ \text{and } L_\chi^{(i,j)}$$

could be extruded in the complementary dimension creating restriction planes given by $\Phi_\alpha^{(i,j)}$. Next, the face cells, ridges, and vertices of the polytope are found using a simple and fast implementation for *vertex enumeration* [19], see Fig. 15.

At this stage, it could be conveniently established whether the current configuration is valid, in other words, determine if the point x_{conf} belongs to the polytope.

This assertion is formally given by

$$\mathbf{A}x_{\text{conf}} < b.$$

In case this assertion is held, there is no need to go through the following optimization phase because of the spheres meeting on their surface, resulting the maximal density

$$\widehat{f}\left(x_{\overline{\text{conf}}}\right) = 1.$$

The opposite situations represent the degenerated configurations resulting from noisy measurements and previously discussed errors. For instance, the point $x_{\underline{\text{conf}}}$ represents an invalid configuration, outside of the polytope where no intersection of spheres exists, see Fig. 15.

The target solution for the latter cases necessarily implies a decay in the density, because at least one of the vector components has to be modified for the point $x_{\underline{\text{conf}}}$ in order to become a valid configuration $x_{\overline{\text{conf}}}$. This offset signifies a dilatation or relative contraction of the sphere(s) depending on the magnitude and direction of the displacement V_{opt},

$$x_{\overline{\text{conf}}} = x_{\underline{\text{conf}}} + V_{\text{opt}},$$

which transforms the degenerated configuration into a valid one, see Fig. 15. Here, the optimal criterion to accomplish is to calculate the minimal-length offset vector transformation V_{opt},

$$V_{\text{opt}} := [v_{r_1}, \ldots, v_{r_n}] \in \mathbf{S}^n,$$

retaining as much density as possible by eluding degradation of the spheres, reducing the radial deviation within (12).

The geometric intuitive way of finding such a vector is to find the closest point from $x_{\underline{\text{conf}}}$ on the cells or ridges of the polytope that could be efficiently computed by perpendicularly projecting the point $x_{\underline{\text{conf}}}$ to each hyperplane,

$$x^{(i,j)}_{\overline{\text{conf}}} = x_{\underline{\text{conf}}} - \left(V_\alpha^{(i,j)} \cdot x_{\underline{\text{conf}}}\right) V_\alpha^{(i,j)}, \tag{17}$$

and selecting the closest one from those points holding the assertion given by (16). Although this technique is computationally efficient and geometrically correct, the outcoming solution is not optimal, because within this space only the absolute directed distance is considered. No contribution effects of different deviations are assessed, producing nonminimal density decay.

This limitation could be vanquished by considering a *homothety* transformation $\mathbf{H}(\mathbf{S}^n)$, i.e., a deviation normalization of the radial configuration space inspired by the concept behind the *Mahalanobis* [4] distance.

The spatial density function of a Gaussian sphere Ω_i given by (12) could be conveniently reformulated in the radial domain as

$$\widehat{f}(\Omega_i, x) = e^{-\frac{1}{2}\left(\frac{r_x}{\sigma_i} - \frac{r_i}{\sigma_i}\right)^2}, \tag{18}$$

so that the deviation of the endowed normal distribution scales the implicitly defined radius r_x and the mean radius r_i of the sphere Ω_i by the factor σ_i^{-1}. This normalization mapping could be generalized for the whole radial configuration space $\mathbf{S}^n$ as

$$\mathbf{H} = \operatorname{diag}\big[\sigma_1^{-1}, \ldots, \sigma_n^{-1}\big]. \tag{19}$$

This matrix actually represents the inverse covariance matrix Σ^{-1} of the total density function given by (20). This could be easily visualized by the alternative expression[12]

$$\widehat{f}_t(x) = e^{-\frac{1}{2}\sum_{i=1}^{n}(\frac{\|x-X_i\|}{\sigma_i} - \frac{r_i}{\sigma_i})^2}. \tag{20}$$

Based on (20) and taking into account the uncorrelated radial distributions, it is clear that the underlying covariance matrix $\mathbf{H}^{-1} = \Sigma$ has zero elements outside its trace. Because of this fact, the proposed normalization $\mathbf{Sd}^n = \mathbf{H}(\mathbf{S}^n)$ could take place by applying the matrix $\mathbf{H}$ as an operator over the orthonormal vector bases of $\mathbf{S}^n$ as

$$\acute{e}_i = \mathbf{H}\hat{e}_i.$$

The Euclidean metric within this resulting space is uniformly isomorphic with the density space. Displacements of the same length arising from the same position imply equal density decay in all directions reflecting different dilatation or contractions of those involved Gaussian spheres. Note that this normalization takes place before the vertex enumeration for the polytope extraction has been computed, reflecting the effects within the affine[13] strata while computing the optimal points in (17), see Fig. 15.

The application of the previous methods within the normalized radial configuration space $\mathbf{Sd}^n$ does not only ensure the optimal solution with minimal decay, but it also benefits from the available certainty provided from the spheres with smaller deviation (higher reliable percepts) by introducing smaller displacements in the corresponding dimension of the displacement vector $V_{\text{opt}}^d \in \mathbf{Sd}^n$.

In other words, the spheres which have a wider deviation easily expand (or contract) their surfaces than those with smaller ones in order to obtain the highest possible density at the meeting operation.

This method delivers the optimal trade-off fusion while performing the management of the modeled uncertainty.

4.5 Duality and Uniqueness

In case the latter method has taken place in $\mathbf{Sd}^3$ (considering three spheres) obtaining the optimal configuration $x_{\overline{\text{conf}}} \in \mathbf{Sd}^3$, there is still a duality to solve while back

[12] By rewriting the exponent as a vector column and arranging in a standard form $x^t\Sigma^{-1}x$.

[13] In the Hesse normal form of the hyperplanes.

mapping this configuration into the physical Euclidean space. This issue is solved straightforward by computing the point pair solution

$$J_{\bigwedge_{i=1}^{3}} = \bigwedge_{i=1}^{3} \Omega_i\big(\sigma_i(x_{\overline{\text{conf}}} \cdot \hat{e}_i), X_i\big).$$

In case both solutions lie within the valid[14] subspace, a simple cross-check against the location of percepts not involved in previous calculations will robustly disambiguate the solution. It is possible to obtain a unique solution by using four spheres for the optimization task, i.e., to represent the setup within $\mathbf{Sd}^4$. In this way $x_{\overline{\text{conf}}} \in \mathbf{Sd}^4$ could be again mapped back into the physical Euclidean space by means of the meet operator unveiling the position of the robot as

$$P_{\bigwedge_{i=1}^{4}} = \bigwedge_{i=1}^{4} \Omega_i\big(\sigma_i(x_{\overline{\text{conf}}} \cdot \hat{e}_i), X_i\big).$$

5 Conclusion

This approach solves the model-based visual self-localization using conformal geometric algebra and Gaussian spheres. The proposed method translates the statistical optimization problem of finding the maximal density location for the robot into a radial normalized density space $\mathbf{Sd}^n$ which allows a very convenient description of the problem. Within this domain, it is not only possible to draw the geometric restrictions which ensure the intersection of spheres, but it also attains the optimal fusion and trade-off of the available information provided from the percepts by incorporating the available information of each landmark according to its uncertainty.

The considered world model of the kitchen consists of 611 rectangular prisms, 124 cylinders, and 18 general polyhedra with 846 faces, all arranged by 1,524 general transformations (rotation, translation, and scaling) with a total of 13,853 vertices and 25,628 normal vectors composed in the scene-graph from the CAD model and verified against real furniture with laser devices, see Fig. 1(a).

The global self-localization of the humanoid robot ARMAR-III within the modeled environment was successfully performed using this approach. The scanning strategy takes 15–20 seconds processing 20 real stereo images. The graph model pruning takes 100–150 ms. The hypotheses generation–validation takes 200–500 ms. Finally, the vertex enumeration takes approximately 15–50 ms depending on the configuration.

Acknowledgements The work described in this book chapter was partially conducted within the German Humanoid Research project SFB588 funded by the German Research Foundation (DFG: Deutsche Forschungsgemeinschaft) and the EU Cognitive Systems project PACO-PLUS (FP6-027657) funded by the European Commission.

---

[14] Above the floor and inside the modeled space.

References

1. Asfour, T., Azad, P., Vahrenkamp, N., Regenstein, K., Bierbaum, A., Welke, K., Schröder, J., Dillmann, R.: Toward humanoid manipulation in human-centred environments. Robot. Auton. Syst. **56**(1), 54–65 (2008)
2. Asfour, T., Regenstein, K., Azad, P., Schroder, J., Bierbaum, A., Vahrenkamp, N., Dillmann, R.: ARMAR-3: an integrated humanoid platform for sensory-motor control. In: 6th International Conference on Humanoid Robots, IEEE-RAS, 4–6 Dec. 2006, pp. 169–175
3. Asfour, T., Welke, K., Azad, P., Ude, A., Dillmann, R.: The Karlsruhe humanoid head. In: IEEE-RAS International Conference on Humanoid Robots (2008)
4. John, H., Horst, S.: Handbook of Mathematics and Computational Science. Springer, Berlin (2006). ISBN: 978-0-387-94746-4
5. Bayro, E., Sobczyk, G.: Geometric Algebra with Applications in Science and Engineering. Birkhäuser, Basel (2001). ISBN: 978-0-8176-4199-3
6. Dorst, L., Fontijne, D., Mann, S.: Geometric Algebra for Computer Science, An Object-Oriented Approach to Geometry. The Morgan Kaufmann Series in Computer Graphics. Morgan Kaufmann, San Mateo (2007). ISBN: 0-123-69465-5
7. Harris, C., Stephens, M.J.: A combined corner and edge detector. In: Alvey Vision Conference, pp. 147–152 (1988)
8. Lowe, D.: Object recognition from local scale-invariant features. In: The Proceedings of the Seventh IEEE International Conference on Computer Vision, Sept. 1999, pp. 1150–1157
9. Gonzalez-Aguirre, D., Asfour, T., Bayro-Corrochano, E., Dillmann, R.: Model-based visual self-localization using geometry and graphs. In: ICPR 2008. 19th International Conference on Pattern Recognition, Tampa, Florida Dic. (2008)
10. Ullman, S.: High-Level Vision. MIT Press, Cambridge (1996). ISBN: 978-0-262-71007-7
11. Se, S., Lowe, D., Little, J.: Vision-based global localization and mapping for mobile robots. IEEE Trans. Robot. Autom. **21**, 364–375 (2005)
12. Duda, R., Hart, P.: Pattern Classification and Scene Analysis. Wiley, New York (1973). ISBN: 0-471-22361-1
13. Comaniciu, D., Meer, P.: Mean shift: a robust approach toward feature space analysis. IEEE Trans. Pattern. Anal. Mach. Intell. **24**, 603–619 (2002)
14. Gonzalez-Aguirre, D., Bayro-Corrochano, E.: A geometric approach for an intuitive perception system of humanoids. In: International Conference on Intelligent Autonomous Systems, Proceedings IAS-9, March 2006
15. Pollefeys, M., Gool, L., Vergauwen, M., Verbiest, F., Cornelis, K., Tops, J., Koch, R.: Visual modeling with a hand-held camera. Int. J. Comput. Vis. **59**(3), 207–232 (2004)
16. Hartley, R., Zisserman, A.: Multiple View Geometry in Computer Vision. Cambridge University Press, Cambridge (2004). ISBN: 0521540518
17. Kallenberg, O.: Foundations of Modern Probability. Springer, New York (1997). ISBN: 0387953132
18. Korn, T., Granino, A.: Mathematical Handbook for Scientists and Engineers. Dover, New York (2000). ISBN: 0-486-41147-8
19. Avis, D., Fukuda, K.: A pivoting algorithm for convex hulls and vertex enumeration of arrangements and polyhedra. Int. J. Discrete Comput. Geom. **8**, 295–313 (1992)

Part V
Conformal Mapping and Fluid Analysis

Geometric Characterization of $\mathcal{M}$-Conformal Mappings

K. Gürlebeck and J. Morais

Abstract In this paper we study the description of monogenic functions in $\mathbb{R}^3$ by their geometric mapping properties. The monogenic functions are considered at first as general quasi-conformal mappings. We consider the local mapping properties of a monogenic function and show that this class of functions can be defined as a special subclass of quasiconformal mappings. It is proved that monogenic functions with nonvanishing Jacobian determinant map infinitesimal balls onto special ellipsoids. The condition on the semiaxes of these ellipsoids is explicitly given.

1 Introduction

The application of complex-analytic methods plays an important role in the theory of partial differential equations in the plane. Because many practical problems lead to models in the three- or four-dimensional spaces, there are different approaches to generalize complex analysis to the higher-dimensional real Euclidean space (see, e.g., [10, 11, 13, 14, 20]). An important issue in this paper is to study spatial generalizations of holomorphic functions in the setting of quaternionic analysis. In the complex case, these functions define, under certain conditions, two-dimensional conformal mappings between geometrical objects and transform partial differential equations to other differential equations. Therefore, in this paper we are interested in the construction of spatial generalizations of holomorphic functions (they are called *monogenic* in Quaternionic and Clifford analysis) and the description of their geometric mapping properties. We have in mind to find out in which sense the well-known properties of complex conformal mappings can be transferred to the higher-dimensional case. The frequent use of quaternions in the study of three-dimensional problems motivates the restriction to monogenic functions defined in domains of

K. Gürlebeck (✉)
Institut für Mathematik/Physik, Bauhaus-Universität Weimar, Coudraystr. 13B, 99421 Weimar, Germany
e-mail: klaus.guerlebeck@uni-weimar.de

E. Bayro-Corrochano, G. Scheuermann (eds.), *Geometric Algebra Computing*,
DOI 10.1007/978-1-84996-108-0_16, © Springer-Verlag London Limited 2010

$\mathbb{R}^3$ with values in the reduced quaternions. A serious study of the geometric properties of monogenic functions requires to consider at first the local behavior of such mappings. This is the main goal of the paper.

The extension of theoretical and practical conformal mapping methods in complex analysis to the higher-dimensional real Euclidean space, particularly in the setting of quaternionic analysis, has originated many questions. Several attempts have been made to study monogenic functions in $\mathbb{R}^{n+1}$ by a corresponding differentiability concept or by the existence of a well-defined hypercomplex derivative (see, e.g., [6, 7, 12, 16, 20]). However, it is not commonly realized that monogenic functions can be characterized via a generalized conformality concept. It is shown by examples in [4] and later on in [3] that these functions can preserve some of the geometrical properties like length, distance, or special angles while mapping special domains onto the ball. Both papers [3, 4] are mainly concerned with generalizations of the Bergman reproducing kernel approach to numerical mapping problems analogously to the complex Bergman kernel method of constructing the conformal mapping from a domain onto the disk.

At this point it is important to remark that, in contrast to the planar case, in spaces $\mathbb{R}^{n+1}$ of dimension $n \geq 2$ the set of conformal mappings is restricted to the set of Möbius tranformations only, at first shown by Liouville in 1850 [15] for the three-dimensional case. We state the result:

Theorem 1 (Liouville's theorem) *Let $\Omega \subset \mathbb{H}$. A C^1-function $f : \Omega \rightarrow \mathbb{H}$ is a conformal mapping if and only if f is a Möbius transformation.*

It is well known that the theory of monogenic functions does not cover the set of Möbius tranformations in higher dimensions and that the Möbius transformations are not monogenic. One can only expect that monogenic functions represent certain quasiconformal mappings. On the other hand, the class of all quasiconformal mappings is much bigger than the class of monogenic functions. The question arises if monogenic functions correspond to a special subclass of quasiconformal mappings.

In [17], Malonek introduced the concept of monogenic-conformal mappings realized by Clifford-valued functions defined in $\mathbb{R}^{n+1}$. The main tool in his paper is the study of relations between special surface and volume differential forms. Together with the geometric interpretation of the hypercomplex derivative [7], dilatations and distortions of these mappings can be estimated. This includes the description of the interplay between the Jacobian determinant and the hypercomplex derivative of monogenic functions (see [5]).

In this contribution we will characterize monogenic functions by more visible geometric mapping properties. It is clear that the local mapping properties of a monogenic function or of a real analytic function are mainly determined by the behavior of the linear part of their Taylor expansions. In this line of reasoning, we start by presenting the geometric behavior of the linear part of a monogenic function with values in the reduced quaternions. As a consequence, we show that monogenic functions can be defined as mappings which map infinitesimal balls onto explicitly characterized ellipsoids and vice versa.

2 Preliminaries

Let $\{\mathbf{e}_0, \mathbf{e}_1, \mathbf{e}_2, \mathbf{e}_3\}$ be an orthonormal basis of the Euclidean vector space $\mathbb{R}^4$. We introduce the (quaternionic-)product subject to the following multiplication rules:

$$\mathbf{e}_0^2 = \mathbf{e}_0; \qquad \mathbf{e}_i \mathbf{e}_0 = \mathbf{e}_0 \mathbf{e}_i = \mathbf{e}_i, \quad i = 1, 2, 3;$$
$$\mathbf{e}_1^2 = \mathbf{e}_2^2 = \mathbf{e}_3^2 = -1; \qquad \mathbf{e}_1 \mathbf{e}_2 = -\mathbf{e}_2 \mathbf{e}_1 = \mathbf{e}_3.$$

This noncommutative product generates the algebra of real quaternions denoted by $\mathbb{H}$. The real vector space $\mathbb{R}^4$ will be identified with $\mathbb{H}$ by means of

$$a := (a_0, a_1, a_2, a_3) \in \mathbb{R}^4 \quad \longleftrightarrow \quad \mathbf{a} := a_0\mathbf{e}_0 + a_1\mathbf{e}_1 + a_2\mathbf{e}_2 + a_3\mathbf{e}_3 \in \mathbb{H},$$

where $a_i \in \mathbb{R}$ $(i = 0, 1, 2, 3)$.

We remark that the vector $\mathbf{e}_0$ is the multiplicative unit element of $\mathbb{H}$. From now on we will identify $\mathbf{e}_0$ with 1, and we simply write

$$\mathbf{a} := a_0 + a_1\mathbf{e}_1 + a_2\mathbf{e}_2 + a_3\mathbf{e}_3.$$

We shall denote by $\mathbf{Sc}(\mathbf{a}) := a_0$ the scalar part of $\mathbf{a}$ and by $\mathbf{Vec}(\mathbf{a}) := a_1\mathbf{e}_1 + a_2\mathbf{e}_2 + a_3\mathbf{e}_3$ its vector part. Analogously to the complex case, the (quaternionic-) conjugate element of $\mathbf{a}$ is

$$\overline{\mathbf{a}} := \mathbf{Sc}(\mathbf{a}) - \mathbf{Vec}(\mathbf{a}) = a_0 - a_1\mathbf{e}_1 - a_2\mathbf{e}_2 - a_3\mathbf{e}_3.$$

The norm of $\mathbf{a}$ is given by $|\mathbf{a}| = \sqrt{\mathbf{a}\overline{\mathbf{a}}} = \sqrt{a_0^2 + a_1^2 + a_2^2 + a_3^2}$ and coincides with its corresponding Euclidean norm as a vector in $\mathbb{R}^4$.

Let us now consider the subset

$$\mathcal{A} := \text{span}_{\mathbb{R}}\{1, \mathbf{e}_1, \mathbf{e}_2\}$$

of the algebra $\mathbb{H}$. Its elements are called reduced quaternions. The real vector space $\mathbb{R}^3$ is to be identified with $\mathcal{A}$ via

$$x := (x_0, x_1, x_2) \in \mathbb{R}^3 \quad \longleftrightarrow \quad \mathbf{x} := x_0 + x_1\mathbf{e}_1 + x_2\mathbf{e}_2 \in \mathcal{A}.$$

We emphasize that $\mathcal{A}$ is a real vectorial subspace, but not a subalgebra, of $\mathbb{H}$.

Now, let us consider an open subset Ω of $\mathbb{R}^3$ with a piecewise smooth boundary. A quaternion-valued function or, briefly, an $\mathbb{H}$-valued function is a mapping $\mathbf{f} : \Omega \to \mathbb{H}$. From now on we will restrict the latter to functions with values in $\mathcal{A}$, i.e., such that

$$\mathbf{f}(x) = [\mathbf{f}(x)]_0 + \sum_{i=1}^{2} [\mathbf{f}(x)]_i \mathbf{e}_i,$$

where the coordinate-functions $[\mathbf{f}]_i$ $(i = 0, 1, 2)$ are real-valued functions in Ω. Properties such as continuity, differentiability, or integrability are ascribed coordinate-wisely.

For continuously real-differentiable functions $\mathbf{f} : \Omega \to \mathcal{A}$, the operator D defined by

$$D\mathbf{f} = \frac{\partial \mathbf{f}}{\partial x_0} + \mathbf{e}_1 \frac{\partial \mathbf{f}}{\partial x_1} + \mathbf{e}_2 \frac{\partial \mathbf{f}}{\partial x_2} \tag{1}$$

is called the generalized Cauchy–Riemann operator on $\mathbb{R}^3$. It can be understood as a three-dimensional extension of the classical Cauchy–Riemann operator $\partial_{\bar{z}}$. In the same way, we define the conjugate quaternionic Cauchy–Riemann operator $\overline{D}$ by

$$\overline{D}\mathbf{f} = \frac{\partial \mathbf{f}}{\partial x_0} - \mathbf{e}_1 \frac{\partial \mathbf{f}}{\partial x_1} - \mathbf{e}_2 \frac{\partial \mathbf{f}}{\partial x_2}, \tag{2}$$

which is a generalization of the operator ∂_z.

Definition 1 A continuously real-differentiable $\mathcal{A}$-valued function $\mathbf{f}$ is called monogenic (resp., antimonogenic) in Ω if $D\mathbf{f} = 0$ (resp., $\overline{D}\mathbf{f} = 0$) in Ω.

Remark 1 It is not necessary to distinguish between left- and right-monogenic (resp., antimonogenic) functions in the case of $\mathcal{A}$-valued functions because $D\mathbf{f} = 0$ (resp., $\overline{D}\mathbf{f} = 0$) implies that $\mathbf{f}D = 0$ (resp., $\mathbf{f}\overline{D} = 0$) and vice versa.

Analogously as in the complex one-dimensional case, $\frac{1}{2}\overline{D}$ defines a derivative of monogenic functions. This was shown for arbitrary dimensions in [7], where $(\frac{1}{2}\overline{D})\mathbf{f}$ was called hypercomplex derivative of $\mathbf{f}$. In the four-dimensional case we find analogous results in [18] and [20].

The generalized Cauchy–Riemann operator (1) and its conjugate (2) factorize the three-dimensional Laplace operator, that is,

$$\Delta_3 \mathbf{f} = D\overline{D}\mathbf{f} = \overline{D}D\mathbf{f} \tag{3}$$

for $\mathbf{f} \in C^2$, which implies that any monogenic function is also a harmonic function. This means that quaternionic analysis gives us a refinement of harmonic analysis.

For future use, we also need the connection between quaternionic analysis and Clifford analysis. We restrict ourselves to the case of $\mathcal{C}\ell_{0,3}$, the universal real Clifford algebra of signature $(0, 3)$, constructed over the canonical orthonormal basis $\{\boldsymbol{\varepsilon}_0, \boldsymbol{\varepsilon}_1, \boldsymbol{\varepsilon}_2, \boldsymbol{\varepsilon}_3\}$ of the Euclidean vector space $\mathbb{R}^4$. The basis elements satisfy the following multiplication rules:

$$\boldsymbol{\varepsilon}_i \boldsymbol{\varepsilon}_j + \boldsymbol{\varepsilon}_j \boldsymbol{\varepsilon}_i = -2\delta_{i,j}\boldsymbol{\varepsilon}_0, \quad i, j = 1, 2, 3;$$

$$\boldsymbol{\varepsilon}_0 \boldsymbol{\varepsilon}_i = \boldsymbol{\varepsilon}_i \boldsymbol{\varepsilon}_0 = \boldsymbol{\varepsilon}_i, \quad i = 1, 2, 3,$$

where $\delta_{i,j}$ stands for the Kronecker symbol.

The skew-field of the quaternions can be identified with the even subalgebra $\mathcal{C}\ell_{0,3}^{+}$ by setting

$$\mathbf{e}_1 \cong -\boldsymbol{\varepsilon}_1\boldsymbol{\varepsilon}_2, \qquad \mathbf{e}_2 \cong -\boldsymbol{\varepsilon}_1\boldsymbol{\varepsilon}_3, \qquad \mathbf{e}_3 \cong +\boldsymbol{\varepsilon}_2\boldsymbol{\varepsilon}_3.$$

As a consequence, starting with the Dirac operator in $\mathcal{C}\ell_{0,3}^{+}$,

$$\mathcal{D} = \varepsilon_1 \frac{\partial}{\partial y_0} + \varepsilon_2 \frac{\partial}{\partial y_1} + \varepsilon_3 \frac{\partial}{\partial y_2},$$

we obtain the generalized Cauchy–Riemann operator via the identification

$$\mathcal{C}\ell_{0,3}^{+} \ni y = y_0\boldsymbol{\varepsilon}_1 + y_1\boldsymbol{\varepsilon}_2 + y_2\boldsymbol{\varepsilon}_3 \quad \longleftrightarrow \quad \mathbf{x} = y_0 - y_1\mathbf{e}_1 - y_2\mathbf{e}_2 \in \mathcal{A}$$

and

$$-\boldsymbol{\varepsilon}_1\mathcal{D} \cong D.$$

Moreover, a vector-valued function $\tilde{f}$ in $\mathcal{C}\ell_{0,3}^{+}$ is identified with the $\mathcal{A}$-valued function $\mathbf{f}$ by

$$\tilde{f} = [\tilde{f}]_0\varepsilon_1 + [\tilde{f}]_1\varepsilon_2 + [\tilde{f}]_2\varepsilon_3 \quad \longleftrightarrow \quad \mathbf{f} = [\mathbf{f}]_0 - [\mathbf{f}]_1\mathbf{e}_1 - [\mathbf{f}]_2\mathbf{e}_2, \tag{4}$$

where $[\tilde{f}(y)]_i = [\mathbf{f}(x)]_i$ $(i = 0, 1, 2)$.

These identifications define a link between null solutions of the Dirac equation in $\mathcal{C}\ell_{0,3}^{+}$ and of the Cauchy–Riemann equations in $\mathbb{H}$. It is well known that the solutions of the Dirac equations are invariant under rotations. These two facts allow us, under certain conditions, to transpose some mapping properties of our original $\mathcal{A}$-valued functions under rotations. More precisely, we have:

Lemma 1 *Let* $\mathbf{f}$ *be an* $\mathcal{A}$*-valued monogenic function defined in a domain* Ω *of* $\mathbb{R}^3$. *Then,*

$$\mathcal{D}\tilde{f} = \mathcal{D}\big(U\tilde{f}U^{-1}\big),$$

where U *is an orthogonal transformation in* $\mathbb{R}^3$, *and* $\tilde{f} = [\mathbf{f}]_0\varepsilon_1 - [\mathbf{f}]_1\varepsilon_2 - [\mathbf{f}]_2\varepsilon_3$.

3 Monogenic-Conformal Mappings and Their Relation to Monogenic Functions

Nowadays the concept of conformal mappings is used in complex analysis, including applications to elliptic partial differential equations, differential geometry, potential theory, mathematical physics, and calculus of variations. Conformal mappings are closely related to the differentiability and analyticity of complex functions in the way that every conformal mapping is holomorphic.

The extension of conformal mapping methods in $\mathbb{C}$ to the higher-dimensional real Euclidean space, particularly in the setting of quaternionic analysis, has originated many questions in the last decades. It is not commonly realized that monogenic functions can be characterized via a generalized conformality concept. In [17] the concept of monogenic-conformal mappings described by paravector-valued real differentiable functions in $\Omega \subset \mathbb{R}^{n+1}$ with values in the Clifford algebra $\mathcal{C}\ell_{0,n}$ has been introduced for the first time. These mappings are called in [17] $\mathcal{M}$-conformal mappings. The relation of this concept with the geometric interpretation of the hypercomplex derivative $\frac{1}{2}\overline{D}$ allowed it to complete the theory of monogenic functions by providing a still missing geometric characterization of those functions. It was shown that monogenic-conformal mappings are generalizations of conformal mappings in the plane. However, they are different from conformal mappings (in the sense of Gauss) in higher dimensions.

In the following, we will use the notation of [17]. We consider in $\Omega \subset \mathbb{R}^{n+1}$ a positively oriented differentiable n-dimensional hypersurface $\mathcal{S}$ with coherent oriented boundary $\partial\mathcal{S}$. Let $z^* \in \mathcal{S}$ be a fixed point, and $(\mathcal{S}_m)$ a regular sequence of subdomains which is shrinking to z^* as m tends to infinity and whereby z^* belongs to all $\mathcal{S}_m$.

From [17] we take the following theorem:

Theorem 2 *Let* $\mathbf{f}$ *be a paravector-valued real differentiable function in* $\Omega \subset \mathbb{R}^{n+1}$. *This function realizes locally in the neighborhood of a fixed point* $z = z^*$ *a left* $\mathcal{M}$*-conformal mapping if and only if* $\mathbf{f}$ *is left monogenic, and its left hypercomplex derivative is different from zero.*

The proof of the previous theorem is based in the following integral representation of the hypercomplex derivative of a monogenic function $\mathbf{f}$:

$$\left(\frac{1}{2}\overline{D}\right)\mathbf{f} = (-1)^n \lim_{m\to\infty}\left[\int_{\mathcal{S}_m} d\mu \wedge dz\right]^{-1} \int_{\partial\mathcal{S}_m} (d\mu\mathbf{f})$$

in case the limit exists for all possible regular sequences. To be more precise, this integral is described by the limit of the "quotient" of a 2-form (surface area) and a 3-form (volume) (Sudbery [20]). For more details on this representation, we refer the reader to [7] and [17].

However, the geometric properties of such a result are not directly visible. In the next section we will show that the description of monogenic functions can be also formulated by more accessible geometric mapping properties.

4 Influence of the Linear Part of a Monogenic Function

Due to the equivalence between hypercomplex differentiability and monogenicity, the question arises if from this point of view $\mathcal{A}$-valued monogenic functions in $\mathbb{R}^3$ can also be defined by their geometric mapping properties.

It is well known from complex analysis that conformal mappings (and also quasi-conformal mappings, cf. Ahlfors [1]) in the plane are realized in the neighborhood of a point by all complex functions for which the complex derivative exists and is not equal to zero. This can be easily shown by looking to the differential of a nonconstant real differentiable function $f : \Omega \to \mathbb{C}$ defined in a domain $\Omega \subset \mathbb{R}^2$ (see, e.g., [17]). This usual treatment includes the description of the connection between the Jacobian determinant and the derivative of such functions.

On the other hand, it is also known that the Riemann mapping theorem allows one to deform an (almost) arbitrary domain by a conformal mapping onto a disk (or vice versa). In particular, it also allows one to prescribe the direction in which the real axis should be mapped. This behavior is related to conformal (local) properties of the complex derivative of a complex-valued function. For that, one usually requires the knowledge of the argument of the derivative at the origin.

To be more precise, assume a general function $f : \Omega \subset \mathbb{R}^2 \to B \subset \mathbb{C}$. We suppose also that the domain B denotes a ball centered at the origin. The first derivative in the origin will give the first (linear) approximation of the mapping function f in terms of a linear mapping represented in the form $w = f(0) + zf'(0)$ $(z \in \mathbb{C})$, valid in a neighborhood of $z = 0$. In other words, for a first approximation, we require nothing else than the identity function. Also, it implies that a small ball is (locally) mapped onto a small ball. However, in the 3D-case the situation is completely different, since the class of conformal functions is now reduced to the class of Möbius transformations. A particular important example is the case of $\mathbf{f}(z) = \mathbf{z}$ for $\mathbf{z} = x_0 + x_1\mathbf{e}_1 + x_2\mathbf{e}_2 \in \mathcal{A}$. This can be viewed as the counterpart of the monomial z in $\mathbb{C}$, but unfortunately it is not a monogenic function.

In this line of reasoning we discuss briefly some general aspects concerning a general real analytic $\mathcal{A}$-valued function $\mathbf{f}$ defined in $\mathbb{R}^3$. Clearly, one has the decomposition

$$\mathbf{f}(x) := \mathbf{f}(0) + \mathrm{lin}\mathbf{f}(x) + \mathbf{R}(x), \tag{5}$$

where $\mathrm{lin}\mathbf{f}$ is the linear part (of the Taylor expansion) of $\mathbf{f}$, and $\mathbf{R}$ denotes the remaining part (degree $k \geq 2$).

Under the above hypotheses, we claim that the linear part of $\mathbf{f}$ is also an $\mathcal{A}$-valued function naturally related to the hypercomplex derivative of $\mathbf{f}$ at the origin. On the other hand, $\mathrm{lin}\mathbf{f}$ is a general linear function or, more explicitly, a general $\mathcal{A}$-valued linear function of three real variables and nine real parameters. Therefore, its coordinates $[\mathrm{lin}\mathbf{f}]_i$ $(i = 0, 1, 2)$ are given by

$$\begin{aligned} \big[\mathrm{lin}\mathbf{f}(x)\big]_0 &= a_0x_0 + a_1x_1 + a_2x_2, \\ \big[\mathrm{lin}\mathbf{f}(x)\big]_1 &= b_0x_0 + b_1x_1 + b_2x_2, \\ \big[\mathrm{lin}\mathbf{f}(x)\big]_2 &= c_0x_0 + c_1x_1 + c_2x_2, \end{aligned} \tag{6}$$

where, as described, the coefficients a_i, b_i, c_i $(i = 0, 1, 2)$ are real valued.

Remark 2 From now on we will restrict ourselves only to mappings from the interior of a ball into the interior of a special domain of $\mathbb{R}^3$.

Remark 3 If the function is vanishing at the origin, then each monogenic approximation polynomial begins with the linear expression lin**f**. It is clear then that the invariance of the origin guarantees that the interior of the ball will be mapped to the interior of the image domain.

We will show in this section that monogenic functions are closely related to special ellipsoids in the neighborhood of a given point. As is well known, ellipsoids play an important role in many disciplines as simple effective object models. Among their applications which enjoy a great variety, there are the following two closely related classes: ellipsoidal models are used both directly to capture shape and indirectly as bounding approximations. Nowadays, in medical areas, ellipsoids are also used to model the shape, volume, or area of anatomical parts, such as the heart, spine, and blood vessels.

Ellipsoids can be represented by its center and the lengths of the semiaxes together with the orientation. An ellipsoid $\mathcal{E}^*$ with semiaxes parallel to the coordinate directions is then generated by the quadratic form

$$\mathcal{E}^* := \left\{(x_0, x_1, x_2) : \frac{x_0^2}{\alpha^2} + \frac{x_1^2}{\beta^2} + \frac{x_2^2}{\gamma^2} = 1\right\}, \tag{7}$$

where α, β, and γ are the lengths of the semiaxes.

On the other hand, if the axes are rotated to a general position, then the equation for an ellipsoid $\mathcal{E}$ can be written as

$$ax_0^2 + bx_1^2 + cx_2^2 + dx_0x_1 + ex_0x_2 + fx_1x_2 = 1, \tag{8}$$

where a, b, c, d, e, and f are new constants. Those six constants indicate the shape and the size of the ellipsoid $\mathcal{E}$ and the directions of its axes. After a rotation it is clear that the directions of the axes remain orthogonal.

Remark 4 The absence of linear terms in (8) means that we keep the origin at the center of the ellipsoid. We use this simplification because the displacement of its center will not add anything essential.

Later on we will also use the representation of an ellipsoid by a symmetric matrix. Denoting by E the referred matrix, an equivalent definition of a general ellipsoid $\mathcal{E}$ is given by

$$\mathcal{E} := \left\{X : X^T E X = 1,\ X = (x_0\ x_1\ x_2)^T\right\}.$$

In this sense, the geometric properties of $\mathcal{E}$ are reflected in algebraic properties of the corresponding symmetric matrix E. If λ_i and ν_i are the eigenvalues and eigenvectors of E, then the principal axes of the ellipsoid are parallel to ν_i, and the corresponding lengths of the semiaxes are given by $|\lambda_i|^{-1}$. For convenience, in what follows, we will restrict ourselves to ellipsoids centered at the origin. There is no loss of generality in this assumption because we are interested only in the shape of our ellipsoids.

After the above preparations, we are now ready to characterize the mapping properties of the linear part of an arbitrary $\mathcal{A}$-valued monogenic function. In the following we consider the image of a ball under the mapping $\mathbf{linf}$, being a positively oriented differentiable surface with coherent oriented boundary.

Theorem 3 *Let* $\mathbf{linf}$ *be an* $\mathcal{A}$*-valued linear function defined in a domain* Ω *of* $\mathbb{R}^3$ *with nonvanishing Jacobian determinant. Then, the function* $\mathbf{linf}$ *is monogenic if and only if it maps a ball onto an ellipsoid with the property that the reciprocal of the length of one semiaxis is equal to the sum of the reciprocals of the lengths of the other two semiaxes.*

Proof Taking into account identification (4), we start to write the linear function $\mathbf{linf}$ in the form

$$\mathbf{linf}(x) = \big[\mathbf{linf}(x)\big]_0 - \big[\mathbf{linf}(x)\big]_1\mathbf{e}_1 - \big[\mathbf{linf}(x)\big]_2\mathbf{e}_2, \tag{9}$$

where the coordinates $[\mathbf{linf}]_i$ $(i = 0, 1, 2)$ are given as in (6). We begin to prove the necessary condition. By the monogenicity of $\mathbf{linf}$, it holds

$$\begin{cases} a_0 = -(b_1 + c_2), \\ b_0 = a_1, \\ c_0 = a_2, \\ c_1 = b_2. \end{cases} \tag{10}$$

For $X = (x_0\ x_1\ x_2)^T$, let A be the matrix which represents $\mathbf{linf}$, i.e., such that

$$(1,\ -\mathbf{e}_1,\ -\mathbf{e}_2)\,AX = \mathbf{linf}(x).$$

The matrix A is symmetric and is given by

$$A = \begin{pmatrix} -(b_1 + c_2) & a_1 & a_2 \\ a_1 & b_1 & b_2 \\ a_2 & b_2 & c_2 \end{pmatrix}.$$

Now, as is well known from Linear Algebra, there exists a real orthogonal matrix C such that

$$C^T AC = D = \operatorname{diag}(\lambda_1, \lambda_2, \lambda_3),$$

where λ_1, λ_2, and λ_3 are the eigenvalues of A (real-valued and different from zero). Furthermore, Newton's identities (Vieta's formulae) (see [21]) relate the eigenvalues of A in the following way:

$$\lambda_1 + \lambda_2 + \lambda_3 = \operatorname{tr}(A) = 0,$$

$$\lambda_1\lambda_2\lambda_3 = \det(A).$$

Hereby $\operatorname{tr}(A)$ and $\det(A)$ denote the trace and determinant of the matrix A, respectively.

In the following, let us assume without loss of generality that $\det(A) > 0$. For $\det(A) < 0$, the proof is similar. Hence, we have to consider two different cases:

1. If $\lambda_1 > 0$, then from the previous conditions we have that λ_2 and λ_3 are both negative numbers.
2. If $\lambda_1 < 0$, one can easily see that λ_1 and λ_2 must have different signs.

Then, from the relations

$$\lambda_1 = (-\lambda_2) + (-\lambda_3) > 0,$$

or

$$\lambda_2 = (-\lambda_1) + (-\lambda_3) > 0,$$

or

$$\lambda_3 = (-\lambda_1) + (-\lambda_2) > 0,$$

respectively, we get the conditions for the lengths of the semiaxes α, β, and γ. This means that one can express the reciprocal of the length of one semiaxis as the sum of the reciprocals of the lengths of the other two semiaxes. To end with the proof of the necessary condition, notice that the maximum of the eigenvalues gives the minimum value of the lengths of the semiaxes.

For the sufficient condition, having in mind Lemma 1, the main idea is to rotate an arbitrary ellipsoid (with the described property of the semiaxes) such that the directions of its semiaxes $y^{(0)}$, $y^{(1)}$, $y^{(2)}$ coincide with the directions of the standard coordinate system (y_0, y_1, y_2). Such an ellipsoid is given by (7). The referred property of the semiaxes is invariant under rotations. It is a direct consequence that $D\mathbf{linf} = 0$ and that $\mathbf{linf}$ must be monogenic. For more details, we refer the reader to [8]. □

Remark 5 The composition of a linear function with a translation allows one to extend the result to an ellipsoid centered at an arbitrary point $\tilde{x}$. This composition preserves the monogenicity.

The result of Theorem 3 allows us to describe monogenic functions as a special class of quasiconformal mappings. If we visualize quasiconformal mappings in $\mathbb{R}^3$ by points, given by the lengths of the semiaxes of the associated ellipsoids, then the monogenic functions (with nonvanishing Jacobian determinant) can be seen as a two-dimensional manifold in $\mathbb{R}^3$. This manifold is defined by the relation between the semiaxes as in Theorem 3 and visualized in Fig. 1.

Next, one has to show that Theorem 3 can be generalized to arbitrary real-analytic functions which have the described local mapping properties. A monogenic function with nonvanishing linear part will map in the small balls to the special class of ellipsoids. Nonvanishing linear part means that all directional first derivatives of the function are different from zero. Equivalently this can be characterized by the Jacobian determinant and the hypercomplex derivative:

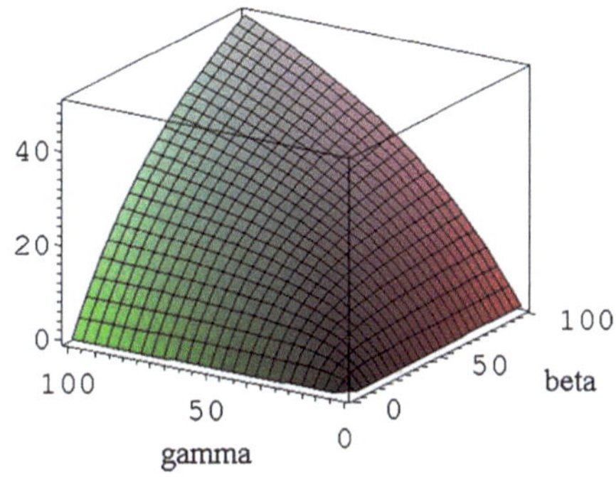

Fig. 1 Geometric characterization of monogenic functions

Theorem 4 (See [5]) *Let* $\mathbf{f}$ *be an* $\mathbb{H}$*-valued monogenic function defined in a domain* Ω *of* $\mathbb{R}^4$*. Then we have*

$$\det J_{\mathbf{f}}(0) = 0 \quad \Longleftrightarrow \quad \exists \mathbf{p} : |\mathbf{p}| = 1 \wedge \overline{D}\mathbf{f}(\overline{\mathbf{p}}z)\big|_{z=0} = 0.$$

This theorem can be considered as a higher-dimensional analogue of the condition in complex analysis, saying that $f'(z_0) = 0$ for a holomorphic function f. For more details and further relations to the hypercomplex derivative and quasi-conformal mappings, see the paper [5].

Theorem 5 *Let* $\mathbf{f}$ *be an* $\mathcal{A}$*-valued real-analytic function defined in a domain* Ω *of* $\mathbb{R}^3$ *with nonvanishing Jacobian determinant. Then, the function* $\mathbf{f}$ *is monogenic if and only if it maps locally a ball onto an ellipsoid with the property that the reciprocal of the length of one semiaxis is equal to the sum of the reciprocals of the lengths of the other two semiaxes.*

Proof The necessary condition can be found in [8]. For the sufficient condition, we suppose that the real-analytic function $\mathbf{f}$ maps the unit ball to an ellipsoid with the referred property. It is clear that the function $\mathbf{f}$ can be expanded in a real Taylor series in a neighborhood of an arbitrary point $a \in \Omega$:

$$\begin{aligned}\mathbf{f}(x) &= \mathbf{f}(a) + \sum_{n=1}^{+\infty} \sum_{|\alpha|=n} (x_0 - a_0)^{\alpha_0}(x_1 - a_1)^{\alpha_1}(x_2 - a_2)^{\alpha_2} \frac{\partial^{\alpha_0+\alpha_1+\alpha_2}}{\partial x_0^{\alpha_0}\partial x_1^{\alpha_1}\partial x_2^{\alpha_2}}\mathbf{f}(a) \\ &= \mathbf{f}(a) + \operatorname{lin}\mathbf{f}(x-a) + \mathbf{R}(x-a).\end{aligned}$$

For simplicity, we take $a = (0, 0, 0)$. We then have

$$\mathbf{f}(x) = \mathbf{f}(0) + \operatorname{lin}\mathbf{f}(x) + \mathbf{R}(x).$$

Now, we deal with the property of monogenicity of the mapping $\mathbf{f}$. We discuss this property only locally for a neighborhood of the point $x = 0$. Without loss of generality (removing higher-order terms from the Taylor expansion described above), we can consider

$$\mathbf{f}(x) \cong \mathbf{f}(0) + \mathrm{lin}\mathbf{f}(x),$$

where $\mathbf{f}(0)$ is to be understood as a translation of the ellipsoid generated by the linear part. Since, by assumption, $\mathrm{lin}\mathbf{f}$ maps balls to our special ellipsoids, we can apply Theorem 3, and for an arbitrary rotation R, its associated symmetric matrix has the form

$$R\tilde{A}R^T,$$

where $\tilde{A}$ is a diagonal matrix, and its elements satisfy the following relation:

$$\frac{\partial}{\partial x_0}[\mathrm{lin}\mathbf{f}]_0 + \frac{\partial}{\partial x_1}[\mathrm{lin}\mathbf{f}]_1 + \frac{\partial}{\partial x_2}[\mathrm{lin}\mathbf{f}]_2 = 0. \tag{11}$$

On the other hand, having in mind identification (4), we write

$$\begin{aligned}\mathrm{lin}\mathbf{f}(x) = {} & x_0\frac{\partial}{\partial x_0}[\mathbf{f}]_0(0) + x_1\frac{\partial}{\partial x_1}[\mathbf{f}]_0(0) + x_2\frac{\partial}{\partial x_2}[\mathbf{f}]_0(0)\\ & - \mathbf{e}_1\left(x_0\frac{\partial}{\partial x_0}[\mathbf{f}]_1(0) + x_1\frac{\partial}{\partial x_1}[\mathbf{f}]_1(0) + x_2\frac{\partial}{\partial x_2}[\mathbf{f}]_1(0)\right)\\ & - \mathbf{e}_2\left(x_0\frac{\partial}{\partial x_0}[\mathbf{f}]_2(0) + x_1\frac{\partial}{\partial x_1}[\mathbf{f}]_2(0) + x_2\frac{\partial}{\partial x_2}[\mathbf{f}]_2(0)\right).\end{aligned}$$

Then, it comes directly from the structure of the matrix associated to $\mathrm{lin}\mathbf{f}$ together with the relation (11) that the conditions of $\mathbf{f}$ at the origin satisfy the following system:

$$\begin{cases}\frac{\partial}{\partial x_0}[\mathbf{f}]_0(0) + \frac{\partial}{\partial x_1}[\mathbf{f}]_1(0) + \frac{\partial}{\partial x_2}[\mathbf{f}]_2(0) = 0,\\ \frac{\partial}{\partial x_1}[\mathbf{f}]_0(0) + \frac{\partial}{\partial x_0}[\mathbf{f}]_1(0) = 0,\\ \frac{\partial}{\partial x_2}[\mathbf{f}]_0(0) + \frac{\partial}{\partial x_0}[\mathbf{f}]_2(0) = 0,\\ \frac{\partial}{\partial x_1}[\mathbf{f}]_2(0) + \frac{\partial}{\partial x_2}[\mathbf{f}]_1(0) = 0,\end{cases}$$

which means, in a compact form, that $D\mathbf{f}(0) = 0$.

Extending this argument to an arbitrary point $a \in \Omega$ by translation of our Taylor series, we obtain that the function $\mathbf{f}$ satisfies at each point of the domain the Cauchy–Riemann equation (i.e., $D\mathbf{f}(a) = 0$). Hence, $\mathbf{f}$ is monogenic, and this completes our proof. □

According to the fact that the function $\mathbf{f}$ has rank 3, it is natural to ask for the properties of the inverse mapping. We first again restrict ourselves to linear functions and work with ellipsoids having the property that the reciprocal of the length of one semiaxis is equal to the sum of the reciprocals of the lengths of the other two semiaxes.

By the next corollary, we state that a linear map between special regions in $\mathbb{R}^3$ has an inverse function that is monogenic.

Corollary 1 *Let* linf *be an $\mathcal{A}$-valued linear function defined in a domain Ω of $\mathbb{R}^3$ with nonvanishing Jacobian determinant. Then,* **linf** *maps an ellipsoid (with the properties described above) onto a ball if and only if its inverse function* $\mathrm{lin}\mathbf{f}^{-1}$ *is monogenic.*

After this direct consequence of Theorem 3 we show now that the previous corollary can be generalized to arbitrary real-analytic functions which have the described local mapping properties. A function with nonvanishing linear part that maps (in the small) the special class of ellipsoids onto balls has an inverse function that is monogenic.

Corollary 2 *Let* **f** *be an $\mathcal{A}$-valued real-analytic function defined in a domain Ω of $\mathbb{R}^3$ with nonvanishing Jacobian determinant. Then,* **f** *maps locally an ellipsoid (with the properties described above) onto a ball if and only if its inverse function* $\mathbf{f}^{-1}$ *is monogenic.*

The noncommutative nature of the quaternionic algebra raises new obstacles in the study of inverses of $\mathcal{M}$-conformal mappings. In general, the inverse of a monogenic function is not monogenic. In the following we will study at least some properties of the linear part of the inverse of an $\mathcal{A}$-valued monogenic function.

Theorem 6 *Let* **f** *be an $\mathcal{A}$-valued real-analytic function defined in a domain Ω of $\mathbb{R}^3$ with nonvanishing Jacobian determinant and such that it maps locally an ellipsoid (with the properties described above) onto a ball. Then, its linear part satisfies the following properties:*

1. $\mathrm{curl}(\mathrm{lin}\mathbf{f}) = 0$;
2. $|\,\mathrm{div}(\mathrm{lin}\mathbf{f}(x))| = \begin{cases} -\alpha+\beta+\gamma, & \alpha \text{ minimum},\\ \alpha-\beta+\gamma, & \beta \text{ minimum},\\ \alpha+\beta-\gamma, & \gamma \text{ minimum};\end{cases}$
3. $|\,\mathrm{div}(\mathrm{lin}\mathbf{f}(x))| \in \begin{cases} [\alpha,\beta+\gamma], & \alpha \text{ minimum},\\ [\beta,\alpha+\gamma], & \beta \text{ minimum},\\ [\gamma,\alpha+\beta], & \gamma \text{ minimum};\end{cases}$
4. $\overline{D}\,\mathrm{lin}\mathbf{f} = 0$ *if*
   a. $\gamma = \frac{1}{2}(3 \pm \sqrt{5})\beta$ *or* $\beta = \frac{1}{2}(3 \pm \sqrt{5})\gamma$, α *minimum*;
   b. $\gamma = \frac{1}{2}(\sqrt{5} \pm 1)\alpha$, β *minimum*;
   c. $\beta = \frac{1}{2}(\sqrt{5} \pm 1)\alpha$, γ *minimum*.

Proof We have seen in Corollary 1 that the inverse of **linf**, which maps a ball onto the special ellipsoid, is monogenic. Also, it is easily seen that the matrix associated to $\mathrm{lin}\mathbf{f}^{-1}$, here denoted by A, is symmetric. From linear algebra it is well known that the matrix that represents the original function **linf**, A^{-1}, is also symmetric. This means clearly that curl(**linf**) $= 0$.

Denoting by μ_1, μ_2, μ_3 the eigenvalues of A^{-1}, it is clear that

$$\operatorname{div}(\mathbf{linf}(x)) = \operatorname{tr}(A^{-1}) = \mu_1 + \mu_2 + \mu_3 = \frac{1}{\lambda_1} + \frac{1}{\lambda_2} + \frac{1}{\lambda_3},$$

where $\lambda_1, \lambda_2, \lambda_3$ stand for the eigenvalues of A.

Now, assume without loss of generality that div(**linf**(x)) is positive and the parameter α assumes the minimum value of the semiaxes, i.e.,

$$\frac{1}{\alpha} = \frac{1}{\beta} + \frac{1}{\gamma}.$$

Then, Conditions 2 and 3 are immediate since

$$\frac{1}{\lambda_1} + \frac{1}{\lambda_2} + \frac{1}{\lambda_3} = -\alpha + \beta + \gamma = \frac{\beta^2 + \gamma^2 + \beta\gamma}{\beta + \gamma}. \tag{12}$$

For Property 4, we assume again without loss of generality that α is the minimum value. Then, from the previous equality it follows

$$\overline{D}\,\mathbf{linf} = \frac{\lambda_2^2 + \lambda_3^2 + 3\lambda_2\lambda_3}{(\lambda_2 + \lambda_3)\lambda_2\lambda_3}. \qquad \square$$

Remark 6 Taking into account the expression on the right-hand side of (12), it is clear that div(**linf**(x)) cannot be zero, and consequently **linf** cannot be monogenic.

Remark 7 In Property 4 it is shown that for some special choices of the parameters α, β, and γ, the linear part of the inverse of an $\mathcal{A}$-valued monogenic function is antimonogenic.

5 Observations and Perspectives

With the obtained results, we are able to characterize monogenic mappings completely by their local geometric properties. Having in mind the desired analogy to conformal mappings in the plane, one can ask for more special properties. In the here presented approach, we relate the shapes of 3D-objects in the domain of definition and in the image of a mapping. In the paper [17] a certain ratio between surfaces and volumes was the background for the definition of $\mathcal{M}$-conformal mappings. Nothing has been said about the transformation of angles. We want to study by examples here only the transformation of right angles under monogenic transformations. The idea behind is to map orthogonal grids from the sphere to the surface of more general domains. Looking for the moment again only at local properties, we study the problem for linear monogenic mappings. We have seen before that each

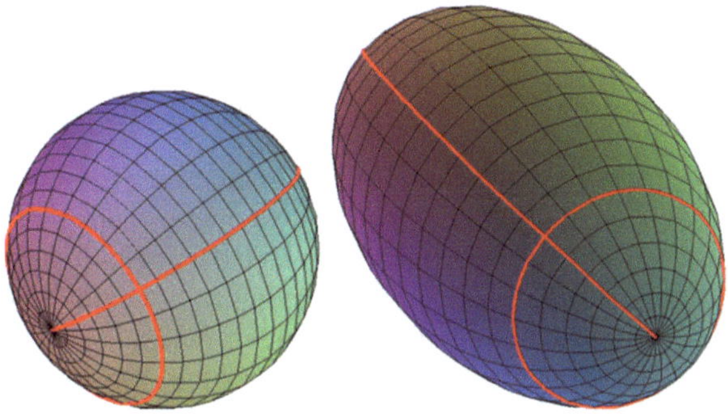

Fig. 2 Images of curves on the unit ball under the linear monogenic transformation $\mathbf{f}(x) = 2x_0 + x_1\mathbf{e}_1 + x_2\mathbf{e}_2$

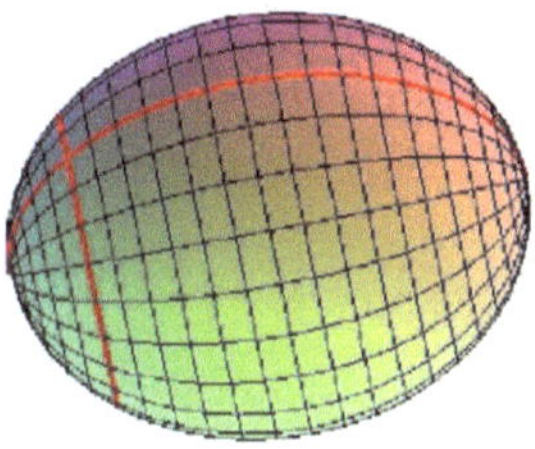

Fig. 3 Images of curves on the unit ball under the linear monogenic transformation $\mathbf{f}(x) = 4x_0 + 5x_1\mathbf{e}_1 - x_2\mathbf{e}_2$

monogenic linear function can be transformed by a rotation such that the representing matrix has diagonal form. All angles are invariant under this rotation. Studying now the question under which conditions the images of meridians and latitude circles intersect orthogonally, we find the additional condition that $b_1^2 - c_2^2 = 0$. The first solution $b_1 = c_2$ defines a special nondegenerate linear and monogenic function. For $b_1 = 1$ and $c_2 = 1$, Fig. 2 visualizes the images of a meridian and a circle with latitude $\pi/4$. Indeed, we can see that the right angles are preserved.

To get an impression that this preservation is not automatically satisfied, we consider a second example. Here we see that the angles are changing. The exact angle in Fig. 3 is about 122°.

Looking now at the second solution of the extra equation, we obtain $b_1 = -c_2$. This implies immediately that $a_0 = 0$ and our mapping degenerates because it does not depend on x_0. Such a mapping is a so-called *hyperholomorphic constant* (see [7]). A hyperholomorphic constant is a monogenic function with identically vanishing hypercomplex derivative. Moreover, taking into account that the set of complex numbers is isomorphic to the (commutative) even subalgebra $\mathcal{C}\ell_{0,2}$, it is easy to prove that such a hyperholomorphic constant is isomorphic to an antiholomorphic function in the (x_1, x_2)-plane (see [19], pp. 10–11). This means that such a mapping will preserve all angles in the (x_1, x_2)-plane (not necessarily the orientation).

Analyzing the condition $b_1 = c_2$ again, we find that this is exactly the condition for a linear monogenic function to be orthogonal to the subspace of hyperholomorphic constants in $L_2(B)$. Moreover, in [9] it was shown that each monogenic

$\mathcal{A}$-valued function $\mathbf{f}$ permits an orthogonal decomposition

$$\mathbf{f}(x) := \mathbf{f}(0) + \mathbf{g}(x) + \mathbf{h}(x), \tag{13}$$

where $\mathbf{h}$ is a hyperholomorphic constant and $\mathbf{g}$ is the so-called principal part of $\mathbf{f}$.

Collecting all these properties, we understand from the simple example that locally each monogenic function can be decomposed as an orthogonal sum of a principal monogenic mapping, preserving right angles as described, and a hyperholomorphic constant, preserving all angles in a plane.

At the moment it is still an open problem how the described local mapping properties can be connected with the global mapping properties of $\mathcal{M}$-conformal mappings. There is already one paper [2], where the authors study the global behavior of a generalized Joukowski mapping in the three-dimensional space.

Acknowledgement The second author is sponsored by *Fundação para a Ciência e a Tecnologia—FCT* via the Ph.D./grant SFRH/BD/19174/2004.

References

1. Ahlfors, L.V.: Lectures on Quasiconformal Mappings. Van Nostrand, Princeton (1966). Reprinted by Wadsworth Inc., Belmont (1987)
2. De Almeida, R., Malonek, H.R.: On a higher dimensional analogue of the Joukowski transformation. Num. Anal. Appl. Math., AIP Conf. Proc. **1048**, 630–633 (2008)
3. Bock, S., Falcão, M.I., Gürlebeck, K., Malonek, H.R.: A 3-dimensional Bergman kernel method with applications to rectangular domains. J. Comput. Appl. Math. **189**, 6–79 (2006)
4. Boone, B.: Bergmankern en conforme albeelding: een excursie in drie dimensies. Proefschrift, Universiteit Gent Faculteit van de Wetenshappen (1991–1992)
5. Cerejeiras, P., Gürlebeck, K., Kähler, U., Malonek, H.R.: A quaternionic Beltrami-type equation and the existence of local homeomorphic solutions. ZAA **20**, 17–34 (2001)
6. Delanghe, R., Kraußhar, R.S., Malonek, H.R.: Differentiability of functions with values in some real associative algebras: approaches to an old problem. Bull. Soc. R. Sci. Liège **70**(4–6), 231–249 (2001)
7. Gürlebeck, K., Malonek, H.R.: A hypercomplex derivative of monogenic functions in $\mathbb{R}^{n+1}$. Complex Var. **39**, 199–228 (1999)
8. Gürlebeck, K., Morais, J.: On mapping properties of monogenic functions. CUBO Math. J. **11**(01), 73–100 (2009)
9. Gürlebeck, K., Morais, J.: Bohr type theorems for monogenic power series. Comput. Methods Funct. Theory **9**(2), 633–651 (2009)
10. Gürlebeck, K., Sprössig, W.: Quaternionic Analysis and Elliptic Boundary Value Problems. Math. Research, vol. 56. Akademieverlag, Berlin (1989)
11. Gürlebeck, K., Sprössig, W.: Quaternionic Calculus for Engineers and Physicists. Wiley, New York (1997)
12. Kraußhar, R.S., Malonek, H.R.: A characterization of conformal mappings in $\mathbb{R}^4$ by a formal differentiability condition. Bull. Soc. R. Sci. Liège **70**, 35–49 (2001)
13. Kravchenko, V.V.: Applied Quaternionic Analysis. Research and Exposition in Mathematics, p. 28. Heldermann, Berlin (2003)
14. Kravchenko, V.V., Shapiro, M.V.: Integral Representations for Spatial Models of Mathematical Physics. Research Notes in Mathematics. Pitman, London (1996)
15. Liouville, L.: Extension au cas de trois dimensions de la question du tracé géographique. Application de l'analyse à la géometrie, G. Monge, Paris, pp. 609–616 (1850).

16. Malonek, H.R.: Power series representation for monogenic functions in $\mathbb{R}^{m+1}$ based on a permutational product. Complex Var. Theory Appl. **15**, 181–191 (1990)
17. Malonek, H.R.: Monogenic functions and M-conformal mappings. In: Brackx, F., et al. (eds.) Clifford Analysis and its Applications. Kluwer, Dordrecht (2001)
18. Mitelman, I.M., Shapiro, M.V.: Differentiation of the Martinelli–Bochner integrals and the notion of hyperderivability. Math. Nachr. **172**, 211–238 (1995)
19. Morais, J.: Approximation by homogeneous polynomial solutions of the Riesz system in $\mathbb{R}^3$. Ph.D. thesis, Bauhaus-Universität Weimar (2009)
20. Sudbery, A.: Quaternionic analysis. Math. Proc. Cambr. Philos. Soc. **85**, 199–225 (1979)
21. Van der Waerden, B.L.: Algebra. Springer, Berlin (1966)

Fluid Flow Problems with Quaternionic Analysis—An Alternative Conception

K. Gürlebeck and W. Sprößig

Abstract This article deals with some classes of fluid flow problems under given initial-value and boundary-value conditions. Using a quaternionic operator calculus, representations of solutions are constructed. For the case of a bounded velocity, a numerically stable semi-discretization procedure for the solution of the problem is presented.

1 Introduction

The study of hydrodynamic problems leads J. d'Alembert (1752) and L. Euler (1777) to the notion of a holomorphic function in the plane, long before A. Cauchy (1814) started with his comprehensive work. B. Riemann used in his papers the connection between holomorphic functions and planar flows of fluids and heat. On the other hand, complex analysis initiated new solution methods for boundary value problems in several research areas of mathematical physics, in particular, in planar fluid- and aerodynamics (N.E. Joukowski 1911) and elasticity theory (G.W. Kolossov 1909).

Although real quaternions were available since Sir W.R. Hamilton's invention in 1843, further 60 years should lapse until the first paper in quaternionic analysis could appear. This paper by A.C. Dixon from the University of Belfast was entitled *On the Newtonian Potential.* In the thirties of the last century, G.N. Moisil and N. Teodorescu from the University of Cluj, on the one hand, and the group around the Swiss mathematician R. Fueter, on the other hand, published a series of fundamental papers on quaternionic analysis. The first closed representation of real Clifford analysis (= generalization of complex analysis in $\mathbb{R}^n$), which can also be

W. Sprößig (✉)
Fakultaet fuer Mathematik und Informatik, TU Bergakademie Freiberg, Prueferstraße 9, 09596 Freiberg, Germany
e-mail: sproessig@math.tu-freiberg.de

E. Bayro-Corrochano, G. Scheuermann (eds.), *Geometric Algebra Computing*,
DOI 10.1007/978-1-84996-108-0_17,

seen as a generalization of quaternionic analysis, was written by the Belgian group around R. Delanghe in 1982 [2].

The solution of boundary-value problems of partial differential equations by function theoretic methods needs a corresponding knowledge on the behavior of Cauchy's parameter integral near to the boundary. The Russian Y.V. Sokhotzki (1873) and the Slovenian mathematician J. Plemelj (1908) found independently of each other jump formulae for Cauchy's integral. In 1912 D. Pompeiu proved the formula

$$u(z) = \frac{1}{2\pi i} \int_{\partial G} \frac{u(\zeta)}{\zeta - z} d\zeta - \frac{1}{\pi} \int_G \partial_{\bar{z}} f(\zeta) \frac{1}{\zeta - z} d\xi\, d\eta, \tag{1.1}$$

where $G \subset \mathbb{R}^3$ is a bounded domain with piecewise smooth boundary ∂G, and $\zeta = \xi + i\eta$. This formula bears today the name formula of Borel–Pompeiu. G.C. Moisil proved in 1930 a corresponding analogue in R^3. The weakly singular integral operator in the second term of Borel–Pompeiu's formula is called Teodorescu transform. The Russian mathematician A.W. Bitzadse found in 1953 a quaternionic analogue to Cauchy's integral in the plane for compact Lyapunov surfaces and quaternionic Hölder continuous functions. T.G. Gegelia proved in 1968 its L_p-boundedness. This result was generalized in 1982 by R. Coifman, A. McIntosh, and Y. Meyer on Lipschitz surfaces in $\mathbb{R}^n$.

Based on three operators: a Cauchy–Riemann-type operator, a Teodorescu transform, and a generalized Cauchy-type integral operator in the books [7, 8], an operator calculus was developed, acting on quaternion-valued or Clifford algebra-valued functions. A new alternative approach for the treatment of fluid flow problems in the space could be established using Plemelj–Sokhotzki-type formulae and Borel–Pompeiu's formula by the help of the mentioned operators. Roughly speaking, a calculus for the spatial fluid flow problems was developed which is analogous to the known complex analytic approach in the plane, advantageously used for a long time.

In this paper we will give a survey on the state of the art. For this reason we will omit the most proofs and refer for that to special research papers. We describe the basics of the applied operator calculus, construct first representation formulas for the steady-state case, and generalize them to the time-dependent case. Finally, we discuss a semi-discretization method for the numerical solution of the initial-boundary-value problems for the Navier–Stokes equations and certain coupled problems involving the Navier–Stokes equations. Our discussion is based on the assumption that a well-adapted numerical method for the remaining elliptic boundary-value problems is available. For a discussion of this problem, we refer to [6]. If the velocity is bounded, then the presented semi-discretization is numerically stable.

2 Operator Calculus

In this section we introduce a general operator conception. By suitable examples we intend to demonstrate the general ideas of this approach.

2.1 Operator Triple

Let X, Y, Z be Banach spaces, and let T, Tr, P be bounded linear operators such that:

(i) $T : X \to \operatorname{im} T \subset Y$ is injective.
(ii) $\operatorname{Tr} : Y \to Z$ is a generalized trace operator.
(iii) The operator $P : \operatorname{im}\operatorname{Tr} \to Y$ satisfies $P \operatorname{Tr} P u = P u$.

We assume:

(iv) $\operatorname{im}\operatorname{Tr} T \subset \ker P$.
(v) $\operatorname{im} T \cap \ker \operatorname{Tr} = \{0\}$.

It is easy to see that $\operatorname{im} T \cap \operatorname{im} P = \{0\}$. The reader can find the proofs in [12].

Theorem 2.1 (Mean value formula [12]) *Set* $\operatorname{im} T \oplus \operatorname{im} P =: Y_1 \subset Y$. *There is a unique linear operator* L *with* $\mathbf{D}(L) = Y_1$ *and* $L : \mathbf{D}(L) \to X$ *such that*

$$u = P \operatorname{Tr} u + T L u. \tag{2.1}$$

Corollary 2.2 [12] *The following relations between the operators* L, P, *and* T *are valid*:

(i) *The operator* L *is the left inverse to the operator* T, *i.e.*, $LT = I$.
(ii) *Set* $R := TL$. *Then* R *is a projection onto* Y_1 *with* $\operatorname{im} R = \operatorname{im} T$.
(iii) $\ker L = \operatorname{im} P \operatorname{Tr}$.

Definition 2.3 (L-Holomorphy) Elements $u \in \ker L \cap Y$ are called L-holomorphic. L is called algebraic derivative. $P \operatorname{Tr}$ is called the initial projection, and T is denoted as general Teodorescu transform.

Remark 2.4 From the point of view of a general operator theory T is also called *algebraic integral*.

2.2 Plemelj-Type Projections

Set $P_r := \operatorname{Tr} P : \operatorname{im}\operatorname{Tr} \to Z$ and $Q_r := I - P_r$. Then the following properties are valid:

(i) The operators P_r, Q_r are idempotent, i.e., we have $P_r^2 = P_r$ and $Q_r^2 = Q_r$ and furthermore $Q_r P_r = P_r Q_r = 0$.
(ii) An element $\xi \in Z$ is the generalized trace of an element u from $\ker L$ if and only if $P_r \xi = \xi$.
(iii) We have $Q_r \xi = \operatorname{Tr} T L u$.

The operators P_r, Q_r are called general Plemelj projections.

The condition $\operatorname{im} TL \cap \ker \operatorname{Tr} = \{0\}$ can be seen as a very general formulation of a maximum principle. In the original papers this condition is called uniqueness condition. We mention her only shortly that there is another approach to this kind of operator calculus, based on the theory of right-invertible operators (see, e.g., [14, 16] for the general theory and [20] for many applications).

2.3 Examples of L-Holomorphy

2.3.1 Fractional Calculus

We consider the absolutely continuous function

$$\big(I_{a+}^{\alpha} u\big)(t) := \frac{1}{\Gamma(\alpha)} \int_a^t \frac{1}{(t-\tau)^{1-\alpha}} u(\tau)\, d\tau, \tag{2.2}$$

which has almost everywhere a derivative in $L_1[a,1]$. Take now for the operator Tr the restriction of a function to the point t and define P by

$$P := \sum_{k=0}^{n-1} \frac{(t-a)^{\alpha-k-1}}{\Gamma(\alpha-k)} \frac{d^{n-k-1}}{dt^{n-k-1}}. \tag{2.3}$$

Furthermore, we introduce

$$(Lu)(t) := \frac{1}{\Gamma(1-\alpha)} \frac{d}{dt} \big(I_{a+}^{1-\alpha} u\big)(t), \qquad (Tu)(t) := \big(I_{a+}^{\alpha} u\big)(t), \tag{2.4}$$

and with $n = [\alpha] + 1$ we recognize

$$(P \operatorname{Tr} u)(t) := \sum_{k=0}^{n-1} \frac{(t-a)^{\alpha-k-1}}{\Gamma(\alpha-k)} \frac{d^{n-k-1}}{dt^{n-k-1}} I_{a+}^{n-\alpha} u(t). \tag{2.5}$$

$(I_{a+}^{\alpha} u)(t)$ is called Riemann–Liouville fractional integral, and $(D_{a+}^{\alpha} u)(t)$ is denoted by Riemann–Liouville fractional derivative.

2.4 Quaternionic Analysis

2.4.1 Real and Complex Quaternions

Let $\mathbb{H}$ be the algebra of real quaternions and $a \in \mathbb{H}$; then $a = \sum_{k=0}^{3} \alpha_k e_k$. Further, let $e_k^2 = -e_0 = -1$, $e_1 e_2 = -e_2 e_1 = e_3$, $e_2 e_3 = -e_3 e_2 = e_1$, $e_3 e_1 = -e_1 e_3 = e_2$. The natural addition (defined as in $\mathbb{R}^4$) and the above-defined multiplication turn $\mathbb{H}$

to a skew-field. A quaternionic conjugation, a norm, and a multiplicative inverse are given by

$$\overline{e_0} = e_0, \qquad \overline{e_k} = -e_k \quad (k = 1, 2, 3), \qquad \overline{a} = a_0 - \sum_{k=1}^{3} \alpha_k e_k = a_0 - \mathbf{a},$$

$$\overline{a}a = a = |a|^2_{\mathbb{R}^4} =: |a|^2_{\mathbb{H}},$$

$$a^{-1} := \frac{1}{|a|^2}\overline{a}, \qquad \overline{ab} = \overline{b}\,\overline{a}.$$

We denote by $\mathbb{H}(\mathbb{C})$ the set of quaternions with complex coefficients, i.e.,

$$a = \sum_{k=0}^{3} \alpha_k e_k \quad (\alpha_k \in \mathbb{C}). \tag{2.6}$$

For $k = 0, 1, 2, 3$, we have the commutator relation $ie_k = e_k i$. Any complex quaternion a has the decomposition $a = a^1 + ia^2$ ($a^j \in \mathbb{H}$). Therefore, also the notation $\mathbb{CH}$ is used. We have three possible conjugations:

1. $\overline{a}^{\mathbb{C}} := a^1 - ia^2$,
2. $\overline{a}^{\mathbb{H}} := \overline{a}^1 + i\overline{a}^2$,
3. $\overline{a}^{\mathbb{CH}} := \overline{a}^1 - i\overline{a}^2$.

2.4.2 D_a-Holomorphy

Let $a := ia_0$ with $a_0 \in \mathbb{R}$. Then

$$D_a := \sum_{k=1}^{3} e_k \partial_k + a \tag{2.7}$$

is called Dirac operator, where e_k ($k = 1, 2, 3$) are the quaternionic units. Functions $u \in C^1(G)$ are called D_a-holomorphic iff $D_a u = 0$.

2.4.3 Spaces of $\mathbb{H}$ ($\mathbb{HC}$)-Valued Functions

We will say that $u \in X$ iff $u_i \in X$ and $u = \sum_{i=0}^{3} u_i e_i$. Let $G \subset \mathbb{R}^3$ be a bounded domain with sufficiently smooth boundary. $C^{k,\lambda}(G)$ denotes the space of all k-times continuously differentiable functions such that the kth derivatives are λ-Hölder-continuous. A norm is given by

$$\|u\|_{k,\lambda} = \sup_{|s| \le k, x \in G} \left|D^s u(x)\right| + \sup_{|s| = k, x \in G, x \ne y} \frac{|D^s u(x) - D^s u(y)|}{|x - y|^\lambda}. \tag{2.8}$$

Let $k = 0, 1, \ldots$ and $p \geq 1$; then the quaternionic (Clifford algebra-valued) Sobolev space is defined by

$$W_p^k(G) = \{u \in L_p(G) : D^s u \in L_p(G); |s| \leq k\}. \tag{2.9}$$

A norm is given by $\|u\|_{p,k} = (\sum_{|s|\leq k} \|D^s u\|_p^p)^{\frac{1}{p}}$. Let

$$I_\lambda(u) = \int_{G\times G} \frac{|u(x) - u(y)|^p}{|x - y|^{k+\lambda p}}\, dx\, dy \tag{2.10}$$

with $k = [k] + \lambda$; $\lambda \in (0, 1)$. Then we can define

$$W_p^k(G) = \{u \in L_p(G) : D^s u \in L_p; |s| \leq k; I_\lambda(D^s u) < \infty\}. \tag{2.11}$$

A norm is given by

$$\|u\|_{p,k} = \left(\|u\|_{p,[k]}^p + \sum_{|s|=[k]} I_\lambda(D^s u)\right)^{\frac{1}{p}}. \tag{2.12}$$

Such spaces are called to be of quaternionic (Clifford algebra-valued) Sobolev–Slobodetzkij type.

2.5 Bergman–Hodge Decomposition

Consider the real Clifford algebra $C\ell_{0,n}$ generated by $e_1, e_2, \ldots, e_n$ with basis

$$e_0; e_1, \ldots, e_n; e_1e_2, \ldots, e_{n-1}e_n; \ldots; e_1 \ldots e_n.$$

The operator $D = \sum_{i=1}^n e_i \partial_i$ is called Dirac operator with zero mass. Functions of the class

$$\ker D(G) \cap C^{m,\lambda}(\overline{G}) \tag{2.13}$$

are called D-holomorphic or simply holomorphic.

Theorem 2.5 [7] *The sets* $\ker D(G) \cap C^{m,\lambda}(\overline{G})$ *and* $\ker D(G) \cap W_p^k(G)$ *are closed subsets in* $C^{m,\lambda}(G)$, $W_p^k(G)$, *respectively, and are called* Bergman spaces.

Proof Two facts are important: We need a mean value formula and Weierstrass' theorem for sequences of holomorphic functions. □

A very important tool is a suitable decomposition of the quaternionic Hilbert space. This can be obtained in the following way. We introduce in $L_2(G)$ the inner product

$$(u,v)_2 = \int_G \overline{u}(x)v(x)\,dx \in \mathbb{H}, \tag{2.14}$$

which fulfils the relations

$$\lambda(u,v)_2 = (\overline{\lambda}u, v)_2, \quad (u,v)_2\lambda = (u, v\lambda)_2 \quad (\lambda \in \mathbb{H}). \tag{2.15}$$

We then find the orthogonal decomposition

$$L_2(G) = L_2(G) \cap \ker D \oplus \overline{D}\mathring{W}_2^1(G) \quad \text{(Bergman–Hodge decomposition).} \tag{2.16}$$

The orthoprojection

$$\mathbf{P} : L_2(G) \overset{\text{onto}}{\to} \ker D \cap L_2(G) \tag{2.17}$$

is called Bergman-type projection relative to the inner product $(u,v)_2$. We denote by

$$\mathbf{Q} := I - \mathbf{P} : L_2(G) \overset{\text{onto}}{\to} D\mathring{W}_2^1(G). \tag{2.18}$$

It can be proved that

$$\mathbf{P} := F_\Gamma(\operatorname{tr} T F_\Gamma)^{-1} T.$$

Remark For sufficiently smooth $\mathbb{H}$-holomorphic functions, the operator $\operatorname{tr} T F_\Gamma$ coincides with the single-layer potential $V_\Gamma u := \frac{1}{4\pi}\int_\Gamma \frac{un}{|y-x|}\,d\Gamma$.

Theorem 2.6 (Trace theorem) *We have the following isomorphism*:

$$\operatorname{tr} V_\Gamma : W_p^{k-\frac{1}{p}}(\Gamma) \cap \operatorname{im} P_\Gamma \to W_p^{k+1-\frac{1}{p}}(\Gamma) \cap \operatorname{im} Q_\Gamma. \tag{2.19}$$

This result can be found in [8]. Analogous results can be proved for the generalized Cauchy–Riemann operator $\partial = \partial_0 + D$.

2.6 Quaternionic Operator Calculus

Let $X = W_p^k(G)$, $Y = W_p^{k+1}(G)$, $Z = W_p^{k+1-\frac{1}{p}}(\Gamma)$, $k = 0,1,2,\ldots; 1 < p < \infty$. Further, let

$$L := D + i\alpha \quad \text{(Dirac operator with mass),} \tag{2.20}$$

$$(Tu)(x) := -\frac{1}{\sigma_3}\int_G e_{i\alpha}(x-y)u(y)\,dy \quad \text{(Teodorescu type transform),} \tag{2.21}$$

$$(Pu)(x) := \frac{1}{\sigma_3}\int_\Gamma e_{i\alpha}(x-y)n(y)u(y)\,d\Gamma_y \quad \text{(CF-type operator)}, \tag{2.22}$$

$$(\mathrm{Tr}\,u)(\xi) := \text{n.-t.-}\lim_{z\to\xi\in\Gamma, z\in G} u(z). \tag{2.23}$$

n.-t.-lim stands for the nontangential limit, and the abbreviation “CF” means Cauchy–Fueter. The disturbed Cauchy–Fueter kernel is given by

$$e_{i\alpha(x)} := -\left(\frac{i\alpha}{2\pi}\right)^{(3/2)}\left[|x|^{-1/2}K_{3/2}(\alpha|x|)\omega - K_{1/2}(\alpha|x|)\right], \tag{2.24}$$

where $\omega \in S^2$, and $K_\mu(t)$ denotes Macdonald's function.

2.7 Discrete Quaternionic Analysis

For an introduction of a discrete calculus, we have to define the following lattice sets:

$$\mathbb{R}_h^3 := \{(ih, jh, kh) : i, j, k \text{ integer},\ h > 0\}, \qquad G_h := G \cap \mathbb{R}_h^3,$$
$$\Gamma_h := \{x \in G_h : \mathrm{dist}(x, \mathrm{co}\, G_h) \le \sqrt{3}h\}.$$

Let $V_{i,h}^{\pm}x$ be the shift of x by $\pm h$ in the x_i-direction. Then

$$\begin{aligned}
\Gamma_{h,\ell(r)} &:= \{x \in \Gamma_h : \exists i : V_{i,h}^{\pm}x \notin G_h\} \quad \text{(left (right) side planes)},\\
\Gamma_{h,\ell(r);i} &:= \{x \in \Gamma_h : V_{i,h}^{\pm}x \notin G_h\},\\
\Gamma_{h,\ell(r);i,j} &:= \Gamma_{h,\ell(r);i} \cap \Gamma_{h,\ell(r);j} \quad \text{(left (right) edges)},\\
\Gamma_{h,\ell(r);i,j,k} &:= \Gamma_{h,\ell(r);i,j} \cap \Gamma_{h,\ell(r);k} \quad \text{(left (right) corners)}.
\end{aligned}$$

Let $X = W_{2,h}^1(G_h)$, $Y = L_{2,h}(G_h)$, $Z = W_{2,h}^{\frac{1}{2}}(G_h)$. Then

$$(Lu)(x) := (D_h^{\pm}u)(x) = \sum_{i=1}^{3} e_i\left[u(V_{i,h}^{\pm}x) - u(x)\right]\frac{1}{h}, \tag{2.25}$$

$$\begin{aligned}
(Tu)(x) &:= (T_h^{\pm}u)(x)\\
&= \left(\sum_{\mathrm{int}\,G_h\cup\Gamma_{h,\ell(r)}} + \sum_{\text{left (right) corners}} - \sum_{\text{left (right) edges}}\right)\\
&\quad \times e_h^{\pm}(x-y)u(y)h^3,
\end{aligned} \tag{2.26}$$

where $e_h^{\pm}$ are the discrete fundamental solutions of $D_h^{\pm}$.

We have

$$(P\,\mathrm{Tr}\,u)(x) := \big(F_h^{\pm}u\big)(x) = \sum_{i=1}^{3}\Big(-\sum_{s_i}+\sum_{s_{ij}}-\sum_{s_{ijk}}\Big)e_h^{\pm}\big(x - V_{i,h}^{\mp}y\big)n(y)u(y)h^2$$

$$+\sum_{i=1}^{3}\sum_{\substack{y\in\Gamma_{h,\ell(r);m,j,k}\\ m\neq j\neq k}} h^{\pm}(x-y)e_i u(y)h^2,$$

where $s_i = \Gamma_{h,\ell;i}\cup\Gamma_{h,r;i}$, $s_{ij} := \Gamma_{h,\ell;j} - V_{i,h}^{+}\Gamma_{h,\ell}$, $s_{ijk} := \Gamma_{h,\ell;j,k} - V_{i,h}^{+}\Gamma_{h,\ell;i,k}$.

In [7] the following mean value theorem is proved:

Theorem 2.7

$$u(x) = \big(F_h^{\pm}u\big)(x) + T_h^{\pm}D_h^{\pm}u(x). \tag{2.27}$$

These formulae are also called (generalized or discrete) Borel–Pompeiu formulae.

To be constructive we need a representation for the fundamental solutions of the discrete Dirac operators. Following [5] *and* [8], $E_h(x)$ *is the solution of the difference equation*

$$-\Delta_h E_h(x) = -\sum_{i=1}^{3} D_{i,h}^{-}D_{i,h}^{+}E_h(x) = \delta_h(x) = \begin{cases} h^{-3}, & x=0,\\ 0, & x\in\mathbb{R}_h^3\setminus\{0\}.\end{cases}$$

We get, with the Fourier transform F,

$$E_h(x) = \frac{1}{\sqrt{2\pi}^3}R_h F\left(\frac{1}{d^2}\right). \tag{2.28}$$

The function d *is defined by*

$$d^2 = \frac{4}{h^2}\left(\sin^2\frac{h\xi_1}{2} + \sin^2\frac{h\xi_2}{2} + \sin^2\frac{h\xi_3}{2}\right), \tag{2.29}$$

and $R_h u$ *is the restriction of the continuous function* u *onto the lattice* $\mathbb{R}_h^3$. *Due to the factorization* $D_{j,h}^{\pm}D_{j,h}^{\mp} = \Delta_h$, *we get*

$$e_h^{\pm}(x) := D_{j,h}^{\mp}E_h(x). \tag{2.30}$$

3 Fluid Flow Problems

3.1 A Brief History of Fluid Dynamics

In 1730 Daniel Bernoulli made the observation that the pressure p of a fluid decreases if its speed increases. Nowadays, this is called Bernoulli's principle. Eleven years later, Leonard Euler, who was invited by Frederick the Great to Potsdam, was charged by him with the construction of a water fountain. In 1755, after thorough studies of the motion of the fluid, he formulated Newton's law for the rate of change of the momentum of a fluid element. This is a set of equations that exactly represents the flow of a fluid as long as one can suppose that the fluid is inviscid:

$$\partial_t \mathbf{u} + (\mathbf{u} \cdot \nabla)\mathbf{u} = -\frac{1}{\rho}\nabla p \quad \text{(Euler's equations)}. \tag{3.1}$$

So he derived the equations of an ideal fluid (no viscous effects included). In 1822 Claude-Louis Navier derived an equation which also considers the inner friction of a flowing fluid without understanding the character of a viscous fluid. His derivation was based on a molecular theory of attraction and repulsion between molecules. J.D. Anderson wrote in his A History of Aerodynamics: *The irony is that although Navier had no conception of shear stress and did not set out to obtain equations that would describe motion involving friction, he nevertheless arrived at the proper form for such equations.* Navier's equations were several times rediscovered (Cauchy 1828, Denise Poisson 1828, and Barré de Saint-Venant 1843). Saint-Venant's model even includes the turbulence case. George Stokes published in 1845 a strong mathematical derivation and explained the equations in the sense that is currently understood. Therefore these equations are called nowadays Navier–Stokes equations (NSE) given by

$$\partial_t \mathbf{u} - \Delta \mathbf{u} + \frac{\rho}{\eta}(u \cdot \nabla)\mathbf{u} + \frac{1}{\eta}\nabla p = f \quad \text{in } G \tag{3.2}$$

for some bounded domain in $\mathbb{R}^3$.

3.2 Stationary Linear Stokes Problem

Let $G \subset \mathbb{R}^3$ be a bounded domain with sufficiently smooth boundary Γ. The linear Stokes system reads as follows:

$$-\Delta \mathbf{u} + \frac{1}{\eta}\nabla p = \frac{\rho}{\eta}\mathbf{f} \quad \text{in } G, \tag{3.3}$$

$$\operatorname{div} \mathbf{u} = f_0 \quad \text{in } G, \tag{3.4}$$

$$\mathbf{u} = \mathbf{g} \quad \text{on } \Gamma. \tag{3.5}$$

Here η is the viscosity, and ρ the density of the fluid. We have to look for the velocity u and the hydrostatical pressure p. Between f_0 and $\mathbf{g}$, we have to fulfil the relation

$$\int_G f_0 \, dx = \int_\Gamma \mathbf{n}\mathbf{g} \, d\Gamma. \tag{3.6}$$

For $g = 0$, the measure of the compressibility f_0 satisfies the identity

$$\int_G f_0 \, dx = 0. \tag{3.7}$$

For all such real functions f_0, the unique solution (p is unique up to a real constant) can be represented as follows:

Theorem 3.1 [7] *Let* $f := f_0 + \mathbf{f} \in W_p^k(G, \mathbb{H})$ $(k \geq 0, 1 < p < \infty)$. *Then we have*

$$\mathbf{u} = \frac{\rho}{\eta} T_G \operatorname{Vec} T_G \mathbf{f} - \frac{\rho}{\eta} T_G \operatorname{Vec} F_\Gamma (\operatorname{tr}_\Gamma T_G \operatorname{Vec} F_\Gamma)^{-1} \operatorname{tr}_\Gamma T_G \operatorname{Vec} T_G \mathbf{f} - T_G f_0, \tag{3.8}$$

$$p = \rho \operatorname{Sc} T_G \mathbf{f} - \rho \operatorname{Sc} F_\Gamma (\operatorname{tr}_\Gamma T_G \operatorname{Vec} F_\Gamma)^{-1} \operatorname{tr}_\Gamma T_G \operatorname{Vec} T_G \mathbf{f} + \eta f_0. \tag{3.9}$$

In that way we strongly separate velocity and pressure!

3.3 Nonlinear Stokes Problem

If the compressibility depends on the velocity and the nonlinear outer forces, then the equations read as follows:

$$-\Delta \mathbf{u} + \frac{1}{\eta} \nabla p = \Lambda f(u) \quad \text{in } G, \tag{3.10}$$

$$\operatorname{div}\left(\eta^{-1} \mathbf{u}\right) = 0 \quad \text{in } G, \tag{3.11}$$

$$\mathbf{u} = 0 \quad \text{on } \Gamma. \tag{3.12}$$

The viscosity η $(\eta > 0)$ depends on the position. Suppose that $f \in L_2(G, \mathbb{H})$, $p \in W_2^1(G)$, $\eta \in C^\infty(G)$.

Theorem 3.2 (Representation formula [7]) *Every solution of the nonlinear Stokes problem permits the representation*

$$u = \Lambda R B f - R B D p,$$
$$0 = \operatorname{Sc} \Lambda \mathbf{Q} T_G B f - \operatorname{Sc} \mathbf{Q} T_G B D p.$$

Here the operator B *stands for the multiplication by* η^{-1}, *and* $R := T_G \mathbf{Q} T_G$.

Theorem 3.3 (Iteration procedure [7]) *Suppose that* $f(u)$, B, *and* Λ *satisfy the following estimates*:

(i) $\|f(u) - f(v)\|_2 \leq L\|u - v\|_{2,1}$ *for* $\|u\|_{2,1}, \|v\|_{2,1} \leq 1$,
(ii) $\|B\|_2 \leq K$ *for positive constants* K, L,
(iii) $\Lambda < \{\|T\|_{\operatorname{im}\mathbf{Q} \cap L_2, L_2} \|T\|_{L_2, L_2} K L\}^{-1}$.

Then the iteration procedure

$$u_n = \Lambda RBf(u_{n-1}) - RBDp_n,$$
$$\Lambda \operatorname{Sc} DBRf(u_{n-1}) = \operatorname{Sc} DBRDp_n,$$

with $\|u_0\|_{2,1} \leq 1$ $(u_0 \in \overset{\circ}{W}{}_2^1(G, \mathbb{H}))$, *converges to the unique solution* $\{u, p\} \in \overset{\circ}{W}{}_2^1(G, \mathbb{H}) \cap \ker(\operatorname{div} B) \times L_2(G)$ *of* (3.10)–(3.12), *where* p *is unique up to a real constant.*

3.4 Stationary Navier–Stokes Problem

As we have already seen, the stationary NSE are given by

$$-\Delta\mathbf{u} + \frac{\rho}{\eta}(u \cdot \nabla)\mathbf{u} + \frac{1}{\eta}\nabla p = f \quad \text{in } G, \tag{3.13}$$
$$\operatorname{div}\mathbf{u} = 0 \quad \text{in } G, \tag{3.14}$$
$$\mathbf{u} = 0 \quad \text{on } \Gamma. \tag{3.15}$$

In dependence on the special application, terms can be dropped out:

- $\nu\Delta\mathbf{u} \approx 0 \ldots$ supersonic motion around an airfoil,
- $\partial_t\mathbf{u} \approx 0 \ldots$ geostrophic flow (steady-state case),
- $(\mathbf{u} \cdot \nabla)\mathbf{u} \approx 0 \ldots$ creeping flow in ground.

Theorem 3.4 [8] *The main result is now the following*:

1. *Let* $f \in L_2(G, \mathbb{H})$ *and* $p \in W_2^1(G)$. *Every solution permits the operator integral representation*

$$u = -\frac{\rho}{\eta}RM(u) - \frac{1}{\eta}T_G\mathbf{Q}p, \tag{3.16}$$
$$0 = \frac{\rho}{\eta}\operatorname{Sc}\mathbf{Q}T_GM(u) - \frac{1}{\eta}\operatorname{Sc}\mathbf{Q}p. \tag{3.17}$$

2. *The above system has a unique solution* $\{u, p\} \in \overset{\circ}{W}{}_2^1(G, \mathbb{H}) \cap \ker(\operatorname{div} B) \times L_2(G)$, *where* p *is unique up to a real constant, if*
   (i) $\|f\|_p \leq (18K^2C_1)^{-1}$ *with* $K := \frac{\rho}{\eta}\|T_G\|_{[L_2\cap \operatorname{im}\mathbf{Q}, W_2^1]}\|T\|_{[L_p, L_2]}$,
   (ii) $u_0 \in \overset{\circ}{W}{}_2^1(G, \mathbb{H}) \cap \ker(\operatorname{div} B)$ *with* $\|u_0\|_{2,1} \leq \min(V, \frac{1}{4KC_1} + W)$ *hold.*

Corollary 3.5 *Setting* $V := (2KC_1)^{-1}$, $W := [(4KC_1)^{-2} - \frac{\rho\|f\|_p}{\eta C_1}]^{\frac{1}{2}}$, *and* $C_1 = 9^{\frac{1}{p}}C$, *where* C *is the embedding constant from* W_2^1 *in* L_2, *the iteration procedure*

(*starting with* u_0)

$$u_n = \frac{\rho}{\eta} RM(u_{n-1}) - \frac{1}{\eta} RDp_n,$$

$$\frac{\rho}{\eta} \operatorname{Sc} \mathbf{Q} T_G M(u_{n-1}) = -\frac{1}{\eta} \operatorname{Sc} \mathbf{Q} p_n, \qquad u_0 \in \overset{\circ}{W}{}_2^1(G, \mathbb{H}) \cap \ker \operatorname{div}$$

converges in $W_2^1(G, \mathbb{H}) \times L_2(G)$.

3.5 Navier–Stokes Equations with Heat Conduction

We will now consider the flow of a viscous fluid under the influence of temperature. The corresponding equations read as follows:

$$-\Delta \mathbf{u} + \frac{\rho}{\eta}(u \cdot \nabla)\mathbf{u} + \frac{1}{\eta}\nabla_p + \frac{\gamma}{\eta}\mathbf{g}w = f \quad \text{in } G, \tag{3.18}$$

$$-\nabla w + \frac{m}{\kappa}(u \cdot \nabla)w = \frac{1}{\kappa}h \quad \text{in } G, \tag{3.19}$$

$$\operatorname{div} \mathbf{u} = 0 \quad \text{in } G, \tag{3.20}$$

$$\mathbf{u}, w = 0 \quad \text{on } \Gamma. \tag{3.21}$$

Here, γ is the Grashof number, κ the temperature conductivity, and m stands for the Prandtl number. Applying the quaternionic operator calculus, we can rewrite the previous system equivalently:

$$u = -R\left[M(u) - \frac{\gamma}{\eta} e_3 w\right] - \frac{1}{\eta} T_G \mathbf{Q} p, \tag{3.22}$$

$$0 = \operatorname{Sc} DR\left[M(u) - \frac{\gamma}{\eta} e_3 w\right] - \frac{1}{\eta}\mathbf{Q} p, \tag{3.23}$$

$$w = -\frac{m}{\kappa} R \operatorname{Sc}(uD)w + Rg, \tag{3.24}$$

where $M(u) := \frac{\rho}{\eta}(\mathbf{u} \cdot \operatorname{grad})u + f(u) - F$.

Remark The Grashof number approximates the ratio of the buoyancy forces to the viscous forces in a fluid,

$$\gamma = \operatorname{Gr} = \frac{g\alpha \Delta w L^3}{\mu^2} = \frac{\text{buoyancy forces}}{\text{viscous forces}}, \tag{3.25}$$

where g is the gravitational acceleration constant, α is the (thermal) volume expansion coefficient of the fluid, Δw is the temperature difference between the fluid and the wall, L is the characteristic length, and μ is the kinematic viscosity of the fluid.

Remark The Prandtl number is a dimensionless parameter of a convecting system that gives the regime of convection. It has the formula

$$m = \mathrm{Pr} = \frac{\mu}{a} = \frac{\text{viscous diffusion rate}}{\text{thermal diffusion rate}},$$

where μ is again the kinematic viscosity, and a is the thermal diffusivity of the fluid.

Remark The Reynolds number is named after Osborne Reynolds, who proposed it in 1883. It is the ratio of inertial forces to viscous forces in a fluid. The expression for this dimensionless measure is

$$\mathrm{Re} = \frac{uL}{\mu} = \frac{\text{inertial forces}}{\text{viscous forces}},$$

where L is the characteristic length, u is the average velocity of the flow, and μ is the kinematic viscosity of the fluid. A fluid flow in a pipe is laminar for Reynolds numbers less than 2000; for values greater than 4000, the flow is turbulent. The Reynolds number is, roughly speaking, the product of Grashof number and Prandtl number.

We consider the following iteration procedure:

$$u_n = -R\left[M(u_{n-1}) - \frac{\gamma}{\eta} e_3 w_{n-1}\right] - \frac{1}{\eta} T_G \mathbf{Q} p_n, \tag{3.26}$$

$$0 = \mathrm{Sc}\, DR\left[M(u_{n-1}) - \frac{\gamma}{\eta} e_3 w_{n-1}\right] - \frac{1}{\eta} \mathbf{Q} p_n, \tag{3.27}$$

$$w_n = -\frac{m}{\kappa} R\, \mathrm{Sc}(u_n D) w_n + Rg. \tag{3.28}$$

The computation of w_n will be done by the inner iteration

$$w_n^j = \frac{m}{\kappa} R\, \mathrm{Sc}(u_n D) w_n^{j-1} + Rg. \tag{3.29}$$

The initial values of the iterations are u_0 ($u_0 = 0$ is possible) and w_n^0 ($w_n^0 = 0$ is possible, $w_n^0 = w_{n-1}$ is better).

Theorem 3.6 (Convergence result)

1. *Let $u_n \in \overset{\circ}{W}{}_2^1$. Further, let $m \neq 4\kappa$ and $\|u_n\| < \kappa/mKC$. The sequence $\{w_n^{\{j\}}\}_{j\in N}$ converges in $W_2^1(G)$.*
2. *Let $F \in L_2(G, \mathbb{H})$, $g \in L_2(G)$, $f : W_2^1(G, \mathbb{H}) \to L_2(G, \mathbb{H})$ with $\|f(u) - f(v)\|_2 \le L\|u - v\|_{2,1}$, and $f(0) = 0$. Under the additional conditions*

$$\text{(i)} \quad \frac{\rho}{\eta}\|F\|_2 + \frac{\gamma}{\eta} K|d|^{-1}\|g\|_2 < \frac{1}{16K^2C} \quad \left(d := \left(4 - \frac{m}{\kappa}\right)\kappa\right), \tag{3.30}$$

$$\text{(ii)} \quad \|g\|_2 < \left(1 - \frac{1}{\sqrt{2}}\right)\eta d^2 \left(\frac{1}{32K^3Cm}\right), \tag{3.31}$$

$$\text{(iii)} \quad m < 4\kappa, \tag{3.32}$$

the sequence $\{u_n, w_n, p_n\}_{\{n \in \mathbb{N}\}}$ *converges in* $W_2^1 \times W_2^1 \times L_2$ *to the unique solution* $(u, w, p) \in \mathring{W}_2^1(G, \mathbb{H}) \times \mathring{W}_2^1(G, \mathbb{H}) \times L_2(G)$ *of the originally boundary-value problem, where p is unique up to a real constant.*

Remark We note that conditions (i) and (ii) can always be realized for fluids with big enough viscosity number.

3.6 Continuous and Discrete Teodorescu Transforms

Let be G_h a lattice G_h with meshwidth h. We study the operators $T_h^{\pm}$. All basic relations (Borel–Pompeiu formula, Hilbert-space decomposition) have a corresponding discrete version, too. This enables us to find strong relations between both the continuous and discrete transforms under weak conditions.

The discrete Teodorescu transform is suitably chosen. One can estimate the "distance" between the continuous operator and the discrete operator.

Let f be a Riemann-integrable function that belongs to $L_\infty(G)$. Then

$$\left\|T_h^+ f - Tf\right\|_{C_h(G)} \to 0 \quad \text{as } h \to 0.$$

If $f \in C^{0,\beta}(\overline{G}),\ 0 < \beta \le 1$, then we have

$$\left\|T_h^+ f - Tf\right\|_{C_h(G)} \le C\big(G, \|f\|_{q,h}, \|f\|_q\big)h^{-2+\frac{3}{p}}|\ln h| \tag{3.33}$$

$$+ K_{p,G}|h|^\beta \|f\|_{C_h(G)} \tag{3.34}$$

for $p < 3/2, 1/p + 1/q = 1$, where $\|\cdot\|_q$ denotes the norm in L_q, and $\|\cdot\|_{q,h}$ is the corresponding discrete norm. Restricting the range for p, we get for $f \in \mathbf{R} \cap L_\infty(G)$ and $p \in (\frac{6}{5}, \frac{3}{2})$,

$$\left\|T_h^+ f - Tf\right\|_{p,G} \to 0 \quad \text{as } h \to 0.$$

Under the assumptions that $f \in C^{0,\beta}(\overline{G}),\ p \in (\frac{6}{5}, \frac{3}{2}),\ 0 < \beta \le 1$, we obtain

$$\left\|T_h^+ f - Tf\right\|_{p,G} \le C\|f\|_{\infty,G}h^{-2+\frac{3}{p}}$$

$$+ C_1\big(\|f\|_{q',h,G}, \|f\|_{q',G}\big)h^{-2+\frac{3}{p'}}|\ln h| \tag{3.35}$$

$$+ K_{p',G}\|f\|_{C^{0,\beta}}(G)h^\beta. \tag{3.36}$$

Here $\mathbf{R}$ describes the class of Riemann-integrable functions.

The proof of these properties requires some work, but there are just technical difficulties. The proof can be found in [7].

3.7 Discrete Version of Navier–Stokes Equations

A corresponding discrete version of Navier–Stokes problem on a lattice is given by

$$-\Delta_h u + \frac{1}{\eta}\nabla_h^+ p + \frac{\rho}{\eta}(u, \nabla_h^-)u = \frac{\rho}{\eta} f \quad \text{in } \operatorname{int} G_h, \tag{3.37}$$

$$\operatorname{div}_h^- u = 0 \quad \text{in } \operatorname{int} G_h, \tag{3.38}$$

$$u = 0 \quad \text{on } \partial G_h, \tag{3.39}$$

where $\nabla_h^\pm := (D_{1,h}^\pm, D_{2,h}^\pm, D_{3,h}^\pm)$, $\operatorname{div}_h^- v := \sum_{i=1}^3 D_{i,h}^-$, and

$$D_{k,h}^\pm u_{i_1\dots i_n} := \pm\frac{1}{h}(u_{i_1\dots i_k\pm 1,\dots i_n} - u_{i_1\dots i_k\dots i_n}). \tag{3.40}$$

It can be proved that

$$\left\| \frac{\rho}{\eta} M_h^{*,-}(u) \right\|_{p,h}^p := \left\| \frac{\rho}{\eta}(u, \nabla_h^-)u \right\|_{p,h}^p \le 9C^p \|u\|_{2,1,h}^{2p} \quad \big(q < 6,\ p = 2q/(2+q)\big). \tag{3.41}$$

Applying the described discrete operator calculus as described in [7], we get the equivalent system

$$u = -T_h^- \mathbf{Q}_h^+ T_h^+ M_h^-(u) - \frac{1}{\eta} T_h^- \mathbf{Q}_h^+ p \quad \text{in } G_h, \tag{3.42}$$

$$\operatorname{Sc} \mathbf{Q}_h^+ T_h^+ M_h^-(u) = \frac{1}{\eta} \operatorname{Sc} \mathbf{Q}_h^+ p \quad \text{in } G_h. \tag{3.43}$$

The operator $\mathbf{Q}_h^+$ denotes the orthoprojection onto the (discrete) Hilbert subspace $D_h^-(\mathring{W}_{2,h}^1(G_h))$, and $M_h^-(u) := M_h^{*,-}(u) - \frac{\rho}{\eta} f$.

It can be proved that this discrete problem has a unique solution u_h in G_h under the same conditions as in the continuous case. If $f \in C^{0,\beta}(\overline{G})$ with $0 < \beta < 1$, then we have the estimate

$$\|u - u_h\|_{2,1;h} \le C\big(h^\beta + h|\ln h|\big) \quad \text{for } h < h_0. \tag{3.44}$$

3.8 Stationary Magneto-Hydromechanics

Magnetic fluids are very important from the technical point of view. These equations read as follows:

$$-\Delta\mathbf{u}+\frac{\rho}{\eta}(\mathbf{u}\cdot\nabla)\mathbf{u}+\frac{1}{\eta}\nabla p-\frac{1}{\mu\eta}\mathbf{B}\times\operatorname{rot}\mathbf{B}=\frac{\rho}{\eta}f \quad \text{in } G, \tag{3.45}$$

$$\Delta\mathbf{B}+\mu\sigma\operatorname{rot}(\mathbf{B}\times\mathbf{u})=\mathbf{g} \quad \text{in } G, \tag{3.46}$$

$$\operatorname{div}\mathbf{B}=0 \quad \text{in } G, \tag{3.47}$$

$$\operatorname{div}\mathbf{u}=0 \quad \text{in } G, \tag{3.48}$$

$$\mathbf{B}=\mathbf{u}=0 \quad \text{on } \Gamma, \tag{3.49}$$

where μ is the permeability, and σ is a measure of the electric charge. $\mathbf{B}$ describes the field induction.

In quaternionic notation, we obtain

$$DD\mathbf{u}+\frac{1}{\eta}D\,p+M(\mathbf{u})=\frac{1}{\mu\eta}\mathbf{B}\times(D\times\mathbf{B}) \quad \text{in } G, \tag{3.50}$$

$$DD\mathbf{B}+\mu\sigma[\mathbf{u},\mathbf{B}]=\mathbf{g} \quad \text{in } G, \tag{3.51}$$

$$\operatorname{Sc} D\mathbf{u}=0 \quad \text{in } G, \tag{3.52}$$

$$\operatorname{Sc} D\mathbf{B}=0 \quad \text{in } G, \tag{3.53}$$

$$\mathbf{B}=\mathbf{u}=0 \quad \text{on } \Gamma. \tag{3.54}$$

We add the following trivial problems:

$$\Delta u_0=0, \quad \Delta B_0=0 \quad \text{in } G, \tag{3.55}$$

$$u_0=0, \quad B_0=0 \quad \text{on } \Gamma. \tag{3.56}$$

This allows us to formulate the equivalent problem for quaternion-valued functions $u=u_0+\mathbf{u}$ and $B=B_0+\mathbf{B}$ instead of vector-valued functions $\mathbf{u}$ and $\mathbf{B}$. The latter class is not adapted to the structure of the algebra $\mathbb{H}$. We introduce the abbreviations

$$M(u):=\frac{\rho}{\eta}\big[M^*(u)-f\big],\ M^*(u):=(\mathbf{u}\cdot D)\mathbf{u}, \tag{3.57}$$

$$[u,B]:=D\times(\mathbf{B}\times\mathbf{u})=(\mathbf{u}\cdot D)\mathbf{B}-(\mathbf{B}\cdot D)\mathbf{u}. \tag{3.58}$$

Notice that $[u,B]$ is just a Lie bracket, $f=\mathbf{f}$, and $g=\mathbf{g}$.

We get the following operator integral representation:

$$u=-\frac{\rho}{\eta}T_G\mathbf{Q}T_G\,M(u)-\frac{1}{\eta}T_G\mathbf{Q}p+\frac{1}{\mu\eta}T_G\mathbf{Q}T_G\big(\mathbf{B}\times(D\times\mathbf{B})\big), \tag{3.59}$$

$$B=T_G\mathbf{Q}T_Gg-\mu\sigma T_G\mathbf{Q}T_G[u,B], \tag{3.60}$$

$$0=\operatorname{Sc}\rho\mathbf{Q}T_GM(u)+\operatorname{Sc}\mathbf{Q}p-\frac{1}{\mu}\operatorname{Sc}QT_G\big(\mathbf{B}\times(D\times\mathbf{B})\big), \tag{3.61}$$

$$0=\operatorname{Sc}\mathbf{Q}T_Gg-\mu\sigma\operatorname{Sc}\mathbf{Q}T_G[u,B]. \tag{3.62}$$

We know that any element $\mathbf{Q}u$ of $\operatorname{im}\mathbf{Q}$ has the representation $\mathbf{Q}u = Dv$ with a suitable $v \in \overset{\circ}{W}{}_2^1(G, \mathbb{H})$. Then it follows that

$$\operatorname{tr}_\Gamma T_G \mathbf{Q}u = \operatorname{tr}_\Gamma T_G Dv = \operatorname{tr}_\Gamma v - \operatorname{tr}_\Gamma F_\Gamma \operatorname{tr} v = 0. \tag{3.63}$$

Let $g = 0$. The solution $\{u, p, 0\}$ solves the corresponding Navier–Stokes equations. We look for solutions with $\mathbf{B} \not\equiv 0$ and consider the following iteration method. For $n = 0, 1, 2, \ldots,$ we calculate

$$u_n = -T_G \mathbf{Q} T_G \left[\frac{\rho}{\eta} M(u_{n-1}) - \frac{1}{\mu\eta} B_{n-1} \times (D \times B_{n-1}) \right] - \frac{1}{\eta} T_G \mathbf{Q} p_n, \tag{3.64}$$

$$0 = \operatorname{Sc}[\rho \mathbf{Q} T_G M(u_{n-1}) + \mathbf{Q} p_n - \frac{1}{\mu} \mathbf{Q} T_G \big[B_{n-1} \times (D \times B_{n-1}) \big], \tag{3.65}$$

$$B_n^{(j)} = -\mu\sigma T_G \mathbf{Q} T_G \big[u_n, B_n^{(j-1)} \big] + T_G \mathbf{Q} T_G g \quad (j = 1, 2, \ldots), \tag{3.66}$$

$$0 = -\mu\sigma \operatorname{Sc} \mathbf{Q} T_G \big[u_n, B_n^{(j-1)} \big] + \operatorname{Sc} \mathbf{Q} T_G g. \tag{3.67}$$

Here B_n will be computed by using again an “inner” iteration.

We underline that at each step of the iteration one has to solve only a linear Stokes problem.

3.8.1 Estimations

Let $u, B \in \overset{\circ}{W}{}_2^1(G, \mathbb{H})$ and $1 < p < 3/2$. Then we obtain

$$\big\| [u, B] \big\|_p^p \leq 2 \sum_{j,i=1}^{3} C^p \| u_i \|_{2,1}^p, \tag{3.68}$$

$$\| B_j \|_{2,1}^p \leq 18 C^p \| u \|_{2,1}^p \| B \|_{2,1}^p. \tag{3.69}$$

The embedding constant C can be calculated by the estimate

$$\| u \|_q = \| T_G D u \|_q \leq \| T_G \|_{[L_2, L_q]} \| D u \|_2 \leq \| T_G \|_{[L_2, L_q]} \| u \|_{2,1}, \tag{3.70}$$

which leads to $\| T_G \|_{[L_2, L_q]} = C$ for $q < 6$, $p = 2q/(2+q)$. Set now $C_1 = 9^{1/p} C$. We get

$$\big\| [u, B] \big\|_p \leq 2 C_1 \| u \|_{1,2} \| B \|_{2,1}. \tag{3.71}$$

Next, we consider the inner iteration

$$B_n^{(j)} = \mu\sigma T_G \mathbf{Q} T_G\big[\big(B_n^{(j-1)} \cdot D\big)u_n - B_n^{(j-1)}(u_n \cdot D)\big] + T_G \mathbf{Q} T_G g, \tag{3.72}$$

$$B_n^{(0)} = 0. \tag{3.73}$$

We abbreviate:

$$K := \|T_G\|_{[L_2 \cap \operatorname{im}\mathbf{Q}, W_2^1]} \|T_G\|_{[L_p, L_2]}. \tag{3.74}$$

It follows that

$$\big\|B_n^{(j)}\big\|_{2,1} \leq \mu\sigma K 2C_1 \|u_n\|_{2,1} \|B_n^{(j-1)}\|_{2,1} + K\|g\|_p. \tag{3.75}$$

Because of

$$\|u_n\|_{2,1} < \frac{1}{4C_1\mu\sigma K}, \tag{3.76}$$

we have

$$\big\|B_n^{(j)}\big\|_{2,1} \leq 2K\|g\|_p. \tag{3.77}$$

3.8.2 Inner Iteration

Hence, $(B_n^{(j)})$ is bounded in $W_2^1(G, \mathbb{H})$, and there exists a weakly convergent subsequence. On the other hand, we have

$$\big\|B_n^{(j)} - B_n^{(j-1)}\big\|_{2,1} = 2C_1\mu\sigma \big\|T_G \mathbf{Q} T_G\big[u_n, B_n^{(j-1)} - B_n^{(j-2)}\big]\big\|_{2,1} \tag{3.78}$$

$$\leq 2C_1 K\mu\sigma \|u_n\|_{2,1} \big\|B_n^{(j-1)} - B_n^{(j-2)}\big\|_{2,1}$$

$$< \frac{1}{2}\big\|B_n^{(j-1)} - B_n^{(j-2)}\big\|_{2,1}. \tag{3.79}$$

From these relations we get the strong convergence of $(B_n^{(j)})$ in $W_2^1(G, \mathbb{H})$ to the limit function B_n which satisfies the estimate

$$\|B_n\|_{2,1} \leq 2K\|g\|_p. \tag{3.80}$$

3.8.3 A Priori Estimates

It is easy to see that

$$\|Du\|_2 \geq \frac{1}{\|T_G\|_{[\operatorname{im}\mathbf{Q} \cap L_2, W_2^1]}} \|u\|_{2,1}. \tag{3.81}$$

Furthermore, one can obtain

$$\frac{1}{\|T_G\|_{[\mathrm{im}\,\mathbf{Q},W_2^1]}}\|u\|_{2,1}+\frac{1}{\eta}\|\mathbf{Q}p\|_2\le 2^{1/2}\|T_G\|_{[L_p,L_2]}\left[\frac{\rho}{\eta}\left\|M^*(u)\right\|_p\right. \tag{3.82}$$

$$\left.+\frac{\rho}{\eta}\|f\|_p+\frac{2}{\mu\eta}\|B\|_{2,1}^2C_1\right], \tag{3.83}$$

and so

$$\frac{1}{\|T_G\|_{[\mathrm{im}\,\mathbf{Q},W_2^1]}}\|u\|_{2,1}\le 2\|T_G\|_{[L_p,L_2]}$$

$$\times\left[\frac{\rho}{\eta}C_1\|u\|_{2,1}^2+\frac{\rho}{\eta}\|f\|_p+\frac{2}{\mu\eta}\|B\|_{2,1}^2C_1\right], \tag{3.84}$$

$$\|u\|_{2,1}\le 2K\frac{\rho}{\eta}C_1\|u\|_{2,1}^2+2K\frac{\rho}{\eta}\|f\|_p+\frac{4K}{\mu\eta}C_1\|B\|_{2,1}^2. \tag{3.85}$$

We have the following quadratic inequality for $\|u\|_{2,1}$:

$$2K\frac{\rho}{\eta}C_1\|u\|_{2,1}^2-\|u\|_{2,1}+2K\frac{\rho}{\eta}\|f\|_p+\frac{4K}{\mu\eta}C_1\|B\|_{2,1}^2\ge 0, \tag{3.86}$$

$$\|u\|_{2,1}^2-\frac{\eta}{2K\rho C_1}\|u\|_{2,1}+\frac{1}{C_1}\|f\|_p+\frac{2}{\rho\mu C_1}\|B\|_{2,1}^2=0, \tag{3.87}$$

so that

$$\frac{\eta}{4K\rho C_1}-\sqrt{\frac{\eta^2}{16K^2\rho^2C_1^2}-\frac{1}{C_1}\|f\|_p-\frac{2}{\rho\mu}\|B\|_{2,1}^2} \tag{3.88}$$

$$\le\|u\|_{2,1}\le\frac{\eta}{4K\rho C_1}+\sqrt{\frac{\eta^2}{16K^2\rho^2C_1^2}-\frac{1}{C_1}\|f\|_p-\frac{2}{\rho\mu}\|B\|_{2,1}^2}. \tag{3.89}$$

There arises the necessary condition

$$\frac{1}{C_1}\|f\|_p+\frac{2}{\rho\mu}\|B\|_{2,1}^2<\frac{\eta^2}{16K^2\rho^2C_1^2}. \tag{3.90}$$

On the other hand, we have to realize for any u_n the condition

$$\|u_n\|_{2,1}\le\frac{1}{4C_1\mu\sigma K}. \tag{3.91}$$

3.8.4 Convergence Results

Now we are able to estimate

$$\|u_n - u_{n-1}\|_{2,1} \le \left\| T_G Q T_G \frac{\rho}{\eta} \left[M(u_{n-1}) - M(u_{n-2}) \right] \right\|_{2,1} \tag{3.92}$$

$$+ \left\| T_G Q T_G \frac{1}{\mu\eta} \left[B_{n-1} \times (D \times B_{n-1}) - B_{n-2} \right. \right.$$

$$\left. \left. \times (D \times B_{n-2}) \right] \right\|_{2,1} \tag{3.93}$$

$$+ \frac{1}{\eta} \left\| T_G Q (p_n - p_{n-1}) \right\|_{2,1}. \tag{3.94}$$

From the a priori estimate we get

$$\sqrt{2} \|T_G\|_{[\operatorname{im} Q, W_2^1]} \left\| T_G N(u, B) \right\|_2 \ge \|u\|_{2,1} + \frac{1}{\eta} \|T_G\|_{[\operatorname{im} Q, W_2^1]} \|Qp\| \tag{3.95}$$

with

$$N(u, B) = \frac{1}{\eta} \left[\rho f + \frac{1}{\mu} M^*(B) - \rho M^*(u) - \frac{1}{2\mu} D|B|^2 \right]. \tag{3.96}$$

For vector fields B_1 and B_2, we have

$$\begin{aligned} B_1 \times (D \times B_2) &= (B_1 \cdot D) B_2 - D(B_1 \cdot B_2), \\ \|D B_1 \cdot B_2\|_p &\le C_1 \|B_1\|_{1,2} \|B_2\|_{1,2}. \end{aligned} \tag{3.97}$$

Further, we obtain

$$\|u_n - u_{n-1}\|_{2,1} \le 3 \|T_G Q T_G\| \left\| \frac{\rho}{\eta} \left[M^*(u_{n-1}) - M^*(u_{n-2}) \right] \right\|_{2,1} \tag{3.98}$$

$$+ 2 \left\| T_G Q T_G \frac{1}{\mu\eta} \left[B_{n-1} \times \big(D \times (B_{n-2} + B_{n-1}) \big) \right] \right\|_{2,1} \tag{3.99}$$

$$+ \left\| (B_{n-1} - B_{n-2}) \times (D \times B_{n-2}) \right\|_{2,1} \tag{3.100}$$

$$\le \frac{3KC_1}{\eta} \Big\{ \rho \|u_{n-1} - u_{n-2}\|_{2,1} \left[\|u_{n-1}\|_{2,1} + \|u_{n-2}\|_{2,1} \right] \tag{3.101}$$

$$+ \frac{4}{\mu} \|B_{n-1} - B_{n-2}\|_{2,1} \left[\|B_{n-1}\|_{2,1} + \|B_{n-2}\|_{2,1} \right] \Big\}. \tag{3.102}$$

On the other hand,

$$\|B_n - B_{n-1}\|_{2,1} \le 8 \mu \sigma K C_1 \|g\|_p \|u_n - u_{n-1}\|_{2,1}. \tag{3.103}$$

Now we have

$$\frac{\eta}{\rho} < \frac{1}{\mu\sigma} \tag{3.104}$$

and

$$\|u_n\| \leq \frac{1}{12\mu\sigma K C_1}, \tag{3.105}$$

$$\|u_n - u_{n-1}\|_{2,1} \leq \left(384K^3C_1^2\|g\|2_p + \frac{1}{2}\right)\|u_{n-1} - u_{n-2}\|_{2,1}. \tag{3.106}$$

A sufficient condition should be

$$\|f\|_p + \frac{8K^2}{\rho\mu}\|g\|_p^2 < \frac{63}{64}\frac{\eta^2}{K^2\rho^2C_1}. \tag{3.107}$$

By straightforward calculation one will find that the sequence $\{\|u_n\|_{2,1}\}$ is decreasing and separated from zero. We can now formulate the following theorem:

Theorem 3.7 *Let $f, g \in L_p(G, \mathbb{H})$. We assume that*

(i) $\frac{\mu\sigma\eta}{\rho} < 1$,
(ii) $\|f\|_p + \frac{8K^2}{\rho\mu}\|g\|_p^2 < \frac{63}{64}\frac{\eta^2}{K^2\rho^2C_1}$,
(iii) $\|g\|_p < \frac{\eta}{768K^3C_1^2\sigma}$ *with*

$$K := \|T_G\|_{[L_2\cap \operatorname{im}\mathbf{Q}, W_2^1]}\|T_G\|_{[L_p, L_2]},$$

$$C_1 := 9^{\frac{1}{p}}\|T\|_{[L_2, L_q]} \quad \left(q < 6,\ p = \frac{2q}{2+q}\right).$$

Then the operator integral equation has the unique solution $\{u, B, p\} \in \mathring{W}_2^1(G, \mathbb{H}) \times \mathring{W}_2^1(G, \mathbb{H}) \times L_2(G, \mathbb{R})$, where u, B are uniquely defined, and p is unique up to an additive real constant. Our iteration method converges in $\mathring{W}_2^1(G, \mathbb{H}) \times \mathring{W}_2^1(G, \mathbb{H}) \times L_2(G, \mathbb{R})$ to this solution if $B_0, u_0 \in \mathring{W}_2^1(G, \mathbb{H}) \cap \ker \operatorname{div}$ and are sufficiently small.

4 Time-Dependent Fluid Flow Problems

4.1 Characterization of Fluids

The so-called deviatoric stress tensor $(\tau_{k\ell})$ determines the character of the fluid flow. One has to distinguish the following fluids:

Ideal fluids: Let $\tau_{k\ell} = 0$, i.e., the stress only depends on the hydrostatical pressure. Then we have

$$\partial_t \mathbf{u} + (\mathbf{u} \cdot \nabla)\mathbf{u} + \frac{1}{\rho}\nabla p = 0 \quad \text{(Euler equations)}, \tag{4.1}$$

$$\nabla \cdot \mathbf{u} = 0 \quad \text{(incompressibility)}. \tag{4.2}$$

Newtonian fluids: Let $\tau_{k\ell} = 2\mu\dot{\gamma}_{k\ell}$ with $\dot{\gamma}_{k\ell} := \frac{1}{2}(\partial_k u_\ell + \partial_\ell u_k)$; μ denotes the dynamic viscosity, and $\dot{\gamma}_{k\ell}$ is the so-called shear rate tensor. The incompressibility condition reads now

$$\operatorname{div} \mathbf{u} = \sum_{\ell=1}^{3} \dot{\gamma}_{k\ell} = 0. \tag{4.3}$$

Fluid flow is now governed by the Navier–Stokes equations

$$\partial_t \mathbf{u} + (\mathbf{u} \cdot \nabla)\mathbf{u} + \frac{1}{\rho}\nabla p - \nu\Delta\mathbf{u} = \frac{f}{\rho}. \tag{4.4}$$

The expression $\nu\nabla\mathbf{u}$ with kinematic viscosity $\nu := \mu/\rho$ is called viscous term, which is responsible for the dissipation effect (loss of energy). The expression $(\mathbf{u} \cdot \nabla)\mathbf{u}$ is called convection term.

Newtonian compressible fluids: The deviatoric stress tensor is now more complicated and given by

$$\tau_{k\ell} = 2\mu\dot{\gamma}_{k\ell} + \lambda\dot{\gamma}_{k\ell}\delta_{k\ell}. \tag{4.5}$$

λ, μ are called Lamé coefficients, which are not independent. Saint Venant found in 1843 that $\lambda = -\frac{2}{3}\mu$. The Navier–Stokes equations transmute to

$$\rho\partial_t \mathbf{u} - \mu\Delta\mathbf{u} - (\mu + \lambda)\nabla \operatorname{div} \mathbf{u} + \nabla\rho = \rho f - \rho(\mathbf{u} \cdot \nabla)\mathbf{u}, \tag{4.6}$$

$$\nabla \cdot (\rho\mathbf{u}) = 0 \quad \text{(Stokes–Poisson equations)}, \tag{4.7}$$

$$\Phi(p, \rho) = 0. \tag{4.8}$$

Now we assume the following nonlinear relation:

$$\tau_{k\ell} = \tau_{k\ell}(\mathbf{u}) =: \mathbf{M}\mathbf{u}. \tag{4.9}$$

Further we assume that $\nabla \cdot \mathbf{u} = 0$. Then we have the so-called generalized Navier–Stokes equations [10, 15]

$$\rho\partial_t \mathbf{u} - \operatorname{div} \mathbf{M}(\mathbf{u})\nabla\mathbf{u} + \nabla p + \rho(\mathbf{u} \cdot \nabla)\mathbf{u} = \rho\mathbf{f}. \tag{4.10}$$

Let $\mathbf{M}\mathbf{u} := \nu(\alpha + |\dot{\gamma}_{k\ell}(u)|^{p-2})\dot{\gamma}_{k\ell}$ (ν, α are constants) The case $p = 2$, $\nu\alpha =: \mu$ describes a Newtonian fluid. For arbitrary p, these fluids are called non-Newtonian fluids with p-structure.

Fluids can be for a fixed time structure viscous (= convex curve in the shear-rate–shear-stress system (sss)), Newtonian (= straight line in (sss)), or dilatant (concave curve in (sss)). Structure viscous means shear-thinning. For instance, solid color becomes liquid while stiring. Dilatant fluids mean shear-thickening. An example would be walking in wet sands at the beach. The depth of sinking depends on the velocity inverse proportionally.

In the case of a constant shear-rate we have a convex curve (increasing viscosity in the time). Such fluids are called rheopex. Usually, for a fluid which has a decreasing viscosity in the time, one uses the notation thixotrop. Thixotrop fluids correspond with structure viscous fluids.

5 Rothe's Method of Semi-Discretization

5.1 Time-Dependent Stokes Problem

The time-dependent Stokes' problem reads as follows:

$$\frac{1}{\mu}\partial_t \mathbf{u} - \Delta \mathbf{u} + \frac{1}{\mu}\nabla p = \frac{1}{\mu} f \quad \text{in } (0,T)\times G, \tag{5.1}$$

$$\operatorname{div} \mathbf{u} = 0 \quad \text{in } (0,T)\times G, \tag{5.2}$$

$$u(0,\cdot) = u_0 \quad \text{in } \{0\}\times G, \tag{5.3}$$

$$u = g \quad \text{on } (0,T)\times \Gamma. \tag{5.4}$$

Let now be

$$T = n\tau, \quad T>0, \tau \text{ meshwidth}, \quad \text{and} \quad u_k := u(k\tau,\cdot), \qquad p_k := p(k\tau,\cdot). \tag{5.5}$$

We approximate the partial derivative with respect to the time t by forward differences:

$$\partial_t u \sim \frac{u_{k+1}-u_k}{\tau}. \tag{5.6}$$

5.1.1 Semi-Discretization

We obtain from (5.1) the following discretized equation:

$$\frac{\rho}{\mu\tau} u_{k+1} + DDu_{k+1} + \frac{1}{\mu} Dp_{k+1} = \frac{f}{\mu} + \frac{\rho}{\mu\tau} u_k \quad (k=0,1,\ldots,n-1). \tag{5.7}$$

A suitable factorization leads to

$$(D+a)(D-a)u_{k+1} + 1/\mu\, D\, p_{k+1} = f/\mu + a_0^2 u_k \quad (k=0,1,\ldots,n-1). \tag{5.8}$$

This equation permits the operator integral representation

$$u_{k+1} = -\frac{1}{\mu} T_{-a} \mathbf{Q}_a T_a D\, p_{k+1} + T_{-a} \mathbf{Q}_a T_a \left(\frac{f}{\mu} + a_0^2 u_k \right) + \mathbf{H}_k. \tag{5.9}$$

$\mathbf{H}_k$ is explicitly described. We have

$$\mathbf{H}_k := T_{-a} F_{\Gamma,a} (\mathrm{tr}_\Gamma\, V_a)^{-1} Q_{\Gamma,-a} g + F_{\Gamma,-a} g, \tag{5.10}$$

$$u_{k+1} = -\frac{1}{\mu} T_{-a} \mathbf{Q}_a \tilde{p}_{k+1} + T_{-a} \mathbf{Q}_a T \left(\frac{f}{\mu} + a_0^2 u_k \right) + \mathbf{H}_k, \tag{5.11}$$

$$\tilde{p}_{k+1} = p_{k+1} - a\, T_a p_{k+1}. \tag{5.12}$$

Moreover,

$$(D_a \tilde{p}_{k+1} = D p_{k+1}!), \tag{5.13}$$

which yields the well-known form.

5.2 Oseen's Equation

$$\frac{\mathbf{f}}{\mu} = 1/\nu \partial_t \mathbf{u} - \Delta \mathbf{u} + 1/\nu (\mathbf{v} \cdot \nabla \mathbf{u}) + 1/\mu \nabla p$$

$$\text{in } (0, T] \times G \text{ (Oseen equation)}, \tag{5.14}$$

$$\operatorname{div} \mathbf{u} = 0 \quad \text{(Continuity condition)}, \tag{5.15}$$

$$u = g \quad \text{(Boundary condition)}, \tag{5.16}$$

$$\mathbf{u}(0, \cdot) = \mathbf{u}_0 \quad \text{(Initial value condition)}, \tag{5.17}$$

where $\mathbf{v}$ is a dominant flow vector (suitably chosen), ν is the kinematic viscosity, μ is the dynamic viscosity, and $\mathbf{n}$ is the unit vector of the outer normal.

5.3 A Special Discretization Method

Let $T > 0$, $T = n\tau$ with meshwidth τ. We abbreviate $u_k := u(k\tau, \cdot)$, $p_k := p(k\tau, \cdot)$, $f_k = 1/\tau \int_{k\tau}^{(k+1)\tau} f(t, x)\, dt$. Using the forward differences

$$(\partial_t u)(k\tau, \cdot) \sim \frac{u_{k+1} - u_k}{\tau} \quad (k = 0, 1, \ldots, n-1), \tag{5.18}$$

we get

$$u_{k+1} - u_k = -\tau(\mathbf{v} \cdot D)u_k - \tau D p_{k+1} + \tau \nu \Delta u_{k+1} + \frac{\tau f_k}{\rho} \quad (*). \tag{5.19}$$

Using the decomposition idea of Kalthoff/Schwarzer/Herrmann (Stuttgart), we introduce the function u^* by

$$\begin{aligned}(u_{k+1} - u^*) - (u_k - u^*) &= -\tau(v \cdot D)u_k - \tau D p_{k+1} \\ &\quad + \frac{\tau}{\rho} f_k + \Delta u_{k+1}(\nu\tau).\end{aligned} \tag{5.20}$$

It follows that

$$\begin{aligned}u_{k+1} - u^* &= -\tau D p_{k+1} + \nu\tau \Delta u_{k+1}; \\ u_k - u^* &= -\tau(v \cdot D)u_k - \frac{\tau}{\rho} f_k.\end{aligned} \tag{5.21}$$

We have now to consider the problem

$$\operatorname{div} u_{k+1} = \operatorname{div} u^* - \tau \Delta p_{k+1} + \nu\tau \Delta \operatorname{div} u_{k+1} = 0 \tag{5.22}$$

with $u^* = u_k + \tau(v \cdot D)u_k + \frac{\tau}{\rho} f_k$.

It remains to deal with the following equations:

$$\Delta(-\nu \operatorname{div} u_{k+1} + p_{k+1}) = +\frac{\operatorname{div} u^*}{\tau}, \tag{5.23}$$

$$\Delta\eta = +\frac{\operatorname{div} u^*}{\tau}. \tag{5.24}$$

Borel–Pompeiu's formula yields

$$D p_{k+1} = T \frac{\operatorname{div} u^*}{\tau} + \phi_{k+1}, \quad \phi_{k+1} \in \ker D. \tag{5.25}$$

Inserting this into $(*)$, we can conclude

$$u_{k+1} - \nu\tau \Delta u_{k+1} = -\tau(v \cdot D)u_k + T \operatorname{div} u^* + \tau \phi_{k+1} + \frac{\tau}{\rho} f_k + u_k, \tag{5.26}$$

and with $a_0^2 = \frac{1}{\nu\tau}$, we arrive at

$$\begin{aligned}(a_0^2 - \Delta)u_{k+1} &= -\left(\frac{1}{\nu}(v \cdot D)u_k + a_0^2 u_k\right) \\ &\quad + \left(\frac{f_k}{\mu} + a_0^2 T \operatorname{div} u^*\right) + \frac{\phi_{k+1}}{\nu}.\end{aligned} \tag{5.27}$$

Remark Each $\tilde{\phi}_{k+1} := \frac{\phi_{k+1}}{\nu}$ defines another approximation!

With $S := a_0^2(-\frac{1}{\nu}(v \cdot D) + I)$ and $R_k := \frac{f_k}{\mu} + a_0^2 T \operatorname{div} u^*$, we obtain

$$D_{ia} D_{-ia} u_{k+1} = S u_k + R_k + \tilde{\phi}_{k+1}. \tag{5.28}$$

Borel–Pompeiu's formula leads to

$$\begin{aligned} u_{k+1} &= T_{-ia} T_{ia}(S u_k + R_k) + T_{-ia} T_i a \tilde{\phi}_{k+1} \\ &\quad + T_{-ia} \phi_{+ia,k+1} + \phi_{-ia,k+1}, \end{aligned} \tag{5.29}$$

where $\phi_{\pm ia,k+1} \in \ker D_{\pm a}$.

It remains to determine $\phi_{ia,k+1}$ and $\phi_{-ia,k+1}$.

$$\bullet F_{\Gamma,-ia} u_{k+1} = \phi_{-ia,k+1} = F_{\Gamma,-ia} g_{k+1}; \quad g_{k+1} = g\big((k+1)\tau, \cdot\big). \tag{5.30}$$

5.4 $(D + ia)$-Holomorphic Functions

Let us now compute $(D + ia)$-holomorphic functions. As for $\operatorname{tr}_\Gamma u_{k+1} = g_{k+1}$, we have

$$\begin{aligned} &\operatorname{tr}_\Gamma \big[+T_{-ia} T_{ia}(S u_k + R_k) + T_{-ia} T_{ia} \tilde{\phi}_{k+1} + T_{-ia} \phi_{ia,k+1} \big] \\ &\quad = Q_{\Gamma,-ia} g_{k+1}. \end{aligned} \tag{5.31}$$

We know that

$$\operatorname{tr}_\Gamma : (I - F_{\Gamma,-ia}) g_{k+1} \to Q_{\Gamma,-ia} g_{k+1}. \tag{5.32}$$

Therefore,

$$\operatorname{tr}_\Gamma \phi_{ia,k+1} = (\operatorname{tr}_\Gamma T_{-ia} F_{\Gamma,ia})^{-1} \big(-T_{-ia} T_{ia} \big(S u_k + \big(R_k - \tilde{\phi}_{k+1} \big) \big) \big) \tag{5.33}$$

$$+ (\operatorname{tr}_\Gamma T_{-ia} F_{\Gamma,ia})^{-1} Q_{\Gamma,-ia} g_{k+1}. \tag{5.34}$$

Consequently, we get

$$\phi_{ia,k+1} = F_{\Gamma,ia} (\operatorname{tr}_\Gamma T_{-ia} F_{\Gamma,ia})^{-1} \big(-T_{-a} T_{ia} \big(S u_k + \tilde{R}_k \big) \big) \tag{5.35}$$

$$+ F_{\Gamma,ia} (\operatorname{tr}_\Gamma T_{-ia} F_{\Gamma,ia})^{-1} Q_{\Gamma,-ia} g_{k+1}. \tag{5.36}$$

5.4.1 Final Representation Formula

After substitution and straight forward calculation it follows

$$u_{k+1} = T_{-ia} \big(T_{ia} \big(S u_k + \tilde{R}_k \big) \tag{5.37}$$

$$- F_{\Gamma,ia} (\operatorname{tr}_\Gamma T_{-ia} F_{\Gamma,ia})^{-1} T_{-ia} T_{ia} \big(S u_k + \tilde{R}_k \big) \big) \tag{5.38}$$

$$+ T_{-ia}F_{\Gamma,ia}(\mathrm{tr}_\Gamma\, T_{ia}F_{\Gamma,ia})^{-1}Q_{\Gamma,-ia}g_{k+1} + F_{\Gamma,-ia}g_{k+1}, \tag{5.39}$$

$$u_{k+1} = T_{-ia}\mathbf{Q}_{ia}T_{ia}\big(Su_k + \tilde{R}_k\big) + T_{-ia}F_{\Gamma,ia}(\mathrm{tr}_\Gamma\, T_{-ia}F_{\Gamma,ia})^{-1}Q_{\Gamma,-ia}g_{k+1} + F_{\Gamma,-ia}g_{k+1},$$

with the Bergman projection $\mathbf{P}_{ia} = F_{\Gamma,a}(\mathrm{tr}_\Gamma\, T_{-ia}F_{\Gamma,ia})^{-1}T_{-ia}$ and $\mathbf{Q}_{ia} = I - \mathbf{P}_{ia}$.

5.4.2 Influence of the Boundary Condition $\mathbf{H}_{k+1}$

We have

$$\mathbf{H}_{k+1} = F_{\Gamma,-ia}g_{k+1} + T_{-ia}\mathbf{P}_i a D_{-ia}H_{k+1} \quad (k = 0,\dots,n-1), \tag{5.40}$$

where H_{k+1} is a smooth continuation into the domain G. If $\tilde{H}_{k+1}$ is another extension, then it follows

$$T_{-ia}\mathbf{P}_{ia}D_{-ia}\big(H_{k+1} - \tilde{H}_{k+1}\big) = 0 \tag{5.41}$$

and

$$D_{-ia}\mathring{W}_2^1(G) = \mathrm{im}\,\mathbf{Q}_{ia}. \tag{5.42}$$

Notice that $\mathrm{tr}_\Gamma\, H_{k+1} = \mathrm{tr}_\Gamma\, \tilde{H}_{k+1}$.

5.4.3 Boundary-Value Problem for $\mathbf{H}_{k+1}$

$\mathbf{H}_{k+1}$ solves the following boundary-value problem

$$(D + ia)(D - ia)v_{k+1} = 0 \quad \text{in } G, \tag{5.43}$$

$$v_{k+1} = g_{k+1} \quad \text{on } \Gamma. \tag{5.44}$$

Using Borel–Pompeiu's formula and the formulae of Plemelj–Sokhotzkij type, we obtain

$$\mathrm{tr}_\Gamma\, T_{-ia}\mathbf{Q}_{ia}T_{ia}\big(Su_k + \tilde{R}_k\big) = 0. \tag{5.45}$$

5.4.4 Relation to the Initial-Value Condition

Let us abbreviate: $T_{-ia}\mathbf{Q}_{ia}T_{ia} =: X_a$. Then we obtain

$$\begin{aligned} u_{k+1} &= X_a S u_k + \mathbf{H}_{k+1} + X_a\tilde{R}_k \\ &= X_a S\big(X_a S u_{k-1} + \mathbf{H}_k + X_a\tilde{R}_{k-1}\big) + \mathbf{H}_{k+1} + X_a\tilde{R}_k \\ &= (X_a S)^2 u_{k-1} + X_a S\mathbf{H}_k + X_a S X_a\tilde{R}_{k-1} + \mathbf{H}_{k+1} + X_a\tilde{R}_k \end{aligned}$$

$$= (X_aS)^2 u_{k-1} + (X_aS)^1 \big(\mathbf{H}_k + X_a\tilde{R}_{k-1}\big) + (X_aS)^0 \big(\mathbf{H}_{k+1} + X_a\tilde{R}_k\big)$$

$$= (X_aS)^{k+1} u_0 + \sum_{l=1}^{k} (X_aS)^l \big(X_a\tilde{R}_{k-l} + \mathbf{H}_{k-l+1}\big) + X_a\tilde{R}_k + \mathbf{H}_{k+1}.$$

The boundary condition remains fulfilled, i.e.,

$$\operatorname{tr}_\Gamma X_a \ldots = 0. \tag{5.46}$$

6 Approximation and Stability

6.1 Approximation Property

$$\begin{aligned}
Lu &:= \frac{1}{\nu}\partial_t u - \frac{1}{\nu}(v \cdot D)u + \frac{1}{\nu} + DDu, \quad t = k\tau;\\
L_\tau u &:= \frac{1}{\nu\tau}\big(u(t+\tau)\big) - u(t,\cdot) + DDu(t+\tau,\cdot)\\
&\quad + \frac{1}{\nu}Dp(t+\tau,\cdot) - \frac{1}{\nu}(v\cdot D)u(t,\cdot);\\
|L_\tau u_k - Lu_k| &\le \bigg|\frac{1}{\nu\tau}(u_{k+1} - u_k) + DD\,u_{k+1} + \frac{1}{\nu}Dp_{k+1} - \frac{1}{\nu}(v\cdot D)u_k\\
&\quad - \frac{1}{\nu}\partial_t u_k + \frac{1}{\nu}(v\cdot D)u_k - \frac{1}{\nu}Dp_k - DD\,u_k\bigg|\\
&= \frac{1}{\nu\tau}\big(u_k + \tau\partial_t u_k + \tau^2\partial_t^2 u(k\tau + \theta\tau,\cdot)\big) - \frac{1}{\nu\tau}u_k - \frac{1}{\nu}\partial_t u_k \qquad (6.1)\\
&\quad - \left(\frac{1}{\nu}\partial_t u_{k+1} - \frac{1}{\nu}\partial_t u_k\right) - \frac{1}{\mu}(f_{k+1} - f_k)\\
&\le \max_G\left|\frac{\tau}{2}\frac{1}{\nu}\partial_{tt}^2 u(k\tau+\theta\tau,\cdot)\right| + \max_G\left|\frac{1}{\nu}\partial_{tt}u(k\tau+\theta_1\tau,\cdot)\right|\\
&\quad + \max_G\left|\frac{\tau}{\mu}\partial_t u(k\tau+\theta_2\tau,\cdot)\right|\\
&\le 2\frac{\tau}{\nu}\max_G\left|\partial_{tt}^2 u\big((k+\overline{\theta})\tau,\cdot\big)\right| + \max_G\left|\frac{\tau}{\mu}\partial_t u\big((k+\theta_2)\tau,\cdot\big)\right|\\
&\le C\tau \to 0.
\end{aligned}$$

6.2 Stability

We follow now the results in [1], where the following estimates of the norm of the operator T_{ia_0} were obtained:

$$\|T_{\pm ia_0}\|_{L(C)} \leq \frac{c}{a_0}\|u\|_C. \tag{6.2}$$

It remains to understand that $X_a S$ is bounded in a suitable Sobolev space. First, we have to state that X_a is smoothing, and therefore $X_a(v \cdot D)$ has to be studied. Let $a = ia_0$. Then we have

$$\begin{aligned} D(v\,u) &= -vDu + (v \cdot D)u + \quad \text{and} \quad D = D_a - a + (Dv)u, \\ T_a(v \cdot D)u &= T_a D_a u - aT_a u + T_a v D_a u + T_a v\, au - T_a(Dv)u \\ &= (I - F_a - aT_a + T_a v D_a + T_a va)u - T_a(Dv)u. \end{aligned}$$

As for $\text{Sc}\, v = 0$, it follows that

$$T_a(v \cdot D)u = (1 + v)(I - F_a) + a(v-1)T_a - T_a(Dv)u,$$

and this means that $T_a S = T_a(I + \tau(v \cdot D))$ is uniformly bounded with respect to τ.

6.3 Representation Formulae

From

$$u_{k+1} - \tau\nu\Delta u_{k+1} = -\tau(v \cdot D)u_k + \frac{\tau}{\rho}f_k + u_k - \frac{1}{\nu}Dp_{k+1},$$

$$F(u_k) = \big(I - \tau(v \cdot D)\big)u_k a_0^2 + \frac{f_k}{\mu}$$

it follows that

$$u_{k+1} = -\frac{1}{\nu}T_{-a}\mathbf{Q}_a T_a D p_{k+1} + T_{-a}\mathbf{Q}_a T_{-a}F(u_k) + \mathbf{H}_k$$

and

$$\begin{aligned} T_a D p_{k+1} &= T_a(D + a)p_{k+1} - aT_a p_{k+1} = p_{k+1} - F_a p_{k+1} - aT_a p_{k+1} \\ &= \tilde{p}_{k+1} - F_a p_{k+1} \\ \mathbf{Q}_a T_a D p_{k+1} &= \mathbf{Q}_a \tilde{p}_{k+1} \quad (F_a p_{k+1} \in \operatorname{im} \mathbf{P}_a). \end{aligned}$$

Because of

$$\begin{aligned} \tilde{p}_{k+1} &= p_{k+1} - aT_a p_{k+1}, \\ D_a \tilde{p}_{k+1} &= D p_{k+1}, \end{aligned}$$

we obtain

$$u_{k+1} = -\frac{1}{\nu}\, T_{-a}\mathbf{Q}_a \tilde{p}_{k+1} + T_{-a}\mathbf{Q}_a T_a F(u_k).$$

7 More Relevant Problems in Fluid Dynamics

7.1 Magnetic Benard's Problem

Governing equations:

$$\partial_t \mathbf{u} + (\mathbf{u}\cdot\nabla)\mathbf{u} - \frac{1}{\mu\rho}(\mathbf{B}\cdot\nabla)\mathbf{B} + \frac{1}{2\rho\mu}\nabla\big(|\mathbf{B}|^2\big) + \frac{1}{\rho}\nabla p - \eta\Delta\mathbf{u} = \mathbf{f}, \tag{7.1}$$

$$\partial_t \mathbf{B} + (\mathbf{u}\cdot\nabla)\mathbf{B} - (\mathbf{B}\cdot\nabla)\mathbf{u} = \frac{1}{\mu\sigma}\Delta\mathbf{B}. \tag{7.2}$$

Divergence free fields:

$$\operatorname{div}\mathbf{u} = 0, \tag{7.3}$$

$$\operatorname{div}\mathbf{B} = 0. \tag{7.4}$$

Boundary conditions:

$$\mathbf{u} = 0 \quad \text{on } \Gamma, \tag{7.5}$$

$$\mathbf{B} = 0 \quad \text{on } \Gamma. \tag{7.6}$$

7.1.1 Time-Discretization of the Magnetic Benard Problem

Let $T > 0$, $T = n\tau$, and τ the meshwidth. We abbreviate again $\mathbf{u}_k := \mathbf{u}(k\tau,\cdot)$, $p_k := p(k\tau,\cdot)$, $B_k = B(k\tau,\cdot)$. Further we set $\mathbf{f}_k = (1/\tau)\int_{k\tau}^{(k+1)\tau} \mathbf{f}(t,x)\,dt$ and $g_k = (1/\tau)\int_{k\tau}^{(k+1)\tau} g(t,x)\,dt$ and approximate

$$(\partial_t \mathbf{u})(k\tau,\cdot) \sim \frac{\mathbf{u}_{k+1} - \mathbf{u}_k}{\tau} \quad (k = 0,1,\ldots,n-1). \tag{7.7}$$

Then we obtain for $(k = 0,1,\ldots,n-1)$:

$$u_{k+1} - u_k = \tau(u_k\cdot D)u_k + \frac{\tau}{\mu\rho}(B_k\cdot D)B_k$$

$$- \frac{\tau}{2\rho\mu} D|B_k|^2 - \frac{\tau}{\rho} Dp_{k+1} \tag{7.8}$$

$$+ \tau\eta\Delta u_{k+1} + \tau f_k, \tag{7.9}$$

$$B_{k+1} - B_k = -\tau(u_k \cdot D)B_k + B_k \cdot D)u_k + \frac{\tau}{\mu\sigma}\Delta B_{k+1} + \tau g_k, \tag{7.10}$$

$$u_{k+1} - \eta\tau\Delta u_{k+1} + \frac{\tau}{\rho}Dp_{k+1} = -\tau(u_k \cdot D)u_k - \frac{\tau}{\mu\rho}(B_k \cdot D)B_k \tag{7.11}$$

$$-\frac{\tau}{2\rho\mu}D|B_k|^2 + \tau f_k =: \tau\mathbf{F}(u_k, B_k) + \tau f_k, \tag{7.12}$$

$$B_{k+1} - \frac{\tau}{\mu\sigma}\Delta B_{k+1} = B_k - \tau(u_k \cdot D)B_k$$

$$+ (B_k \cdot D)u_k + \tau g_k \tag{7.13}$$

$$=: \tau\mathbf{G}(u_k, B_k) + \tau g_k. \tag{7.14}$$

We introduce virtual functions u^* and B^*. It then follows that

$$\left(u_{k+1} - u^*\right) + \left(u^* - u_k\right) = \tau\mathbf{F}(u_k, B_k) + \tau f_k - \frac{\tau}{\rho}Dp_{k+1} + \tau\eta\Delta u_{k+1}, \tag{7.15}$$

$$\left(B_{k+1} - B^*\right) + \left(B^* - B_k\right) = \tau\mathbf{G}(u_k, B_k) + \tau g_k \frac{\tau}{\mu\sigma}\Delta B_{k+1}. \tag{7.16}$$

We work with the ansatz

$$B_{k+1} - B^* = \frac{\tau}{\mu\sigma}\Delta B_{k+1}, \qquad B^* - B_k = \tau\mathbf{G}(u_k, B_k) + \tau g_k, \tag{7.17}$$

$$\begin{aligned} u_{k+1} - u^* &= -\frac{\tau}{\rho}Dp_{k+1} + \tau\eta\Delta u_{k+1}, \\ u^* - u_k &= \tau\mathbf{F}(u_k, B_k) + \tau f_k. \end{aligned} \tag{7.18}$$

Setting $\operatorname{div} u_{k+1} =: U_{k+1}$ and $\operatorname{div} B_{k+1} =: b_{k+1}$, it then follows:

$$U_{k+1} - \operatorname{div} u^* = \Delta\left(-\frac{\tau}{\rho}p_{k+1} + \tau\eta U_{k+1}\right), \tag{7.19}$$

$$b_{k+1} - \operatorname{div} B^* = -\frac{\tau}{\mu\sigma}\Delta b_{k+1}. \tag{7.20}$$

7.1.2 Divergence and Uniqueness

The Poisson equation

$$\Delta\left(\frac{1}{\rho}p_{k+1} + \eta U_{k+1}\right) = \frac{\operatorname{div} u^*}{\tau} \tag{7.21}$$

is always solvable. If we take this result as the hydrostatical pressure multiplied by $(1/\rho)$, then the scalar function U_{k+1} has to be 0. Uniqueness results are obtained under the following growth and boundedness conditions:

$$\left|\nabla_x u(x,t)\right|, \left|\nabla_x B(x,t)\right| \leq \text{const}, \tag{7.22}$$

$$\left|p(x,t)\right| \leq \text{const}\big(1+|x|\big)^{-1/2}. \tag{7.23}$$

Meanwhile several examples for nonuniqueness are known (cf. [11]) if the mentioned conditions are not satisfied.

7.1.3 Velocity Calculation

It is easy to see that

$$\frac{\tau}{\rho} D p_{k+1} = -T \operatorname{div} u^* + \phi_{k+1}, \quad \phi_{k+1} \in \ker D. \tag{7.24}$$

This leads to

$$u_{k+1} - u^* = -\eta\tau D D u_{k+1} - T \operatorname{div} u * + \phi_{k+1}. \tag{7.25}$$

Thus we obtain with $b^2 = 1/(\eta\tau)$

$$D D u_{k+1} + b^2 u_{k+1} = F\big(u^*, \phi_{k+1}\big) =: F_{k+1}, \tag{7.26}$$

and so

$$(D - ib)(D + ib) u_{k+1} = F_{k+1}. \tag{7.27}$$

Factorization methods with an estimation of the truncation error and the approximation properties can be found under the assumption that the velocity is bounded at all times (cf. [18] and [19]).

7.2 Boussinesq's Formulation of Poisson–Stokes' Problem

Let be the body-force **f** equal to the gravity acceleration **g**. Then we have

$$\rho_0 \partial_t \mathbf{u} + \rho_0 (\mathbf{u} \cdot \nabla)\mathbf{u} = -\nabla p + \mu \Delta \mathbf{u} + \rho_0 \mathbf{g}\big[1 - \alpha(w - w_0)\big], \tag{7.28}$$

$$\rho_0 c \partial_t w + \rho_0 c (\mathbf{u} \cdot \nabla) w = \operatorname{div}(k \nabla w) + \rho_0 \beta + \Phi, \tag{7.29}$$

$$\operatorname{div} \mathbf{u} = 0, \tag{7.30}$$

where the last term in the first equation is the buoyancy force, β is the heat consumption (loss) through chemical reactions, c is the difference between the two heat capacities, i.e., $c = c_p - c_v$, and k is the coefficient of thermal conductivity. Φ is the so-called dissipation function, which is, roughly speaking, connected with the long-time average of entropy production, and α is the volume expansion coefficient.

7.3 Shallow Water Equations

The French mathematician Barré de Saint-Vainant described the viscous, rotating shallow water equations as follows:

$$\partial_t u + (u \cdot \nabla_S)u = -2a \wedge u + \nu \Delta_S u - g\nabla_S h, \tag{7.31}$$

$$\partial_t H = -(u \cdot \nabla_S)H + H\nabla_S \cdot u, \tag{7.32}$$

$$H := h(t,x) - h_B(t,x), \tag{7.33}$$

$$u(0,x) = u_0, \tag{7.34}$$

$$h(0,x) = h_0. \tag{7.35}$$

Here H denotes the total depth of the fluid, $h_B(t,x)$ determines the bottom surface topography, $2a \wedge u$ is the Coriolis force, g is the gravity acceleration, ν is the dynamic viscosity, and u is the velocity. A good reference is the book [13].

7.3.1 Assumptions for the Formulation of St. Venant's Equations

- The direction of the rotation axis coincides with the e_3 axis.
- The angular velocity $|\omega|$ is induced by Coriolis.
- The external forces reduce here to the gravity.
- $\partial_\tau p = -\rho g$, i.e., the vertical pressure gradient equals the buoyancy forces.
- The scale of horizontal motion is much larger than the scale of vertical motion.
- The fluid is incompressible, homogeneous, has a constant density (here normalized to 1).

A good reference is the dissertation by Fengler [4].

7.4 Forecasting Equations

Shallow water equations are similar to a model of forecasting equations restricted on the sphere

$$\partial_t u + (v \cdot \nabla_S)u = \nu \Delta_S u - \frac{1}{\rho}\nabla_S p - 2a \wedge u + F \quad \text{in } G, \tag{7.36}$$

$$c_v \partial_t T = -c_v(u \cdot \nabla_S)T + Q, \tag{7.37}$$

$$\partial_t \rho = -(u \cdot \nabla_S)\rho, \tag{7.38}$$

with the vector derivative (Günter's gradient) ∇_S, the Beltrami operator Δ_S, the surface gradient $\nabla_S p$, the heating rate Q, and the surface divergence $\nabla_S \cdot u$. The vector of outer forces F also contained the so-called apparent gravity that is gravity reduced by the centrifugal force. Moreover, it is assumed that $\nabla_S \cdot u = 0$ with boundary conditions on $\partial G =: \Gamma$.

8 Numerical Examples

Our goal is now to show by a numerical example how good the performance of the described methods is. For simplicity, we consider only a stationary Navier–Stokes equation. As we have seen above, the solution of such problems is the "kernel" of the iteration procedures. This means that it is of greatest importance to solve these problems efficiently. This numerical example is taken from the paper [3], where also Navier–Stokes equations in unbounded domains were studied. The calculations were done by N. Faustino.

Table 1 Error estimates and convergence for Example 1

h	Grid	Number of iterations	$\|u_{n+1} - u_n\|$ (last iteration)
1	$5 \times 5 \times 5$	3	4.6875×10^{-7}
2/3	$7 \times 7 \times 7$	3	1.8243×10^{-7}
1/2	$9 \times 9 \times 9$	3	1.0317×10^{-7}
1/3	$13 \times 13 \times 13$	3	4.2204×10^{-8}
1/4	$17 \times 17 \times 17$	3	2.7564×10^{-8}

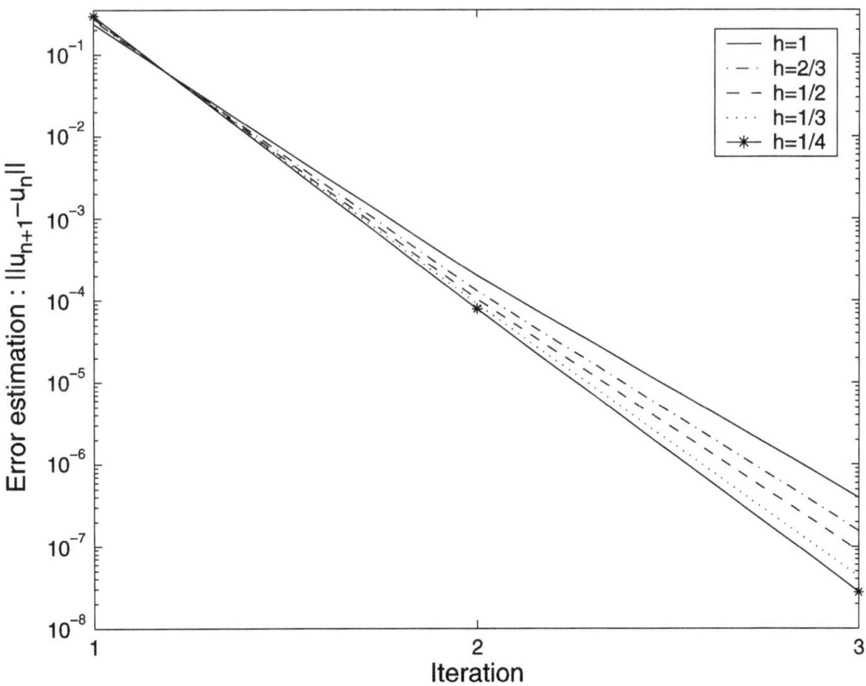

Fig. 1 Error estimation for different mesh sizes during the iteration

For the domains $\Omega = [-2,2]^3$, we use a cubic equidistant grid of $(N+1) \times (N+1) \times (N+1)$, N even, points with mesh size $h = \frac{4}{N}$ corresponding to

$$\Omega_h = \{mh = (m_1 h, m_2 h, m_3 h)^T : -N/2 \le m_1, m_2, m_3 \le N/2\}.$$

For the stopping criteria, we use the discrete Sobolev norm, $\|.\| := \|.\|_{w^1_{p+3/\alpha}}$ with $p = 4$ and $\alpha = 1/2$, and fix $\delta = 10^{-6}$ as error tolerance.

Example 1 Consider the stationary Navier–Stokes in $\Omega = [-2,2]^3$

$$\hat{f}(x) = \big(|x_1|, |x_2|, |x_3|\big)^T.$$

Note that the right-hand side is not differentiable.

Table 1 shows the error after the third iteration for different grids. Figure 1 illustrates the development of the error during the iteration process. We can see that the error decreases quickly. This is in accordance to our theoretical results. The iteration was based on a fixed-point principle. We can also observe in Example 1 that the lack of regularity at the origin is not an obstacle for the method.

9 Conclusions

The presented methods were also applied to the solution theory of some Galpern–Sobolev equations, an alternative model to the three-dimensional generalized Korteweg–de Vries equations (we refer to [9]). Meanwhile, results are published also on a three-dimensional analogue to Airy‘s equation for waves in rigid materials [17].

References

1. Bahmann, H., Gürlebeck, K., Shapiro, M., Sprößig, W.: On a modified Teodorescu transform. Integral Transform. Spec. Funct. **12**(3), 213–226 (2001)
2. Brackx, F., Delanghe, R., Sommen, F.: Clifford Analysis. Pitman Research Notes Math., vol. 76. Pitman, London (1982)
3. Faustino, N., Gürlebeck, K., Hommel, A., Kähler, U.: Difference potentials for the Navier–Stokes equations in unbounded domains. J. Differ. Equ. Appl. **12**(6), 577–596 (2006)
4. Fengler, M.J.: Vector spherical harmonic and vector wavelet based non-linear Galerkin schemes for solving the incompressible Navier–Stokes equation on the sphere. Shaker, D386 (2005)
5. Gürlebeck, K.: Grundlagen einer diskreten räumlich verallgemeinerten Funktionentheorie und ihrer Anwendungen. Habilitationsschrift, TU Karl-Marx-Stadt (1988)
6. Gürlebeck, K., Hommel, A.: On finite difference potentials and their applications in a discrete function theory. Math. Methods Appl. Sci. **25**(16–18), 1563–1576 (2002)
7. Gürlebeck, K., Sprößig, W.: Quaternionic Analysis and Elliptic Boundary Value Problems. Birkhäuser, Basel (1990)

8. Gürlebeck, K., Sprößig, W.: Quaternionic and Clifford Calculus for Physicists and Engineers. Mathematical Methods in Practice. Wiley, New York (1997)
9. Gürlebeck, K., Sprößig, W.: Representation theory for classes of initial value problems with quaternionic analysis. Math. Methods Appl. Sci. **25**, 1371–1382 (2002)
10. Hung, L.U.: Finite Element Analysis of Non-Newtonian Flow. Springer, Berlin (1989)
11. Ishimura, N., Nakamura, M.: Uniqueness for unbounded classical solutions of the MHD equations. MMAS **20**, 617–623 (1997)
12. Le, T.H., Sprößig, W.: On a new notion of holomorphy and its applications. Cubo Math. J. **11**(1), 145–162 (2008)
13. Majda, A.: Introduction to PDE's and Waves for Atmosphere and Ocean. Courant-Lecture Notes, vol. 9. Courant Institute of Mathematical Sciences, New York (2003)
14. Przeworska-Rolewicz, D.: Algebraic theory of right invertible operators. Stud. Math. **48**, 129–144 (1973)
15. Rajagapol, K.R.: Mechanics of Non-Newtonian Fluids. Pitman, London (1998)
16. Ryabenskij, V.S.: The Method of Difference Potentials for Some Problems of Continuum Mechanics. Nauka, Moscow (1987) (in Russian)
17. Schlichting, A., Sprößig, W.: Norm estimations of the modified Teodorescu transform with application to a multidimensional equation of Airy's type. In: Simos, Th.E., Psihoyios, G., Tsitouras, Ch. (eds.) Numerical Analysis and Applied Mathematics. American Institute of Physics (AIP) Conference Series, vol. 1048 (2008)
18. Sprößig, W., Gürlebeck, K.: Representation theory for classes of initial value problems with quaternionic analysis. Math. Methods Appl. Sci. **25**, 1371–1382 (2002)
19. Sprößig, W.: Fluid flow equations with variable viscosity in quaternionic setting. In: Advances in Applied Clifford Algebras, vol. 17, pp. 259–272. Birkhäuser, Basel (2007)
20. Tasche, M.: Eine einheitliche Herleitung verschiedener Interpolationsformeln mittels der Taylorschen Formel der Operatorenrechnung. Z. Angew. Math. Mech. **61**, 379–393 (1981) (in German)

Part VI
Crystallography, Holography and Complexity

Interactive 3D Space Group Visualization with CLUCalc and Crystallographic Subperiodic Groups in Geometric Algebra

Eckhard M.S. Hitzer, Christian Perwass, and Daisuke Ichikawa

Abstract The Space Group Visualizer (SGV) for all 230 3D space groups is a standalone PC application based on the visualization software CLUCalc. We first explain the unique geometric algebra structure behind the SGV. In the second part we review the main features of the SGV: The GUI, group and symmetry selection, mouse pointer interactivity, and visualization options. We further introduce the joint use with Hahn (Space-group Symmetry, 5th edn., International Tables of Crystallography, vol. A, Springer, Dordrecht, 2005). In the third part we explain how to represent the 162 so-called subperiodic groups of crystallography in geometric algebra. We construct a new compact geometric algebra group representation symbol, which allows us to read off the complete set of geometric algebra generators. For clarity, we moreover state explicitly which generators are chosen. The group symbols are based on the representation of point groups in geometric algebra by versors.

1 Introduction

Crystals are fundamentally periodic geometric arrangements of molecules. The directed distance between two such elements is a Euclidean vector in $\mathbb{R}^3$. Intuitively all symmetry properties of crystals depend on these vectors. Indeed, the geometric product of vectors [13], combined with the conformal model of 3D Euclidean space [2, 15, 25–27, 38, 39, 42], yields an algebra fully expressing crystal point and space groups [19, 22, 28–30, 33, 45]. Two successive reflections at (non)parallel planes express (rotations) translations, etc. [8, 10]. This leads to a one-to-one correspondence of geometric objects and symmetry operators [35] with vectors and their products, ideal for creating a suite of interactive visualizations using CLUCalc [44] and OpenGL [28, 29, 31, 45].

E.M.S. Hitzer (✉)
Department of Applied Physics, University of Fukui, Bunkyo 3-9-1, 910-8507 Fukui, Japan
e-mail: hitzer@mech.u-fukui.ac.jp

E. Bayro-Corrochano, G. Scheuermann (eds.), *Geometric Algebra Computing*,
DOI 10.1007/978-1-84996-108-0_18,

For crystallographers, the subperiodic space groups [37] in 2D and 3D with only one or two degrees of freedom for translations are also of great interest.

We begin in Sect. 2 by explaining the representation of point and space groups in conformal geometric algebra. Next we explain the basic functions of the software implementation, called Space Group Visualizer [46] in Sect. 3. Finally in Sect. 4 we show how to construct a new compact geometric algebra group representation symbol for subperiodic space groups (Frieze groups, rod groups, and layer groups), which allows us to read off the complete set of geometric algebra generators. For clarity, we moreover state explicitly which generators are chosen.

2 Point Groups and Space Groups in Clifford Geometric Algebra

2.1 Cartan–Dieudonné and Geometric Algebra

Clifford's associative geometric product [13] of two vectors simply adds the inner product to the outer product of Grassmann

$$\boldsymbol{a}\boldsymbol{b} = \boldsymbol{a} \cdot \boldsymbol{b} + \boldsymbol{a} \wedge \boldsymbol{b}. \tag{1}$$

Under this product, parallel vectors commute, and perpendicular vectors anti-commute:

$$\boldsymbol{a}\boldsymbol{x}_{\|} = \boldsymbol{x}_{\|}\boldsymbol{a}, \qquad \boldsymbol{a}\boldsymbol{x}_{\perp} = -\boldsymbol{x}_{\perp}\boldsymbol{a}. \tag{2}$$

This allows us to write the reflection of a vector $\boldsymbol{x}$ at a hyperplane through the origin with normal $\boldsymbol{a}$ as

$$\boldsymbol{x}' = -\boldsymbol{a}^{-1}\boldsymbol{x}\boldsymbol{a}, \qquad \boldsymbol{a}^{-1} = \frac{\boldsymbol{a}}{\boldsymbol{a}^2}. \tag{3}$$

The composition of two reflections at hyperplanes whose normal vectors $\boldsymbol{a}, \boldsymbol{b}$ subtend the angle $\alpha/2$ yields a rotation around the intersection of the two hyperplanes by α:

$$\boldsymbol{x}' = (\boldsymbol{a}\boldsymbol{b})^{-1}\boldsymbol{x}\boldsymbol{a}\boldsymbol{b}, \qquad (\boldsymbol{a}\boldsymbol{b})^{-1} = \boldsymbol{b}^{-1}\boldsymbol{a}^{-1}. \tag{4}$$

Continuing with a third reflection at a hyperplane with normal $\boldsymbol{c}$ according to the Cartan–Dieudonné theorem [4, 6, 12, 18] yields the rotary reflections and inversions

$$\boldsymbol{x}' = -(\boldsymbol{a}\boldsymbol{b}\boldsymbol{c})^{-1}\boldsymbol{x}\boldsymbol{a}\boldsymbol{b}\boldsymbol{c}, \qquad \boldsymbol{x}'' = -i^{-1}\boldsymbol{x}i, \qquad i \doteq \boldsymbol{a} \wedge \boldsymbol{b} \wedge \boldsymbol{c}, \tag{5}$$

where $\doteq$ means equality up to nonzero scalar factors (which cancel out in (6)). In general the geometric product S of k normal vectors corresponds to the composition of reflections to all symmetry transformations [22] of 2D and 3D crystal cell point groups:

$$\boldsymbol{x}' = (-1)^k S^{-1}\boldsymbol{x}S = \widehat{S}^{-1}\boldsymbol{x}S = S^{-1}\boldsymbol{x}\widehat{S}, \tag{6}$$

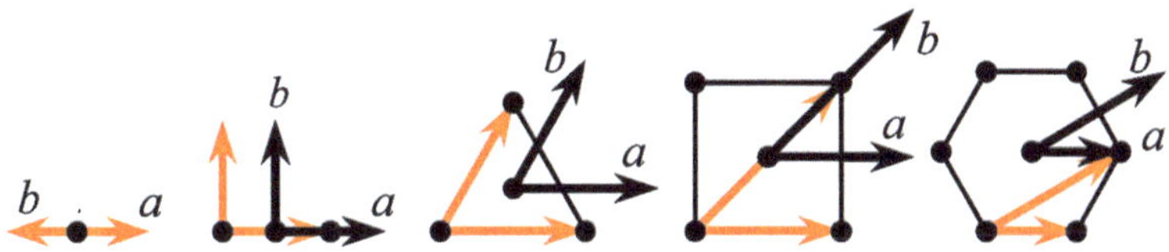

Fig. 1 Regular polygons ($p = 1, 2, 3, 4, 6$) and point group generating vectors $\boldsymbol{a}, \boldsymbol{b}$ subtending angles π/p shifted to center

Table 1 Geometric [22, 24] and international [20] notation for 2D point groups

Crystal	Oblique		Rectangular		Trigonal		Square		Hexagonal	
Geometric	$\bar{1}$	$\bar{2}$	1	2	3	$\bar{3}$	4	$\bar{4}$	6	$\bar{6}$
International	1	2	m	mm	$3m$	3	$4m$	4	$6m$	6

where $\widehat{S} = (-1)^k S$ is the *grade involution* or *main involution* in Clifford geometric algebra. We call the product of invertible vectors S in (6) *versor* [14, 22, 24, 35, 38], but the names *Clifford monomial* of invertible vectors, *Clifford group element*, or *Lipschitz group element* are equally in use [38, 41].

2.2 Two-Dimensional Point Groups

2D point groups [22] are generated by multiplying vectors selected [28, 29, 45] as in Fig. 1. The index p can be used to denote these groups as in Table 1. For example, the hexagonal point group is given by multiplying its two generating vectors $\boldsymbol{a}, \boldsymbol{b}$:

$$6 = \{\boldsymbol{a}, \boldsymbol{b}, R = \boldsymbol{ab}, R^2, R^3, R^4, R^5, R^6 = -1, \boldsymbol{a}R^2, \boldsymbol{b}R^2, \boldsymbol{a}R^4, \boldsymbol{b}R^4\}. \quad (7)$$

The rotation subgroups are denoted with bars, e.g., $\bar{6}$. The identities $\boldsymbol{a}^2 = \boldsymbol{b}^2 = 1$ and $R^6 = -1$ directly correspond to relations in the group presentation [47] of 6.

2.3 Three-Dimensional Point Groups

The selection of three vectors $\boldsymbol{a}, \boldsymbol{b}, \boldsymbol{c}$ from each crystal cell [22, 28, 29, 45] for generating all 3D point groups is indicated in Fig. 2. Using $\angle(\boldsymbol{a}, \boldsymbol{b})$ and $\angle(\boldsymbol{b}, \boldsymbol{c})$, we can denote all 32 3D point groups (alias crystal classes) as in Table 2. For example, the monoclinic point groups are then (international symbols of Hermann–Maugin: $2/m$, m, and 2, respectively)

$$2\bar{2} = \{\boldsymbol{c}, R = \boldsymbol{a} \wedge \boldsymbol{b} = i\boldsymbol{c}, i = \boldsymbol{c}R, 1\}, \qquad 1 = \{\boldsymbol{c}, 1\}, \qquad \bar{2} = \{i\boldsymbol{c}, 1\}. \quad (8)$$

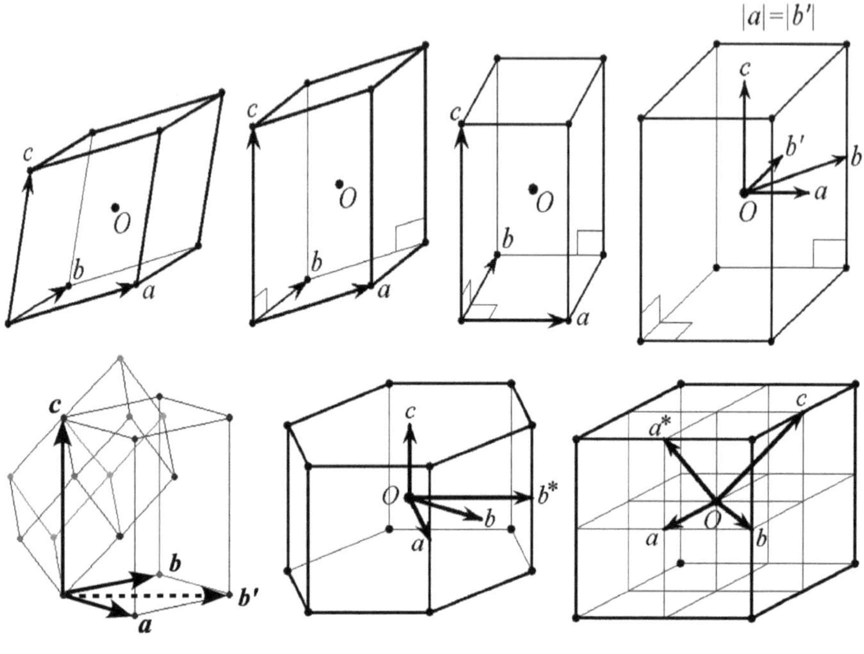

Fig. 2 7 crystal cells with vector generators $\boldsymbol{a}, \boldsymbol{b}, \boldsymbol{c}$: triclinic, monoclinic, orthorhombic, tetragonal, trigonal (rhombohedral), hexagonal, cubic

Table 2 Geometric 3D point group symbols [10, 22] and generators with $\theta_{\boldsymbol{a},\boldsymbol{b}} = \pi/p$, $\theta_{\boldsymbol{b},\boldsymbol{c}} = \pi/q$, $\theta_{\boldsymbol{a},\boldsymbol{c}} = \pi/2$, $p, q \in \{1, 2, 3, 4, 6\}$

Symbol	$p=1$	$p \neq 1$	$\bar{p}$	pq	$\bar{p}q$	$p\bar{q}$	$\bar{p}\bar{q}$	$\overline{pq}$
Generators	$\boldsymbol{a}$	$\boldsymbol{a}, \boldsymbol{b}$	$\boldsymbol{ab}$	$\boldsymbol{a}, \boldsymbol{b}, \boldsymbol{c}$	$\boldsymbol{ab}, \boldsymbol{c}$	$\boldsymbol{a}, \boldsymbol{bc}$	$\boldsymbol{ab}, \boldsymbol{bc}$	$\boldsymbol{abc}$

2.4 Space Groups

The smooth composition with translations is best done in the conformal model [1–3, 5, 7, 9, 11, 14, 15, 17, 21, 25–27, 42, 43] of Euclidean space (in the GA of $\mathbb{R}^{4,1}$), which adds two null-vector dimensions for the origin $\boldsymbol{e}_0$ and infinity $\boldsymbol{e}_\infty$.

$$X = \boldsymbol{x} + \frac{1}{2}\boldsymbol{x}^2\boldsymbol{e}_\infty + \boldsymbol{e}_0, \qquad \boldsymbol{e}_0^2 = \boldsymbol{e}_\infty^2 = X^2 = 0, \qquad X \cdot \boldsymbol{e}_\infty = -1. \tag{9}$$

The $+\boldsymbol{e}_0$ term integrates projective geometry, and the $+\frac{1}{2}\boldsymbol{x}^2\boldsymbol{e}_\infty$ term ensures $X^2 = 0$. The inner product of two conformal points gives their Euclidean distance and therefore a plane m equidistant from two points A, B as

$$X \cdot A = -\frac{1}{2}(\boldsymbol{x} - \boldsymbol{a})^2 \quad \Longrightarrow \quad X \cdot (A - B) = 0, \qquad m = A - B \propto \boldsymbol{p} - d\boldsymbol{e}_\infty, \tag{10}$$

Table 3 Computing with reflections and translations The vectors $\boldsymbol{a}, \boldsymbol{b}$ are pictured in Fig. 1

$\angle(\boldsymbol{a}, \boldsymbol{b})$	180°	90°	60°	45°	30°
$T_{\boldsymbol{a}}\boldsymbol{b} =$	$\boldsymbol{b}T_{-\boldsymbol{a}}$	$\boldsymbol{b}T_{\boldsymbol{a}}$	$\boldsymbol{b}T_{\boldsymbol{a}-\boldsymbol{b}}$	$\boldsymbol{b}T_{\boldsymbol{a}-\boldsymbol{b}}$	$\boldsymbol{b}T_{\boldsymbol{a}-\boldsymbol{b}}$
$T_{\boldsymbol{b}}\boldsymbol{a} =$	$\boldsymbol{a}T_{-\boldsymbol{b}}$	$\boldsymbol{a}T_{\boldsymbol{b}}$	$\boldsymbol{a}T_{\boldsymbol{b}-\boldsymbol{a}}$	$\boldsymbol{a}T_{\boldsymbol{b}-2\boldsymbol{a}}$	$\boldsymbol{a}T_{\boldsymbol{b}-3\boldsymbol{a}}$

Table 4 Monoclinic space group versor generators, $T^A = T_{\boldsymbol{b}+\boldsymbol{c}}^{1/2}$, int. = international [20], geo. = geometric, alt. = alternative, columns 3 and 4: [24]. $T_{\boldsymbol{a}}, T_{\boldsymbol{b}}, T_{\boldsymbol{c}}$ suppressed

Int.#	Int. name	Geo. name	Geo. generators	Int. generators	Alt. generators
3	$P2$	$P\bar{2}$	$i\boldsymbol{c} = \boldsymbol{a} \wedge \boldsymbol{b}$		
4	$P2_1$	$P\bar{2}_1$	$i\boldsymbol{c}T_{\boldsymbol{c}}^{1/2}$		
5	$C2$	$A\bar{2}$	$i\boldsymbol{c}, T^A$		
6	Pm	$P1$	$\boldsymbol{c}$		
7	Pc	P_a1	$\boldsymbol{c}T_{\boldsymbol{a}}^{1/2}$		
8	Cm	$A1$	$\boldsymbol{c}, T^A$		
9	Cc	A_a1	$\boldsymbol{c}T_{\boldsymbol{a}}^{1/2}, T^A$		
10	$P2/m$	$P2\bar{2}$	$\boldsymbol{c}, i\boldsymbol{c}$	$i, i\boldsymbol{c}$	$i, \boldsymbol{c}$
11	$P2_1/m$	$P2\bar{2}_1$	$\boldsymbol{c}, i\boldsymbol{c}T_{\boldsymbol{c}}^{1/2}$	$i, i\boldsymbol{c}T_{\boldsymbol{c}}^{1/2}$	$i, \boldsymbol{c}T_{\boldsymbol{c}}^{1/2}$
12	$C2/m$	$A2\bar{2}$	$\boldsymbol{c}, i\boldsymbol{c}, T^A$	$iT^A, i\boldsymbol{c}T^A, T^A$	$i, \boldsymbol{c}, T^A$
13	$P2/c$	$P_a2\bar{2}$	$\boldsymbol{c}T_{\boldsymbol{a}}^{1/2}, i\boldsymbol{c}$	$i, i\boldsymbol{c}T_{\boldsymbol{a}}^{1/2}$	$i, \boldsymbol{c}T_{\boldsymbol{a}}^{1/2}$
14	$P2_1/c$	$P_a2\bar{2}_1$	$\boldsymbol{c}T_{\boldsymbol{a}}^{1/2}, i\boldsymbol{c}T_{\boldsymbol{c}}^{1/2}$	$i, i\boldsymbol{c}T_{\boldsymbol{a}+\boldsymbol{c}}^{1/2}$	$i, \boldsymbol{c}T_{\boldsymbol{a}+\boldsymbol{c}}^{1/2}$
15	$C2/c$	$A_a2\bar{2}$	$\boldsymbol{c}T_{\boldsymbol{a}}^{1/2}, i\boldsymbol{c}, T^A$	$i, i\boldsymbol{c}T_{\boldsymbol{a}}^{1/2}, T^A$	$i, \boldsymbol{c}T_{\boldsymbol{a}}^{1/2}, T^A$

where $\boldsymbol{p}$ is a unit normal to the plane, and d is its signed scalar distance from the origin. Reflecting at two parallel planes m, m' with distance $\boldsymbol{t}/2$, we get the so-called *translator* (*transla*tion operat*or* by $\boldsymbol{t}$)

$$X' = m'mXmm' = T_{\boldsymbol{t}}^{-1}XT_{\boldsymbol{t}}, \qquad T_{\boldsymbol{t}} = 1 + \frac{1}{2}\boldsymbol{t}\boldsymbol{e}_{\infty}. \tag{11}$$

Reflection at two nonparallel planes m, m' yields the rotation around the m, m'-intersection by twice the angle subtended by m, m'.

Group-theoretically the conformal group $C(3)$ is isomorphic to $O(4, 1)$, and the Euclidean group $E(3)$ is the subgroup of $O(4, 1)$ leaving infinity $\boldsymbol{e}_{\infty}$ invariant [22, 24, 38]. Now general translations and rotations are represented by geometric products of vectors. To study combinations of versors, it is useful to know that (cf. Table 3)

$$T_{\boldsymbol{t}}\boldsymbol{a} = \boldsymbol{a}T_{\boldsymbol{t}'}, \quad \boldsymbol{t}' = -\boldsymbol{a}^{-1}\boldsymbol{t}\boldsymbol{a}. \tag{12}$$

Applying these techniques, one can compactly tabulate geometric space group symbols and generators [24]. Table 4 implements this for the 13 monoclinic space groups. All this is interactively visualized by the Space Group Visualizer [46].

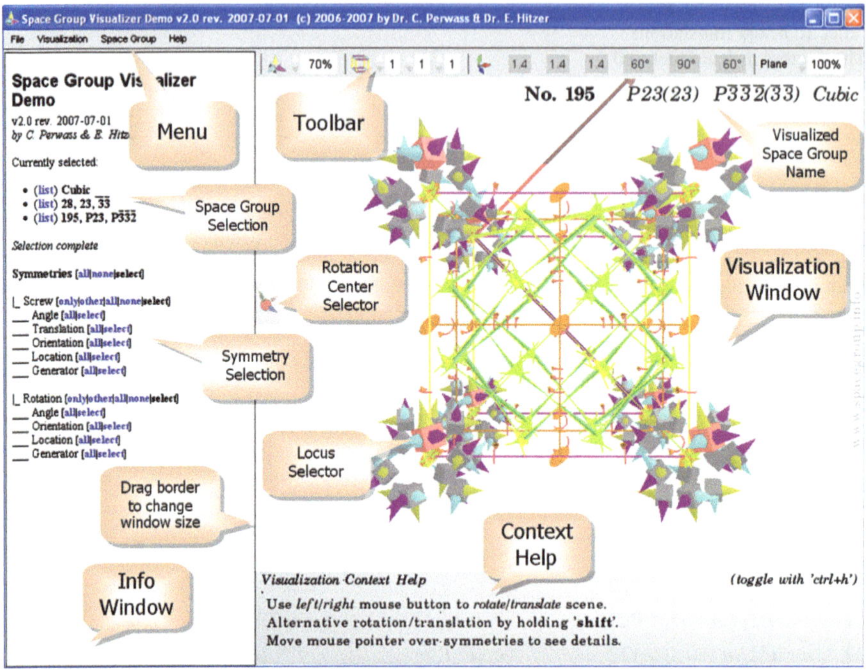

Fig. 3 GUI of the Space Group Visualizer

3 Interactive Software Implementation

The realization in software relies on the visual multivector software CLUCalc [44]. The excellent graphics rendering is based on OpenGL graphics. The space group symmetry definitions described in the previous sections are denoted for each space group in the form of an XML input file. The XML files serve as input for a CLUCalc script named Space Group Visualizer (SGV) [46].

3.1 The Space Group Visualizer GUI

Figure 3 shows the SGV GUI. The SGV toolbar is magnified and annotated in Fig. 4. Depending on the displayed space group, basis vector lengths and (or) angles may be fixed (i.e., they may not be changed by the user). This is indicated by toolbar elements shaded in gray.

3.2 Space Group and Symmetry Selection

Figure 5 shows the interactive (hyperlink like) space group selection. Clicking blue text elements in the browser panel on the left of the GUI allows us to access crystal systems, crystal classes (point groups), and individual space groups.

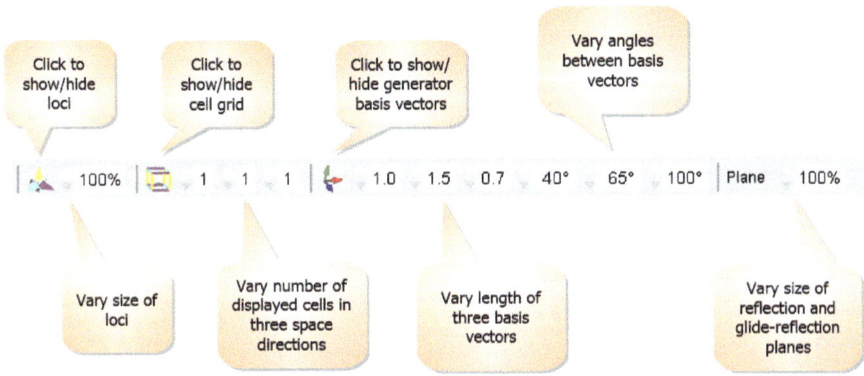

Fig. 4 Toolbar of the Space Group Visualizer

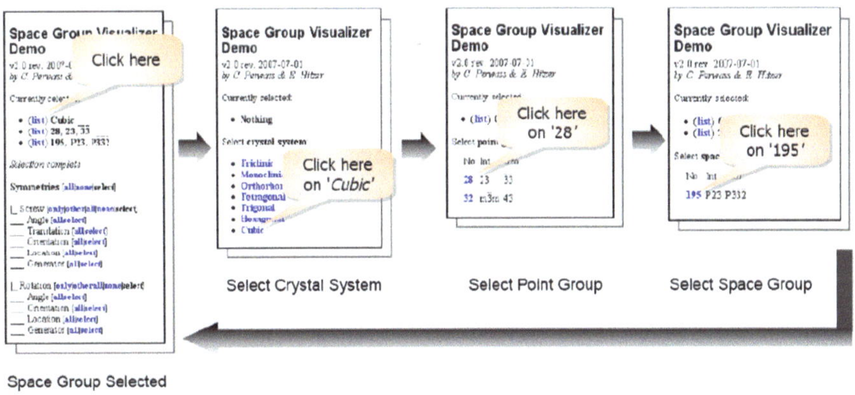

Fig. 5 Space group selection from the Space Group Visualizer browser panel

Figure 6 illustrates the selection of symmetries from the complete list of *Symmetries* (left SGV GUI browser panel), which are present in the currently selected space group. Symmetries that are to be displayed can be selected according to their properties (angle, orientation, location, translation component). Several properties selected together will display only those symmetries that satisfy all properties. Another way is to open the generator product list of a certain type of symmetry and select individual geometric algebra generator products to be displayed (or to be removed from the display).

3.3 Mouse Pointer Interactivity

The mouse pointer allows a variety of visual interactions and animations, depending over which part of the visualization it is placed. Moving the mouse pointer over a

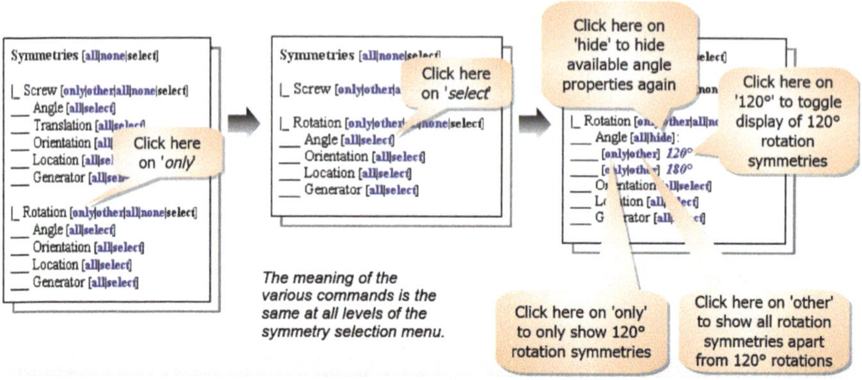

Fig. 6 Space group selection of individual symmetries or groups of symmetries to be displayed

symmetry element visualization both animates the symmetry and displays detailed information about this symmetry group element in the lower right corner. Animation means dynamic color and size changes and the motion of general elements along a trajectory tracing the symmetry operation incrementally. Placement of the mouse pointer over a general element (locus) selector activates it (blinking). The mouse pointer over the rotation center (of view) selector allows us to change the rotation center of the mouse activated view rotation (described below).

The mouse pointer can be placed anywhere inside the visualization window. Holding down the left (right) mouse button and moving the mouse will rotate (translate) the visualization. Alternative rotation axes (translation directions) are activated by additionally holding the SHIFT key. With a 3D-mouse (3D connexion) one can rotate and translate the view along all axes simultaneously.

First placing the mouse pointer over a general element (locus) selector permits one to translate and rotate it (together with all its symmetric partners). This provides an excellent way to grasp how one general element and the 3D symmetry represented in the space group determine the whole crystal structure.

A special feature of the SGV is the direct 3D graphics interaction. Simply placing the mouse pointer over a symmetry activates it and allows one to:

- Select only the activated symmetry (left mouse button). All other symmetries disappear from the view.
- Holding the CTRL key at the same time (while pressing the left mouse button) shows all symmetries (and only these) of the same type.
- Clicking the right mouse button removes an activated symmetry from the view.
- Holding the CTRL key at the same time (while pressing the right mouse button) removes all symmetries of the same type.

3.4 Visualization Options in Detail

The visualization drop-down menu allows one to toggle (activate and deactivate) the following visual functions

- Full screen mode.
- Orthographic view. The orthographic view allows the most direct comparison with ITA orthographic projections [20].
- Animation of the origin locus when a symmetry is activated (animated).
- Rotation animation of the whole view when it is *pushed* with the (left) mouse button.
- Reset the crystal view to visualizer default values.
- Reset general element (loci) positions.

The special visualization lighting menu provides a relative position light source. It is positioned relative to the visualization coordinate frame and moves with the visualization. Deselecting this option fixes the light source relative to the observer. The light source can optionally be positioned at the center of the coordinate frame, which is relative (or absolute) depending on the (de)selection of the relative position option. The ambient light submenu allows one to adjust the brightness of the ambient light, leading to more dramatic effects for darker settings.

The color scheme menu item allows one to select the current color scheme. For example, a scheme with black background is more suitable for use in presentations, while a white background is better for publications, etc. It is possible via an XML file to individually define further color schemes. A color stereo option allows one to specify cinema type stereo colors, which are best viewed with corresponding cinema color glasses in order to perform the full spatial 3D effect akin to virtual reality.

The cell-type menu allows one to select between different cell choices in the IT, vol. A [20], and (if different) a special geometric algebra type cell, which has the generating vectors $\boldsymbol{a}, \boldsymbol{b}, \boldsymbol{c}$ as cell axis attached to the cell origin. Details are given in [34].

3.5 Integration with the Online International Tables of Crystallography

Through the window menu an additional window can be opened for displaying the pages of a Space group from the online version of the International Tables of Crystallography, vol. A (ITA) [20]. For this, the user must hold a valid user ID and password. When the online ITA can be accessed, the SGV and the online ITA window will always show the *same* space group. The user can synchronously navigate from space group to space group either in the SGV or in the online ITA window (cf. Fig. 7).

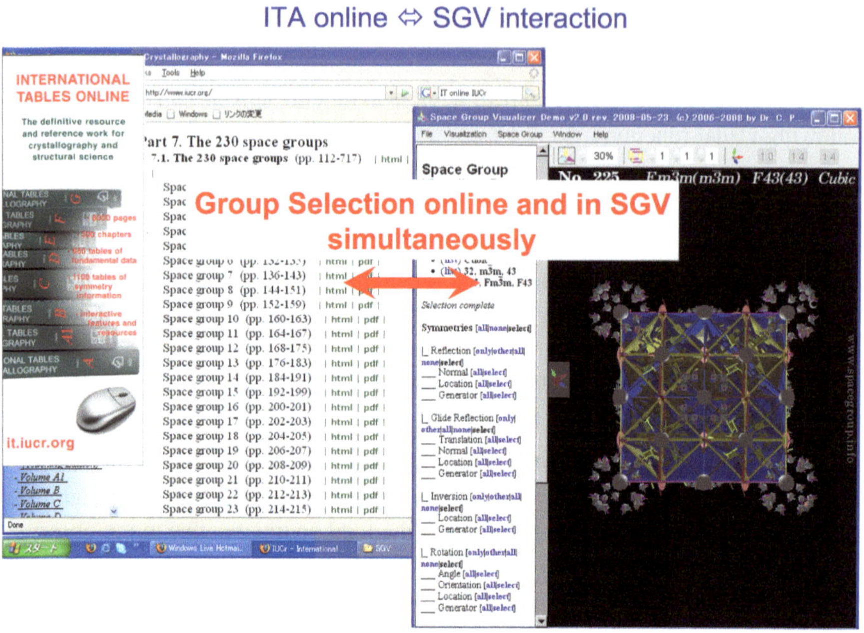

Fig. 7 Synchronous space group selection in the SGV window and the online ITA [20] space group window

4 Subperiodic Groups Represented in Clifford Geometric Algebra

Now we begin to explain the details of the new geometric algebra-based representation of so-called *subperiodic* space groups. These include the seven *frieze* groups (in 2D space, 1 DOF for translation), the 75 *rod* groups (in 3D space, 1 DOF for translation), and the 80 *layer* groups (in 3D space, 2 DOF for translations).

Compared to the geometric 2D and 3D space group symbols in [24], we have introduced dots: If one or two dots occur between the Bravais symbol ($\not{p}$, p, c) and index 1, the vector b or c, respectively, is present in the generator list. If one or two dots appear between the Bravais symbol and the index 2 (without or with bar), then the vectors b, c or a, c, respectively, are present in the generator list.

In agreement [24] the indices a, b, c, n (and g for frieze groups) in the first, second, or third position after the Bravais symbol indicate that the reflections a, b, c (in this order) become glide reflections. The index n indicates diagonal glides. The dots also serve as symbolic a, b, c position indicators. For example, rod group 5: $\not{p}_c1$ has glide reflection $aT_c^{1/2}$, rod group 19: $\not{p}_{.c}2$ has $bT_c^{1/2}$, and layer group 39: $p_b2_a2_n$ has $aT_b^{1/2}$, $bT_a^{1/2}$, and $cT_{a+b}^{1/2}$.

The notation $\overline{n}_p$ indicates a right-handed screw rotation of $2\pi/n$ around the $\overline{n}$-axis, with pitch $T_t^{p/n}$, where t is the shortest lattice translation vector parallel to the axis, in the screw direction. For example, the layer group 21: $p\bar{2}\bar{2}_1\bar{2}_1$ has the screw generators $bcT_a^{1/2}$ and $acT_b^{1/2}$.

In the following we discuss specific issues for frieze groups, rod groups, and layer groups. In the current publication we restrict ourselves to the new symbols for triclinic and monoclinic rod and layer groups. The full tables will be published elsewhere.

4.1 Frieze Groups

Figure 8 shows the generating vectors a, b of oblique and rectangular cells for 2D frieze groups. The only translation direction is a. Table 5 lists the seven frieze groups with new geometric symbols and generators. The abbreviations SG# and SGN mean space group number and space group name (symbol), respectively.

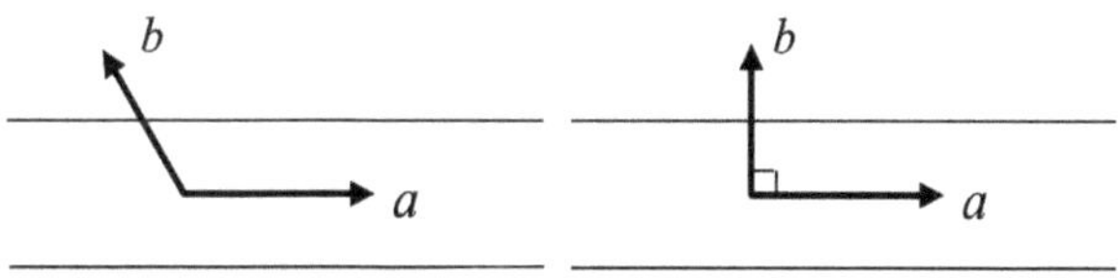

Fig. 8 Generating vectors a, b of oblique and rectangular cells for 2D frieze groups

Table 5 Table of frieze groups. Group number (col. 1), international frieze group notation [37] (col. 2), related international 3D space group numbers [20] (col. 3), and notation [20] (col. 4), geometric 3D space group notation [24] (col. 5), related international 2D space group numbers [20] (col. 6), and notation [20] (col. 7), related geometric 2D space group notation [24] (col. 8), *geometric frieze group notation* (col. 9), *geometric algebra frieze group versor generators* (col. 10). The pure translation generator T_a is omitted

Frieze group #	Intern. notat.	3D SG#	Intern. 3D SGN	Geom. 3D SGN	2D SG#	Intern. 2D SGN	Geom. 2D SGN	Geom. notat.	Frieze group generators
Oblique									
F_1	$\mathcal{p}1$	1	$P1$	$P\bar{1}$	1	$p1$	$p\bar{1}$	$\mathcal{p}\bar{1}$	
F_2	$\mathcal{p}211$	3	$P2$	$P\bar{2}$	2	$p2$	$p\bar{2}$	$\mathcal{p}\bar{2}$	$a \wedge b$
Rectangular									
F_3	$\mathcal{p}1m1$	6	Pm	$P1$	3	$pm(p1m1)$	$p1$	$\mathcal{p}1$	a
F_4	$\mathcal{p}11m$	6	Pm	$P1$	3	$pm(p11m)$	$p1$	$\mathcal{p}.1$	b
F_5	$\mathcal{p}11g$	7	Pc	P_a1	4	$pg(p11g)$	p_g1	$\mathcal{p}._g1$	$bT_a^{1/2}$
F_6	$\mathcal{p}2mm$	25	$Pmm2$	$P2$	6	$p2mm$	$p2$	$\mathcal{p}2$	a, b
F_7	$\mathcal{p}2mg$	28	$Pma2$	$P2_a$	7	$p2mg$	$p2_g$	$\mathcal{p}2_g$	$a, bT_a^{1/2}$

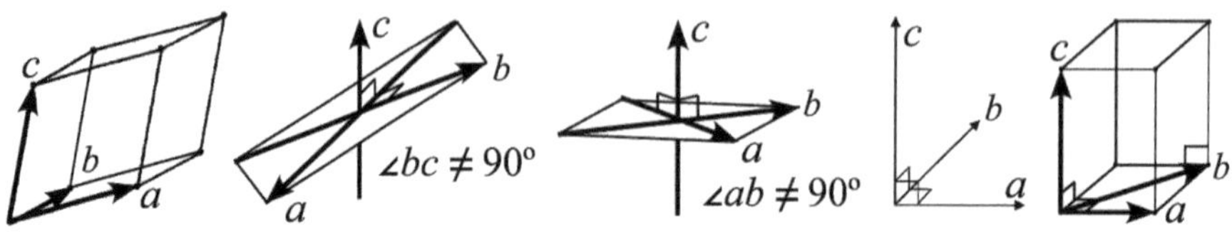

Fig. 9 From left to right: Triclinic, monoclinic inclined, monoclinic orthogonal, orthorhombic, and tetragonal cell vectors a, b, c for rod and layer groups

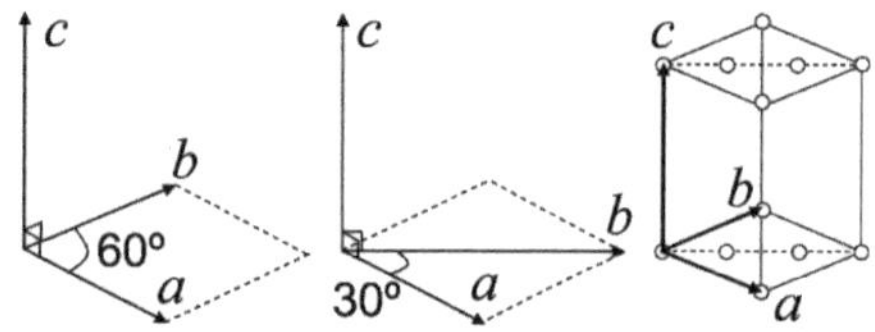

Fig. 10 Generating vectors a, b, c of trigonal (*left*), hexagonal (*center*), and hexagonally centered (*right*, Bravais symbol: H or h) cells for 3D rod and layer groups

4.2 Rod Groups

Figure 9 shows the generating vectors a, b, c of triclinic, monoclinic, orthorhombic, and tetragonal cells for 3D rod and layer groups. Figure 10 shows the same for trigonal and hexagonal cells. For rod groups, the only translation direction is c. There is a total of 75 rod groups in all 3D crystal systems. Table 6 lists the triclinic and monoclinic rod groups with new geometric symbols and generators: Rod group number (col. 1), international rod group notation [37] (col. 2), related international 3D space group numbers [20] (col. 3), and notation [20] (col. 4), related geometric 3D space group notation [24] (col. 5), *geometric rod group notation* (col. 6), *geometric algebra generators* (col. 7).

4.3 Layer Groups

For layer groups, the two translation directions are a, b. There is a total of 80 layer groups. Tables 7 list the triclinic and monoclinic 3D layer groups with new geometric symbols and generators: Layer group number (col. 1), international layer group notation [37] (col. 2), related international 3D space group numbers [20] (col. 3), and notation [20] (col. 4), related geometric 3D space group notation [24] (col. 5), *geometric layer group notation* (col. 6), *geometric algebra generators* (col. 7). The layer groups are classified according to their 3D crystal system/2D Bravais system. The monoclinic/oblique(rectangular) system corresponds to the monoclinic/orthogonal(inclined) system of Fig. 9. Figure 10 shows the hexagonally centered cell with Bravais symbols H (space group) and h (layer group).

Table 6 Table of triclinic and monoclinic rod groups. The pure translator T_c is omitted

Rod group #	Intern. notat.	3D space group #	Intern. 3D SGN	Geom. 3D SGN	Geom. notat.	Rod group generators
Triclinic						
R_1	$\not{p}1$	1	$P1$	$P\bar{1}$	$\not{p}\bar{1}$	
R_2	$\not{p}2$	2	$P\bar{1}$	$P\overline{22}$	$\not{p}\overline{22}$	$a \wedge b \wedge c$
Monoclinic/inclined						
R_3	$\not{p}211$	3	$P112$	$P\bar{2}$	$\not{p}..\bar{2}$	$b \wedge c$
R_4	$\not{p}m11$	6	Pm	$P1$	$\not{p}1$	a
R_5	$\not{p}c11$	7	Pc	P_c1	$\not{p}_c1$	$aT_c^{1/2}$
R_6	$\not{p}2/m11$	10	$P2/m$	$P2\bar{2}$	$\not{p}2\bar{2}$	$a, b \wedge c$
R_7	$\not{p}2/c11$	13	$P2/c$	$P_a2\bar{2}$	$\not{p}_c2\bar{2}$	$aT_c^{1/2}, b \wedge c$
Monoclinic/orthogonal						
R_8	$\not{p}112$	3	$P112$	$P\bar{2}$	$\not{p}\bar{2}$	$a \wedge b$
R_9	$\not{p}112_1$	4	$P2_1$	$P\bar{2}_1$	$\not{p}\bar{2}_1$	$(a \wedge b)T_c^{1/2}$
R_{10}	$\not{p}11m$	6	Pm	$P1$	$\not{p}..1$	c
R_{11}	$\not{p}112/m$	10	$P2/m$	$P\bar{2}2$	$\not{p}\bar{2}2$	$a \wedge b, c$
R_{12}	$\not{p}112_1/m$	11	$P2_1/m$	$P\bar{2}_12$	$\not{p}\bar{2}_12$	$(a \wedge b)T_c^{1/2}, c$

5 Conclusion

We have briefly reviewed the geometric algebra representation of three-dimensional Euclidean space $\mathbb{R}^3$ in the so-called conformal model in the GA of $\mathbb{R}^{4,1}$, and its use for the representation of 2D and 3D point groups and space groups. The key point is to only use physical crystal lattice vectors for the group generation. The second part introduced the interactive software visualization of 3D space group symmetries based on the established geometric algebra representation. This implementation uses the conformal model both for generating the graphics itself and for internally computing with space group transformations.

Future options are the visualization of *noncharacteristic space group orbits* [16] and *magnetic space groups* [36]. The latter seems particularly attractive as it may nicely integrate the bivector representation of spin [23] in the real Dirac–Hestenes equation of relativistic quantum physics. Based on CLUCalc [44], a first rudimentary geometric algebra *protein visualizer* has been programmed recently for proteins of several thousand (up to 10 000) atoms. A possible future molecule (or ion group) toolbox may therefore be able to display complex biomolecule crystals as well.

We have further devised a new Clifford geometric algebra representation for the 162 subperiodic space groups using versors. In the future this may also be extended to *magnetic subperiodic space groups* [40]. We expect that the present work forms a suitable foundation for interactive visualization software of subperiodic space groups similar to the SGV visualization of the 3D space groups of Sect. 3. Fig-

Table 7 Table of triclinic and monoclinic layer groups. The pure translators T_a, T_b are omitted

Layer group #	Intern. notat.	3D space group #	Intern. 3D SGN	Geom. 3D SGN	Geom. notat.	Layer group generators
Triclinic/oblique						
L_1	$p1$	1	$P1$	$P\bar{1}$	$p\bar{1}$	
L_2	$p\bar{1}$	2	$P\bar{1}$	$P\overline{22}$	$p\overline{22}$	$a \wedge b \wedge c$
Monoclinic/oblique						
L_3	$p112$	3	$P2$	$P\bar{2}$	$p\bar{2}$	$a \wedge b$
L_4	$p11m$	6	Pm	$P1$	$p..1$	c
L_5	$p11a$	7	Pc	P_a1	$p.._a1$	$cT_a^{\frac{1}{2}}$
L_6	$p112/m$	10	$P2/m$	$P\bar{2}2$	$p\bar{2}2$	$a \wedge b, c$
L_7	$p112/a$	13	$P2/c$	$P_a2\bar{2}$	$p\bar{2}2_a$	$a \wedge b, cT_a^{\frac{1}{2}}$
Monoclinic/rectangular						
L_8	$p211$	3	$P2$	$P\bar{2}$	$p.\bar{2}$	$b \wedge c$
L_9	$p2_111$	4	$P2_1$	$P\bar{2}_1$	$p.\bar{2}_1$	$(b \wedge c)T_a^{\frac{1}{2}}$
L_{10}	$c211$	5	$C2$	$A\bar{2}$	$c.\bar{2}$	$b \wedge c, T_{a+b}^{1/2}$
L_{11}	$pm11$	6	Pm	$P1$	$p1$	a
L_{12}	$pb11$	7	Pc	P_a1	p_b1	$aT_b^{\frac{1}{2}}$
L_{13}	$cm11$	8	Cm	$A1$	$c1$	$a, T_{a+b}^{1/2}$
L_{14}	$p2/m11$	10	$P2/m$	$P2\bar{2}$	$p2\bar{2}$	$a, b \wedge c$
L_{15}	$p2_1/m11$	11	$P2_1/m$	$P2\bar{2}_1$	$p2\bar{2}_1$	$a, (b \wedge c)T_a^{\frac{1}{2}}$
L_{16}	$p2/b11$	13	$P2/c$	$P_a2\bar{2}$	$p_b2\bar{2}$	$aT_b^{\frac{1}{2}}, b \wedge c$
L_{17}	$p2_1/b11$	14	$P2_1/c$	$P_a2\bar{2}_2$	$p_b2\bar{2}_1$	$aT_b^{\frac{1}{2}}, (b \wedge c)T_a^{\frac{1}{2}}$
L_{18}	$c2/m11$	12	$C2/m$	$A2\bar{2}$	$c2\bar{2}$	$a, b \wedge c, T_{a+b}^{1/2}$

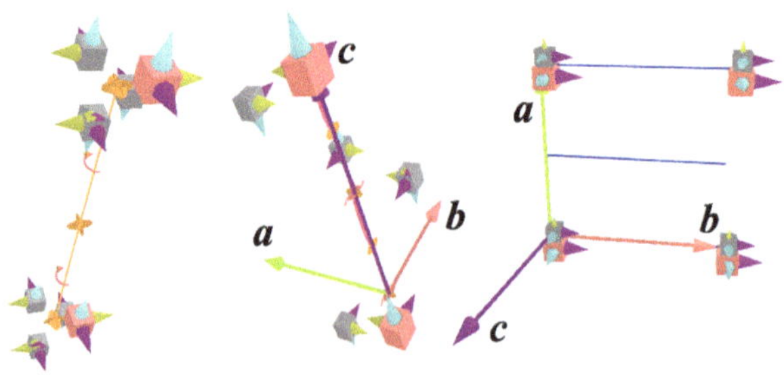

Fig. 11 How a future subperiodic space group viewer software might depict rod groups 13: $\wp\bar{2}\bar{2}\bar{2}$ and 14: $\wp\bar{2}_1\bar{2}\bar{2}$, and the layer group 11: $p1$, based on [32, 46]

ure 11 shows how the rod groups 13: $\wp\bar{2}\bar{2}\bar{2}$ and 14: $\wp\bar{2}_1\bar{2}\bar{2}$, and the layer group 11: $p1$ might be visualized in the future, based on [32, 46].

Acknowledgements E. Hitzer wishes to thank God: *It (Jerusalem) shone with the glory of God, and its brilliance was like that of a very precious jewel, like a jasper, clear as crystal* [48].

He thanks his family, his students M. Sakai, K. Yamamoto, T. Ishii, his colleagues G. Sommer, D. Hestenes, J. Holt, H. Li, T. Matsumoto, D. Litvin, H. Wondratschek, M. Aroyo, H. Fuess, N. Ashcroft, M. Tribelski, A. Hayashi, N. Onoda, Y. Koga, and the organizers of AGACSE 2008.

References

1. Ahlfors, L.: Clifford numbers and Moebius transformations in $\mathbb{R}^n$. In: Chisholm, J., Common, A. (eds.) Clifford Algebras and Their Applications in Mathematical Physics. Reidel, Dordrecht (1986)
2. Angles, P.: Construction de revêtements du groupe conforme d'un espace vectoriel muni d'une métrique de type (p,q). Ann. IHP Sect. A **33**(1), 33–51 (1980)
3. Angles, P.: Conformal Groups in Geometry and Spin Structures. Progress in Mathematical Physics. Birkhäuser, Basel (2007)
4. Aragón-González, G., Aragón, J.L., Rodríguez-Andrade, M.A., Verde-Star, L.: Reflections, rotations, and Pythagorean numbers. Adv. Appl. Clifford Algebr. Online First (2008)
5. Bonola, R.: La Geometria Non-euclidia: Exposizione Storico-Critico del Suo Sviluppo. Zanichelli, Bologna (1906). Translated by H.S. Carslaw, Non-Euclidean Geometry. Dover, New York (1955). References to Friedrich Ludwig Wachter's horosphere on pp. 62, 63 and 88. Download version: http://www.archive.org/details/noneuclideangeom00bonorich
6. Cartan, E.: La Théorie des Spineurs I, II. Actualités Scientifiques et Industrielles, vol. 643. Hermann, Paris (1938)
7. Cox, H.: Application of Grassmann's Ausdehnungslehre to properties of circles. Q. J. Pure Appl. Math. **25**, 1–71 (1891)
8. Coxeter, H.S.M.: Discrete groups generated by reflections. Ann. Math. **35**, 588–621 (1934)
9. Coxeter, H.S.M.: Non-Euclidean Geometry, Toronto, 1942, 3rd edn. Oxford University Press, Cambridge (1957). References to Friedrich Ludwig Wachter's horosphere on pp. 7 and 220
10. Coxeter, H.S.M., Moser, W.O.J.: Generators and Relations for Discrete Groups, 4th edn. Springer, Berlin (1980)
11. Crumeyrolle, A.: Chap. 12 of Orthogonal and Symplectic Clifford Algebras. Kluwer, Dordrecht (1990)
12. Dieudonné, J.: Sur les Groupes Classiques. Actualités Scientifiques et Industrielles, vol. 1040. Hermann, Paris (1948)
13. Doran, C., Lasenby, A.: Geometric Algebra for Physicists. Cambridge University Press, Cambridge (2003)
14. Dorst, L., Fontijne, D., Mann, S.: Geometric Algebra for Computer Science: An Object-oriented Approach to Geometry. Morgan Kaufmann Series in Computer Graphics. Morgan Kaufmann, San Mateo (2007)
15. Dress, A.W.M., Havel, T.F.: Distance geometry and geometric algebra. Found. Phys. **23**(10), 1357–1374 (1993)
16. Engel, P., Matsumoto, T., Steinmann, G., Wondratschek, H.: The non-characteristic orbits of the space groups. Z. Kristallogr., Supplement Issue No. 1. Raumgruppen-Projektionen. In Abhandlungen der Mathematisch-Physikalischen Klasse der Sächsischen Akademie (1984)
17. Forder, H.G.: The Calculus of Extension. Cambridge University Press, Cambridge (1941)
18. Gallier, J.: Geometric Methods and Applications. For Computer Science and Engineering. Springer, New York (2001). Chap. 7
19. Gutierrez, J.D.M.: Operaciones de simitria mediante algebra geometrica aplicadas a grupos cristalograficos. Thesis, UNAM, Mexico (1996)
20. Hahn, T. (ed.) Space-group Symmetry, 5th edn. International Tables of Crystallography, vol. A. Springer, Dordrecht (2005). Online: it.iucr.org
21. Hestenes, D.: Old wine in new bottles. In: Bayro-Corrochano, E., Sobczyk, G. (eds.) Geometric Algebra: A Geometric Approach to Computer Vision, Quantum and Neural Computing, Robotics, and Engineering, pp. 498–520. Birkhäuser, Basel (2001)

22. Hestenes, D.: Point groups and space groups in geometric algebra. In: Dorst, L., et al. (ed.) Applications of Geometric Algebra in Computer Science and Engineering, pp. 3–34. Birkhäuser, Basel (2002)
23. Hestenes, D.: Spacetime physics with geometric algebra. Am. J. Phys. **71**(7), 691–714 (2003)
24. Hestenes, D., Holt, J.: The crystallographic space groups in geometric algebra. J. Math. Phys. **48**, 023514 (2007)
25. Hestenes, D., Li, H., Rockwood, A.: New algebraic tools for classical geometry. In: Sommer, G. (ed.) Geometric Computing with Clifford Algebras, pp. 4–26. Springer, Berlin (2001)
26. Hitzer, E.: Euclidean Geometric Objects in the Clifford Geometric Algebra of {Origin, 3-Space, Infinity}. Bull. Belg. Math. Soc. Simon Stevin **11**(5), 653–662 (2004)
27. Hitzer, E.: Conic sections and meet intersections in geometric algebra, computer algebra and geometric algebra with applications. Lecture Notes in Computer Science, vol. 3519, pp. 350–362. Springer, Berlin (2005). GIAE 2004, Revised Selected Papers
28. Hitzer, E., Perwass, C.: Crystal cells in geometric algebra. In: Proc. Int. Symp. on Adv. Mech. Eng. (ISAME), Fukui, pp. 290–295 (2004)
29. Hitzer, E., Perwass, C.: Full geometric description of all symmetry elements of crystal space groups by the suitable choice of only three vectors for each bravais cell or crystal family. ISAME, Busan, pp. 19–25 (2005)
30. Hitzer, E., Perwass, C.: Crystal cell and space lattice symmetries in Clifford geometric algebra. In: Simos, T.E., et al. (ed.) Int. Conf. on Numerical Analysis and Applied Mathematics, Rhodes, pp. 937–941. Wiley, VCH, New York, Weinheim (2005)
31. Hitzer, E., Perwass, C.: Space group visualizer for monoclinic space groups. Bull. Soc. Sci. Form **21**, 38–39 (2006)
32. Hitzer, E., Perwass, C.: The space group visualizer. In: Proceedings of ISAMPE, pp. 172–181 (2006)
33. Hitzer, E., Perwass, C.: Three vector generation of crystal space groups in geometric algebra. Bull. Soc. Sci. Form **21**, 55–56 (2006)
34. Hitzer, E., Perwass, C.: Interactive 3D space group visualization with CLUCalc and the Clifford geometric algebra description of space groups. Proc. of ICCA8, Las Campinas, Brazil (2008, accepted)
35. Hitzer, E., Tachibana, K., Buchholz, S., Yu, I.: Carrier method for the general evaluation and control of pose, molecular conformation, tracking, and the like. Adv. Appl. Clifford Algebr. Online First (2009)
36. Kopcik, V.A.: Subnikovskie Gruppy: Spravochnik po Simmetrii i Fizicheskim Svoistvam Kristallicheskikh Struktur. Izd. Moskovskogo Univ., Moskva (1966). 722 p. (in Russian)
37. Kopsky, V., Litvin, D.B. (eds.): Subperiodic Groups, 1st edn. International Tables for Crystallography, vol. E. Kluwer, Dordrecht (2002). Online: it.iucr.org
38. Li, H.: Invariant Algebras and Geometric Reasoning. World Scientific, Singapore (2008)
39. Lie, S.: On a Class of Geometric Transformations. University of Oslo Press, Oslo (1872)
40. Litvin, D.B.: Tables of properties of magnetic subperiodic groups. Acta Crystallogr. Sect. A, Found. Crystallogr. A **61**, 382–385 (2005)
41. Lounesto, P.: Clifford Algebras and Spinors, 2nd edn. Cambridge University Press, Cambridge (2001)
42. Lounesto, P., Latvamaa, E.: Conformal transformations and Clifford algebras. Proc. Am. Math. Soc. **79**(4), 533–538 (1980)
43. Maks, J.: Modulo (1, 1) periodicity of Clifford algebras and generalized (anti-) Moebius transformations. Thesis, Delft University of Technology, The Netherlands (1989)
44. Perwass, C.: CLUCalc—interactive visualization. www.clucalc.info
45. Perwass, C., Hitzer, E.: Interactive visualization of full geometric description of crystal space groups. ISAME, Busan, pp. 276–282 (2005)
46. Perwass, C., Hitzer, E.: Space group visualizer. www.spacegroup.info (2005)
47. Souvignier, B.: Space groups. Syllabus for MaThCryst summer school, 15–20 July 2007, Havanna, Cuba
48. The Holy Bible, New International Version. International Bible Society, Colorado Springs (1984)

Geometric Algebra Model of Distributed Representations

Agnieszka Patyk

Abstract Formalism based on GA is an alternative to distributed representation models developed so far: Smolensky's tensor product, Holographic Reduced Representations (HRR), and Binary Spatter Code (BSC). Convolutions are replaced by geometric products interpretable in terms of geometry, which seems to be the most natural language for visualization of higher concepts. This paper recalls the main ideas behind the GA model and investigates recognition test results using both inner product and a clipped version of matrix representation. The influence of accidental blade equality on recognition is also studied. Finally, the efficiency of the GA model is compared to that of previously developed models.

1 Introduction

Since the early 1980s, a new idea of representing knowledge has emerged by the name of distributed representation. It has been the answer to the problems of recognition, reasoning, and language processing—people accomplish these everyday tasks effortlessly, often with only noisy and partial information, while computational resources required for these assignments are enormous. To this day many models have been built, in which arbitrary variable bindings, short sequences of various lengths, and predicates are all usually represented as fixed-width high-dimensional vectors that encode information throughout the elements. In 1990 Smolensky [14] described how tensor product algebra provides a framework for the distributed representation of recursive structures. Unfortunately, Smolensky's tensor product does

A. Patyk (✉)
Faculty of Applied Physics and Mathematics, Gdańsk University of Technology, 80-233 Gdańsk, Poland
e-mail: patyk@mif.pg.gda.pl

A. Patyk
Centrum Leo Apostel (CLEA), Vrije Universiteit Brussel, 1050 Brussels, Belgium

E. Bayro-Corrochano, G. Scheuermann (eds.), *Geometric Algebra Computing*,
DOI 10.1007/978-1-84996-108-0_19, © Springer-Verlag London Limited 2010

not meet all criteria of reduced representations as the size of the tensor increases with the size of the structure. Nevertheless, Smolensky and Dolan [15] have shown that tensor product algebra can be used in some architectures as long as the size of a tensor is restricted. In 1994 Plate [13] worked up his Holographic Reduced Representation (HRR) that uses circular convolution and vector addition to combine vectors representing elements of a domain in hierarchical structures. Elements are represented by randomly chosen high-dimensional vectors. A vector representing a structure is of the same size as the vectors representing the elements it contains. In 1997 Pentti Kanerva [10, 11] introduced Binary Spatter Code (BSC) that is very similar to HRR and is often referred to as a form of HRR. In BSC objects are represented by binary vectors, and the boolean exclusive OR is used instead of convolution. The clean-up memory is an important part of any distributed representation model as an auto-associative collection of all atomic objects and complex statements produced by that system. Given a noisy extracted vector, such a structure must be able to recall the most similar item stored or indicate that no matching object had been found.

The geometric algebra (GA) model, which is the focus of this paper, is an alternative to models developed so far. It has been inspired by the well-known fact that most people think in pictures, i.e., two- and three-dimensional shapes, not by using sequences of ones and zeroes. As far as brain functions are concerned, geometric computing has been applied thus far only in the context of primate visual system [5, Chaps. 1 and 2].

In the GA model convolutions are replaced by geometric products, and superposition is performed by ordinary addition. Sentences are represented by multivectors—superpositions of blades. The concept of GA first appeared in the 19th century works of Grassmann and Clifford but was abandoned for almost a century until Hestenes brought up the subject in [8] and [9]. The Hestenes system has recently found applications in quantum computation (Czachor et al. [1–4, 6]), which appears to be a promising leap from cognitive systems based on traditional computing.

Section 2 of this paper recalls basic operations that can be performed on blades and multivectors, using the example Kanerva [11] gave to illustrate BSC. For further details on multivectors and interesting exercises, the reader may refer to [7, 12]. Section 3 gives rise to discussion about various ways of asking questions and investigates the percentage of correctly recognized items under two possible constructions. Section 4 introduces measures of similarity based not on only the inner product of a multivector but also on its matrix representation. Finally, Sect. 5 studies the influence of accidental blade equality on recognition, and Sect. 6 compares the performance of the GA model with HRR and BSC.

2 Geometric Algebra Model

Distributed representation models developed so far were based on long binary or real vectors. However, most people tend to think by pictures, not by sequences of numbers. Therefore geometric algebra with its ability to describe shapes is the most natural language to mimic human thought process and to represent atomic objects and complex sentences. Furthermore, geometric product of two multivectors is geometrically meaningful, unlike the convolution or a binary exclusive OR operation performed on two vectors.

In this paper we consider the $C\ell_n$ algebra generated by the orthonormal vectors $b_i = \{0,\ldots,0,1,0\ldots,0\}$ for $i \in \{1,\ldots,n\}$. The inner product used throughout the paper is an extension of the inner product $\langle\cdot|\cdot\rangle$ from the Euclidean space $\mathbb{R}^n$. For blades $X_{\langle k\rangle} = x_1 \wedge \cdots \wedge x_k$ and $Y_{\langle l\rangle} = y_1 \wedge \cdots \wedge y_l$, the inner product $\cdot : C\ell_n \times C\ell_n \to \mathbb{R}$ is defined as

$$\langle X_{\langle k\rangle}|Y_{\langle l\rangle}\rangle = \begin{vmatrix} \langle x_1|y_l\rangle & \langle x_1|y_{l-1}\rangle & \cdots & \langle x_1|y_1\rangle \\ \langle x_2|y_l\rangle & \langle x_2|y_{l-1}\rangle & \cdots & \langle x_2|y_1\rangle \\ \vdots & & \ddots & \vdots \\ \langle x_k|y_l\rangle & \langle x_k|y_{l-1}\rangle & \cdots & \langle x_k|y_1\rangle \end{vmatrix} \quad \text{for } k = l, \tag{1}$$

$$\langle X_{\langle k\rangle}|Y_{\langle l\rangle}\rangle = 0 \quad \text{for } k \neq l \tag{2}$$

and is extended by linearity to the entire algebra.

Originally, the GA model was developed as a geometric analogue of BSC and HRR and was described by Czachor, Aerts, and De Moor in [1] and [4]. Before switching from geometric product to BSC and HRR, one has to realize that geometric product is a projective representation of boolean exclusive OR. Let $x_1\ldots x_n$ and $y_1\ldots y_n$ be binary representations of two n-bit numbers x and y, and let $c_x = c_{x_1\ldots x_n} = b_1^{x_1}\ldots b_n^{x_n}$ and $c_y = c_{y_1\ldots y_n} = b_1^{y_1}\ldots b_n^{y_n}$ be their corresponding blades, b_i^0 being equal to $\mathbf{1}$. The following examples show that the geometric product of two blades c_x and c_y equals, up to a sign, $c_{x\oplus y}$:

$$b_1b_1 = c_{10\ldots0}c_{10\ldots0} = \mathbf{1} = c_{0\ldots0} = c_{(10\ldots0)\oplus(10\ldots0)}, \tag{3}$$

$$b_1b_{12} = c_{10\ldots0}c_{110\ldots0} = b_1b_1b_2 = b_2 = c_{010\ldots0} = c_{(10\ldots0)\oplus(110\ldots0)}, \tag{4}$$

$$\begin{aligned} b_{12}b_1 &= c_{110\ldots0}c_{10\ldots0} = b_1b_2b_1 = -b_2b_1b_1 = -b_2 \\ &= -c_{010\ldots0} = -c_{(110\ldots0)\oplus(10\ldots0)} = (-1)^D c_{(110\ldots0)\oplus(10\ldots0)}, \end{aligned} \tag{5}$$

the number D being calculated as

$$D = y_1(x_2+\cdots+x_n) + y_2(x_3+\cdots+x_n) + \cdots + y_{n-1}x_n = \sum_{k<l} y_k x_l. \tag{6}$$

The original BSC is illustrated by an example taken from [4, 11]: atomic objects are represented by randomly chosen strings of bits, "$\oplus$" is a componentwise addition mod 2, and "$\boxplus$" represents a thresholded sum producing a binary vector—the threshold is set at one half of sentence chunks, and a random string of bits is added in case of an even number of sentence chunks to break the tie. The encoded record is

$$PSmith = (name \oplus Pat) \boxplus (sex \oplus male) \boxplus (age \oplus 66), \tag{7}$$

and the decoding of *name* uses the involutive nature of XOR:

$$\begin{aligned} Pat' &= name \oplus PSmith \\ &= name \oplus \big[(name \oplus Pat) \boxplus (sex \oplus male) \boxplus (age \oplus 66)\big] \\ &= Pat \boxplus (name \oplus sex \oplus male) \boxplus (name \oplus age \oplus 66) \\ &= Pat \boxplus \text{noise} \to Pat. \end{aligned} \tag{8}$$

In order to switch from BSC to HRR, the $x \mapsto c_x$ map described in [4] is used.

In the GA model, roles and fillers are represented by randomly chosen blades

$$PSmith = name * Pat + sex * male + age * 66. \tag{9}$$

The "+" is an ordinary addition, and "$*$" written between clean-up memory items denotes the geometric product—this notation will be traditionally omitted when writing down operations performed directly on blades and multivectors. The "+" written in the superscript denotes the reversion of a blade or a multivector. The whole record now corresponds to the multivector

$$PSmith = c_{a_1\ldots a_n} c_{x_1\ldots x_n} + c_{b_1\ldots b_n} c_{y_1\ldots y_n} + c_{c_1\ldots c_n} c_{z_1\ldots z_n}, \tag{10}$$

and the decoding operation "$\sharp$" of *name* with respect to *PSmith* is defined as follows:

$$PSmith \sharp name = name^{+} * PSmith \tag{11}$$

$$= c^{+}_{a_1\ldots a_n}[c_{a_1\ldots a_n} c_{x_1\ldots x_n} + c_{b_1\ldots b_n} c_{y_1\ldots y_n} + c_{c_1\ldots c_n} c_{z_1\ldots z_n}]$$

$$= c_x \pm c_{a\oplus b\oplus y} \pm c_{a\oplus c\oplus z} \tag{12}$$

$$= Pat + \text{noise}. \tag{13}$$

It remains to employ the cleanup memory to find the element closest to *Pat'*—similarity is computed by means of the inner (scalar) product. When using the decoding symbol "$\sharp$", we assume that the reader knows which model is used at the time. Therefore, there will be no variations of the "$\sharp$" symbol in BSC, HRR, or in two possible GA models (depending on the way of asking a question).

For an actual example, let us choose the following representation for roles and fillers of *PSmith*:

$$\left.\begin{aligned} Pat &= c_{00100}, \\ male &= c_{00111}, \\ 66 &= c_{11000}, \end{aligned}\right\} \text{fillers} \tag{14}$$

$$\left.\begin{aligned} name &= c_{00010}, \\ sex &= c_{11100}, \\ age &= c_{10001}. \end{aligned}\right\} \text{roles} \tag{15}$$

The whole record then reads

$$\begin{aligned} PSmith &= name * Pat + sex * male + age * 66 \\ &= c_{00010}c_{00100} + c_{11100}c_{00111} + c_{10001}c_{11000} \\ &= -c_{00110} + c_{11011} + c_{01001}. \end{aligned} \tag{16}$$

The decoding of *PSmith*'s *name* will produce the following result:

$$\begin{aligned} name^{+} * PSmith &= c_{00100} + c_{11001} - c_{01011} \\ &= Pat + \text{noise} = Pat'. \end{aligned} \tag{17}$$

At this point, inner products between Pat' and the elements of the clean-up memory need to be compared. Item in the clean-up memory yielding the highest inner product will be the most likely candidate for *Pat*:

$$\langle Pat|Pat' \rangle = c_{00100} \cdot (c_{00100} + c_{11001} - c_{01011}) = 1 \neq 0, \tag{18}$$

$$\langle male|Pat' \rangle = 0, \tag{19}$$

$$\langle 66|Pat' \rangle = 0, \tag{20}$$

$$\langle name|Pat' \rangle = 0, \tag{21}$$

$$\langle sex|Pat' \rangle = 0, \tag{22}$$

$$\langle age|Pat' \rangle = 0, \tag{23}$$

$$\langle PSmith|Pat' \rangle = 0. \tag{24}$$

A question arises as to how to extract information from a multivector, should a question be asked on the left-hand-side of a multivector

$$name \ * \ PSmith \tag{25}$$

or on the right-hand-side

$$PSmith * name. \tag{26}$$

Furthermore, should we use *name* or rather $name^{+}$? Since we can ask about both the role and the filler, we should be able to ask both right-hand-side and left-hand-side questions according to the principles of geometric algebra. Such an approach, however, would make the rules of decomposition unclear, which is against the philosophy of distributed representations. The problem of asking reversed questions on the appropriate side of a sentence is that we should be able to distinguish roles from fillers. This implies that atomic objects should be partly hand-generated, which is not a desirable property of a distributed representation model. If we decide that a question should always be asked on one fixed side of a sentence, there is no point in reversing the blade since there is no certainty that the fixed side is the appropriate one. Independently of the hand-sidedness of questions, in test results the moduli of inner products are compared instead of their actual (possibly negative) values. For right-hand-side questions, we can reformulate (11)–(13) in the following way:

$$PSmith\sharp name = PSmith * name \tag{27}$$

$$= [c_{a_1\ldots a_n} c_{x_1\ldots x_n} + c_{b_1\ldots b_n} c_{y_1\ldots y_n}$$
$$+ c_{c_1\ldots c_n} c_{z_1\ldots z_n}] c_{a_1\ldots a_n} \tag{28}$$

$$= \pm c_x \pm c_{b\oplus y\oplus a} \pm c_{c\oplus z\oplus a} \tag{29}$$

$$= \pm Pat + \text{noise} = Pat'. \tag{30}$$

The decoding of *PSmith*'s *name* will then take the form

$$PSmith * name = -c_{00100} - c_{11001} - c_{01011} = Pat', \tag{31}$$

resulting in $|\langle PSmith|Pat'\rangle| = |-1| = 1$. We will study the effects of asking questions in various ways in the next section.

3 Recognition

Before we investigate the percentage of correctly recognized items, we need to introduce the following definitions. Let S and Q denote the *sentence* and the *question*, respectively. Let $\mathcal{A}$ be the set of all clean-up memory items A for which $\langle S \sharp Q|A\rangle \neq 0$. We will call $\mathcal{A}$ a set of *potential answers*. Let $m = \max\{|\langle S \sharp Q|A\rangle| : A \in \mathcal{A}\}$ and $T = \{A \in \mathcal{A} : |\langle S \sharp Q|A\rangle| = m\}$. A *pseudo-answer* is an answer belonging to the set T but different from the correct answer to $S \sharp Q$—even if the difference is only in the meaning and not in the multivector. Of course, the set T might also include the correct answer; therefore, it is called the set of *(pseudo-)correct* answers and is actually the set of answers leading to the highest modulus of the inner product. We assume that a noisy statement has been recognized correctly if its counterpart in the clean-up memory is among the (pseudo-)correct answers.

There are some doubts concerning how the sentences should be built—Plate [13] adds an additional vector denoting action id (usually a verb) to a sentence, e.g.,

$$(\underline{eat} + eat_{\text{agt}} \circledast Mark + eat_{\text{obj}} \circledast theFish)/\sqrt{3}, \tag{32}$$

where "$\circledast$" denotes circular convolution. We will distinguish between two types of sentence constructions

- Plate construction, e.g., $eat + eat_{\text{agt}} * Mark + eat_{\text{obj}} * theFish$,
- agent-object construction, e.g., $eat_{\text{agt}} * Mark + eat_{\text{obj}} * theFish$.

The agent–object construction will often be denoted as "A–O" for short, especially in table headings. Preliminary tests conducted on the GA model were designed to investigate which type of construction suits GA better. The vocabulary set and the sentence set for these tests are included in Table 1. The sentence set is especially

Table 1 Contents of the clean-up memory used in tests described throughout this paper

Number of blades	Contents	
1	A total of 42 atomic objects: 19 fillers, 7 single-feature roles, and 8 double-feature roles	
2	(1a)	$bite_{agt} * Fido + bite_{obj} * Pat$
	(2a)	$flee_{agt} * Pat + flee_{obj} * Fido$
3	(3a)	$see_{agt} * John + see_{obj} * (1a)$
	(*PSmith*)	$name * Pat + sex * male + age * 66$
4	(1b)	$bite_{agt} * Fido + bite_{obj} * PSmith$
	(2c)	$flee_{agt} * PSmith + flee_{obj} * Fido$
	(4a)	$cause_{agt} * (1a) + cause_{obj} * (2a)$
5	(3b)	$see_{agt} * John + see_{obj} * (1b)$
	(5a)	$see_{agt} * John + see_{obj} * (4a)$
6	(4c)	$cause_{agt} * (1b) + cause_{obj} * (2a)$
7	(*DogFido*)	$class * animal + type * dog + taste * chickenlike + name * Fido + age * 7 + sex * male + occupation * pet$
8	(1c)	$bite_{agt} * DogFido + bite_{obj} * Pat$
	(2b)	$flee_{agt} * Pat + flee_{obj} * DogFido$
	(4b)	$cause_{agt} * (1b) + cause_{obj} * (2c)$
9	(3c)	$see_{agt} * John + see_{obj} * (1c)$
	(5b)	$see_{agt} * John + see_{obj} * (4b)$
10	(1d)	$bite_{agt} * DogFido + bite_{obj} * PSmith$
	(2d)	$flee_{agt} * PSmith + flee_{obj} * DogFido$
11	(3d)	$see_{agt} * John + see_{obj} * (1d)$

Each sentence carries a number (e.g., "(3a)") to make further equations more readable

filled with similar sentences to test sensitivity of the GA model to confusing data. Each sentence carries a number (e.g., "(3a)") to make further equations more compact and readable.

3.1 Right-Hand-Side Questions

In the previous section we commented on the use of reversions and the choice of the side of a statement that a question should be asked on. The argument for right-hand-side (generally: fixed-hand-side) questions without a reversion was that rules of decomposition of a statement should be clear and unchangeable. However, the use of right-hand-side questions poses a problem best described by the following example. Let the clean-up memory contain seven roles and fillers

$$\begin{aligned} see_{\text{agt}} &= c_{00101}, & John &= c_{00101}, \\ see_{\text{obj}} &= c_{01010}, & Pat &= c_{10000}, \\ bite_{\text{agt}} &= c_{10110}, & Fido &= c_{10001}, \\ bite_{\text{obj}} &= c_{00001}, \end{aligned} \tag{33}$$

and two sentences mentioned in Table 1,

$$\text{(1a)} \quad bite_{\text{agt}} * Fido + bite_{\text{obj}} * Pat = c_{00111} - c_{10001}, \tag{34}$$

$$\text{(3a)} \quad see_{\text{agt}} * John + see_{\text{obj}} * \text{(1a)} = -c_{00000} - c_{01101} - c_{11011}. \tag{35}$$

Let us now ask the question (3a) $\sharp$ $see_{\text{obj}} = \text{(3a)} * see_{\text{obj}}$. The decoded answer

$$\begin{aligned} \text{(3a)} * see_{\text{obj}} &= (-c_{00000} - c_{01101} - c_{11011})c_{01010} \\ &= -c_{01010} + c_{00111} + c_{10001} && (36) \\ &= \text{noise} + bite_{\text{agt}} * Fido - bite_{\text{obj}} * Pat && (37) \end{aligned}$$

results in one noisy chunk and two chunks resembling sentence (1a) but having a partially different sign than the original (1a). Furthermore,

$$\begin{aligned} \text{(1a)} \cdot \big(\text{(3a)} * see_{\text{obj}}\big) &= (c_{00111} - c_{10001}) \cdot (-c_{01010} + c_{00111} + c_{10001}) \\ &= c_{00111} \cdot c_{00111} - c_{10001} \cdot c_{10001} && (38) \\ &= -\mathbf{1} + \mathbf{1} = 0. && (39) \end{aligned}$$

Such a situation would not have happened if we asked differently

$$see^{+}_{\text{obj}} * \text{(3a)}, \tag{40}$$

since $see^{+}_{\text{obj}} see_{\text{obj}} = \mathbf{1}$ for normalized atomic objects.

The similarity of (1a) and (3a) $* \, see_{\text{obj}}$ equals zero because the nonzero similarities of blades (i.e., **1**s) belonging to these statements cancelled each other out. Cancellation could be most likely avoided if sentence (1a) had an odd number of blades. This observation has been backed up by test results comparing the performance of Plate construction and the agent–object construction.

As expected, the agent–object construction seems to work better for sentences from which a rather simple information is to be derived, e.g., *PSmith* $\sharp$ *name* or (5a) $\sharp$ see_{obj} (Figs. 1 and 2, respectively). However, when the information asked was more complex, e.g., (5a) $\sharp$ see_{obj}, the Plate construction seemed more appropriate (Fig. 3). This might be so for at least two reasons:

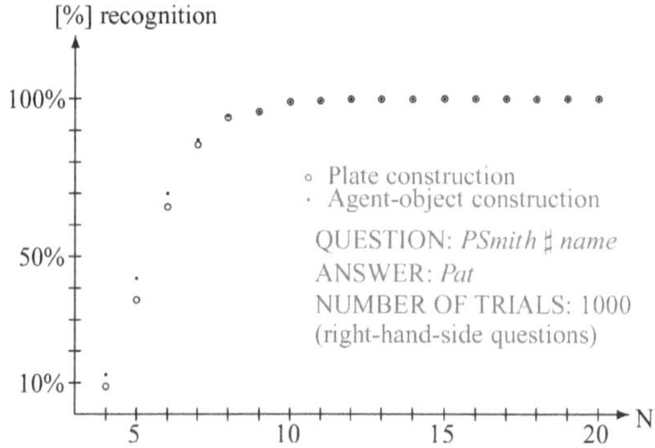

(a) N	(b) Plate	(c) A-O	$\lvert(b)-(c)\rvert$
4	8.9%	12.6%	3.7%
5	36.3%	43.1%	6.8%
6	65.6%	70.0%	4.4%
7	85.5%	87.0%	1.5%
8	94.2%	94.6%	0.4%
9	96.0%	96.0%	0.0%
10	99.0%	99.0%	0.0%
11	99.4%	99.4%	0.0%
12	99.9%	99.9%	0.0%
13	99.9%	99.9%	0.0%
14	99.9%	99.9%	0.0%
15	100.0%	100.0%	0.0%
16	100.0%	100.0%	0.0%
17	100.0%	100.0%	0.0%
18	99.9%	99.9%	0.0%
19	100.0%	100.0%	0.0%
20	100.0%	100.0%	0.0%

Fig. 1 Recognition test results for $PSmith \sharp name$

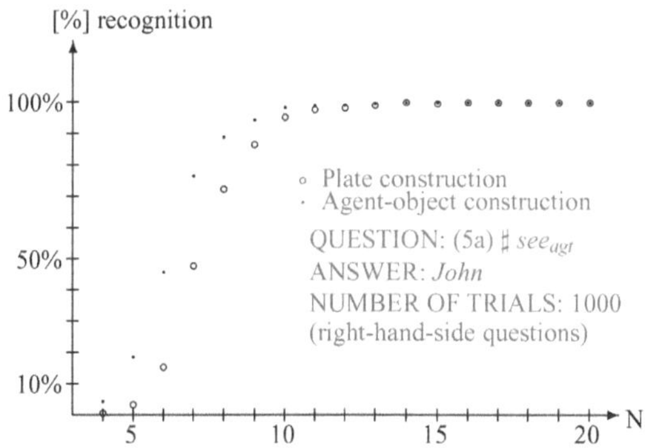

(a) N	(b) Plate	(c) A-O	$\lvert(b)-(c)\rvert$
4	0.6%	4.4%	3.8%
5	3.3%	18.5%	15.2%
6	15.3%	45.8%	30.5%
7	47.8%	76.4%	28.6%
8	72.2%	88.9%	16.7%
9	86.4%	94.3%	7.9%
10	95.1%	98.2%	3.1%
11	97.7%	98.9%	1.2%
12	98.1%	99.2%	1.1%
13	99.1%	99.6%	0.5%
14	99.9%	99.9%	0.0%
15	99.6%	99.9%	0.3%
16	100.0%	100.0%	0.0%
17	100.0%	100.0%	0.0%
18	100.0%	100.0%	0.0%
19	100.0%	100.0%	0.0%
20	100.0%	100.0%	0.0%

Fig. 2 Recognition test results for (5a) $\sharp$ see_{agt}

- Some of the blades belonging to the answer of (5a) $\sharp$ see_{obj} appear in numerous entries in the cleanup memory listed in Table 1, causing the GA model to misinterpret the answers it receives after the inner products have been computed.
- The number of blades in (5a) $\sharp$ see_{obj} is uneven when using Plate construction; hence the possibility that blades' similarities cancel each other out is smaller than in case of the agent–object construction—such hypothesis would be backed up by test results depicted in Fig. 4.

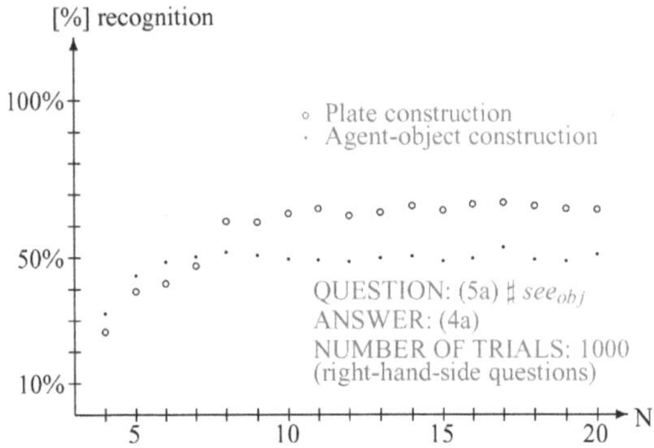

(*a*) N	(*b*) Plate	(*c*) A-O	$\|(b)-(c)\|$
4	26.4%	32.2%	5.8%
5	39.2%	44.2%	5.0%
6	41.8%	48.6%	6.8%
7	47.3%	50.1%	2.8%
8	61.4%	51.7%	9.7%
9	61.2%	50.6%	10.6%
10	64.0%	49.5%	14.5%
11	65.4%	49.3%	16.1%
12	63.2%	48.7%	14.5%
13	64.3%	49.9%	14.4%
14	66.4%	50.4%	16.0%
15	64.9%	48.8%	16.1%
16	66.9%	49.8%	17.1%
17	67.3%	53.3%	14.0%
18	66.4%	49.4%	17.0%
19	65.4%	48.8%	16.6%
20	65.2%	51.1%	14.1%

Fig. 3 Recognition test results for (5a) ♯ see_{obj}

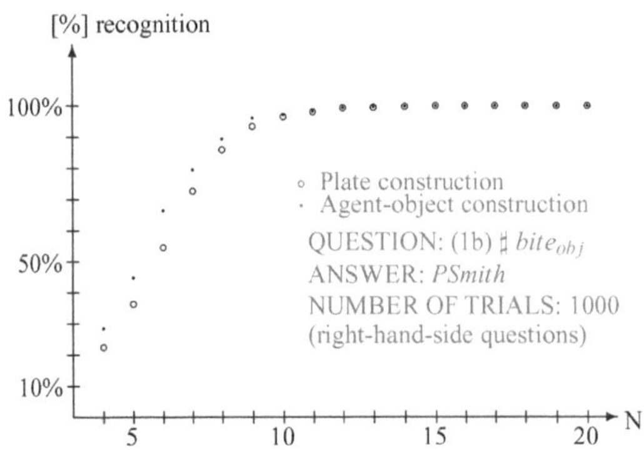

(*a*) N	(*b*) Plate	(*c*) A-O	$\|(b)-(c)\|$
4	22.5%	28.6%	6.1%
5	36.3%	44.6%	8.3%
6	54.6%	66.5%	11.9%
7	72.6%	79.5%	6.9%
8	85.8%	89.4%	3.6%
9	93.5%	96.1%	2.6%
10	96.4%	97.2%	0.8%
11	98.0%	98.3%	0.3%
12	99.4%	99.5%	0.1%
13	99.5%	99.7%	0.2%
14	99.7%	99.8%	0.1%
15	100.0%	100.0%	0.0%
16	99.9%	100.0%	0.1%
17	100.0%	100.0%	0.0%
18	100.0%	100.0%	0.0%
19	99.9%	100.0%	0.1%
20	100.0%	100.0%	0.0%

Fig. 4 Recognition test results for (1b) ♯ $bite_{\text{obj}}$

These hypotheses led to a conclusion that perhaps a random blade should be added to those sentences that have an even number of blades, similarly to BSC.

A correct answer might not be recognized for two reasons:

- The correct answer has an even number of blades, and their similarities cancelled each other out completely because of having opposite signs; hence such an answer does not even appear within the set of potential answers.

- There are some pseudo-answers leading to a higher inner product because the similarities of blades of a correct answer cancelled each other partially.

Adding random extra blades that make the number of blades in a multivector odd (for short: *odding blades*) is a solution to the first reason why a correct answer is not recognized. Further, an odding blade acts as a distinct marker belonging only to one sentence (for sufficiently large data size) distinguishing it from other sentences, unlike the extra blade representing action in Plate construction which may appear in numerous sentences. Unfortunately, to address the second problem, we need to employ some other measurement of similarity than the inner product. We will show in Sect. 4 that Hamming and Euclidean measures perform very well in that case.

Observation of preliminary recognition test results led to a conclusion that sentences with an even number of blades behave quite differently than sentences with an odd number of blades. In the following tests we inspected the average number of times that blades' similarities cancelled each other out completely during the computation of similarity via inner product.

The complete cancellation of similarities takes place only when exactly half of blades of the correct answer carry a plus sign and the other half carry a minus sign. If the correct answer has $2K \geq 2$ blades, then the probability of exactly half of blades having the same sign is

$$\frac{\binom{2K}{K}}{2^{2K}} \tag{41}$$

under the assumption that the sentence set is chosen completely at random without the interference of the experimenter. Figures 5, 6, 7 show three examples of questions yielding an even-number blade answer and the average number of times their blades' similarities cancelled each other out completely.

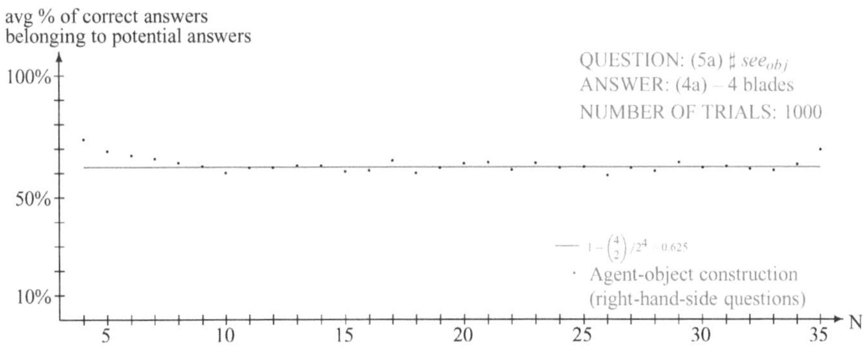

Fig. 5 The average number of times a correct answer appears within the set of potential answers ((5a) ♯ see_{obj})

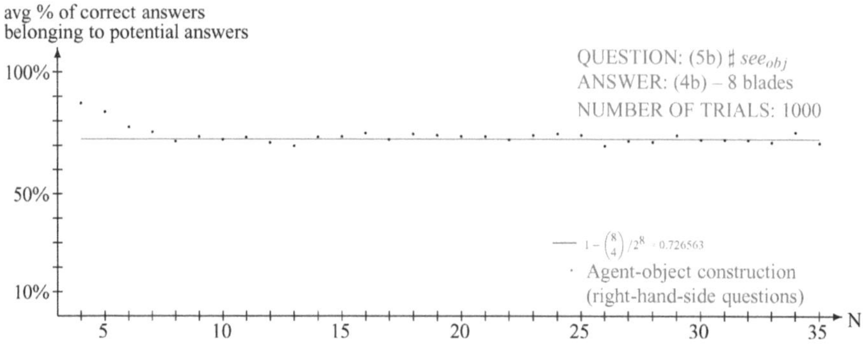

Fig. 6 The average number of times a correct answer appears within the set of potential answers ((5b) ♯ see_{obj})

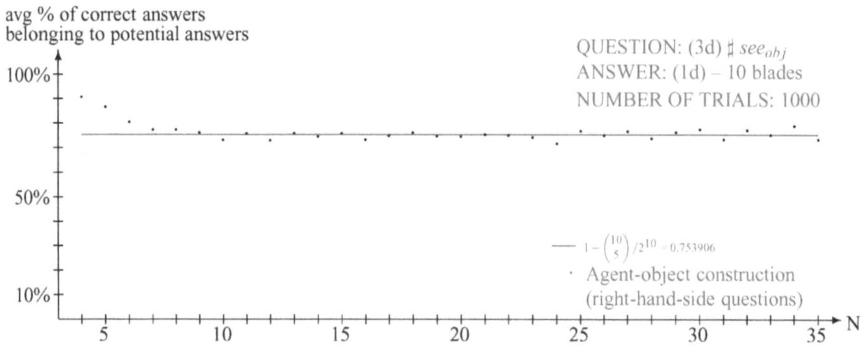

Fig. 7 The average number of times a correct answer appears within the set of potential answers ((3d) ♯ see_{obj})

3.2 Appropriate-Hand-Side Reversed Questions

Let us recall some roles and fillers,

$$\begin{aligned} see_{\mathrm{agt}} &= c_{00101}, & John &= c_{00101}, \\ see_{\mathrm{obj}} &= c_{01010}, & Pat &= c_{10000}, \\ bite_{\mathrm{agt}} &= c_{10110}, & Fido &= c_{10001}, \\ bite_{\mathrm{obj}} &= c_{00001}, \end{aligned} \tag{42}$$

and two sentences mentioned in Table 1,

$$\text{(1a)} \quad bite_{\mathrm{agt}} * Fido + bite_{\mathrm{obj}} * Pat = c_{00111} - c_{10001}, \tag{43}$$

$$\text{(3a)} \quad see_{\mathrm{agt}} * John + see_{\mathrm{obj}} * \text{(1a)} = -c_{00000} - c_{01101} - c_{11011}. \tag{44}$$

The answers to questions (3a) $\sharp$ see_{obj} and (3a) $\sharp$ *John* should be computed in different ways:

$$(3\mathrm{a}) \sharp\, see_{\mathrm{obj}} = see^{+}_{\mathrm{obj}} * (3\mathrm{a}) \approx (1\mathrm{a}), \tag{45}$$

$$(3\mathrm{a}) \sharp\, John = (3\mathrm{a}) * John^{+} \approx see_{\mathrm{agt}}. \tag{46}$$

We will concentrate only on the first question

$$\begin{aligned} see^{+}_{\mathrm{obj}} * (3\mathrm{a}) &= c^{+}_{01010}(-c_{00000} - c_{01101} - c_{11011}) \\ &= c_{01010}(c_{00000} + c_{01101} + c_{11011}) \\ &= c_{01010} + c_{00111} - c_{10001} \qquad (47) \\ &= \text{noise} + (1\mathrm{a}) = (1\mathrm{a})'. \qquad (48) \end{aligned}$$

Only two elements of the clean-up memory are similar to $(1\mathrm{a})'$,

$$\left|\langle see_{\mathrm{obj}} | (1\mathrm{a})' \rangle\right| = 1, \tag{49}$$

$$\begin{aligned} \left|\langle (1\mathrm{a}) | (1\mathrm{a})' \rangle\right| &= \left|(c_{00111} - c_{10001}) \cdot (c_{01010} + c_{00111} - c_{10001})\right| \\ &= |c_{00111} \cdot c_{00111} + c_{10001} \cdot c_{10001}| \\ &= |-\mathbf{1} - \mathbf{1}| = 2, \qquad (50) \end{aligned}$$

where see_{obj} is similar to the noise term only by accident.

Asking reversed questions on the appropriate side has one huge advantage over fixed-hand-side questions: no similarities cancel each other out, neither completely nor partially, while similarity is being computed, and hence there is no need for adding odding vectors. For small data size, blades may cancel each other out at the moment a sentence is created. Nevertheless, in all cases recognition will quickly reach 100%, and the only problem that might appear is that several items of the clean-up memory might be equally similar.

4 Other Measures of Similarity

The inner product is not the only way to measure the similarity of concepts stored in the clean-up memory. This section comments on the use of matrix representation and its advantages in the unavoidable presence of similarity cancellation and many equally probable answers. We will show that comparison by Hamming and Euclidean measures gives promising results in such cases.

4.1 Matrix Representation

Matrix representations of GA, although not efficient, are useful for performing cross-checks of various GA constructions and algorithms. An arbitrary n-bit record can be encoded into the matrix algebra known as Cartan representation of Clifford algebras as follows:

$$b_{2k} = \underbrace{\sigma_1 \otimes \ldots \otimes \sigma_1}_{n-k} \otimes \sigma_2 \otimes \underbrace{1 \otimes \ldots \otimes 1}_{k-1}, \tag{51}$$

$$b_{2k-1} = \underbrace{\sigma_1 \otimes \ldots \otimes \sigma_1}_{n-k} \otimes \sigma_3 \otimes \underbrace{1 \otimes \ldots \otimes 1}_{k-1}, \tag{52}$$

using Pauli's matrices

$$\sigma_1 = \begin{pmatrix} 0 & 1 \\ 1 & 0 \end{pmatrix}, \qquad \sigma_2 = \begin{pmatrix} 0 & -i \\ i & 0 \end{pmatrix}, \qquad \sigma_3 = \begin{pmatrix} 1 & 0 \\ 0 & -1 \end{pmatrix}. \tag{53}$$

Let us once again consider the roles and fillers of *PSmith*:

$$\left.\begin{aligned} Pat &= c_{00100}, \\ male &= c_{00111}, \\ 66 &= c_{11000}, \end{aligned}\right\} \text{fillers} \tag{54}$$

$$\left.\begin{aligned} name &= c_{00010}, \\ sex &= c_{11100}, \\ age &= c_{10001}, \end{aligned}\right\} \text{roles} \tag{55}$$

as described in Sect. 2. Their explicit matrix representations are

$$Pat = c_{00100} = b_3 = \sigma_1 \otimes \sigma_1 \otimes \sigma_1 \otimes \sigma_3 \otimes 1, \tag{56}$$

$$\begin{aligned} male = c_{00111} &= b_3 b_4 b_5 \\ &= (\sigma_1 \otimes \sigma_1 \otimes \sigma_1 \otimes \sigma_3 \otimes 1)(\sigma_1 \otimes \sigma_1 \otimes \sigma_1 \otimes \sigma_2 \otimes 1)(\sigma_1 \otimes \sigma_1 \otimes \sigma_3 \otimes 1 \otimes 1) \\ &= \sigma_1 \otimes \sigma_1 \otimes \sigma_3 \otimes (-i\sigma_1) \otimes 1, \end{aligned} \tag{57}$$

$$\begin{aligned} 66 = c_{11000} &= b_1 b_2 = (\sigma_1 \otimes \sigma_1 \otimes \sigma_1 \otimes \sigma_1 \otimes \sigma_3)(\sigma_1 \otimes \sigma_1 \otimes \sigma_1 \otimes \sigma_1 \otimes \sigma_2) \\ &= 1 \otimes 1 \otimes 1 \otimes 1 \otimes (-i\sigma_1), \end{aligned} \tag{58}$$

$$name = c_{00010} = b_4 = \sigma_1 \otimes \sigma_1 \otimes \sigma_1 \otimes \sigma_2 \otimes 1, \tag{59}$$

$$\begin{aligned} sex = c_{11100} &= b_1 b_2 b_3 \\ &= (\sigma_1 \otimes \sigma_1 \otimes \sigma_1 \otimes \sigma_1 \otimes \sigma_3)(\sigma_1 \otimes \sigma_1 \otimes \sigma_1 \otimes \sigma_1 \otimes \sigma_2) \\ &\quad \times (\sigma_1 \otimes \sigma_1 \otimes \sigma_1 \otimes \sigma_3 \otimes 1) \\ &= \sigma_1 \otimes \sigma_1 \otimes \sigma_3 \otimes 1 \otimes (-i\sigma_1), \end{aligned} \tag{60}$$

$$\begin{aligned} age = c_{10001} &= b_1 b_5 = (\sigma_1 \otimes \sigma_1 \otimes \sigma_1 \otimes \sigma_1 \otimes \sigma_3)(\sigma_1 \otimes \sigma_1 \otimes \sigma_3 \otimes 1 \otimes 1) \\ &= 1 \otimes 1 \otimes (-i\sigma_2) \otimes \sigma_1 \otimes \sigma_3. \end{aligned} \tag{61}$$

Figure 8 shows six blades making up *PSmith* for $n = 5$, and Fig. 9 shows the matrix representation of *PSmith* for $n \in \{6, 7\}$; black dots indicate nonzero matrix entries.

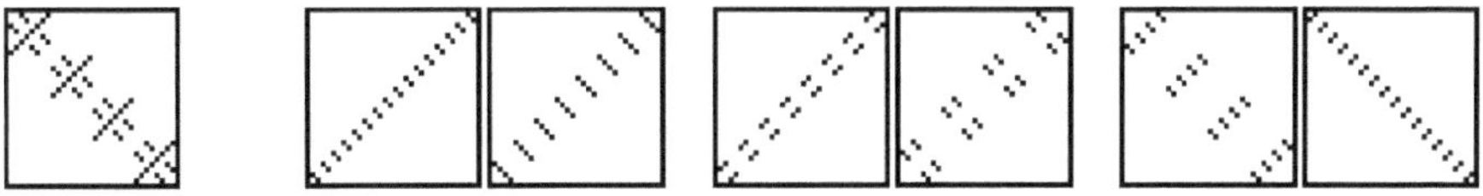

Fig. 8 *PSmith* and its blades for $n = 5$

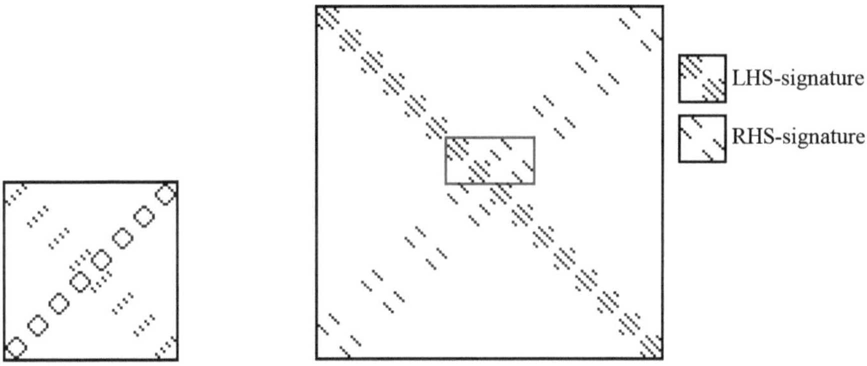

(a) $n = 6$

$name = 001001$ $Pat = 000001$
$sex = 110011$ $male = 110101$
$age = 010001$ $66 = 001011$

(b) $n = 7$

$name = 1010100$ $Pat = 1000001$
$sex = 0000110$ $male = 1010110$
$age = 0011010$ $66 = 0000100$

Fig. 9 An example of matrix representation of *PSmith* for $n \in \{6, 7\}$

The regularity of patterns placed along the diagonals is not accidental. Consider Cartan representation of blades b_{2k} and b_{2k-1}—the shortest sequence of $n - k$ σ_1's will occur for $k = \lceil \frac{n}{2} \rceil$ (in other words, in blade b_n). Therefore, each blade $b_1, \dots, b_n$ has at least $\lfloor \frac{n}{2} \rfloor$ of σ_1's placed at the beginning of the formula describing its representation. Hence, there are exactly $2^{\lfloor \frac{n}{2} \rfloor}$ "boxes" of patterns placed along one of the diagonals, each one of dimension $2^{\lceil \frac{n}{2} \rceil} \times 2^{\lceil \frac{n}{2} \rceil}$. To extract a part individual for a given blade, one needs to consider only the last $\lceil \frac{n}{2} \rceil + 1$ of σ's or unit matrices belonging to its representation—the extra σ_1 bearing the number $n - \lceil \frac{n}{2} \rceil$ is needed to preserve the direction of the diagonal—either "top left to bottom right" or "top right to bottom left." If $c = b_{\alpha_1} \dots b_{\alpha_m}$ is a blade representing an atomic object in the clean-up memory, say *Pat*, then such an object has a "top left to bottom right" orientation if and only if $m \equiv 0 \pmod 2$. Therefore, we can reformulate (51) and (52) in the following way:

$$b_{2k} = \underbrace{\sigma_1 \otimes \dots \otimes \sigma_1}_{\lceil \frac{n}{2} \rceil - k + 1} \otimes \sigma_2 \otimes \underbrace{\mathbf{1} \otimes \dots \otimes \mathbf{1}}_{k-1}, \tag{62}$$

$$b_{2k-1} = \underbrace{\sigma_1 \otimes \dots \otimes \sigma_1}_{\lceil \frac{n}{2} \rceil - k + 1} \otimes \sigma_3 \otimes \underbrace{\mathbf{1} \otimes \dots \otimes \mathbf{1}}_{k-1}. \tag{63}$$

Consider once again the representation of *PSmith* depicted in Fig. 9b. To distinguish *PSmith* from other object, we only need to store two of its "boxes," each "box" lying along a different diagonal, Fig. 9b shows such two parts. We will call the two different "boxes" *left-hand-side signatures* and *right-hand-side signatures* depending on the corner the diagonal is anchored to at the top of the matrix. It is worth noticing that signatures for $n = 2k - 1$ and $n = 2k$ are of the same size, which causes some test results diagrams to resemble step functions.

Note that the use of tensor products in GA bears no resemblance to Smolensky's model, as the rank of a tensor does not increase with the growing complexity of a sentence.

4.2 The Hamming Measure of Similarity

The first most obvious method of comparing two matrices or their signatures would be to compute the number of entries they have in common and the number of entries they differ by. Let $X = [x_{ij}]$ and $Y = [y_{ij}]$ be signatures of matrices, i.e., $i \in \{1, \dots, 2^{\lceil \frac{n}{2} \rceil + 1}\}$, $j \in \{1, \dots, 2^{\lceil \frac{n}{2} \rceil}\}$. Let

$$c(x_{ij}, y_{ij}) = \begin{cases} 1 & \text{if } x_{ij} \neq 0 \text{ and } y_{ij} \neq 0, \\ 0 & \text{otherwise,} \end{cases} \tag{64}$$

$$u(x_{ij}, y_{ij}) = 1 - c(x_{ij}, y_{ij}). \tag{65}$$

Now let us count the numbers of *common points* and *uncommon points*

$$C(X, Y) = \sum_{i,j} c(x_{ij}, y_{ij}), \tag{66}$$

$$U(X, Y) = \sum_{i,j} u(x_{ij}, y_{ij}). \tag{67}$$

Finally, the Hamming measure for comparing the signatures of matrices computes the ratio of common and uncommon points

$$H(X, Y) = \begin{cases} \frac{C(X,Y)}{U(X,Y)} & \text{if } U(X, Y) \neq 0, \\ \infty & \text{otherwise.} \end{cases} \tag{68}$$

Such a measure of similarity is fairly fast to calculate since it does not involve computing any mathematical operations except addition and the final division of $C(X, Y)$ and $U(X, Y)$.

4.3 The Euclidean Measure of Similarity

The second most obvious method for computing matrix similarity is via Euclidean distance. Again, let $X = [x_{ij}]$ and $Y = [y_{ij}]$ be signatures of matrices for

$i \in \{1, \ldots, 2^{\lceil \frac{n}{2} \rceil + 1}\}$, $j \in \{1, \ldots, 2^{\lceil \frac{n}{2} \rceil}\}$. Let

$$E(X, Y) = \begin{cases} \frac{1}{\sum_{i,j} \sqrt{||x_{ij}|^2 - |y_{ij}|^2|}} & \text{if } \sum_{i,j} \sqrt{||x_{ij}|^2 - |y_{ij}|^2|} \neq 0, \\ \infty & \text{otherwise.} \end{cases} \tag{69}$$

This kind of measure uses more mathematical operations requiring greater time to compute: the modulus of a complex number, multiplication, and the square root. The Hamming measure involved calculating only addition and the ratio of common and uncommon points. Calculating the ratio in both measures results in those measures taking on a role of "probability" that the matrices are alike rather than describing the distance between them; therefore, one should avoid calling those measures "metrics."

4.4 Performance of Hamming and Euclidean Measures

In this section we present some test results comparing the effectiveness of Hamming and Euclidean measures against the computation of similarity by inner product. These tests are conducted on the data set presented in Table 1. Once the inner product test indicates more than one potential answer, Hamming and Euclidean measures are employed upon the subset of the potential answers—not upon the whole clean-up memory. Figures 10, 11, 12 show test results for sentences with various numbers of blades using two types of construction, agent–object construction and agent–object construction with odding blades.

There was no significant difference between results obtained using the agent–object construction with odding blades and those obtained with the help of Plate construction; therefore, the results for Plate construction are not presented in the diagrams. Nevertheless, it is more in the spirit of distributed representations to use agent–object construction with odding blades since the additional blade is drawn at random, whereas the use of Plate construction makes data more predictable. Poor recognition in case of the agent–object construction without odding blades results from complete or partial similarity cancellation.

It becomes apparent that the best types of construction of sentences for GA are agent–object construction with odding blades and the Plate construction, as they ensure that sentences have an odd number of blades. Further, it is advisable to compute similarity by the means of Hamming measure or the Euclidean measure instead of the inner product. The Euclidean measure recognizes 100% of items much faster (i.e., for smaller data size), but for large data size, both measures behave identically. Therefore Hamming measure should be used to calculate similarity since its computation requires less time. The success of those measures is due to the fact that the differences between matrices or their signatures lessen the similarity, whereas differences in blades did not lessen the value of the inner product considerably.

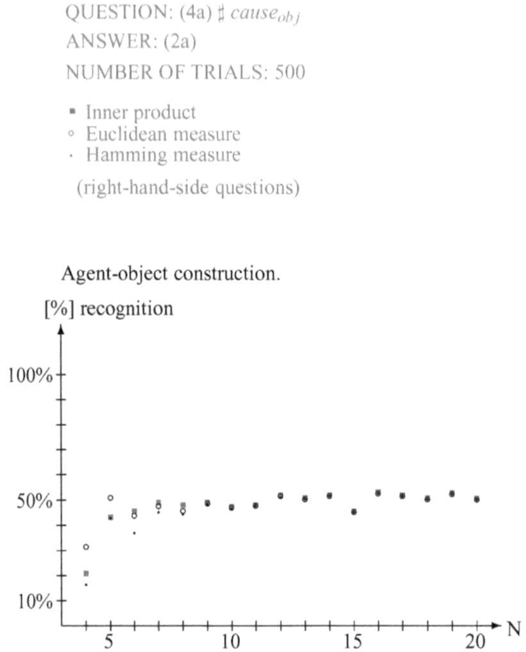

N	Hamming	Euclidean	Inner pr.
4	16.2%	31.4%	20.2%
5	42.8%	51.0%	42.8%
6	37.0%	43.8%	45.2%
7	45.2%	47.6%	48.6%
8	44.4%	45.8%	47.6%
9	48.4%	48.6%	48.8%
10	46.6%	47.0%	47.0%
11	47.8%	47.8%	47.6%
12	51.0%	51.6%	51.6%
13	50.4%	50.4%	50.4%
14	51.6%	51.6%	51.6%
15	45.2%	45.2%	45.2%
16	52.6%	52.6%	52.6%
17	51.6%	51.6%	51.6%
18	50.4%	50.4%	50.4%
19	52.4%	52.4%	52.4%
20	50.2%	50.2%	50.2%

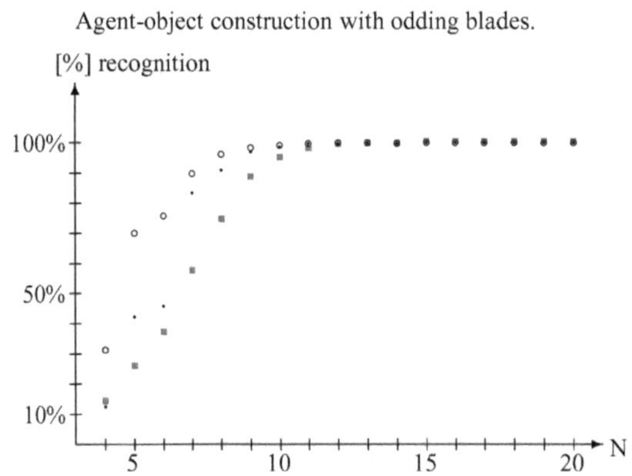

N	Hamming	Euclidean	Inner pr.
4	12.6%	31.2%	14.0%
5	42.2%	70.0%	25.6%
6	45.8%	75.8%	36.8%
7	83.4%	89.8%	57.4%
8	91.0%	96.2%	74.4%
9	97.0%	98.4%	88.4%
10	98.6%	99.2%	94.8%
11	99.8%	99.8%	97.8%
12	99.6%	100.0%	99.2%
13	100.0%	100.0%	99.4%
14	99.8%	99.8%	99.4%
15	100.0%	100.0%	100.0%
16	100.0%	100.0%	100.0%
17	100.0%	100.0%	100.0%
18	100.0%	100.0%	100.0%
19	100.0%	100.0%	100.0%
20	100.0%	100.0%	100.0%

Fig. 10 Recognition test via inner product, Hamming and Euclidean measure for (4a) ♯ $cause_{\text{obj}}$

5 The Average Number of Potential Answers

Our point of interest in this section will be analyzing the influence of accidental blade equality on the number of potential answers under agent–object construction with appropriate-hand-side reversed questions. The following estimates assume *ideal conditions*, i.e., no two chunks of a sentence are identical up to a constant at

QUESTION: (3b) ♯ see_{obj}
ANSWER: (1b)
NUMBER OF TRIALS: 500

- ▪ Inner product
- ∘ Euclidean measure
- · Hamming measure

(right-hand-side questions)

Agent-object construction.

[%] recognition

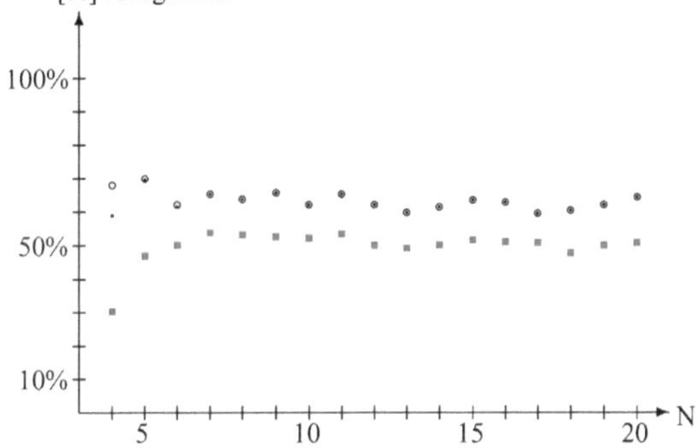

N	Hamming	Euclidean	Inner pr.
4	59.0%	68.0%	29.8%
5	69.6%	70.0%	46.4%
6	61.6%	62.2%	49.8%
7	65.4%	65.4%	53.4%
8	63.8%	63.8%	52.8%
9	65.8%	65.8%	52.2%
10	62.2%	62.2%	51.8%
11	65.4%	65.4%	53.0%
12	62.2%	62.2%	49.6%
13	59.8%	59.8%	48.8%
14	61.6%	61.6%	49.8%
15	63.6%	63.6%	51.2%
16	63.0%	63.0%	50.6%
17	59.6%	59.6%	50.4%
18	60.6%	60.6%	47.4%
19	62.2%	62.2%	49.6%
20	64.4%	64.4%	50.4%

Agent-object construction with odding blades.

[%] recognition

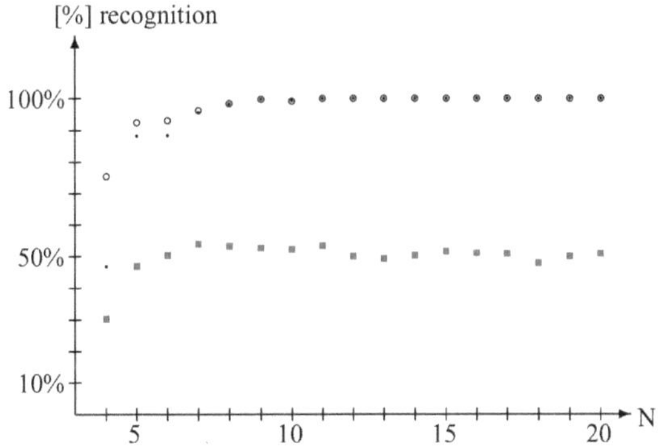

N	Hamming	Euclidean	Inner pr.
4	46.8%	75.4%	26.8%
5	88.2%	92.4%	40.4%
6	88.4%	93.0%	45.6%
7	95.6%	96.2%	66.4%
8	98.2%	98.4%	78.0%
9	99.8%	99.8%	83.8%
10	99.4%	99.2%	90.6%
11	100.0%	100.0%	93.6%
12	100.0%	100.0%	91.6%
13	100.0%	100.0%	93.6%
14	100.0%	100.0%	94.2%
15	100.0%	100.0%	95.0%
16	100.0%	100.0%	92.6%
17	100.0%	100.0%	95.0%
18	100.0%	100.0%	93.6%
19	100.0%	100.0%	92.8%
20	100.0%	100.0%	93.4%

Fig. 11 Recognition test via inner product, Hamming and Euclidean measure for (3b) ♯ see_{obj}

any time. Intuitively, such conditions could be met for sufficiently large lengths of the input vectors, whereas vectors of short length will be linearly dependent with high probability. We will estimate the average number of times that a nonzero inner product comes up when a noisy output is compared with items stored in the clean-up memory. We will deal with the following issues:

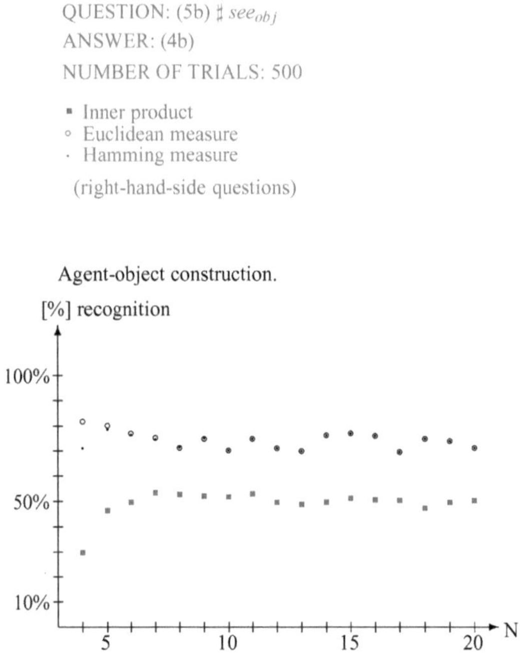

N	Hamming	Euclidean	Inner pr.
4	71.0%	81.8%	40.0%
5	78.6%	80.2%	47.0%
6	76.4%	77.0%	55.8%
7	74.6%	75.2%	57.0%
8	71.6%	71.2%	56.0%
9	75.0%	74.8%	59.2%
10	70.2%	70.2%	54.6%
11	74.8%	74.8%	60.8%
12	71.0%	71.0%	57.0%
13	70.0%	70.0%	54.6%
14	76.2%	76.2%	61.8%
15	77.0%	77.0%	60.0%
16	76.0%	76.0%	62.6%
17	69.6%	69.6%	52.2%
18	74.8%	74.8%	59.8%
19	74.0%	74.0%	55.2%
20	71.2%	71.2%	57.0%

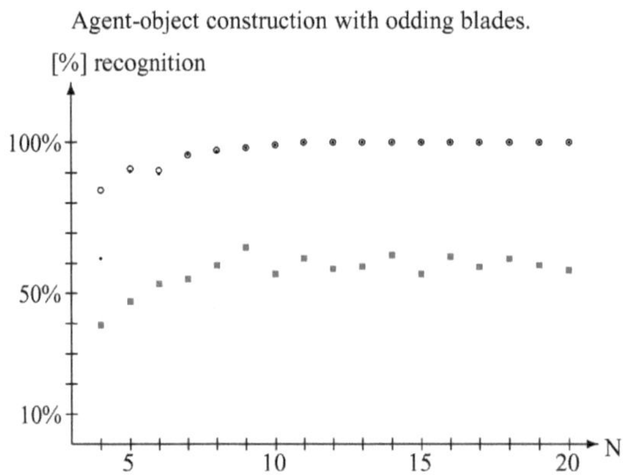

N	Hamming	Euclidean	Inner pr.
4	61.6%	84.2%	39.0%
5	90.4%	91.2%	46.8%
6	89.6%	90.8%	52.6%
7	96.0%	95.8%	54.2%
8	97.0%	97.4%	58.8%
9	98.2%	98.2%	64.8%
10	99.2%	99.2%	55.8%
11	100.0%	100.0%	61.2%
12	100.0%	100.0%	57.6%
13	100.0%	100.0%	58.4%
14	100.0%	100.0%	62.2%
15	100.0%	100.0%	55.8%
16	100.0%	100.0%	61.8%
17	100.0%	100.0%	58.2%
18	100.0%	100.0%	61.0%
19	100.0%	100.0%	58.8%
20	100.0%	100.0%	57.2%

Fig. 12 Recognition test via inner product, Hamming and Euclidean measure for (5b) ♯ see_{obj}

- How often does the system produce identical blades representing atomic objects?
- How often does the system produce identical sentence chunks from different blades?
- How do the two above problems affect the number of potential answers?

Let V be the set of multivectors over $\mathbb{R}^N$ stored in the clean-up memory, and let $\omega(V)$ be the maximum number of blades stored in a multivector in V. The set of all multivectors having the number of blades equal to k is denoted by S_k (S_1 being the set of atomic objects). Naturally, $V = S_1 \cup \cdots \cup S_{\omega(V)}$. Let $\tilde{n}$ be a noisy answer to some question. Under ideal conditions, for every $c \in V$,

$$|\langle \tilde{n} | c \rangle| \neq 0 \quad \Longleftrightarrow \quad \tilde{n} \text{ and } c \text{ share a common blade.} \tag{70}$$

We will begin with a simple example of a multivector with one meaningful blade and L noisy blades. Let $r_0, \dots, r_L$ be roles and $f_0, \dots, f_L$ be fillers for some $L > 0$. Consider the question

$$(r_0 * f_0 + r_1 * f_1 + \cdots + r_L * f_L) \sharp r_0 \tag{71}$$

which results in the following noisy answer:

$$f_0 + \tilde{n}_1 + \cdots + \tilde{n}_L, \qquad \tilde{n}_i = r_0^+ * r_i * f_i, \quad 0 < i \leq L. \tag{72}$$

Surely, the original answer f_0 belongs to S_1. Let $s \in V$ be an arbitrary element of the clean-up memory.

Case 1. Let $s \in S_1$ and $s \neq f_0$, in a sense that s might have the same blade as f_0 but is remembered under a different meaning in the clean-up memory. Using basic probability methods, we obtain

$$\left|\left\langle s \middle| (f_0 + \tilde{n}_1 + \cdots + \tilde{n}_L)\right\rangle\right| \neq 0 \quad \Longleftrightarrow \quad s = f_0 \text{ or } s = \tilde{n}_1 \text{ or } \dots \text{ or } s = \tilde{n}_L, \tag{73}$$

$$\mathrm{P}[s = f_0 \text{ or } s = \tilde{n}_1 \text{ or } \dots \text{ or } s = \tilde{n}_L] = \frac{L+1}{2^N}. \tag{74}$$

Since all blades in S_1 are chosen independently, the following is true:

$$\sum_{s \in S_1, s \neq f_0} \mathrm{P}[s = f_0 \text{ or } s = \tilde{n}_1 \text{ or } \dots \text{ or } s = \tilde{n}_L] = \frac{(|S_1| - 1)(L+1)}{2^N}. \tag{75}$$

Case 2. Let $s \in S_k$ for some $1 < k \leq \omega(V)$ be a multivector made of k blades, $s = s_1 + \cdots + s_k$. Since

$$\mathrm{P}\big[s \text{ does not contain any of } \{f_0, \tilde{n}_1, \dots, \tilde{n}_l\}\big] = \left(1 - \frac{L+1}{2^N}\right)^k, \tag{76}$$

we receive the following formula:

$$\sum_{s \in S_k} \mathrm{P}\big[s \text{ contains at least one of } \{f_0, \tilde{n}_1, \dots, \tilde{n}_l\}\big] = |S_k|\left(1 - \left(1 - \frac{L+1}{2^N}\right)^k\right). \tag{77}$$

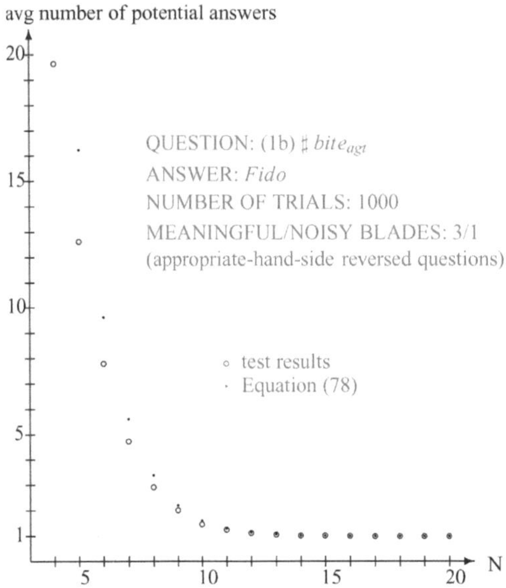

(a) N	(b) Eq. (78)	(c) test results	$\|(b)-(c)\|$
4	25.9406	19.626	6.31459
5	16.2396	12.626	3.61357
6	9.62587	7.8	1.82587
7	5.61888	4.736	0.882881
8	3.39396	2.93	0.463961
9	2.2152	2.014	0.205202
10	1.6153	1.473	0.142299
11	1.30909	1.254	0.055092
12	1.15491	1.125	0.029909
13	1.07755	1.057	0.0205455
14	1.0388	1.026	0.0127955
15	1.0194	1.021	0.00159653
16	1.0097	1.006	0.00370316
17	1.00485	1.0	0.00485194
18	1.00243	1.003	0.00057394
19	1.00121	1.0	0.00121305
20	1.00061	1.002	0.00139347

Fig. 13 Average number of potential answers per 1000 trials with a 1:3 meaningful-to-noisy blades ratio

Thus, when probing for an answer of $f_0 + \tilde{n}_1 + \cdots + \tilde{n}_L,\ L > 0$, we are likely to receive an average of

$$1 + \frac{(|S_1|-1)(L+1)}{2^N} + \sum_{k=2}^{\omega(V)} |S_k| \left(1 - \left(1 - \frac{L+1}{2^N}\right)^k\right) \tag{78}$$

potential answers.

Figure 13 shows test results compared with exact values given by (78) for noisy answers containing one meaningful blade and three noisy blades. Note that (78) is also valid for right-hand-side questions.

The situation becomes more complex when we are to deal with answers having more than one blade. Although items in S_1 are always chosen independently, we cannot say the same about items belonging to $S_i,\ 1 < i \le \omega(V)$, since the sentence set is chosen by the experimenter. Let us consider the question

$$(bite_{\mathrm{agt}} * Fido + bite_{\mathrm{obj}} * PSmith) \sharp\ bite_{\mathrm{obj}} \tag{79}$$

yielding an answer of four blades,

$$name * Pat + sex * male + age * 66 + bite^{+}_{\mathrm{obj}} * bite_{\mathrm{agt}} * Fido. \tag{80}$$

Clearly, the correct answer (*PSmith*) belongs to S_3, but there is one other element of the clean-up memory listed in Table 1 that contains a portion of *PSmith*'s blades,

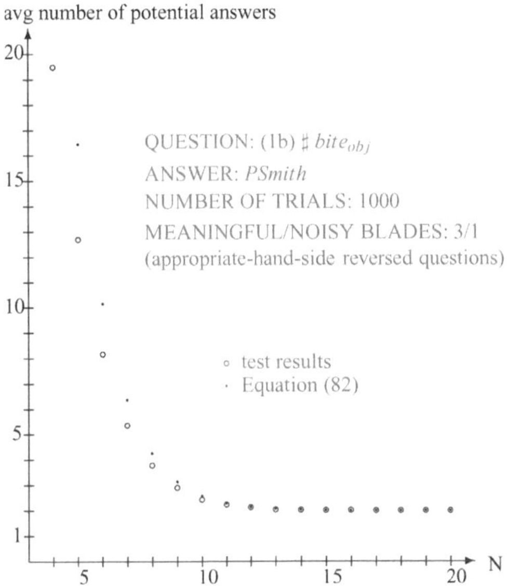

(*a*) N	(*b*) Eq. (82)	(*c*) test results	$\lvert(b)-(c)\rvert$
4	25.7459	19.479	6.26695
5	16.4272	12.698	3.72919
6	10.1488	8.175	1.97385
7	6.36	5.368	0.992003
8	4.25906	3.785	0.47406
9	3.15034	2.901	0.249337
10	2.58051	2.441	0.139507
11	2.29161	2.249	0.0426052
12	2.14614	2.136	0.0101427
13	2.07316	2.045	0.0281567
14	2.0366	2.032	0.00459971
15	2.01831	2.021	0.0026948
16	2.00915	2.015	0.00584606
17	2.00458	2.004	0.00057730
18	2.00229	2.003	0.00071126
19	2.00114	2.0	0.00114439
20	2.00057	2.0	0.00057219

Fig. 14 Average number of potential answers per 1000 trials with a 3:1 meaningful-to-noisy blades ratio

DogFido:

$$\begin{aligned} &\mathit{class} * \mathit{animal} + \mathit{type} * \mathit{dog} + \mathit{taste} * \mathit{chickenlike} \\ &\quad + \mathit{name} * \mathit{Fido} + \mathit{age} * 7 + \mathit{sex} * \mathit{male} + \mathit{occupation} * \mathit{pet}, \end{aligned} \tag{81}$$

the common blade being $\mathit{sex} * \mathit{male}$. We have two answers that, under ideal conditions, will surely result in a nonzero inner product: the correct answer in S_3 and a potential answer in S_7. By calculations analogous to those leading to (78), the average number of answers giving a nonzero inner product for the above example is

$$\begin{aligned} &2 + \frac{4|S_1|}{2^N} + \sum_{k\in\{3,7\}} \big(|S_k|-1\big)\left(1-\left(1-\frac{4}{2^N}\right)^k\right) \\ &\quad + \sum_{k=2,k\notin\{3,7\}}^{\omega(V)} |S_k|\left(1-\left(1-\frac{4}{2^N}\right)^k\right). \end{aligned} \tag{82}$$

The number of meaningful blades in this example is odd, and therefore (82) is also valid for right-hand-side questions. Figure 14 shows test results compared with exact values computed by (82).

Let us consider another example. Now the question is

$$(4a) \ \sharp \ \mathit{cause}_{\mathrm{obj}}, \tag{83}$$

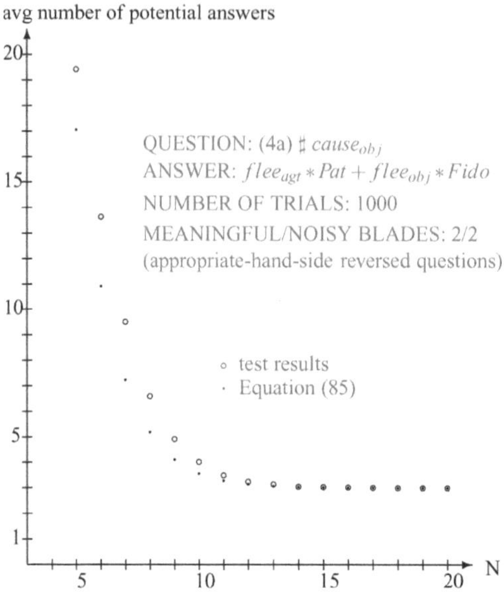

(a) N	(b) Eq. (85)	(c) test results	$\lvert (b)-(c) \rvert$
4	26.1696	26.908	0.738392
5	17.06	19.432	2.37202
6	10.9365	13.65	2.71353
7	7.24505	9.513	2.26795
8	5.19916	6.601	1.40184
9	4.11975	4.924	0.804253
10	3.56505	4.018	0.452952
11	3.28383	3.488	0.204166
12	3.14225	3.246	0.103753
13	3.07121	3.142	0.0707938
14	3.03562	3.06	0.0243762
15	3.01782	3.038	0.0201829
16	3.00891	3.017	0.00809016
17	3.00446	3.005	0.00054476
18	3.00223	3.003	0.00077229
19	3.00111	3.0	0.00111387
20	3.00056	3.001	0.00044306

Fig. 15 Average number of potential answers per 1000 trials with a 2:2 meaningful-to-noisy blades ratio

and the answer has a 2:2 meaningful-to-noisy blades ratio,

$$flee_{\mathrm{agt}} * Pat + flee_{\mathrm{obj}} * Fido + cause^{+}_{\mathrm{obj}} * (bite_{\mathrm{agt}} * Fido + bite_{\mathrm{obj}} * Pat). \tag{84}$$

Apart from the correct answer in S_2 ($flee_{\mathrm{agt}} * Pat + flee_{\mathrm{obj}} * Fido$), there are also two potential answers belonging to the clean-up memory listed in Table 1

- sentence (2b) in S_8—the common blade is $flee_{\mathrm{agt}} * Pat$,
- sentence (2c) in S_4—the common blade is $flee_{\mathrm{obj}} * Fido$.

Therefore, the equation for calculating the estimated number of potential answers for this example takes the following form:

$$3 + \frac{4|S_1|}{2^N} + \sum_{k\in\{2,4,8\}} (|S_k| - 1)\left(1 - \left(1 - \frac{4}{2^N}\right)^k\right) + \sum_{k=3,k\notin\{4,8\}}^{\omega(V)} |S_k|\left(1 - \left(1 - \frac{4}{2^N}\right)^k\right), \tag{85}$$

which is illustrated by Fig. 15.

In this example test results for right-hand-side questions (see Fig. 16) will differ from those obtained by formula (85) by about 0.5. This is because the scalar product of (4a) ♯ $cause_{\mathrm{obj}}$ and the correct answer will produce two **1**s that, with probability 0.5, will have opposite signs and will cancel each other out. Potential answers (2b) and (2c) do not cause such problems, since the number of their blades is odd. In half

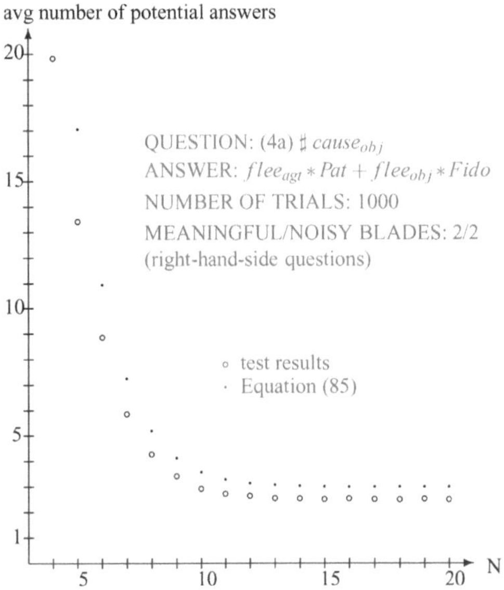

(a) N	(b) Eq. (85)	(c) test results	$\vert(b)-(c)\vert$
4	26.1696	19.815	6.35461
5	17.06	13.42	3.63998
6	10.9365	8.854	2.08247
7	7.24505	5.854	1.39105
8	5.19916	4.265	0.934156
9	4.11975	3.411	0.708747
10	3.56505	2.914	0.651048
11	3.28383	2.727	0.556834
12	3.14225	2.645	0.497247
13	3.07121	2.556	0.515206
14	3.03562	2.543	0.492624
15	3.01782	2.513	0.504817
16	3.00891	2.532	0.47691
17	3.00446	2.508	0.496455
18	3.00223	2.495	0.507228
19	3.00111	2.525	0.476114
20	3.00056	2.496	0.504557

Fig. 16 Average number of potential answers per 1000 trials with a 2:2 meaningful-to-noisy blades ratio (right-hand-side questions)

the cases the number of potential answers will be 2 (sentences (2b) and (2c)), and in half the cases it will be 3 (sentences (2a), (2b), and (2c)), achieving the average of 2.5 potential answers.

We are now ready to work out a more general formula describing the average number of potential answers for noisy statements with multiple meaningful blades. Let S and Q denote the sentence and the question, respectively. Let p_k be the number of potential answers to $S \sharp Q$ in the subset S_k of the clean-up memory V, denote by L the number of blades in $S \sharp Q$, and let $p = p_1 + \cdots + p_{\omega(V)}$. The formula for calculating the estimated number of potential answers to $S \sharp Q$ then reads

$$p + \frac{(|S_1| - p_1)L}{2^N} + \sum_{k=2}^{\omega(V)} \left(|S_k| - p_k\right)\left(1 - \left(1 - \frac{L}{2^N}\right)^k\right), \tag{86}$$

provided that we use appropriate-hand-side reversed questions. As far as right-hand-side questions are concerned, this equation may be regarded only as the upper bound due to cancellation; for a closer estimate, one should investigate elements of the clean-up memory that have an even number of blades.

6 Comparison with Previously Developed Models

The most important performance measure of any new distributed representation model is the comparison of its efficiency in relation to previously developed models. This section comments on test results performed on GA, BSC, and HRR.

QUESTION: *PSmith* ♯ *name*
ANSWER: *Pat*
NUMBER OF TRIALS: 1000
GA CONSTRUCTION: Agent-object with odding blades, right-hand-side questions
MEANINGFUL/NOISY BLADES: 1/2
▪ GA, Hamming measure
∘ BSC
· HRR

All vector lengths: *N*.

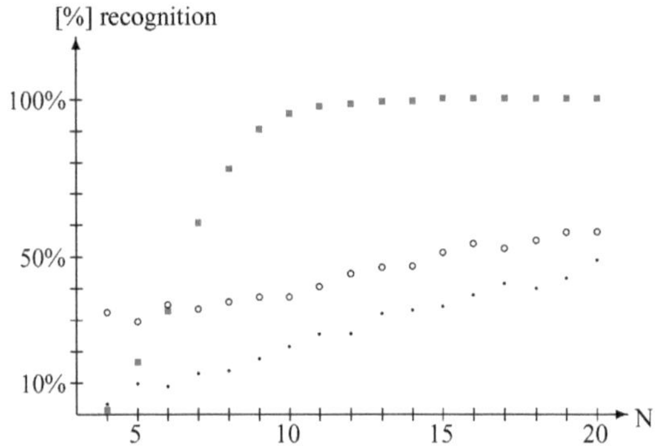

N	Hamming	HRR	BSC
4	0.8%	3.2%	32.5%
5	16.0%	9.7%	29.5%
6	32.4%	8.9%	34.9%
7	60.4%	12.9%	33.5%
8	77.6%	13.8%	35.8%
9	90.2%	17.7%	37.3%
10	95.2%	21.4%	37.3%
11	97.4%	25.5%	40.6%
12	98.2%	25.7%	44.7%
13	99.0%	32.0%	46.7%
14	99.2%	33.2%	47.1%
15	100.0%	34.3%	51.4%
16	100.0%	37.9%	54.3%
17	100.0%	41.7%	52.7%
18	100.0%	40.1%	55.3%
19	100.0%	43.3%	57.8%
20	100.0%	49.0%	57.9%

Vector lengths: *N* for GA, 13*N* for HRR and BSC.

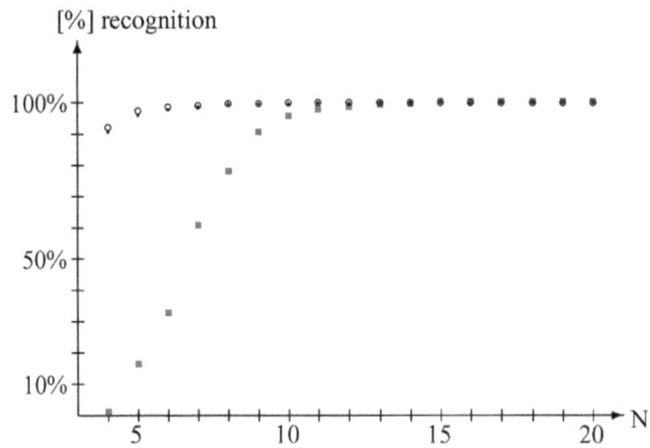

N	Hamming	HRR	BSC
4	0.8%	90.6%	92.2%
5	16.0%	95.9%	97.4%
6	32.4%	97.7%	98.7%
7	60.4%	98.6%	99.1%
8	77.6%	99.5%	99.7%
9	90.2%	99.6%	99.7%
10	95.2%	99.6%	99.9%
11	97.4%	99.9%	100.0%
12	98.2%	99.8%	100.0%
13	99.0%	100.0%	100.0%
14	99.2%	100.0%	100.0%
15	100.0%	100.0%	100.0%
16	100.0%	99.9%	100.0%
17	100.0%	100.0%	100.0%
18	100.0%	100.0%	100.0%
19	100.0%	100.0%	100.0%
20	100.0%	100.0%	100.0%

Fig. 17 Comparison of recognition for GA, BSC, and HRR—*PSmith* ♯ *name*

Naturally, the question of data size arises as a GA clean-up memory item may store information in more than one vector (blade), unlike in architectures known so far. Further, the preferred way of recognition for GA requires the usage of matrix signatures comprising up to $2^{1+2\lceil \frac{N}{2} \rceil}$ entries. However, since one only needs blades to calculate the matrix signatures, it has been assumed that tests comparing efficiency of various models should be conducted using the following sizes of data:

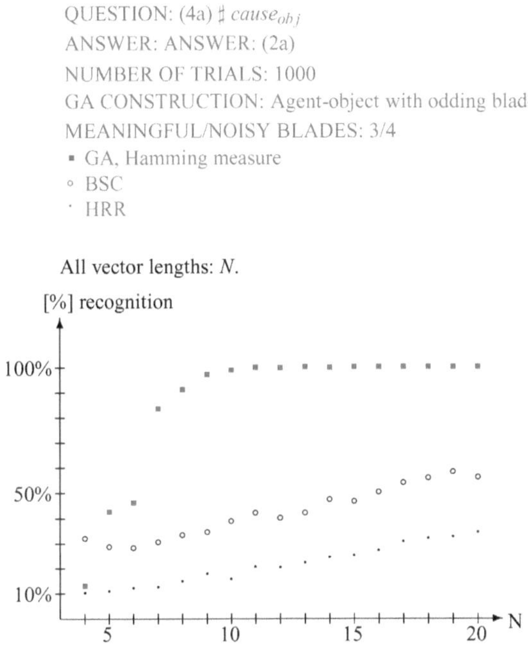

N	Hamming	HRR	BSC
4	12.6%	10.4%	32.0%
5	42.2%	11.1%	28.8%
6	45.8%	12.3%	28.4%
7	83.4%	12.7%	30.7%
8	91.0%	14.9%	33.5%
9	97.0%	18.0%	34.6%
10	98.6%	15.9%	39.0%
11	99.8%	20.7%	42.3%
12	99.6%	20.5%	40.2%
13	100.0%	22.4%	42.3%
14	99.8%	24.6%	47.7%
15	100.0%	25.3%	46.9%
16	100.0%	27.3%	50.6%
17	100.0%	30.9%	54.3%
18	100.0%	32.2%	56.2%
19	100.0%	32.9%	58.7%
20	100.0%	34.6%	56.5%

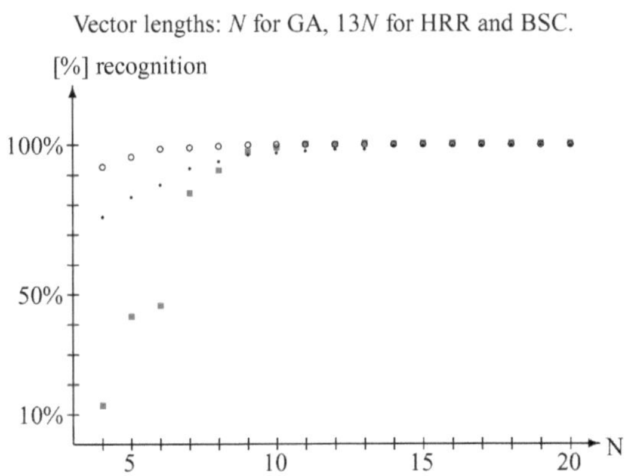

N	Hamming	HRR	BSC
4	12.6%	75.9%	92.7%
5	42.2%	82.6%	95.9%
6	45.8%	86.7%	98.6%
7	83.4%	92.1%	99.1%
8	91.0%	94.5%	99.5%
9	97.0%	96.6%	99.9%
10	98.6%	97.2%	100.0%
11	99.8%	97.9%	100.0%
12	99.6%	98.3%	100.0%
13	100.0%	98.4%	100.0%
14	99.8%	99.5%	100.0%
15	100.0%	99.5%	100.0%
16	100.0%	99.5%	100.0%
17	100.0%	99.7%	100.0%
18	100.0%	99.8%	100.0%
19	100.0%	99.9%	100.0%
20	100.0%	99.7%	100.0%

Fig. 18 Comparison of recognition for GA, BSC, and HRR—(4a) ♯ $cause_{\mathrm{obj}}$

- N bits for a single blade in GA,
- KN bits for a single vector in BSC and HRR,

where K is the maximum number of blades stored in a complex sentence belonging to GA's clean-up memory under agent–object construction with odding blades. For the data set presented in Table 1, the maximum number of blades is stored in items (3d) and (5b) and is equal to 13. Such an approach to the test data size will certainly

QUESTION: (5a) ♯ see_{obj}
ANSWER: (4a)
NUMBER OF TRIALS: 1000
GA CONSTRUCTION: Agent-object with odding blades, right-hand-side questions
MEANINGFUL/NOISY BLADES: 7/2
▪ GA, Hamming measure
∘ BSC
· HRR

All vector lengths: N.

[%] recognition

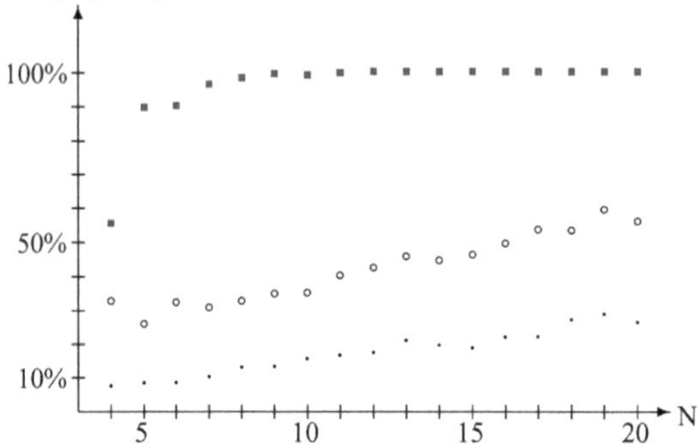

N	Hamming	HRR	BSC
4	55.2%	7.6%	32.9%
5	89.4%	8.4%	26.1%
6	90.0%	8.6%	32.5%
7	96.2%	10.2%	31.0%
8	98.0%	13.0%	32.9%
9	99.4%	13.4%	35.1%
10	99.0%	15.6%	35.3%
11	99.6%	16.6%	40.4%
12	100.0%	17.5%	42.6%
13	100.0%	21.1%	46.0%
14	100.0%	19.7%	44.8%
15	100.0%	18.8%	46.5%
16	100.0%	22.2%	49.9%
17	100.0%	22.3%	53.8%
18	100.0%	27.3%	53.6%
19	100.0%	29.0%	59.7%
20	100.0%	26.6%	56.4%

Vector lengths: N for GA, $13N$ for HRR and BSC.

[%] recognition

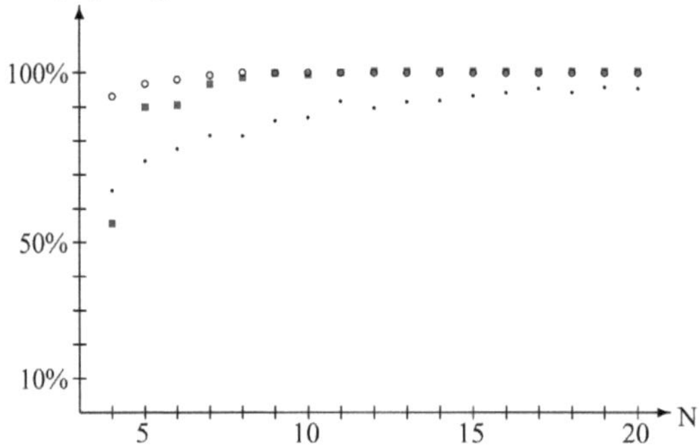

N	Hamming	HRR	BSC
4	55.2%	65.3%	93.0%
5	89.4%	74.0%	96.8%
6	90.0%	77.6%	98.0%
7	96.2%	81.6%	99.3%
8	98.0%	81.5%	100.0%
9	99.4%	85.9%	99.9%
10	99.0%	86.9%	100.0%
11	99.6%	91.6%	100.0%
12	100.0%	89.7%	100.0%
13	100.0%	91.5%	100.0%
14	100.0%	91.8%	100.0%
15	100.0%	93.3%	100.0%
16	100.0%	94.1%	100.0%
17	100.0%	95.4%	100.0%
18	100.0%	94.4%	100.0%
19	100.0%	95.8%	100.0%
20	100.0%	95.5%	100.0%

Fig. 19 Comparison of recognition for GA, BSC, and HRR—(5a) ♯ see_{obj}

prove redundant for GA sentences having a lesser number of blades; nevertheless, it is only fair to provide relatively the same data size for all compared models.

Figures 17, 18, 19 show comparison of performance for GA, BSC, and HRR, and tested sentences range in meaningful-to-noisy blades ratio from 1:2 to 7:2. Clearly, GA with the use of Hamming and Euclidean measure ensures quite a remarkable recognition percentage for sentences of great complexity and therefore great noise, whereas the HRR model works better for statements of low complexity. There is no significant difference in performance of the BSC model as far as complexity of tested sentences is concerned. BSC does remain the best model, provided that vector lengths for BSC are sufficiently longer than that of GA. Under uniform length of vectors and blades, GA recognizes sentences better than HRR or BSC, regardless of their complexity.

7 Conclusion

We have presented a new model of distributed representation that is based on the way humans think, while models developed so far were designed to use arrays of numbers mainly in order to be easily simulated by computers.

After a brief recollection of the main ideas behind the GA model, we investigated three types of sentence constructions, namely the Plate construction, the agent–object construction, and the agent–object construction with odding blades. Two methods of asking questions were also investigated. As a result, in face of shortcomings of recognition based solely on the inner product, matrix representation has been employed as a recognition tool for the GA model. Using test results computed on a toy model, we have shown that Hamming and Euclidean measures of similarity perform very well under the agent–object construction with odding blades.

We also studied the ways in which the number of potential answers is affected by situations in which the system draws at random identical blades denoting different atomic objects or in which identical sentence chunks are produced from different blades. A formula estimating the number of potential counterparts of a noisy piece of information has been derived. Finally, the performance of the GA model has been compared with that of BSC and HRR models using sentences of various complexity.

Acknowledgements I would like to thank Rafał Abłamowicz and Ian Bell for many inspiring conversations that took place during the AGACSE conference in Grimma in August 2008. Furthermore, my participation in that conference would not have been possible without the support from Centrum Leo Apostel (CLEA) at the Vrije Universiteit in Brussels. I also acknowledge support from the LFPPI network.

References

1. Aerts, D., Czachor, M., De Moor, B.: On geometric-algebra representation of binary spatter codes. Preprint (2006). arXiv:cs/0610075 [cs.AI]
2. Aerts, D., Czachor, M.: Cartoon computation: Quantum-like algorithms without quantum mechanics. J. Phys. A **40**, F259 (2007)
3. Aerts, D., Czachor, M.: Tensor-product vs. geometric-product coding. Phys. Rev. A **77**, 012316 (2008). arXiv:0709.1268 [quant-ph]
4. Aerts, D., Czachor, M., De Moor, B.: Geometric analogue of holographic reduced representation. Preprint (2007). arXiv:0710.2611

5. Bayro-Corrochano, E.: Handbook of Geometric Computing. Springer, Berlin (2005)
6. Czachor, M.: Elementary gates for cartoon computation. J. Phys. A **40**, F753 (2007)
7. Dorst, L., Fontijne, D., Mann, S.: Geometric Algebra for Computer Science. Morgan-Kauffman, San Mateo (2007)
8. Hestenes, D.: Space-Time Algebra. Gordon and Breach, New York (1966)
9. Hestenes, D., Sobczyk, G.: Clifford Algebra to Geometric Calculus: A Unified Language for Mathematics and Physics. Reidel, Dordrecht (1984)
10. Kanerva, P.: Binary spatter codes of ordered k-tuples. In: von der Malsburg, C., et al. (eds.) Artificial Neural Networks ICANN Proceedings. Lecture Notes in Computer Science, vol. 1112, pp. 869–873. Springer, Berlin (1996)
11. Kanerva, P.: Fully distributed representation. In: Proc. 1997 Real World Computing Symposium (RWC'97, Tokyo), Real World Computing Partnership, Tsukuba-City, Japan, pp. 358–365 (1997)
12. Lounesto, P.: Clifford Algebras and Spinors, 2nd edn. Cambridge University Press, Cambridge (2001)
13. Plate, T.: Holographic Reduced Representation: Distributed Representation for Cognitive Structures. CSLI Publications, Stanford (2003)
14. Smolensky, P.: Tensor product variable binding and the representation of symbolic structures in connectionist systems. Artif. Intell. **46**, 159–216 (1990)
15. Smolensky, P., Dolan, C.: Tensor product production system: a modular architecture and representation. Connect. Sci. **1**(1), 53–68 (1989)

Computational Complexity Reductions Using Clifford Algebras

René Schott and G. Stacey Staples

Abstract Given a computing architecture based on Clifford algebras, a natural context for determining an algorithm's time complexity is in terms of the number of geometric (Clifford) operations required. In this paper the existence of such a processor is assumed, and a number of graph-theoretical problems are considered. This paper is an extension of previous work, in which the authors defined the "nilpotent adjacency matrix" associated with a finite graph and showed that a number of graph problems of complexity class NP are polynomial in the number of Clifford operations required. Previous results are recalled and illustrated with Mathematica examples. New results are obtained, and old results are improved by the development of new techniques. In particular, a matrix-free approach is developed to count matchings, compute girth, and enumerate proper cycle covers of finite graphs. These new results and techniques are also illustrated with Mathematica examples.

1 Introduction

This paper is an extension of earlier work (cf. [19]), in which the current authors investigated complexity reductions for a number of combinatorial and graph-theoretic problems. The graph-theoretic results of that paper were based primarily on nilpotent adjacency matrix methods developed in a number of earlier publications [17, 18, 20–22].

Combinatorial properties of Clifford algebras provide the underlying context for this work. All algebras used herein can be realized within Clifford algebras of appropriate signature and grow with the size of the graph being considered. An ideal architecture would be able to perform computations in Clifford algebras of arbitrary dimension.

G.S. Staples (✉)
Department of Mathematics and Statistics, Southern Illinois University Edwardsville, Edwardsville, IL 62026-1653, USA
e-mail: sstaple@siue.edu

E. Bayro-Corrochano, G. Scheuermann (eds.), *Geometric Algebra Computing*, DOI 10.1007/978-1-84996-108-0_20, © Springer-Verlag London Limited 2010

The previous graph results are recalled in summary in Sect. 3, where additional examples are computed using Mathematica. In Sect. 4, new results are obtained, and some previous results are improved by eliminating adjacency matrices, thereby removing the consideration of matrix multiplication in determining computational complexity.

Moreover, results on graph and vertex coverings by disjoint "proper" cycles are obtained by introducing the "three-nil algebra" constructed within a Clifford algebra of appropriate signature. The generators $\{\xi_j\}_{1\le i\le n}$ of this algebra satisfy $\xi_i\,\xi_j = \xi_j\,\xi_i$ and $\xi_j^3 = 0$ for $1 \le i, j \le n$.

Given a computing architecture based on Clifford algebras, the natural context for determining an algorithm's time complexity is in terms of the number of geometric (Clifford) operations required. This paper assumes the existence of such a processor and examines the number of Clifford operations needed for a number of combinatorial problems known to be of NP time complexity.

While Clifford algebra computations can be performed on general purpose processors through the use of software libraries like CLU [14], GluCat [11], GAIGEN [7], and the Maple package CLIFFORD [1], direct hardware implementations of data types and operators is the best way to exploit the computational power of Clifford algebras. To this end, a number of hardware implementations have been developed.

To our knowledge, the first such hardware implementation was a Clifford coprocessor design developed by Perwass, Gebken, and Sommer [15]. Another was the color edge detection hardware developed by Mishra and Wilson [12, 13], whose work focused on the introduction of a hardware architecture for applications involving image processing.

More recently, Gentile, Segreto, Sorbello, Vassallo, Vitabile, and Vullo have developed a parallel embedded coprocessing core that directly supports Clifford algebra operators (cf. [8–10]). The prototype was implemented on a Field Programmable Gate Array, and initial tests showed a $4\times$ speedup for Clifford products over the analogous operations in GAIGEN.

Also of interest is the work of Aerts and Czachor [2], who have shown that quantum-like computations can be performed within Clifford algebras without the associated problem of noise and need for error correction.

2 Preliminaries

Definition 1 (Clifford algebra of signature (p,q)) For fixed $n \ge 1$, the 2^n-dimensional algebra $\mathcal{C}\ell_{p,q}$ $(p+q=n)$ is defined as the associative algebra generated by the collection $\{\mathbf{e}_i\}$ $(1 \le i \le n)$ along with the unit scalar $\mathbf{e}_0 = \mathbf{e}_\emptyset = 1 \in \mathbb{R}$, subject to the following multiplication rules:

$$\mathbf{e}_i\,\mathbf{e}_j = -\mathbf{e}_j\,\mathbf{e}_i \quad \text{for } i \ne j, \tag{1}$$

$$\mathbf{e}_i^2 = \begin{cases} 1, & 1 \le i \le p, \\ -1, & p+1 \le i \le n. \end{cases} \tag{2}$$

Products are multi-indexed by subsets of $[n] = \{1, \ldots, n\}$ (in canonical order) according to

$$\mathbf{e}_{\underline{i}} = \prod_{\iota \in \underline{i}} \mathbf{e}_\iota, \tag{3}$$

where $\underline{i}$ is an element of the power set $2^{[n]}$.

Define $f_i = (\mathbf{e}_i + \mathbf{e}_{2n+i}) \in \mathcal{C}\ell_{2n,2n}$ for each $1 \leq i \leq 2n$. Then by defining $\zeta_j = f_{2j-1} f_{2j}$ for each $1 \leq j \leq n$, the following useful algebra is obtained.

Definition 2 Let $\mathcal{C}\ell_n^{\text{nil}}$ denote the real Abelian algebra generated by the collection $\{\zeta_i\}$ $(1 \leq i \leq n)$ along with the scalar $1 = \zeta_0$ subject to the following multiplication rules:

$$\zeta_i \zeta_j = \zeta_j \zeta_i \quad \text{for } i \neq j, \text{ and} \tag{4}$$

$${\zeta_i}^2 = 0 \quad \text{for } 1 \leq i \leq n. \tag{5}$$

It is evident that a general element $u \in \mathcal{C}\ell_n^{\text{nil}}$ can be expanded as

$$u = \sum_{\underline{i} \in 2^{[n]}} u_{\underline{i}}\, \zeta_{\underline{i}}, \tag{6}$$

where $\underline{i} \in 2^{[n]}$ is a subset of $[n] = \{1, 2, \ldots, n\}$ used as a multi-index, $u_{\underline{i}} \in \mathbb{R}$, and $\zeta_{\underline{i}} = \prod_{\iota \in \underline{i}} \zeta_\iota$.

Given $u \in \mathcal{C}\ell_n^{\text{nil}}$, define the *grade-k part* of u by

$$\langle u \rangle_k = \sum_{|\underline{i}|=k} u_{\underline{i}}\, \zeta_{\underline{i}}, \tag{7}$$

where $|\underline{i}|$ denotes the cardinality of the multi-index $\underline{i}$.

Recalling the Clifford algebra $\mathcal{C}\ell_{p,q,r}$ defined by Porteous [16], in which ${\mathbf{e}_i}^2 = 0$ for $p+q+1 \leq i \leq p+q+r$, one could simply define $\zeta_i = \mathbf{e}_{2i-1}\, \mathbf{e}_{2i}$ for $1 \leq i \leq n$ in the Clifford algebra $\mathcal{C}\ell_{0,0,2n}$. Equivalently, one could use disjoint bivectors of the Grassmann algebra.

Note that defining $\xi_i = \zeta_{2i-1} + \zeta_{2i} \in \mathcal{C}\ell_{2n}^{\text{nil}}$ for $1 \leq i \leq n$ gives ${\xi_i}^3 = 0$ and $\xi_i \xi_j = \xi_j \xi_i$. This construction leads to another useful commutative algebra.

Definition 3 Let $n > 0$ be an integer, and let the collection $\{\xi_1, \ldots, \xi_n\}$ satisfy the following: ${\xi_i}^k = 0$ if and only if $k \geq 3$ for each $1 \leq i \leq n$ and $\xi_i \xi_j = \xi_j \xi_i$ for $1 \leq i, j \leq n$. The *three-nil algebra* is the 3^n-dimensional algebra generated by the collection $\{\xi_i\}$ along with the unit scalar and is denoted by $\mathcal{C}\ell_n^{\text{3nil}}$.

Note that elements of the three-nil algebra have canonical expansion of the following form:

$$u = \sum_{\mathbf{v} \in \mathbb{Z}_3{}^n} \alpha_{\mathbf{v}} \xi_1^{v_1} \cdots \xi_n^{v_n}. \tag{8}$$

Given vector $\mathbf{v} \in \mathbb{Z}_3^n$, let diag($\mathbf{v}$) denote the $n \times n$ diagonal matrix whose main diagonal is $\mathbf{v}$. Given $u \in \mathcal{C}\ell_n^{3\text{nil}}$, define the *grade-k part* of u by

$$\langle u \rangle_k = \sum_{\substack{\mathbf{v} \in \mathbb{Z}_3^n \\ \text{rank}(\text{diag}(\mathbf{v}))=k}} \alpha_{\mathbf{v}} \xi_1^{v_1} \cdots \xi_n^{v_n}. \tag{9}$$

In other words, the grade-k part of u is the expansion over terms with k distinct generators. This definition extends naturally to the grade-(j,k) part of an element in $\mathcal{C}\ell_n^{3\text{nil}} \otimes \mathcal{C}\ell_n^{\text{nil}}$.

Remark 1 Letting $\varepsilon_i = \frac{1}{2}(1 + \mathbf{e}_i \mathbf{e}_{n+i}) \in \mathcal{C}\ell_{n,n}$ for each $1 \le i \le n$ gives another useful commutative algebra whose generators are idempotent, i.e., elements of $\{\varepsilon_i\}_{1 \le i \le n}$ satisfy the following multiplication rules:

$$\varepsilon_i \varepsilon_j = \varepsilon_j \varepsilon_i \quad \text{for } i \neq j, \quad \text{and} \tag{10}$$

$$\varepsilon_i^2 = \varepsilon_i \quad \text{for } 1 \le i \le n. \tag{11}$$

Combinatorial properties of this algebra make it applicable to graph theory as well [17, 19].

Letting u denote an arbitrary element of $\mathcal{C}\ell_n^{\text{nil}}$, the *scalar sum* of coefficients will be denoted by

$$\langle\langle u \rangle\rangle = \sum_{\underline{i} \in 2^{[n]}} \langle u, \zeta_{\underline{i}} \rangle = \sum_{\underline{i} \in 2^{[n]}} u_{\underline{i}}. \tag{12}$$

The definitions of scalar sum and grade-k part extend naturally to $\mathcal{C}\ell_n^{3\text{nil}}$.

A number of norms can be defined on $\mathcal{C}\ell_n^{\text{nil}}$. One that will be used later is the *infinity norm* defined by

$$\left\| \sum_{\underline{i} \in 2^{[n]}} u_{\underline{i}} \zeta_{\underline{i}} \right\|_\infty = \max_{\underline{i} \in 2^{[n]}} |u_{\underline{i}}|. \tag{13}$$

An algorithm's time complexity is typically determined by counting the number of operations required to process a data set of size n in worst-, average-, and best-case scenarios. The operation of multiplying two integers is typical. Multiplying a pair of integers in classical computing is assumed to require a constant interval of time, independent of the integers. The architecture of a classical computer makes this assumption natural.

The existence of a processor whose registers accommodate storage and manipulation of elements of $\mathcal{C}\ell_n^{\diamondsuit}$ is assumed through the remainder of this paper.

The $\mathcal{C}\ell$ complexity of an algorithm will be determined by the required number of $\mathcal{C}\ell_n^{\diamondsuit}$ operations or *$\mathcal{C}\ell$ops* required by the algorithm. In other words, multiplying (or adding) a pair of elements $u, v \in \mathcal{C}\ell_n^{\diamondsuit}$ will require one $\mathcal{C}\ell$op in $\mathcal{C}\ell_n^{\diamondsuit}$, where $\diamondsuit$ can be replaced by either "nil" or "3nil".

Evaluation of the infinity norm is another matter. In one possible model of such an evaluation, the scalar coefficients in the expansion of $u \in \mathcal{C}\ell_n^{\diamondsuit}$ are first paired off, and all pairs are then compared in parallel. In this way, evaluation of the infinity norm has complexity $O(\log 2^n) = O(n)$ (cf. [19]).

2.1 Graph Preliminaries

The reader is referred to [24] for graph theory beyond the essential notation and terminology found here. A *graph* $G = (V, E)$ is a collection of vertices V and a set E of unordered pairs of vertices called *edges*. Two vertices $v_i, v_j \in V$ are *adjacent* if there exists an edge $e_{ij} = \{v_i, v_j\} \in E$. In this case, the vertices v_i and v_j are said to be *incident* with e_{ij}.

The number of edges incident with a vertex is referred to as the *degree* of the vertex. A graph is said to be *regular* if all its vertices are of equal degree. A graph is *finite* if V and E are finite sets, that is, if $|V|$ and $|E|$ are finite numbers.

A *k-walk* $\{v_0, \ldots, v_k\}$ in a graph G is a sequence of vertices in G with *initial vertex* v_0 and *terminal vertex* v_k such that there exists an edge $(v_j, v_{j+1}) \in E$ for each $0 \leq j \leq k-1$. A k-walk contains k edges. A *k-path* is a k-walk in which no vertex appears more than once. A *closed k-walk* is a k-walk whose initial vertex is also its terminal vertex. A *k-cycle* is a closed k-path with $v_0 = v_k$.

For convenience, two-cycles (which have a repeated edge) will be allowed in the current work. The term *proper cycle* will refer to any cycle of length three or greater.

A *Hamiltonian cycle* is an n-cycle in a graph on n vertices, i.e., it contains V. Given a graph G, the *circumference* and *girth* of G are defined as the lengths of the longest and shortest cycles in G, respectively.

Given a graph $G = (V, E)$ on n vertices, a *cycle cover* of G is a collection of cycles $\{C_1, \ldots, C_k\}$ contained as subgraphs of G such that each vertex of G is contained in exactly one of the cycles.

A graph G is said to be *connected* if for every pair of vertices v_i, v_j in G, there exists a k-walk on G with initial vertex v_i and terminal vertex v_j for some positive integer k.

A *connected component* of a graph G is a connected subgraph G' of maximal size. In other words, $V(G') \subseteq V(G)$, $E(G') \subseteq E(G)$, and there is no connected subgraph G'' with the property $V(G') \subsetneq V(G'')$.

A *tree* is a connected graph that contains no cycles.

Given a graph $G = (V, E)$, a *matching* of G is a subset $E_1 \subset E$ of the edges of G having the property that no pair of edges in E_1 shares a common vertex. The largest possible matching on a graph with n vertices consists of $n/2$ edges, and such a matching is called a *perfect matching*.

The following graph-theoretic results will be useful in later sections.

Lemma 1 *Let G be a connected graph on $n \geq 2$ vertices. Then G is a tree if and only if G contains $n-1$ edges.*

Proof Proof is by induction on the number of vertices n. When $n=2$, the graph G contains one edge and is a tree by definition. Assuming the lemma is true for some positive integer $n \geq 2$, let G be a connected graph on n vertices, and let the graph H be constructed by appending one vertex v to G. In other words, $V(H) = V(G) \cup \{v\}$. In order to make H connected, one edge must be appended, joining v to some existing vertex u of G. Now H is a connected graph on $n+1$ vertices and is a tree, since v is incident with only one edge.

It remains to be seen that appending two edges incident with v prevents H from being a tree. Suppose a second edge incident with v is appended to H. This edge is incident with some vertex $w \neq u$ of G. Since G is connected, there exists a walk in G having initial vertex u and terminal vertex w. Appending vertex v and its two incident edges to G yields a cycle in H. Thus, H cannot be a tree.

Hence, at most one edge can be appended to G in constructing H. The $(n+1)$-vertex connected graph H consists of n edges, and the proof is complete. □

Lemma 2 *Let G be a connected graph on $n \geq 3$ vertices. Then G is a cycle if and only if G is regular of degree 2.*

Proof Proof is by induction on the number of vertices n. Note that when $n=3$, the only connected graph on three vertices of degree 2 is the 3-cycle. Assume that the lemma is true for some $n \geq 3$ and let G be a connected graph on n vertices containing n edges.

Let H be a connected graph constructed from G by appending one vertex v and an edge incident with v. The edge incident with v must also be incident with a vertex u of G, which is now of degree 3. To correct this, one edge incident with u and another vertex w must be removed, lowering the degree of w to 1. In order to make H regular of degree 2, a new edge incident with v and w is appended. This makes H a cycle on $n+1$ vertices. □

Lemma 3 (Handshaking Lemma) *If G is any graph of e edges, then*

$$\sum_{v \in V_G} \deg(v) = 2e. \tag{14}$$

Proof Since each edge is incident with exactly two vertices, summing degrees over all vertices counts each edge exactly twice. □

3 Complexity Reduction for Graph Problems: Nilpotent Adjacency Matrix Approach

In this section, methods and results of the initial work [19] are recalled, and a number of Mathematica examples are presented. In the subsequent section, new methods are developed, some results are improved, and some new results are obtained.

Definition 4 Let G be a graph on n vertices, and let $\{\zeta_i\}$, $1 \le i \le n$ denote the null-square generators of $\mathcal{C}\ell_n^{\text{nil}}$. Define the *nilpotent adjacency matrix* associated with G by

$$\Lambda_{ij} = \begin{cases} \zeta_j & \text{if } (v_i, v_j) \in E(G), \\ 0 & \text{otherwise.} \end{cases} \tag{15}$$

Thus, Λ defined over $\mathcal{C}\ell_n^{\text{nil}}$ implies that Λ^k is an $n \times n$ zero matrix for all $k > n$.

Theorem 1 *Let Λ be the nilpotent adjacency matrix of an n-vertex graph G. For any $m > 1$ and $1 \le i \le n$, summing the coefficients of $(\Lambda^m)_{ii}$ yields the number of m-cycles based at v_i occurring in G.*

Proof Proof is by induction on m. Entries of $(\Lambda^m)_{ii}$ are sums of products of ζ_js, with each product representing the vertices contained in a closed walk of length m based at the ith vertex. Because ${\zeta_j}^2 = 0$ for each j, terms involving revisited vertices reduce to zero. □

Example 1 To count the 7-cycles in the 16-vertex graph of Fig. 1, the nilpotent adjacency matrix is constructed over $\mathcal{C}\ell_{16}^{\text{nil}}$. Computations are performed with Mathematica procedures available online at http://www.siue.edu/~sstaple.

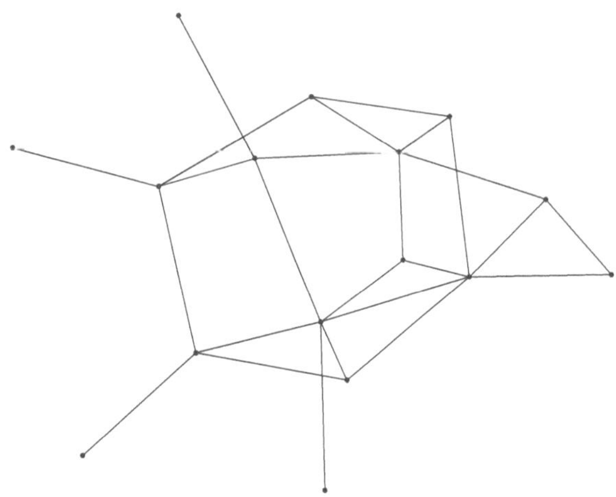

Fig. 1 Randomly generated graph on 16 vertices

$$\begin{pmatrix}
0 & 0 & 0 & 0 & e_{\{5\}} & 0 & 0 & 0 & e_{\{9\}} & 0 & 0 & 0 & 0 & 0 & e_{\{15\}} & 0 \\
0 & 0 & 0 & 0 & 0 & 0 & 0 & e_{\{8\}} & 0 & e_{\{10\}} & 0 & 0 & 0 & 0 & e_{\{15\}} & 0 \\
0 & 0 & 0 & 0 & 0 & 0 & 0 & e_{\{8\}} & 0 & 0 & 0 & 0 & 0 & 0 & 0 & 0 \\
0 & 0 & 0 & 0 & 0 & 0 & 0 & 0 & e_{\{9\}} & e_{\{10\}} & 0 & 0 & 0 & 0 & e_{\{15\}} & 0 \\
e_{\{1\}} & 0 & 0 & 0 & 0 & 0 & 0 & 0 & 0 & 0 & 0 & 0 & 0 & 0 & e_{\{15\}} & 0 \\
0 & 0 & 0 & 0 & 0 & 0 & 0 & 0 & 0 & 0 & 0 & e_{\{12\}} & 0 & 0 & 0 & 0 \\
0 & 0 & 0 & 0 & 0 & 0 & 0 & 0 & e_{\{9\}} & 0 & e_{\{11\}} & 0 & 0 & 0 & e_{\{15\}} & 0 \\
0 & e_{\{2\}} & e_{\{3\}} & 0 & 0 & 0 & 0 & 0 & 0 & e_{\{10\}} & 0 & e_{\{12\}} & 0 & 0 & 0 & 0 \\
e_{\{1\}} & 0 & 0 & e_{\{4\}} & 0 & 0 & e_{\{7\}} & 0 & 0 & 0 & e_{\{11\}} & 0 & e_{\{13\}} & 0 & 0 & 0 \\
0 & e_{\{2\}} & 0 & e_{\{4\}} & 0 & 0 & 0 & e_{\{8\}} & 0 & 0 & 0 & 0 & e_{\{13\}} & e_{\{14\}} & e_{\{15\}} & 0 \\
0 & 0 & 0 & 0 & 0 & 0 & e_{\{7\}} & 0 & e_{\{9\}} & 0 & 0 & e_{\{12\}} & 0 & 0 & 0 & 0 \\
0 & 0 & 0 & 0 & 0 & e_{\{6\}} & 0 & e_{\{8\}} & 0 & 0 & e_{\{11\}} & 0 & e_{\{13\}} & 0 & 0 & 0 \\
0 & 0 & 0 & 0 & 0 & 0 & 0 & 0 & e_{\{9\}} & e_{\{10\}} & 0 & e_{\{12\}} & 0 & 0 & 0 & e_{\{16\}} \\
0 & 0 & 0 & 0 & 0 & 0 & 0 & 0 & 0 & e_{\{10\}} & 0 & 0 & 0 & 0 & 0 & 0 \\
e_{\{1\}} & e_{\{2\}} & 0 & e_{\{4\}} & e_{\{5\}} & 0 & e_{\{7\}} & 0 & 0 & e_{\{10\}} & 0 & 0 & 0 & 0 & 0 & 0 \\
0 & 0 & 0 & 0 & 0 & 0 & 0 & 0 & 0 & 0 & 0 & 0 & e_{\{13\}} & 0 & 0 & 0
\end{pmatrix}$$

```
In[78]:= (* Count 7 cycles*)
         ScalarSum[Tr[ClNilMatrixPower[B, 7]]] / 14

Out[78]= 40
```

Notation To simplify the notation, $\mathrm{tr}(\Lambda^m)$ is replaced by τ_m in the remainder of the paper.

Using the Coppersmith–Winograd algorithm, multiplying two $n \times n$ matrices can be done in $O(n^{2.376})$ time [5]. It is not clear that the same asymptotic speedup can be accomplished for the $\mathcal{C}\ell$ case. However, in the remainder of the paper, β will represent the exponent associated with matrix multiplication. In the worst case, multiplication of two $n \times n$ matrices with entries in $\mathcal{C}\ell_n^{\text{nil}}$ requires n^3 $\mathcal{C}\ell$ops, so $\beta \le 3$.

Corollary 1 *Enumerating the k-cycles in a finite graph on n vertices requires* $O(n^\beta \log k)$ $\mathcal{C}\ell$*ops in* $\mathcal{C}\ell_n^{\text{nil}}$.

Corollary 2 *Enumerating the Hamiltonian cycles in a finite graph on n vertices requires* $O(n^\beta \log n)$ $\mathcal{C}\ell$*ops in* $\mathcal{C}\ell_n^{\text{nil}}$.

Corollary 3 *Let* Λ *be the nilpotent adjacency matrix of an n-vertex graph G. Let* $X_{m,\ell}$ *denote the number of* ℓ*-tuples of pairwise disjoint m-cycles appearing in the graph G, where* $m \ge 3$ *and* $1 \le \ell \le \lfloor n/m \rfloor$*. Then*

$$\left\langle\!\left\langle (\tau_m)^\ell \right\rangle\!\right\rangle = (2m)^\ell \ell! X_{m,\ell}. \tag{16}$$

The following proposition is an immediate corollary of Theorem 1.

Proposition 1 (Graph circumference) *Let G be a graph on n vertices with nilpotent adjacency matrix Λ. The length of the longest cycle in G is the largest integer k such that*

$$\tau_k \neq 0. \tag{17}$$

Corollary 4 *Computing the circumference of a graph on n vertices requires $O(n^\beta \log n)$ $\mathcal{C}\ell$ops in $\mathcal{C}\ell_n^{\mathrm{nil}}$.*

Proof Computing τ^n requires $O(n^\beta \log n)$ $\mathcal{C}\ell$ops. □

Example 2 The circumference and girth of the graph in Fig. 2 are computed using Mathematica.

```
In[55]:= Girth = 0;
         Circumference = 0;
         Mx = ClNilMatrixPower[B, 2] // ClNilExpand;
         For[k = 3, k ≤ 16, k++,
          Mx = ClNilMatrixProduct[Mx, B] // ClNilExpand;
          If[ScalarSum[Tr[Mx]] ≠ 0 ∧ Girth == 0, Girth = k];
          Circumference = If[ScalarSum[Tr[Mx]] ≠ 0, k, Circumference];]
         Print["Girth : ", Girth]
         Print["Circumference : ", Circumference]

Girth : 3

Circumference : 13
```

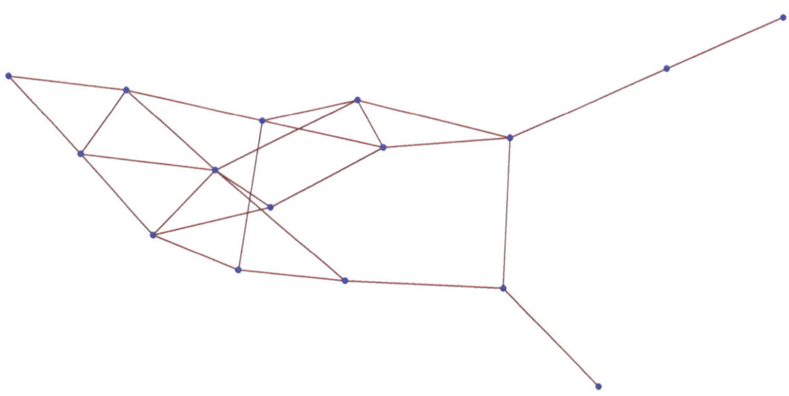

Fig. 2 A randomly generated graph on 16 vertices

Corollary 5 (Graph girth) *Let G be a graph on n vertices with nilpotent adjacency matrix Λ. The length of the shortest cycle in G is the smallest integer k such that*

$$\tau_k \neq 0. \tag{18}$$

The complexity of computing a graph's girth can now be determined in the same manner as the complexity of computing circumference.

Corollary 6 *Computing the girth of a graph on n vertices requires $O(n^\beta \log n)$ $\mathcal{C}\ell ops$ in $\mathcal{C}\ell_n^{\text{nil}}$.*

In the next proposition, C denotes the diagonal matrix $\text{Diag}(\zeta_1, \dots, \zeta_n)$. It is used to account for the initial vertices of paths in G.

Proposition 2 (Longest path) *Let G be a graph on n vertices with nilpotent adjacency matrix Λ. The length of the longest path in G is the largest integer k such that*

$$C\Lambda^k \neq \mathbf{0}. \tag{19}$$

Here, $\mathbf{0}$ denotes the $n \times n$ zero matrix.

Corollary 7 *Computing the length of the longest path in a graph on n vertices requires $O(n^\beta (\log n)^2)$ $\mathcal{C}\ell ops$ in $\mathcal{C}\ell_n^{\text{nil}}$.*

Proof The maximum possible path length is n. For each $1 \leq k \leq n$, computing $\mathcal{C}\Lambda^k$ requires $O(n^\beta \log k + n^2) = O(n^\beta \log n)$ $\mathcal{C}\ell$ops. Using binary search then requires testing $O(\log n)$ values of k in Proposition 2. □

4 Matrix-Free Approach to Representing Graphs

Some of the complexity results recalled in the previous section can be improved by eliminating the nilpotent adjacency matrix. Moreover, the matrix-free approach facilitates additional results such as the enumeration of matchings.

Let $G = (V, E)$ be a graph on n vertices. The adjacency structure of G is represented uniquely within $\mathcal{C}\ell_n^{\text{nil}}$ by

$$\Gamma = \sum_{\{v_i, v_j\} \in E(G)} \zeta_{\{v_i, v_j\}}. \tag{20}$$

In particular, note that Γ is a sum of bivectors representing edges of G by using each edge's incident vertices as indices.

Denote by $\mathcal{C}\ell_n^{\text{nil}} \otimes \mathbb{R}[t]$ the ring of polynomials in the unknown t with $\mathcal{C}\ell_n^{\text{nil}}$ coefficients.

Proposition 3 *Let G be a graph on n vertices with $\Gamma \in \mathcal{C}\ell_n^{\text{nil}}$ as defined in* (20). *Let M denote the number of edges in a maximal matching of G. Then, $e^{t\Gamma}$ is a polynomial in $\mathcal{C}\ell_n^{\text{nil}} \otimes \mathbb{R}[t]$, and*

$$M = \deg_t e^{t\Gamma}. \tag{21}$$

Proof Note that each edge of G is uniquely identified by the pair of vertices with which it is adjacent. Let k be a nonnegative integer such that $k \leq \frac{n}{2}$. From construction of Γ it follows that Γ^k is a sum of blades representing k-subsets of edges of G. Moreover, by the null-square property of the vertex labels ζ_k, such k-subsets of edges must represent k-matchings of G, since two edges incident with a common vertex would result in a blade with a squared generator.

Let V_k denote the collection of subsets of V representing the vertices incident with edges in a k-matching of G and note that $|V_k| = 2k$. For $\underline{i} \in V_k$, let $\alpha_{\underline{i}}$ denote the number of k-matchings on the vertex set $\underline{i}$, observe that $\Gamma^\ell = 0$ for all $\ell > \frac{n}{2}$, and that $\Gamma^k = k! \sum_{\underline{i} \in V_k} \alpha_{\underline{i}} \zeta_{\underline{i}}$. Hence,

$$e^{t\Gamma} = \sum_{k=0}^{\infty} \frac{(t\Gamma)^k}{k!} = \sum_{k=0}^{n/2} \frac{(t\Gamma)^k}{k!} = \sum_{k=0}^{n/2} t^k \sum_{\underline{i} \in V_k} \alpha_{\underline{i}} \zeta_{\underline{i}}. \tag{22}$$

It follows immediately that $\deg_t e^{t\Gamma} = k$ if and only if k is the greatest integer for which a k-matching of G exists. □

Example 3 Figure 3 shows Mathematica code used to generate the random bipartite graph on 14 vertices pictured below and compute the size of a maximal matching.

```
n = 14;
Z = DiagonalMatrix[Table[ς{i}, {i, 1, n}]];
SeedRandom;
A = GenRandGraph[n];
While[Not[BipartiteQ[FromAdjacencyMatrix[A]]],
 A = GenRandGraph[n];]

GraphPlot[A]
B = A.Z;
ψ = Sum[ς{j} B[[j]][[k]], {j, 1, n}, {k, 1, j}];
exptψ = Sum[Expand[t^k ψ^k / k!], {k, 0, Floor[n / 2]}];
Print["Size of Maximal Matching: ",
 Length[BipartiteMatching[FromAdjacencyMatrix[A]]]]
Print["Degree of e^tψ: ", Exponent[exptψ, t]]
```

Fig. 3 Mathematica code for matchings example

Out[77]=

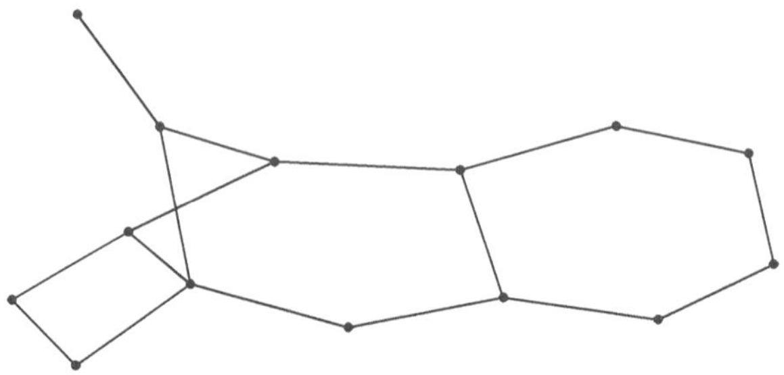

```
Size of Maximal Matching: 7

Degree of e^{tψ}: 7
```

The Mathematica package *Combinatorica* is used to corroborate the result of Proposition 3.

Corollary 8 *The problem of computing the size of a maximal matching in a graph on n vertices is of complexity $O(n)$ in $\mathcal{C}\ell_n^{\text{nil}}$.*

Proof Evaluating $e^{t\Gamma}$ is accomplished by computing $t^{\lfloor n/2 \rfloor}\Gamma^{\lfloor n/2 \rfloor}$, which requires $O(\log(n/2))$ $\mathcal{C}\ell$ops in $\mathcal{C}\ell_n^{\text{nil}}$ when successive squaring is used. Computing all intermediate powers, summing and multiplying scalars results in $O(n)$ $\mathcal{C}\ell$ops. □

Recalling that a perfect matching of a graph on an even number n of vertices is a matching containing $n/2$ vertices, the following result is immediate.

Corollary 9 *Counting the perfect matchings of a bipartite graph on n vertices is of complexity $O(\log n)$ in $\mathcal{C}\ell_n^{\text{nil}}$.*

In light of this result, an improvement on the matrix permanent is expected. It is known that counting the perfect matchings of a bipartite graph is of the same complexity as computing the permanent of its adjacency matrix.

The problem of computing the permanent of a matrix is known to be ♯P-complete [3, 23]. Methods of approximating the permanent using Clifford algebras have also been discussed [4].

Proposition 4 *Let $M = (m_{ij})_{n\times n}$ denote an arbitrary $n \times n$ matrix. Let $\{\zeta_i\}_{1\le i\le n}$ denote commutative null-square generators of $\mathcal{C}\ell_n^{\text{nil}}$, and define $a_i = \sum_{j=1}^n m_{ij}\zeta_j$ for each $i = 1, 2, \ldots, n$. Then,*

$$\prod_{i=1}^{n} a_i = \operatorname{per}(M)\zeta_{[n]}. \tag{23}$$

Proof Recall the definition of the matrix permanent:

$$\operatorname{per}(M) := \sum_{\sigma \in S_n} \prod_{i=1}^{n} m_{i\sigma(i)}. \tag{24}$$

Note that $i \neq j \Rightarrow \sigma(i) \neq \sigma(j)$.

Now consider the product

$$\begin{aligned}\prod_{i=1}^{n} a_i &= \prod_{i=1}^{n} (m_{i\,1}\zeta_1 + \cdots + m_{i\,n}\zeta_n) \\ &= \sum_{(k_1,\ldots,k_n)\in[n]^n} m_{1\,k_1} m_{2\,k_2} \cdots m_{n\,k_n}\, \zeta_{k_1}\zeta_{k_2}\cdots\zeta_{k_n}. \end{aligned} \tag{25}$$

The null-square property of the collection $\{\zeta_i\}$ implies that the sum is over n-tuples of distinct integers:

$$\begin{aligned} &\sum_{(k_1,\ldots,k_n)\in[n]^n} m_{1\,k_1} m_{2\,k_2} \cdots m_{n\,k_n}\, \zeta_{k_1}\zeta_{k_2}\cdots\zeta_{k_n} \\ &\quad = \sum_{\substack{(k_1,\ldots,k_n)\in[n]^n \\ i\neq j \Rightarrow k_i \neq k_j}} m_{1\,k_1} m_{2\,k_2} \cdots m_{n\,k_n}\, \zeta_{k_1}\zeta_{k_2}\cdots\zeta_{k_n} \\ &\quad = \sum_{\sigma\in S_n} m_{1\sigma(1)} m_{2\sigma(2)} \cdots m_{n\sigma(n)}\, \zeta_{[n]} = \operatorname{per}(M)\zeta_{[n]}. \end{aligned} \tag{26}$$

□

An immediate corollary gives the complexity of computing the permanent of an $n \times n$ matrix.

Corollary 10 *Computing the permanent of an arbitrary* $n \times n$ *matrix requires* $O(n)$ $\mathcal{C}\ell$*ops in* $\mathcal{C}\ell_n^{\mathrm{nil}}$.

Example 4 A randomly generated binary matrix is generated, and its permanent is computed in Fig. 4.

The corresponding elements of $\mathcal{C}\ell_{16}^{\mathrm{nil}}$ appear in Fig. 5.

The product $\prod_{i=1}^{16} b_i$ is computed in Fig. 6. Note that the loop performs $15 = n - 1$ multiplications in $\mathcal{C}\ell_{16}^{\mathrm{nil}}$.

Allowing cycles of length two, the permanent of a graph's adjacency matrix counts the number of cycle covers of the graph. This is clear from the definition of permanent (24) when one recalls that every permutation $\sigma \in S_n$ can be written as a product of disjoint cycles.

```
n = 16;
SetSignature[n, 0]
SeedRandom;
A = GenRandGraph[n];
Print["A = ", A // MatrixForm]
Print["Permanent(A) = ", Permanent[A]]
```

$$
A = \begin{pmatrix}
0 & 0 & 1 & 0 & 0 & 1 & 0 & 0 & 0 & 0 & 0 & 0 & 1 & 0 & 0 & 1 \\
0 & 0 & 0 & 0 & 0 & 0 & 0 & 0 & 1 & 0 & 0 & 0 & 0 & 0 & 0 & 0 \\
1 & 0 & 0 & 0 & 1 & 1 & 0 & 0 & 1 & 0 & 0 & 0 & 0 & 1 & 1 & 1 \\
0 & 0 & 0 & 0 & 1 & 0 & 1 & 0 & 0 & 1 & 0 & 0 & 0 & 0 & 0 & 0 \\
0 & 0 & 1 & 1 & 0 & 0 & 0 & 0 & 0 & 0 & 0 & 0 & 0 & 0 & 0 & 0 \\
1 & 0 & 1 & 0 & 0 & 0 & 1 & 0 & 0 & 1 & 0 & 0 & 0 & 0 & 1 & 0 \\
0 & 0 & 0 & 1 & 0 & 1 & 0 & 0 & 0 & 0 & 1 & 0 & 0 & 0 & 1 & 0 \\
0 & 0 & 0 & 0 & 0 & 0 & 0 & 0 & 0 & 0 & 0 & 0 & 0 & 0 & 1 & 0 \\
0 & 1 & 1 & 0 & 0 & 0 & 0 & 0 & 0 & 0 & 0 & 0 & 1 & 0 & 0 & 0 \\
0 & 0 & 0 & 1 & 0 & 1 & 0 & 0 & 0 & 0 & 0 & 0 & 1 & 1 & 0 & 0 \\
0 & 0 & 0 & 0 & 0 & 0 & 1 & 0 & 0 & 0 & 0 & 0 & 0 & 0 & 0 & 0 \\
0 & 0 & 0 & 0 & 0 & 0 & 0 & 0 & 0 & 0 & 0 & 0 & 1 & 0 & 0 & 0 \\
1 & 0 & 0 & 0 & 0 & 0 & 0 & 0 & 1 & 1 & 0 & 1 & 0 & 0 & 1 & 0 \\
0 & 0 & 1 & 0 & 0 & 0 & 0 & 0 & 0 & 1 & 0 & 0 & 0 & 0 & 0 & 0 \\
0 & 0 & 1 & 0 & 0 & 1 & 1 & 1 & 0 & 0 & 0 & 0 & 1 & 0 & 0 & 0 \\
1 & 0 & 1 & 0 & 0 & 0 & 0 & 0 & 0 & 0 & 0 & 0 & 0 & 0 & 0 & 0
\end{pmatrix}
$$

```
Permanent(A) = 9
```

Fig. 4 Random matrix and its permanent

Corollary 11 (Complexity of cycle covers) *Counting the cycle covers of a finite graph on n vertices requires $O(n)$ $\mathcal{C}\ell$ops in $\mathcal{C}\ell_n^{\text{nil}}$.*

Example 5 The graph associated with the matrix appearing in Example 4 is shown in Fig. 7. The number of cycle covers of this graph is 9.

We now introduce a method for computing the girth and counting the proper cycle covers of a graph. The adjacency structure of a graph G on n vertices is represented within $\mathcal{C}\ell_n^{3\text{nil}} \otimes \mathcal{C}\ell_{|E|}^{\text{nil}}$ by

$$\Xi = \sum_{\{v_i,v_j\}\in E(G)} \xi_{\{v_i,v_j\}} \zeta_{\overline{v_i v_j}}. \tag{27}$$

Here, each edge $\{v_i, v_j\}$ of G is associated with a single generator $\zeta_{\overline{v_i v_j}}$ of $\mathcal{C}\ell_{|E|}^{\text{nil}}$, while each vertex is associated with a generator of $\mathcal{C}\ell_n^{3\text{nil}}$.

$b\_1 = e_{\{3\}} + e_{\{6\}} + e_{\{13\}} + e_{\{16\}}$ $b\_2 = e_{\{9\}}$

$b\_3 = e_{\{1\}} + e_{\{5\}} + e_{\{6\}} + e_{\{9\}} + e_{\{14\}} + e_{\{15\}} + e_{\{16\}}$ $b\_4 = e_{\{5\}} + e_{\{7\}} + e_{\{10\}}$

$b\_5 = e_{\{3\}} + e_{\{4\}}$ $b\_6 = e_{\{1\}} + e_{\{3\}} + e_{\{7\}} + e_{\{10\}} + e_{\{15\}}$

$b\_7 = e_{\{4\}} + e_{\{6\}} + e_{\{11\}} + e_{\{15\}}$ $b\_8 = e_{\{15\}}$

$b\_9 = e_{\{2\}} + e_{\{3\}} + e_{\{13\}}$ $b\_10 = e_{\{4\}} + e_{\{6\}} + e_{\{13\}} + e_{\{14\}}$

$b\_11 = e_{\{7\}}$ $b\_12 = e_{\{13\}}$

$b\_13 = e_{\{1\}} + e_{\{9\}} + e_{\{10\}} + e_{\{12\}} + e_{\{15\}}$ $b\_14 = e_{\{3\}} + e_{\{10\}}$

$b\_15 = e_{\{3\}} + e_{\{6\}} + e_{\{7\}} + e_{\{8\}} + e_{\{13\}}$ $b\_16 = e_{\{1\}} + e_{\{3\}}$

Fig. 5 Elements of $\mathcal{C}\ell_{16}^{\text{nil}}$ associated with random binary matrix of Fig. 4

```
p = b1;
For[k = 2, k ≤ n, k++,
  p = p⊖bk // ClNilExpand]
Print["Product of b_ell = ", p]
```

Product of b_ell = $9\, e_{\{1,2,3,4,5,6,7,8,9,10,11,12,13,14,15,16\}}$

Fig. 6 Product $\prod_{i=1}^{1} 6 b_i = \text{per}(A)\zeta_{[16]}$

Fig. 7 Graph associated with matrix A of Example 4

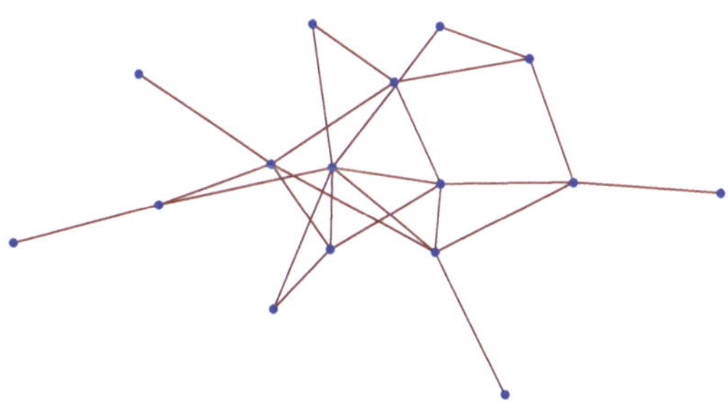

Observe that a path of length m in a connected graph consists of m edges and $m+1$ vertices. Hence, any path is also a tree. However, in order for a tree to be a path, each vertex must be incident with no more than two vertices. Labeling vertices with elements ξ_i satisfying $\xi_i^3 = 0$ allows us to "sieve out" paths.

Recall that the term *proper cycle* refers to any cycle of length greater than two.

It is now possible to obtain an upper bound on the number of Hamiltonian paths in a graph by considering the dimension of the smallest subspace containing $\langle \Xi^n \rangle_{(2(n-1),n)}$. For convenience, the following notation is defined.

Definition 5 Let $\mathcal{A}$ be an algebra, and let $u \in \mathcal{A}$. Then the *dimension* of u is defined as the dimension of the smallest linear subspace S of $\mathcal{A}$ such that $u \in S$. In other words,

$$\dim(u) = \min_{\{S:\mathcal{A}\supseteq S\ni u\}} \dim(S). \tag{28}$$

Proposition 5 *Let G be a graph on n vertices with $\Xi \in \mathcal{C}\ell_n^{3\text{nil}} \otimes \mathcal{C}\ell_{|E|}^{\text{nil}}$ as defined in* (27). *Let X denote the number of coverings of vertices of G by a Hamiltonian path or exactly one path and one or more disjoint proper cycles. Then,*

$$\dim\bigl(\langle \Xi^n \rangle_{(n,n-1)}\bigr) = X. \tag{29}$$

Proof Each nonzero coefficient of the expansion of $\langle \Xi^n \rangle_{(n,n-1)}$ corresponds to a unique subgraph G' of G having n vertices and $n-1$ edges. Moreover, each vertex has maximum degree two. In light of Lemmas 1 and 2, it follows that either G' is a Hamiltonian path or G' consists of exactly one path and one or more disjoint proper cycles as connected components. □

An immediate consequence of Proposition 5 is the following result concerning Hamiltonian paths.

Corollary 12 *Let G be a graph on n vertices with $\Xi \in \mathcal{C}\ell_n^{3\text{nil}} \otimes \mathcal{C}\ell_{|E|}^{\text{nil}}$ as defined in* (27). *Let H denote the number of Hamiltonian paths in G. Then,*

$$\dim\bigl(\langle \Xi^n \rangle_{(n,n-1)}\bigr) = 0 \quad\Longrightarrow\quad H = 0, \tag{30}$$

and

$$H \le \dim\bigl(\langle \Xi^n \rangle_{(n,n-1)}\bigr). \tag{31}$$

Proposition 6 *Let G be a graph on n vertices with $\Xi \in \mathcal{C}\ell_n^{3\text{nil}} \otimes \mathcal{C}\ell_{|E|}^{\text{nil}}$ as defined in* (27). *Let k be a positive integer. Then each term in the expansion of $\langle \Xi^k \rangle_{(k,k)}$ represents a covering of G with proper cycles; that is, each term represents a subgraph G' whose connected components are disjoint cycles of minimum length* 3.

Proof By properties of $\Xi \in \mathcal{C}\ell_n^{3\text{nil}} \otimes \mathcal{C}\ell_{|E|}^{\text{nil}}$, each term of $\langle \Xi^k \rangle$ corresponds to a subgraph G' of G having k vertices and k edges. Moreover, the degree of each vertex is less than or equal to two. By Lemma 2, the connected components of G' are cycles. Since edges are labeled with null-square generators of $\mathcal{C}\ell_{|E|}^{\text{nil}}$, these cycles contain no repeated edges and are therefore proper. □

An immediate consequence of Proposition 6 is the following result concerning Hamiltonian cycles.

Corollary 13 *Let G be a graph on n vertices with $\Xi \in \mathcal{C}\ell_n^{3\text{nil}} \otimes \mathcal{C}\ell_{|E|}^{\text{nil}}$ as defined in* (27). *Let Z_H denote the number of Hamiltonian cycles in G. Then,*

$$\dim\left(\left\langle \Xi^n \right\rangle_{(n,n)}\right) = 0 \quad \Longrightarrow \quad Z_H = 0, \tag{32}$$

and

$$Z_H \leq \dim\left(\left\langle \Xi^n \right\rangle_{(n,n)}\right). \tag{33}$$

Example 6 For each $k = 3, 4, \ldots, 8$, the grade (k,k) part of Ξ^k is computed using Mathematica for a randomly generated graph on eight vertices. Each nonzero term of $\langle \Xi^k \rangle_{(k,k)}$ corresponds to a covering of a k-vertex subgraph using disjoint cycles of length $j \leq k$. The fact that $\langle \Xi^8 \rangle_{(8,8)} = 0$ implies that the graph contains no Hamiltonian cycles.

In the Mathematica implementation of products of null-cubes, the multiindex is defined by a vector in $\mathbb{Z}_3^8$ according to

$$\xi_{\mathbf{v}} = \prod_{j=1}^{8} \xi_j^{v_j}. \tag{34}$$

```
(* Overload Times operator to handle null-squares ζ_{j} and null-cubes ξ_{j} *)
Unprotect[Times];
ζ = Symbol["ζ"];
ζ_{a_} ζ_{b_} := If[Length[a ⋂ b] > 0, 0, ζ_{a⋃b}];
ξ = Symbol["ξ"]
ξ_{v_} ξ_{w_} := If[Max[v + w] < 3, ξ_{v+w}, 0];
Protect[Times];
Unprotect[Power];
(x_ /; ! FreeQ[x, ζ_{a_}])^(n_Integer) := Module[{y, f}, y = Expand[x]; Switch[EvenQ[n],
   True, If[n == 0, Return[1], Composition[Expand] [Distribute[f[y, y]] /.f → Times]^(n/2)],
   False, If[n == 1, Return[x], Composition[Expand] [Distribute[f[y, y]] /.f → Times]^((n-1)/2) x]]];
(x_ /; ! FreeQ[x, ξ_{a_}])^(n_Integer) := Module[{y, f}, y = Expand[x]; Switch[EvenQ[n],
   True, If[n == 0,Return[1],Composition[Expand][Distribute[f[y, y]]/.f → Times]^(n/2)],
   False, If[n == 1,Return[x],Composition[Expand][Distribute[f[y, y]]/.f → Times]^((n-1)/2) x]]];
Protect[Power]; Unprotect[Expand];
Expand[x_/;! FreeQ[x,ζ_]]:=DeleteCases[Distribute[x, Plus, Times], 0.` ζ_];
Expand[x_/;! FreeQ[x,ξ_]]:=DeleteCases[Distribute[x, Plus, Times], 0.` ξ_];Protect[Expand];
```

```
n = 8;
Z = DiagonalMatrix[Table[ξ_{e_i}, {i, 1, n}]];
SeedRandom;
A = GenRandGraph[n];

GraphPlot[A]
B = A.Z;
ψ = Sum[ξ_{e_j} B[[j]][[k]], {j, 1, n}, {k, 1, j}];
```

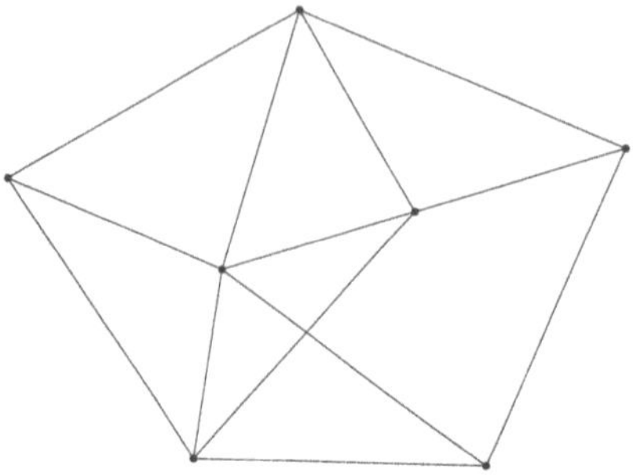

•

```
Grade[b_ ξ_a_] := MatrixRank[DiagonalMatrix[a]];
Grade[ξ_a_] := MatrixRank[DiagonalMatrix[a]];
Grade[ζ_a_] := Length[a];
Grade[b_ ζ_a_] := Length[a];

GradeJK[u_] := {Grade[u /. {ζ_ → 1}], Grade[u /. {ξ_ → 1}]}

In[208]:= GradeJKPart[u_, {j_, k_}] :=
  Sum[If[GradeJK[u[[ell]]]=={j,k},u[[ell]],0],{ell,1,Length[u]}]

In[326]:= Ξ = Sum[ζ_{k} ψ[[k]], {k, 1, Length[ψ]}]

Out[326]= ζ_{1}ξ_{0,0,0,0,0,0,1,1}+ζ_{2}ξ_{0,0,0,0,1,0,0,1}+ζ_{3}ξ_{0,0,0,0,1,0,1,0}+
  ζ_{4}ξ_{0,0,0,0,1,1,0,0}+ζ_{5}ξ_{0,0,0,1,0,0,1,0} +ζ_{6}ξ_{0,0,0,1,0,1,0,0}+
  ζ_{7}ξ_{0,0,0,1,1,0,0,0}+ζ_{8}ξ_{0,0,1,0,0,0,0,1}+ζ_{9} ξ_{0,0,1,0,0,1,0,0}+
  ζ_{10}ξ_{1,0,0,0,0,0,0,1}+ζ_{11}ξ_{1,0,0,0,1,0,0,0}+ζ_{12}ξ_{1,0,0,1,0,0,0,0}+
  ζ_{13}ξ_{1,0,1,0,0,0,0,0}

In[335]:= For[k = 3, k ≤ n, k++,
  Print["k = ", k, ". Grade (k,k) part of Ξ^k : ",
  GradeJKPart[Expand[Ξ^k],{k, k}]]]

k = 3.  Grade (k,k) part of Ξ^k : 6 ζ_{1,2,3} ξ_{0,0,0,0,2,0,2,2} +
  6 ζ_{3,5,7} ξ_{0,0,0,2,2,0,2,0} + 6 ζ_{4,6,7} ξ_{0,0,0,2,2,2,0,0} +
  6 ζ_{2,10,11} ξ_{2,0,0,0,2,0,0,2} + 6 ζ_{7,11,12} ξ_{2,0,0,2,2,0,0,0} +
  6 ζ_{8,10,13} ξ_{2,0,2,0,0,0,0,2}

k = 4.  Grade (k,k) part of Ξ^k : 24 ζ_{1,2,5,7} ξ_{0,0,0,2,2,0,2,2} +
  24 ζ_{3,4,5,6} ξ_{0,0,0,2,2,2,2,0}+ 24 ζ_{2,4,8,9} ξ_{0,0,2,0,2,2,0,2} +
  24 ζ_{1,3,10,11} ξ_{2,0,0,0,2,0,2,2} + 24 ζ_{1,5,10,12} ξ_{2,0,0,2,0,0,2,2} +
  24 ζ_{2,7,10,12} ξ_{2,0,0,2,2,0,0,2} + 24 ζ_{3,5,11,12} ξ_{2,0,0,2,2,0,2,0} +
  24 ζ_{4,6,11,12} ξ_{2,0,0,2,2,2,0,0} + 24 ζ_{2,8,11,13} ξ_{2,0,2,0,2,0,0,2} +
  24 ζ_{4,9,11,13} ξ_{2,0,2,0,2,2,0,0} + 24 ζ_{6,9,12,13} ξ_{2,0,2,2,0,2,0,0}
```

k = 5. Grade (k,k) part of Ξ^k :

$120\,\zeta_{\{1,2,4,5,6\}}\,\xi_{\{0,0,0,2,2,2,2,2\}} + 120\,\zeta_{\{1,3,4,8,9\}}\,\xi_{\{0,0,2,0,2,2,2,2\}} +$
$120\,\zeta_{\{1,5,6,8,9\}}\,\xi_{\{0,0,2,2,0,2,2,2\}} + 120\,\zeta_{\{2,6,7,8,9\}}\,\xi_{\{0,0,2,2,2,2,0,2\}} +$
$120\,\zeta_{\{1,2,5,11,12\}}\,\xi_{\{2,0,0,2,2,0,2,2\}} + 120\,\zeta_{\{1,3,7,10,12\}}\,\xi_{\{2,0,0,2,2,0,2,2\}} +$
$120\,\zeta_{\{1,5,7,10,11\}}\,\xi_{\{2,0,0,2,2,0,2,2\}} + 120\,\zeta_{\{2,3,5,10,12\}}\,\xi_{\{2,0,0,2,2,0,2,2\}} +$
$120\,\zeta_{\{2,4,6,10,12\}}\,\xi_{\{2,0,0,2,2,2,0,2\}} + 120\,\zeta_{\{1,3,8,11,13\}}\,\xi_{\{2,0,2,0,2,0,2,2\}} +$
$120\,\zeta_{\{2,4,9,10,13\}}\,\xi_{\{2,0,2,0,2,2,0,2\}} + 120\,\zeta_{\{4,8,9,10,11\}}\,\xi_{\{2,0,2,0,2,2,0,2\}} +$
$120\,\zeta_{\{1,5,8,12,13\}}\,\xi_{\{2,0,2,2,0,0,2,2\}} + 120\,\zeta_{\{6,8,9,10,12\}}\,\xi_{\{2,0,2,2,0,2,0,2\}} +$
$120\,\zeta_{\{2,7,8,12,13\}}\,\xi_{\{2,0,2,2,2,0,0,2\}} + 120\,\zeta_{\{4,7,9,12,13\}}\,\xi_{\{2,0,2,2,2,2,0,0\}} +$
$120\,\zeta_{\{6,7,9,11,13\}}\,\xi_{\{2,0,2,2,2,2,0,0\}}$

k = 6. Grade (k,k) part of Ξ^k :

$720\,\zeta_{\{1,3,6,7,8,9\}}\,\xi_{\{0,0,2,2,2,2,2,2\}} + 720\,\zeta_{\{1,4,5,7,8,9\}}\,\xi_{\{0,0,2,2,2,2,2,2\}} +$
$720\,\zeta_{\{2,3,5,6,8,9\}}\,\xi_{\{0,0,2,2,2,2,2,2\}} + 720\,\zeta_{\{1,3,4,6,10,12\}}\,\xi_{\{2,0,0,2,2,2,2,2\}} +$
$720\,\zeta_{\{1,4,5,6,10,11\}}\,\xi_{\{2,0,0,2,2,2,2,2\}} + 720\,\zeta_{\{1,3,4,9,10,13\}}\,\xi_{\{2,0,2,0,2,2,2,2\}} +$
$720\,\zeta_{\{1,5,6,9,10,13\}}\,\xi_{\{2,0,2,2,0,2,2,2\}} + 720\,\zeta_{\{1,3,7,8,12,13\}}\,\xi_{\{2,0,2,2,2,0,2,2\}} +$
$720\,\zeta_{\{1,5,7,8,11,13\}}\,\xi_{\{2,0,2,2,2,0,2,2\}} + 720\,\zeta_{\{2,3,5,8,12,13\}}\,\xi_{\{2,0,2,2,2,0,2,2\}} +$
$720\,\zeta_{\{3,5,7,8,10,13\}}\,\xi_{\{2,0,2,2,2,0,2,2\}} + 720\,\zeta_{\{2,4,6,8,12,13\}}\,\xi_{\{2,0,2,2,2,2,0,2\}} +$
$720\,\zeta_{\{2,6,7,9,10,13\}}\,\xi_{\{2,0,2,2,2,2,0,2\}} + 720\,\zeta_{\{2,6,8,9,11,12\}}\,\xi_{\{2,0,2,2,2,2,0,2\}} +$
$720\,\zeta_{\{4,6,7,8,10,13\}}\,\xi_{\{2,0,2,2,2,2,0,2\}} + 720\,\zeta_{\{4,7,8,9,10,12\}}\,\xi_{\{2,0,2,2,2,2,0,2\}} +$
$720\,\zeta_{\{6,7,8,9,10,11\}}\,\xi_{\{2,0,2,2,2,2,0,2\}} + 720\,\zeta_{\{3,4,5,9,12,13\}}\,\xi_{\{2,0,2,2,2,2,2,0\}} +$
$720\,\zeta_{\{3,5,6,9,11,13\}}\,\xi_{\{2,0,2,2,2,2,2,0\}}$

k = 7. Grade (k,k) part of Ξ^k :

$5040\,\zeta_{\{1,2,3,6,9,12,13\}}\,\xi_{\{2,0,2,2,2,2,2,2\}} +$
$5040\,\zeta_{\{1,2,4,5,9,12,13\}}\,\xi_{\{2,0,2,2,2,2,2,2\}} +$
$5040\,\zeta_{\{1,2,5,6,9,11,13\}}\,\xi_{\{2,0,2,2,2,2,2,2\}} +$
$5040\,\zeta_{\{1,3,4,6,8,12,13\}}\,\xi_{\{2,0,2,2,2,2,2,2\}} +$
$5040\,\zeta_{\{1,3,6,7,9,10,13\}}\,\xi_{\{2,0,2,2,2,2,2,2\}} +$
$5040\,\zeta_{\{1,3,6,8,9,11,12\}}\,\xi_{\{2,0,2,2,2,2,2,2\}} +$
$5040\,\zeta_{\{1,4,5,6,8,11,13\}}\,\xi_{\{2,0,2,2,2,2,2,2\}} +$
$5040\,\zeta_{\{1,4,5,7,9,10,13\}}\,\xi_{\{2,0,2,2,2,2,2,2\}} +$
$5040\,\zeta_{\{1,4,5,8,9,11,12\}}\,\xi_{\{2,0,2,2,2,2,2,2\}} +$
$5040\,\zeta_{\{2,3,5,6,9,10,13\}}\,\xi_{\{2,0,2,2,2,2,2,2\}} +$
$5040\,\zeta_{\{3,4,5,6,8,10,13\}}\,\xi_{\{2,0,2,2,2,2,2,2\}} +$
$5040\,\zeta_{\{3,4,5,8,9,10,12\}}\,\xi_{\{2,0,2,2,2,2,2,2\}} +$
$5040\,\zeta_{\{3,5,6,8,9,10,11\}}\,\xi_{\{2,0,2,2,2,2,2,2\}}$

k = 8. Grade (k,k) part of Ξ^k : 0

Definition 6 Given $\psi \in \mathcal{C}\ell_n^{3\text{nil}} \otimes \mathcal{C}\ell_{|E|}^{\text{nil}}$, the *grade-balanced exponential* of ψ is defined by

$$\exp_{\text{gb}}(u) = \sum_{k=0}^{\infty} \left\langle \frac{\psi^k}{k!} \right\rangle_{(k,k)}. \tag{35}$$

Proposition 7 *Let G be a graph on n vertices with $\Xi \in \mathcal{C}\ell_n^{3\text{nil}} \otimes \mathcal{C}\ell_{|E|}^{\text{nil}}$ as defined in* (27). *Let* Girth(G) *denote the length of the smallest nontrivial cycle in G. Then,* *as a polynomial in* $\mathcal{C}\ell_n^{3\text{nil}} \otimes \mathcal{C}\ell_{|E|}^{\text{nil}} \otimes \mathbb{R}[t]$,

$$\deg_t\left(t^n\left(\exp_{\text{gb}}\left(\frac{\Xi}{t}\right) - 1\right)\right) = n - \text{Girth}(G). \tag{36}$$

Proof Note first that $t^n \exp_{\text{gb}}(\frac{\Xi}{t})$ is a polynomial in t of degree at most n. The degree of the polynomial is equal to the smallest exponent k appearing among nonzero terms in the expansion

$$\exp_{\text{gb}}\left(\frac{\Xi}{t}\right) - 1 = \sum_{k=1}^{n}\left\langle\frac{\Xi^k}{k!t^k}\right\rangle_{(k,k)}. \tag{37}$$

By Proposition 6, nonzero terms in the expansion of $\langle\Xi^k\rangle_{(k,k)}$ represent disjoint cycle covers of k-vertex subgraphs in G. The smallest positive integer k_0 for which such a cover exists must correspond to k_0-cycles in G, i.e., $k_0 = \text{Girth}(G)$. Then $\frac{t^n}{t^{k_0}} = t^{n-\text{Girth}(G)}$, from which the result is obtained. □

Corollary 14 *Computing the girth of a graph with n vertices and* $|E|$ *edges requires* $O(n \log n)$ $\mathcal{C}\ell$*ops in* $\mathcal{C}\ell_n^{3\text{nil}} \otimes \mathcal{C}\ell_{|E|}^{\text{nil}}$.

Proof Computing $\exp_{\text{gb}}(\frac{\Xi}{t})$ requires computing $\langle\frac{\Xi^k}{k!t^k}\rangle_{(k,k)}$ for $0 \le k \le n$. □

Proposition 8 *Let G be a graph on n vertices with* $\Xi \in \mathcal{C}\ell_n^{3\text{nil}} \otimes \mathcal{C}\ell_{|E|}^{\text{nil}}$ *as defined in* (27). *Let Z denote the number of proper cycle covers of G. Then,*

$$\dim\left(\left\langle\Xi^n\right\rangle_{(n,n)}\right) = Z. \tag{38}$$

Proof Note that Ξ^n represents n-edge subsets taken from the graph G. Recalling that any collection of n edges incident with n vertices in a connected graph is a cycle if and only of every vertex has degree 2. By construction of the three-nil algebra, the terms of Ξ^n represent subgraphs in which the maximum vertex degree is two and the minimum vertex degree is one.

All that remains is to show no vertex of this subgraph can have degree one. By the Handshaking Lemma (Lemma 3), since the subgraph G' contains n vertices and n edges, we have

$$\sum_{v\in V_{G'}} \deg(v) = 2n. \tag{39}$$

Since the maximum vertex degree is two, this sum can be written in the form

$$\begin{aligned}\sum_{v\in V_{G'}} \deg(v) &= \sum_{\substack{v\in V_{G'}\\ \deg(v)=1}} \deg(v) + \sum_{\substack{v\in V_{G'}\\ \deg(v)=2}} \deg(v)\\ &= \left|\left\{v \in V_{G'} : \deg(v) = 1\right\}\right| + 2\left|\left\{v \in V_{G'} : \deg(v) = 2\right\}\right|\end{aligned}$$

$$= \left(n - \left|\left\{v \in V_{G'} : \deg(v) = 2\right\}\right|\right) + 2\left|\left\{v \in V_{G'} : \deg(v) = 2\right\}\right|$$
$$= n + \left|\left\{v \in V_{G'} : \deg(v) = 2\right\}\right| = 2n. \qquad (40)$$

Hence, $|\{v \in V_{G'} : \deg(v) = 2\}| = n$. It follows that G' is a graph whose connected components are nontrivial cycles. □

Corollary 15 *Computing the number of proper cycle covers of a graph on n vertices and* $|E|$ *edges requires* $O(\log n)$ $\mathcal{C}\ell$*ops in* $\mathcal{C}\ell_n^{\text{3nil}} \otimes \mathcal{C}\ell_{|E|}^{\text{nil}}$.

5 Conclusion

The advantages of a computer architecture capable of dealing naturally with geometric objects are numerous. If one assumes the existence of such a machine, a natural measure of algorithmic complexity is the number of Clifford operations ($\mathcal{C}\ell$ops) required by the algorithm.

In terms of numbers of $\mathcal{C}\ell$ops required, a number of problems of complexity class NP are of polynomial complexity. We assert that a Clifford computer would have natural advantages for solving an assortment of combinatorial and graph-theoretic problems.

Using currently available technology and techniques, the computations performed are not truly geometric operations. Multiplying two multivectors still requires keeping track of nonzero coefficients of blades of all grades. In fact, handling homogeneous grade-k multivectors requires keeping track of up to $\binom{n}{k}$ coefficients

n	Time
1	0.001696
2	0.002168
3	0.004618
4	0.011810
5	0.038795
6	0.130727
7	0.484850
8	1.856420
9	7.242576
10	28.576006

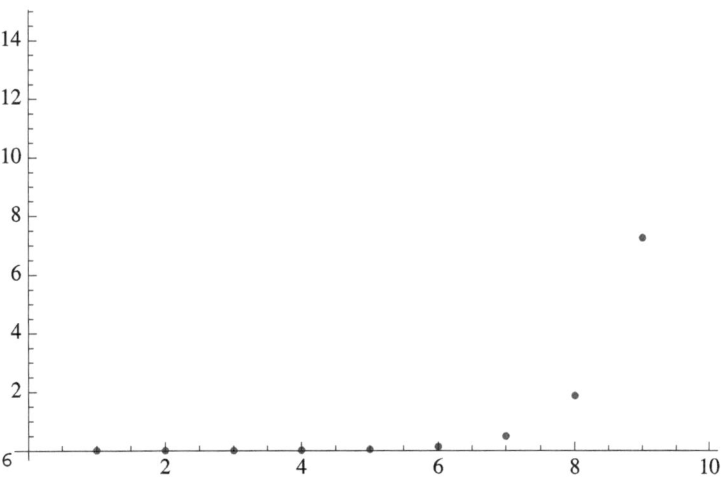

Fig. 8 Time (in secs) required for product of random multivectors in $\mathcal{C}\ell_n^{\text{nil}}$

using the methods employed here. It is worth noting however that this might be dramatically improved by using a multiplicative representation of k-blades such as that found in the work of Fontijne [6].

Examples were computed on a 2.4 GHz MacBook Pro with 4 GB of 667 MHz DDR2 SDRAM running Mathematica 6 for MAC OS X with the packages *Cliffmath08* and *CliffSymNil08* available online at www.siue.edu~sstaple. Multivector products in $\mathcal{C}\ell_n^{\text{nil}}$ are computed combinatorially. The times of computing products of arbitrary randomly generated multivectors in $\mathcal{C}\ell_n^{\text{nil}}$ appear in Fig. 8.

The true power of a geometric computing architecture is yet to be realized. In an ideal architecture, multiplying two multivectors would require one operation and an appropriate measurement (projection) to recover the relevant information. However, an architecture in which the multivector product is of polynomial complexity would be sufficient for dramatic reductions in complexity.

Acknowledgement The authors thank the referees for a number of helpful comments.

References

1. Abłamowicz, R., Fauser, B.: CLIFFORD—a Maple package for Clifford algebra computations. http://math.tntech.edu/rafal/
2. Aerts, D., Czachor, M.: Cartoon computation: quantum-like computing without quantum mechanics. J. Phys. A: Math. Theor. **40**, F259–F263 (2007)
3. Ben-Dor, A., Halevi, S.: Zero-one permanent is ♯P-complete, a simpler proof. In: Proceedings of the 2nd Israel Symposium on the Theory and Computing Systems, pp. 108–117 (1993)
4. Chien, S., Rasmussen, L., Sinclair, A.: Clifford algebras and approximating the permanent. J. Comp. Syst. Sci. **67**, 263–290 (2003)
5. Coppersmith, D., Winograd, S.: Matrix multiplication via arithmetic progressions. J. Symb. Comput. **9**, 251–280 (1990)
6. Fontijne, D.: Efficient implementation of geometric algebra. Ph.D. thesis, University of Amsterdam (2007)
7. Fontijne, D., Bouma, T., Dorst, L.: GAIGEN: a geometric algebra implementation generator, University of Amsterdam, NL, July 2002. http://www.science.uva.nl/ga/gaigen
8. Franchini, S., Gentile, A., Grimaudo, M., Hung, C.A., Impastato, S., Sorbello, F., Vassallo, G., Vitabile, S.: A sliced coprocessor for native Clifford algebra operations. In: Proceedings of the 10th Euromicro Conference on Digital System Design Architectures, Methods and Tools (DSD 2007), pp. 436–439 (2007)
9. Gentile, A., Segreto, S., Sorbello, F., Vassallo, G., Vitabile, S., Vuollo, V.: CliffoSor, an innovative FPGA-based architecture for geometric algebra. In: Proceedings of the International Conference on Engineering of Reconfigurable Systems and Algorithms (ERSA 2005), pp. 211–217
10. Gentile, A., Segreto, S., Sorbello, F., Vassallo, G., Vitabile, S., Vuollo, V.: CliffoSor: a parallel embedded architecture for geometric algebra and computer graphics. In: Proceedings of the IEEE International Workshop on Computer Architecture for Machine Perception (CAMP 2005), pp. 90–95. IEEE Comput. Soc., Los Alamitos (2005)
11. Leopardi, P.: The GluCat home page, http://glucat.sourceforge.net/
12. Mishra, B., Wilson, P.: Color edge detection hardware based on geometric algebra. http://eprints.ecs.soton.ac.uk/13188/
13. Mishra, B., Wilson, P.: Hardware implementation of a geometric algebra processor core. In: Proceedings of ACA 2005, IMACS, Int. Conf. on Advancement of Computer Algebra, Nara, Japan (2005). http://eprints.ecs.soton.ac.uk/10957/

14. Perwass, C.: The CLU Project web page, http://www.perwass.de/cbup/clu.html
15. Perwass, C., Gebken, C., Sommer, G.: Implementation of a Clifford algebra co-processor design on a field programmable gate array. In: Clifford Algebras Applications to Mathematics, Physics, and Engineering. Progress in Mathematical Physics, vol. 34. Birkhäuser, Basel (2004)
16. Porteous, I.: Lecture 2: Mathematical structure of Clifford algebras. In: Abłamowicz, R., Sobczyk, G. (eds.) Lectures on Clifford (Geometric) Algebras and Applications. Birkhäuser, Basel (2003)
17. Schott, R., Staples, G.S.: Nilpotent adjacency matrices and random graphs. Ars Comb. (2010, to appear)
18. Schott, R., Staples, G.S.: Nilpotent adjacency matrices, random graphs, and quantum random variables. J. Phys. A: Math. Theor. **41**, 155205 (2008)
19. Schott, R., Staples, G.S.: Reductions in computational complexity using Clifford algebras. Adv. Appl. Clifford Algebr. (2010, to appear)
20. Staples, G.S.: Clifford-algebraic random walks on the hypercube. Adv. Appl. Clifford Algebr. **15**, 213–232 (2005)
21. Staples, G.S.: Graph-theoretic approach to stochastic integrals with Clifford algebras. J. Theor. Probab. **20**, 257–274 (2007)
22. Staples, G.S.: Norms and generating functions in Clifford algebras. Adv. Appl. Clifford Algebr. **18**, 75–92 (2008)
23. Valiant, L.: The complexity of computing the permanent. Theor. Comput. Sci. **8**, 189–201 (1979)
24. West, D.: Introduction to Graph Theory, 2nd edn. Prentice Hall, New York (2001)

Part VII
Efficient Computing with Clifford (Geometric) Algebra

Efficient Algorithms for Factorization and Join of Blades

Daniel Fontijne and Leo Dorst

Abstract Subspaces are powerful tools for modeling geometry. In geometric algebra, they are represented using blades and constructed using the outer product. Producing the actual geometrical intersection (meet) and union (join) of subspaces, rather than the simplified linearizations often used in Grassmann–Cayley algebra, requires efficient algorithms when blades are represented as a sum of basis blades. We present an efficient blade factorization algorithm and use it to produce implementations of the join that are approximately 10 times faster than earlier algorithms.

1 Introduction

Blades, defined as outer products of vectors, are used in geometric algebra to represent geometric objects such as directions, points, planes, and circles. In implementations of geometric algebra [9, 11, 12], blades are commonly represented as weighted sums of orthogonal basis blades, which we call the *additive representation* [10]. A basis blade is the outer product of basis vectors. 2^n basis blades are required to form a basis for an n-dimensional geometric algebra (2^n is the size of the set of all combinations of up to n basis vectors). This basis of blades allows for straightforward and efficient implementation of many bilinear products such as the outer product, at least for $n < 10$.

However, the additive representation makes the implementation of the true subspace union (join) and intersection (meet) substantially more involved and expensive than bilinear products, because they are nonlinear products [6]. To implement the join of blades in additive representation, one of the input blades must be (partially) factored in terms of the outer product.

Unfortunately, the terms meet and join have historically been used to denote simplified linearizations of the true subspace intersection and union. For example, [8]

D. Fontijne (✉)
University of Amsterdam, Kruislaan 403, 1098 SJ, Amsterdam, The Netherlands
e-mail: fontijne@science.uva.nl

E. Bayro-Corrochano, G. Scheuermann (eds.), *Geometric Algebra Computing*, DOI 10.1007/978-1-84996-108-0_21,

defines the join as the outer (wedge) product (i.e., $\mathsf{join}(\mathbf{A},\mathbf{B}) = \mathbf{A}\wedge\mathbf{B}$) and the meet as $\mathsf{meet}(\mathbf{A},\mathbf{B}) = (\mathbf{A}^*\wedge\mathbf{B}^*)^*$, where the dual * is with respect to the whole vector space. As a result, software packages based on these definitions (such as `Bigebra/CLIFFORD` [1]) require the dimension of the vector space to be set (`dim_V := ...`) before computing the meet. For example, in `CLIFFORD`:

```
> dim_V:=3: M3:=meet(e1we2,e2we3);
          M3 := -e2
> dim_V:=4: M4:=meet(e1we2,e2we3);
          M4 := 0
```

This shows that in `CLIFFORD` the meaning of intersection (meet) depends on the dimension of the full space, which we find rather strange. So once again, we would like to stress that this paper is about computing the join (and indirectly the meet) of blades in any geometric (possibly degenerate) situation. That is, the join is the true geometric subspace union, e.g., $\mathsf{join}(\mathbf{e}_1,\mathbf{e}_1) = \mathbf{e}_1$ and not $\mathsf{join}(\mathbf{e}_1,\mathbf{e}_1) = 0$.

In this paper, we present a new blade factorization algorithm for blades represented as a sum of based blades. We also use the new factorization algorithm to construct a new algorithm to compute the join. Both new algorithms are improvements of previous work [2, 3, 6]. Our main contributions are the FastFactorization algorithm, which factors blades by simply rearranging coordinates, and the StableFastJoin algorithm, which efficiently computes the join in a numerically stable way.

Having a fast and stable join implementation is especially important because it encourages people to make proper use of geometric algebra. One of the great advantages of geometric algebra is that many equations are universal and that the join allows for universal subspace union. However, if it is perceived as slow in practice, people will rewrite their equations to avoid the join and replace it with linear inner and outer products. This will often break the universality of their equations, splitting one join into many different cases (e.g., if the variable is a line, then do this; if the variable is a plane, then do that; and so on).

We only briefly consider computing the meet (true subspace intersection) in Sect. 5.3. Experimentation showed that adjusting our FastJoin algorithm to compute the meet directly is somewhat slower than computing it from the join using [6]

$$\mathsf{meet}(\mathbf{A},\mathbf{B}) = \big(\mathbf{B}\rfloor\mathsf{join}(\mathbf{A},\mathbf{B})^{-1}\big)\rfloor\mathbf{A} \tag{1}$$

and also leads to more generated code. This is why the topic is ignored in the main text.

We assume a Euclidean metric in all computations, as both factorization and join are metric-independent operations (i.e., we use the LIFT described in [4]). Throughout the paper, n is the dimension of the vector space V^n, and k is used to denote the grade of the blade in the current context.

2 Blade Factorization

The problem of factorizing a blade is the following. Given a k-blade $\mathbf{B}$ (in the additive representation), find k vectors $\mathbf{b}_i$ such that

$$\mathbf{B} = \mathbf{b}_1 \wedge \mathbf{b}_2 \wedge \cdots \wedge \mathbf{b}_k,$$

where for reasons of numerical stability, we prefer the factors $\mathbf{b}_i$ to be "sufficiently nonparallel."

Not all k-vectors (homogeneous multivectors of grade k) are blades, i.e., not all k-vectors can be factored (for example, $\mathbf{e}_1 \wedge \mathbf{e}_2 + \mathbf{e}_3 \wedge \mathbf{e}_4$). Such nonfactorizable k-vectors are invalid input to our factorization algorithm. However, the algorithm will not attempt to detect such input, since feeding it a nonblade does not have catastrophic consequences. If required, a separate bladedness test as described in [6] can be employed to make sure that the input to the algorithm is a blade.

To find factors, one may project "probing vectors" $\mathbf{p}_i$ onto the blade. In geometric algebra, this is done by the projection operator [5, 6]

$$\mathbf{q}_i = \left(\mathbf{p}_i \rfloor \mathbf{B}^{-1}\right) \rfloor \mathbf{B}. \tag{2}$$

If $\mathbf{q}_i$ is not zero, it is a factor of $\mathbf{B}$. By finding k linearly independent vectors $\mathbf{q}_i$ a factorization of $\mathbf{B}$ has been found (up to scale). A straightforward choice for the probing vectors $\mathbf{p}_i$ are basis vectors $\mathbf{e}_i$. To obtain the factorization, one selects a total of k independent projected vectors $\mathbf{q}_i$. This procedure is the essence of the outer factorization algorithm presented in detail in [6], based on ideas in [3]. It works, but the projection is computationally rather expensive.

2.1 New Algorithm for Blade Factorization

Figure 1 inspired our new factorization algorithm. The figure illustrates the steps involved in the projection of a vector onto a 2-blade using geometric algebra by means of the two contractions in (2). The first inner product $(\mathbf{p}_i \rfloor \mathbf{B}^{-1})$ computes the orthogonal complement in $\mathbf{B}$ of the projected vector onto $\mathbf{B}$ (with a proportionality factor of $1/\|\mathbf{B}\|^2$). The second inner product in $(\mathbf{p}_i \rfloor \mathbf{B}^{-1}) \rfloor \mathbf{B}$ computes the orthogonal complement of the result within $\mathbf{B}$, and this rotates the previous result in the $\mathbf{B}$-plane, to become the actual projection.

If we are only looking for factors, we are more interested in the fact that the final result $\mathbf{q}_i$ is in $\mathbf{B}$ than in it being the precise projection of the original probing vector $\mathbf{p}_i$. However. this "being in $\mathbf{B}$" is already guaranteed by the second contraction in the projection equation. Hence, in our fast outer factorization algorithm, we save on the first contraction by using the orthogonal complement of the probing vector $\mathbf{p}_i$ with respect to the largest basis blade in $\mathbf{B}$. The whole operation then becomes merely the selection of appropriate coordinates. We turn this idea into an algorithm as follows:

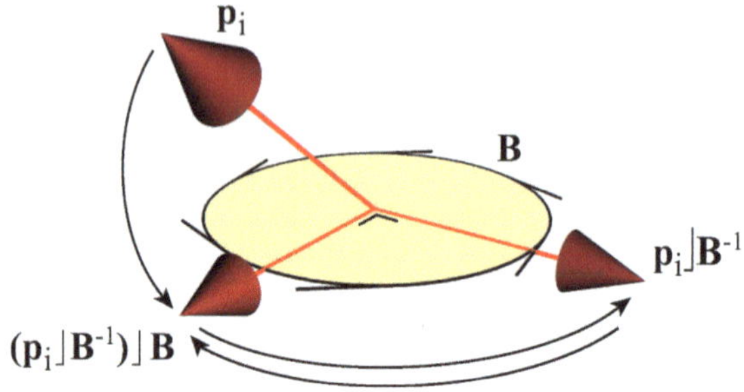

Fig. 1 Orthogonal projection of a vector onto a 2-blade, displayed as the undoing of the orthogonal-complement-step of the inner product

Algorithm FastFactorization($\mathbf{B}$):

Let $\mathbf{B}$ be a k-blade with $1 < k < n$ (to exclude trivial cases). The algorithm computes a factorization $\mathbf{B} = \beta\, \mathbf{b}_1 \wedge \mathbf{b}_2 \wedge \cdots \wedge \mathbf{b}_k$, where β is a scalar:

1. Find the basis blade $\mathbf{F}$ to which the absolute largest coordinate of $\mathbf{B}$ refers. Let $\mathbf{f}_i$ be the basis vectors in $\mathbf{F}$ (i.e., $\mathbf{F} = \mathbf{f}_1 \wedge \mathbf{f}_2 \wedge \cdots \wedge \mathbf{f}_k$), and let β be the coordinate that refers to $\mathbf{F}$.
2. Compute $\mathbf{B}_s = \mathbf{B}/\beta$.
3. For each $\mathbf{f}_i$, compute: $\mathbf{b}_i = (\mathbf{f}_i \rfloor \mathbf{F}^{-1}) \rfloor \mathbf{B}_s$.

The independency of the vectors $\mathbf{b}_i$ guarantees that they form a factorization of $\mathbf{B}_s$.

Theorem *The factors* $\mathbf{b}_i$ *as computed by the* FastFactorization *algorithm are linearly independent.*

Proof $\mathbf{B}_s$ is represented as a sum of grade k basis blades $\mathbf{E}_j$: $\mathbf{B}_s = \sum_{j=1}^{\binom{n}{k}} \frac{\beta_j}{\beta} \mathbf{E}_j$. By distributivity, each $\mathbf{b}_i = \sum_{j=1}^{\binom{n}{k}} \frac{\beta_j}{\beta} (\mathbf{f}_i \rfloor \mathbf{F}^{-1}) \rfloor \mathbf{E}_j$. Let us analyze the contribution of each $\mathbf{E}_j$ to each $\mathbf{b}_i$:

- If $\mathbf{E}_j$ is equal to $\mathbf{F}$, then $(\mathbf{f}_i \rfloor \mathbf{F}^{-1}) \rfloor \mathbf{E}_j = \mathbf{f}_i$.
- Else if $\mathbf{E}_j$ is orthogonal to $\mathbf{F}$, we find $(\mathbf{f}_i \rfloor \mathbf{F}^{-1}) \rfloor \mathbf{E}_j = 0$.
- Else $\mathbf{E}_j = \sigma_j (\mathbf{f}_i \rfloor \mathbf{F}^{-1}) \wedge \mathbf{e}_j$ for some basis vector $\mathbf{e}_j$, where the sign $\sigma_j = \pm 1$ depends on the order of basis vectors in $\mathbf{E}_j$. This $\mathbf{e}_j$ cannot be equal to $\mathbf{f}_i$, for then $\mathbf{E}_j$ is equal to $\mathbf{F}$, and also $\mathbf{e}_j$ cannot be any of the other basis vectors in $\mathbf{F}$, for then $\mathbf{E}_j$ would be 0 as it would contain the same basis vector twice. Thus, in this case,

$$(\mathbf{f}_i \rfloor \mathbf{F}^{-1}) \rfloor \mathbf{E}_j = (\mathbf{f}_i \rfloor \mathbf{F}^{-1}) \rfloor \big(\sigma_j (\mathbf{f}_i \rfloor \mathbf{F}^{-1}) \wedge \mathbf{e}_j\big) = (\mathbf{f}_i \rfloor \mathbf{F}^{-1})^2 \sigma_j\, \mathbf{e}_j = \pm \sigma_j \mathbf{e}_j,$$

  where $\mathbf{e}_j$ is not a factor of $\mathbf{F}$.

Thus each $\mathbf{b}_i$ equals

$$\mathbf{b}_i = \mathbf{f}_i + \sum_{\mathbf{e}_j \wedge \mathbf{F} \neq 0} (\pm \sigma_j) \frac{\beta_j}{\beta} \mathbf{e}_j,$$

so that it does not contain any other factor of $\mathbf{F}$ than $\mathbf{f}_i$ itself. Since the $\mathbf{f}_i$ are linearly independent, so are the $\mathbf{b}_i$. □

To illustrate the proof, let us compute the factors of the example $\mathbf{B}$ given above:

$$\mathbf{B} = -0.8\,\mathbf{e}_1 \wedge \mathbf{e}_2 \wedge \mathbf{e}_3 + 0.4\,\mathbf{e}_1 \wedge \mathbf{e}_3 \wedge \mathbf{e}_4 - 0.2\,\mathbf{e}_2 \wedge \mathbf{e}_3 \wedge \mathbf{e}_4 + 0.6\,\mathbf{e}_1 \wedge \mathbf{e}_2 \wedge \mathbf{e}_4.$$

Then $\mathbf{F} = \mathbf{e}_1 \wedge \mathbf{e}_2 \wedge \mathbf{e}_3$ and $\beta = -0.8$, and after scaling $\mathbf{B}$s becomes

$$\mathbf{B}_s = \mathbf{e}_1 \wedge \mathbf{e}_2 \wedge \mathbf{e}_3 - 0.5\,\mathbf{e}_1 \wedge \mathbf{e}_3 \wedge \mathbf{e}_4 + 0.25\,\mathbf{e}_2 \wedge \mathbf{e}_3 \wedge \mathbf{e}_4 - 0.75\,\mathbf{e}_1 \wedge \mathbf{e}_2 \wedge \mathbf{e}_4.$$

The factors are:

$$\begin{aligned} \mathbf{b}_1 &= (\mathbf{e}_1 \rfloor \mathbf{F}^{-1}) \rfloor \mathbf{B}_s = \mathbf{e}_1 \quad\quad\quad + 0.25\,\mathbf{e}_4, \\ \mathbf{b}_2 &= (\mathbf{e}_2 \rfloor \mathbf{F}^{-1}) \rfloor \mathbf{B}_s = \quad \mathbf{e}_2 \quad\quad + 0.5\,\mathbf{e}_4, \\ \mathbf{b}_3 &= (\mathbf{e}_3 \rfloor \mathbf{F}^{-1}) \rfloor \mathbf{B}_s = \quad\quad \mathbf{e}_3 \; - 0.75\,\mathbf{e}_4. \end{aligned} \tag{3}$$

The diagonal typesetting of $\mathbf{e}_1$, $\mathbf{e}_2$, $\mathbf{e}_3$ should make it obvious that the $\mathbf{b}_i$ are linearly independent, and because the $\mathbf{b}_i$ are all contained in $\mathbf{B}$, and there are grade($\mathbf{B}$) of them, they must form a factorization of $\mathbf{B}$.

Even though the factors are linearly independent, they are not orthogonal in general. For an estimation of how nonorthogonal the factors can be in the worst case, let us assume that the $\mathbf{B}$ is placed so skewly relative to the basis that it has an equal weight for each basis blade (it is not guaranteed that such an element $\mathbf{B}$ is indeed a blade). For a unit blade $\mathbf{B}$, this gives, as the worst case,

$$\mathbf{B} = \frac{1}{\sqrt{\binom{n}{k}}} \sum_{j=1}^{\binom{n}{k}} \pm \mathbf{E}_j, \quad \text{so} \quad \mathbf{B}_s = \sum_{j=1}^{\binom{n}{k}} \pm \mathbf{E}_j.$$

For such an element $\mathbf{B}_s$, the largest possible absolute value of an inner product between a pair of factors is the number of nonzero coordinates:

$$\|\mathbf{b}_i \cdot \mathbf{b}_j\| \le n - k,$$

and the largest absolute value of an inner product between a pair of normalized factors is

$$\big\|\text{unit}(\mathbf{b}_i) \cdot \text{unit}(\mathbf{b}_j)\big\| \le \frac{n-k}{n-k+1}.$$

This last equation shows that the factorization method could be less usable in, for example, 10-D space (e.g., if $k = 5$ and $n = 10$, then $\|\text{unit}(\mathbf{b}_i) \cdot \text{unit}(\mathbf{b}_j)\| \le \frac{5}{6}$) than it is in, for example, 3-D space, because the factors are potentially less and less orthogonal as the dimension of space n increases. However, in such spaces a factorized blade representation becomes more attractive than the additive representation

[10], and issues of efficient factorization of blades in the additive representation are then less crucial.

We still have to prove that the outer product of the computed factors $\mathbf{b}_i$ actually is equal to $\mathbf{B}_s$, i.e., that the scale and orientation of the computed factors are correct.

Theorem *The outer product of the factors* $\mathbf{b}_i$ *as computed by the* FastFactorization *algorithm is equal to* $\mathbf{B}_s$:

$$\mathbf{B}_s = \mathbf{b}_1 \wedge \mathbf{b}_2 \wedge \cdots \wedge \mathbf{b}_k.$$

Proof We know that the computed factors $\mathbf{b}_i$ are actually in $\mathbf{B}_s$ and are linearly independent. So the proof can be simplified to proving that the largest (unit) basis blade $\mathbf{F}$ of $\mathbf{B}_s$ is correctly computed, because then the other basis blades will have the right scale automatically . Now, $\mathbf{F} = \mathbf{f}_1 \wedge \mathbf{f}_2 \wedge \cdots \wedge \mathbf{f}_k$, and the contribution to this basis blade of each computed factor $\mathbf{b}_i$ is $(\mathbf{f}_i \rfloor \mathbf{F}^{-1}) \rfloor \mathbf{F}$. Hence, if we can prove that

$$(\mathbf{f}_i \rfloor \mathbf{F}^{-1}) \rfloor \mathbf{F} = \mathbf{f}_i,$$

the full theorem holds. This is trivial because $\mathbf{f}_i$ is contained in $\mathbf{F}$, allowing us to rewrite

$$(\mathbf{f}_i \rfloor \mathbf{F}^{-1}) \rfloor \mathbf{F} = (\mathbf{f}_i\, \mathbf{F}^{-1})\, \mathbf{F} = \mathbf{f}_i\, (\mathbf{F}^{-1}\, \mathbf{F}) = \mathbf{f}_i. \qquad \square$$

3 Algorithms for Computing the Join of Blades

Given the FastFactorization algorithm of the previous section, it is straightforward to formulate fast algorithms for the join based on [2, 6].

3.1 Fast Join Algorithm

Given two blades $\mathbf{A}$ and $\mathbf{B}$, the following algorithm computes the join $\mathbf{J} = \mathsf{join}(\mathbf{A}, \mathbf{B})$. A small constant threshold value ε is required.

Algorithm FastJoin($\mathbf{A}, \mathbf{B}, \varepsilon$):

1. Filter out trivial cases:

   - If $\mathbf{A}$ and/or $\mathbf{B}$ is zero, return $\mathbf{J} = 0$.
   - Else if $\mathbf{A}$ is of grade 0, return $\mathbf{J} = \mathrm{unit}(\mathbf{B})$.
   - Else if $\mathbf{B}$ is of grade 0, return $\mathbf{J} = \mathrm{unit}(\mathbf{A})$.
   - Else if $\mathbf{A}$ and/or $\mathbf{B}$ is of grade n, return $\mathbf{J} = \mathbf{I}_n$.
   - Otherwise continue with Step 2.

2. If required, swap **A** and **B** so that $\text{grade}(\mathbf{A}) \geq \text{grade}(\mathbf{B})$. If the swap is applied, remember the "swapping-sign" $\sigma = (-1)^{\text{grade}(\mathbf{A})\,\text{grade}(\mathbf{B})}$, otherwise set $\sigma = 1$. The swap is for reasons of efficiency.
3. Set $\mathbf{J} \leftarrow \text{unit}(\mathbf{A})$.
4. Find the largest basis blade term **F** in **B**.
5. While $\text{grade}(\mathbf{J}) \neq n$ and not all basis vectors $\mathbf{f}_i$ in **F** have been tried:
   a. Take any basis vector $\mathbf{f}_i$ in **F** which has not been tried yet.
   b. Compute $\mathbf{b}_i = (\mathbf{f}_i \rfloor \mathbf{F}^{-1}) \rfloor \mathbf{B}$.
   c. Compute $\mathbf{H} = \mathbf{J} \wedge \text{unit}(\mathbf{b}_i)$.
   d. If $(\|\mathbf{H}\| \geq \varepsilon)$, set $\mathbf{J} \leftarrow \text{unit}(\mathbf{H})$.
6. Return $\sigma\,\mathbf{J}$.

The implementation of this algorithm can be made very efficient by generating optimized code for each combination of arguments in Steps 5b and 5c, see Sect. 4.

The algorithm constantly forces blades and vectors to unit size in Steps 5c and 5d. In an optimized version of the algorithm, one may remove these normalizations and only normalize the final outcome. To apply this optimization, modify the following steps of the algorithm:

Step 5(c). Compute $\mathbf{H} = \mathbf{J} \wedge \mathbf{b}_i$.
Step 5(d). If $(\frac{\|\mathbf{H}\|}{\|\mathbf{J}\|\,\|\mathbf{b}_i\|} \geq \varepsilon)$, set $\mathbf{J} \leftarrow \mathbf{H}$.
Step 6. Return $\sigma\,\text{unit}(\mathbf{J})$.

However, if this optimization is employed, one should be careful that floating point precision does not underflow, as **J** may be scaled by a factor of $\varepsilon\,\|\mathbf{b}_i\|$ in each loop.

3.2 Computational Example

As an illustration, lets us compute the $\text{join}(\mathbf{A}, \mathbf{B})$ using the FastJoin algorithm using

$$\mathbf{A} = \mathbf{e}_1 \wedge \mathbf{e}_2,$$

$$\mathbf{B} = \mathbf{e}_2 \wedge (0.8\mathbf{e}_1 + 0.3\mathbf{e}_3 + 0.5\mathbf{e}_4) = 0.3\mathbf{e}_2 \wedge \mathbf{e}_3 - 0.8\mathbf{e}_1 \wedge \mathbf{e}_2 + 0.5\mathbf{e}_2 \wedge \mathbf{e}_4.$$

Note that in this example, values are rounded to two decimals.

Since the input is nontrivial and the grades of **A** and **B** are equal, Steps 1 and 2 of the algorithm have no effect. We continue with Step 3.

Step 3: $\mathbf{J} \leftarrow \mathbf{A}$.
Step 4: $\mathbf{F} = -0.8\mathbf{e}_1 \wedge \mathbf{e}_2$.

Step 5a, loop 1: $\mathbf{f}_1 = \mathbf{e}_1$.
Step 5b, loop 1: $\mathbf{b}_1 = 1.00 * \mathbf{e}_1 + 0.38 * \mathbf{e}_3 + 0.63 * \mathbf{e}_4$.
Step 5c, loop 1: $\mathbf{H} \leftarrow 0.51 * \mathbf{e}_1 \wedge \mathbf{e}_2 \wedge \mathbf{e}_4 + 0.30 * \mathbf{e}_1 \wedge \mathbf{e}_2 \wedge \mathbf{e}_3$.
Step 5d, loop 1: $\mathbf{J} \leftarrow 0.86 * \mathbf{e}_1 \wedge \mathbf{e}_2 \wedge \mathbf{e}_4 + 0.51 * \mathbf{e}_1 \wedge \mathbf{e}_2 \wedge \mathbf{e}_3$.

Step 5a, loop 2: $\mathbf{f}_2 = \mathbf{e}_2$.
Step 5b, loop 2: $\mathbf{b}_2 = \mathbf{e}_2$.
Step 5c, loop 2: $\mathbf{H} \leftarrow 0$.
Step 5d, loop 2: $\mathbf{H} < eps$, so $\mathbf{J}$ is not overwritten.

Hence the returned result is the grade 3 blade $0.86 * \mathbf{e}_1 \wedge \mathbf{e}_2 \wedge \mathbf{e}_4 + 0.51 * \mathbf{e}_1 \wedge \mathbf{e}_2 \wedge \mathbf{e}_3$.

3.3 Grade Stability of Fast Join Algorithm

The main loop of the above join algorithm tries to increase the grade of the current $\mathbf{J}$ by taking the outer product with factors of $\mathbf{B}$ in Step 5c and accepts the result if its norm is above some threshold value ε. This leads to a problem in stability of the grade of result: the factorization of $\mathbf{B}$ is dependent on the basis, and hence so is the grade of $\mathbf{J}$ as computed by the algorithm.

This is best illustrated using a simple example. Let $\mathbf{A}$ and $\mathbf{B}$ be the blades

$$\mathbf{A} = \mathbf{e}_1 \wedge \mathbf{e}_2, \qquad \mathbf{B} = \mathbf{e}_1 \wedge \mathbf{e}_2 + 0.2\,\mathbf{e}_1 \wedge \mathbf{e}_3 + 0.2\,\mathbf{e}_2 \wedge \mathbf{e}_3,$$

and compute the factors

$$\mathbf{b}_1 = \mathbf{e}_1 - 0.2\,\mathbf{e}_3, \qquad \mathbf{b}_2 = \mathbf{e}_2 + 0.2\,\mathbf{e}_3.$$

If we now compute Step 5c of the FastJoin algorithm for both $\mathbf{b}_1$ and $\mathbf{b}_2$, we find

$$\mathbf{H}_1 = \mathbf{A} \wedge \text{unit}(\mathbf{b}_1) = -0.196\,\mathbf{e}_1 \wedge \mathbf{e}_2 \wedge \mathbf{e}_3,$$
$$\mathbf{H}_2 = \mathbf{A} \wedge \text{unit}(\mathbf{b}_2) = 0.196\,\mathbf{e}_1 \wedge \mathbf{e}_2 \wedge \mathbf{e}_3.$$

Therefore, $\|\mathbf{H}_1\| = \|\mathbf{H}_2\| = 0.196$. If ε is taken as 0.2, the algorithm would accept neither $\mathbf{H}_1$ nor $\mathbf{H}_2$, and hence return a join of grade 2.

Now suppose that we rotate the input over $\pi/4$ in the $\mathbf{e}_1 \wedge \mathbf{e}_2$-plane, producing

$$\mathbf{A}' = \mathbf{e}_1 \wedge \mathbf{e}_2, \qquad \mathbf{B}' = \mathbf{e}_1 \wedge \mathbf{e}_2 + 0.283\,\mathbf{e}_2 \wedge \mathbf{e}_3.$$

This leads to a different factorization,

$$\mathbf{b}'_1 = \mathbf{e}_1 - 0.283\,\mathbf{e}_3, \qquad \mathbf{b}'_2 = \mathbf{e}_2,$$

resulting in

$$\mathbf{H}'_1 = \mathbf{A}' \wedge \text{unit}\big(\mathbf{b}'_1\big) \approx -0.272\,\mathbf{e}_1 \wedge \mathbf{e}_2 \wedge \mathbf{e}_3,$$

Hence, with $\varepsilon = 0.2$, the algorithm would now return a $\mathbf{J}$ of grade 3. Only the position of the blades relative to the basis changed relative to the previous situation, not their mutual position, so we would have expected the grade of their geometrical join to have been the same in both cases.

In some application where speed is essential and precision matters less, this behavior may be acceptable, but in general one would prefer the grade of the computed join to be independent of the choice of basis. Hence modifications to the FastJoin algorithm are required.

3.4 Improved Fast Join Algorithm

If the required grade of the join $\mathbf{J}$ could somehow be computed in advance, this knowledge could be used to guide the FastJoin algorithm. For instance, we could first run the FastJoin algorithm as is, and if the resulting $\mathbf{J}$ does not have the required grade, we could "lower our standards" and selectively accept some $\mathbf{b}_i$ that were rejected earlier, until the grade of the join is correct. To compute the required grade of the join, we use the following equation [6]:

$$\mathrm{grade}\big(\mathrm{join}(\mathbf{A},\mathbf{B})\big) = \frac{\mathrm{grade}(\mathbf{A}) + \mathrm{grade}(\mathbf{B}) + \mathrm{grade}(\mathbf{A}\Delta\mathbf{B})}{2}. \tag{4}$$

The delta product Δ is the geometric symmetric difference [4]. It can be computed as the highest-grade part of the geometric product $\mathbf{A}\,\mathbf{B}$ which is nonzero. In practice, however, we have to deal with floating point round-off errors. To quantify this, when selecting the top-grade part of the geometric product, we use a small threshold δ and select as the value of $\mathbf{A}\Delta\mathbf{B}$ the highest-grade part of $(\mathrm{unit}(\mathbf{A})\mathrm{unit}(\mathbf{B}))$ which has a norm $\geq\delta$. The grade of the delta product can be determined efficiently using lazy evaluation; details follow in Sect. 4.4, for now we call this function $\mathrm{FastDeltaGrade}(\mathbf{A},\mathbf{B},\delta)$.

We use this function to improve our FastJoin algorithm as follows:

Algorithm $\mathrm{StableFastJoin}(\mathbf{A},\mathbf{B},\varepsilon,\delta)$:

Start with Steps 1–5 of $\mathrm{FastJoin}(\mathbf{A},\mathbf{B},\varepsilon)$.

6. If $(\mathrm{grade}(\mathbf{J}) = n)$ or $(\mathrm{grade}(\mathbf{J}) = \mathrm{grade}(\mathbf{A}) + \mathrm{grade}(\mathbf{B}))$, return $\sigma\mathbf{J}$. (σ was computed in Step 2 of the FastJoin algorithm.) Otherwise:
7. Compute $\mathrm{grade}(\mathrm{join}(\mathbf{A},\mathbf{B}))$ using (4) and threshold δ for the delta product.
8. While $(\mathrm{grade}(\mathbf{J}) < \mathrm{grade}(\mathrm{join}(\mathbf{A},\mathbf{B})))$
   a. For all valid i, compute $\mathbf{b}_i = (\mathbf{f}_i \rfloor \mathbf{F}^{-1}) \rfloor \mathbf{B}$.
      Set $\mathbf{b}_m$ to that $\mathbf{b}_i$ which leads to the largest $\|\mathbf{J}\wedge\mathbf{b}_i\|$.
   b. Update $\mathbf{J} \leftarrow \mathbf{J}\wedge\mathbf{b}_m$.
9. Return $\sigma\mathbf{J}$.

Note that one should set $\varepsilon \geq \delta$, or else in Step 7 blade $\mathbf{J}$ may already have a grade which is larger than the grade required by $\mathrm{FastDeltaGrade}(\mathbf{A},\mathbf{B},\delta)$. Also note that in Step 8a it makes no sense to try any of the $\mathbf{b}_i$ which were already accepted in Step 5d or in an earlier iteration of Step 8a.

3.5 Numerical Stability of the Fast Join Algorithms

Another issue in the FastJoin algorithm is numerical stability due to the use of floating point values which causes round off errors.

The main thing we can do to improve numerical stability of both join algorithms is to choose good factors $\mathbf{b}_i$ in Step 5c of the algorithms. Each factor $\mathbf{b}_i$ potentially extending the current iteration value of the join $\mathbf{J}$ can be written as the sum

$$\mathbf{b}_i = \mathbf{b}_i^{\parallel} + \mathbf{b}_i^{\perp},$$

where $\mathbf{b}_i^{\parallel}$ is parallel to the current $\mathbf{J}$, and $\mathbf{b}_i^{\perp}$ is orthogonal to it. The outer product rejects $\mathbf{b}_i^{\parallel}$ and uses only $\mathbf{b}_i^{\perp}$, i.e.,

$$\mathbf{H} = \mathbf{J} \wedge \mathbf{b}_i = \mathbf{J} \wedge \mathbf{b}_i^{\perp}.$$

However, when $\|\mathbf{b}_i^{\parallel}\| \gg \|\mathbf{b}_i^{\perp}\|$, it is likely that the floating point precision of $\mathbf{J} \wedge \mathbf{b}_i$ is low. Hence we would prefer to select those $\mathbf{b}_i$ which result in the largest $\|\mathbf{J} \wedge \mathbf{b}_i\|$ because that $\mathbf{b}_i$ is most orthogonal to the current $\mathbf{J}$.

One solution is trying every $\mathbf{b}_i$ in each loop and use the one which results in the largest norm, but this is inefficient.

Fortunately, it is trivial to adjust the StableFastJoin algorithm to be more precise, with only a minimal performance impact. By running Steps 1 through 5 (its FastJoin part) using a relatively large threshold like $\varepsilon = 10^{-2}$, we only accept factors in Step 5d which are reasonably orthogonal to $\mathbf{J}$. When the input blades are in a nondegenerate configuration, the computed $\mathbf{J}$ will have the required grade (as verified by Step 6 or 7). Otherwise, we will find the best factors in Step 8. This step is more expensive, but it rarely needs to execute.

It is important to realize that in the StableFastJoin algorithm, δ only affects the correctness of the grade of result, while ε has an effect on the efficiency of computation and the numerical precision of the result. That is, a larger ε will let the algorithm run more slowly on average, but it will also increase numerical precision of the outcome.

4 Implementation

Below we describe the interesting parts of the implementation of each algorithm. First, we motivate why code generation was used to write the implementations. Then we describe the implementation in some more detail, also showing a small piece of code from the core of each, in order to convey the essence of each operation at the level of coordinates. Because these functions are used internally in the implementation, coordinates are passed in the form of arrays of floating point values, instead of being encapsulated in classes as is usual in `Gaigen 2`.

We also give a complexity order of the size of the generated code. This is relevant because the amount of generated code is large enough to limit the practicality of the code generation approach, especially for the join algorithms.

4.1 Code Generation

After the FastFactorization algorithm has found the largest basis blade **F**, it only has to scale, copy, and selectively negate certain coordinates from the input blade to the output factors. Exactly what coordinates have to be copied where depends on the grade of the input blade and which basis blade was the largest. This can be implemented very efficiently by explicitly spelling out the code for each possible case. However, writing that amount of code by hand is tedious and error-prone.

Thus, to implement the FastFactorization and the FastJoin algorithms, we have written a code generator on top of the `Gaigen 2` code generation framework [10]. The code generator generates a `C++` implementation of the algorithms for a specific n and for a specific order and orientation of the basis elements. It generates the code of the factorization code without any conditionals (`if`, `else`, `switch`) for each valid case. For the join, it also writes out the outer products. We have also written a code generator for evaluation of the grade of the delta product.

4.2 Implementation of the Fast Factorization Algorithm

The generated implementation of the FastFactorization algorithm first filters out trivial cases (0, scalars, vectors, pseudoscalars). These are handled separately.

If the input blade is not one of these trivial cases, the implementation finds the absolute largest coordinate of the input blade and the respective basis blade **F**. The proper scale β of the factorization is computed, and the coordinates are "normalized," so that the absolute largest coordinate is 1.

The implementation then calls an optimized factorization function which implements the actual factorization. One such generated function is available for each possible largest basis blade **F**, and they are called via a lookup table. Figure 2 shows an example of such a function. It is clearly visible that the function just copies (and possibly negates) coordinates of the input blade to the coordinates of a factor. Also, a number of coordinates of the factors are set to one or zero, as was already apparent in (3).

The delightful computational simplicity of this way of factorization is the main idea of this paper. It makes our factorization fast and, with it, the algorithms for the join that employ it.

In the order of $\sum_k \binom{n}{k} = 2^n$ of these functions are generated, each with code size proportional to $k\,n$, for a total code size of $O(\sum_k nk\binom{n}{k}) = O(n^2\,2^{n-1})$.

4.3 Implementation of the Fast Join Algorithm

The generated implementation of the FastJoin algorithm closely follows its description in Sect. 3.1.

```
void factorE234grade3(const float *B, float **b) {
  b[2][0] = B[0];
  b[1][0] = -B[1];
  b[0][0] = B[2];
  b[0][1] = b[1][2] = b[2][3] = 1.0f; // B[3];
  b[2][4] = B[6];
  b[1][4] = -B[8];
  b[0][4] = B[9];
  b[0][2] = b[0][3] = b[1][1] = b[1][3] = b[2][1] = b[2][2] = 0.0f;
}
```

Fig. 2 Example of a generated factorization function which computes the factors $\mathbf{b}_i$ of a normalized blade $\mathbf{B}$. This function implements the core of the FastFactorization algorithm for $n = 5$, $k = 3$, and $\mathbf{F} = \mathbf{e}_2 \wedge \mathbf{e}_3 \wedge \mathbf{e}_4$. The indices refer to a particular ordering of the basis blades in this implementation. For instance, the first line `b[2][0] = B[0]` corresponds to $(\mathbf{e}_4 \rfloor (\mathbf{e}_2 \wedge \mathbf{e}_3 \wedge \mathbf{e}_4)^{-1}) \rfloor (\beta_{123}\, \mathbf{e}_1 \wedge \mathbf{e}_2 \wedge \mathbf{e}_3) = \beta_{123}\, \mathbf{e}_1$

```
void factorAndOuterProductE35G3(const float *J, const float *B, float *H) {
  H[0] = J[3] * B[5] - J[2] * B[6] + J[0] * B[9];
  H[1] = J[6] * B[5] - J[5] * B[6];
  H[2] = J[8] * B[5] - J[7] * B[6] - J[4] * B[9];
  H[3] = J[9] * B[5] - J[5] * B[9];
  H[4] = J[9] * B[6] - J[6] * B[9];
  return B[5] * B[5] + B[6] * B[6] + B[9] * B[9];
}
```

Fig. 3 Example of a generated function which combines the extraction of a single factor of $\mathbf{B}$ with computing the outer product of $\mathbf{J}$ with that factor. The function stores the outcome of the outer product in $\mathbf{H}$ and returns the squared norm of the factor. This particular function computes $\mathbf{H} = \mathbf{J} \wedge ((\mathbf{e}_i \rfloor \mathbf{F}) \rfloor \mathbf{B})$ for $(\mathbf{e}_i \rfloor \mathbf{F} = \mathbf{e}_3 \wedge \mathbf{e}_5)$, grade$(\mathbf{J}) = 3$, and $n = 5$

The most significant optimization is the main loop of the algorithm: the implementation combines the factorization of Step 5b with the outer product of Step 5c. This allows it to take advantage of the zero coordinates which are in the factors due to the FastFactorization algorithm. One factor-and-outer-product function is generated for each valid combination of $\mathbf{e}_i \rfloor \mathbf{F}$ and grade$(\mathbf{J})$ in Steps 5b, c of the algorithm. These functions are called via a lookup table. Figure 3 shows an example of such a function.

There are in the order of $O(\sum_j \sum_k \binom{n}{k}) = O(n\, 2^n)$ of these functions, each with code size proportional to $n\binom{n}{j}$, for a total code size of $O(\sum_j \sum_k \binom{n}{k} n \binom{n}{j}) = O(n 2^{2n})$.

The functions do not normalize the factors; instead, they return the squared norm of the factors. The main loop uses this norm to correct the threshold check (Step 5d). This optimization was described in Sect. 3.1.

After the main loop of the algorithm has been completed, the function may compute the grade of the delta product. This is only done if its outcome is required to verify that $\mathbf{J}$ is of the correct grade, as was described in Sect. 3.4. The delta product is implemented by an optimized function described next.

4.4 Implementation of the Delta Product

The delta product is the highest-nonzero-grade part of the geometric product. In the context of the StableFastJoin algorithm, we are only interested in its grade, but we need to compute (part of) the actual delta product to establish what this grade is. Fortunately, it is possible to limit the candidate grades, so that we can avoid computing the full geometric product. Our FastDeltaGrade($\mathbf{A}, \mathbf{B}, \delta$) algorithm evaluates the grade of the delta product using the following optimizations:

- First of all, we compute the grade parts from high-to-low and abort early when we find that the norm of the grade part is above the threshold δ. We implement this early-abort strategy per coordinate. So only as many coordinates of the geometric product are computed as are required to determine which grade part has a norm exceeding δ.
- Secondly, no grade part of the geometric product $\mathbf{A}\,\mathbf{B}$ above n needs to be computed, as such grade parts do not exist in the geometric algebra of V^n. Also, no grade part above $2n - (\text{grade}(\mathbf{A}) + \text{grade}(\mathbf{B}))$ needs to be computed; these grade parts cannot be nonzero for then (4) would have grade(join($\mathbf{A}, \mathbf{B}$)) exceeding n, which is obviously impossible.
- The lowest possible grade part that may be nonzero is (grade($\mathbf{A}$) − grade($\mathbf{B}$)) (recall that grade($\mathbf{A}$) $\geq$ grade($\mathbf{B}$) after Step 2 of the FastJoin algorithm). But this grade part actually never has to be computed explicitly: if all grade parts above it are zero, then $\mathbf{B}$ must be fully contained in $\mathbf{A}$, and then grade($\mathbf{A}\Delta\mathbf{B}$) = grade($\mathbf{A}$) − grade($\mathbf{B}$).

To summarize, the grade parts k of the geometric product which must actually be evaluated to implement the delta product for use with the FastJoin algorithm are

$$\text{grade}(\mathbf{A}) - \text{grade}(\mathbf{B}) < k \leq \min\big\{n, 2n - \text{grade}(\mathbf{A}) - \text{grade}(\mathbf{B})\big\}.$$

The code for the implementation of the FastDeltaGrade($\mathbf{A}, \mathbf{B}, \delta$) algorithm is generated automatically. One function is generated for each valid combination of grade($\mathbf{A}$) and grade($\mathbf{B}$). Figure 4 shows an example of such a function.

If a stands for the grade of $\mathbf{A}$ and b stands for the grade of $\mathbf{B}$, then in the order of $O(\sum_b b^2) = O(n^3)$ such functions are generated, each with a code size in the order of $O(\sum_b \sum_{a\leq b} \binom{n}{b}\binom{n}{a}) < O(\sum_b \binom{n}{b} 2^{n-1}) = O(2^{2n-1})$. The total code size is $O(\sum_b \sum_{a\leq b} ba\binom{n}{b}\binom{n}{a}) < O((\sum_b b\binom{n}{b})(\sum_a a\binom{n}{a})) = O(n^2 2^{2n-2})$.

4.5 Benchmarks

We performed our benchmarks on a 1.83 GHz Core2 Duo notebook, using a single thread (i.e., one CPU). The programs were compiled with Visual C++ 2005, using standard optimization settings. 32 bit floating point arithmetic was used.

We ran benchmarks for $3 \leq n \leq 6$. Above 6-D, our particular implementation starts to make less sense because the amount of generated code becomes too large

```
int deltaProductG2G2(const float *A, const float *B, const float EPSILON2) {
  float c, n2;

  // check if norm of grade 2 is above threshold:
  c = -A[1]*B[2] + A[2]*B[1];
  n2 = c * c;
  if (n2 > EPSILON2) return 2; // return: delta product is grade 2
  c = A[0]*B[2] - A[2]*B[0];
  n2 += c * c;
  if (n2 > EPSILON2) return 2; // return: delta product is grade 2
  c = -A[0]*B[1] + A[1]*B[0];
  n2 += c * c;
  if (n2 > EPSILON2) return 2; // return: delta product is grade 2

  // grade 2 is near-zero, so delta product must be of grade 0:
  return 0; // return: delta product is grade 0
}
```

Fig. 4 Example of a function that computes $\mathbf{A}\Delta\mathbf{B}$ for $\text{grade}(\mathbf{A}) = 2$, $\text{grade}(\mathbf{B}) = 2$, and $n = 3$. Note how the squared norm of the grade 2 part is built up, allowing for early-abort after the evaluation of each coordinate

```
inline blade3 op(const vector& x, const blade2& y){
  return blade3(
    -x.c[1]*y.c[1] + x.c[2]*y.c[0] + x.c[0]*y.c[2],
     x.c[0]*y.c[4] - x.c[1]*y.c[3] + x.c[3]*y.c[0],
    -x.c[2]*y.c[3] + x.c[3]*y.c[1] + x.c[0]*y.c[5],
     x.c[3]*y.c[2] - x.c[2]*y.c[4] + x.c[1]*y.c[5]);
}
```

Fig. 5 Example of a generated outer product function from `Gaigen 2`. Our benchmarks are relative to this type of functions. This particular example computes the outer product of a vector and a 2-blade for $n = 4$

(one may then switch to a conventional hand-written implementation which will be somewhat less efficient, see the discussion in Sect. 5). Besides giving absolute values, such as the number of factorizations that can be performed per second, we also list benchmarks relative to the outer product of a vector and a 2-blade in the same algebra. This gives a fair impression of how expensive a factorization or join is relative to a straightforward bilinear product. The efficiency of such bilinear products as generated by `Gaigen 2` is comparable to optimized hand-coded linear algebra [10]. Figure 5 shows an example of an outer product.

4.5.1 Factorization Benchmarks

To benchmark the FastFactorization($\mathbf{A}$, $\mathbf{B}$) algorithm, we generated a number of random blades of random grades. The grades of the blades were uniformly distributed over the range $[0, n]$. A random k-blade was generated by computing the outer product of k random vectors. A random vector was generated by setting the n coordinates of the vector to random values, uniformly distributed in the range $[-1, 1]$. Table 1 shows the results for our implementation.

Table 1 Benchmarks of our implementation of the FastFactorization algorithm. The first column ("*without orthogonalization*") lists the cost of a factorization relative to the outer product of a vector and a 2-blade, without using Gram–Schmidt to orthogonalize resulting the factors. The third column "*with orthogonalization*" lists those values for the case where the Gram–Schmidt algorithm *is* applied after factorization. The columns "*factorizations per second*" lists absolute values (in millions), with and without using orthogonalization

n	Without orthogonalization	Factorizations per second (w/o)	With orthogonalization	Factorizations per second (with)
3	5.1×	15 M	8.5×	9.1 M
4	5.2×	9.2 M	9.9×	4.8 M
5	3.4×	5.2 M	6.5×	2.8 M
6	3.8×	2.8 M	7.1×	1.5 M

For reference, we also benchmarked the implementation of the factorization algorithm as described in [6], which is the direct predecessor the FastFactorization algorithm. This algorithm uses the regular projection operator to find factors, and the implementation does not use code generation as extensively as the implementation of our new algorithm. We measured 1.5 M factorizations per second in 3-D, 0.47 M factorizations per second in 4-D, and 0.25 M factorizations per second in 5-D (a 6-D implementation is not provided with the software of [6]). Thus we achieved a more than tenfold improvement in performance with our new algorithm.

4.5.2 Join Benchmarks

To benchmark the FastJoin and StableFastJoin algorithms, we generated pairs of random blades **A** and **B** using the same method as described for the factorization benchmark. The grades of the blades **A** and **B** were uniformly distributed over the range $[0, n]$. However, the pairs of random blades were generated such that they shared a common factor. The grade of the common factor was uniformly distributed over the range $[0, \min\{\text{grade}(\mathbf{A}), \text{grade}(\mathbf{B})\}]$.

We measured the time it took to compute the join and the time it took to compute the meet as well (i.e., the meet is computed from the join using (1)). We benchmarked both the FastJoin algorithm ($\varepsilon = 10^{-6}$) and the StableFastJoin algorithm ($\varepsilon = 10^{-2}$, $\delta = 10^{-6}$).

Table 2 shows the relative results, while Table 3 shows the absolute results.

In Table 2 also shows figures for computing the Gram–Schmidt orthogonalization of the factors of each pair of random blades. For this, we retained the factors that generated the blades, performed a standard Gram–Schmidt orthogonalization, and discarded the dependent factors (using the same ε threshold as for the join). This algorithm was implemented using the same principles (i.e., optimizing and unrolling the inner loop of the algorithm, but without using code generation) as the FastJoin algorithm. Hence, the last column allows us to compare the FastJoin algorithms to

Table 2 Relative benchmarks of our implementation of the FastJoin algorithm. The values are relative to the outer product of a vector and a grade 2 blade in the same algebra (i.e., lower values are better)

n	FastJoin	FastJoin + meet	StableFastJoin	StableFastJoin + meet	Gram–Schmidt
3	9.8×	12×	9.8×	12×	12×
4	8.7×	11×	9.1×	11×	12×
5	5.8×	7.7×	7.0×	8.8×	7.9×
6	6.4×	9.5×	6.8×	10×	8.0×

Table 3 Absolute benchmarks of our implementation of the FastJoin algorithm (millions of products per second). See Table 2 for relative benchmarks

n	FastJoin	FastJoin + meet	StableFastJoin	StableFastJoin + meet
3	7.4 M	6.0 M	7.4 M	6.0 M
4	5.4 M	4.1 M	5.2 M	4.0 M
5	3.1 M	2.4 M	2.6 M	2.1 M
6	1.8 M	1.2 M	1.6 M	1.1 M

a classical linear algebra approach for computing a minimal basis set which spans a subspace union.

As with the factorization algorithm, we also performed a benchmark on the join algorithm described in [6], which was the predecessor to our new algorithm. This algorithm does not use fast factorization and uses less extensive code generation. It also computes the join and meet simultaneously, a strategy which actually paid off for that algorithm (as opposed to our new algorithm, see Sect. 5.3). We measured 0.58 M joins per second in 3-D, 0.46 M joins per second in 4-D, and 0.29 M joins per second in 5-D. Thus we have achieved approximately a 10-times performance improvement.

4.5.3 Code Size

Table 4 lists the size of the generated code for our factorization and join implementation. The code size grows in approximate agreement with the computed complexities of $O(n^2 2^{n-1})$ and $O(n 2^{2n})$, respectively. We state "approximate" because for low-dimensional spaces, the constant code size of the algorithm (which is included in the table) can be relatively large compared to the amount of generated code, especially for the factorization algorithm.

Table 4 Source code size of our FastFactorization and StableFastJoin implementations

n	FastFactorization	FastJoin
3	3.74 kB	14.7 kB
4	5.85 kB	26.4 kB
5	11.4 kB	75.9 kB
6	25.8 kB	321 kB
7	62.1 kB	1.47 MB

5 Discussion

5.1 Fast Factorization Algorithm

Our benchmarks show that the FastFactorization algorithm is in the order of 5 times slower than a regular outer product in the same space. Adding a Gram–Schmidt orthogonalization to orthogonalize the factors approximately doubles the cost of the function.

The time complexity of the FastFactorization algorithm is $O(\binom{n}{n/2}) = O(n^{-1/2}2^n)$ (using the Stirling approximation of factorials) due to the step which finds the largest coordinate of the input blade. The fact that this step uses conditional statements makes it extra expensive on modern processors. The outer product of a vector and a 2-blade relative to which we presented the benchmarks in Table 1 has a time complexity of $O(n^3)$ and uses no conditional statements (an outer product of arbitrary blades has a time complexity around $O(2^n)$). As a result, FastFactorization is about five times more expensive than such an outer product. The benchmarks in Fig. 1 suggest that the FastFactorization algorithm becomes less expensive compared to the outer product as n becomes larger, but if one plots $\binom{n}{n/2}/n^3$ for $1 \leq n \leq 20$, it becomes clear that $n = 6$ is in fact the turning point beyond which the FastFactorization should become exceedingly expensive relative to the outer product. So our figure of five times slower is only valid for the limited range of n for which we benchmarked.

Note that it is rather remarkable (but logical) that in general the FastFactorization algorithm does not use all coordinates of the input blade once it has found which coordinate is the largest one: the k factors of a k-blade have $k\,n$ coordinates, which in many cases is less than the $\binom{n}{k}$ coordinates of the blade. For example, in Fig. 2, the coordinates `B[4]`, `B[5]`, and `B[7]` are not used (`B[3]` is known to be `1`, and so it is not used either). This demonstrates the redundant encoding of blades in the additive representation, and the implications of the Plücker relations [7].

The code size (Table 4) of the generated implementation is acceptable (less than 100 kB) up to 7-D, but extrapolation of the figures suggests that a 10-D implementation would about 1 MB in size. This is confirmed by the theoretical figure that code size should be in the order of $O(n^2 2^{n-1})$. Thus, in high-dimensional spaces, we recommend using a more conventional implementation approach. Our initial implementation of the FastFactorization algorithm was implemented without using code generation and was about two times slower than the generated implementation.

The FastFactorization algorithm is a useful building block for other algorithms. In this paper we used it for computing the join. Another useful application may be a fast "blade manifold projection" function which projects a nonblade onto the blade manifold in Grassmann space (the elements satisfying the Plücker relations). This may be implemented by naively "factoring" the nonblade and using the factors thus obtained to compute a valid blade as their outer product. We have also used the FastFactorization algorithm to factor conformal point pairs into individual points.

5.2 FastJoin Algorithms

The benchmarks show that our implementation of the FastJoin algorithms is slightly faster than an implementation of Gram–Schmidt orthogonalization applied to the factors of the input blades. This is quite remarkable, as it means that even if the only geometry you need is computing the join, you may be better off using the basis-of-blades representation rather than a factorized representation in terms of basis sets (at least for such low-dimensional spaces).

To make sure the grade of the join which is computed by our FastJoin algorithm is independent of the (arbitrary) basis, use of the delta product is required, invoking additional computational cost. However, the delta product needs to be invoked only when the algorithm cannot determine that it has computed a join of the right grade. As a result, the cost of the StableFastJoin (which uses the delta product) is only about 10% higher than that of the straightforward FastJoin algorithm.

The time complexity of the FastJoin algorithms is $O(n^2\binom{n}{n/2}) = O(n^{3/2}2^n)$, as we need to compute in the order of n the outer product of vectors with blades (in Step 5c), and each of these outer products has a time complexity of order $n\binom{n}{n/2}$. This means that the cost FastJoin algorithm relative to a vector-2-blade outer product should increase right from $n = 3$. The fact that the benchmarks in Fig. 2 do not agree with this is likely due to the decreasing relative cost of the overhead (filtering out special cases, and such) of the algorithm.

We implemented our join algorithms using code generation. Starting around 7-D, this is no longer tractable. The generated code for 6-D is 0.32 MB, while the code for 7-D is 1.47 MB in size; generating and compiling the 7-D code took several minutes. The size of the code is in the order of $n2^{2n}$. Hence, for $n \geq 7$, we recommend using a more conventional implementation which does not explicitly spell out the functions used in the inner loop for all possible arguments.

5.3 Simultaneous Computation of Meet and Join Costs More

It is possible to compute the meet directly, instead of computing it from the join using (1). In [6, 10], we factorize the dual of the delta product

$$(\mathbf{A}\Delta\mathbf{B})^* = \mathbf{s}_1 \wedge \mathbf{s}_2 \wedge \cdots \wedge \mathbf{s}_k$$

and project the factors $\mathbf{s}_i$ onto either $\mathbf{A}$ or $\mathbf{B}$. If those projections are not zero, they are factors of $\mathsf{meet}(\mathbf{A}, \mathbf{B})$. This method is due to [2].

Using this approach, one may also compute both the join and the meet simultaneously. Since factors of the dual of the join may be obtained as the rejection of the $\mathbf{s}_i$ from $\mathbf{A}$ or $\mathbf{B}$ (i.e., the operation $(\mathbf{s}_i \wedge \mathbf{A})\mathbf{A}^{-1}$, see [6]), such an algorithm is able to compute the join and the meet simultaneously and terminate as soon as either is fully known.

We implemented this idea using the same code generation techniques as used for the FastJoin algorithms, expecting it to be somewhat faster than the StableFastJoin algorithm. However, it turned out to be slightly slower (5% to 15%) and required about 1.5 times as much code to be generated.

One of the reasons for it being slower is that a full evaluation of the (dual of the) delta product is always required. By contrast, the StableFastJoin algorithm merely computes the grade of the delta product (not its numerical value), and only when "in doubt." One of the reasons for the larger code size is that one needs both a meet-from-join and a join-from-meet function.

Besides being slower, the implementation of the simultaneous meet and join algorithm also takes more effort because the algorithm is more complex. For all these reasons, we did not include a detailed description of it in this paper.

6 Conclusion

Using our FastFactorization algorithm, the outer factorization of a k-blade in V^n is a computationally trivial operation. It amounts to copying and possibly negating selected coordinates of the input blade into the appropriate elements of the factors. Implemented as such, factorization is only about five times slower than an outer product in the same algebra, at least in the low-dimensional spaces (less than 7-D) for which we benchmarked. The $O(n^{-1/2}2^n)$ time complexity of the algorithms is determined by the number of coordinates of the input k-blade, which becomes exceedingly large in high-dimensional spaces.

The join and meet of blades are relatively expensive products, due to their nonlinearity. However, when efficiently implemented through our FastJoin algorithm, their cost is only in the order of 10 times that of an outer product in the same algebra, compared to 100 times in previous research. Again, these figures are valid only for low-dimensional spaces. The $O(n^{3/2}2^n)$ time complexity makes clear that in high-dimensional spaces, one should use a multiplicative presentation of blades and use classical linear algebra algorithms like QR (which has $O(n^3)$ time complexity) to implement the join, as in [10]. Our StableFastJoin algorithm, which takes both grade and numerical stability into account, is just 10% slower than the FastJoin algorithm.

These speeds are obtained at the expense of generating efficient code that spells out the operations for certain combinations of the basis blades and grades in the arguments. While efficient, this is only truly possible for rather low-dimensional spaces, since the amount of code scales as $O(n^2 2^{n-1})$ for FastFactorization and $O(n2^{2n})$ for the FastJoin algorithm.

References

1. Ablamowicz, R., Fauser, B.: CLIFFORD—a Maple package for Clifford algebra computations with bigebra, http://math.tntech.edu/rafal/cliff12/, March 2009
2. Bell, I.: Personal communication (2004–2005)
3. Bouma, T.: Projection and factorization in geometric algebra. Unpublished paper (2001)
4. Bouma, T., Dorst, L., Pijls, H.: Geometric algebra for subspace operations. Acta Math. Appl. **73**, 285–300 (2002)
5. Dorst, L.: The inner products of geometric algebra. In: Dorst, L., Doran, C., Lasenby, J. (eds.) Applications of Geometric Algebra in Computer Science and Engineering, pp. 35–46. Birkhäuser, Basel (2002)
6. Dorst, L., Fontijne, D., Mann, S.: Geometric Algebra for Computer Science: An Object Oriented Approach to Geometry. Morgan Kaufmann, San Mateo (2007)
7. Eastwood, M., Michor, P.: Some remarks on the Plücker relations. Rendiconti del Circolo Matematico di Palermo, pp. 85–88 (2000)
8. Fauser, B.: A treatise on quantum Clifford algebras. Habilitation, Uni. Konstanz, Jan. 2003
9. Fontijne, D.: Gaigen 2: a geometric algebra implementation generator. GPCE 2006 Proceedings
10. Fontijne, D.: Efficient Implementation of geometric algebra. Ph.D. thesis, University of Amsterdam (2007). http://www.science.uva.nl/~fontijne/phd.html
11. Hildenbrand, D., Koch, A.: Gaalop—high performance computing based on conformal geometric algebra. In: AGACSE 2008 Proceedings. Springer, Berlin (2008)
12. Perwass, C.: The CLU project, Clifford algebra library and utilities. http://www.perwass.de/cbup/programs.html

Gaalop—High Performance Parallel Computing Based on Conformal Geometric Algebra

Dietmar Hildenbrand, Joachim Pitt, and Andreas Koch

Abstract We present Gaalop (Geometric algebra algorithms optimizer), our tool for high-performance computing based on conformal geometric algebra. The main goal of Gaalop is to realize implementations that are most likely faster than conventional solutions. In order to achieve this goal, our focus is on parallel target platforms like FPGA (field-programmable gate arrays) or the CUDA technology from NVIDIA. We describe the concepts, current status, and future perspectives of Gaalop dealing with optimized software implementations, hardware implementations, and mixed solutions. An inverse kinematics algorithm of a humanoid robot is described as an example.

1 Introduction

In recent years, geometric algebra, and especially the 5D Conformal geometric algebra, has proved to be a powerful tool for the development of geometrically intuitive algorithms in a lot of engineering areas like robotics, computer vision, and computer graphics. However, runtime performance of these algorithms was often a problem. In this chapter, we present our approach for the automatic generation of high-performance implementations with a focus on parallel target platforms like FPGA or CUDA. In Sects. 2 and 3, we present some related work and the basics of conformal geometric algebra. Our main goal with Gaalop is to realize implementations that are most likely faster than conventional solutions. The main concepts combining both approaches for the optimization of software and of hardware implementations are presented in Sect. 4. The corresponding architecture of Gaalop is described in Sect. 5. An inverse kinematics algorithm for the leg of a humanoid robot is presented in Sect. 6 as a complex example for the use of Gaalop. The current status of Gaalop and its future perspectives can be found in Sect. 7.

D. Hildenbrand (✉)
Interactive Graphics Systems Group, University of Technology Darmstadt, Darmstadt, Germany
e-mail: dhilden@gris.informatik.tu-darmstadt.de

E. Bayro-Corrochano, G. Scheuermann (eds.), *Geometric Algebra Computing*,
DOI 10.1007/978-1-84996-108-0_22,

2 Related Work

Despite the tremendous expressive power of geometric algebra, it has only seen limited use in practical applications. One of the reasons for this might be that the actual processing of geometric algebra algorithms requires significant computational effort. Related tools with the intention of optimizing geometric algebra implementations focus either on pure software or pure hardware solutions.

2.1 Software Implementations

The most advanced pure software solution is Gaigen developed at the university of Amsterdam (see [2] and [3]). You can find some benchmarks comparing Gaigen with other software implementations in [3].

2.2 Hardware Implementations

To resolve the above mentioned quandary, it is promising to look at dedicated hardware architectures for the acceleration of geometric algebra algorithms. Current integrated circuit technology offers a means to achieve this in the form of field-programmable gate arrays (FPGAs). These *reconfigurable* devices allow the implementation of a wide variety of digital logic circuits without the need for a very expensive photochemical circuit fabrication. Furthermore, the same device is able to realize different logic circuits by reconfiguring them onto the same silicon area.

2.2.1 Prior Attempts

The first serious approach is described by Perwass et al. [16]. That accelerator realizes the geometric product implemented on a 20-MHz FPGA connected via the PCI bus to the host computer. Due to the limited capacity of the FPGA employed, techniques such as wide parallel or pipelined processing and the use of fast on-chip memories were not exploited. Similarly, subspace coefficients consist only of 24-bit integers, and other fixed or floating point formats are not supported. The architecture is able to process multivectors of up to eight dimensions, with smaller vectors being processed faster. While the resulting accelerator does achieve a speedup over a conventional software programmable processor when counting clock cycles, comparisons with actual clock cycles lead to a practical *slow-down* when using the FPGA-based solution over simple software running on a conventional computer.

A different approach was presented by Gentile et al. [5]: This accelerator supports functions beyond the geometric product, namely, the outer product, contractions, etc., each being implemented on a dedicated hardware unit. The architecture is limited to multivectors of three to four dimensions. As before, the coefficients are limited to integers, in this case 16-bit wide. The FPGA implementation requires a lot of communication with the host computer over the PCI bus. Additionally, when taking the different clock frequencies into account to compute the real-world execution times, this approach does not lead to a speedup compared to a software implementation.

An update of this work is given by Franchini et al. [4]: the operation-specific hardware units have now been replaced by a variable number of so-called slices. Each slice is able to compute all operations of the four-dimensional geometric algebra. The coefficients have now been extended to 32 bit integers. In terms of hardware, a slice consists of a 32-bit wide arithmetic logic unit capable of addition, subtraction, multiplication, and logical computations. The geometric algebra operations are decomposed into these primitive calculations, with their execution being orchestrated step-by-step by on-chip software (microcode). The FPGA implementation achieves a clock frequency of 45 MHz and runs by a factor 3× to 4× faster than a software-programmable processor when counting cycles. When actually considering the 2 GHz clock frequency of the reference processor, the actual execution time again slows down by a factor of 9× to 12× versus software.

The first coprocessor to lift the integer limitation on coefficients is the custom-fabricated integrated circuit (ASIC) implementation introduced by Mishra and Wilson [12], which allows two-dimensional multivectors with double precision floating-point coefficients. At its core, it consists of a floating point adder and multiplier each, supported by smaller hardware units to compute the product of basis blades. While pipeline-parallel execution is employed within these compute units, actual geometric algebra operations (geometric product, rotor, etc.) are again computed sequentially by decomposing them into primitive calculations controlled by microcode. The experimental evaluation of the system in [13] shows a real *wall-clock speed-up* of 3× over a software programmable processor.

2.3 Our Proof-of-Concept Approach

In [9] we could show that an approach with symbolic simplification of geometric algebra algorithms is able to lead to an implementation which is three times faster than a conventional solution. In a second stage we implemented this inverse kinematics algorithm also on hardware and got an additional speedup of more than 100 times (see [10]).

When studying all of the prior hardware attempts, it is obvious that most of them lead to an application slowdown instead of the hoped-for acceleration. The major reason for this disappointing result is due to the architectural choices made. The discrepancy in achievable clock frequencies of conventional processors (which are now into multiple gigahertz), and that of FPGAs (which currently top out at 500–600 MHz), implies the need for massive parallelism in the FPGA to achieve better performance.

As a proof-of-concept, we implemented an accelerator [10] for a specific geometric algebra algorithm, namely the inverse kinematics of the arm of a virtual human. It is a completely different architectural approach compared to the approaches described above. Instead of coarse granular computation units capable of handling entire geometric algebra operators, we decomposed the geometric algebra description into the underlying scalar equations. These equations are optimized symbolically and employ only basic arithmetic operators. The resulting set of equations was then implemented one arithmetic operator at a time. For each of these arithmetic operators, we carefully examined the range of values to be processed for the specific problem. With this data, and external requirements on computational precision (in this case, the positional accuracy of the hand), we determined for each operator the optimal numerical representation (e.g., values in the range of 0 to 100 with $1/16$ mm of accuracy would be represented as 11-bit unsigned fixpoint numbers). The circuits of the operators were then optimally matched to their representation and to one of their operands being the constant.

The resulting accelerator, which exploits parallelism between multivector components, between fine-grained arithmetic operators, and in a pipelined fashion over the entire computation, achieves currently a real-world speedup in execution time of $185\times$ over a conventional processor with a 1.5-GHz clock frequency. The compute pipeline consists of 363 stages with an average of 12 arithmetic operators per stage. This extreme degree of parallelism allows the real-world acceleration even though the FPGA device (which is by now two generations out of date) only runs at 100-MHz clock frequency.

One of the aims of the Gaalop project is to develop a tool flow for automatically executing the optimization and hardware generation which we had to perform manually for our reference design. Before giving an overview of the planned flow, we will first give a brief introduction into geometric algebra.

3 Conformal Geometric Algebra

While points and vectors are normally used as basic geometric entities, in the 5D conformal geometric algebra we have a wider variety of basic objects. For example, spheres and circles are simply represented by algebraic objects. To represent a circle, you only have to intersect two spheres, which can be done with a basic algebraic operation. Alternatively, you can simply combine three points to obtain the circle through these three points.

Table 1 lists the two representations of the geometric entities in conformal geometric algebra. In this table, **x** and **n** are marked bold to indicate that they represent 3D entities as linear combination of the 3D base vectors e_1, e_2, and e_3:

$$\mathbf{x} = x_1 e_1 + x_2 e_2 + x_3 e_3. \tag{1}$$

The additional two base vectors are indicated by

- e_0 representing the 3D origin
- e_∞ representing the point at infinity

Table 1 Representations of the conformal geometric entities

Entity	Standard representation	Direct representation
Point	$P = \mathbf{x} + \frac{1}{2}\mathbf{x}^2 e_\infty + e_0$	
Sphere	$s = P - \frac{1}{2}r^2 e_\infty$	$s^* = x_1 \wedge x_2 \wedge x_3 \wedge x_4$
Plane	$\pi = \mathbf{n} + d e_\infty$	$\pi^* = x_1 \wedge x_2 \wedge x_3 \wedge e_\infty$
Circle	$z = s_1 \wedge s_2$	$z^* = x_1 \wedge x_2 \wedge x_3$
Line	$l = \pi_1 \wedge \pi_1$	$l^* = x_1 \wedge x_2 \wedge e_\infty$
Point Pair	$P_p = s_1 \wedge s_2 \wedge s_3$	$P_p^* = x_1 \wedge x_2$

The $\{s_i\}$ represent different spheres, and the $\{\pi_i\}$ different planes.

The two representations are dual to each other. In order to switch between the two representations, the dual operator which is indicated by '*', can be used. For example, in the standard representation, a sphere is represented with the help of its center point P and its radius r, while in the direct representation, it is constructed by the outer product '$\wedge$' of four points x_i that lie on the surface of the sphere ($x_1 \wedge x_2 \wedge x_3 \wedge x_4$). In standard representation, the dual meaning of the outer product is the intersection of geometric entities. For example, a circle is defined by the intersection of two spheres ($s_1 \wedge s_2$).

Blades are the basic computational elements and the basic geometric entities of geometric algebras. The 5D conformal geometric algebra consists of blades with *grades* 0, 1, 2, 3, 4, and 5, whereby a scalar is a 0-*blade* (blade of grade 0). The element of grade five is called the pseudoscalar. A linear combination of blades is called a *k-vector*. So a bivector is a linear combination of blades with grade 2. Other k-vectors are vectors (grade 1), trivectors (grade 3), and quadvectors (grade 4). Furthermore, a linear combination of blades of different grades is called a *multivector*. Multivectors are the general elements of a geometric algebra. Table 2 lists all the 32 blades of conformal geometric algebra. The indices indicate 1: scalar, 2...6: vector, 7...16: bivector, 17...26: trivector, 27...31: quadvector, 32: pseudoscalar.

A point $P = x_1 e_1 + x_2 e_2 + x_3 e_3 + \frac{1}{2}\mathbf{x}^2 e_\infty + e_0$ (see Table 1 and (1)), for instance, can be written in terms of a multivector as the following linear combination of blades:

$$P = x_1 * blade[2] + x_2 * blade[3] + x_3 * blade[4] + \frac{1}{2}\mathbf{x}^2 * blade[5] + blade[6]. \quad (2)$$

For more details, refer, for instance, to the book [2] and to the tutorials [8] and [6].

4 Concepts

The main goal of Gaalop is the combination of the elegance of algorithms using geometric algebra with generation of implementations that are most likely faster than conventional implementations. Depending on the application, these can be either optimized software implementations, or optimized hardware implementations, or a mixture between them.

Table 2 The 32 blades of the 5D conformal geometric algebra

Index	Blade	Grade	Index	Blade	Grade
1	1	0	17	$e_1 \wedge e_2 \wedge e_3$	3
2	e_1	1	18	$e_1 \wedge e_2 \wedge e_\infty$	3
3	e_2	1	19	$e_1 \wedge e_2 \wedge e_0$	3
4	e_3	1	20	$e_1 \wedge e_3 \wedge e_\infty$	3
5	e_∞	1	21	$e_1 \wedge e_3 \wedge e_0$	3
6	e_0	1	22	$e_1 \wedge e_\infty \wedge e_0$	3
			23	$e_2 \wedge e_3 \wedge e_\infty$	3
7	$e_1 \wedge e_2$	2	24	$e_2 \wedge e_3 \wedge e_0$	3
8	$e_1 \wedge e_3$	2	25	$e_2 \wedge e_\infty \wedge e_0$	3
9	$e_1 \wedge e_\infty$	2	26	$e_3 \wedge e_\infty \wedge e_0$	3
10	$e_1 \wedge e_0$	2			
11	$e_2 \wedge e_3$	2	27	$e_1 \wedge e_2 \wedge e_3 \wedge e_\infty$	4
12	$e_2 \wedge e_\infty$	2	28	$e_1 \wedge e_2 \wedge e_3 \wedge e_0$	4
13	$e_2 \wedge e_0$	2	29	$e_1 \wedge e_2 \wedge e_\infty \wedge e_0$	4
14	$e_3 \wedge e_\infty$	2	30	$e_1 \wedge e_3 \wedge e_\infty \wedge e_0$	4
15	$e_3 \wedge e_0$	2	31	$e_2 \wedge e_3 \wedge e_\infty \wedge e_0$	4
16	$e_\infty \wedge e_0$	2	32	$e_1 \wedge e_2 \wedge e_3 \wedge e_\infty \wedge e_0$	5

For that purpose, we propose a two-stage approach with

- symbolic optimization
- use of the inherent fine-grained parallel structure

of geometric algebra algorithms. Algorithms can vary from just a set of formulas to complex control flows.

4.1 Symbolic Optimization

We use the symbolic computation functionality of Maple (together with a library for geometric algebras [1]) in order to optimize the geometric algebra algorithm. Algorithms can be developed visually with CLUCalc [15] and afterwards be optimized with Gaalop. Gaalop parses and translates the CLUCalc code to Maple code. A small Maple library provided with Gaalop implements the corresponding CLUCalc functions in Maple. Maple computes the coefficients of the desired variable symbolically, returning an efficient implementation depending just on the input variables.

As an example, the following CLUCalc code computes the intersection circle C of two spheres $S1$ and $S2$. While CLUCalc requires the definition of the variables $x1, x2, x3, y1, y2, y3, r1$, and $r2$, we do not want to compute with fixed values for these variables. So just the second part is needed for Gaalop.

```
DefVarsN3();
:IPNS;

x1 = 0.2; x2 = 0.3; x3 = 0.5; r1 = 0.7;
y1 = 0.7; y2 = 1.1; y3 = 1.3; r2 = 0.9;

// Gaalop uses the code below

P1 = x1*e1 +x2*e2 +x3*e3 +1/2*(x1*x1+x2*x2+x3*x3)*einf +e_0;
P2 = y1*e1 +y2*e2 +y3*e3 +1/2*(y1*y1+y2*y2+y3*y3)*einf +e_0;

S1 = P1 - 1/2 * r1*r1 * einf;
S2 = P2 - 1/2 * r2*r2 * einf;

?C = S1 ^ S2;
```

A question mark in CLUCalc at the beginning of a line prints the result after evaluation of the corresponding line in the output window. Gaalop interprets these question marks almost the same, as it computes and prints out the coefficients of the following variable symbolically, depending on the previous input.

The computation of the conformal points $P1$ and $P2$ and the spheres $S1$ and $S2$ correspond to Table 1.

The resulting C code generated by Gaalop for the intersection circle C is as follows and only depends on the variables $x1$, $x2$, $x3$, $y1$, $y2$, $y3$, $r1$, and $r2$:

```
float C [32] = {0.0};

C[7] = x1*y2-x2*y1;
C[8] = x1*y3-x3*y1;

C[9] = -0.5*y1*x1*x1-0.5*y1*x2*x2-0.5*y1*x3*x3+0.5*y1*r1*r1
       +0.5*x1*y1*y1+0.5*x1*y2*y2+0.5*x1*y3*y3-0.5*x1*r2*r2;

C[10] = -y1+x1;
C[11] = -x3*y2+x2*y3;

C[12] = -0.5*y2*x1*x1-0.5*y2*x2*x2-0.5*y2*x3*x3+0.5*y2*r1*r1
        +0.5*x2*y1*y1+0.5*x2*y2*y2+0.5*x2*y3*y3-0.5*x2*r2*r2;

C[13] = -y2+x2;

C[14] = -0.5*y3*x1*x1-0.5*y3*x2*x2-0.5*y3*x3*x3+0.5*y3*r1*r1
        +0.5*x3*y1*y1+0.5*x3*y2*y2+0.5*x3*y3*y3-0.5*x3*r2*r2;

C[15] = -y3+x3;

C[16] = -0.5*y3*y3+0.5*x3*x3+0.5*x2*x2+0.5*r2*r2
        -0.5*y1*y1-0.5*y2*y2+0.5*x1*x1-0.5*r1*r1;
```

Gaalop always computes optimized 32-dimensional multivectors. Since a circle is described with the help of a bivector, only the blades 7 to 16 (see Table 2) are used. As you can see, all the corresponding coefficients of this multivector are very simple expressions with basic arithmetic operations. A more complex example is described in Sect. 6.

4.2 Use of Inherent Fine-Grained Parallel Structure

With the help of symbolic optimization, the geometric algebra algorithm is transformed into an algorithm computing the coefficients of 32D multivectors using only basic arithmetical operations. This can be implemented very efficiently in digital logic on silicon devices such as FPGAs using parallel computation of coefficients of multivectors, deeply pipelined processing, and the exploitation of constant values by propagating them directly into the circuit. These techniques are described in Sect. 2.3 and in more detail in [10] for an inverse kinematics example.

5 The Architecture of Gaalop

Figure 1 shows an overview over the architecture of Gaalop. Its input is a geometric algebra algorithm written in CLUCalc (see [15]). Via symbolic simplification it is transformed into a generic intermediate representation (IR) that can be used for generation of different output formats. Gaalop supports sequential platforms, such as C and Java, and parallel platforms, such as CUDA [14] or FPGA descriptions (as a structural hardware description, currently written in the Verilog language). CLUCalc can also be used as an output format in order to visualize the optimized results (see [15]).

The basis of the mapping the IR, which is expressed on an abstract mathematical/behavioral level, to a hardware accelerator is the technology already used in the COMRADE compiler [11]. COMRADE is designed to translate from ANSI C (complete language, no additional user annotations required) into hybrid hardware/software applications, with the hardware parts being executed on an FPGA. Since geometric algebra algorithms are far more abstract than C (which contains, e.g., pointers and gotos), they are considerably easier to optimize and translate efficiently to an FPGA-based accelerator.

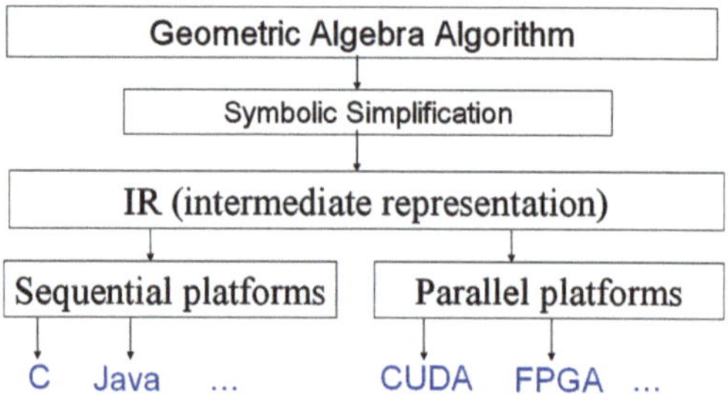

Fig. 1 Architecture of Gaalop

Fig. 2 The robot "Mr. DD Junior 2" of the RoboCup team, the Darmstadt Dribbling Dackels (DDD)

6 Inverse Kinematics of the Leg of a Humanoid Robot

In this section we present an inverse kinematics algorithm for the leg of the humanoid robot "Mr. DD Junior 2" (see Fig. 2). We use CLUCalc code as an example for the input language of Gaalop. Parts of the generated C code are shown as an example for a target implementation of Gaalop.

"Mr. DD Junior 2" is a humanoid robot of about 38-cm total height and was used for the 2005 RoboCup competition [17] where robots play soccer completely autonomously. Its legs have six degrees of freedom each: from hip to foot, the first joint rotates about the forward oriented axis, the next three joints about the sidewards oriented axis, and the last two joints about the forward and upward oriented axis. This is different from the standard configuration of humanoid robot legs, where the joint that rotates about the upward oriented axis usually is located in the hip. The robot is equipped with a camera for vision, a pocket PC for computation and servo motors for actuation. The robot must localize itself on the field which has color-coded landmarks, identify other players, the location of the ball, and the goal. For walking, "Mr. DD Junior 2" uses the following inverse kinematics approach: The motion of the hips and feet are given by smooth trajectories that are described by several parameters. and the joint angle trajectories of the legs are computed from

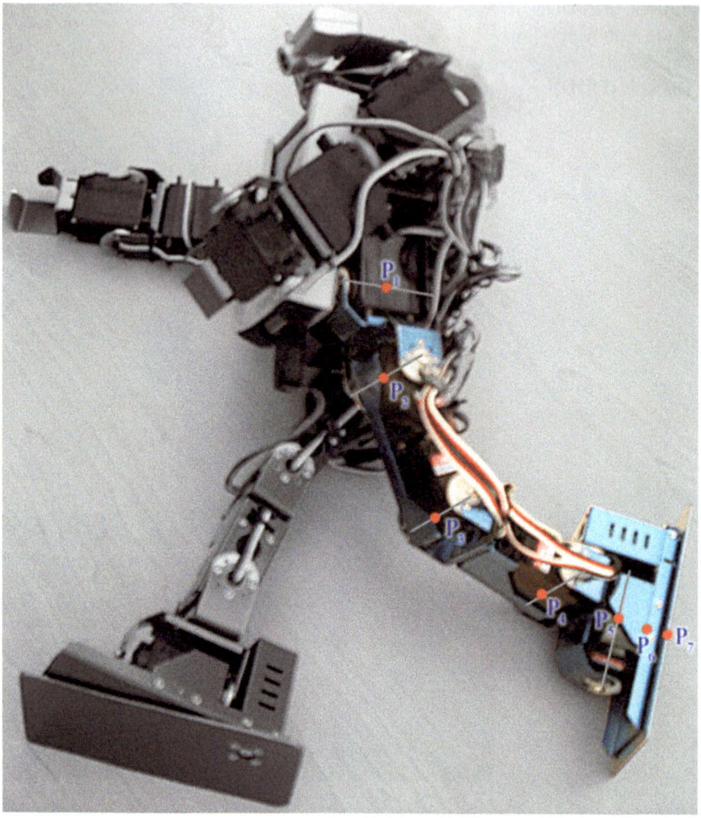

Fig. 3 The leg of the robot "Mr. DD Junior 2" with the indication of the points P_1 to P_7

the hips and feet trajectories by inverse kinematics. This computation is done online on the pocket PC, which also is used for image processing.

6.1 Solving the Inverse Kinematics Algorithm

The following inverse kinematics algorithm has been developed using conformal geometric algebra to solve the 6-DOF kinematic chain for the leg of the humanoid robot "Mr. DD Junior 2" (see Fig. 3). The leg consists of joints with one degree of freedom each. The hip (P_1) defines the first joint and lies in the origin, rotating about the x-axis. Three joints rotating about the y-axis, one rotating about the x-axis and a final one rotating about the z-axis follow, leading to the foot-point (P_7). The coordinates of the foot (P_x, P_y, P_z), the normal of the foot (n), and the lengths of the links ($l_1, l_2, l_3, l_4, l_5, l_6$) are needed to solve the inverse kinematics chain. Refer to Table 3 for a list of the input and output parameters of the algorithm.

Table 3 Input/Output of the inverse kinematics algorithm

Input		Output	
Var	Description	Var	Description
P_x	Foot x-value	ang_1	angle at P_1
P_y	Foot y-value	ang_2	angle at P_2
P_z	Foot z-value	ang_3	angle at P_3
n	Normal of foot x-value	ang_4	angle at P_4
	Normal of foot y-value	ang_5	angle at P_5
	Normal of foot z-value	ang_6	angle at P_6
l_1	Length of 1st link		
l_2	Length of 2nd link		
l_3	Length of 3rd link		
l_4	Length of 4th link		
l_5	Length of 5th link		
l_6	Length of 6th link		

6.1.1 Computation of the Positions of Link 5 and 6 in the Kinematics Chain

Since there is only a rotation about the z-axis in P_6, links 5 and 6 (in P_5 and in P_6) are on the normal (n) of the foot. The translator T_1 is needed to translate P_7 about l_6 in direction of n:

$$\begin{aligned} T_1 &= 1 - \left(\frac{1}{2} n l_6\right) e_\infty, \\ P_6 &= T_1 P_7 \tilde{T}_1. \end{aligned} \tag{3}$$

Please notice that a translator is defined by the expression $T = 1 - \frac{1}{2} t_{\mathrm{vec}} e_\infty$ with t_{vec} being the 3D translation vector and that a translation is defined by a multiplication of the translator from the left and of its reverse from the right. Another translator (T_2) is necessary to compute P_5, where the distance to P_7 is $l_5 + l_6$:

$$\begin{aligned} T_2 &= 1 - \left(\frac{1}{2} n (l_5 + l_6)\right) e_\infty, \\ P_5 &= T_2 P_7 \tilde{T}_2. \end{aligned} \tag{4}$$

See Fig. 4.

6.1.2 Computation of the Position of Link 4

By taking a closer look at the kinematic chain, one will notice that P_1, P_2, P_3, P_4, P_5 define a plane π_3, which includes the x-axis. Since the joints in P_2, P_3, and P_4 all rotate about the y-axis, these and the joints directly connected to them (P_1 and P_5)

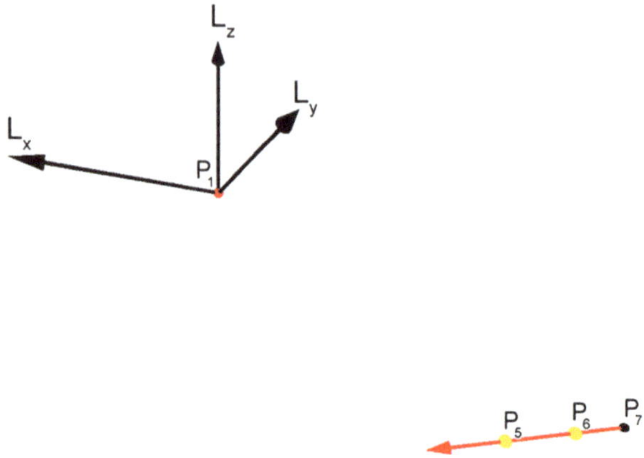

Fig. 4 Step A. Translating P_7 by l_6 and l_5 in direction of n to get P_6 and P_5

are all in one plane. Every plane can be defined by three points, which here are P_1, P_5, and the auxiliary point P_{H2} on the x-axis. Another plane, defined by the points P_4, P_5, P_6, P_7, is orthogonal to plane π_3. So the projection of the line through P_5 and P_7 onto the plane π_3 yields L_{Proj}, which intersects the sphere S_5 (with center P_5 and radius l_4), resulting in a point pair. Function `pp_get2nd` selects link 4 (P_4) from it.

$$\begin{aligned} S_5 &= \text{Sphere}(P_5, l_4), \\ \pi_3 &= (P_{H2} \wedge P_1 \wedge P_5 \wedge e_\infty)^*, \\ \pi_3 &= \frac{\pi_3}{|\pi_3|}, \\ L_{P5P7} &= (P_5 \wedge P_7 \wedge e_\infty)^*, \\ L_{\text{Proj}} &= \frac{(\pi_3 \cdot L_{P5P7})}{\pi_3}, \\ P_4 &= \text{pp\_get2nd}\big((S_5 \wedge L_{\text{Proj}})^*\big). \end{aligned} \tag{5}$$

See Fig. 5.

Sphere(x, r) generates a sphere around x with the radius r:

$$\text{Sphere}(x, r) = x - \frac{1}{2} r^2 e_\infty. \tag{6}$$

The functions pp_get1st and pp_get2nd each pick one point out of a point pair:

$$\text{pp\_get1st}(x) = \frac{\sqrt{|x \cdot x|} - x}{e_\infty \cdot x}, \tag{7}$$

$$\text{pp\_get2nd}(x) = \frac{-\sqrt{|x \cdot x|} + x}{e_\infty \cdot x}. \tag{8}$$

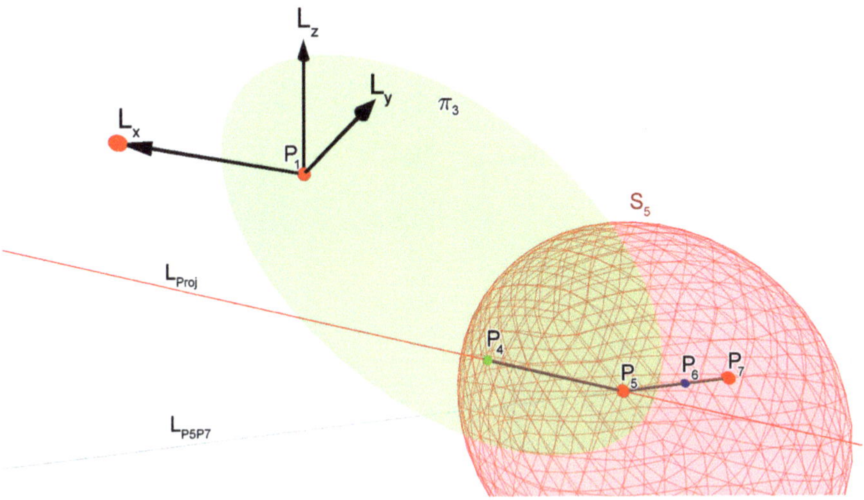

Fig. 5 Step B. Projection of the line through P_7 and P_5 onto the green plane defined by P_1, P_5, and an auxiliary point on the x-axis. Intersection of the sphere around P_5 with radius l_4 and the projected line returns P_4

6.1.3 Computation of the Position of Link 2

Links 1 and 2 are located on the yz-plane π_1, so the intersection of planes π_1 and π_3 results in a line, with P_1 and P_2 on it. The distance between P_2 and P_1 is l_1; hence the intersection of the sphere S_1 around P_1 with radius l_1 results in a point pair, from which P_2 can be selected

$$\begin{aligned} \pi_1 &= e_1, \\ S_1 &= \text{Sphere}(P_1, l_1), \\ L_1 &= \pi_1 \wedge \pi_3, \\ P_2 &= \text{pp\_get1st}\big((L_1 \wedge S_1)^*\big). \end{aligned} \tag{9}$$

See Fig. 6.

6.1.4 Computation of the Position of Link 3

The intersection of the two spheres S_2 and S_4 results in a circle Z_3. P_3 must be located on circle Z_3 and on plane π_3 as well; thus the intersection of Z_3 and π_3

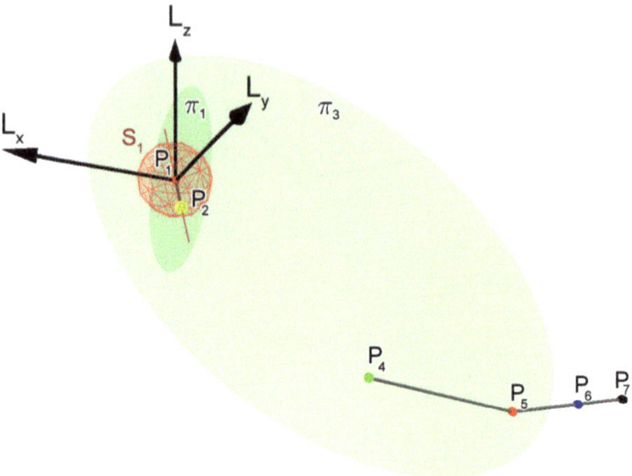

Fig. 6 Step C. Intersecting the sphere around P_1 with radius l_1 with the intersection of the plane π_3 and the yz-plane returns P_2

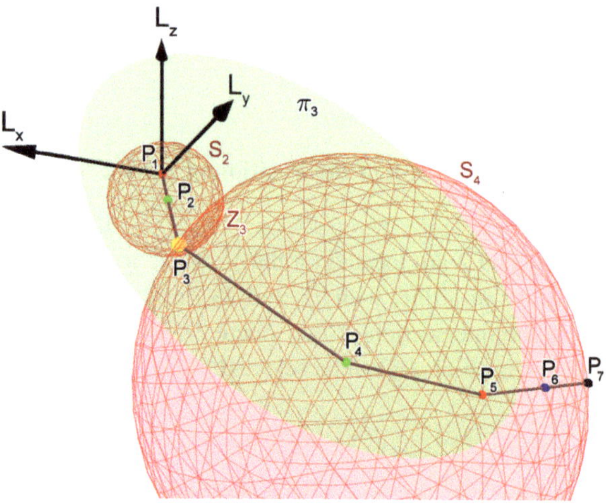

Fig. 7 Step D. The intersection of the spheres around P_2 with radius l_2 and around P_4 with radius l_3 results in *the red circle*. Intersecting the circle with the plane π_3 returns P_3

results in a point pair again, from which P_3 can be selected

$$\begin{aligned} S_2 &= \text{Sphere}(P_2, l_2), \\ S_4 &= \text{Sphere}(P_4, l_3), \\ Z_3 &= S_2 \wedge S_4, \\ P_3 &= \text{pp\_get1st}\big((Z_3 \wedge \pi_3)^*\big). \end{aligned} \tag{10}$$

See Fig. 7.

6.1.5 Compute the Angles of the Links

Since a line is defined by two points, three points are necessary to generate two intersecting lines and to compute the angle in between. To compute the angle of the first link, a point above the origin (0,0,1) is used as a parameter.

$$\text{angle}(x, y, z) = \pi - \arccos\left(\frac{(x \wedge y \wedge e_\infty) \cdot (z \wedge y \wedge e_\infty)}{|x \wedge y \wedge e_\infty||z \wedge y \wedge e_\infty|}\right). \tag{11}$$

6.2 *Symbolic Optimization of the Kinematic Chain*

In this section, we present the CLUCalc code for the just described inverse kinematics algorithm as an example for the input language of Gaalop. Parts of the generated C code are presented as an example for a target implementation of Gaalop.

CluCalc models the first part of the inverse kinematics algorithms as follows:

```
PH3 = VecN3(hx-1,hy,hz);
:P1 = VecN3(hx,hy,hz);
:P7 = VecN3(-px,py,pz);

norm1 = FNorm[1]*e1 - FNorm[2]*e2 - FNorm[3]*e3;
?norm2 = 1/abs(norm1);
norm3 = norm1 * norm2;

//generate translator and compute P6
tvec = (norm3*(len[6]))/2;
T1 = 1 - tvec * einf;
?P6 = T1 * P7 * ~T1;

tvec = (norm3*(len[5]+len[6]))/2;
T2 = 1 - tvec * einf;
?P5 = T2 * P7 * ~T2;

S5 = Sphere(P5,len[4]);
Pi3 = *(PH3 ^ e0 ^ P5 ^ einf);
?Pi3a = 1/abs(Pi3);
?Pi3b = Pi3 * Pi3a; // Pi3/abs(Pi3)
```

Gaalop optimizes the above CLUCalc code (see Sect. 4.1 for details about the optimization approach) and generates the following C code:

```
float norm2_opt[32] = {0.0};
norm2_opt[1]=1/sqrt(FNorm[1]*FNorm[1]+FNorm[2]*FNorm[2]
                    +FNorm[3]*FNorm[3]);

float P6_opt[32] = {0.0};
P6_opt[2]=-px+FNorm[1]*norm2_opt[1]*len[6];
P6_opt[3]=py-FNorm[2]*norm2_opt[1]*len[6];
```

```
P6_opt[4]=pz-FNorm[3]*norm2_opt[1]*len[6];
P6_opt[5]=0.5*FNorm[1]*FNorm[1]*norm2_opt[1]*norm2_opt[1]
         *len[6]*len[6]+0.5*py*py+0.5*pz*pz-FNorm[1]
         *norm2_opt[1]*len[6]*px+0.5*FNorm[3]*FNorm[3]
         *norm2_opt[1]*norm2_opt[1]*len[6]*len[6]+0.5
         *px*px+0.5*FNorm[2]*FNorm[2]*norm2_opt[1]
         *norm2_opt[1]*len[6]*len[6]-FNorm[2]*norm2_opt[1]
         *len[6]*py-FNorm[3]*norm2_opt[1]*len[6]*pz;
P6_opt[6]=1;

float P5_opt[32] = {0.0};
P5_opt[2]=-px+FNorm[1]*norm2_opt[1]*len[5]
         +FNorm[1]*norm2_opt[1]*len[6];
P5_opt[3]=-FNorm[2]*norm2_opt[1]*len[5]
         -FNorm[2]*norm2_opt[1]*len[6]+py;
P5_opt[4]=-FNorm[3]*norm2_opt[1]*len[5]
         -FNorm[3]*norm2_opt[1]*len[6]+pz;
P5_opt[5]=0.5*FNorm[2]*FNorm[2]*norm2_opt[1]*norm2_opt[1]
         *len[6]*len[6]+0.5*FNorm[3]*FNorm[3]*norm2_opt[1]
         *norm2_opt[1]*len[6]*len[6]+0.5*FNorm[1]*FNorm[1]
         *norm2_opt[1]*norm2_opt[1]*len[6]*len[6]-FNorm[3]
         *norm2_opt[1]*pz*len[5]-FNorm[1]*norm2_opt[1]*px
         *len[5]-FNorm[2]*norm2_opt[1]*py*len[5]+0.5*py*py
         +0.5*pz*pz+FNorm[2]*FNorm[2]*norm2_opt[1]
         *norm2_opt[1]*len[5]*len[6]+FNorm[1]*FNorm[1]
         *norm2_opt[1]*norm2_opt[1]*len[5]*len[6]+FNorm[3]
         *FNorm[3]*norm2_opt[1]*norm2_opt[1]*len[5]*len[6]
         +0.5*px*px-FNorm[1]*norm2_opt[1]*len[6]*px
         -FNorm[3]*norm2_opt[1]*len[6]*pz-FNorm[2]
         *norm2_opt[1]*len[6]*py+0.5*FNorm[1]*FNorm[1]
         *norm2_opt[1]*norm2_opt[1]*len[5]*len[5]
         +0.5*FNorm[2]*FNorm[2]*norm2_opt[1]*norm2_opt[1]
         *len[5]*len[5]+0.5*FNorm3*FNorm3*norm2_opt[1]
         *norm2_opt[1]*len[5]*len[5];
P5_opt[6]=1;

float Pi3a_opt[32] = {0.0};
Pi3a_opt[1]=1/sqrt(hy*hy*P5_opt[4]*P5_opt[4]-2*hy*P5_opt[4]
           *hz*P5_opt[3]+hz*hz*P5_opt[3]*P5_opt[3]
           +P5_opt[4]*P5_opt[4]*hx*hx-2*P5_opt[4]*hx*hz
           *P5_opt[2]-2*P5_opt[4]*P5_opt[4]*hx+hz*hz
           *P5_opt[2]*P5_opt[2]+2*hz*P5_opt[2]*P5_opt[4]
           +P5_opt[4]*P5_opt[4]+P5_opt[3]*P5_opt[3]*hx*hx
           -2*P5_opt[3]*hx*hy*P5_opt[2]-2*P5_opt[3]
           *P5_opt[3]*hx+hy*hy*P5_opt[2]*P5_opt[2]+2*hy
           *P5_opt[2]*P5_opt[3]+P5_opt[3]*P5_opt[3]);


float Pi3b_opt[32] = {0.0};
Pi3b_opt[2]=-Pi3a_opt[1]*(-P5_opt[4]*hy+hz*P5_opt[3]);
Pi3b_opt[3]=Pi3a_opt[1]*(-P5_opt[4]*hx+hz*P5_opt[2]
           +P5_opt[4]);
Pi3b_opt[4]=-Pi3a_opt[1]*(-P5_opt[3]*hx+hy*P5_opt[2]
           +P5_opt[3]);
```

As you can see, the optimized code is very complex in terms of length. Therefore we only list the CLUCalc code for the second part of the algorithm below.

```
LP5P7 = *(P5 ^ P7 ^ einf);
LProj = (Pi3b.LP5P7)/Pi3b;
:P4 = pick2nd(*(S5 ^ LProj));

Pi1 = e1;
S1 = Sphere(P1,len(1));
L1 = Pi1 ^ Pi3:Red;
:P2 = pick1st(*(L1 ^ S1));

S2 = Sphere(P2,len(2));
S4 = Sphere(P4,len(3));
C3 = S2 ^ S4;
?P3 = pick1st(*(C3 ^ Pi3));

?angle(VecN3(0,1,0),P1,P2);
?angle(P1,P2,P3);
?angle(P2,P3,P4);
?angle(P3,P4,P5);
?angle(P4,P5,P6);
```

From the runtime performance point-of-view, our optimized C code achieved results comparable to the conventional algorithm. This is why the actual speedup that Gaalop can provide for this inverse kinematics application will result from future implementations on parallel platforms.

Because of the similarity of this inverse kinematics algorithms to our proof-of-concept application (see Sect. 2.3 and [9]), we expect for an FPGA implementation a comparable hardware speedup of about 100 times.

7 Current Status and Future Perspectives

Gaalop is currently able to handle sequential conformal geometric algebra algorithms. The algorithm is currently transformed into C code, CLUCalc code, and simple LaTeX formulas. The latest news can be always found on the Gaalop homepage [7].

We are just extending Gaalop in order to handle control flow with loops, conditions, etc. We are also developing generators for additional output formats like Java code, CUDA [14], and FPGA descriptions.

One focus will lie on mixed solutions handling reasonable combinations of software and hardware implementations.

8 Conclusion

Geometric algebra is applicable in many different engineering scenarios and provides a straightforward and intuitive problem solving approach. With the help of our Gaalop tool, these algorithms can be automatically transformed into high runtime performance implementations. With these results, we are convinced that conformal geometric algebra will be able to become more and more fruitful in a great variety of engineering applications.

References

1. Abłamowicz, R., Fauser, B.: Clifford/bigebra, a Maple package for Clifford (co)algebra computations (2009). ©1996–2009, RA&BF
2. Dorst, L., Fontijne, D., Mann, S.: Geometric Algebra for Computer Science, An Object-Oriented Approach to Geometry. Morgan Kaufman, San Mateo (2007)
3. Fontijne, D.: Efficient implementation of geometric algebra. Ph.D. thesis, University of Amsterdam (2007)
4. Franchini, S., Gentile, A., Grimaudo, M., Hung, C.A., Impastato, S., Sorbello, F., Vassallo, G., Vitabile, S.: A sliced coprocessor for native Clifford algebra operations. In: Euromico Conference on Digital System Design, Architectures, Methods and Tools (DSD) (2007)
5. Gentile, A., Segreto, S., Sorbello, F., Vassallo, G., Vitabile, S., Vullo, V.: Cliffosor, an innovative FPGA-based architecture for geometric algebra. In: ERSA 2005, pp. 211–217 (2005)
6. Hildenbrand, D.: Geometric computing in computer graphics using conformal geometric algebra. Comput. Graph. **29**(5), 802–810 (2005)
7. Hildenbrand, D., Pitt, J.: The Gaalop homepage. Available at http://www.gaalop.de (2010)
8. Hildenbrand, D., Fontijne, D., Perwass, C., Dorst, L.: Tutorial geometric algebra and its application to computer graphics. In: Eurographics Conference Grenoble (2004)
9. Hildenbrand, D., Fontijne, D., Wang, Y., Alexa, M., Dorst, L.: Competitive runtime performance for inverse kinematics algorithms using Conformal geometric algebra. In: Eurographics Conference Vienna (2006)
10. Hildenbrand, D., Lange, H., Stock, F., Koch, A.: Efficient inverse kinematics algorithm based on conformal geometric algebra using reconfigurable hardware. In: GRAPP Conference Madeira (2008)
11. Kasprzyk, N., Koch, A.: High-level-language compilation for reconfigurable computers. In: Proceedings International Conference on Reconfigurable Communication-centric SoCs (ReCoSoC) (2005)
12. Mishra, B., Wilson, P.R.: Color edge detection hardware based on geometric algebra. In: European Conference on Visual Media Production (CVMP) (2006)
13. Mishra, B. Wilson, P.R.: VLSI implementation of a geometric algebra parallel processing core. Technical report, Electronic Systems Design Group, University of Southampton, UK (2006)
14. NVIDIA. The CUDA homepage. Available at http://www.nvidia.com/object/cuda_home.html (2009)
15. Perwass, C.: The CLU homepage. Available at http://www.clucalc.info (2010)
16. Perwass, C., Gebken, C., Sommer, G.: Implementation of a Clifford algebra co-processor design on a field programmable gate array. In: Ablamowicz, R. (ed.) CLIFFORD ALGEBRAS: Application to Mathematics, Physics, and Engineering. Progress in Mathematical Physics, pp. 561–575. Birkhäuser, Basel (2003). 6th Int. Conf. on Clifford Algebras and Applications, Cookeville, TN
17. The RoboCup Federation. Robocup official site. Available at http://www.robocup.org

Some Applications of Gröbner Bases in Robotics and Engineering

Rafał Abłamowicz

Abstract Gröbner bases in polynomial rings have numerous applications in geometry, applied mathematics, and engineering. We show a few applications of Gröbner bases in robotics, formulated in the language of Clifford algebras, and in engineering to the theory of curves, including Fermat and Bézier cubics, and interpolation functions used in finite element theory.

1 Introduction

Gröbner bases were introduced in 1965 by Buchberger [5–8]. For an excellent exposition on their theory, see [10, 11, 17], while for a basic introduction with applications, see [9]. These bases gave rise to development of computer algebra systems like muMath, Maple, Mathematica, Reduce, AXIOM, CoCoCA, Macaulay, etc. Buchberger's Algorithm to compute Gröbner bases has been made more efficient [5] or replaced with another approach [12]. Algorithms to compute Gröbner bases have been implemented, for example, in Maple [26], Singular [15, 24], and FGb [12]. For a multitude of applications of Gröbner bases, see [8, 14] and, in particular, an online repository [16]. For a recent new application in geodesy, see [4].

2 Gröbner Basis Theory in Polynomial Rings

We follow presentation and notation from [10]. Let $k[x_1, \dots, x_n]$ be a polynomial ring in n indeterminates over a field k. Let $f_1, \dots, f_s$ be polynomials in

R. Abłamowicz (✉)
Department of Mathematics, Tennessee Technological University, Box 5054, Cookeville, TN 38505, USA
e-mail: rablamowicz@tntech.edu

E. Bayro-Corrochano, G. Scheuermann (eds.), *Geometric Algebra Computing*,
DOI 10.1007/978-1-84996-108-0_23, © Springer-Verlag London Limited 2010

$k[x_1, \dots, x_n]$. Then $\langle f_1, \dots, f_s\rangle$ denotes an ideal finitely generated by the chosen polynomials. We say that the polynomials form a basis of the ideal. Then, by $\mathbf{V}(f_1, \dots, f_s)$ we denote an *affine variety* defined by $f_1, \dots, f_s$, that is, a subset of k^n, possibly empty, consisting of all common zeros of $f_1, \dots, f_s$, namely

$$\mathbf{V}(f_1, \dots, f_s) = \{(a_1, \dots, a_n) \in k^n \mid f_i(a_1, \dots, a_n) = 0 \text{ for all } 1 \le i \le s\}.$$

In order to define Gröbner bases in the ideals of $k[x_1, \dots, x_n]$, we first need a concept of a monomial order.

Definition 1 A *monomial order* on $k[x_1, \dots, x_n]$ is any relation $>$ on $\mathbb{Z}^n_{\ge 0} = \{(\alpha_1, \dots, \alpha_n) \mid \alpha_i \in \mathbb{Z}_{\ge 0}\}$, or equivalently, any relation on the set of monomials x^α, $\alpha \in \mathbb{Z}^n_{\ge 0}$, satisfying: (i) $>$ is a total ordering on $\mathbb{Z}^n_{\ge 0}$ (for any $\alpha, \beta \in \mathbb{Z}^n_{\ge 0}$, $\alpha > \beta$, $\alpha = \beta$, or $\beta > \alpha$); (ii) If $\alpha > \beta$ and $\gamma \in \mathbb{Z}^n_{\ge 0}$, then $\alpha + \gamma > \beta + \gamma$; (iii) $>$ is a well-ordering on $\mathbb{Z}^n_{\ge 0}$ (every nonempty subset has smallest element). We will say that $x^\alpha > x^\beta$ when $\alpha > \beta$. Let $f = \sum_\alpha a_\alpha x^\alpha$ be a nonzero polynomial in $k[x_1, \dots, x_n]$. Then, the *multidegree* of f is

$$\operatorname{multideg}(f) = \max\{\alpha \in \mathbb{Z}^n_{\ge 0} : a_\alpha \ne 0\},$$

where the maximum is taken with respect to $>$. The *leading coefficient* of f is $\operatorname{LC}(f) = a_{\operatorname{multideg}}(f)$; the *leading monomial* of f is $\operatorname{LM}(f) = x^{\operatorname{multideg}(f)}$; and the *leading term* of f is $\operatorname{LT}(f) = \operatorname{LC}(f) \cdot \operatorname{LM}(f)$.

Some of the monomial orders are: (i) lexicographic (lex)[1]: $\alpha >_{\text{lex}} \beta$ if, in the vector difference $\alpha - \beta \in \mathbb{Z}^n$, the left-most nonzero entry is positive; (ii) graded reverse lex: $\alpha >_{\text{grevlex}} \beta$ if either $|\alpha| > |\beta|$, or $|\alpha| = |\beta|$ and in $\alpha - \beta \in \mathbb{Z}^n$ the right-most nonzero entry is negative; (iii) graded inverse lex: $\alpha >_{\text{ginvlex}} \beta$ if either $|\alpha| > |\beta|$, or $|\alpha| = |\beta|$ and in $\alpha - \beta \in \mathbb{Z}^n$ the right-most nonzero entry is positive.

In order to divide any polynomial f by a list of polynomials $f_1, \dots, f_s$, we need a generalized division algorithm.

Theorem 1 (General division algorithm) *Fix a monomial order $>$ on $\mathbb{Z}^n_{\ge 0}$, and let $F = (f_1, \dots, f_s)$ be an ordered s-tuple of polynomials. Then every $f \in k[x_1, \dots, x_n]$ can be written as*

$$f = a_1 f_1 + \cdots + a_s f_s + r, \tag{1}$$

where $a_i, r \in k[x_1, \dots, x_n]$, and either $r = 0$ or r is a linear combination, with coefficients in k, of monomials, none of which is divisible by any of $\operatorname{LT}(f_1), \dots, \operatorname{LT}(f_s)$. *We call r a* remainder *of f on division by F. Furthermore, if $a_i f_i \ne 0$, then we have* $\operatorname{multideg}(f) \ge \operatorname{multideg}(a_i f_i)$.

[1] In the following, $\operatorname{lex}(x_1, \dots, x_n)$ will denote the lex order in which $x_1 > x_2 > \cdots > x_n$.

The remainder r in (1) is not unique as it depends on the order of polynomials in F and on the monomial order. This shortcoming of the Division Algorithm disappears when we divide polynomials by a Gröbner basis.

Definition 2 Let $I \subset k[x_1, \ldots, x_n]$ be a nonzero ideal. Then, $\mathrm{LT}(I)$ is the set of leading terms of elements of I, and $\langle \mathrm{LT}(I) \rangle$ is the ideal generated by the elements of $\mathrm{LT}(I)$.

The ideal $\langle \mathrm{LT}(I) \rangle$ is an example of a monomial ideal. Dickson's Lemma states that every monomial ideal $I \subset k[x_1, \ldots, x_n]$ has a finite basis [10]. Since the polynomial ring $k[x_1, \ldots, x_n]$ is Noetherian, the famous theorem of Hilbert states that every ideal $I \subset k[x_1, \ldots, x_n]$ is finitely generated.

Theorem 2 (Hilbert basis theorem) *Every ideal $I \subset k[x_1, \ldots, x_n]$ has a finite generating set. That is, $I = \langle g_1, \ldots, g_s \rangle$ for some $g_1, \ldots, g_s \in I$.*

Definition 3 Fix a monomial order. A finite subset $G = \{g_1, \ldots, g_t\}$ of an ideal I is said to be a *Gröbner basis* if $\langle \mathrm{LT}(g_1), \ldots, \mathrm{LT}(g_t) \rangle = \langle \mathrm{LT}(I) \rangle$.

As a consequence of Hilbert's theorem, every ideal $I \subset k[x_1, \ldots, x_n]$ other than $\{0\}$ has a Gröbner basis once a monomial order has been chosen.

When dividing f by a Gröbner basis, we denote the remainder as $r = \overline{f}^G$. Due to the uniqueness of r, one gets unique coset representatives for elements in the quotient ring $k[x_1, \ldots, x_n]/I$: The coset representative of $[f] \in k[x_1, \ldots, x_n]/I$ will be $\overline{f}^G$.

Gröbner bases are computed using various algorithms. The most famous one is the Buchberger's algorithm that uses S-polynomials. One of its many modifications is discussed in [10], whereas [12] implements a completely different approach.

Definition 4 The *S-polynomial* of $f_1, f_2 \in k[x_1, \ldots, x_n]$ is defined as $S(f_1, f_2) = \frac{x^\gamma}{\mathrm{LT}(f_1)} f_1 - \frac{x^\gamma}{\mathrm{LT}(f_2)} f_2$, where $x^\gamma = \mathrm{lcm}(\mathrm{LM}(f_1), \mathrm{LM}(f_2))$, and $\mathrm{LM}(f_i)$ is the leading monomial of f_i w.r.t. some monomial order.[2]

Theorem 3 (Buchberger theorem) *A basis $\{g_1, \ldots, g_t\} \subset I$ is a Gröbner basis of I if and only if $\overline{S(g_i, g_j)}^G = 0$ for all $i < j$.*

Buchberger's algorithm for finding a Gröbner basis implements the above criterion: If $F = \{f_1, \ldots, f_s\}$ fails because $\overline{S(f_i, f_j)}^G \neq 0$ for some $i < j$, then add this remainder to F and try again.

Gröbner bases computed with the Buchberger's algorithm are usually too large: A standard way to reduce them is to replace any polynomial f_i with its remainder on division by $\{f_1, \ldots, f_{i-1}, f_{i+1}, \ldots, f_t\}$, removing zero remainders, and for

[2]Here, $\mathrm{lcm}(\mathrm{LM}(f_1), \mathrm{LM}(f_2))$ denotes the least common multiple of the leading monomials $\mathrm{LM}(f_1)$ and $\mathrm{LM}(f_2)$.

polynomials that are left, making their leading coefficient equal to 1. This produces a *reduced Gröbner basis*. For a fixed monomial order, it is well known that any ideal in $k[x_1, \ldots, x_n]$ has a *unique* reduced Gröbner basis. See, for example, [10] and references therein.

2.1 Examples of Using Gröbner Bases

There are many problems, in many different areas of mathematics and applied sciences, that can be solved using Gröbner bases. Here we just list a few applied problems:

- Solving systems of polynomial equations, e.g., intersecting surfaces and curves, finding closest point on a curve or on a surface to the given point, Lagrange multiplier problems (especially those with several multipliers), etc. Solutions to these problems are based on the so-called Extension Theory [10].
- Finding equations for equidistant curves and surfaces to curves and surfaces defined in terms of polynomial equations, such as conic sections, Bézier cubics; finding syzygy relations among various sets of polynomials, for example, symmetric polynomials, finite group invariants, interpolating functions, etc. Solutions to these problems are based on the so-called Elimination Theory [10].
- Finding equidistant curves and surfaces as envelopes to appropriate families of curves and surfaces, respectively [2, 10].
- The implicitization problem, i.e., eliminating parameters and finding implicit forms for curves and surfaces.
- The forward and the inverse kinematic problems in robotics [7, 10].
- Automatic geometric theorem proving [7, 8, 10].
- Expressing invariants of a finite group in terms of generating invariants [10].
- Finding relations between polynomial functions, e.g., interpolation functions (syzygy relations).
- For recent applications in geodesy, see [4].
- See also bibliography on Gröbner bases at Johann Radon Institute for Computational and Applied Mathematics (RICAM) [16].

Our first example is an inverse kinematics problem consisting of finding an elbow point of a robot arm on the circle of intersection between two spheres. This problem is elegantly formulated in the language of conformal algebra CGA in [18, 19].

Example 1 (Elbow of a robot arm) We model CGA as a Clifford algebra of a five-dimensional real vector space V which is an extension of 3D Euclidean space by an origin-infinity plane. Let $\{\mathbf{e}_1, \mathbf{e}_2, \mathbf{e}_3, \mathbf{e}_4, \mathbf{e}_5\}$ be basis vectors for V which satisfy the following relations in CGA:

$$\begin{aligned} &\mathbf{e}_i^2 = 1, \qquad \mathbf{e}_i \cdot \mathbf{e}_j = \mathbf{e}_j \cdot \mathbf{e}_i = 0, \qquad \mathbf{e}_i \cdot \mathbf{e}_4 = \mathbf{e}_i \cdot \mathbf{e}_5 = 0, \\ &\mathbf{e}_4^2 = \mathbf{e}_5^2 = 0, \qquad \mathbf{e}_4 \cdot \mathbf{e}_5 = -1, \end{aligned} \tag{2}$$

for $i, j = 1, 2, 3$ and $i \neq j$.[3] Euclidean points, spheres, and planes are modeled, respectively, in CGA by the following 5D vectors:

$$P = \mathbf{p} + \frac{1}{2}\mathbf{p}^2 e_\infty + e_0, \qquad S = \mathbf{s} + \frac{1}{2}(\mathbf{s}^2 - r^2)e_\infty + e_0, \qquad \pi = \mathbf{n} + d e_\infty, \quad (3)$$

where $\mathbf{p}$ is the 3D point location, $\mathbf{s}$ is the 3D sphere center, and r is the sphere radius,[4] $\mathbf{n}$ is the 3D unit normal vector of the plane, and d is the distance of the plane from the origin. In particular, for a sphere, we have $S^2 = r^2$.[5] Two spheres S_1 and S_2 intersect in a 3D circle (resp., a single point, or do not intersect) when $S_1 \cdot S_2 + r_1 r_2 > 0$ (resp., $S_1 \cdot S_2 + r_1 r_2 = 0$, or otherwise). The circle is represented in CGA by the element $C = S_1 \wedge S_2$.

It is shown in [18] that when two spheres intersect in a circle, the bivector C equals the following quantity:

$$Z = \mathbf{c} \wedge \mathbf{n_c} - \mathbf{n_c} \wedge e_0 - (\mathbf{c} \cdot \mathbf{n_c})E + \left[(\mathbf{c} \cdot \mathbf{n_c})\mathbf{c} - \frac{1}{2}(\mathbf{c}^2 - r^2)\mathbf{n_c}\right] \wedge e_\infty, \quad (4)$$

where $E = e_\infty \wedge e_0$, $\mathbf{c}$ is the circle center, r is its radius, and vector $\mathbf{n_c}$ is normal to the plane π in 3D containing the circle. We will solve a system of polynomial equations resulting from the condition $Z = C$ for the components of $\mathbf{c}$, $\mathbf{n_c}$ and for the radius r with a Gröbner basis. For example, let

$$f_1 = 4(x_1 - 1)^2 + 4x_2^2 + 4x_3^2 - 9 \quad \text{and} \quad f_2 = (x_1 + 1)^2 + x_2^2 + x_3^2 - 4 \quad (5)$$

be in $\mathbb{R}[x_1, x_2, x_3]$. Then, $\mathbf{V}(f_1)$ and $\mathbf{V}(f_2)$ are the two spheres viewed as varieties. These two spheres are represented in CGA as these 1-vectors:

$$S_1 = \mathbf{e}_1 - \frac{5}{8}e_\infty + e_0, \qquad S_2 = -\mathbf{e}_1 - \frac{3}{2}e_\infty + e_0. \quad (6)$$

Since $S_1 \cdot S_2 + r_1 r_2 = \frac{33}{8} > 0$, the spheres intersect in a circle C given as

$$C = S_1 \wedge S_2 = -\frac{17}{8}\mathbf{e}_1 \wedge e_\infty + 2\mathbf{e}_1 \wedge e_0 - \frac{7}{8}e_0 \wedge e_\infty. \quad (7)$$

By letting $C = Z$ and $\mathbf{n_c} = (n_1, n_2, n_3)$, $\mathbf{c} = (c_1, c_2, c_3)$, and by equating symbolic coefficients at corresponding Grassmann basis monomials, we obtain the following system of polynomial equations:

---

[3] We identify $\mathbf{e}_4$ with the origin vector e_0 and $\mathbf{e}_5$ with the infinity vector e_∞ from [18]. Thus, CGA is isomorphic to the Clifford algebra $C\ell_{4,1}$. The dot $\cdot$ denotes the inner product in V.

[4] Of course, we can recover the analytic equation of the sphere by setting $(\mathbf{x} - \mathbf{s})^2 = S^2$ where $\mathbf{x}$ is a vector in 3D from the origin to a surface point on the sphere S.

[5] To be precise, $S^2 = r^2 1$ where 1 is the identity element of CGA. When $r = 0$, the sphere becomes a point.

$$
\begin{aligned}
f_1 &= c_1n_3 - c_3n_1 = 0,\\
f_2 &= -c_3n_2 + c_2n_3 = 0,\\
f_3 &= c_1n_2 - c_2n_1 = 0,\\
f_4 &= 8c_3n_3 + 8c_2n_2 + 8c_1n_1 + 7 = 0,\\
f_5 &= -n_1 - 2 = 0,\\
f_6 &= -n_2 = 0,\\
f_7 &= -n_3 = 0,\\
f_8 &= 8c_1c_2n_2 + 4c_1^2n_1 + 17 + 4n_1r^2 - 4n_1c_2^2 + 8c_1c_3n_3 - 4n_1c_3^2 = 0,\\
f_9 &= c_2^2n_2 + 2c_2c1n_1 - n_2c_1^2 + 2c_2c_3n_3 + n_2r^2 - n_2c_3^2 = 0,\\
f_{10} &= 2c_3c_2n_2 + c_3^2n_3 + 2c_3c_1n_1 - n_3c_1^2 - n_3c_2^2 + n_3r^2 = 0.
\end{aligned}
\tag{8}
$$

The reduced Gröbner basis for the ideal I generated by the above ten polynomials in lex order with $n_1 > n_2 > n_3 > c_1 > c_2 > c_3 > r$ is

$$\{-495 + 256r^2, c_3, c_2, -7 + 16c_1, n_3, n_2, n_1 + 2\}, \tag{9}$$

from which we get, as expected, that $\mathbf{n_c} = (-2, 0, 0)$, $\mathbf{c} = (0, 0, 0)$, and $r = \frac{\sqrt{5}}{2}$. In the same way, one can handle the degenerate cases where the spheres just touch at a single point, where one is included in the other, and where they do not intersect.

Our second example is related to the above and shows how to visualize the circle of intersection of two spheres $C = S_1 \cap S_2$ as an intersection of a cylinder and a plane. Often such visualizations simplify the picture and especially when one considers an additional constraint.

Example 2 Let S_1 and S_2 be the spheres defined by the polynomials f_1 and f_2 given in (5), that is, $S_1 = \mathbf{V}(f_1)$ and $S_2 = \mathbf{V}(f_2)$. Then, a reduced Gröbner basis for the ideal J generated by these two polynomials for the *lex* order $x_1 > x_2 > x_3$ is

$$G = \{256x_2^2 + 256x_3^2 - 495,\ 16x_1 - 7\}, \tag{10}$$

where the first polynomial c gives the cylinder $\mathbf{V}(c)$, and the second of course is the plane $\mathbf{V}(\pi)$. Thus, the circle $C = S_1 \cap S_2 = \mathbf{V}(c) \cap \mathbf{V}(\pi)$ can be visualized in two different ways: As the intersection of the two spheres Fig. 1 or as the intersection of the cylinder and the plane Fig. 2. By adding an additional constraint consisting, for example, of an additional plane $\mathbf{V}(\pi_2)$ defined by the polynomial $\pi_2 = x_3 - x_1 - \frac{1}{4}$, we can identify two points on the circle C and the plane π_2. See Fig. 3. To find their coordinates, it is enough to solve the system of polynomial equations $f_1 = 0$, $f_2 = 0$, $\pi_2 = 0$. We can employ the Gröbner basis approach once more by computing a reduced basis for the ideal J generated by f_1, f_2, π_2 for the lex order $x_1 > x_2 > x_3$,

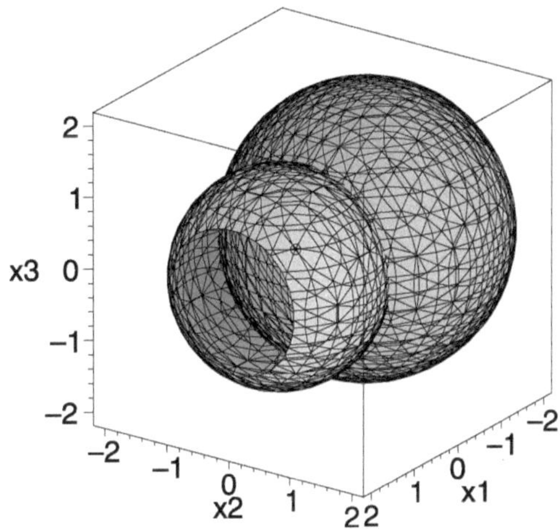

Fig. 1 Circle C as the intersection of two spheres $\mathbf{V}(S_1) \cap \mathbf{V}(S_2)$

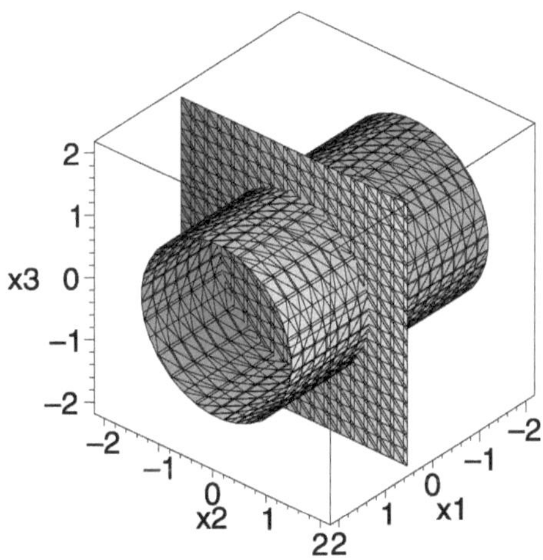

Fig. 2 Circle C as the intersection of the cylinder and the plane $\mathbf{V}(c) \cap \mathbf{V}(\pi_1)$

and we get

$$G = \left\{16x_3 - 11, 128x_2^2 - 187, 16x_1 - 7\right\}, \tag{11}$$

which gives the two points $P_1,\ P_2 = (x_1 = \frac{7}{16},\ x_2 = \pm\frac{1}{16}\sqrt{374},\ x_3 = \frac{11}{16})$.

Our third example is a classical problem of finding equidistant curves (envelopes) of various polynomial curves. Here we show how a general envelope of a parabola can be computed in a general case. Such problems also appear in engineering in designing cam mechanisms (cf. [22] and [25]).

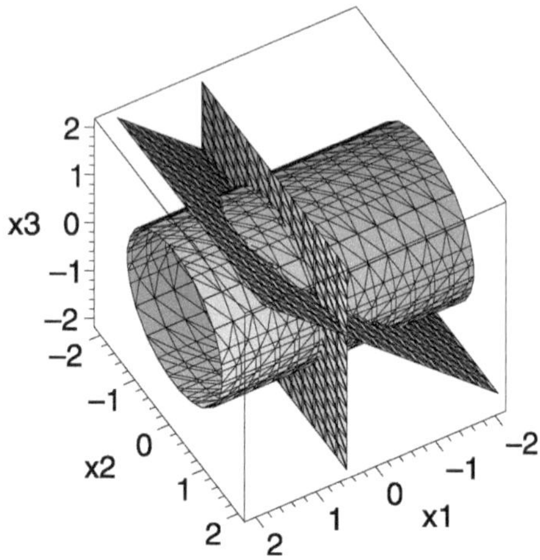

Fig. 3 Two points as the intersection of the circle C and the additional plane $\mathbf{V}(\pi_2)$

Example 3 (Equidistant curves to a parabola) We compute equidistant curves to the parabola defined by the polynomial

$$f_1 = 4py_0 - x_0^2 = 0, \tag{12}$$

where $|p|$ denotes the distance between the focus $F = (0, p)$ and the vertex $V = (0, 0)$. Polynomial f_2 defines the circle of radius (offset) r centered at a point (x_0, y_0) on the parabola f_1,

$$f_2 = (y - y_0)^2 + (x - x_0)^2 - r^2 = 0, \tag{13}$$

while the polynomial f_3,

$$f_3 = 2xp - 2x_0p + x_0y - x_0y_0 = 0 \tag{14}$$

gives the condition that a point $P(x, y)$ on the circle f_2 lies on a line perpendicular to the parabola f_1 at the point (x_0, y_0). There are two such points for any given point (x_0, y_0), one on each side of the parabola. All these points P belong to an affine variety $\mathbf{V} = \mathbf{V}(f_1, f_2, f_3)$—the *envelope* of the family of circles—and define two equidistant curves at the distance r from the parabola. To find a single polynomial equation for this envelope, we compute a reduced Gröbner basis for the ideal $I = \langle f_1, f_2, f_3\rangle \subset \mathbb{R}[x_0, y_0, x, y, p, r]$ for a suitable elimination order. Then, eliminating variables x_0 and y_0 gives a single polynomial $g \in \mathbb{R}[x, y, p, r]$ that defines the envelope. Polynomial g provides a Gröbner basis for the second elimination ideal $I_2 = I \cap \mathbb{R}[x, y, p, r]$.

The reduced Gröbner basis G for the ideal I for $lex(y_0, x_0, x, y, r, p)$ order consists of fourteen homogeneous polynomials while $I_2 = \langle g\rangle$ where g is as follows:

$$\begin{aligned} g = &-2pr^2yx^2 + 8pr^2y^3 + 8p^2r^2y^2 - 32yp^3r^2 + 16p^4r^2 - 16y^4p^2 + 32y^3p^3 \\ &- 16p^4y^2 + 3r^2x^4 + 8p^2r^4 + 20p^2r^2x^2 - y^2x^4 + 10ypx^4 - x^6 - x^4p^2 \end{aligned}$$

$$+8py^3x^2 - 32x^2y^2p^2 + 8x^2yp^3 - 3r^4x^2 + 2r^2x^2y^2$$
$$+r^6 - r^4y^2 - 8pr^4y. \tag{15}$$

It is possible now to analytically analyze singularities of the envelope by finding points on $\mathbf{V}(g)$ where $\nabla g = 0$. This gives a critical value $r_{\text{crit}} = 2|p|$ of r that determines whether $\mathbf{V}(I_2)$ has one or three singular points. For more details, as well as for a complete treatment of other conics, see [2].

In a manner similar to Example 3, it is possible to analyze envelopes and their singularities of other curves defined via polynomial equations like Fermat curves, Bézier cubics, etc., and surfaces, like quadrics, Bézier surfaces. This aids in studying the so-called *caustics* [3], shell structures through the finite element analysis [2], and in designing machinery [22].

3 Fermat Curves and Bézier Cubics

In this section we briefly discuss other curves such as the Fermat curves and the Bézier cubics. We begin with the Fermat curves.

3.1 Fermat Curves

The Fermat curves are defined as

$$f_1 = x_0^n + y_0^n - c^n, \quad n \in \mathbb{Z}^+,\ c > 0.$$

Define f_2 to be the circle of radius r centered at (x_0, y_0) on f_1, and let f_3 give a normal line to f_1 at (x_0, y_0).

Let $n = 3$ and $I = \langle f_1, f_2, f_3\rangle \subset \mathbb{R}[x_0, y_0, x, y, r, c]$, where

$$f_1 = x_0^3 + y_0^3 - c^3, \qquad f_2 = (x - x_0)^2 + (y - y_0)^2 - r^2,$$
$$f_3 = xy_0^2 - x_0y_0^2 - x_0^2y + x_0^2y_0.$$

For the elimination order lexdeg($[x_0, y_0], [x, y, r, c]$),[6] the reduced Gröbner basis for I consists of 129 polynomials of which only one $g \in \mathbb{R}[x, y, r, c]$. Polynomial g has 266 terms and is homogeneous of degree 18. The variety $\mathbf{V}(I)$ contains our offset curves shown in Fig. 4.

---

[6] In the degree lexicographic order lexdeg($[x_0, y_0], [x, y, r, c]$) monomials involving only x_0 and y_0 are compared using the total degree tdeg(x_0, y_0) (tdeg is also known as the graded reverse lexicographic order grevlex); monomials involving only x, y, r, c are compared using the term order tdeg(x, y, r, c); a monomial involving x_0 or y_0 is higher than another monomial involving only x, y, r, c. Such a term order is usually used to eliminate the indeterminates listed in the first list, namely x_0, y_0 [10].

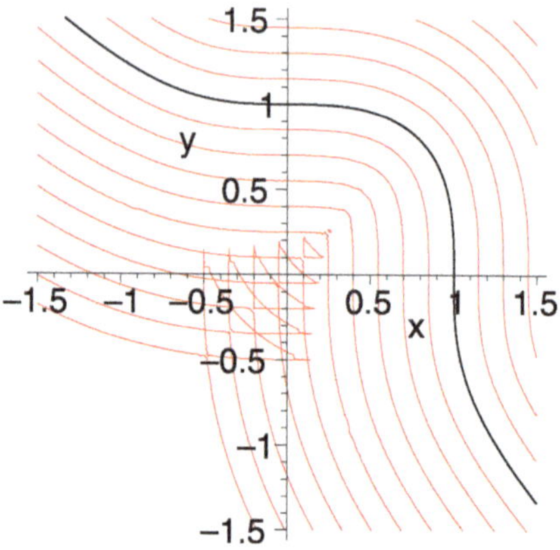

Fig. 4 Fermat cubic $n = 3$ with equidistant curves and growing singularities

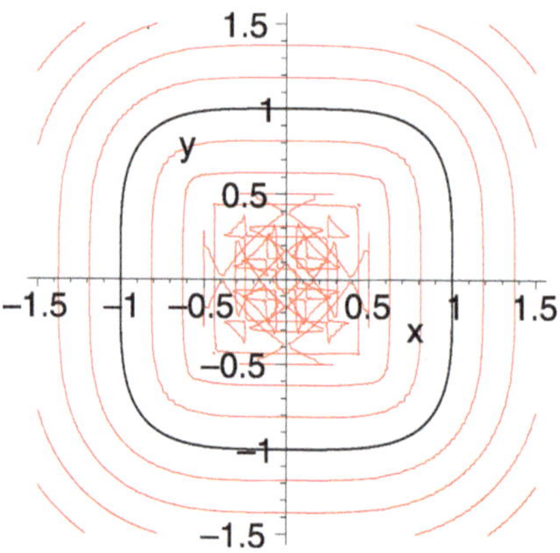

Fig. 5 Fermat cubic $n = 4$ with equidistant curves and growing singularities

Let $n = 4$ and $I = \langle f_1, f_2, f_3 \rangle \subset \mathbb{R}[x_0, y_0, x, y, r, c]$, where

$$f_1 = x_0^4 + y_0^4 - c^4, \qquad f_2 = (x - x_0)^2 + (y - y_0)^2 - r^2,$$
$$f_3 = xy_0^3 - x_0y_0^3 - x_0^3y + x_0^3y_0.$$

A reduced Gröbner basis for I for the same order consists of 391 polynomials of which only one $g \in \mathbb{R}[x, y, r, c]$. Polynomial g has 525 terms and is homogeneous of degree 32. The variety $\mathbf{V}(I)$ contains our offset curves shown in Fig. 5.

3.2 *Bézier Cubics*

A Bézier cubic is defined parametrically as

$$\begin{aligned} X &= (1-t)^3x_1 + 3t(1-t)^2x_2 + 3t^2(1-t)x_3 + t^3x_4, \\ Y &= (1-t)^3y_1 + 3t(1-t)^2y_2 + 3t^2(1-t)y_3 + t^3y_4, \end{aligned} \tag{16}$$

Fig. 6 Bézier cubic in implicit form with control points and points where curvature is maximum or minimum

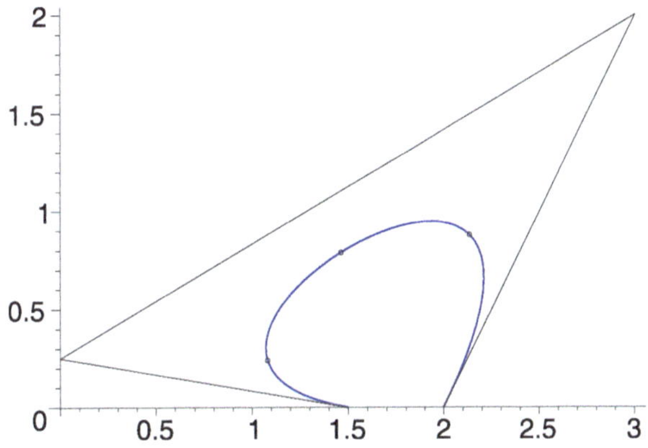

where (x_i, y_i), $i = 1, \ldots, 4$, are the coordinates of four control points, and $0 \le t \le 1$.

3.2.1 First Application of Gröbner Bases to Bézier Cubics

In our first application of Gröbner basis we show how to eliminate the parameter t from the defining polynomials (16) and find a single polynomial that defines the cubic implicitly. We compute a reduced Gröbner basis G for the ideal

$$I = \langle x - X, y - Y \rangle \subset \mathbb{R}[x, y, x_i, y_i, t]$$

for the elimination order lexdeg$([t], [x, y, x_i, y_i])$. The basis G has 12 polynomials of which only one g does not contain t, or $g \in \mathbb{R}[x, y, x_i, y_i]$. Then, g is homogeneous of degree 6, and it contains 460 terms.

Let the control points be $(\frac{3}{2}, 0)$, $(0, \frac{1}{4})$, $(3, 2)$, $(2, 0)$. Then, the Bézier cubic in Fig. 6 has this implicit form:

$$\begin{aligned} g = &-213948x + 66420y - 214164yx - 145656y^2x + 89964x^2y \\ &+ 110079x^2 + 135756 + 78608y^3 - 18522x^3 + 219456y^2 = 0. \end{aligned} \quad (17)$$

3.2.2 Second Application of Gröbner Bases to Bézier Cubics

As our second application, we find a parameterization $(x(t), y(t))$ for a variety of equidistant curves to a Bézier cubic by computing a reduced Gröbner basis G for a suitable ideal I in elimination order lexdeg$([y], [x, t])$.

Let the control points be $(2, 0)$, $(3, 3)$, $(4, 1)$, $(3, 0)$; then

$$X = 2 + 3t - 2t^3, \qquad Y = 9t - 15t^2 + 6t^3,$$

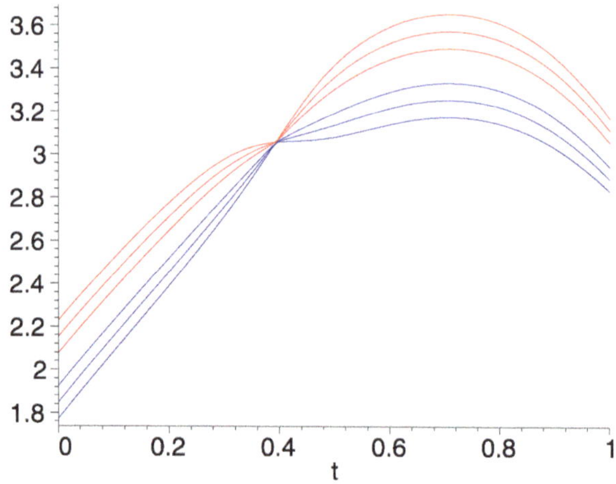

Fig. 7 Graphs of gx_{1i} and gx_{2i} for $i = 1, 2, 3$

and consider an ideal $I = \langle x - X, y - Y, f_3, f_4\rangle \in \mathbb{R}[x, y, r, t]$, where f_3 gives an equation of the circle of radius r at $(X(t), Y(t))$ on the cubic

$$f_3 = (y - Y)^2 + (x - X)^2 - r^2, \tag{18}$$

while f_4 gives an equation of a normal at $(X(t), Y(t))$ to the cubic

$$\begin{aligned} f_4 = 3x - 6xt^2 - 6 + 417t^2 - 90t - 642t^3 \\ - 120t^5 + 9y - 30yt + 18yt^2 + 450t^4 \end{aligned} \tag{19}$$

for the offset values $r = \frac{2}{25}, \frac{4}{25}, \frac{6}{25}$. For each r, the basis G has four polynomials $I = \langle h_1, h_2, h_3, h_4\rangle$: Polynomial h_1 depends only on x, t and is quadratic in x, while polynomials h_2, h_3, h_4 depend on x, y, t. The discriminant of h_1 is always positive or zero when $t_s =$ for $0 \le t \le 1$. This means that x can always be parameterized in terms of t by solving $h_1 = 0$ for x with radicals [10].

Let gx_{1i}, gx_{2i} be the solutions of $h_1 = 0$ for x for three offset values of r, $i = 1, 2, 3$. Then each gx_{1i}, gx_{2i} is continuous but not smooth. Furthermore, the first elimination ideal is $I_1 = \langle h_1\rangle = I \cap \mathbb{R}[x, t]$. Here, indices 1 (red) and 2 (blue) refer to opposite sides of the Bézier cubic in Fig. 7. The graphs intersect at $t = t_s =$ given above. Polynomials h_2, h_3 are linear in y, while h_4 is quadratic in y, and $h_2, h_3, h_4 \in \mathbb{R}[x, t][y]$. Since one of their leading coefficients is a nonzero constant, by the Extension Theorem [10], any partial solution of $h_1 = 0$ in terms of x and t is extendable to the y-solution.

To find a parameterization for $y = y(t)$, substitute gx_{1i}, gx_{2i} into h_3 for each $i = 1, 2, 3$ and solve for y: We obtain discontinuous functions gy_{1i}, gy_{2i} that belong to the variety $\mathbf{V}(h_2, h_3, h_4)$. The single discontinuity appears at $t = t_s$. See Fig. 8.

For the offset values r smaller than r_{crit}, the equidistant curves are smooth curves in the Euclidean plane since, as shown later, they can be defined globally by an equation $g(x, y) = 0$ where g is a polynomial of two variables, and the partial derivatives

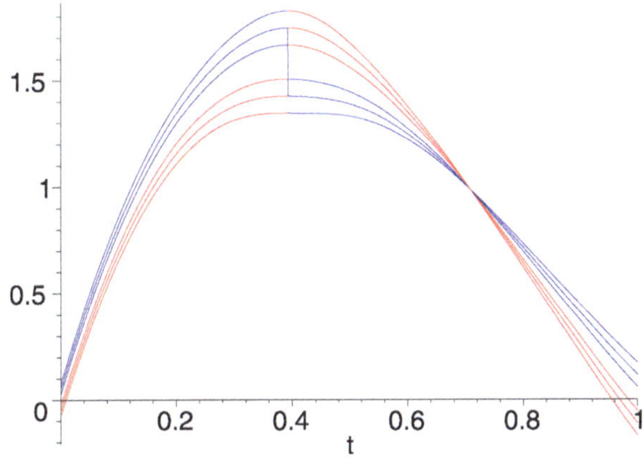

Fig. 8 Graphs of gy_{1i} and gy_{2i} for $i = 1, 2, 3$

g_x and g_y are not simultaneously equal to 0. That is, equidistant curves to the Bézier cubic form a nonsingular variety when $r < r_{\text{crit}}$. See Fig. 9.

For the Bézier cubic, we conjecture that $r_{\text{crit}} = \frac{1}{\kappa_{\max}} = \rho_{\min}$. In this example we chose the three values of r to satisfy this condition. Thus, we can parameterize the equidistant curves for the Bézier cubic in our example as, for one side,

$$Gx_{1i}(t) = \begin{cases} gx_{1i}(t), & t < t_s, \\ gx_{2i}(t), & t \geq t_s, \end{cases} \qquad Gy_{1i}(t) = \begin{cases} gy_{1i}(t), & t < t_s, \\ y_{1i}, & t = t_s, \\ gy_{2i}(t), & t > t_s, \end{cases} \tag{20}$$

where $y_{1i} = \lim_{t \to t_s^-} gy_{1i}(t) = \lim_{t \to t_s^+} gy_{2i}(t)$, $i = 1, 2, 3$. Likewise, for the other side,

$$Gx_{2i}(t) = \begin{cases} gx_{2i}(t), & t < t_s, \\ gx_{1i}(t), & t \geq t_s, \end{cases} \qquad Gy_{2i}(t) = \begin{cases} gy_{2i}(t), & t < t_s, \\ y_{2i}, & t = t_s, \\ gy_{1i}(t), & t > t_s, \end{cases} \tag{21}$$

where $y_{2i} = \lim_{t \to t_s^-} gy_{2i}(t) = \lim_{t \to t_s^+} gy_{1i}(t)$, $i = 1, 2, 3$.

3.2.3 Third Application of Gröbner Bases to Bézier Cubics

In this section we will find a general polynomial $g \in \mathbb{R}[x, y, r]$ so that the polynomial equation $g = 0$ will give equidistant curves to a Bézier cubic at an arbitrary offset r. We first find a reduced Gröbner basis for a suitable ideal in the elimination order lexdeg($[t, X, Y], [x, y, r]$).

Let the control points be $(2, 0), (3, 3), (4, 1), (3, 0)$; then define

$$f_1 = X - 2 - 3t + 2t^3, \qquad f_2 = Y - 9t + 15t^2 - 6t^3, \tag{22}$$

and consider an ideal $I = \langle f_1, f_2, f_3, f_4 \rangle \subset \mathbb{R}[t, X, Y, x, y, r]$ where f_3 as in (18) gives an equation of the circle of radius r at $(X(t), Y(t))$ on the cubic, while f_4

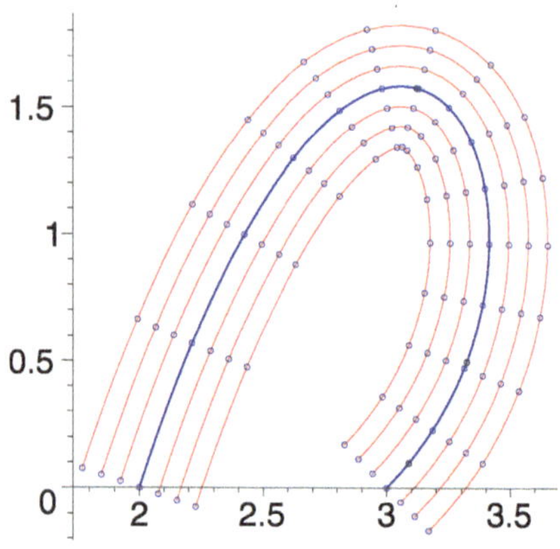

Fig. 9 Bézier cubic with parametric equidistant curves, nodes, points of max/min curvature, and switch points where $t = t_s$

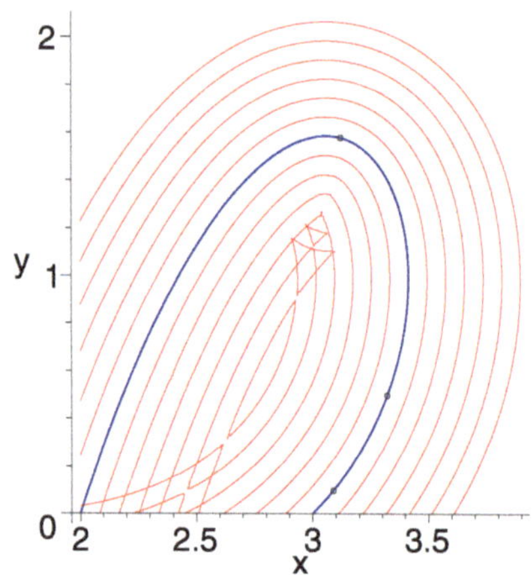

Fig. 10 Bézier cubic with implicit equidistant curves showing growing singularities as the offset r increases

gives an equation of a normal at $(X(t), Y(t))$ to the cubic

$$f_4 = -x + 2xt^2 + X - 2Xt^2 - 3y + 10yt - 6yt^2 + 3Y \tag{23}$$

for an arbitrary (nonnegative) offset r.

The reduced Gröbner basis for I contains twenty seven polynomials, of which only one belongs to $\mathbb{R}[x, y, r]$. That is,

$$I_3 = I \cap \mathbb{R}[x, y, r] = \langle g \rangle,$$

where g is of total degree ten in x, y, r, and it has 161 terms (but only 66 terms for any specific value of r).

In Fig. 10 we plot a few equidistant curves for various offsets with growing singularities of which one is again of the dove-tail type and it appears across the point of maximum curvature.

By analyzing solutions to $\nabla(g) = 0$ on $g = 0$ for a general offset r, one can search for singularities, if any, of equidistant curves to this specific Bézier cubic.

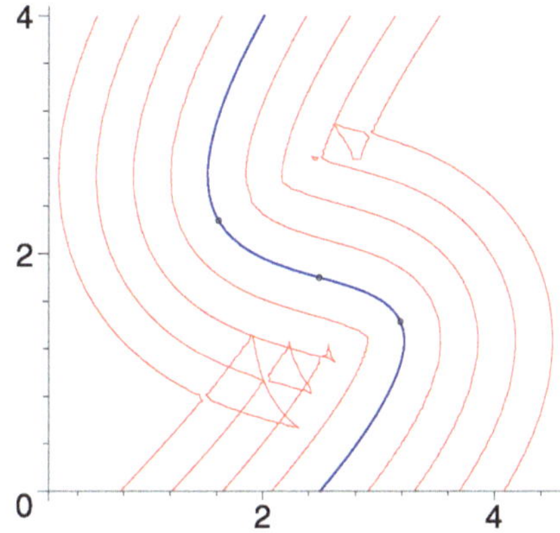

Fig. 11 Bézier cubic with implicit equidistant curves showing growing singularities as the offset r increases

As an another example, let us define an S-shaped Bézier cubic

$$X = 2 - 6t + 21t^2 - \frac{29}{2}t^3, \qquad Y = 4 - \frac{21}{2}t + 18t^2 - \frac{23}{2}t^3. \tag{24}$$

Its graph along with a few equidistant offset curves is shown in Fig. 11.

Let us summarize advantages and disadvantages of this approach.

- *Advantages*:
  - Equidistant curves to any Bézier cubic are given globally as one single polynomial of total degree ten in x, y, r for any offset r.
  - This permits their analytic analysis, including analysis of their singularities, if any, as well as finding the critical value of the offset r_{crit}.
  - Ease of graphing.
  - Ease of finding nodes for finite elements (to any desired accuracy).
- *Disadvantages*:
  - Computational complexity although computation of g for the given cubic, i.e., for the chosen control points, takes only a few seconds.
  - Computation of g for a general Bézier cubic when

    $$I = \langle f_1, f_2, f_3, f_4 \rangle \subset \mathbb{R}[X, Y, x, y, t, r, x_i, y_i], \quad i = 1, 2, 3, 4,$$

    is beyond the computational ability of present-day PC.

Example 4 (Distance to ellipse) In this example we find a point (or points) on ellipse $f_1 = \frac{x^2}{a^2} + \frac{y^2}{b^2} - 1$ that minimizes distance from the ellipse to a given point $P = (x_0, y_0)$ not on the ellipse and such that $x_0 \neq 0$.[7] Thus, one needs first to find points Q on the ellipse such that a line T tangent to the ellipse at Q is orthogonal to the vector $\overrightarrow{QP}$. Let $f_2 = a^2 y(x - x_0) - b^2 x(y - y_0)$. Then, the condition $f_2 = 0$ assures

[7] When $x_0 = 0$, then the solution is obvious.

that the vector $\overrightarrow{QP} \perp T$. We will use values $x_0 = 4,\ y_0 = \frac{3}{2},\ a = 2,\ b = 1$. Thus, we must to solve the system of equations

$$f_1 = 4y^2 + x^2 - 4 = 0 \quad \text{and} \quad f_2 = 6yx - 32y + 3x = 0 \tag{25}$$

for x and y. We will find the reduced Gröbner basis for the ideal $I = \langle f_1, f_2 \rangle$ that defines $V = \mathbf{V}(f_1, f_2)$ for lex(x, y) order. The basis contains two polynomials

$$\begin{aligned} g_1 &= -18y^3 + 9 - 9y^2 + 12x - 110y, \\ g_2 &= -9 - 36y + 229y^2 + 36y^3 + 36y^4. \end{aligned} \tag{26}$$

Observe that g_2 belongs to $I_2 = I \cap \mathbb{R}[y]$. Observe also that the leading coefficient in g_1 w.r.t. lex(x, y) is 12; hence by the Extension Theorem [10], every partial solution to the system $\{g_1 = 0, g_2 = 0\}$ on the variety $\mathbf{V}(g_2)$ can be extended to a complete solution of (25) on the variety V. Since polynomial g_2 is of degree 4, its solutions are expressible in radicals. When approximated, two real values of y are $y_1 = 0.2811025120$ and $y_2 = -0.1354474035$. Each of the exact values of y, when substituted into the equation $g_1 = 0$ yields the exact value of x. Thus, we have two points Q on the ellipse whose approximate coordinates are $Q_1 = (1.919355494, 0.2811025085)$ and $Q_2 = (-1.981569077, -0.1354473991)$. Checking the distances, one finds $\|\overrightarrow{Q_1P}\| = 2.411388118 < \|\overrightarrow{Q_2P}\| = 6.201117385$, or, that the point Q_1 is closest to the given point P.

Repeating this example in the purely symbolic case where a, b, x_0, y_0 remain unassigned, returns again a two-polynomial reduced Gröbner basis for I:

$$\begin{aligned} G = \big[& a^4y^4 - a^4y^2b^2 + 2a^2y^2b^4 - 2a^2b^2y^4 + a^2y^2x_0^2b^2 + 2a^2b^2y^3y_0 \\ & - 2a^2yb^4y_0 - b^6y_0^2 - 2y^3y_0b^4 - y^2b^6 + 2yb^6y_0 + y^4b^4 + y^2y_0^2b^4, \\ & a^2b^4y_0 - b^6y_0 - a^2b^2y^2y_0 + b^4y^2y_0 + a^4yb^2 - 2a^2yb^4 + yb^6 \\ & - a^2x_0^2yb^2 - a^4y^3 + 2a^2b^2y^3 - b^4y^3 + x_0b^4xy_0\big], \end{aligned} \tag{27}$$

where the first polynomial is of degree 4 in y and is, in principle, solvable with radicals. The second polynomial is again of degree 1 in the variable x. Thus, in general, this problem is solvable in radicals.

In [4] a similar problem is studied: It consists of finding the distance between a point P on the Earth's topographic surface and the closest to it point p on the *international reference ellipsoid* $\mathbb{E}^2_{a,a,b}$. This is another example of a constrained minimization problem which is set up with the help of a constrained Lagrangian, while the resulting system of four polynomial equations is solved with Gröbner basis.

Example 5 (Rodrigues matrix) Recall that the trigonometric form of a quaternion $a = a_0 + \mathbf{a} \in \mathbb{H}$ is $a = \|a\|(\cos\alpha + \mathbf{u}\sin\alpha)$, where $\mathbf{u} = \mathbf{a}/|\mathbf{a}|$, $|\mathbf{a}|^2 = {a_1}^2 + {a_2}^2 + {a_3}^2$, and α is determined by $\cos\alpha = a_0/\|a\|$, $\sin\alpha = |\mathbf{a}|/\|a\|$, $0 \leq \alpha < \pi$. Then, any quaternion can be written as

$$a = \|a\|\big(\cos\alpha + |\mathbf{a}|^{-1}(a_1\mathbf{i} + a_2\mathbf{j} + a_3\mathbf{k})\sin\alpha\big). \tag{28}$$

The following theorem can be found in [20, 21].

Theorem 4 *Let a and r be quaternions with nonzero vector parts where $\|a\| = 1$, so $a = \cos\alpha + \mathbf{u}\sin\alpha$ where $\mathbf{u}$ is a unit vector. Then, the norm and the scalar part of the quaternion $r' = ara^{-1}$ equal those of r, that is, $\|r'\| = \|r\|$ and $\mathrm{Re}(r') = \mathrm{Re}(r)$. The vector component $\mathbf{r}' = \mathrm{Im}(r')$ gives a vector $\mathbf{r}' \in \mathbb{R}^3$ resulting from a finite rotation of the vector $\mathbf{r} = \mathrm{Im}(r)$ by the angle 2α counter-clockwise about the axis $\mathbf{u}$ determined by a.*

Let $a = a_0 + \mathbf{a}$, $b = b_0 + \mathbf{b} \in \mathbb{H}$. Let $\mathbf{v}_a$, $\mathbf{v}_b$, and $\mathbf{v}_{ab}$ be vectors in $\mathbb{R}^4$ whose coordinates equal those of $a, b, ab \in \mathbb{H}$.

Then, the vector representation of the product ab is

$$ab \mapsto \mathbf{v}_{ab} = G_1(a)\mathbf{v}_b = G_2(b)\mathbf{v}_a, \tag{29}$$

where

$$G_1(a) = \begin{bmatrix} a_0 & -\mathbf{a}^T \\ \mathbf{a} & a_0 I + K(\mathbf{a}) \end{bmatrix}, \qquad G_2(b) = \begin{bmatrix} b_0 & -\mathbf{b}^T \\ \mathbf{b} & a_0 I - K(\mathbf{b}) \end{bmatrix}, \tag{30}$$

and

$$K(\mathbf{a}) = \begin{bmatrix} 0 & -a_3 & a_2 \\ a_3 & 0 & -a_1 \\ -a_2 & a_1 & 0 \end{bmatrix}, \qquad K(\mathbf{b}) = \begin{bmatrix} 0 & -b_3 & b_2 \\ b_3 & 0 & -b_1 \\ -b_2 & b_1 & 0 \end{bmatrix}, \tag{31}$$

are skew-symmetric matrices determined by the vector parts $\mathbf{a}$ and $\mathbf{b}$ of the quaternions a and b, respectively. For properties of matrices $G_1(a)$ and $G_2(b)$, see [20, 21]. Theorem 4 implies that mapping $r \mapsto r' = ara^{-1}$, $\|a\| = 1$, gives the rotation $\mathbf{r} \mapsto \mathbf{r}'$ in $\mathbb{R}^3$. Using 4×4 matrices, it can be written as

$$\mathbf{v}_r \mapsto \mathbf{v}_r' = G_1(a)G_2\big(a^{-1}\big)\mathbf{v}_r = G_1(a)G_2^T(a)\mathbf{v}_r, \tag{32}$$

where

$$G_1(a)G_2^T(a) = \begin{bmatrix} 1 & 0 \\ 0 & \underbrace{(2a_0^2 - 1)I + 2\mathbf{a}\mathbf{a}^T + 2a_0 K(\mathbf{a})}_{R(a)} \end{bmatrix}. \tag{33}$$

The 3×3 matrix $R(a)$ in the product $G_1(a)G_2^T(a)$ is the well-known Rodrigues matrix of rotation. [13] The Rodrigues matrix has this form in terms of the components of a:

$$R(a)=\begin{bmatrix} a_0^2+a_1^2-a_2^2-a_3^2 & 2a_1a_2-2a_0a_3 & 2a_1a_3+2a_0a_2 \\ 2a_1a_2+2a_0a_3 & a_0^2-a_1^2+a_2^2-a_3^2 & 2a_2a_3-2a_0a_1 \\ 2a_1a_3-2a_0a_2 & 2a_2a_3+2a_0a_1 & a_0^2-a_1^2-a_2^2+a_3^2 \end{bmatrix}. \tag{34}$$

Entries of $R(a)$ are homogeneous polynomials of degree 2 in $\mathbb{R}[a_0,a_1,a_2,a_3]$. Separating the scalar and the vector parts of the quaternion r in the 4D representation (32), we get

$$\mathrm{Re}(r')=\mathrm{Re}(r), \qquad \mathrm{Im}(r')=\boxed{\mathbf{r}'=R(a)\mathbf{r}}=R(a)\,\mathrm{Im}(r). \tag{35}$$

The first relation shows that the scalar part of r remains unchanged, while the vector part $\mathbf{r}'$ of r' is a result of rotation of the vector part $\mathbf{r}$ of r about the axis $\mathbf{a}=a_1\mathbf{i}+a_2\mathbf{j}+a_3\mathbf{k}$, and the angle of counter-clockwise rotation is 2α. Observe that $\det R(a)=\|a\|^6$ and $R(a)^T R(a)=\|a\|^4 I$. Since $\det R(a)=\|a\|^6$ and $R(a)^T R(a)=\|a\|^4 I$, the Rodrigues matrix $R(a)$ gives a rotation if and only if $\|a\|=1$. We intend to find the rotation axis $\mathbf{a}$ and the rotation angle 2α by expressing the quaternionic entries (a_0,a_1,a_2,a_3) in terms of the entries of an orthogonal matrix M of determinant 1. For that purpose, we will use a technique of Gröbner basis and the theory of elimination [10]. Let $M=(m_{ij})$ be an orthogonal 3×3 matrix, that is, $M^T M=I$. Since $M^T M$ is symmetric, this one constraint gives us six polynomial constraints on the entries of M:

$$c_1=m_{11}^2+m_{21}^2+m_{31}^2-1, \qquad c_2=m_{12}^2+m_{22}^2+m_{32}^2-1,$$
$$c_3=m_{13}^2+m_{23}^2+m_{33}^2-1, \qquad c_4=m_{11}m_{12}+m_{21}m_{22}+m_{31}m_{32},$$
$$c_5=m_{11}m_{13}+m_{21}m_{23}+m_{31}m_{33}, \qquad c_6=m_{12}m_{13}+m_{22}m_{23}+m_{32}m_{33}.$$

We add one more constraint, namely, that $\det M=1$:

$$\begin{aligned} c_7={}&m_{11}m_{22}m_{33}-m_{11}m_{23}m_{32}-m_{21}m_{12}m_{33}\\ &+m_{21}m_{13}m_{32}+m_{31}m_{12}m_{23}-m_{31}m_{13}m_{22}-1. \end{aligned}$$

A Gröbner basis G_J for the syzygy ideal $J=\langle c_1,c_2,\ldots,c_7\rangle$ with respect to the order $\mathrm{lex}(m_{11},m_{12},\ldots,m_{33})$ contains twenty polynomials including five polynomials from the original set. This means that the seven constraint polynomials are not algebraically independent. Define nine polynomials $f_k\in\mathbb{R}[a_0,a_1,a_2,a_3,m_{ij}]$:

$$[f_1,f_2,f_3,f_4,f_5,f_6,f_7,f_8,f_9]=\big[m_{ij}-R(a)_{ij}\big]. \tag{36}$$

Our goal is to express the four parameters a_0,a_1,a_2,a_3 in terms of the nine matrix entries m_{ij} that are subject to the seven constraint (syzygy) relations $c_s=0$, $1\le s\le 7$. This should be possible up to a sign since for any rotation in $\mathbb{R}^3$ given by an orthogonal matrix M, $\det M=1$, there are two unit quaternions a and $-a$ such that $R(a)=R(-a)=M$.

We compute a Gröbner basis G_I for the ideal $I = \langle f_1, \ldots, f_9, c_1, \ldots, c_7\rangle$ for $\mathrm{lex}(a_0, a_1, a_2, a_3, m_{11}, m_{12}, \ldots, m_{33})$ order. G_I contains fifty polynomials of which twenty polynomials are in $\mathbb{R}[m_{ij}]$: thus they provide a basis G_J for the syzygy ideal J. We need to solve the remaining thirty polynomial equations for a_0, a_1, a_2, a_3, so we divide them into a set S_l of twenty linear polynomials in a_0, a_1, a_2, a_3, and a set S_{nl} of ten nonlinear polynomials in a_0, a_1, a_2, a_3. The first four polynomials in S_{nl} are:

$$
\begin{aligned}
a_0^2 &= \frac{1}{4}(1 + m_{11} + m_{22} + m_{33}), & a_1^2 &= \frac{1}{4}(1 + m_{11} - m_{22} - m_{33}),\\
a_2^2 &= \frac{1}{4}(1 - m_{11} + m_{22} - m_{33}), & a_3^2 &= \frac{1}{4}(1 - m_{11} - m_{22} + m_{33}),
\end{aligned} \tag{37}
$$

which easily shows that $\|a\| = 1$, the quaternion a defined by the orthogonal matrix M is a unit quaternion.

The remaining six polynomials in S_{nl} are:

$$
\begin{aligned}
a_0a_1 &= \frac{1}{4}(m_{32} - m_{23}), & a_0a_2 &= \frac{1}{4}(m_{13} - m_{31}),\\
a_1a_2 &= \frac{1}{4}(m_{12} + m_{21}), & a_0a_3 &= \frac{1}{4}(m_{21} - m_{12}),\\
a_1a_3 &= \frac{1}{4}(m_{13} + m_{31}), & a_2a_3 &= \frac{1}{4}(m_{23} + m_{32}).
\end{aligned} \tag{38}
$$

The remaining twenty polynomials from S_l are linear in a_0, a_1, a_2, a_3. Let A be the coefficient matrix of that linear homogeneous system. Matrix A is 20×4, but it can be easily reduced to 14×4 by analyzing its submatrices and normal forms of their determinants modulo the Gröbner basis G_J. It can be shown that this symbolic matrix is of rank 3. That is, there is always a one-parameter family of solutions. Once that one-parameter family of solutions is found, two unit quaternions $\pm a$ such that $R(\pm a) = M$ can be found from remaining ten nonlinear equations.

For example, let

$$
M = \begin{bmatrix} 1 & 0 & 0\\ 0 & 0 & -1\\ 0 & 1 & 0 \end{bmatrix}.
$$

Then, the linear system $A(a_0, a_1, a_2, a_3)^T = 0$ has the one-parameter solution $a_0 = a_0$, $a_1 = a_0$, $a_2 = 0$, $a_3 = 0$. The system of nonlinear equations reduces to just $4{a_0}^2 = 2$, which gives $a_0 = \pm\frac{1}{2}\sqrt{2}$, and one unit quaternion is: $a = \frac{1}{2}\sqrt{2} + \frac{1}{2}\sqrt{2}\mathbf{i} = a_0 + \mathbf{a}$, $\cos\alpha = \frac{1}{2}\sqrt{2}$, $\sin\alpha = |\mathbf{a}| = \frac{1}{2}\sqrt{2}$. The Rodrigues matrix gives $R(\pm a) = M$, $\alpha = \frac{1}{4}\pi$, so the rotation angle is $2\alpha = \frac{1}{2}\pi$, and the rotation axis $\mathbf{u}$ is just $\mathbf{i}$, as expected.

For another example, consider the following orthogonal matrix:

$$M = \begin{bmatrix} 0 & \frac{\sqrt{210}-5\sqrt{14}}{35} & \frac{-2\sqrt{35}-5\sqrt{21}}{35} \\ \frac{\sqrt{210}+5\sqrt{14}}{35} & \frac{11}{35} & \frac{-7\sqrt{6}+5\sqrt{10}}{35} \\ \frac{-2\sqrt{35}+5\sqrt{21}}{35} & \frac{-7\sqrt{6}-5\sqrt{10}}{35} & \frac{4}{35} \end{bmatrix}$$

with $\det M = 1$. Then, solution to the linear system is $a_0 = a_0$, $a_1 = \frac{-\sqrt{10}}{5}a_0$, $a_2 = \frac{-\sqrt{21}}{5}a_0$, $a_3 = \frac{\sqrt{14}}{5}a_0$. Upon substitution into the nonlinear equations, we find $a_0 = \pm\frac{\sqrt{70}}{14}$, which eventually gives $a = \frac{\sqrt{70}}{14} + (-\frac{\sqrt{7}}{7}\mathbf{i} - \frac{\sqrt{30}}{10}\mathbf{j} + \frac{\sqrt{5}}{5}\mathbf{k})$, $\cos\alpha = \frac{\sqrt{70}}{14}$, $\sin\alpha = |\mathbf{a}| = \frac{3\sqrt{14}}{14}$. It can be verified again that $R(\pm a) = M$ and $\alpha \approx 0.9302740142$ rad.

For our last example, we need the following result from [10, Proposition 3, p. 339] on the so-called *ideal of relations* I_F for a set of polynomials $F = (f_1, \ldots, f_m)$. It is usually used to derive the syzygy relations among homogeneous invariants of finite groups. We will use it to show how one can systematically derive relations among interpolation functions used in finite element theory. Although these relations are well known [23], their systematic derivation with the help of Gröbner basis is less known.

Consider the system of equations

$$y_1 = f_1(x_1, \ldots, x_n), \ldots, y_m = f_m(x_1, \ldots, x_n).$$

Then, the syzygy relations among the polynomials $f_1, \ldots, f_m$ can be obtained by eliminating $x_1, \ldots, x_n$ from these equations. Let $k[x_1, \ldots, x_n]^G$ denote the ring of invariants of a finite group $G \subset \mathrm{GL}(n, k)$. The following result is proven in [10].

Proposition 1 *If* $k[x_1, \ldots, x_n]^G = k[f_1, \ldots, f_m]$, *consider the ideal*

$$J_F = \langle f_1 - y_1, \ldots, f_m - y_m, \rangle \subset k[x_1, \ldots, x_n, y_1, \ldots, y_m].$$

(i) I_F *is the* n*th elimination ideal of* J_F. *Thus,* $I_F = J_F \cap k[y_1, \ldots, y_m]$. (ii) *Fix a monomial order in* $k[x_1, \ldots, x_n, y_1, \ldots, y_m]$, *where any monomial involving one of* $x_1, \ldots, x_n$ *is greater than all monomials in* $k[y_1, \ldots, y_m]$ *and let GB be a Gröbner basis of* J_F. *Then* $GB \cap k[y_1, \ldots, y_m]$ *is a Gröbner basis for* I_F *in the monomial order induced on* $k[y_1, \ldots, y_m]$.

The ideal I_F is known to be a prime ideal of $k[y_1, \ldots, y_m]$.[8] Furthermore, the Gröbner basis for I_F may not be minimal: This is because the original list F of polynomials may contain polynomials which are algebraically dependent. Thus, in

[8] An ideal $I \subset k[x_1, \ldots, x_n]$ is *prime* if whenever $f, g \in k[x_1, \ldots, x_n]$ and $fg \in k[x_1, \ldots, x_n]$, then either $f \in I$ or $g \in I$.

order to obtain a minimal Gröbner basis for I_F or the smallest number of syzygy relations, one needs to assure first that the list F is independent [10].

We will use this proposition encoded in a procedure `SyzygyIdeal` from the SP package [1] to compute syzygy relations among interpolation functions for orders $k=2$ and $k=3$ for triangular elements in finite element method. These elements are referred to as *quadratic* and *cubic* as their interpolation functions are, respectively, quadratic and cubic polynomials, and they contain, respectively, three and four equally spaced nodes per side. For all definitions, see [23, Chap. 9].

Example 6 We derive relations between the interpolation functions for the higher-order Lagrange family of triangular elements with the help of the so-called *area coordinates* L_i. For triangular elements, there are three nondimensional coordinates L_i, $i=1,2,3$, such that

$$L_i = A_i/A, \qquad A = A_1 + A_2 + A_3, \qquad L_1 + L_2 + L_3 = 1 \tag{39}$$

(see [23, Fig. 9.3, p. 408]). Here, A_i is the area of a triangle formed by the nodes j and k (i.e., $i \neq j, i \neq k$) and an arbitrary point P in the element, and A is the total area of the element. Each L_i is a function of the position of the point P. For example, if point P is positioned on a line joining nodes 2 and 3 (or, at the nodes 2 or 3), then $L_1 = 0$ and $L_1 = 1$ when P is at the node 1. Thus, L_1 is the *interpolation function* ψ_1 associated with the node 1 and likewise for L_2 and L_3, that is,

$$\psi_1 = L_1, \qquad \psi_2 = L_2, \qquad \psi_3 = L_3$$

for any triangular element. Functions (polynomials) L_i are used to construct interpolation functions for higher-order triangular elements with k nodes per side.

For $k=3$, we have three equally spaced nodes per side of a triangular element, and the total number of nodes in the element is $n = \frac{1}{2}k(k+1) = 6$. Then, the degree d of the interpolation functions (polynomials) L_i is $d = k - 1 = 2$. The triangular elements are then called *quadratic*.

We define six interpolation functions ψ_s, $s = 1, \dots, 6$, in terms of L_i as in formulas (9.16a), (9.16b), and (9.16c) in [23, p. 410]:

$$\begin{aligned}
&\psi_1 = 2L_1^2 - L_1, \qquad \psi_2 = 4L_1L_2, \qquad \psi_3 = 2L_2^2 - L_2,\\
&\psi_4 = -4L_1L_2 - 4L_2^2 + 4L_2,\\
&\psi_5 = 2L_1^2 + 4L_1L_2 - 3L_1 + 2L_2^2 - 3L_2 + 1,\\
&\psi_6 = -4L_1^2 - 4L_1L_2 + 4L_1.
\end{aligned} \tag{40}$$

The procedure `SyzygyIdeal` yields seven polynomial relations between the functions ψ_s:

$$\begin{aligned}
r_1 &= \psi_4^2 + 4\psi_4\psi_5 + 4\psi_5^2 + 2\psi_4\psi_6 + 4\psi_5\psi_6 + \psi_6^2 - \psi_4 - 4\psi_5 - \psi_6,\\
r_2 &= 2\psi_2\psi_5 + \psi_2\psi_6 + 2\psi_3\psi_6,\\
r_3 &= 4\psi_3\psi_5 + 4\psi_4\psi_5 + 4\psi_5^2 - \psi_2\psi_6 - 2\psi_3\psi_6 + \psi_4\psi_6 + 4\psi_5\psi_6\\
&\quad + \psi_6^2 - 4\psi_5 - \psi_6,\\
r_4 &= \psi_1 - 1 + \psi_2 + \psi_3 + \psi_4 + \psi_5 + \psi_6,\\
r_5 &= \psi_2\psi_4 - \psi_2\psi_6 - 4\psi_3\psi_6 - \psi_4\psi_6,\\
r_6 &= 2\psi_3\psi_4 - 6\psi_4\psi_5 - 8\psi_5^2 + 2\psi_2\psi_6 + 6\psi_3\psi_6 - \psi_4\psi_6 - 8\psi_5\psi_6\\
&\quad - 2\psi_6^2 + 8\psi_5 + 2\psi_6,\\
r_7 &= \psi_2^2 + 4\psi_2\psi_3 + 4\psi_3^2 + 8\psi_4\psi_5 + 12\psi_5^2 - 2\psi_2\psi_6 - 4\psi_3\psi_6\\
&\quad + 2\psi_4\psi_6 + 12\psi_5\psi_6 + 3\psi_6^2 - \psi_2 - 4\psi_3 - 12\psi_5 - 3\psi_6.
\end{aligned} \tag{41}$$

It can be verified by direct substitution of (40) into (41) that the latter well-known relations [23] are satisfied.

When there are $k = 4$ equally spaced nodes per side of a triangular element, then the total number of nodes per element is $n = 10$, and the degree d of the interpolation functions (polynomials) is $d = k - 1 = 3$. In a similar manner as above one can derive twenty nine relations among the ten interpolation polynomials.

4 Conclusions

Our goal was to show a few applications of Gröbner basis technique to engineering problems extending from robotics through curve theory to finite-element method. While applications to the inverse kinematics are well known, formulation of the problem of finding the circle of two intersecting spheres, that is, the plane of the circle, a normal to the plane, and then the radius and the center of the circle in the language of Clifford (geometric) algebra, gave another opportunity to apply the Gröbner basis technique. Since the method is fast and efficient, it can be employed when solving the elbow problem of the robot arm also when formulated in that language.

References

1. Abłamowicz, R., Fauser, B.: SP—a maple package for symmetric polynomials. http://math.tntech.edu/rafal/, 18 March 2009

2. Abłamowicz, R., Liu, J.: On the parallel lines for nondegenerate conics. Department of Mathematics. TTU, Tech. Report No. 2006-1, January 2006, http://www.math.tntech.edu/techreports/techreports.html, 18 March 2009
3. Arnold, V.I.: Wave Fronts and Caustics. Kluwer Academic, Dordrecht (1990)
4. Awange, J.L., Grafarend, E.W.: Solving Algebraic Computational Problems in Geodesy and Geoinformatics. Springer, Berlin (2005)
5. Becker, T., Weispfenning, V.: Gröbner Bases. Springer, Berlin (1993)
6. Buchberger, B.: In: Bose, N.K. (ed.) Multidimensional Systems Theory, p. 184. Reidel, Dordrecht (1985)
7. Buchberger, B.: In: Proc. Marktoberdorf Summer School 1995. Springer, Berlin (1997)
8. Buchberger, B., Winkler, F. (eds.): Gröbner Bases and Applications. Cambridge University Press, Cambridge (1998)
9. Cox, D.: Introduction to Gröbner bases. Proc. Symp. Appl. Math. **53**, 1–24 (1998)
10. Cox, D., Little, J., O'Shea, D.: Ideals, Varieties, and Algorithms: An Introduction to Computational Algebraic Geometry and Commutative Algebra. Springer, New York (2008)
11. Eisenbud, D.: Commutative Algebra with a View Toward Algebraic Geometry. Springer, New York (2004)
12. Faugère, J.-Ch.: FGb, http://fgbrs.lip6.fr/jcf/Software/FGb/, 18 March 2009
13. Goldstein, H.: Classical Mechanics. Addison-Wesley, Reading (1980)
14. Grabmeier, J., Kaltofen, E., Weispfenning, V. (eds.): Computer Algebra Handbook. Springer, Berlin (2003)
15. Greuel, G.-M., Pfister, G.: A *Singular* Introduction to Commutative Algebra. Springer, Berlin, Heidelberg (2002)
16. Gröbner: Basis Bibliography at Johann Radon Institute for Computational and Applied Mathematics (RICAM). www.ricam.oeaw.ac.at/Groebner-Bases-Bibliography/, 18 March 2009
17. Hartshorne, R.: Algebraic Geometry. Springer, New York (1997)
18. Hildenbrand, D., Hitzer, E.: Analysis of point clouds: using conformal geometric algebra. In: GRAPP Conference Madeira
19. Hildenbrand, D., Lange, H., Stock, F., Koch, A.: Efficient inverse kinematics algorithms based on conformal geometric algebra: using reconfigurable hardware. In: GRAPP Conference Madeira
20. Meister, L.: In: Bayro Corrochano, E., Sobczyk, G. (eds.) Geometric Algebra with Applications in Science and Engineering, p. 387. Birkhäuser, Basel (2001)
21. Meister, L., Schaeben, H.: Math. Methods Appl. Sci. **28**, 101–126 (2005)
22. Norton, R.L.: Design of Machinery: An Introduction to the Synthesis and Analysis of Mechanisms and Machines. McGraw-Hill, New York (1999)
23. Reddy, J.N.: An Introduction to the Finite Element Method, 2nd edn. McGraw-Hill, New York (1993)
24. SINGULAR:PLURAL, www.singular.uni-kl.de/, 18 March 2009
25. Uicker, J.J., Pennock, G.R., Shigley, J.E.: Theory of Machines and Mechanisms. Oxford University Press, London (2003)
26. Waterloo Maple Incorporated: Maple, a general purpose computer algebra system, Waterloo, http://www.maplesoft.com (2007)

Index

E. Bayro-Corrochano, G. Scheuermann (eds.), *Geometric Algebra Computing*,
DOI 10.1007/978-1-84996-108-0, © Springer-Verlag London Limited 2010

D

MIX
Papier aus verantwortungsvollen Quellen
Paper from responsible sources
FSC® C105338

If you have any concerns about our products,
you can contact us on
ProductSafety@springernature.com

In case Publisher is established outside the EU,
the EU authorized representative is:
Springer Nature Customer Service Center GmbH
Europaplatz 3, 69115 Heidelberg, Germany

Printed by Libri Plureos GmbH
in Hamburg, Germany